Outstanding Contributions to Logic

Volume 16

More information about this series at http://www.springer.com/series/10033

Janusz Czelakowski
Editor

Don Pigozzi on Abstract Algebraic Logic, Universal Algebra, and Computer Science

 Springer

Editor
Janusz Czelakowski
Institute of Mathematics and Informatics
University of Opole
Opole
Poland

ISSN 2211-2758 ISSN 2211-2766 (electronic)
Outstanding Contributions to Logic
ISBN 978-3-319-74771-2 ISBN 978-3-319-74772-9 (eBook)
https://doi.org/10.1007/978-3-319-74772-9

Library of Congress Control Number: 2017964448

This Springer imprint is published by Springer Nature
The registered company is Springer International Publishing AG
The registered company address is: Gewerbestrasse 11, 6330 Cham, Switzerland

Ad multos annos, Don!

Preface

1. In a most surprising way logic is connected with algebra. Each language, whether natural or artificial, possesses a complex algebraic structure. In the simplest case this structure is revealed in the languages of propositional logics as certain absolutely free algebras. The above linguistic perspective establishes the first bridge between logic and algebra. There are more links between these two domains and they exist on a deeper level. George Boole (1854) proved that the "laws of thought" can be framed algebraically as identities of an algebra. Thus, thinking is also "algebraizable". Boole's work was extended in various directions by a number of researchers, beginning with William Stanley Jevons. Charles Sanders Peirce integrated his work with Boole's during the 1870s. Other significant figures were Augustus de Morgan, Platon Sergeevich Poretskii, William Ernest Johnson, and Ernst Schröder. The conception of a Boolean algebra structure on equivalent statements of a propositional calculus is credited to Hugh MacColl (a four part article from 1877–1879, see (MacColl, 1906), preceding Gottlob Frege's *Begriffschrifft*). The Fregean principle of compositionality, that the meaning of any complex expression is a function of the meanings of its constituents, together with other principles is central in formal semantics. Roughly speaking, these principles establish a homomorphism between the algebraic structure of each language and the algebraic structure constituted by meanings of the expressions of this language.

The above discoveries gave rise to algebraic logic. As a result the link between logic and algebra is cemented and becomes inseparable. Studying the dependence existing between them is still one of the vital areas of scientific activity. It is also in this very context that the differentiation of the professional belonging of the individual researcher—a logician or an algebraist—becomes blurred.

2. Professor Don Pigozzi, together with his colleagues who cooperated with him at the turn of the 1980s and the 1990s, managed to effect a change of the paradigm of algebraic logic. This new situation can best be illustrated with this analogy: at the beginning of the 20th century, mathematicians carried out extensive studies in the area of function spaces, e.g., the spaces of continuous functions on compact topological spaces, the spaces of absolutely integrable functions on measure spaces, etc. A series of deep and detailed results was obtained regarding this sphere. Still, there was a lack of the right key to the general theory which would allow ordering the research field and deriving well-known particular cases from a few notions and theorems. Obviously, the notion of a Hilbert space did order a certain section of the field. However, what was obtained then was rather a set of facts loosely connected with one another, though each separately was an original and deep mathematical theorem. It was not until the theory of Banach spaces was put forward, along with its apparatus of notions and key theorems, that the gathered research material could be ordered and adequate mathematical arsenal was provided to make it possible to lay foundations of functional analysis. We came to deal with a similar situation in algebraic logic in the second half of the last century. The continuum of logical systems, grouped into some categories (modal systems, temporal systems, system of dynamic logic, relevant logics, etc.), entered the stage for good. The then literature of the subject abounded in difficult and sophisticated metalogical results which characterized various aspects of individual systems. In the study of the systems, new semantic tools, like relational semantics, neighborhood semantics, and the like, are made use of. Still, some types of semantics have limitations, since not every system can be adequately semantically characterized in terms of an appropriate completeness theorem; there appears the phenomenon of semantical incompleteness— we are familiar with, e.g., normal modal systems, not possessing an adequate Kripke-style semantics. It was known that the path from a logical system to algebraic semantics leads through Lindenbaum-Tarski algebras (Lindenbaum (1929), Tarski (1930), Łukasiewicz and Tarski (1930)). That was a well-marked and reliable route for classical logic, intuitionistic logic, and a series of other logics. Helena Rasiowa, in her pioneering monograph (Rasiowa, 1974), introduced implicative systems, thought to be a broad class of logical systems for which the generalization of the Lindenbaum-Tarski method, which she investigated, did "work". Here, the key was the notion of an implication, viewed as a binary connective satisfying natural assumptions, analogous to the properties of the implication of classical or intuitionistic logics. Nevertheless, the class of S-logics investigated by Rasiowa does not encompass many elementary intensional systems, like the main modal logics. If one goes through the relevant literature of the 1970s, the scenery was saturated with millions of logical systems, each of which being somehow important, individually examined and described. However, there were no methodological tools available of sufficient generality which would allow framing theses systems from a uniform research perspective. Again, the key which allowed introducing an order were the notions of the Leibniz operator and of a protoalgebraic logics, the latter employing the above-mentioned operator. Both notions were explicitly defined by Wim Blok and Don Pigozzi (1986), (1989). Although the both notions had been known in the literature under other names (e.g., Wójcicki (1988) in his earlier works wrote about the largest matrix congruences which are the same objects as Leibniz congruences, and Czelakowski (1985, 1986) isolated the class of *non-pathological logics*, the studies launched by Wim Blok and Don Pigozzi introduced a universal notional network, as well as gave rise to systematic investigations conducted in the new language of abstract algebraic logic (AAL, in short). It is considered that the most important and indisputable achievements of them are the introduction of the notion of an algebraizable logic in the pioneering monograph entitled *Algebraizable Logics* and also giving the key properties characterizing this class. Algebraizability is a rigorous mathematical notion systematically investigated by many logicians since then. (We omit its definition here.)

3. It can be argued that studies in the field of logic, which were carried out in the 1960s and the 1970s, whose nucleus was the notion of a consequence operation, did not meet with a broader interest at that time. The consequence theory, whose fundamentals had already been laid by Alfred Tarski after the World War I (see Tarski (1956) remained rather underestimated. It was mainly Tarski's concept of truth in formalized languages which attracted interest as his first-rate achievement, especially in the context of later model theoretic and algebraic applications. Apparently it was so. A few causes of such a state of things can be indicated. Firstly, the then trends in the world's logic were different and the foundations of mathematics laid, in particular, by Kurt Gödel and his discoveries developed recursion theory and set theory, especially upon introducing the notion of forcing by Paul Cohen. In many universities the methods of relational semantics for intensional logics were developed. As it was mentioned, in the Warsaw school of Andrzej Mostowski, Helena Rasiowa and Roman Sikorski the algebraic foundations were created for a broad class of non-classical logics, viz. the extended implicative calculi. Roman Suszko, Ryszard Wójcicki and their disciples were also active at that time.

What is a logical system then? In the literature, we will find several definitions. Here, we will limit ourselves to two of them only. Logic (on the propositional level) is most often understood as a set of formulas closed with respect to substitutions and certain rules of inference. For example, an array of modal logics is defined in this way. In another approach, the basic notional frame is composed by structural and finitary consequence operations defined on pertinent languages. This is a more general approach than the former one. The following problem is connected with it: if one accepts the notion of logic, viewed as a certain invariant set of formulas, to be the basis, what consequence operation should be attached to this set? On the level of normal modal systems, as e.g. **S4**, one can associate, in a natural way, *two* different consequence operations with each such a system: the first—the so-called *weak* consequence, determined by the given system, and the detachment rule, as a primitive rule of inference, as well as the so-called *strong* consequence, formed out of the weak one by adjoining the Gödel rule as a primitive rule (and thereby applicable in all derivations!). Which consequence should be chosen? This depends on the research preferences of the given logician. For instance, within the framework of the rapidly developing unification theory for propositional and equational logics, one investigates structurally complete or almost structurally complete systems. Naturally, there are solely strong modal consequences in sight there.

Comprehensive studies of logical systems, chiefly intensional ones, such as modal or temporal logics, etc., which have been conducted in recent years, have caused the consequence theory to be placed in the focal point. The creation of the theory of *entailment relation* (this term can be translated as a multi-conclusion consequence relation), hybrid logics, substructural logics and the growing interest in non-monotonic reasonings has exercised a strong impact on consequence theory. The existence of a continuum of logical systems, of various references and applications, be it in informatics, philosophy, or theory of language or others, resulted in the need of ordering the research field, finding a few common principles, according to which one can notionally embrace the complex matter of contemporary logic.

The introduction of the Leibniz and Suszko operators marked a new stage for metalogic. It was discovered that a uniform classification scheme encompassing all logical systems can be based on some plausible properties of these operators. As a result, abstract algebraic logic (AAL) has emerged. AAL offers a transparent and natural hierarchy of logical systems (and not only propositional ones); each level of the hierarchy is determined by a simple and concrete property of the above-listed operators, such as monotonicity or order-continuity. Let us underline that the notion of a consequence operation is of key importance in the AAL.

4. The central issue that underlay AAL was to get the gist of the dependence between logic and algebra from the mathematical viewpoint; the point is about the question which for over 150 years has been permeating the history of the both disciplines.

AAL takes a more abstract and general approach than the traditional algebraic logic. In contrast to algebraic logic, where the focus is on the algebraic forms of specific deductive systems, AAL is concerned with the *process* of algebraization itself. AAL investigates *degrees* of algebraizability of deductive systems, making use of rigorous mathematical tools. The degree of algebraizability of a system is determined by the place of the system in the hierachy of logics based on the Leibniz operator.

The problem area of AAL and the issue of algebraizability of logical systems in particular, have set new tasks to algebra and logic, viz. to describe the algebraic semantics for deductive systems. The relation between algebra and logic is the strongest in the case of algebraizable systems. Speaking in the most general way, algebraizability of a system L consists in determining the conditions which allow replacing the process of deduction of logical formulas by the process of congruence generation on appropriate algebras in an equivalent manner on the ground of this equational system.

We find a development of the idea of linking logic and algebra present in the prewar Warsaw School, in the works of Adolf Lindenbaum and Alfred Tarski, the result of which was the notion of *Lindenbaum-Tarski algebras* of the given logical system. Not entering into technical and definition-related details, it can be said that the Lindenbaum-Tarski algebras of classical logic are Boolean algebras; by analogy—the Lindenbaum-Tarski algebras of intuitionistic logic are Heyting algebras. As it was mentioned, further studies in that direction were conducted after the World War II in Warsaw. The following key ideas should be mentioned: (1) Language viewed as an abstract algebra. (2) Quantifiers treated as infinite conjunctions and disjunction, that is, suprema and infima in ordered models (Andrzej Mostowski), (3) The theory of consequence operations as a foundation of deductive systems, originated from the works of Alfred Tarski and then supplemented with the notion of structurality by Jerzy Łoś and Roman Suszko. (Structurality reflects the invariance of inference patterns with respect to substitutions in the language.) (4) Introduction of Boolean methods to the theory of models. Many of theses ideas were presented in Rasiowa and Sikorski's book (1963) and in Rasiowa's monograph (1974).

Rasiowa's book shows, at last in a clear way, the relations between logic and the theory of equational classes of algebras and, more generally, between logic and quasivarieties of algebras. (Maltsev, 1971) is the first systematic exposition of the theory of quasivarieties).

As mentioned, a certain problem regarding Rasiowa's framework are *intensional* systems, such as modal logics, temporal logics, dynamic logic, etc., which do not fall under the definitions and notional apparatus introduced in her monograph. How to overcome this basic obstacle? The Leibniz operator Ω turned out the key to solving the problem. It is Wim Blok and Don Pigozzi who must be credited with perceiving the significance of this operator in metalogic. The Leibniz operator is the tool that enables one to generalize the approach based on the construction of Lindenbaum-Tarski algebras by providing a uniform conceptual framework abstracting from the presence of definite logical connectives like an implication or equivalence in the language. Following Lindenbaum, each sentential language S is identified with an absolutely free algebra—is a commonly-accepted paradigm. In the simplest case of sentential logics, the operator Ω assigns to each closed theory T of logic L a certain congruence ΩT on the sentential language. ΩT is, by definition, the largest congruence on S that is compatible with the theory T, i.e., the largest congruence which does not "glue" together formulas belonging to T with those being outside T. The congruence ΩT exists; it is also called the relation of (absolute) *synonymy* on the language with respect to T. As a result, to each logical system L there corresponds the mapping Ω assigning the Leibniz congruence ΩT to each logically closed theory T of L. Let $Th(L)$ be the set of closed theories of L. If the mapping Ω is monotone on the set $Th(L)$, i.e., it assigns a larger congruence to a larger theory, then the logic L is called *protoalgebraic*. One can impose further order conditions on the run of the operator Ω on the lattice $Th(\boldsymbol{L})$. In consequence, one obtains proper subclasses of the class of protoalgebraic logics: equivalential logics, weakly algebraizable logics, algebraizable logics, regularly algebraiable logics, Fregean logics, etc. They form a hierarchy of protoalgebraic systems—the higher the position in the hierarchy, the narrower the class of logics. The largest class is formed by protoalgebraic systems, while Fregean systems make the smallest one (after the name of the German logician Gottlob Frege). Nowadays, extensive literature on protoalgebraic logics and—more broadly—the abstract algebraic logic (e.g., (Czelakowski, 2001), (Font, 2016)) is available. The hierarchy under consideration builds a bridge between logic *sensu largo* and the theory of quasivarieties of algebras, as well as provides most sophisticated methods of description of its individual rungs. Furthermore, the notional apparatus, which has been worked out, has contributed to a substantial broadening of the algebraic discourse with new elements. This influence is evident, especially with reference to studies on various algebraic aspects of equational logic.

5. The class of protoalgebraic logics, vital as it is, since all the "non-pathological" logics belong to it, including all known intensional logics, does not cover all logical systems: some systems "fall" outside the above-mentioned hierarchy. The simplest example is the conjunctive-disjunctive fragment of classical propositional logic. The approach which is based on the Leibniz operator does not work here. However, the new operator Σ, which is called the *Suszko operator* in honor of Roman Suszko comes in aid. This operator was implicitly introduced in unpublished notes by Suszko. Its theory was subsequently developed in a systematic way (see e.g. (Czelakowski, 2001, 2003)), and the name *the Suszko operator* has been coined. The operator is nowadays extensively investigated by other researchers.

The Suszko operator Σ coincides with the Leibniz operator Ω on protoalgebraic systems; moreover, it is always monotone. Properties of the Suszko operator gave rise to researching the hierarchy of *all* logical systems, not only protoalgebraic ones, from the uniform algebraic viewpoint. If we take into account the algebraic aspects of operator Σ, the classes of algebras defined by it do not have to be either varieties or quasivarieties. They are classes closed with respect to the formation of subdirect products and (sometimes) under ultraproducts; they are also called *Lyndon* classes, after the name of the outstanding American logician and mathematician.

6. We present the reader with the *Festschrift* dedicated to Professor Don Pigozzi on the occasion of his 80th birthday. Professor Don Pigozzi was born in June 1935; thus, this book comes out three years after his jubilee. Always, when a round occasion to celebrate the birthday of an outstanding scholar draws closer, there arises the question: In what way should the scholars' community, acquaintances and friends act to express their appreciation? The series entitled *Outstanding Contributions to Logic*, which has been in existence for a few years now, allows choosing a natural and reliable formula: editing an occasional anthology dedicated to the scholar. The formula realized by the

Outstanding Contributions to Logic differs, however, from others: it is more flexible and richer by new motifs. Beside the occasional articles included in the *Festschrift*, which were written by outstanding specialists and disciples, there is also a place in it to present the scientific output of the Jubilarian and discuss to a broad extent his impact on the development of science, to catalogue his works, and—first and foremost—to bring his person closer in the form of an extensive profile which sketches the most important events, not only related to the scientific activity, but also from his personal life. And it is this formula that has been applied in the present volume dedicated to Professor Don Pigozzi.

The publication of the volume is the merit of many people—primarily the authors, many of whom being disciples of Don Pigozzi, who were willing to write occasional articles.

We do hope that the present book will be useful for scholars who are interested in Don Pigozzi's output, as well as in the areas, to the rise and development of which his work contributed, that is AAL, universal algebra and computer science. The published works contain new scientific results. Some of the papers also present chronologically ordered facts relating to the development of the disciplines, to the rise of which Don Pigozzi again contributed considerably, especially the abstract algebraic logic. The papers published in the volume will certainly offer valuable source material for historians of science, especially those who deal with the history of mathematics and logic.

Acknowledgements. It is self-evident that, in the first place, it is Don Pigozzi who should be thanked the most. His life and scholarly output have provided the basic motivation which justifies the publication of this *Festschrift* in the series of *Outstanding Contributions to Logic*. I am personally grateful to Don for his agreeing to undertake to work on editing this volume and also for his invaluable help in realization of a number of editorial and substantial aspects of the projects.

I wish to thank all the authors who decided to submit their papers for publication in the present volume: they did not only respond to the initial invitation to participate in the project, but—first of all—have contributed their original papers and constructively cooperated with the editor. All the submitted works were reviewed. Accordingly, here, I would like to thank the Reviewers, who—through their anonymous work—have also contributed to the appearance of the volume.

Special thanks are due to Professor Jan Zygmunt for his checking the factographic data relating to the prewar works of the logicians mentioned in the papers.

I am grateful to Sven Ove Hansson, the editor of the series. It is thanks to his support that this volume could be brought out. He secured the indispensable documentation relating to the principles of editing it and was greatly helpful in various other matters which required explaining.

I want to thank Christi Lou of Springer Science for the support on the part of the Publisher. She was the one to provide all forms, guidelines and advice, necessary in the implementation of the project.

The book has been typeset using the the program TEX, originally designed by D. Knuth, from the source files submitted by the authors. The book uses the LATEX format, a class file provided by Springer, as well as several packages that were developed under the auspices of the American Mathematical Society.

Janusz Czelakowski
The editor of the volume

References

Blok, W. and Pigozzi, D. (1986). Protoalgebraic logics, *Studia Logica* 45, 337–369.

Blok, W. and Pigozzi, D. (1989). *Algebraizable Logics*, Memoirs of AMS 396, Providence.

Boole, G. (1854). *An Investigation of the Laws of Thought, on which are Founded the Mathematical Theories of Logic and Probabilities*, London, Walton and Maberly; reprinted 1951.

Czelakowski, J. (1985). Algebraic aspects of deduction theorems, *Studia Logica* 44, 369–387.

Czelakowski, J. (1986). Local deduction theorems, *Studia Logica* 45, 377–391.

Czelakowski, J. (2001). *Protoalgebraic Logics*, Kluwer, Dordrecht.

Czelakowski, J. (2003). The Suszko operator, Part 1, *Studia Logica* 74, 181–231.

Font, J. M. (2016). *Abstract Algebraic Logic. An Introductory Textbook*, College Publications, London.

Font, J. M. and Jansana, R. (1996). *General Algebraic Semantics for Sentential Logics*, Springer Verlag, Berlin.

Frege, G. (1879). *Begriffsschrift: eine der arithmetischen nachgebildete Formelsprache des reinen Denkens*, Halle.

Lindenbaum, A. (1929). Méthodes mathématiques dans les researchers sur le systeme de la théorie déduction, *Księga Pamiątkowa Pierwszego Polskiego Zjazdu Matematycznego, Lwów, 7–10 IX 1927*, Kraków.

Łukasiewicz, J. and Tarski, A. (1930). Untersuchungen über des Aussagenkalkül, *Comptes Rendus des Séances de la Société des Sciences et des Lettres de Varsovie* 23, 30–50.

MacColl, H. (1906). *Symbolic Logic and Its Applications*, Longmans, Green.

Maltsev, A. (1971). *The Metamathematics of Algebraic Systems. Collected Papers: 1936–1967*, Studies in Logic and the Foundations of Mathematics 66, North-Holland, Amsterdam.

Monk, D. (1965). Substitutionless Predicate Logic with Identity, *Archiv für Mathematische Logik und Grundlagenforschung* 7, 103–121.

Rasiowa, H. (1974). *An Algebraic Approach to Non-Classical Logics*, PWN and North-Holland, Warsaw/Amsterdam.

Rasiowa, H. and Sikorski, R. (1963). *The Mathematics of Metamathematics*, Monografie Matematyczne 41, PWN, Warsaw.

Tarski, A. (1930). Über einige fundamentale Begriffe der Metamathematik, *Comptes Rendus des Séances de la Société des Sciences et des Lettres de Varsovie* 23, 22–29. English translation in Tarski (1956) On some fundamental concepts of metamathematics, 30–37.

Tarski, A. (1956). *Logic, Semantics, Metamathematics. Papers from 1923 to 1938*, Clarendon Press, Oxford.

Wójcicki, R. (1988). *Theory of Logical Calculi. Basic Theory of Consequence Operations*, Kluwer, Dordrecht.

Contents

Boolean product representations of algebras via binary polynomials 297

Antonino Salibra, Antonio Ledda, and Francesco Paoli

Paraconsistent constructive logic with strong negation as a contraction-free relevant logic ... 323

Matthew Spinks and Robert Veroff

Possible classification of finite-dimensional compact Hausdorff topological algebras ... 381

Walter Taylor

List of Contributors

Hugo Albuquerque
Departament de Filosofia, Universitat de Barcelona (UB), C. Montalegre 7,
E-08001 Barcelona, Spain, e-mail: `hugo.albuquerque@ua.pt`

Josep Maria Font
Departament de Matemàtiques i Informàtica, Universitat de Barcelona
(UB), Gran Via de les Corts Catalanes 585, E-08007 Barcelona, Spain,
e-mail: `jmfont@ub.edu`

Ramon Jansana
Departament de Filosofia, Universitat de Barcelona (UB), C. Montalegre 7,
E-08001 Barcelona, Spain, e-mail: `jansana@ub.edu`

Tommaso Moraschini
Institute of Computer Science, Czech Academy of Sciences, Pod Vodárenskou
věží 2, 182 07 Prague 8, Czech Republic, e-mail: `tommaso.moraschini@`
`gmail.com`

Sergey Babenyshev
Siberian Fire-Rescue Academy, Severnaya Str 1, Zheleznogorsk, 662971,
Russia, e-mail: `sergey.babenyshev@gmail.com`

Clifford Bergman
Department of Mathematics, Iowa State University, Ames, Iowa, 50011,
USA, e-mail: `cbergman@iastate.edu`

Janusz Czelakowski
Insitute of Mathematics and Informatics, Opole University, Oleska 48,
45-052 Opole, Poland, e-mail: `jczel@math.uni.opole.pl`

Isabel Ferreirim
Department of Mathematics, University of Lisbon, Portugal, e-mail:
`imferreirim@ciencias.ulisboa.pt`

Manuel A. Martins

Department of Mathematics and CIDMA, University of Aveiro, Portugal,
e-mail: martins@ua.pt

Alexandr Kazda
IST Austria, Am Campus 1, 3400 Klosterneuburg, Austria, e-mail:
alex.kazda@gmail.com

Marcin Kozik
Theoretical Computer Science, Faculty of Mathematics and Computer
Science, Jagiellonian University, Cracow, Poland, e-mail: kozik@tcs.uj.
edu.pl

Ralph McKenzie
Department of Mathematics, Vanderbilt University, Nashville, TN 37240,
USA, e-mail: ralph.n.mckenzie@vanderbilt.edu

Matthew Moore
Department of Mathematics and Statistics, McMaster University, Hamilton,
Ontario, Canada L8S 4K1, e-mail: matthew.moore@math.mcmaster.ca

Keith A. Kearnes
Department of Mathematics, University of Colorado, Boulder, CO 80309-
0395, USA, e-mail: Keith.Kearnes@Colorado.EDU

George F. McNulty
Department of Mathematics, University of South Carolina, Columbia, South
Carolina 29208, United States of America e-mail: mcnulty@math.sc.edu

Katarzyna Pałasińska
Cracow University of Technology Faculty of Physics, Mathematics and
Computer Science, Cracow, Poland, e-mail: kpalasinska@gmail.com

Anna B. Romanowska
Faculty of Mathematics and Information Science, Warsaw University of
Technology, Warsaw, Poland, e-mail: aroman@mini.pw.edu.pl

Jonathan D.H. Smith
Department of Mathematics, Iowa State University, Ames, Iowa, 50011,
USA, e-mail: jdhsmith@iastate.edu

Antonino Salibra
Università Ca'Foscari Venezia, Italy, e-mail: salibra@dsi.unive.it

Antonio Ledda
Università di Cagliari, Italy, e-mail: antonio.ledda@unica.it

Francesco Paoli
Università di Cagliari, Italy, e-mail: paoli@unica.it

Matthew Spinks
Department of Philosophy, University of Cagliari, 09123 Cagliari, Italy,
e-mail: mspinksau@yahoo.com.au

Robert Veroff
Department of Computer Science, University of New Mexico, Albuquerque
NM 87131, U.S.A., e-mail: `veroff@cs.unm.edu`

Walter Taylor
Mathematics Department, University of Colorado, Boulder, Colorado
80309–0395, USA, e-mail: `walter.taylor@colorado.edu`

George Voutsadakis
School of Mathematics and Computer Science, Lake Superior State
University, Sault Sainte Marie, MI 49783, USA, e-mail: `gvoutsad@lssu.edu`

About the Contributors

The Editor

Janusz Czelakowski is a Polish logician and mathematician. He graduated from the University of Wrocław in 1972. In 1975 he defended his PhD thesis in Mathematics at the University of Warsaw. He worked in the Polish Academy of Sciences till 1992. He is Full Professor at Opole University, Poland, and the head of Chair of Logic and Algebra in the Institute of Mathematics and Informatics. He was Visiting Professor in the Department of Mathematics of the Iowa State University in Ames, U.S.A. (1990/1991 and 1994/1995). His main interests include: abstract algebraic logic, universal algebra, action theory, infinitistic methods in the theory of definitions, algebraic methods in set theory. He is the author of almost 100 publications, including the following monographs: *Protoalgebraic Logics*, Kluwer 2001; *Freedom and Enforcement in Action. Elements of Formal Action Theory*, Springer 2015; *The Equationally Defined Commutator. A Study in Equational Logic and Algebra*, Birkhäuser 2015.

Contributors

Hugo Albuquerque graduated in Applied Mathematics and Computer Science at the University of Lisbon (Instituto Superior Técnico) in 2007, followed by a MSc from University of Aveiro in 2010. In 2016 he defended his PhD thesis in Pure and Applied Logic at the University of Barcelona. His main research interests are abstract algebraic logic and non-classical logics. He is a co-author of the papers *Compatibility operators in abstract algebraic logic*, *The strong version of a sentential logic* and *Note on the full generalized models of the extensions of a logic*, published in The Journal of Symbolic Logic, Studia Logica and Reports on Mathematical Logic, respectively.

Sergey Babenyshev is a Russian logician and mathematician. He graduated from the Krasnoyarsk State University in 1992. In 2004 he defended his PhD thesis *Metatheories of deductive systems* in Mathematics at the Iowa State University in Ames, USA, under the direction of Don Pigozzi. He is Assistant Professor at the Siberian Fire-Rescue Academy, Russia. His main interests include: modal propositional logics, abstract algebraic logic and universal algebra. He is the author of almost 20 publications, including the following papers: *Fully Fregean Logics*, Reports on Mathematical Logic (2003); *Unification in Linear Temporal Logic LTL*, Annals of Pure and Applied Logic (2011); *Linear Temporal Logic LTL: Basis for Admissible Rules*, Journal of Logic and Computation (2011) (both latter papers are in collaboration with V. Rybakov).

Clifford Bergman is the Barbara J. Janson Professor and Chair of the Department of Mathematics at Iowa State University in Ames, Iowa, U.S.A. He received his Ph.D. from the University of California at Berkeley in 1982 under the direction of Ralph McKenzie. (That makes him "first cousin twice removed from Don Pigozzi".) His primary interests are universal algebra, computational complexity, and cryptography. His work has been funded in the U.S. by the National Science Foundation, the National Institute of Justice, and other agencies. He is the author of the 2012 textbook *Universal Algebra: Fundamentals and Selected Topics*, CRC Press.

Isabel Ferreirim is a Portuguese mathematician. She graduated in Mathematics from the University of Lisbon in 1984. In 1992 she defended her PhD thesis in Mathematics at the University of Illinois at Chicago, USA. She has been an Assistant Professor at the University of Lisbon since 1993. In the period 1996-99 she was the Portuguese representative in COST Action 15 (Many-valued logics for Computer Science Applications), an international project funded by the EC. Until 2006 she was a member of the Algebra Center of the University of Lisbon (CAUL). Her main interests include abstract algebraic logic, universal algebra and lattice theory. Her main publications include (with W. J. Blok) *On the structure of hoops* (2000); and (with P. Aglianò and F. Montagna) *Basic hoops: an algebraic study of continuous t-norms* (2007).

Josep Maria Font is a Catalan mathematician and logician. He obtained his Ph. D. in Mathematics from the University of Barcelona in 1981, and he has been affiliated with this university throughout all his academic career, becoming a full professor of the Department of Mathematics and Computer Engineering in 2003. His research is in the areas of algebraic logic and non-classical logics. Among his other scientific interests are the applications of lattice theory and universal algebra to logic, history of logic, foundations of mathematics, history of mathematics, programming in (La)TeX, and editing and publishing mathematics. He has published over 70 papers and 2 books, *A General Algebraic Semantics for Sentential Logics* (Springer, 1996) written jointly with Ramon Jansana, and *Abstract Algebraic Logic — An Introductory Textbook* (College Publications, 2016).

Ramon Jansana is a Spanish logician. He graduated from the Autonomous University of Barcelona in 1974. In 1981 he defended his PhD thesis in Philosophy at the University of Barcelona. He works at this university since 1975 and became a full professor of Logic in 2003. His main interests include: abstract algebraic logic, non-classical logics, and topological dualities for classes of algebras related to logic. He is the author of over 50 papers and 3 books, including the monograph *A General Algebraic Semantics for Sentential Logics* (second revised edition, Lecture Notes in Logic 7, Association for Symbolic Logic, 2009) coauthored with Josep Maria Font. He is also a member of the Steering Committee of the meetings series TACL (Topology, Algebra, and Categories in Logic).

Alexandr Kazda is a Czech algebraist and complexity theorist. He graduated from Charles University in Prague in 2013. He was a postdoc at Vanderbilt University from 2013 to 2015. Since 2013, he works as a postdoc in the group of Vladimir Kolmogorov at IST Austria. Alexandr Kazda is interested in variants of the constraint satisfaction problem and the interplay between universal algebra and complexity theory.

Keith Kearnes is an American mathematician. He received his PhD from the University of California at Berkeley in 1988. He held visiting positions at the University of Hawaii, Vanderbilt University, Harvey Mudd College, and the University of Arkansas. He visited the Technical University at Darmstadt as a Humboldt Fellow (1993/95) and the University of Szeged as a Fulbright Fellow (2010). He held a permanent position at the University of Louisville (1998/00) until moving to the University of Colorado at Boulder, where he is Professor. His academic interests are in algebra, logic and combinatorics. He is the author or coauthor of more than 80 publications, including the monograph *The Shape of Congruence Lattices*, a Memoir of the American Mathematical Society, coauthored with Emil W. Kiss.

Marcin Kozik is a Polish mathematician. In 2004 he obtained a Ph.D. from Vanderbilt University with a dissertation supervised by Ralph McKenzie. Marcin Kozik is working in Universal Algebra and Constraint Satisfaction Problem; his the most important publications include: *Constraint Satisfaction Problems Solvable by Local Consistency Methods*, JACM 2014, characterizing CSPs of bounded width and *Robust Satisfiability of Constraint Satisfaction Problems*, STOC 2012, characterizing all CSPs with robust algorithms.

Antonio Ledda defended his PhD thesis in Logic in 2008. He is Associate Professor at University of Cagliari, Italy. His main interests include: algebraic logic, universal algebra, quantum logic. He is author of almost 60 publications.

Manuel A. Martins is a Portuguese logician and mathematician. He graduated from the University of Aveiro in 1994. In 2004 he defended his PhD thesis in Mathematics at the University of Lisbon. He has been Assistant Professor at Aveiro University since 2004, and he is a member of the Center for Research & Development in Mathematics and Applications. His main interests include: abstract algebraic logic, universal algebra, modal logic and specification of abstract data types. He is the author of more than 25 publications, including: *Behavioural reasoning for conditional equations* (with Don Pigozzi), published in Mathematical Structures for Computer Science in 2007; *Behavioral algebraization of logics* (with C. Caleiro and R. Gonçalves), published in Studia Logica in 2009; and *Deduction-detachment theorem in hidden k-logics* (with S. Babenyshev), published in Journal of Logic and Computation in 2014.

Ralph Nelson Whitfield McKenzie (born in Cisco, Texas, in October 20, 1941) is an American mathematician, logician, and abstract algebraist. He received his doctorate under the direction of Alfred Tarski from the University of California, Berkeley in 1966. He worked in the University of California, Berkeley and then since 1994 in Vanderbilt University. He is a fellow of the American Mathematical Society.

Selected works:

R. N. McKenzie with David Hobby: *The structure of finite algebras*, AMS 1988.

R. N. McKenzie with Ralph Freese: *Commutator Theory for Congruence Modular Varieties*, London Math.Society Lecture Notes, Cambridge University Press 1987 (after Wikipedia).

George F. McNulty is an American mathematician. He graduated for Harvey Mudd College in 1967 with distinction and honors in mathematics. He attained his doctoral degree under the direction of Alfred Tarski from the University of California, Berkeley in 1972. During this time, Don Pigozzi was a more advanced student of Tarski, fnishing in 1970. After three years of post-doctoral experience, George McNulty took a permanent faculty position at the University of South Carolina in 1975, rising through the ranks to the position of professor in 1986. McNulty was a Fulbright-Hays professor at the University of the Philippines (1982–83), an Alexander von Humboldt Research Fellow in Darmstadt (1983), Visiting Stanisław Ulam Lecturer (1998) at the University of Colorado, and he was named a Fellow of the American Mathematical Society with the original group of fellows in 2011. McNulty's mathematical interests lie at the confuence of algebra, mathematical logic, discrete mathematics, and computer science. He is the author of more than 60 publications.

Matthew Moore is an American mathematician. He graduated from The University of Arizona in 2007. In 2013 he defended his PhD thesis in Mathematics at The University of Colorado at Boulder. He worked as a non-tenure track assistant professor at Vanderbilt University in Nashville Tennessee until 2016. He is currently a Britton fellow at McMaster University in Hamilton Ontario. His main interests include universal algebra, theoretical computer science, and logic.

Tommaso Moraschini graduated in Philosophy at the University of Milano, and defended his PhD thesis in Logic at the University of Barcelona in 2016. Since then, he is a post-doc fellow at the Institute of Computer Science of the Czech Academy of Sciences. His main interests include (abstract) algebraic logic, universal algebra, substructural logic and category theory. He has already published in the best journals in these areas.

Katarzyna Pałasińska graduated with her M. Sc. in Mathematics from the Jagiellonian University in 1984 and with her Ph. D. from Iowa State University in 1994. Her PhD thesis *Deductive systems and finite axiomatization properties* was prepared under the supervision of professor Don Pigozzi. Since 1986 she is on the faculty of the Mathematics Department of the Cracow University of Technology, where she currently serves as a vice-chair for teaching. Some of her publications are: *Three-element nonfinitely based matrices*, Studia Logica 1994; *Finite Basis Theorem for Filter-Distributive Protoalgebraic Deductive Systems and Strict Universal Horn Classes*, Studia Logica 2003, *No matrix term-equivalent to Wroński's 3-element matrix is finitely based*, Studia Logica 2004, *Three-element non-finitely axiomatizable matrices and term-equivalence*, Logic and Logical Philosophy 2014.

Francesco Paoli is a Full Professor of Logic at the University of Cagliari, Italy. His main interests include algebraic logic, universal algebra, quantum logic. He is the author of 5 books and several research papers.

Anna Romanowska received her Ph.D. in Mathematics in 1973, and subsequently her Habilitation at Warsaw University of Technology. Since 1967, she has been based at Warsaw University of Technology, starting as an assistant and finishing as a full professor, additionally fulfilling a number of administrative duties. She has held visiting and research positions at various locations, including the Polish Academy of Sciences, Technische Hochschule Darmstadt, and Iowa State University. She has supervised 12 Ph.D. and more than 20 M.S. students. Her main scientific interests involve general and ordered algebraic systems, and their applications in combinatorics, geometry, logic, and computer science, in particular the theory of idempotent and entropic algebras (modes). She is the author of almost 100 publications, including three monographs (with J. D. H. Smith): *Modal Theory* (Heldermann Verlag, 1985), *Post-Modern Algebra* (Wiley, 1999), and *Modes* (World Scientific, 2002). She has organized a number of algebraic conferences, and has served as the editor of many special issues.

Antonino Salibra is a Full Professor of Computer Science at Ca'Foscari University of Venice, Italy. His main interests include algebraic logic, universal algebra, lambda calculus. He is the author of almost 70 publications.

Jonathan Smith received his Ph.D. at the University of Cambridge in 1975, and his Habilitation at the (then) Technische Hochschule Darmstadt in 1983. Since 1984, he has been based at Iowa State University, having been recruited by Don Pigozzi. He has held visiting positions at various locations, including Temple University, the Institute for Mathematics and its Applications, the University of Rome, Chonnam National University, and Warsaw University of Technology. He has supervised more than 20 Ph.D. students. His scientific interests focus primarily on algebra, combinatorics, and information theory, with applications in computer science, physics, and biology. He has published more than 160 papers, and has been the author or editor of a dozen books, including *Mal'cev Varieties* (Springer, 1976), *Modes* (with A.B. Romanowska, World Scientific, 2002), and *An Introduction to Quasigroups and Their Representations* (Chapman & Hall, 2006).

Matthew Spinks studied computer science and mathematics at Monash University, Australia, from where he obtained his Ph.D. in 2003. His research interests lie in non-Fregean logics, especially logics with strong negation and deductive systems arising from the ternary discriminator construct of general algebra. He currently works in industry.

Walter Taylor received a PhD in mathematics from Harvard University in 1968. Since then, he has been on the mathematics faculty at the University of Colorado, in Boulder, with emeritus status since 2007. His research has been in general algebra and related considerations in logic and foundations. In the last twenty years, his focus has been on algebras whose operations are continuous, in some topology of interest. He has around 900 pages of published research. He has been a visitor at the University of New South Wales (1975) and the University of Hawaii, Manoa (1977–78 and 1985). He is the co-author, with George McNulty and Ralph McKenzie, of *Algebras, Lattices, Varieties* (Wadsworth and Brooks/Cole, 1987). He is the author of *The Geometry of Computer Graphics* (Wadsworth and Brooks/Cole, 1991).

Robert Veroff is Professor Emeritus of Computer Science at the University of New Mexico. His primary research is in the area of automated deduction, including the application of automated deduction methods to a variety of problem domains. His work has led directly to the solution of numerous open questions in mathematics and logic, and in May 2003, he was awarded the University of New Mexico, School of Engineering, Senior Faculty Research Excellence Award. In addition to his research contributions, Professor Veroff has authored (with Paul Helman) a computer science text, *Intermediate Problem Solving and Data Structures: Walls and Mirrors*, Benjamin-Cummings

Publishing Co., Inc. 1986, and has edited a book on automated deduction, *Automated Reasoning and Its Applications: Essays in Honor of Larry Wos*, The MIT Press, 1997.

George Voutsadakis is a Greek and American logician and mathematician. He graduated with a Diploma in Computer Engineering and Informatics from the University of Patras, Patras, Greece in 1993. He obtained a M.Sc. in 1995 and a Ph.D. in 1998, in Mathematics, from Iowa State University, Ames, Iowa, under the supervision of Don Pigozzi with Thesis title *Categorical Abstract Algebraic Logic (CAAL)*. He also holds a Ph.D. in Computer Science from Iowa State University, completed in 2010 under the supervision of Giora Slutzki and Vasant Honavar, with Thesis title *Federated Description Logics for the Semantic Web*. He has held post-doctoral positions at Case Western Reserve University and New Mexico State University and has been teaching at Lake Superior State University since 2002. His main interests include algebraic logic, universal and categorical algebra and ordered structures. He is the author of several scientific publications, including: *CAAL: Equivalent Institutions*, Studia Logica, 74 (2003), 275–311; *CAAL: $(\mathcal{I}, N)$-Algberaic Systems*, Applied Categorical Structures, 13 (2005), 265–280, and *CAAL: Models of π-Institutions*, Notre Dame Journal of Formal Logic, 46 (2005), 439-460.

A Mathematical Life

Don Pigozzi

the best of all possible worlds
Leibniz
but not a perfect one
Tarski

Here is the story of my forty years as a professional algebraist and logician and also as a computer science dilettante. It is divided into seven parts chronologically. Each part begins with a narrative in which I describe my relationship with mentors, colleagues, collaborators, and students, often in very personal terms. I am happy to say that all these relationships have been cordial and many of them deeply satisfying.

I also mention in passing my research during these times, but the more detailed discussion of selected papers is left to a Publication part at the end of most sections for those interested. I have tried here to present at least enough detail so that the reader has a feeling for what I and my coauthors have tried to do. The great majority of my research is in universal algebra and algebraic logic. I expect that most of the likely readers of this autobiography will themselves have worked in these areas and may feel that the space devoted to computer science related topics is out of proportion. It is just because they may be unfamiliar with these topics that I have included this material. In any case it may be skipped if the reader chooses, which is why I have separated the description of individual papers from the narrative.

1 Early Life

I was born in Oakland California on the 29th of June, 1935. My first clear memory was listening to the attack on Pearl Harbor (the beginning of World

© Springer International Publishing AG 2018
J. Czelakowski (eds.), *Don Pigozzi on Abstract Algebraic Logic,
Universal Algebra, and Computer Science*, Outstanding Contributions
to Logic 16, https://doi.org/10.1007/978-3-319-74772-9_1

War II for us) on the radio with my parents. For the next four years I was absorbed with following the Allied Forces' slow but steady island hopping advance in the Pacific and, on the Atlantic front, through North Africa, Italy and then northern Europe.

Both my parents were second generation Americans, my mother of German ancestry and my father of Italian. Neither went beyond grammar school, although they were certainly capable of much higher education. My father became a skilled machinist, and went in business with his brother-in-law (my uncle) at the height of the depression two years before I was born. In the 1940s the business became quite successful, affording the family a comfortable lifestyle.

Although not educated themselves, my parents always stressed its importance. My brother Leo, who was seven years older than me, became an electrical engineer.

When I entered The University of California, Berkeley in 1953 I intended to become an engineer also. I soon realized my interests were more theoretical and graduated four and a half years later with a bachelors degree in physics.

I was always most interested in the courses that applied calculus to solve physical problems, but was uncomfortable about the connection. In high school algebra Cramer's rule was about the only thing I found interesting. What was the connection between this mechanical process and the solution of systems of linear equations? I asked my teacher and he said simply that"it just works". I naively concluded that mathematics was an experimental science, and this view wasn't completely dispelled until my senior year in college when I talked my physics advisor into letting me take the senior course in mathematical analysis in the mathematics department. It was taught by Professor Leo Breiman using Rudin's Mathematical Analysis textbook. It was a revelation to me. What I had naively thought was an empirical science was in fact based on a solid logical foundation.

After a short period satisfying my military obligation, I entered the graduate program in mathematics at Berkeley, with the intention of concentrating on analysis. I spent my first semester taking senior level courses that included mathematical logic from Leon Henkin and Boolean algebra from Robert Vaught. In the second semester I took graduate courses: algebra from Maxwell Rosenlicht and mathematical logic from William Craig. I enjoyed both courses, especially algebra and decided to make this my specialty.

Subsequent courses in algebra were less appealing and I drifted rudderless through the graduate program for a time making little progress and avoiding the PhD qualifying examinations. (At that time Berkeley put no pressure on graduate students to finish who did not require Departmental support.) I was suddenly seized by a sense of urgency and decided on logic as my research area. I passed the qualifying exam (on the second try) with mathematical logic as the optional topic, and asked Leon Henkin if he would take me on as a graduate student. He demurred because he already had several students and suggested I ask Alfred Tarski, who at that time was looking for a student to help him with a book on cylindric algebras that he was writing with Henkin and Donald Monk. Tarski agreed, and thus begun a close relationship that lasted five years.

2 Years 1964–1972

At this time Tarski had become especially interested in what he called *The General Theory of Algebra*, more commonly called *Universal Algebra*, UA (a term he didn't like). Cylindric algebras were introduced in the late 1950s by the three authors as an algebraic model of the first order predicate logic, and thus a natural topic for universal algebra. He agreed to write the introductory chapter for the book on this topic, based on a graduate course he was teaching at the time and which I was taking. The cylindric algebra part of the book was to be written by the other two authors. My job was perceived as assisting Tarski with this introductory chapter.

The finished typescript manuscript had been prepared in Boulder and mailed to Berkeley. It included even the color coding for special fonts; this was in that primitive time before TEX. It arrived with a note from Monk saying he expected that Tarski, with my help, would rewrite the introductory chapter as seen fit, and review the cylindric algebra chapters, making any presumably minor changes Tarski cared to; he expected this would take a few months. In fact it took almost five years, during which the entire book was rewritten.

I feel an obligation at this point to record for historical reference how Tarski worked with his graduate students, at least in my case. He always worked at home in the afternoons; mornings were reserved for working in his garden. He came on campus only for classes and seminars and for administrative duties. I recall meeting him the evening of a university holiday and remarking that Robert Vaught and several other faculty members were in their offices working. He was outraged: in Europe he said only clerks work in their office– academics always work at home.

On the weeks we worked we would meet normally two or three times a week, at his house at about seven or eight in the evening. Although not infrequently I would come earlier and he would take me out to one of his favorite restaurants for dinner before we went to work. We would then work in his downstairs office until about midnight. We would then go upstairs to the kitchen where Mrs. Tarski would have prepared a midnight snack for us. Although visibly tired before this break, he would seem to revive and would ask me if I was up to working more. Of course I would agree, though dreading it. Then, with the help of some kind of stimulus pills, he would be very alert for several hours while I was barely functional. Many times we would see the sun rising before he agreed we should stop.

During this time we would systematically work through the Boulder manuscript, revising it as he saw fit. He was interested only in the formulation of the theorems and especially the interconnecting narrative text. The proofs of the extant theorems he left for me to check and, as he gained more confidence in me, to reformulate at my discretion. He would also leave to me the proofs of the new theorems that he added. He felt that the real understanding of the material is imparted in the interconnecting text, and he would spend a longtime reflecting on this while smoking his cigarillos. He would then dictate the text to me. We would discuss it, modify it as needed, and then I would record it. He always liked to have a native English speaker to help him at this stage because he was not completely confident of his command of the language.

My real work was between these meetings. I would edit the dictation I had taken and give it to the secretary of the Logic and Methodology group to type up. I would also have to prepare for the next meeting by working out the proofs of the modified and new theorems and looking up references for our next meeting. Dale Ogar, the secretary, was an excellent technical typist. She was in large part responsible for preparing the final manuscript for submission to the publisher, and we became good friends. I was happy to have her available to type my thesis and grateful for her offer to move it through the university bureaucracy for me while I was in Indiana on my first academic appointment.

There were long periods when Tarski was away from Berkeley and we did not meet. During these hiatuses I would catch up on my work on the cylindric algebras book and work on my thesis, which not surprisingly was on a topic in cylindric algebras, a problem Don Monk gave me: Which classes of cylindric algebras have the *amalgamation property*, that is, given two algebras of the class with isomorphic subalgebras, under what conditions can they be both isomorphically embedded in a third algebra of the class such that the two isomorphic images of the subalgebras coincide? This algebraic problem is closely related to the Craig Interpolation Theorem of first-order predicate logic.

Don Monk was an enormous help to me—in a sense my defacto thesis advisor; Tarski wasn't much interested in the problem. I finished the first draft of the thesis during the summer of 1969, just after finishing my work on the cylindric algebra book and finally sending the completed manuscript to Boulder. The problem was that I had already accepted a one year appointment at Indiana University under the assumption that my thesis would be completed. But Indiana graciously honored their offer and I spent the fall semester of the 1969-70 academic year in Bloomington.

Tarski had accepted a invitation to give a series of lectures at the Pennsylvania State University prior to the start of the academic year. When the opportunity arose Tarski liked to take long motor trips and was eager to see the Dakota badlands (Leszek Kołakowski had told him how impressed he was by them on a motor trip to Berkeley earlier that summer). So Tarski, Mrs. Tarski, and I drove to State College Pennsylvania in my new Ford Maverick, and after the conference we drove to Columbus Ohio where they took a plane to Berkeley, and I went on to Bloomington to join my friend and fellow Berkeley graduate student Ralph Seifert in a new life.

I spent the year teaching and writing the final version of my thesis, which I sent on to Dale Ogar to be typed and submitted to Tarski. My appointment was extended for the 1970-71 year, but I spent the fall of 1970 in Berkeley on a research assistantship to help in the final production of the cylindric algebra book. Indiana did not extend my appointment again, and I accepted an offer of a position at Iowa State University in Ames for the 1972-73 academic year where I remained for 31 years until my retirement in 2002.

Publications: While working on the introductory chapter in the cylindric algebra book Tarski had considered the three operations on classes of algebras S, H, P forming subalgebras, homomorphic images, and direct products, and all their possible compositions. They form a finite ordered set under set theoretical inclusion. When I had just started working for him, he asked me to describe its structure. I instinctively sensed that this was my real qualifying exam, and with considerable anxiety worked hard on the problem which I was fortunately able to solve. This was my first publication (Pigozzi, 1972b) that appeared the same year as my thesis (Pigozzi, 1972a). My other publication of this period (Pigozzi, 1974) dealt with joins of logical theories. It was known that the first-order theory of a single equivalence relation is decidable while that of two equivalent relations is not. This paper shows that the join of two decidable equational theories with disjoint languages is always decidable. On the other hand, the result fails if the languages are not disjoint, even if the two theories coincide on the common part of their languages.

3 Years 1972–1979

The person mainly responsible for me coming to Iowa State was Alexander Abian. He was an analyst turned set theorist. Federico Sioson, a good friend of Donald Monk, with a Berkeley doctorate in universal algebra from Alfred Foster, had come to Iowa State not long before me. I never met him since he tragically died of cancer after only a few years. He together with Abian were responsible for establishing the courses in foundations then being offered in the mathematics department, and I had the opportunity to teach graduate courses in algebra and logic.

Computer science had been part of the mathematics department, but had been split off into a separate department shortly before I arrived. One of those who went into the new department was George Strawn. He was Sioson's only doctoral student at Iowa State. After writing a thesis in universal algebra his interests had turned to computer science. But our research interests were quite compatible. He was responsible for me spending two years in the computer science department as an adjunct professor quite a bit later.

In 1974 I was promoted to associate professor, and in the summer was invited by John Doner and Richard Büchi to speak in the computer science department seminar at Purdue University. John had also been a student of Tarski, writing a thesis on ordinal numbers that Tarski highly regarded. He finished shortly before me. While on a postdoctoral fellowship at the Rand Corporation he obtained a result in automata theory that complemented a well known result of Büchi, who was responsible for John's appointment in West Lafayette at about the same time I came to Bloomington. We became close friends (I was the best man at his wedding). My talk was on equational logic; it became the basis of a manuscript that was published as a technical report of the Purdue computer science department (Pigozzi, 1975).

In 1978 the Department with my encouragement hired another Tarski graduate student, Roger Maddux. His thesis was on a different algebraic model of predicate logic, relation algebra. It had a much longer history than cylindric algebras and was of more interest to Tarski at that time, partly I think because of its structural elegance. Soon after this we acquired another Berkeley logician, Richard (Dick) Epstein. His thesis was in recursive number theory, but at the time he was developing an interest in philosophical logic.

I had enough of cylindric algebras after working exclusively on them for five years and turned my attention to a quite different topic under the influence, among others, of George McNulty. George was another student of Tarski who was at the University of South Carolina. Combinatorial universal algebra is like combinatorial group theory except that instead of word problems, it deals, among other things, with the base decidability of finitely based equational theories and equivalently their varieties of models. (A *variety* is the set of all models of a set of identities, called a *base* of the variety; the *equational theory* of the variety is its largest base, i.e., the set of all its identities. It can also be thought of as any base together with the deductive apparatus for deriving all identities from the base using Birkhoff's rules.)

The key to my results in this area is the notion of a *universal variety*. Roughly speaking an equational theory is universal if any equational theory, under certain restrictions, can be faithfully interpreted in it. This topic, together with work on the structure of equationally complete (i.e., minimal) varieties, locally finite varieties, and questions about the finite axiomatization of various varieties took up most of my research time until I went on my first sabbatical leave during the 1978-79 academic year at the invitation of George Grätzer and the universal algebra group in Winnipeg Manitoba. It was an invigorating experience in several regards: during one two week stretch in the winter the temperature remained at 40 degrees below zero (Celsius and Fahrenheit). Academically it was even more so. The UA group was very hospitable and I was quickly accepted as a member.

I met two mathematicians who like me were visiting: Wilhelm (Wim) Blok and Peter Köhler with whom I immediately bonded and who brought me back to algebraic logic (AL). Wim Blok had just come to Winnipeg on a postdoctoral fellowship after finishing his thesis on varieties of interior algebras under

Philip Dwinger at Amsterdam and Chicago. Peter Köhler was on leave from Giessen University in Germany with his wife Adelheid and young son; Wim and I were invited to the house they were renting several times. His thesis was on the semigroup of varieties of Brouwerian semilattices, and he had collaborated with Blok on a paper on the semigroup of varieties of generalized interior algebras.

Because Wim Blok and I were both single and together in an unfamiliar environment we became very close friends–frequently going out to dinner together in the evenings and sharing personal things about our lives. One incident with Wim at this time was particularly memorable. During the interval of minus 40 degree temperature in Winnipeg Wim and I drove downtown one evening to attend an opera (the Flying Dutchman). The streets were ice covered when we started home and were incredibly slick. I could only drive a couple of miles an hour and could hardly control the car at that speed. I was just barely able to stop behind a line of cars waiting to turn onto the street back to the university. The next two cars also were able to stop, but the one after that couldn't and collapsed the line like an accordion. The damage to all the cars but mine was minor, but I had stopped behind a tow truck with it's boom down, which pushed through the grill and punctured the radiator and totalled the Maverick, the one I drove to Penn State with the Tarskis. I was sorry to lose that car with all its memories, like the burn on the dashboard that Tarski made when he missed the ashtray with his cigarillo.

So began a personal friendship and professional collaboration that lasted over twenty years until Wim's death in a tragic automobile accident in 2003. After the year in Winnipeg Wim went to Simon Fraser University in Vancouver on another postdoctoral fellowship, and then a visiting instructorship at the University of British Columbia where he met his future wife Mary. He was looking for a permanent position after Vancouver, and I was eager to have him at Iowa State, but he accepted a competing offer from the University of Illinois at Chicago where his thesis advisor Philip Dwinger had a part time professorship. I was disappointed but understood that Chicago was the right place for him. They had a more well formed algebra and foundations program than we did, and he was very comfortable in Chicago, having spent much time there with Dwinger working on his thesis. Besides that, he was a big city guy, having essentially grown up in Amsterdam. I understood he and Mary would never be happy in a small town like Ames.

Research-wise my year in Winnipeg was very fruitful. Blok, Köhler and I wrote or began work on several papers dealing with varieties with equationally definable principal congruences. This was an area of universal algebra that was getting a lot of attention in the literature currently and proved to be important in algebraic logic because of its connection with the deduction theorem.

I was also able to work with other members of the UA group on topics not connected to algebraic logic. Jiřy Sichler was a long time member of the Manitoba UA group, and Michael (Mick) Adams was another visitor like me.

Both became good friends, especially Jiřy. He was an avid outdoors-man and invited me back the following year for a canoe trip to an isolated lake in the wild lake district of Manitoba.

My collaboration with Peter Köhler did not extend beyond our time together in Winnipeg. He left academics and began a successful career in software development. However we did keep in touch by email over the years. I stayed one night with Peter and Adelheid at their home in Weilburg-Kubach north of Frankfurt while I was in the area attending a conference. Recently they stayed with us several days in Oakland when they came to the West Cost to visit their son who was working in Washington State. We took them sightseeing in San Francisco, and they then picked up the large Cruise America van they had rented and drove it up to Washington.

Publications: S.I. Adjan and M.O. Rabin (see (Pigozzi, 1976a) for references) show that for every property of groups satisfying a certain condition called the *Markov condition*, the set of all finite presentations that define a group with the property fails to be recursive. The Markov condition is surprisingly simple: a property satisfies it if (I) there exists at least one group with the property, and (II) there exists a group which not only fails to have the property, but cannot be isomorphically embedded in any group with the property.

In (Pigozzi, 1976a) we give a Markov-like condition for showing a property of arbitrary equational theories is base undecidable. The key is finding a condition on equational theories analogous to (II) of the Markov condition. This is based on the notion of a universal equational theory, more specifically a *normal universal theory*, which is defined and studied in (Pigozzi, 1979b). The universality of the equational theory (equivalently the variety) of quasigroups is established in (Pigozzi, 1976b).

The paper (Pigozzi, 1981b) and its sequel (Pigozzi, 1981c) investigate the structure of equationally complete varieties that either fail to be locally finite or are locally finite but are neither congruence-permutable nor congruence-distributive. It turns out that the structure of such a variety can, with some reservations, be as complex as that of an arbitrary variety of the same kind. In contrast the structure of locally finite varieties having at least one of the two congruence conditions had previously been shown to be much simpler to describe.

Two other papers at this time, (Pigozzi, 1979a) and (Pigozzi, 1981a) investigate the problem of the existence of a finite basis for the identities of equationally complete locally finite varieties and of finite groupoids. For example, one of the results of the second paper is that the category of all non-finitely based groupoids and all homomorphisms between them included a subcategory isomorphic to the category of all finite groupoids.

The two papers (Adams et al., 1981) and (Pigozzi and Sichler, 1985) I wrote with Sichler and Adams were in a different vein: endomorphisms of direct products of bounded lattices and homomorphisms of partial and of complete Steiner triple systems.

A variety of algebras has *Equationally Definable Principal Congruences* (EDPC) if there are finitely many equations $p_i(x,y,z.w) \approx q_i(x,y,z,w)$, $i = 1,\ldots,n$, such that, for each algebra in the variety and every quadruple a,b,c,d of its elements, c is congruent to d by the congruence generated by the pair a,b, in symbols $c \equiv d\,\Theta(a,b)$, if and only if $p_i(a,b,c,d) = q_i(a,b,c,d)$ for each i. It is not difficult to show that for all a,b,c,d the join of the congruences $\Theta\bigl(p_i(a,b,c,d), q_i(a,b,c,d)\bigr)$ is the smallest congruence Ψ such that $\Theta(c,d) \subseteq \Theta(a,b) \vee \Psi$, in other words, it is the *dual pseudo-complement* of $\Theta(a,b)$ *relative* to $\Theta(c,d)$ in the lattice of congruences of the algebra. It follows without difficulty that if a variety has EDPC, the join-semilattices of finitely generated, i.e., compact congruences of its members forms a dually pseudo-complemented join semilattice, and in fact these two properties are equivalent.

Dually pseudo-complemented join-semilattices are distributive, consequently, varieties with EDPC are congruence-distributive. This had been an open problem of some interest in the UA literature. These results are contained in (Köhler and Pigozzi, 1980), coauthored with Peter Köhler.

A comprehensive list of varieties with EDPC is given in (Blok and Pigozzi, 1982). There are two principal kinds: discriminator varieties and varieties arising from the algebraization of deductive systems of classical and non-classical logic that satisfy some reasonable version of the deduction theorem. In fact every algebraizable logic of this kind gives rise to a variety with EDPC. Discriminator varieties are both congruence-permutable and semisimple and comprehend all EDPC varieties with these two properties. In the two papers (Blok and Pigozzi, 1982), with Blok, and (Blok et al., 1984), with Blok together with Köhler, two questions are investigated: (I) Can the characterization of congruence-permutable and semisimple EDPC varieties by means of the discriminator function be extended in a useful way to classes of congruence-permutable EDPC varieties that are not semisimple? (II) How comprehensive is the class of EDPC varieties that arise from logic–can it be characterized by some natural algebraic conditions?

A *discriminator variety* has a ternary polynomial function P, called a *discriminator function* such that $p(a,b,c)$ equals c if $a = b$, and a otherwise. Blok and I wrote two more papers, (Blok and Pigozzi, 1994a) and (Blok and Pigozzi, 1994b), over the next ten years investigating various generalizations of the discriminator function, for example, the *quaternary discriminator*, or *if-then-else function*, p where $p(a,b,c,d)$ equals c if $a = b$ and d otherwise.

4 Years 1981–1988

In the fall of 1980 Leszek Szczerba came to Ames as a visiting professor for the academic year at the invitation of Dick Epstein. He came with his wife Bożena and their two sons Stanisław and Włodek (at Epstein's suggestion we called them Stan and Wally). Leszek's field was the foundations of geometry. He had been in Berkeley visiting Tarski where Epstein and Roger Maddux got to know him well, and so did I while he was in Ames. Bożena made great pierogi (a kind of filled dumpling very popular in Poland), and I always looked forward to an invitation to dinner with the family. During his time in Ames the Solidarity movement exploded in Poland and martial law was declared. Leszek considered applying for refugee status to remain in the US, but the family was reluctant so they returned home at the end of the academic year.

The following year we got the okay to hire another person in universal algebra and we brought in Clifford (Cliff) Bergman, and a few years later Jonathan Smith. Cliff was another Berkeley graduate and a student of Ralph McKenzie, and had just completed a postdoctoral year in Hawaii. Jonathan came to us from Darmstadt, Germany where he had been for several years after getting his doctorate from Cambridge. A two semester graduate universal algebra sequence was added to the curriculum in 1993. I taught a rather traditional course the first semester based on my notes from Tarski's course at Berkeley. The topics I chose in the second semester were influenced by my research.

Cliff and Roger Maddux, like me, were interested in computer science and the three of us had held informal joint seminars with George Strawn and other members of the CS Department. I asked George if CS would be interested in joining Math in funding a joint position in the two departments. He said no, but they would like to offer me a halftime position. CS's problem was that having just split off from Math they had very few senior positions filled in their new department. I was intrigued with the idea, but told him I knew very little basic computer science–I didn't ever know how to program. George said that if Math agreed, I could have a halftime adjunct position to teach through their undergraduate curriculum, one course a semester, and they would send a graduate student over to Math to take my undergraduate course. At the end of that time I could decide if I wanted a halftime professorship in CS. Our department agreed, and starting in the fall of 1984 I taught in CS for five semesters, beginning with the most basic programming course for non-majors and ending with the senior level compiler course. I also audited more advanced CS courses.

It was a very difficult two and a half years; the classes were large and I was always just a step ahead of the students. The most stressful times came when programming assignments were due. Occasionally the teaching assistant couldn't figure out what was wrong with a student's program and would send him or her over to me to debug it. I always broke out into a cold sweat when this happened, but thank goodness it didn't happen often and much to my surprise I was always able to solve the problem. The compiler course was a particular challenge. I found the textbook almost uncomprehensible, but I was able to manage things on my own pretty well until one night, near the end of the term, while I was in the student union trying to prepare the next morning's lecture on a critical point, I realized I had no idea what to say.

Thank goodness, a former colleague in the CS department, who was back on a visit and happened to be in the union, was able to explain the problem to me. The CS department had given me two excellent graduate students and thanks to them I was able to get through the course successfully.

In spite of the time and stress I appreciated the opportunity to learn a lot of basic computer science in a very short time, and was gratified to know I was now able to write an actual compiler. But I had been unable to do any research during this time and was greatly relieved to know I would be able to escape gracefully by going on sabbatical leave in Chicago for the 1986-87 academic year.

I spent the summer before Chicago in Europe. Wim Blok and I were invited by Hajnal Andréka, Istvan Németi, and Ildiko Sain to participate in the summer algebra seminar in Budapest. Following that I went on to Warsaw, at the invitation of Anna Romanowska, to speak in the theoretical computer science seminar of the Technical University of Warsaw. I gave the first public presentation of Blok and my definition of an algebraizable logic. Jeszek Szczerba graciously offered to drive me through western Poland to meet a number of logicians and algebraists whose work I knew well but had never met. The highlights of the tour were a visit to Andrzej Wroński in Kraków, Piotr Wojtylak in Katowice, and to Ryszard Wójcicki, Janusz Czelakowski and Wiesław Dziobiak at the Section of Logic of the Polish Academy of Sciences in Łódź. Czelakowski would in time become one of my closest collaborators in research on algebraic logic. Before returning to the US I visited Aldo Ursini in Siena, another universal algebraist whose work I knew well.

Chicago is a great place to visit and The University of Illinois at Chicago (UIC) a great place too for a logician and algebraist to work. At the time the Department included, besides Wim Blok, also John Baldwin, David Marker, and Joel Berman; in addition the Philosophy Department contained several prominent people working in philosophical logic. Berman was kind enough to let me have his apartment, which was available at the time, for the year, and within easy walking distance from the campus.

During my time in Chicago Wim and I were finally able to complete our long delayed paper on a general framework for investigating the precise connection between algebra and logic. This was finally submitted to the Memoirs of the AMS in April of 1987 and appeared in January of 1989 under the title *Algebraizable Logics* (Blok and Pigozzi, 1989). We were gratified to learn that the first edition sold out shortly after publication. The monograph has recently appeared in the Classic Reprints series published by the *Advanced Reasoning Forum*, a series edited by my former colleague at Iowa State Dick Epstein. Wim and I had already started work on it while in Winnipeg in 1978. The earliest version was intended to be the first part of our paper on the deduction theorem in algebraic logic, which itself was still only in manuscript form when the EDPC I paper (Blok and Pigozzi, 1982) was submitted in July of 1981. (A reference to this manuscript can be found in the (Blok and Pigozzi, 1982) bibliography.)

The deduction theorem paper, not to be confused with *Local Deduction Theorems in Algebraic Logic* (Blok and Pigozzi, 1991b), also has a long and troubled history. It was rewritten several times and finally submitted to the *Bulletin of Symbolic Logic* of the Association for Symbolic Logic. It was conditionally accepted, but the referee's suggestions were so cogent that we determined the last part had to be completely re-written. I worked for a long time expanding the discussion of the history of the subject and adding many new references. I sent the manuscript to Wim in 1995 with the recommendation to make any additional changes he saw fit and then send it on to the editor without having to consult me. But he died before he could complete his work. Thanks to Joel Berman I got his files and with some effort could see that the paper was near completion. I was able to interpret the notes he had made and trace most of the new but frequently incomplete bibliographic references he intended to insert. But these revisions have not yet been completely implemented. Considering the time that has elapsed since the paper was first submitted, I have been reluctant to resubmit it to the Bulletin of Symbolic Logic or any other journal. The editors have graciously offered to include it in this volume, but because it still requires substantial work before it is ready for publication I have decided to decline the offer in order not to further delay the appearance of the present volume.

Following my sabbatical I remained in Chicago in the autumn of 1987 as a visiting professor (teaching linear algebra and computer programming). During the intervening summer Walter Carnielli invited me to teach a short course on algebraic logic at the State University of Campinas, Brazil. I was thankful to Dick Epstein, who had just returned from an extended stay in Brazil, for arranging this invitation. This was the first of many trips I would make to Latin America over the years. Research-wise, Wim and I continued working on the EDPC papers in which various generalizations of the discriminator function are investigated. This in turn led to my first foray into computer science research: *Data Types Over Multi-valued Logics* (Pigozzi, 1990) and *Equality-Test and If-Then-Else Algebras: Axiomatization and Specification* (Pigozzi, 1991). I also obtained an analog for quasivarieties of Baker's famous result that every finite member of a congruence-distributive variety is finitely based (Pigozzi, 1988a).

Because Wim was now married our relationship wasn't as close as in Winnipeg. But I was soon just as comfortable around Mary as I was with him. The three of us frequently enjoyed dinner together, at their home or a restaurant. Wim loved good food and wine and kept up to date on the best restaurants in the Chicago area. For many years I would drive the 350 miles from Ames to Chicago for Thanksgiving dinner with them and then go with Wim to a concert or opera. Wim was an excellent violinist and violaist, and played the viola regularly in a string quartet among friends. He also played the viola in a local orchestra organized and conducted by his colleague at UIC David Tartakoff. For several years he and Mary rented a cottage in Lake Michigan in Indiana for a week or two and would often invite me to join them for a

weekend. We would take short trips in the nearby countryside: to the Indiana
Dunes and the fall nesting area of the Sand Hill Cranes. My times with Wim
and Mary are some of my fondest memories. I find myself often thinking of
Wim and realizing how much I miss him.

Publications Algebraic logic in the modern sense began with Tarski's
1935 paper *The foundations of the calculus of systems*. The sentences of *classical sentential logic* (CSL) form the universe of an algebra of sentences whose
operations correspond to the connectives of the logic. Two sentences φ and
ψ are equivalent if the two implications $\varphi \to \psi$ and $\psi \to \varphi$ are theorems in
CSL (i.e. tautologies). This defines a congruence relation Θ on the sentence
algebra and the resulting quotient algebra is a Boolean algebra, the *Tarski-
Lindenbaum algebra* (the TL) of CSL. A sentence φ is a theorem of CSL if and
only if the equation $\varphi/\Theta \approx \top/\Theta$ is an identity in the TL algebra. Conversely
a deductive system for CSL can be constructed from any set of axioms for
Boolean algebra, together with the Birkhoff laws for equational logic.

A number of nonclassical sentential logics can be algebraized in this
way: the intuitionistic logic of Heyting, the multi-valued logic of Post and
Łukasiewicz, and the modal logics **S4** and **S5** of Lewis. There is a correspondence between theorems and algebraic identities that allows the deductive
apparatus of each algebraizable logic to be interpreted in the equational theory of the TL algebra of the logic. Many higher order metalogical properties
also have natural algebraic interpretations, for example the deduction theorem as EDPC.

There are however important sentential logics to which this method does
not apply, such as the Lewis systems **S1, S2, S3** which do not have the
Rule of Necessitation $\varphi/\Box\varphi$. In this case the equivalence relation Θ defined
as for CSL is not a congruence since $\vdash \varphi \to \psi$ does not imply $\vdash \Box\varphi \to \Box\psi$.
Other examples arise from considerations of strict (non-material) forms of
implication: the systems **R** and **E** of *relevance* and *entailment* introduced
by Anderson and Belnap where there exist theorems φ and ψ for which
$\varphi \to \psi$ is not a theorem. The question naturally arises if any of these can be
algebraized by some method other than the classical one, or if they are in
some sense inherently non-algebraizable. What is needed is a precise, formal
definition of algebraizability that is generally accepted, such as the formal
definition of a recursive function is accepted as capturing the intuitive notion
of a constructible function. This is what Wim and I set out to do in the
monograph *Algebraizable Logics* (Blok and Pigozzi, 1989). But we must leave
it to the reader to decide if we have been successful.

A class K algebras is an *equivalent algebraic semantics* for a sentential
logic $\mathcal{S}$ if the consequence relation $\vdash_{\mathcal{S}}$ can be interpreted in the (semantical)
equational consequence $\models_{\mathsf{K}}$ of K and vice-versa, and moreover the two interpretations are inverses of one another in a natural sense. By definition $\mathcal{S}$ is
algebraizable if it has an equivalent algebraic semantics. Equivalent algebraic
semantics are unique in the sense that they all generate the same quasivariety.
This definition is not very useful from a practical point of view since it is not

intrinsic in the sense it requires a priori knowledge of the equivalent algebraic semantics. Two intrinsic characterizations are given in the monograph.

Let T be a *theory* of a sentential logic $\mathcal{S}$, i.e., a subset of sentences closed under the consequence relation of $\mathcal{S}$. It is natural to think of sentences φ and ψ as being equivalent with respect to T if either one can be replaced by the other in any third sentence without affecting that sentence's truth or falsity relative to T, i.e., whether or not the sentence is contained in T. This is a congruence relation on the sentence algebra, called the *Leibniz congruence* of T (because of the connection with Leibniz's definition of truth in second order logic), and is denoted by ΩT. It turns out to be the largest congruence compatible with T in the sense that if φ is in T then so is every sentence equivalent to φ. In order for $\mathcal{S}$ to be algebraizable it is necessary that ΩT properly includes $\Omega T'$ whenever T properly includes T', i.e., the Leibniz operator is one-one and order preserving on the lattice of theories. The condition is also sufficient provided a certain other natural condition holds. Although natural the condition is not easy to establish, but the characterization is useful to showing a given logic fails to be algebraizable.

The second intrinsic characterization is useful for establishing algebraizability. A logic is algebraizable if there exists a finite set of sentences in two variables that collectively have many of the properties of the biconditional $x \leftrightarrow y$ of classical logic. It is obtained by adding one additional condition to the definition of *equivalential logic* introduced and studied by J. Czelakowski, T. Prucnal, and A. Wroński. Using these two characterizations we show that there is a large class of modal logics, including **S1, S2, S3**, that are not algebraizable in our sense. Entailment logic **E** also fails to be algebraizable, but Relevance logic **R** is algebraizable.

There is also a large class of logics that, like **S1, S2, S3**, need not be algebraizable, but are amenable to most of the standard methods of AL. A logic $\mathcal{S}$ is *protoalgebraic* if for every $\mathcal{S}$-theory T, every pair of ΩT-equivalent sentences φ and ψ are interderivable relative to T, i.e., $T, \varphi \vdash_\mathcal{S} \psi$ and $T, \psi \vdash_\mathcal{S} \varphi$. It turns out that $\mathcal{S}$ is protoalgebraic if and only if the Leibniz operator is monotonic on the lattice of theories. Their algebraic counterpart are not algebras but the *matrix models* of the logic, that is pairs $\langle \boldsymbol{A}, F \rangle$ where $\boldsymbol{A}$ is an algebra over the same language as the sentence algebra of $\mathcal{S}$ and F is a subset of the universe A of $\boldsymbol{A}$ closed under the interpretation in $\boldsymbol{A}$ of the consequence relation of $\mathcal{S}$. F is called an $\mathcal{S}$-*filter* of $\boldsymbol{A}$. An $\mathcal{S}$-theory is a filter of the algebra of sentences. Protoalgebraic logics are studied in detail in (Blok and Pigozzi, 1986b) and in the monograph *Protoalgebraic Logics* by J. Czelakowski.

The three publications just discussed have been followed by a large literature, by a number of different authors, on a new kind of algebraic logic where the focus is on the nature of the connection between logic and algebra in general rather than on specific logical systems. The term *Abstract Algebraic Logic* (AAL) has been suggested for this new line of research.

I returned to universal algebra with the paper *Finite Basis Theorems for Relatively Congruence-Distributive Quasivarieties* (Pigozzi, 1988a). Consider a quasivariety Q. A congruence relation Θ on an algebra A (not necessarily in Q) is said to be *relative* to Q, a Q-*congruence*, if the quotient algebra A/Θ is a member of Q. The set of relative congruences $Con_Q A$ forms a complete lattice that contains the identity congruence only if $A \in$ Q. Q is *relatively congruence-distributive* if $Con_Q A$ is distributive for every A in Q. A subquasivariety R of Q is called a *relative subvariety* of Q if it is of the form V ∩ Q for some variety V, i.e., a base for R can be obtained by adjoining only identities to a base for Q.

THEOREM I. *Every finitely generated and relatively congruence-distributive quasivariety is finitely based.*

THEOREM II. *Let Q be a relatively congruence-distributive quasivariety. Then every finitely generated relative subvariety of Q is finitely based.*

These are the main results of (Pigozzi, 1988a). The first theorem generalizes Baker's theorem for congruence-distributive varieties in the sense that the latter is an easy corollary.

The property of being relatively congruence-distributive need not be inherited by subquasivarieties. In particular, a quasivariety may generate a congruence-distributive variety without itself having the property, and vice-versa; examples of both kind are easy to find. Some problems pertinent to finite basis theorem for relatively congruence-distributive quasivarieties have been characterized by J. Czelakowski and W. Dziobiak in their paper *Congruence Distributive quasivarieties whose finitely subdirectly irreducible members form a universal class.*

Theorems I and II are closely related, in fact each is a corollary of the other. But they can apply in quite different situations. Let K be any class of algebras and let Qv K be the quasivariety generated by K. When K is included in a quasivariety Q, let Va_Q K be the relative subvariety of Q generated by K. In general Qv K is strictly smaller than Va_Q K, and one may be relatively congruence-distributive while the other is not. The difference between the two theorems can be clearly seen by comparing their syntactical forms. To simplify matters let us consider a single finite algebra A. If A is contained in a relatively congruence-distributive quasivariety Q, then by Theorem II the identities of A are logical consequences of the finite set of quasi-identities of A. However, if Qv A itself is relatively congruence-distributive, then the quasi-identities of A are finitely based by Theorem I.

The applicability of the various finite basis results is limited, at least in comparison with analogous results for varieties, by the difficulty encountered in establishing relative congruence-distributivity in concrete situations. No condition like the existence of Jónsson terms is known to characterize the property. Such a condition, if it existed would likely be radically different from the familiar Mal'cev-style conditions since relative congruence-distributivity is not inherited by subquasivarieties. However some general methods are known to work in limited situations. We refer the reader to (Pigozzi, 1988a) for references.

In Section 7 of (Pigozzi, 1988a) we show how any algebra $\boldsymbol{A}$ with at least two elements can be transformed into an algebra $\boldsymbol{Et\,A}$ by adjoining a (Boolean) equality-test operation to $\boldsymbol{A}$. The quasivariety generated by $\boldsymbol{Et\,A}$ is relatively congruence-distributive, but the reduct obtained by discarding the equality-test operation need not be. The single-sorted algebra $\boldsymbol{Et\,A}$ is admittedly an artificial construction. But this type of construction is natural for the multi-sorted algebras that play a prominent role in software development, which is the topic of the two papers *Equality Test* and *If-Then-Else Algebras: Axiomatization and Specification* (Pigozzi, 1991) and *Data Types Over Multi-Valued Logics* (Pigozzi, 1990).

A *data structure* $\boldsymbol{S}$ is a multi-sorted algebra in which every element is denoted by a *ground term*, i.e., a term without variables. An (*abstract*) *data type* is an isomorphism class of data structures. An sxiom set Γ is an *initial* (*final*) *specification* of the data type of $\boldsymbol{S}$ if Γ is a set of sentences in some formal language describing $\boldsymbol{S}$, and $\boldsymbol{S}$ is the initial (final) object in the category of models of Γ. Alternatively, Γ is an initial specification of $\boldsymbol{S}$ if $\boldsymbol{S}$ is a model of Γ and every ground identity of $\boldsymbol{S}$ is a logical consequence of Γ; final specifications can be similarly characterized. A specification is called *universal, conditional,* or *equational* if Γ is, respectively, a set of universal first-order sentences, conditional equations (i.e., quasi-equations), or equations. In (Pigozzi, 1991) the conditional theory of multi-sorted algebras with equality tests and the equational theory of data structures with equality tests and if-then-else operations are investigated. The main results of the paper are the following: An *equality test* (*multi-sorted*) *algebra* has a two element Boolean sort with a separate binary operation eq_S for each non-Boolean sort S where $eq_S(a,b)$, for all a,b in sort S, takes the value TRUE if $a = b$ and FALSE otherwise. The *if-then-else algebras* are obtained from the equality-test algebras by adjoining the *if-then-else operation* $[\ ,\ ,\ ]_S$ for each non-Boolean sort S where $[\text{TRUE},a,b]_S = a$ and $[\text{FALSE},a,b]_S = b$ for all a,b in sort S.

A simple algorithm is given that converts any universal initial or final specification of an equality-test data type $\boldsymbol{S}$ into a conditional specification of $\boldsymbol{S}$: moreover, the new specification is *complete* in the sense that it is both initial and final. As a consequence every semicomputable or cosemicomputable equality test data type is computable. If $\boldsymbol{S}$ is an arbitrary data type, essentially the same algorithm can be used to convert a universal initial specification of it into a conditional specification with equality tests as hidden operations. In this case the new specification will be complete only if the original one is. Thus an arbitrary data type is computable if and only if its equality enrichment is semicomputable, or equivalently, cosemicomputable.

5 Years 1988–1995

While still on leave from ISU I returned to Ames for the 1988 spring semester, where I finished my work on the quasivariety paper (Pigozzi, 1988a). But most of the time I was occupied with organizing together with Cliff Bergman and Roger Maddux a conference: Algebraic Logic and Universal Algebra in Computer Science. It was held at ISU in June of 1988, and was to the best of my knowledge the first national conference to bring logicians, universal algebraists, and computer scientists together for the purpose of exploring the connections between these disciplines. It attracted a large number of participants from all over the United States and Canada. Thanks to Dana Scott, who enthusiastically supported multi-disciplinary efforts of this kind and agreed to be one of the principal speakers, we were able to obtain financial support from the National Science Foundation, the Office of Naval Research, and the Institute for Applied Mathematics in Minneapolis The invited speakers in order of their position on the program: Eric Wagner, Dana Scott, István Németi, Dexter Kozen, H.P.Gumm, Vaughan Pratt, and Bjarni Jónsson. The proceedings of the conference, edited by the three organizers, appeared as a volume of Lecture Notes in Computer Science (Bergman et al., 1990).

The organization of such a large conference in a small town in agricultural heart of the country proved challenging, especially to novices like the three organizers. The nearest good size airport was in Des Moines, thirty miles from Ames, which meant that we had to arrange transportation for the participants who arrived by air at various times of the day. The most suitable accommodation for them was a hotel abut five miles from the site of the conference on campus. Transportation also had to be provided between them twice a day. The University provided the vans and graduate students happily agreed to be drivers. All in all the conference turned out to be a big success, much to the satisfaction of the organizers and their helpers.

The summer of 1988 turn out to be another busy one for me. In July Walter Taylor and I were invited to speak at the Ninth Latin American School of Mathematics (ELAM) in Santiago Chile. We both owed this invitation of Renato Lewin, of the Pontifical Catholic University of Chile, who was a former student of Walter and one of the organizers of the conference. I knew Renato through Wim Blok. Wim had helped him with some questions about the algebras of modal logic that arose in connection with his thesis. I was greatly impressed that Walter was able to give his talk at the ELAM conference in Spanish. I could only speak English. This was true of all my subsequent visits to Chile to my increasing embarrassment. At one point I promised I would give my talk in Spanish on my next visit to Chile, which I thought would give me lots of time. Sadly I was not able to keep this promise. Renato later became interested in **AAL** and our collaboration deepened and resulted in several more visits to Santiago as a visiting scholar at the Catholic University. It was also due to his influence while a member of the organizing committees of two different meetings of the Latin American Symposium of

Mathematical Logic that I was invited to present hour addresses. The first was the IX meeting in July of 1991 in Bahia Blanca, Argentina where I spoke on *An Introdution to Lambda Abstractions Algebras* (Pigozzi and Salibra, 1993a), joint work with Antonino Salibra. The second was the X meeting in July of 1995 in Bogotá, Colombia. Here I spoke on my joint paper with Janusz Czelakowski *Amalgamation and Interpolation in Abstract Algebraic Logic* (Czelakowski and Pigozzi, 1999).

Shortly after the ELAM meeting, in August of 1988, there was a large conference on algebraic logic presented by the Janos Bolyai Mathematical Society in Budapest and organized by Hajnal Andréka, Istvan Németi, and Ildiko Sain. Wim Blok and I were invited to jointly present a series of talks our work that began with the Algebraizable Logics monographic (that had recently appeared) amd was evolving into what would eventually be called abstract algebraic logic. It was valuable exposure for us because it attracted algebraists and logician from all over the world. I met Josep Maria Font, and my future graduate student Katarzyna Pałasińska, for the first time here.

My university service during my time at ISU was limited, but during the 1988-89 academic years I was a member of the search committee for the new dean of the College of Science and Humanities. I tried to promote candidates sensitive to the needs of the mathematical disciplines and emailed John Addison, who at the time was chairman of the Department of Mathematics at Berkeley, inviting him to apply. He thanked me and asked for some time to consider the application, but eventually declined. I thought he would have made a very good dean.

My one other significant university service was during the academic years 1990-93 when I served on the Promotion and Tenure Committee of the College. The work of the Committee was all done during the Christmas break between the Fall and Spring semesters, but it was intense. Every committee member had to read the promotion documents for each candidate that were submitted by the twenty-three departments of the colleges. The Committee would then met to decide which candidates to recommend to the dean; the Committee was only advisory. At a meeting with the dean it was decided which ones the dean would approve. I was chairman during my last year on the committee and had the responsibility of explaining to the general faculty the process by which the tenure decisions were made and defending them. I had purposely avoided any involvement in university politics up to this time, and was totally unprepared for the resentment some of the humanity departments, in particular the English Department, felt towards the sciences because of a perceived bias in their favor in the tenure process. I was one of the principal participants at an open meeting about the process and made some off-the-cuff and admittedly banal remarks on the committee's role in it. In the following questions and comments period I was washed over by a wave of resentment and frustration that left me speechless and running for cover.

In the summer of 1989 the Soviet government's xenophobia had weakened to the point that the academic community in Siberia was able to organize an international conference in Novosibirsk in honor of A.I. Mal'cev. I owe my invitation to Victor Gorbunov I think because of the finite basis results for quasivarieties paper that had recently appeared. Gorbunov was a specialist on quasivarieties. I got special treatment that did not correspond to my position in the hierarchy of participants. For example, although I was not one of the plenary speakers, I was given a room to myself in the guest hotel, rather than being placed in a student dormitory like many of my colleagues. The site of the conference was Akademgorodok (academic town) a part of Novosibirsk about thirty km from the center. It was the home of the Siberian division of the Soviet Academy of Sciences, the Novosibirsk State University, and a large number of research institutes. It is surrounded by a pine and birch forest on the shore of the OB sea, an artificial reservoir of the River OB. The social highlight of the conference for me was an invitation from Victor Gorbunov to join him, Ralph McKenzie, and George McNulty for what turned out to be a bacchanalian dinner one evening in his cabin on the OB reservoir. The night ended with a searing banya followed by a naked dive in the reservoir by Victor and all three guests. I was saddened to find out a few years later that Victor had died prematurely.

Janusz Czelakowski was invited to Ames as visiting professor for the 1989-90 academic year and again in 1994-95. The origin of AAL can be traced back to the work of Polish logicians during the 1970's, chiefly Helena Rasiowa in Warsaw and Ryszard Wójcicki in Łódź, with the latter working in the wider context of matrices. Janusz, Wójcicki's student, greatly advanced these investigations by bringing to them a thorough understanding of universal algebra. A series of his papers in the 1980s significantly influenced Wim and my work, and I was eager to have him in Ames to work with me and participate in our seminars. Janusz's wife Bożena came with him, and she was a pleasure to have with us. First of all, she spoke almost perfect English, being an English teacher back home. Like Bożena Szczerba, she was an excellent cook, and I eagerly looked forward to invitations to dinner. I found out much later that she was very lonely during her first few days in Ames. She was in a foreign land, knowing no one and terribly homesick. She told me she was on the verge of going back to Poland, when a group of neighbors showed up at her door to welcome her with a basket of gifts and an invitation to join the community. (This is a tradition in the United States, especially in small towns, called a *Welcome Wagon.*) From this time on she was a happy citizen of our small town.

In 1993 George Voutsadakis enrolled in the graduate program at Iowa State and eventually turned out to be my second Iowa State doctoral student (the first was Katarzyna Pałasińska). He came from Greece where he had graduated from the University of Patras. George took my graduate course in algebra and asked to work towards a masters degree under me. His intention after getting it was to go on for a doctorate in mainstream algebra, possibly

at another university. But while working on his masters thesis he became
more and more attracted to the kind of logic oriented algebra I was doing.
About this time I was reading about the theory of *Institutions*, a category
description of logical systems in computer science that had been developed
by Joseph Gougen and Rodney Burstall. George had taken Jonathan Smith's
category course and I proposed that George investigate reformulating the
algebraization of logic within the framework of institutions to see what new
insight this might give. He attacked the problem with great energy and soon
knew far more about categories than I. He wrote a fine thesis and received
his doctorate in 1998.

He got a one-year visiting position at Western Reserve University where
he became a friend and collaborator of Charles Wells. At the end of the
year he was offered a tenure-track position at WRU. But Charles Wells was
retiring and for this and other reasons he decided to accept another offer
from the Physical Sciences Laboratory in Las Cruces New Mexico to work
on what was essentially artificial intelligence, developing intelligence agents
for computer-based Army simulations. There he joined a group in the area of
finite dynamical systems. He learned a lot there that led to several publications. The upper level administration at the lab was not easy to deal with and
after several years he returned to academic life by taking a position at Lake
Superior State University in Upper Michigan where he teaches and maintains
an active research program

During this period I began professional relationships with several people
that took my research in three different directions: one was the lambda calculus, more specifically its algebraization. The other two were into the realm
of computer science: Logic Programming and Software Development.

Antonino Salibra came to Ames to visit me for three months in 1992. At
that time he had a position at the University of Bari, but eventually moved
to Venice. I had gotten to know Antonino from several conferences we had
both attended. He was working at this time on the algebraization of the
untyped lambda calculus and wanted to discuss this with me. I was skeptical
about the project at first. I had some knowledge of the formalism of the
calculus but had not studied it in any detail, and had never considered the
possibility it might be algebraizable. Antonino described to me a variety of
algebras, that he called *Lambda Abstraction Algebras* (LAA). It was clear to
me almost at once that this was the natural algebraization of the untyped
lambda calculus in the same sense the varieties of cylindric and polyadic
algebras were algebraizations of first order predicate logic. Antonino and I
then began to investigate the theory of LAAs that resulted in five joint papers
over the next six years (Pigozzi and Salibra, 1994, 1993c,a, 1995, 1998).

During the years I was teaching in the CS Department I became interested
in logic programming. This was about the time I recall that the Japanese
were promoting logic programming as the "fourth-generation programming
language". I never really understood what this meant, but was aware that,
whatever it mesnt, Prolog was its prototype. I read J.W. Lloyd's book *Foun-*

dations of Logic Programming, which had recently appeared, and was happy to see that Prolog was very much like the classical propositional logic. I joined two members of CS, Professor Giora Slutzki and Bamshad Mobasher, a graduate student, in investigating the subject. We were joined part of the time by a Polish computer scientist, Jacek Leszczyłowski, who was visiting the CS Department at the time. Our research resulted in four papers and extended conference abstracts (Mobasher et al., 1993b, 1997a,b, 2000) and a technical report (Mobasher et al., 1993a). It also led to Mobasher's doctoral dissertation for which Sluzki and I were co-chairs of his committee.

I met Gary Leavens for the first time at a conference on algebraic methods in Iowa City in May of 1991. He had recently joined the CS Department at ISU; his research area was software development. The significance of algebraic methods had been growing in this area, a fact I was already aware of. Gary and I soon became good friends and collaborators. I had received a National Science Foundation (NSF) research grant in 1989 whose scope included algebraic data types, and had published two papers in the area. So it was natural that Gary and I would begin collaborating, and in fact we became good friends. The CISE division of NSF was eager to promote collaboration among computer scientists and mathematicians. Gary and I were able to get three consecutive research grants between 1995 and 2002, the year I retired, with Gary as the Principal Investigator. We published four papers during this period (Leavens and Pigozzi, 1991, 1997, 1998, 2000) and a technical report (Leavens and Pigozzi, 2002).

Publications: The correlation between interpolation theorems of logic and certain properties of the class of models related to the amalgamation property is well known. In classical sentential and first-order logic it takes the form of a correspondence between Craig's interpolation theorem and Robinson's joint consistency lemma. In the algebraic versions of these logics the joint consistency property can be replaced by the amalgamation property for Boolean algebras and locally finite cylindric algebras, respectively. The connection between interpolation and amalgamation has also been explored in the context of intermediate and modal logics, equational logic, and general deductive systems in the sense of Tarski. More recently, logical interpolation results have been shown to have interesting applications for the specification of abstract data types, especially with regard to the important problem of modularization.

In the paper *Amalgamation and Interpolation in Abstract Algebraic Logic* (Czelakowski and Pigozzi, 1999) a unified theory of interpolation, joint consistency, model extension, and amalgamation is presented that comprehends all these results. In this general context the Craig interpolation property ramifies into several different interpolation-like properties, one of which is closely related to the familiar congruence extension property of universal algebra. It is shown how under quite weak conditions on the logical system, each interpolation property is equivalent to an extension or amalgamation-like property of the appropriate model class.

Lambda Abstraction Algebras: The untyped lambda calculus is a formulation of an intensional as opposed to an extensional theory of functions, that is, a theory of functions viewed a "rules" rather than "sets of ordered pairs". Its basic feature is that functions are not distinguished from the elements of the domains on which the functions act. Thus a function can, in theory, take other functions, including itself, as arguments. There are two primitive notions: *application*, the operation of applying a function to an argument, and *lambda (functional) abstraction*, the process of forming a function from the "rule" that defines it.

As one would expect there are no simple models of the untyped lambda calculus, but one can imagine idealized models that are "constructed" in the following way: Start with any set S (possibly empty), and successively form the sets $T_0 = S$, $T_1 = S^S$, $T_2 = (S \cup S^S)^{S \cup S^S}$, $T_3 = (T_1 \cup T_2)^{T_1 \cup T_2}, \ldots$. Iterate the construction until a "fixpoint" is reached, giving a set V satisfying the "domain equation" $V^V = V$. We know of course that if V^V is interpreted as the set of all functions from V to itself in the usual set-theoretical sense, then the above iterative process can never reach a fixpoint since no set can satisfy the domain equation. By restricting the functions we consider to certain admissible ones, and interpreting V^V accordingly, domains satisfying the domain equation, or a somewhat weaker form of it that guarantees there are in some sense enough admissible functions, have been found. (The first such model was constructed by Dana Scott, *Data types as lattices*, SIAM J. Comput. **5** 1976). Such domains are the "natural" models of the untyped lambda calculus. They are called *environment models* in the literature. They can be characterized by means of an injective partial mapping $\lambda : V^V \xrightarrow{\text{p}} V$ whose domain is the set of admissible functions. λ may be thought of as the process of encoding admissible functions as elements of V. With functions encoded this way, application can be viewed as a binary operation on V. Let $\mathbf{V}$ be the domain V enriched by the application operation and the encoding mapping. We will denote the application operation by $\cdot^{\mathbf{V}}$ and the encoding mapping by $\lambda^{\mathbf{V}}$.

Intuitively, each admissible function in V^V has two forms, an *intensional* one and an *extensional* one. In its intensional form it is represented by a term $t(x)$ of the lambda calculus with a free variable x. For each $v \in V$, let $t[\![v]\!]$ be value $t(x)$ takes in V when x is interpreted as v. Then its extensional form is the function $\langle t[\![v]\!] : v \in V \rangle \in V^V$, which is encoded as the element $\lambda^{\mathbf{V}}(\langle t[\![v]\!] : v \in V \rangle)$ of V. It is represented by the term $\lambda x.t(x)$ obtained by applying lambda abstraction to $t(x)$. Note that $t(x)$ and $\lambda x.t(x)$ both represent the same function, but in environment models only the extensional form corresponds to an actual element of the universe of the model; this is an essential difference between the models of lambda calculus and lambda abstraction algebras.

The two forms of the function are connected by the operation of application. Intuitively, the value $t[\![v]\!]$ of the function at a particular argument v is obtained by applying its extensional form to v; symbolically,

$\langle t[\![v]\!] : v \in V \rangle(v) = t[\![v]\!]$. Expressed in the environment model this becomes

$$\lambda^{\mathbf{V}}\big(\langle t[\![v]\!] : v \in V \rangle\big) \cdot^{\mathbf{V}} v = t[\![v]\!].$$

In the lambda calculus itself this relationship is represented by the fundamental axiom of β-*conversion*:

$(\beta)(\lambda x.t)s = t[s/x]$, for all terms t, s and variable x such that s is free for x in t.

Terms of the lambda calculus are constructed as follows: There is an infinite set of variables, each of which is a term; if t and s are terms, so are $t \cdot s$ and $\lambda x(t)$ for each variable x. By convention we write ts for $t \cdot s$ and $\lambda x.t$ for $\lambda x(t)$. An occurrence of a variable x in a term is *bound* if it lies within the scope of a lambda abstraction λx; otherwise it is *free*. s is *free for* x in t if no free occurrence of x in t lies within the scope of a lambda abstraction with respect to a variable that occurs free in s. $t[s/x]$ is the result of substituting s for all free occurrences of x in t.

The other fundamental axiom of the lambda calculus is α-*conversion*:

$(\alpha)\lambda x.t = \lambda y.t[y/x]$,   if y does not occur free in t.

α-conversion says that bound variables can be replaced in a term under the appropriate condition. A *lambda theory* is any set of equations that is closed under α and β conversion and the five axioms of equational logic.

The following *completeness theorem* of A. R. Meyer is a basic result of lambda calculus: Every lambda theory consists of precisely the equations valid in some environment model.

Although the axioms of lambda calculus are all in the form of equations, the lambda calculus is not a true equational theory since the variable-binding properties of lambda abstraction prevent variables in lambda calculus from operating as real algebraic variables. The way in which lambda abstraction theory arises from the lambda calculus parallels the way cylindric algebras are obtained from first-order logic. The axioms of first-order logic are like those of lambda calculus in that the formula-variables can not be substituted without restriction. In both cases the source of the problem is the way substitution for individuals is handled. By dealing with substitution at the level of the object language rather than the metalanguage, i.e., by abstracting it, a pure equational formalization of lambda calculus can be developed giving rise to the theory of lambda abstraction algebras. Like cylindric algebras, and in contrast to the lambda calculus, the axioms of lambda abstraction theory are pure identities (more accurately, they turn out to be equivalent to pure identities). Among the seven axioms, the first six constitute a recursive definition of the abstract substitution operator; they express precisely the metamathematical content of β-conversion. The last axiom is an algebraic translation of α-conversion. The most significant feature of the axioms is that they are true identities in the sense that they continue to hold when arbitrary terms are substituted for the variables. Thus the theory of lambda abstraction algebras gives a pure equational theory of lambda calculus, and lambda abstraction algebras form a variety in the universal-algebraic sense.

There is a notion of a "natural" lambda abstraction algebra—the algebras that the axioms of lambda abstraction theory are intended to characterize. They correspond to functional polyadic algebras and, more loosely, to representable cylindric algebras; we call them *functional lambda abstraction algebras*. They are closely related to the environment models of lambda calculus. The important point here is that, in contrast to environment models, the intensional form of the function corresponds to an actual element of the functional lambda abstraction algebra. In this sense functional lambda abstraction algebras are richer than environment models, and this greater richness translates into a more algebraic theory.

The basic theory of lambda abstraction algebras is developed in (Pigozzi and Salibra, 1993a). The main result there may be viewed as the natural algebraic analogue of the completeness theorem for the lambda calculus. It is the functional representation theorem for locally finite lambda abstraction algebras: every locally finite lambda abstraction algebra is isomorphic to a functional lambda abstraction algebra obtained by coordinatizing an environment model having the same carrier set. This result corresponds to what is called the functional representation theorem for locally finite polyadic Boolean algebras.

However the natural algebraic analogues of the completeness theorem for first-order logic are the stronger representation theorem for simple, locally finite polyadic Boolean algebras of infinite degree and the representation theorem for locally finite cylindric algebras. The representation theorem for locally finite lambda abstraction algebras that corresponds to these results is the main result of (Pigozzi and Salibra, 1993c).

Logic Programming: First-order logic turns out to be unsuitable as a basis for a knowledge representation language in Artificial Intelligence (AI) systems. The problem with first-order logic is its monotonicity: The truth of a inference is determined only by its structure, not by the truth or falsity of its atomic components, and so it can only deal with two truth-values, truth and falsity. An intelligence agent however must deal with information that is uncertain or incomplete. This discussion suggests that such systems must have two common characteristics: they must rely on the expressive power of an underlying multi-valued logic that can deal with contradictory as well as incomplete or uncertain information, and secondly, such systems should be able to interpret statements not only based on their truth or falsity, but also based on some measure of the knowledge or information contained within those statements.

Attention has focused on logics that have a knowledge dimension as well as a truth dimension and thus can be used to model the connection between truth and knowledge in a particular logic program or deductive database. The first logic of this kind originated with N. D. Belnap's four-valued logic. It is based on the idea that information in a database can have both a positive and a negative content with regard to the truth of a particular event. The two situations in which only positive or only negative information is available give rise to two truth values that can be identified with classical `true` and `false`, respectively. But there are two other situations: when the information has both a positive and a negative content, and where there is no information of either kind. These lead to a third and fourth "truth-value" that are denoted respectively by $\top$ and $\bot$. Part of the motivation here is that, in a distributed database, information about a given event is collected from various sources at various times and some of it might be contradictory. So the truth-value of the event can be viewed as representing our state of knowledge about the classical truth or falsity of the event rather than its actual truth or falsity.

In (Mobasher et al., 1993b,a) a knowledge-based procedural semantics based on the 4-valued Belnap logic is developed and its soundness and completeness with respect to M. Fitting's declarative fixpoint semantics is proved. A novel feature of this procedural semantics is the introduction of completely symmetric notions of *proof* and *refutation*. Intuitively, the existence of a proof, respectively refutation, for a given goal corresponds to having positive, respectively negative, information about it.

M. C. Ginsberg has suggested using general *bilattices* as the underlying framework for various AI inference systems including those based on default logics, truth maintenance systems, probabilistic logics, and others. Bilattices are mathematical structures with multiple "truth-values" and two separate orderings, called knowledge and truth orderings, that provide a framework for the study of knowledge-truth interaction. (The Belnap logic is a four-element bilattice.) These ideas were pursued by Fitting in the context of logic programming semantics. More recently, bilattices and their extensions have been used in the literature to model a variety of reasoning mechanisms about uncertainty in the presence of incomplete or contradictory information. For example, a variant of Fitting's extension of logic programming to bilattices has been used to deal with a form of negation as failure (where the inability to prove a statement is true is interpreted as it being false) as well as a second explicit negation in logic programs. Bilattices have also been extended to include a third ordering (called the *precision* ordering) in order to effectively deal with varying degrees of belief and doubt in probabilistic deductive databases.

In (Mobasher et al., 1997b) the procedural semantics in (Mobasher et al., 1993b,a) is first generalized to an arbitrary distributive bilattice. (It could just as easily be generalized to any partially ordered algebra.) We introduce the notion of a *b-derivation* for each element of the bilattice except $\top$ and $\bot$. (In the 4-element case `true`-derivations coincide with proofs and `false`-derivations with refutations.) We prove the soundness and completeness theorems for this procedural semantics, again with respect to Fitting's declarative fixpoint semantics.

Although the resulting logic programming system is quite satisfactory in some respects, for example the symmetry between truth and refutation in the 4-element case is carried over and the mathematical theory is quite smooth, it has some undesirable defects. For a given truth-value b, the search for a b-derivation of a complex goal G may entail searches for c-derivations of the subformulas of G for a large number of truth-values c that are only remotely related to b; moreover, this complexity ramifies as we pass down the parse tree of G. It turns out that for finite distributive bilattices (and, more generally, bilattices with the *descending chain property*), we can restrict our attention to derivations that range over a relatively small subset of special truth-values. These special truth-values turn out to be the so-called *join irreducible* elements of the knowledge part of the bilattice. Ginsberg has discussed the ramifications of reducing the complexity of bilattice based inference systems by focusing on a smaller set of representative elements called *grounded* elements. As we will see, join-irreducible elements provide an even smaller set of representative elements which represent the most "primitive" bits of information. In fact, this difference could be exponential for certain classes of bilattices.

A *join-irreducible* procedural semantics is developed in (Mobasher et al., 1997b) as an alternative to the standard one. The join-irreducible semantics, which represents the most novel feature of this paper, can provide the basis for effective implementation of a family of logic programming languages which, depending on the choice of the underlying logic, can be used for a variety of reasoning tasks in intelligent systems.

Software Development: The main advantage of abstract data types (ADTs) in programming is that they allow reasoning at an appropriate level. In reasoning about code that uses an ADT, clients rely on the ADT's specification, instead of using more complex and overly specific reasoning about the ADT's implementation. The soundness of such an abstract reasoning technique means that if an implementation is certified correct, then its visible behavior will not be surprising. By *visible behavior* we mean, informally, the printed or returned results of programs. By *surprising behavior* we mean visible behavior that would contradict the predictions of the specification. Completeness of an abstract reasoning technique means that if an implementation cannot exhibit surprising behavior, then it can be certified as correct.

We investigate sound and complete model-theoretic techniques for proving that a candidate implementation of an ADT is correct. For reasons discussed below, we are especially interested in specifications that are incomplete and not term-generated. For us, a complete specification is one for which all of its models are behaviorally equivalent, and a specification is not term-generated if there are nonvisible types that fail to have a complete system of constructors. We shall also assume that a candidate implementation has already been adapted to the interface (signature) required.

What is known about the soundness and completeness of techniques for proving that a candidate implementation of an ADT is correct? We shall restrict ourselves here to model-theoretic methods. Previous model-theoretic work on this problem, like our work, is based on comparisons to paradigmatic models. In most work, there is only one paradigmatic model mentioned, and so the ADT's specification must be complete. If the specification is incomplete, there is no way to choose a single paradigm, and the technique must be adapted somehow to deal with the choice of an appropriate paradigm before the comparison. However, it is a simple matter to adapt this technique to incomplete specifications by using a collection of paradigms. These paradigms collectively span the permitted behaviors, and thus to prove the correctness of a candidate implementation, one must first choose a paradigm and then make the comparison.

The paper (Leavens and Pigozzi, 1997) concentrates, therefore, on how to compare an implementation algebra to a paradigm, once a paradigm is selected. Several authors have studied such notions previously. For our purposes the most important technique is that of Oliver Schoett. He casts the problem as one of showing that a partial algebra $\mathbf{A}$ can be used in place of the paradigm, a partial algebra $\mathbf{A}$ that is assumed not to exhibit surprising behavior. Then an arbitrary algebra $\mathbf{B}$ will also fail to exhibit surprising behavior if the two algebras are behaviorally equivalent in the sense that any program that is run in the two algebras has the same output. He makes the natural assumption that only visible data is legitimate input-output for the program. He proves that the existence of a bisimulation between $\mathbf{A}$ and $\mathbf{B}$, i.e., a homomorphic relation that is the identity on visible types, is both necessary and sufficient for the behavior of $\mathbf{A}$ to be equivalent to the behavior of $\mathbf{B}$.

It can be argued however that Schoett's criterion for behavioral equivalence is not restrictive enough. It fails to detect some behavioral differences that an ADT implementor might care about. The main problem with his approach is that programs can take only visible data as input and hence algebras can be compared only with respect to the behavior of visible data. For example, in the context of specifying a parameterized type (e.g., a parameterized priority-queue), consider the specification of its formal type parameter, PO. The only operation that would be specified for PO would be a comparison predicate, leq, taking two POs and returning a Boolean; no constructors would be specified for PO. In this example, the type PO would not be a visible type (i.e., it could not be directly input or output). Hence the only visible type in the example is the Booleans, and the type PO is hidden. Because PO is hidden and there are no constructors for it, programs with visible input-output cannot make any interesting observations. Hence, using Schoett's criterion, even candidate implementations that, say, fail antisymmetry would be certified as correct. In (Leavens and Pigozzi, 1997) we adapt Schoett's technique by considering not just observations with visible inputs, but "procedures" with nonvisible inputs. For example, this allows us to make behavioral distinctions in the PO example. That is, we allow the behavior of nonvisible data to be compared in different models, leading to a stronger notion of implementation which is important in situations where the specification is not term-generated.

ADTs that are not term-generated are even more important for object-oriented programming than they are in more conventional programming with ADTs. For example, a library of object-oriented ADTs typically includes a type `Collection` that is "abstract" in the sense that it has no constructors. Such a type will have subtypes such as `Set`, `Bag`, `List`, and `Array`. Existing objects of one of these subtypes can be treated as if they were collections. This is analogous to the way that objects having the type of a formal type parameter, such as `PO`, are treated in parameterized code. It is also apparent from this example why it is important to be able to compare nonvisible data. It is natural to want to compare the behavior of a bag constructed from the integers 1, 2, and 3, for instance, with that of a set constructed from the same integers. But this cannot always be achieved by simply comparing the behavior of the visible data, such as the integers 1, 2, and 3 in two different models, because a (deterministic) program with only visible input would construct either a set in both $\mathbf{A}$ and $\mathbf{B}$ or a bag in both $\mathbf{A}$ and $\mathbf{B}$, but not a set in $\mathbf{A}$ and a bag in $\mathbf{B}$. This problem is the original motivation for our study of "procedures" with nonvisible inputs.

In (Leavens and Pigozzi, 1997) we give a sound and complete algebraic technique for proving the correctness of an implementation, which need not be term-generated. The technique uses a general notion of simulation, which in turn uses a generalization of the notion of homomorphic relation; such a generalization is necessary because standard homomorphic relations do not give a complete characterization technique for specifications that are not term-generated.

6 Years 1995–1997

This period was the most intense of my professional life. I spent most of the time in various parts of Europe while away from Ames on a combination of sabbatical leave and leave without pay.

In the early 1990s I became aware of a group of researchers in the *Institute of Mathematics* at the University of Barcelona (IMUB) who were beginning to do some serious work on AAL. I began to correspond with Josep Maria Font, a principal member of the group. His graduate student Raimon Elgueta's thesis topic was the algebraic model theory for languages without equality. Josep Maria asked me to become Raimon's co-advisor, which I readily agreed to. Raimon and I consulted frequently about his thesis by email, and he was able to come to Ames once to work directly with me. He eventually obtained his doctorate from the University of Barcelona in 1994.

When I got sabbatical leave for the 1995-96 academic year I asked Josep Maria if I could join his group. He readily agreed and obtained for me a position as a visiting scholar at the *Center for Mathematical Research*, (CRM), of the *Institute of Catalan Studies*. The CRM was located on the campus of the *Autonomous University of Barcelona* (UAB). Moreover I received financial support in the form of a grant from the *Ministry of Science and Education of Spain*. I also had the use of one of the apartments that the CRM maintains for visitors in the nearby town of San Cugat del Vallès. (The origin of its name has an interesting history. San Cugat was a Christian monk. He was born in North Africa and came to Barcelona to evangelize in the area. He was martyred by the Romans about 304 AD. The Benedictine Abbey in San Cugat del Vallès is by tradition the site where he died by decapitation.)

The CRM turned out to be a wonderful place to work, with seminars in all areas of mathematics and visitors from all over the world. Its Director Manuel Castellet was very accommodating and organized daylong sightseeing tours to various parts of Catalonia every month for the visitors. I looked forward to them and was one of the few visitors to go on all of them. The IMUB had a close relationship with the CRM, and the seminars organized by the AAL group were of course of most interest to me. The UAB is twenty-four km from Barcelona but is easily accessible by an efficient rail service and excellent roads. I frequently attended the seminars that were held in Barcelona.

That fall in Barcelona was stimulating and productive in many ways. Apart from the professional stimulus, there was the city itself: the Ramblas, the Gothic Quarter, evidence of Gaudi's work all over. I would often take the train into Barcelona on the weekend to enjoy its pleasures. One of the first things I did was climb all the way to the top of one of towers of the Sagrada Familia for the view of the City. I would return many times over the years.

All the members of the AAL group became good friends: Josep Maria, Ramon Jansana, Ventura Verdú, Jordi Rebagliato, Antoni Torrens, and Ángel Gil. My closest collaborators were Josep Maria and Ramon Jansana, and our collaboration has extended over many years. They took me on many weekends to visit historical sites in Catalonia. The two most impressive are of great importance of Catalonians' sense of national identity: the Monastery of Santa Marie del Poblet near Tarragona, and the Santa Maria de Montserrat Abbey on Montserrat mountain.

The Monastery of Santa Maria del Poblet was founded in 1153, and over the three centuries it rose is size and wealth. Starting in the fourteenth century all the kings of Aragon and Catalonia were buried there. Its architecture was one of the supreme examples of Catalan Gothic, but by the 1890s Poblet was a wreck and its restoration became a top priority. The restoration (base on a plan by a young Gaudi and friend) was completed some time ago and now Poblet has become a major tourist attraction.

Montserrat, whose name means 'serrated mountain', plays an important role in the cultural and spiritual life of Catalonia. It is Catalonia's most important religious retreat and groups of young people from Barcelona and all over Catalonia make overnight hikes at least once in their lives to watch the sunrise from the heights of Montserrat. The Virgin of Montserrat (the black

virgin), is Catalonia's favourite saint, and is located in the sanctuary of the
Mare de Déu de Montserrat next to the Benedictine monastery nestling
in the towers and crags of the mountain. I made two visits to Montserrat's
and climbed to the top of the mouontain both times for the great view of the
surrounding countryside.

We also visited the Penedés area south of Barcelona that is Catalonia's
major wine producing area. We toured the Cordorniu and Torres estates and
caves in Sant Sadurní d'Anoia. We then drove over to Stiges on the coast,
an art center. While the artistic roots of Stiges date back to the late 19th
century, the town became the center for the 1960 counterculture in mainland
Spain.

Maybe our most most memorable trip was to the town of Roses overlooking
Cala Montjoi, a bay on Catalonia's Costa Brava. There we had dinner at El
Bulli, a Michelin 3-star restaurant that has been described as "the most
imaginative generator of haute cuisine on the planet".

In planning my sabbatical it was my intention to go to Kraków af-
ter my stay in Barcelona to visit my former graduate student Katarzyna
Pałasińska and her family, then to Budapest to work with István Németi,
Hajnal Andréka, and Ildikó Sain, and finally to Siena to see Aldo Ursini and
Paolo Aglianó. Moreover I was determined to do it all by car. This meant that
I had to either rent or buy a car in Barcelona, and since I intended to keep
it for almost a year buying seemed to be the better choice. For a foreigner
buying a car and registering it in Catalonia is no easy matter. But thanks to
the very patient help of Ventura Verdóu, Antoni Torrens and their friends I
was able to buy a used Opel sedan in good shape and register it, although the
latter took up more than a day at the motor vehicle department and wasn't
okayed until we made it up all the way to the chief of the department. I will
be eternally grateful to my friends for their effort on my behalf.

I left Barcelona about a week before Christmas 1995 with the intention of
driving to Kraków in three days. The first day's drive to Karlsruhe Germany
would be the longest, about 1200 km. I left San Cugat at 5 AM. I packed
a bag lunch so that I wouldn't have to stop to eat, only for fuel. I crossed
into France just south of Perpignan, skirted around Lyon, and crossed the
Rhine and into Germany at Mulhouse. It was 5 PM. I had been on the road
for twelve hours and was still 165 km from Karlsruhe. It was getting dark.
(I hadn't realized how far north I had come.) On top of it all it started to
rain hard, and the two-lane road I was on was congested with cars going 100
km/hr with little more than a car length between them. I could only go with
the flow and pray that no one in the long line of cars would make a mistake.
I finally got to Karlsruhe and with some effort found a reasonable hotel near
the train station. I fell into bed about nine, without supper, sixteen hours
after leaving San Cugat. I slept soundly for ten hours and was rewarded when
I got up with, what I perceived in my famished state, to be a magnificent
breakfast buffet upon which I gorged myself.

The next leg of the journey to Vienna was shorter, 740 km, and uneventful except for crossing the border into Austria that at that time was not an EU country. I was worried when they took me out of line and went into the administration building with my passport and auto registration papers. But eventually they let me pass, requiring only that I put a blue E sticker (for España) on the back of the car. (This caused some trouble later when I returned to the UAB with its hotbed of student action for Catalonian independence from Madrid.) After spending the night in a hotel on the outskirts of Vienna I drove through Brno to Cieszyn, the closest border point to Kraków where I saw many large trucks transporting wrecked cars from Western Europe into Poland. Crossing the border I drove through Bielsko-Biała to Kraków. I had no problem at either the Czech or Polish borders.

The most interesting, and as it turned out, the most harrowing event of the whole trip occurred in the early morning of the last day. It was very cold. I had just crossed into the Czech Republic and found myself on a large open plain in the bright early morning winter sunlight. There was no sign of human habitation in any direction except for the long straight road that stretched before me to the horizon. I was driving very fast, much beyond the speed limit. I saw a large truck in the distance that I quickly caught up and passed. I noticed something in the distance that I thought might be a turn in the road but I didn't reduce my speed. When I got closer I could see it was a wooden barrier marking where the road made a ninety degree turn to the left. I slammed on the breaks, but unfortunately unknown to me some snow mudt had fallen during the night and the road was very slippery. The car slid off the end of the road and smashed through the barrier embedding itself in a snow bank. The car had no visible damage and I could start it, but it was so firmly embedded I couldn't back it out. I was seized with panic. I could see nothing about me except what appeared to be a small village on a small hill way in the distance. It was very cold, and I knew it would take me at least an hour to walk there. They probably knew no English and I would have difficulty making it understood what had happened. The police would surely be brought in. I had heard they were very tough on foreign drivers and I had obviously been speeding. While I stood there lamenting my carelessness and wondering what to do other than commit suicide, the truck I had passed stopped. The driver was Italian and spoke good English. He offered to try to pull me out, and tied a rope between his bumper and mine. But the rope broke when he tried to back up, and my despair deepened after the brief moment of hope. But he didn't give up. He doubled the rope and tried again and this time it worked and my car was free. I was overwhelmed with relief and gratitude. I offered him all the cash I had, fifty dollars, but he wouldn't take it. He just made me promise to drive slower in the future. He pulled out and I followed, making a point not to pass him.

In the fall of 1989 Katarzyna (Kate) Pałasińska came to Ames to work toward her doctorate under my direction. She came with a solid foundation in logic and algebra as a student of Andrzej Wroński at the Jagiellonian

University in Kraków. She was on leave from her position at the Tadeusz Kościuszko University of Technology. She came with her family: her husband Marek, who himself was on leave from the Jagiellonian, and three young daughters, Magdalena (Magda), Joanna, and Justyna, ages four, two, and seven months, respectively. In anticipation of the cultural shock they would certainly receive, I did what I could to help them begin their new life in a foreign environment. I helped them arrange for an apartment in a student housing complex for families, and with the help of two of my colleagues furnished it as well as I could. I was particularly concerned about the children, and on the advice of my mother bought three large stuffed animals to have waiting for them in the apartment when they arrived. Kate later told me the girls immediatly bonded with the animals. I like to think that these helped the girls cope with the stress of their long journey to what I'm sure seemed to them an alien world. I had no family of my own in Ames and the Pałasiński's soon made me part of theirs. I became like an uncle to the girls. They thrived in Ames and after five years of school became thoroughly Americanized. I grew to greatly admire Marek who had suspended his career to help Kate with hers. He constantly sought employment to supplement Kate's teaching assistantship stipend although the nature of his visa limited his options. And he was a devoted and caring father. Kate received her doctorate in 1994 and the family returned to Kraków. The Pałasińskis invited me for Christmas every year they were in Ames, and I looked forward to celebrating it again in Kraków.

Shortly after the New Year I left for Budapest but would return to Kraków in a month for a longer stay. I drove over the Tatra Mountains into Slovakia. Tarski loved these mountains and frequently hiked in them when he was young. His favorite mountain range in California was the Trinity Alps because they reminded him so much of the Tatras. I then drove on into Hungary and finally to Budapest. Luckily the weather was good although there was evidence of a recent snowfall. Surprisinly there was no problem crossing the borders. I was using a road that must have been used almost exclusively by the locals, for the border crossings were very lightly guarded.

Ildikó kindly let me use her apartment, which she was not using at the time. It was on the Buda side of the Danube, and on most days I would drive over the Danube to in István and Hajnal's flat on the Pest side of the river to work with them and Ildikó on a chapter of a proposed volume of collected articles on the different aspects of algebraic logic. The three of them with Don Monk and Bjarni Jónsson were its editors. The chapter the four of us were working on was intended to be a broadening of algebraic logic by incorporating the important work that the Hungarians had done into the framework of AAL. One night, while driving back to my apartment, I was stopped by the Budapest police and falsely accused of running a red light. (I was exceedingly careful about my driving for fear of just such an event.) I really got frightened when the officer gave me a sobriety test because I had had a large bottle of beer at István and Hajnal's a couple of hours before, and

I knew how hard Hungary was on drunk driving. But I passed the test and to my relief was able to be on my way after paying a small (by western standards) fine directly to the officer. When I left Budapest in late January to return to Kraków we had a substantial preliminary draft of our chapter. We continued to work on it through the following summer, collaborating remotely by email. There were some other draft chapters, but the project lost steam and was eventually abandoned, but I incorporated a part of it in a paper I presented at the Workshop on Combination of Logic and Applications, CombLog'04, in Lisbon seven years later (Pigozzi, 2004).

I remained in Poland as a guest of the Pałasińskis for the rest of the winter. I reluctanly deplaced Magda from her room, but she was happy to move in with her sisters. When Kate had some free time we would work on problems that were left open in her thesis. She arranged talks for me at her university and at the Jagiellonian at the invitation of Andrzej Wroński and Paweł Idziak. In March I drove to Kedzierzyn-Kózle to visit the Czelakowskis. While there Janusz took me to visit Piotr Wotylak and Wojciech Dzik at the Silesian University in Katowice, and also to Wrocław where I gave a talk at the Institute of Mathematics. I also went with Janusz to Opole to see the University where he teaches, and to look at a house for sale. Opole is a long commute from Kedzierzyn-Koźle, 56 km, and he, Bożena, and their daughters were planning to move there. They did eventually, but much later into a lovely new house they built. Janusz took me on a tour of the Opole region. One place we saw was the small town Głogówek. During the Napoleonic wars Ludwig van Beethoven found shelter in the Głogówek castle in the fall of 1806, invited by Count Franz von Oppersdorff. Beethoven worked there on his IV and V symphonies that were commissioned by the count.

My stay in Kedzierzyn-Koźle ended on a sad note. I received a telephone call from my brother that my 95 year old mother had had a heart attack, was in the hospital and not expected to live much longer. I immediately took a train through East Germany to Frankfurt and flew home to Oakland. While I was in route she had rallied and was back home and seemed to be doing pretty well. I stayed with her a couple weeks and returned to Poland. It was only temporary however and within a short time, while I was in Italy, I had to fly back to Oakland, but sadly she died just a couple hours before I arrived home. An academic autobiography is not the place for a eulogy for my mother, but my love for her demands more than these few emotionless words. With the greatest affection she nurtured and protected me in my youth and encouraged and supported me in my adulthood. I owe her more than I can ever hope to express.

I left Poland at the end of March. I retraced my earlier trip from Vienna to Kraków, and then went south into Italy, stopping overnight in Klagenfurt Austria, through Odine, Padova, Bologna, and finally Siena, where Paolo Aglianó met me at the train station and took me home for dinner with his family. The mathematics department of the Univesity of Siena at that time was situated in the center of the city in an old, historic building that had at one time been an officers billet for a calvary unit. The interior had been renovated but the foyer still retained the picturesque evidence of its past.

I had met Aldo Ursini at a conference in the US, and in 1986 when I was in Europe he invited me to Siena for a talk. I reciprocated later when he again was in the US. I was interested in his and Paolo's work on ideals in universal algebra and was curious about its possible consequences for AAL. I had been invited this time to present an eight week graduate course on AAL that was funded by a grant from the National Research Council of Italy. Eight very good students signed up for the course and several faculty members also attended. I enjoyed the opportunity and the time this gave me to present the foundations of this relatively new branch of algebraic logic. I tried to emphasize the ideas that motivated the founders, and I prepared detailed notes for each day's lecture with the idea of using them as a basis of a monograph on the subject. I worked on this the following year for a few months but never could find the time to finish it.

During my time in Siena I got to stay at a medieval monastery that had been renovated and donated to the University for a visitor center. The interior contained a large courtyard, the building on side of which contained cells for the monks that had been cleverly modified into apartments for visitors, with all modern conveniences, but that still retained their original character to an amazing degree. It was the most unique and enjoyable accommodation away from home I had ever experienced. The monastery was out in the countryside in a lovely olive grove. Every morning I would drive the seven kilometers to Siena along a winding country road. I looked forward to turning the last curve and seeing Siena on its hill before me in the early morning spring sunlight.

I was lucky to be in Siena when a special Palio was held. The *Palio di Siena* is a horse race with entries from ten of the fifteen *contrade* or city wards that dates back to the fourteenth century. The jockeys ride bareback and are fiercely competitive, and since the race circles the relatively small *Piazza del Campo* it can be very rough. At the end of my stay Aldo invited me, Paolo and his family to dinner with his family at his home in the country. His house had been an old farmhouse that he had renovated and was in a lovely setting. After dinner he took us for a walk in the surrounding Tuscan countryside. I decided then that there must be no place as beautiful as Tuscany in the early summer. I drove back to Barcelona in one day along the Italian and French Riviera and the Spanish Costa Brava. I put my car in storage and flew to Nashville to attend a big univeral algebra conference at Vanderbilt. I hitched a ride back to Ames with some graduate students, and my sabbatical year was over.

I was however to return to Barcelona the following spring. A special semester on algebraic logic and model theory was to be held at the CRM from May through July of 1997, and I was given another grant from the Ministry of Education and Science of Spain to attend. The highlight of the semester for me was the *Workshop on Abstract Algebraic Logic* that was to be organized by Josep Maria Font and Ramon Jansana for the AAL group of the

IMUB. I was invited to join them on the Organizing Committee. This was the first meeting devoted to AAL. It was attended by specialists throughout the world and highlighted the considerable development of the subject that had occurred during the preceding ten years. This is summarized in *A Survey of Abstract Algebraic Logic* (Font et al., 2000b) which appears in Volume II of the proceedings of the Workshop (Font et al., 2003a). An updated survey can be found in (Font et al., 2009)

I spent the rest of the summer in Barcelona working with Josep Maria and Ramon on joint research projects and preparations for the special issue of Studia Logica to be devoted to papers from the Workshop. I did take some time off for an auto trip to southern Spain with my future wife Judy Casey who joined me shortly after the Workshop. I met Judy a couple of years before through a mutual friend. I was immediately attracted to her, partly because of her gregariousness. In this regard she reminded me of my mother, which is odd since no one else in my family was like this. She was born in Iowa and attended Iowa State. After teaching in elementary schools in California for several years, she obtained a masters degree in childhood development from Stanford, and susequently a masters degree in library science from the University of Northern Iowa. She was working in the Iowa State library when we met. The summer before she was temporarily without a residnce and I asked her to house sit for me while I was away on my years sabbatical. We were married three years later. It was the first marriage for both of us. She was fifty-eight and I was sixty-five. When this autobiography was written we had been married sixteen years, and I enjoyed every day of it. You'll have to ask her how she feels about it.

I was able to get her a nice apartment in San Cugat for a few weeks through the CRM. During this time I was living in a special section of a student dormitory reserved for visiting scholars. The southern Spain trip was a whirlwind six days. I felt I couldn't take more time, although I can't recall now why. We drove from Barcelona to Granada the first day, spent the next day seeing the Alhambra, and went on to Madrid on the third day. We spent half the next day in the Prado and the evening dining on roast sucking pig at the Restaurant Botín near the Plaza Mayor. On the fifth day we went on to Toledo for sightseeing and then back to Madrid. The sixth day was spent in Segovia viewing the best preserved Roman aqueduct in Spain, which was still being used. We also got to see (from a distance) the city ceremoniously greeting the Prince of Asturias, the present King Felipe VI of Spain, who was making an official visit. That evening we drove back to the UAB via Zaragoza, arriving late at night and completely exhausted. I returned to Ames for the start of the Fall Semester, thus finally ending the most eventful two years of my career.

Publications: In 1988, inspired by the work of Roman Suszko, I published a paper (Pigozzi, 1988b) about a concept I called Fregean logic. The concept is an old one, going back to Frege, but I considered it in the framework of abstract algebraic logic. A number of authors have published work on the subject considerably expanding its scope. References to this work can be found in the two papers (Czelakowski and Pigozzi, 2004a,b) coauthored by Janusz Czelakowski and me which are discussed below.

The origin of Fregean logic is Frege's principle of compositionality. Frege's seminal insight, as interpreted by Alonzo Church, was to think of a (declarative) sentence in the same way as one thinks of a proper name. A sentence, like every proper name, must denote or name something (Church's rendering of Frege's *bedeuten*). Church calls the thing it denotes, i.e., its denotation (*Bedeutung*), its *truth-value*. According to Frege a sentence also has a *sense* (*Sinn*), which is also assumed to be compositional. But Frege viewed this concept as extra-linguistic and did not attempt to incorporate it in his formal system.

Frege's analysis of proper names when applied to the denotation of sentences leads to the *principle of compositionality* for truth-values: Assume a constituent part φ of a sentence ϑ is replaced by another sentence φ' to give $\vartheta(\varphi'/\varphi)$. If φ and φ' both have the same truth-value, then so do ϑ and $\vartheta(\varphi'/\varphi)$. Logical systems that uphold the *Frege principle* are sometimes called *truth-functional* or *extensional*. Those that violate it are called *nontruth-functional* or *intensional*. Most modal logics are intensional in this sense.

The first one to formally analyze the Frege principle in a general setting was Roman Suszko. In his view the denotation of a sentence is not its truth-value, but rather something more in keeping with Frege's notion of the sense of a sentence. (Suszko looked to Wittgenstein for support for this view. For him the denotation of a sentence is what the sentence says about a certain "situation". This term was chosen by Suszko to interpret Wittgenstein's *Sachlage*—the state of affairs.) Moreover, he introduced a new binary connective Δ, called the *identity connective*, into the language with the idea that the sentence $\varphi \Delta \psi$ is to be interpreted as the proposition that φ and ψ have the same denotation in this new sense, which for the purposes of this discussion we will view as the proposition that φ and ψ have the same *meaning*. In Suszko's formal system, which he called *logic with identity*, the principal axioms governing the identity-of-meaning connective Δ express its compositionality. Suszko's system also includes all the classical connectives, in particular the biconditional $\leftrightarrow$. As in Frege's system, $\varphi \leftrightarrow \psi$ is to be interpreted as the proposition that φ and ψ have the same truth-value. It is easily shown that the two binary connectives $\leftrightarrow$ and Δ are both compositional only if the sentences $\varphi \leftrightarrow \psi$ and $\varphi \Delta \psi$ are themselves logically equivalent for all sentences φ and ψ. Thus Frege's principle that $\leftrightarrow$ is compositional can be formalized in Suszko's system as the proposition

$$(x \leftrightarrow y) \, \Delta \, (x \, \Delta \, y).$$

Suszko calls this the *Fregean axiom*. When adjoined to the other axioms of logic with identity it gives *Fregean logic*, and extensions of logic with identity in which it fails to hold are called *non-Fregean.*

In the paper (Pigozzi, 1988b) and the two papers (Czelakowski and Pigozzi, 2004a,b) with Janusz Czelakowski the Fregean axiom is investigated within the framework of *abstract algebraic logic*, where a class of algebras is associated with a given logical system based solely on the latter's metalogical properties. We consider a much wider class of deductive systems than those encompassed by Suszko's logic of identity. In particular, we consider deductive systems that are not assumed a priori to have special connectives dedicated to representing identity of meaning and of truth-value.

In (Pigozzi, 1988b) the Fregean axiom is investigated for the assertional logics of classes of algebras with a distinguished constant. Let K be a class of similar algebras with a constant 1. Its assertional logic $\mathcal{S}$ is defined as follows: For all formulas $\varphi_1, \cdots, \varphi_n, \psi$,

$$\varphi_1, \cdots, \varphi_n \models_{\mathcal{S}} \psi$$

iff every homomorhism from the formula algebra into an algebra of K that maps each φ_i into 1 must also map ψ into 1. A class K of algebras with a distingished constant, together with its associated assertional logic, is Fregean if $\Theta(a, 1) = \Theta(b, 1)$ implies $a = b$ for every $\boldsymbol{A}$ in K and every pair of its elements a, b. The main result of (Pigozzi, 1988b): A variety V is Fregean iff it is termwise definitionally equivalent to a variety of Brouwerian (i. e., relatively pseudo-complemented) semilattices with additional operations that are compatible with the congruence relations of the underlying Brouwerian structure.

A deductive system $\mathcal{S}$ by itself is viewed as an "uninterpreted" logic. Its interpretations take the form of matrices $\langle \boldsymbol{A}, F \rangle$ where $\boldsymbol{A}$ is an algebra and $F \subseteq A$, the universe of $\boldsymbol{A}$. An *interpretation of* $\mathcal{S}$ is a matrix $\langle \boldsymbol{A}, F \rangle$ together with a mapping $h: \mathrm{Fm} \to A$ from the set of formulas into A. $h(\varphi)$ is to be thought of as the meaning of the formula φ under the interpretation, and φ is "true" or "false" depending on whether or not $h(\varphi) \in F$. Several natural assumptions are made about interpretations. First of all, the meaning function h is assumed to be a homomorphism from the algebra of formulas $\mathbf{Fm}$ into $\boldsymbol{A}$; this is the *principle of compositionality of meaning*. Secondly, truth and meaning are assumed to be connected by another well-known principle, due to Leibniz. According to the Leibniz principle identity can be characterized in second-order logic by the formula

$$x \approx y \quad \text{iff} \quad \forall P(P(x) \leftrightarrow P(y)),$$

where P ranges over all unary predicates. The principle is adapted to the interpretations of a deductive system $\mathcal{S}$ by restricting attention to predicates that are "definable" in $\mathcal{S}$ by some formula $\vartheta(x)$ with a designated variable x ($\vartheta(x)$ may have other variables that are treated as parameters). Thus we assume that, if the formulas φ and ψ have different meanings in an interpretation, then they can be distinguished by some predicate, i.e., for some formula $\vartheta(x)$, $\vartheta(\varphi/x)$ and $\vartheta(\psi/x)$ have different truth-values. This is also known as the *principle of contextual differentiation*. Finally, we assume the class of interpretations is sound and complete for the consequence relation in the sense that $\Gamma \vdash_S \varphi$ iff φ is true in every interpretation in which each $\psi \in \Gamma$ is true.

A consequence of these assumptions is that the global identity-of-truth-value and identity-of-meaning relations can be characterized entirely in terms of the consequence relation, without direct reference to the interpretations. In fact, the identity-of-truth-value relation of S is given by

$$\Lambda S = \{\, \langle \varphi, \psi \rangle : \forall \Gamma \subseteq \mathrm{Fm}\big(\Gamma \vdash_S \varphi \Leftrightarrow \Gamma \vdash_S \psi\big) \,\},$$

and the identity-of-meaning relation by

$$\Omega S = \{\, \langle \varphi, \psi \rangle : \forall \Gamma \subseteq \mathrm{Fm}\, \forall \vartheta(x) \in \mathrm{Fm}\big(\Gamma \vdash_S \vartheta(\varphi/x) \Leftrightarrow \Gamma \vdash_S \vartheta(\psi/x)\big) \,\}.$$

ΛS and ΩS are called the *Frege relation* and *Leibniz congruence* of S, respectively. The Fregean axiom for S takes the form $\Lambda S = \Omega S$. Arbitrary deductive systems with this property have been identified and investigated in the literature under the name *self-extensional*.

The paradigms for self-extensional deductive systems are the classical and intuitionistic propositional calculi. But these systems have a stronger property: every interpreted classical and intuitionistic logic also satisfies the Fregean axiom, and it is this that is taken to be the defining property of a Fregean deductive system. For any theory T of a deductive system S define:

$$\widetilde{\Lambda}_S T = \{\, \langle \varphi, \psi \rangle : \forall \Gamma \subseteq \mathrm{Fm}\big(T, \Gamma \vdash_S \varphi \Leftrightarrow T, \Gamma \vdash_S \psi\big) \,\},$$

$\widetilde{\Omega}_S T =$

$$\{\, \langle \varphi, \psi \rangle : \forall \Gamma \subseteq \mathrm{Fm}\, \forall \vartheta(x) \in \mathrm{Fm}\big(T, \Gamma \vdash_S \vartheta(x/\varphi) \Leftrightarrow T, \Gamma \vdash_S \vartheta(x/\psi)\big) \,\}.$$

$\widetilde{\Omega}_S T$ is called the *Suszko congruence* of T with respect to S; it can be expressed in the following more perspicuous form by means of the consequence operator Clo_S of S. $\widetilde{\Omega}_S T = \{\, \langle \varphi, \psi \rangle : \forall \vartheta(x) \in \mathrm{Fm}\big(\mathrm{Clo}_S(T, \vartheta(x/\varphi)) = \mathrm{Clo}_S(T, \vartheta(x/\psi))\big) \,\}$. Similarly, $\widetilde{\Lambda}_S T = \{\, \langle \varphi, \psi \rangle : \mathrm{Clo}_S(T, \varphi) = \mathrm{Clo}_S(T, \psi) \,\}$. A deductive system S is *Fregean* if $\widetilde{\Lambda}_S T = \widetilde{\Omega}_S T$ for every theory T of S.

The main result of (Czelakowski and Pigozzi, 2004a) is that the Fregean axiom together with the deduction theorem is the characteristic property of the intuitionistic calculus in the following sense: Every Fregean deductive system with the deduction-detachment theorem (i. e., there exists a formula $x \to y$, not necessary a primitive conective, for which both detachment and the deduction theorem hold) is equivalent in a strong sense to an axiomatic extension of one of the appropriate fragments of the intuitionistic propositional calculus, possibly with arbitrarily many additional connectives that are compatible with intuitionistic logical equivalence in a natural way. Moreover, the same applies to every Fregean protoalgebraic deductive system with conjunction and at least one theorem because these two assumptions guarantee the deduction-detachment theorem holds.

The paper contains some new insights into the problem as to which deductive systems are *strongly algebraizable* (that is whose equivalent algebraic semantics is a variety) both in the form of original results and of elaborations of significant results on this problem due to J. M. Font and R. Jansana. Every Fregean deductive system with the deduction-detachment theorem is strongly algebraizable and its equivalent algebraic semantics is termwise definitionally equivalent to a variety of Hilbert algebras with possibly additional compatible operations. Accordingly, the algebraic counterpart of every protoalgebraic, Fregean deductive system with conjunction and at least one theorem is termwise definitionally equivalent to a variety of Brouwerian semilattices with compatible operations.

The other central topic of the paper is an investigation of the relationship between Fregean deductive systems and their 2nd-order matrix semantics. (These are matrices $\langle A, F \rangle$, where F is a set of subsets of A satisfying certain natural conditions.) A semantic version of the Fregean property is defined and it is proved that, if a protoalgebraic deductive system is Fregean, then every full 2nd-order model of it is Fregean. Conversely, the deductive system determined by any class of Fregean 2nd-order matrices is Fregean. The latter result is used to verify that a particular algebraizable, Fregean deductive system is not strongly algebraizable; the example is due to P. Idziak. A proof is outline showing that the $(\rightarrow, \neg)$-fragment of the intuitionistic propositional calculus is Fregean and algebraizable but not strongly algebraizable. This shows that the behavior of protoalgebraic, Fregean deductive systems that fail to have the deduction-detachment theorem differs strikingly from those that do. A protoalgebraic, Fregean deductive system may be either strongly algebraizable or not, but the deduction-detachment theorem guarantees strong algebraizability in this context. The fact that there is a single binary formula with the deduction-detachment property is essential.

A deductive system has the *multiterm* deduction-detachment theorem if there is a finite set of binary formulas that collectively give the theorem. An example of a Fregean deductive system with the multiterm deduction-detachment theorem is given in (Pigozzi, 1976a) that is algebraizable but not strongly algebraizable. Fregean deductive systems with the multiterm deduction-detachment theorem are studied in detail in this paper.

7 Years 1998 to 2004

Late in 1998 I was invited to speak at the 7th International Conference on Algebraic Methodology and Software Technology (AMAST'98) that was to be held in Amazonia, Brazil in January of the following year. The organizer was Armando Haeberer of Rio De Janeiro. Four years earlier he had invited me to a conference in Rio on Relational and Algebraic Methods in Computer Science (RAMICS 2) at the suggestion of Roger Maddux. I asked Judy if she wanted to go to Amazonia, and as expected she jumped at the chance. We flew to Rio the week before to see the city and then to Manaus on the Amazon River. The conference site was the Aniao Towers, a boutique hotel northwest of Manaus on the Rio Negro, a major tributary whose confluence with the Solimoes forms the Amazon. It consisted of six towers, with all 288 rooms elevated from the rain forest floor by approximately 10-20 meters and connected by approximately 8 km of catwalks that seemed to be the home of a large number of monkeys of various sizes. The hotel organized many interesting excursions into the rain forest for the participants and their guests. Judy with her capacity to make friends with anyone was of course very popular. But what made her the "Belle of AMAST'98" was getting bitten by one of the monkeys on the last evening of the conference. The infirmary nurse assured us that Amazonian monkeys never get rabies, but she had to have rabies shots when she returned to Ames.

During May and June of 1999 I was in Lisbon giving a series of lectures on *Abstract Algebraic Logic and the Specification of Abstract Data Types* at the Center for Algebra of the University of Lisbon. This was at the invitation of Isabel Ferreirim who had been a student of Wim Blok and was now on the faculty of the Center. She said that if I wanted I could use the apartment of a colleague of hers who was on leave. It was in the *Bairro Alto*, which the tour books describe as a "picturesque working class quarter of Lisbon dating from the 16th century that has traditionally been the city's haunt of artists and writers". It's a grid of narrow streets, quiet by day, but transformed at night into the city's vibrant nightlife quarter. I of course was quick to accept the offer, and it certainly lived up to its reputation. During the morning I would prepare detailed notes for the afternoon lecture, which was intended to be the basis of an eventual paper. I would then take the subway across town to the Center for Algebra, returning in the evening to enjoy the offerings of the Bairro. At the end of my stay in Portugal Judy joined me and we rented a car and drove to the Algarve, the southernmost region of Portugal. Sagres is the southwesternmost town in continental Europe with spectacular windswept bluffs overlooking the sea. We splurged and spent the night on top of one of these bluffs in the Pousada de Sagres. The next day we drove to Évora in the Alentejo region. Historically it was a trading and religious center. It has the largest number of national monuments in Portugal apart from Lisbon.

Judy flew back to Ames and I went on to Barcelona for the *5th Barcelona Logic Meeting* at the CRM. I stayed in Barcelona for the rest of the summer working with the AAL group at the IMUB. Judy joined me again in late July for our second motor trip of the summer. This time over the Pyrenees through Andorra into the Cathar country of southern France. And then east to the Rhône delta, the Côte d'Azur to Nice, and as far as Monaco. Heading east again we spent our last and most memorable night of the trip in the small town of Quillan in the foothills of the Pyrenees. We stayed the night in a small provincial hotel and had dinner in the hotel restaurant with a British couple who spend their summer vacation every year in Quillan. It was our only experience of the trip with provincial French cuisine and it was a meal I will never forget. The next day we headed by to Catalonia after visiting the two Cathar mountain castles of Peyrepertuse and Queribus. The whole trip was planned to take only ten days, and except for the last night I decided before we started where we would stop each day and made hotel reservations accordingly. Again I greatly overestimated how much we could do each day and had to push hard to keep on schedule. Consequently we arrived back in San Cugat completely exhausted, especially me because I had to do all the driving.

This was the end of July and I was planning to return to Ames and relax before courses began in late August. But I received an urgent email that I was to be a member of a delegation that was going to Armenia, in a couple of weeks, in return for a delegation from the Mathematics Department of Yerevan State University that came to Ames the year before. I knew this was being contemplated but wasn't expecting it so soon. Their delegation included several logicians whose work I was not familiar with. Their trip to Ames had been arranged for at the college level and I was unaware of it until they arrived. I returned quickly to Ames to send my passport to Chicago for an expedited visa. Our delegation consisted of only three people: myself, Roger Maddux, and Wolfgang Klieman an analyst. It was a long flight: to Frankfurt and then by Armenian Airlines to Yerevan. In 1999 Armenia had only recently become independent and signs of its Soviet past were evident. It is situated in the Caucasus region between Asia and Europe and is among the earliest Christian civilizations.

Our hosts were extraordinarily hospitable. We were lodged in a good hotel, and each of us had our own comfortable room. On the days no scientific activity was scheduled we were taken on a tour of sites in the vicinity of Yerevan. One of particular interest was the large radio telescope that was built by the Russians; Armenia was apparently the center of Soviet astronomical research. It was still operational at the time and we were given a tour of the control center by the director. Sadly it has now been abandoned. On these off days our hosts served us enormous meals, sometimes twice a day. There was far more food and alcoholic beverages than we could possibly consume, but I forced myself to eat more than I should because I was reluctant to disappoint our hosts. The scientific part of the visit consisted of several colloquia at which each participant presented a talk describing a recent research project. In addition smaller groups would meet to discuss the proposed continuing cooperation between the two mathematics departments. Just before leaving Armenia the three of us discussed the potential of establishing a relationship that would be valuable to both parties. Roger and I agreed that the research areas of the logicians at the two schools differed so much that any significant collaboration between them would be unlikely, and Wolfgang felt that this would also be the case for analysis. Thus contact with Yerevan ended with our return to Ames. But I will always have a warm feeling and a tinge of sadness when I recall our friends there, their hospitality and their eagerness to reach out to the West after years of Soviet isolation.

It was about this time that I got my last two graduate students: Manuel Martins and Sergey Babenyshev from Portugal and Russia, respectively. Isabel Ferreirim asked me if I would help her direct the research of a doctoral student that had already advanced to candidacy at the University of Lisbon, but did not want to leave Portugal because, unlike Kate Pałasińska, was not willing to leave his wife and young child or take them to the US, snd did not want to take a extended leave of absence from the position he held at the University of Évora (the second oldest universiy in Portugal, established in 1559). This was similar to the situation with Raimon Elgueta, but Raimon was well advanced on his research while Manuel had not yet chosen a topic. I was reluctant but Isabel told me that Walter Taylor had done this with another of her students, so I agreed to take him on. If I had been aware of the amount of work it would take to direct a student, who was just starting his research, by email correspondence, I probably would not have done it. But Manuel was a good student and well worth the effort.

I had recently been thinking about a more general kind of data structure than I had dealt with in my Lisbon lectures early that summer. Some of the sorts are "hidden" in the sense that the data in them and the methods used to manipulate them are not part of the specification of the program that is intended to "run" on the data structure viewed as a "machine". These data structures with hidden data prove useful in object-oriented programming. I gave Manuel my notes from the Lisbon lectures and asked him to correct the deficiencies, of which there were many, and redevelop the whole theory for data structures with hidden data. He did this with great energy and dedication, and he was able to come to Ames for a month on four separate occasions to confer personally with me. Not too long after we started working together he asked if he should give a talk on his work up to that point at a conference being held in Europe. I had some misgivings and feared that the work of a fledgling student of an outsider might not be received generously by everyone. I agreed though because I admired his confidence and felt the experience even if hard would be useful. My fears were justified, and he was confronted with several caustic comments after he spoke which greatly shook his confidence. Luckily Joseph Goguen was at the conference, and came to him after his talk to speak well of his work and encouraged him to keep at it. This meant a lot to Manuel and I will always be grateful to Joseph for the kindness he showed that day.

Sergey Babenyshev came to me with the recommendation of Vladimir Rybakov. He was well advanced when he arrived and had already fulfilled most of the requirements to advance to candidacy. I gave him a problem Josep Maria Font had raised and he quickly solved it and submitted it for publication. There was another unfinished paper, the result of an attempt of mine to fit Gentzen style calculi into the AAL framework. It contained several open problems, or more precisely, several lines of development that deadened leaving the problem unformulated. This is what I gave Sergey, admittedly not a fair thing to do to a new graduate student, but my energy level was low at the time. I was able to give him a research assistantship during most of his time in Ames that relieved him of the need to teach beyond what the Department required. He rarely came to me for help, and I had to ask him periodically to come and tell me what he had done. But he finished his thesis, got his doctorate and now has a faculty position in Russia with an active research program.

I returned to Ames from Armenia just before classes began. I agreed to teach two sections of discrete mathematics for business and social sciences. The course was required for business majors and with two lectures a week for as many as 105 students and much smaller problem sessions run by graduate students. I thought that with only two lectures to prepare each week I would have more free time, but I was sadly mistaken. The administrative work organizing and managing the problem sessions, quizzes and examinations took all of my time during the semester, and at the end of the semester I was more exhausted than at the beginning. The 2000-01 academic year was easier. I got to teach the graduate abstract algebra sequence, one of my favorites. I was able to take off for a couple of weeks in November to visit the Kanazawa Institute of Science at the invitation of Hiroakira Ono. Judy came with me and we had a great time socially, but I was very disappointed in my performance. I gave a uninspired talk and seemed unable to interact in a meaningful way mathematically with the other participants. I flew home after the meeting, but Judy staid on in Japan touring on her own.

When I got back I attempted to escape the depressing state of my research by becoming a candidate for the Chair of the Department. The only other candidate was Justin Peters, a good friend, and after the Department voted it was a virtual tie. It would be up to the Dean to decide after an interview with both candidates. Mine was delayed because I was attending the Alfred Tarski Centenary Conference in Warsaw where I gave another uninspired talk. But otherwise the conference was quite enjoyable. I got to meet old friends. Anna Romanowska who works closely with Jonathan Smith is a frequent visitor to Ames. I also visited the Szczerba family. Sadly for Leszek it would be the last time. He died several years later. The most entertaining event of the conference was a meeting where people who had known Tarski could relate their personal experiences with him. Roger Maddux and I were called on because we were two of his last students. I told the story about Tarski burning a small hole in the dashboard of my car. I drove that car for many years and cherished that hole for the fond memories of Alfred it made me recall whenever I noticed it. Judy came with me to Warsaw and after the conference we took the train to Kraków to spend some time with the Pałasińskis and then on to Opole to see the Czelakowskis. Judy had met Kate and Janusz in Warsaw but I wanted her to meet their families. They all became instant friends of her's, which always seems to happens when she meets new people.

On our return to Ames I had my interview in the Dean's office and it quickly became evident that I had given little thought to what I planned to do as Chair and that I didn't really want the job (at least not for the right reasons). So Justin became Chair, and I became the Graduate Coordinator for the 2001-02 academic year, which relieved me from teaching. But it took a lot of time and required me to completely rewrite the Department Graduate Student Handbook by the College's mandate. I finally retired in May of 2002, and Judy and I, who were now married, moved to Oakland California and into the family house where I had lived for ten years before moving to the midwest. The house was built by my parents in 1956. It was in the Oakland hills overlooking San Francisco Bay, but had been neglected for many years and was badly deteriorated. Judy and I decided to completely restore and modernize it. This lasted almost three years and took up almost all of our time, but I still had some mathematical responsibilities I couldn't abandon and some unfinished research projects I was reluctant to.

Manuel and Sergey had not yet received their degrees. I helped them via email while they were writing their theses, and in Manuel's case in person when he got a grant to come to Ames a fourth time in February of 2003 shortly before I returned on other business. I was also back in Ames 1n 2004 for Sergey's thesis defence. Manuel's thesis defence was also in 2004, in September, and in Lisbon since he would receive his doctorate from the University of Lisbon. I went to Lisbon for his defence, but this was my last time in Portugal. I had been there the previous summer and had just returned to Oakland when I got a telephone call from Joel Berman that Wim Blok had been killed in an automobile accident. It was a terrible blow and I was in a state of shock for some time.

The last time I had seen Wim was the previous summer at the Annual Meeting of the Association for Symbolic Logic in Chicago in June of 2003. I had the honor of being invited to present one of the plenary addresses. I hesitated accepting for some time because of the doubts I was having about the significance of my research. I finally accepted because I couldn't pass up the opportunity to explain the ideas behind AAL to a wider audience. I spent almost an entire month doing nothing but preparing the lecture, but afterward felt that I had completely failed. The meeting in general was going well and Wim was in a good mood when the two of us had lunch together following the day's talks. But this was dampened when I told him of my doubts about the significance of my research. I remember him responding with a little edge to his voice that it was also his research. This was the last time we talked to each other and I will always regret that we parted on that note.

A short time after Wim's death I was asked by Jacek Malinowski, Chairman of the Editorial Board of Studia Logica to edit a special issue of the journal in memory of Wim. I understood that I could be considered the natural choice for the job, and at any other time would have quickly agreed, but the winter of 2003 was particularly bad for me. We were at a critical stage in the renovation of our house, and I was fully involved in helping Manuel and Sergey finish their theses. But Wiesław Dziobiak contacted me and told me that he, Joel Berman, and James Raftery had agreed to be editors and would like to have me join them, even on a honorary basis. I was uncomfortable sharing credit for the editorial work without actually doing my full share. But I appreciated the editors recognizing the special relationship I had with Wim and wanting me to join them on any basis, so with some reservations I agreed. All the editors were able to meet together in Chicago after Wim's memorial service and plan for the initial phase of our work.

In November of 2003 I heard from Renato Lewin that he had received a grant to bring me to Santiago as a visiting scholar. He had come to Ames on sabbatical leave in the fall of 1994 and we worked on a problem that Renato had raised some time before. It had to do with the so-called annotated logics P_L that were introduced in the late 1980s as a logical framework to deal with deductive databases that contain inconsistent, conflicting or contradictory information. They are non-structural and hence do not have an algebraic semantics in the usual sense. Working together, along with my masters degree student Sara Bowers, we were able to construct a structural and algebraizable logic that simulates the deductive process of P_L in a natural way (Bowers et al., 2001).

Judy was again eager to come with me to Chile, and while Renato and I were working she took off on her own to Patagonia for a week. Renato let her have his cellphone to keep in touch but we didn't hear from her until she returned to Santiago, which caused me, Renato and his wife Carmen a lot of trouble trying to contact her. One Sunday Renato and Carmen took us to see Valparaiso. It was the busiest seaport on the west coast of South America before the Panama Canal was built and is still important. Part of the city sits on the high bluffs that overlook the older commercial part on the coast. After dinner in a restaurant with a spectacular view of the coast we took one of the many funicular railways that climb up and down the bluffs to the center of the city to see the sights, especially the giant flea market that is open every Sunday in the summer. We also went to Viña del Mar, a modern resort city just north of Valparaiso. Irene Mikenberg, a colleague of Renato's, owns a very nice condominium at the top of a high-rise building there and she generously offered it to us for the night when Renato and Carmen would return to Santiago. Irene and a friend of hers joined us in the afternoon and showed us around Viña del Mar and took us to a seafood restaurant on the beach for a good dinner. All-in-all it was a very enjoyable two days; Irene suggested we buy a condominium and spend the North American winters in Chile. The thought is appealing and we are still considering it. Renato invited me twice more. Once in 2005, during which I attended my last conference, and again in 2006. Judy came with me this time and again took off on her own to the Lake Region of southern Chile. The highlight of the trip for me was a weekend with Renato and Carmen at their house in Los Molles, a picturesque village of summer homes on the ocean about 190 km north of Santiago.

Although I have emphasised the social part of these visits to Chile, their real purpose was scientific and in fact most of my time was spend with Renato on campus working on several different projects. It was a disappointment that none of this work led to a publication, and for this I blame myself. This ended the forty years of my professional life. I had authored or coauthored over sixty publications, not a lot compared to many other mathematicians, and many more that were never published because I could never convince myself that they were ready for it. Once Wim reminded me of the old aphorism: "perfect is the enemy of the good".

One final note. In July of 2016 Judy and I flew to Kraków to attend the marriage of Magda Pałasinska, the little four year old girl who came to Ames with her father, mother, and sisters 27 years before.

References

Adams, M. and Pigozzi, D. and Sichler, J. (1981). Endomorphisms of direct unions of bounded lattices, *Archiv der Mathematik* **36**(1), 221–229.

Bergman, C. H. and Maddux, R. D. and Pigozzi D. (1990). *Algebraic logic and universal algebra in computer science*, Volume **425**. Lecture Notes in Computer Science, Conference, Ames Iowa, USA, June 1988.

Blok, W. J. and Köhler, P. and Pigozzi, D. (1984). On the structure of varieties with equationally definable principal congruences, II, *Algebra Universalis* **18**, 334–379.

Blok, W. J. and Pigozzi, D. (1982). On the structure of varieties with equationally definable principal congruences, I, *Algebra Universalis* **13**, 195–227.

Blok, W. J. and Pigozzi, D. (1986a). A finite basis theorem for quasivarieties, *Algebra Universalis* **22**(1), 1–13.

Blok, W. J. and Pigozzi, D. (1986b). Protoalgebraic logics, **45**, 337–369.

Blok, W. J. and Pigozzi, D. (1988). Alfred Tarski's work on general metamathematics, *The Journal of Symbolic Logic* **53** (01), 36–50.

Blok, W. J. and Pigozzi, D. (1989, January). *Algebraizable logics*, Volume **396**, *Memoires of the American Mathematical Society*, Providence.

Blok, W. J. and Pigozzi, D. (1991). *Algebraic Logic*, Volume **50** of Studia Logica, Invited papers.

Blok, W. J. and Pigozzi, D. (1991b). Local deduction theorems in algebraic logic, In J. D. Andréka, H. Monk and I. Németi (Eds.), *Algebraic Logic (Proc. Conf. Budapest 1988)*, Volume **54** of *Colloq. Math. Soc. János Bolyai*, pp. 75–109. Amsterdam: North-Holland.

Blok, W. J. and Pigozzi, D. (1992). Algebraic semantics for universal Horn logic without equality, In A. Romanowska and J. D. H. Smith (Eds.), *Universal Algebra and Quasigroup Theory*, Volume **19** of *Research and Exposition in Mathematics*, pp. 1–56. Berlin: Heldermann Verlag.

Blok, W. J. and Pigozzi, D. (1994a). On the structure of varieties with equationally definable principal congruences, III, *Algebra Universalis* **32**, 545–608.

Blok, W. J. and Pigozzi, D. (1994b). On the structure of varieties with equationally definable principal congruences, IV, *Algebra Universalis* **31**, 1–35.

Blok, W. J. and Pigozzi, D. (1997). On the congruence extension property, *Algebra Universalis* **38**(4), 391–394.

Blok, W. J. and Pigozzi, D. (2003). Abstract algebraic logic and the deduction theorem, Manuscript.

Bowers, S. E. and Lewin, R. A. and Pigozzi, D. (2001). An annotated logic defined by a matrix, In J. Abe and S. Tanaka (Eds.), *Unsolved Problems on Mathematic for the 21th Century–A Trubute to Kioshe Iseki's 80th Birthday*. Amsterdam: IOS Press.

Czelakowski, J. and Pigozzi, D. (1999). Amalgamation and interpolation in abstract algebraic logic, In X. Caicedo and C. M. Montenegro (Eds.), *Models, Algebras, and Proofs*, Volume **203** of *Lecture Notes in Pure and Applied Mathematics*, pp. 187–265. New York, Basel: Marcel Dekker, Inc., Selected papers of the X Latin American symposium on mathematical logic held in Bogotá.

Czelakowski, J. and Pigozzi, D. (2004a). Fregean logics, *Annals of Pure and Applied Logic* **127**, 17–76, Provinces of logic determined. Essays in the memory of Alfred Tarski. Parts IV, V and VI. Eds: Z. Adamowicz, S. Artemov, D. Niwiński, E. Orłowska, A. Romanowska and J. Woleński.

Czelakowski, J. and Pigozzi, D. (2004b). Fregean logics with the multiterm deduction theorem and their algebraization, *Studia Logica* **78**, 171–212.

Font, J. M. and Jansana, R. and Pigozzi, D. (2000a). *Abstract algebraic logic I*, Volume **65**. Studia Logica, Proceedings Workshop on Abstract Algebraic Logic, Centre de Recerca Matemàtica, Bellaterra (Spain), 1–5 July 1997.

Font, J. M. and Jansana, R. and Pigozzi, D. (2000b). Foreward to "Abstract algebraic logic I", *Studia Logica* **65**(1), 1–9, Special issue devoted to the first volume of the Proceedings of the Workshop on Abstract Algebraic Logic. Editors J. M. Font, R. Jansana, and D. Pigozzi. Centre de Recerca Matemàtica, Bellaterra (Spain), 1–5 July 1997.

Font, J. M. and Jansana, R. and Pigozzi, D. (2001). Fully adequate Gentzen systems and the deduction theorem, *Reports on Mathematical Logic*, Volume **35**:115–165.

Font, J. M. and Jansana, R. and Pigozzi, D. (2003a). *Abstract algebraic logic II*, Volume **74**. Studia Logica, Proceedings Workshop on Abstract Algebraic Logic, Centre de Recerca Matemàtica, Bellaterra (Spain), 1–5 July 1997.

Font, J. M. and Jansana, R. and Pigozzi, D. (2003b). Foreward to "Abstract algebraic logic II", *Studia Logica* **74**(1–2), 1–9, Special issue devoted to the second volume of the Proceedings of the Workshop on Abstract Algebraic Logic. Editors J. M. Font, R. Jansana, and D. Pigozzi. Centre de Recerca Matemàtica, Bellaterra (Spain), 1–5 July 1997.

Font, J. M. and Jansana, R. and Pigozzi, D. (2003c). A survey of abstract algebraic logic, *Studia Logica* **74**(1–2), 13–97.

Font, J. M. and Jansana, R. and Pigozzi, D. (2006). Fully adequate Gentzen systems and closure properties of the class of full models, In D. P. J. Berman, W. Dziobiak and J. Raftery (Eds.), *Studia Logica. Special issue in memory of Willem Johannes Blok*, Volume **83**. Springer.

Font, J. M. and Jansana, R. and Pigozzi, D. (2009). Update to "A survey of abstract algebraic logic", *Studia Logica* **91**(1), 125–130.

Köhler, P. and Pigozzi, D. (1980). Varieties with equationally definable principal congruences, *Algebra Universalis* **11**(1), 213–219.

Leavens, G. and Pigozzi, D. (2002). Equational reasoning with subtypes, Technical Report #02–07, Iowa State University Department of Computer Science.

Leavens, G. T. and Pigozzi, D. (1991). Typed homomorphic relations extended with subtypes, In A. M. S. Brooks, M. Main and D. Schmidt (Eds.), *International Conference on Mathematical Foundations of Programming Semantics*, Volume **598** of *Lecture Notes in Computer Science*, pp. 144–167.

Leavens, G. T. and Pigozzi, D. (1997). The behavior-realization adjunction and generalized homomorphic relations, *Theoretical Computer Science* **177**(1), 183–216.

Leavens, G. T. and Pigozzi, D. (1998). Class-based and algebraic models of objects, *Electronic Notes in Theoretical Computer Science* **14**, 214–244.

Leavens, G. T. and Pigozzi, D. (2000). A complete algebraic characterization of behavioral subtyping, *Acta Informatica* **36**(8), 617–633.

Martins, M. A. and Pigozzi, D. (2007). Behavioural reasoning for conditional equations, *Mathematical Structures in Computer Science* **17**(05), 1075–1113.

Mobasher, B. and Leszczyłowski, J. and Slutzki, G. and Pigozzi, D. (1993a, May), A knowledge-based procedural semantics for logic programming, Technical Report #93–15, Iowa State University, Department of Computer Science.

Mobasher, B. and Leszczyłowski, J. and Slutzki, G. and Pigozzi, D. (1993b). Negation as partial failure, In I. M. Pereira and A. Nerode (Eds.), *Logic programming and non-monotonic reasoning. Proceedings 2nd international workshop*, pp. 244–262.

Mobasher, B. and Pigozzi, D. and Slutzki, G. (1997a). Algebraic semantics for knowledge-based logic programs, In L. V. S. Lakshmanan and M. Nivat (Eds.), *Proceedings international workshop on uncertainty in databases and deductive systems*, Ithaca, New York, Switzerland, pp. 77–109. Elsevier.

Mobasher, B. and Pigozzi, D. and Slutzki, G. (1997b). Multi-valued logic programming semantics: an algebraic approach, *Theoretical Computer Science* **171**(1-2), 77–109.

Mobasher, B. and Pigozzi, D. and Slutzki, G. and Voutsadakis, G. (2000). A duality theory for bilattices, *Algebra Universalis* **43**, 109–125.

Pałasińska, K. and Pigozzi, D. (1995a). Gentzen-style axiomatizations in equational logic, *Algebra Universalis* **34**, 128–143.

Pałasińska, K. and Pigozzi, D. (1995b, November). Implication in abstract algebra, Preprint 311, Centre de Recerca Matemàtica, Bellaterra (Spain).

Pałasińska, K. and Pigozzi, D. (2000). Bikraty a programowanie logiczne oparte na wiedzy, In E. Szumakowicz (Ed.), *Granice sztucznej inteligencji: eseje i studia*, pp. 179–211. Kraków: Politechnika Krakowska.

Pigozzi, D. (1972a). Amalgamation, congruence-extension, and interpolation properties of algebras, *Algebra Universalis* **1**, 269–349.

Pigozzi, D. (1972b). On some operations on classes of algebras, *Algebra Universalis* **2**, 346–353.

Pigozzi, D. (1974). The join of equational theories, *Colloquium Mathematicae,* **30**, 15–25, Institute of Mathematics, Polish Academy of Sciences, 1974.

Pigozzi, D. (1975, March). Equational logic and equational classes of algebras, Technical Report CSD 123, Purdue University, Department of Computer Science.

Pigozzi, D. (1976a). Base undecidable properties of universal varieties, *Algebra Universalis* **6**, 193–223.

Pigozzi, D. (1976b). The universality of the variety of quasigroups, *Journal of the Australian Mathematical Society,* **21**, 193–223.

Pigozzi, D. (1979a). Minimal, locally finite varieties that are not finitely axiomatizable, *Algebra Universalis* **9**, 374–390.

Pigozzi, D. (1979b). Universal equational theories and varieties of algebras, *Annals of Mathematical Logic,* **17**, 117–150.

Pigozzi, D. (1980). Varieties of pseudo-Post algebras, In *The Tenth International Symposium on Multiple-valued Logic, Northwestern University, Evanston, Ill., 1980*, pp. 205. IEEE.

Pigozzi, D. (1981a). Finite groupoids without finite bases for their identities, *Algebra Universalis* **13**(1), 329–354.

Pigozzi, D. (1981b). On the structure of equationally complete varieties. I, *Colloquium Mathematicae* **45**:191–201. Institute of Mathematics, Polish Academy of Sciences.

Pigozzi, D. (1981c). On the structure of equationally complete varieties. II, *Transactions of the American Mathematical Society* **264**(2), 301–319.

Pigozzi, D. (1988a).   Finite basis theorems for relatively congruence-distributive quasivarieties, *Transactions of the American Mathematical Society* **310**(2), 499–533.

Pigozzi, D. (1988b). Fregean algebraic logic, In J. D. Andréka, H. Monk and I. Németi (Eds.), *Algebraic Logic*, Number **54** in Colloquia Mathematica Societatis, János Bolyai, pp. 473–502. Amsterdam: North-Holland.

Pigozzi, D. (1990). Data types over multiple-valued logics, *Theoretical Computer Science* **77**(1–2), 161–194.

Pigozzi, Don (1991). Equality-test and if-then-else algebras: Axiomatization and specification, *SIAM Journal on Computing* **20**(4), 766–805.

Pigozzi, D. (1992). A lattice theoretic characterization of equivalent quasivarieties, In O. H. K. L. A. Bokut, Yu. L. Ershov and A. I. Kostríkin (Eds.), *Proceedings of the international conference on algebra honoring A. Mal'cev (Part 3)*, Volume **131** of *Contempory Mathematics*, pp. 187–199.

Pigozzi, D. (1997a, May, June). Lectures in abstract algebraic logic (Draft), Elaboration of notes for a short couse taught at the Department of Mathematics, Univesity of Siena.

Pigozzi, D. (1997b, June). Second-order algebraizable logics (Draft), Centre de Recerca Matemàtica, Bellaterra (Spain).

Pigozzi, D. (1998, July). Abstract algebraic logic: past, present and future, In *Workshop on Abstract Algebraic Logic*, Number 10, pp. 122–139. Bellaterra (Spain): Centre de Recerca Matemàtica.

Pigozzi, D. (1999a).   Abstract algebraic logic, In A. M. Haeberer (Ed.), *Algebraic methods and software technology*, Volume **1548** of *Lecture Notes in Computer Science*, pp. 8–16. Proceedings 7th International Conference, AMAST'98 Amazonia ,Brazil, January 1999: Springer, Verlag.

Pigozzi, D. (1999b, May). Abstract algebraic logic and the specification of abstract data types (Draft), A course of lectures, Center for Algebra, University of Lisbon.

Pigozzi, Don (2002). Abstract algebraic logic, in M. Hazewinkel, editor, *Encyklopaedia of Mathematics, Supplement III*, volume **13** of *Encyklopaedia of Mathematics*, Chapter A, pp. 2–13, Springer Netherlands.

Pigozzi, D. (2004, July 28-30). Combining interpreted languages in abstract algebraic logic, In W. A. Carnielli, F. M. DionÃsio, and P. Mateus (Eds.), *CombLog'04, Workshop on Combination of Logics: Theory and Applications*. CLC, Department of Mathematics, IST, Lisbon, Portugal.

Pigozzi, D. and Salibra, A. (1993a). An introduction to lambda abstraction algebras, In *IX Simposio Latinoamericano de Logica Matematica*, Volume **38**, pp. 93–112. Universidad Nacional del Sur Bahia Bianca, Argentina.

Pigozzi, D. and Salibra, A. (1993b). Polyadic algebras over nonclassical logics, in *Algebraic Methods in Logic and in Computer Science*, volume **28** of *Banach Center Publications*, pp. 51–66, Warszawa, Institute of Polish Academy of Sciences.

Pigozzi, D. and Salibra, A. (1993c). A representation theorem for lambda abstraction algebras, In A. M. Borzyszkowski and S. Sokolowski (Eds.),

Proceedings of the 18th International Symposium on Mathematical Foundations of Computer Science, Volume **7111** of *Lecture Notes in Computer Science*, pp. 629–639. Springer-Verlag.

Pigozzi, D. and Salibra, A. (1994). Dimension-complemented lambda abstraction algebras, In T. R. M. Nivat, C. Rattray and G. Scollo (Eds.), *Algebraic Methodology and Software Technology (AMAST'93)*, Workshops in Computing, pp. 129–136. Springer–Verlag.

Pigozzi, D. and Salibra, A. (1995a). The abstract variable-binding calculus, *Studia Logica* **55**(1), 129–179.

Pigozzi, D. and Salibra, A. (1995). Lambda abstraction algebras: representation theorems, *Theoretical Computer Science* **140**(1), 5–52.

Pigozzi, D. and Salibra, A. (1998). Lambda abstraction algebras: coordinatizing models of lambda calculus, *Fundamenta Informaticae* **33**(2), 149–200.

Pigozzi, D. and Sichler, J. (1985). Homomorphisms of partial and of complete steiner triple systems and quasigroups, In *Universal algebra and lattice theory*, Volume **1149** of *Lecture Notes In Mathematics*, pp. 224–237, Proceedings of Conference in Universal Algebra, Charleston, South Carolina, July 11-14, 1984.

Pigozzi, D and Tardos, G. (1985). The representation of certain lattices as lattices of subvarieties, Manuscript.

Assertional logics, truth-equational logics, and the hierarchies of abstract algebraic logic

Hugo Albuquerque, Josep Maria Font, Ramon Jansana, and Tommaso Moraschini

Dedicated to Professor Don Pigozzi
on the occasion of his 80th birthday

Abstract We establish some relations between the class of truth-equational logics, the class of assertional logics, other classes in the Leibniz hierarchy, and the classes in the Frege hierarchy. We argue that the class of assertional logics belongs properly in the Leibniz hierarchy. We give two new characterizations of truth-equational logics in terms of their full generalized models, and use them to obtain further results on the internal structure of the Frege hierarchy and on the relations between the two hierarchies. Some of these results and several counterexamples contribute to answer a few open problems in abstract algebraic logic, and open a new one.

2010 Mathematics Subject Classification: 03G27, 03B22

Key words: Abstract algebraic logic - Leibniz hierarchy - Frege hierarchy - truth-equational logics - assertional logics - Fregean logics - full generalized models - unital matrices

Hugo Albuquerque
Departament de Filosofia, Universitat de Barcelona (UB), C. Montalegre 7, E-08001 Barcelona, Spain, e-mail: hugo.albuquerque@ua.pt

Josep Maria Font
Departament de Matemàtiques i Informàtica, Universitat de Barcelona (UB), Gran Via de les Corts Catalanes 585, E-08007 Barcelona, Spain, e-mail: jmfont@ub.edu

Ramon Jansana
Departament de Filosofia, Universitat de Barcelona (UB), C. Montalegre 7, E-08001 Barcelona, Spain, e-mail: jansana@ub.edu

Tommaso Moraschini
Institute of Computer Science, Czech Academy of Sciences, Pod Vodárenskou věží 2, 182 07 Prague 8, Czech Republic, e-mail: tommaso.moraschini@gmail.com

© Springer International Publishing AG 2018
J. Czelakowski (eds.), *Don Pigozzi on Abstract Algebraic Logic,
Universal Algebra, and Computer Science*, Outstanding Contributions
to Logic 16, https://doi.org/10.1007/978-3-319-74772-9_2

53

1 Introduction

The contribution of Don Pigozzi to the recent evolution of algebraic logic is enormous. He has shaped the field in such a way that the community has adopted a special name, *abstract algebraic logic*, promoted by him,[1] to refer to the area of algebraic logic that studies in abstract mathematical terms the very process of algebraization of logics, associates with every logic an algebraic counterpart, relates properties of a logic with properties of its algebraic counterpart, and classifies logics according to the type of relation they enjoy with it; all this, with the purpose that once one knows where a logic fits in the classification, the application of the theory built around the classification criteria and their consequences can immediately reveal many of its properties.

Don's fundamental work was developed mostly, but not exclusively, in his long standing collaborations with Willem J. Blok and with Janusz Czelakowski. At its center we find the construction of an impressive edifice, the so-called *Leibniz hierarchy* (Blok and Pigozzi, 1992), based on the notions of *algebraizable logic* (Blok and Pigozzi, 1989) and of *protoalgebraic logic* (Blok and Pigozzi, 1986). Other scholars (such as Janusz Czelakowski, Burghard Herrmann, Ramon Jansana, James Raftery) have also contributed to the enlargement and further study of this hierarchy. The latest addition to the Leibniz hierarchy is the class of *truth-equational logics*, characterized in Raftery (2006); up to now, it is the only class in this hierarchy not contained in the class of protoalgebraic logics (but see below). Don also laid the foundations (Pigozzi, 1991) of the study with algebraic logic tools of the distinction between Fregean and non-Fregean logics (due to Roman Suszko); this gave rise to the technical notion of *Fregean logic* and later on to the construction of a simpler hierarchy, the *Frege hierarchy*,[2] where logics are classified according to replacement properties they (or their models) satisfy. We adress the reader to Czelakowski (2001) for more information on the Leibniz hierarchy, and to Font (2015, 2016) for both hierarchies.

One of the goals of the present paper is to contend that the (already well-known) class of *assertional logics* (also called "1-assertional" in the literature) should be counted among those in the Leibniz hierarchy (notice that it is not included in the class of protoalgebraic logics). We also study the relations between this class and that of truth-equational logics, and between

---

[1] In this Don followed a suggestion of Hajnal Andréka and István Németi, who first used the term, in Section 5.3 of Henkin, Monk, and Tarski (1985), for their abstract model theoretic approach to the algebraization of first-order logic. It was first applied in the present sense (i.e., to the study of sentential logics) in the *Workshop on abstract algebraic logic* organized in Barcelona, under Don's chairmanship, in 1997. In the 2010 version of the *Mathematics Subject Classification*, the term appears with code 03G27. Notice that these initial paragraphs of our Introduction just intend to put the paper in context; a complete exposition of Don's work is found elsewhere in this volume.

[2] See the detailed references given after Definition 12.

each of these and the classes in the Frege hierarchy, in order to further clarify the internal structure of this hierarchy and some relations between the two hierarchies. In particular, we show that the Frege hierarchy becomes considerably simplified inside large portions of the Leibniz hierarchy. Our results, and the construction of some *ad hoc* counterexamples, allow us to answer several open problems on these issues; one of these answers is only partial and opens another new problem.

The structure of the paper is as follows. After summarizing the indispensable preliminaries in Section 2, we introduce assertional logics and truth-equational logics in Section 3. The analysis of several characterizations of the former among the latter supports our claim that the class of assertional logics should be considered as belonging to the Leibniz hierarchy, as it can be characterized by conditions formulated purely in terms of the Leibniz congruence, in the same way as, say, regularly algebraizable logics are characterized among the algebraizable ones. We see that the class of truth-equational logics occupies an intermediate position between the class of assertional logics and the class of logics having an algebraic semantics (the latter not belonging to the Leibniz hierarchy). Then, in Section 4 we introduce the fundamental notion of full generalized model of a logic, present the Frege hierarchy, and establish that Fregean logics with theorems are all assertional, and hence truth-equational, and that for a fully selfextensional logic, to be assertional is the same as to be truth-equational. In Section 5 we give two characterizations, of independent interest, of truth-equational logics in terms of their full generalized models. In Section 6, using these characterizations and the appropriate counterexamples, we prove that for truth-equational logics the Frege hierarchy reduces to exactly three classes, and that for finitary weakly algebraizable logics it reduces to two. Finally, we combine our results in order to answer several open problems on the structure of the Frege hierarchy posed in Font and Jansana (1996) and Font (2003, 2006): we prove that the class of selfextensional logics is not the union of the classes of Fregean and fully selfextensional logics, that there are finitely regularly algebraizable logics that are selfextensional but not fully selfextensional, and that for logics with theorems the class of fully Fregean logics is the intersection of the classes of Fregean logics and of fully selfextensional logics. This last result opens a new problem, that of whether the assumption that the logic has theorems can be deleted from it.

2 Preliminaries

We assume the reader is acquainted with the standard notions, terminology and notations of abstract algebraic logic, as given for instance in Blok and Pigozzi (1989); Czelakowski (2001); Font (2015, 2016); Font and Jansana

(1996); Font, Jansana, and Pigozzi (2003); Raftery (2006); Wójcicki (1988). We recall here just the most central to the paper.

All logics and all algebras we deal with are assumed to share an arbitrary but fixed algebraic language. A (sentential) logic $\mathcal{L}$ is identified with its consequence relation $\vdash_{\mathcal{L}}$.

Three kinds of algebra-based structures play a rôle in this area as models of logics: just plain **algebras** (denoted by $\mathbf{A}, \mathbf{B}$, etc., with universes A, B, resp.), **matrices** in the usual sense (i.e., pairs $\langle \mathbf{A}, F \rangle$ where $F \subseteq A$), and **generalized matrices** (**g-matrices** for short), which are pairs $\langle \mathbf{A}, \mathcal{C} \rangle$ where $\mathcal{C}$ is a closure system of subsets of A. If $\mathbf{A}$ is an algebra, the set of all the $\mathcal{L}$-**filters** of $\mathbf{A}$ is a closure system and is denoted by $\mathcal{F}i_{\mathcal{L}}\mathbf{A}$. A matrix $\langle \mathbf{A}, F \rangle$ is a **model** of a logic $\mathcal{L}$ when $F \in \mathcal{F}i_{\mathcal{L}}\mathbf{A}$, and a g-matrix $\langle \mathbf{A}, \mathcal{C} \rangle$ is a **generalized model** (**g-model** for short) of $\mathcal{L}$ when $\mathcal{C} \subseteq \mathcal{F}i_{\mathcal{L}}\mathbf{A}$. Thus, the largest g-model of $\mathcal{L}$ on $\mathbf{A}$ is the g-matrix $\langle \mathbf{A}, \mathcal{F}i_{\mathcal{L}}\mathbf{A} \rangle$.

Given an algebra $\mathbf{A}$ and a subset F of its universe A, the **Leibniz congruence** of F, denoted by $\Omega^{\mathbf{A}} F$, is the largest congruence of $\mathbf{A}$ that is compatible with F in the sense that it does not identify elements in F with elements of A not in F. Note that this congruence is a purely algebraic object and does not depend on any logic. However, when studying a sentential logic $\mathcal{L}$, the term **Leibniz operator** on $\mathbf{A}$ refers to the map $F \mapsto \Omega^{\mathbf{A}} F$ restricted to $\mathcal{F}i_{\mathcal{L}}\mathbf{A}$. Several classes of logics with particularly well-behaved matrix semantics can be characterized in terms of properties of this operator, constituting the so-called **Leibniz hierarchy** (the part of this hierarchy relevant for the paper is depicted in Figure 1 on page 62). A matrix is **reduced** when its Leibniz congruence is the identity relation. The class of reduced models of $\mathcal{L}$ is denoted by $\mathsf{Mod}^{*}\mathcal{L}$, and the class of its algebraic reducts by $\mathsf{Alg}^{*}\mathcal{L}$. This class of algebras was classically taken to be the most natural algebraic counterpart of the logic $\mathcal{L}$, but in Font and Jansana (1996) it was shown that this may not be the case for some non-protoalgebraic logics, and that a more general algebra-based semantics where generalized matrices replace ordinary matrices seems to yield better results. To introduce it we need a few more definitions.

If $\mathcal{C}$ is a closure system over the universe A of an algebra $\mathbf{A}$, its **Tarski congruence** is $\widetilde{\Omega}^{\mathbf{A}}\mathcal{C} := \bigcap \{\Omega^{\mathbf{A}} F : F \in \mathcal{C}\}$; it is the largest congruence of $\mathbf{A}$ compatible with all $F \in \mathcal{C}$. The map $\mathcal{C} \longmapsto \widetilde{\Omega}^{\mathbf{A}}\mathcal{C}$ is called the **Tarski operator** on $\mathbf{A}$. The **reduction** of a g-matrix $\langle \mathbf{A}, \mathcal{C} \rangle$ is the result of factoring it out by its Tarski congruence, that is, the quotient g-matrix $\langle \mathbf{A}/\widetilde{\Omega}^{\mathbf{A}}\mathcal{C}, \mathcal{C}/\widetilde{\Omega}^{\mathbf{A}}\mathcal{C} \rangle$. A g-matrix is a g-model of a logic if and only if its reduction is. A g-matrix is **reduced** when its Tarski congruence is the identity relation (obviously, the reduction of a g-matrix is always reduced).

The class $\mathsf{Alg}\mathcal{L}$ is defined as the class of the algebraic reducts of the reduced g-models of a logic $\mathcal{L}$. This class of algebras provides another algebraic counterpart of a logic that is useful for all logics, even if they are not protoalgebraic. Moreover, when $\mathcal{L}$ is protoalgebraic, $\mathsf{Alg}\mathcal{L} = \mathsf{Alg}^{*}\mathcal{L}$. In particular, if $\mathcal{L}$ is algebraizable, then $\mathsf{Alg}\mathcal{L}$ coincides with its largest equivalent algebraic semantics introduced by Blok and Pigozzi (1989), and when $\mathcal{L}$ is implicative, $\mathsf{Alg}\mathcal{L}$ coincides with the class of $\mathcal{L}$-algebras as defined by Rasiowa (1974).

A simple construction that plays an important rôle in the paper is the following. If $\mathcal{C}$ is a closure system, for each $F \in \mathcal{C}$ we consider the closure system $\mathcal{C}^F := \{G \in \mathcal{C} : F \subseteq G\}$. Hence, each g-model $\langle \mathbf{A}, \mathcal{C} \rangle$ of a logic $\mathcal{L}$ gives rise to a family of g-models of the form $\langle \mathbf{A}, \mathcal{C}^F \rangle$, one for each for $F \in \mathcal{C}$. In particular, from the largest g-model $\langle \mathbf{A}, \mathcal{F}i_{\mathcal{L}}\mathbf{A} \rangle$ we obtain a g-model of the form $\langle \mathbf{A}, (\mathcal{F}i_{\mathcal{L}}\mathbf{A})^F \rangle$ for each $F \in \mathcal{F}i_{\mathcal{L}}\mathbf{A}$.

If $\langle \mathbf{A}, \mathcal{C} \rangle$ is a g-matrix and $F \in \mathcal{C}$, the ***Suszko congruence*** of F (relative to $\mathcal{C}$) is $\widetilde{\Omega}^{\mathbf{A}}_{\mathcal{C}} F := \widetilde{\Omega}^{\mathbf{A}} \mathcal{C}^F = \bigcap\{\Omega^{\mathbf{A}} G : G \in \mathcal{C}, F \subseteq G\}$. This notion was formally introduced[3] by Czelakowski (2003), in the special case where $\mathcal{C} = \mathcal{F}i_{\mathcal{L}}\mathbf{A}$; in this case, since the relativization is actually determined by $\mathcal{L}$, it makes sense to use the symbol $\widetilde{\Omega}^{\mathbf{A}}_{\mathcal{L}} F$ instead of the more complicated $\widetilde{\Omega}^{\mathbf{A}}_{\mathcal{F}i_{\mathcal{L}}\mathbf{A}} F$, when $F \in \mathcal{F}i_{\mathcal{L}}\mathbf{A}$, and therefore

$$(1) \qquad \widetilde{\Omega}^{\mathbf{A}}_{\mathcal{L}} F := \widetilde{\Omega}^{\mathbf{A}} (\mathcal{F}i_{\mathcal{L}}\mathbf{A})^F = \bigcap\{\Omega^{\mathbf{A}} G : G \in \mathcal{F}i_{\mathcal{L}}\mathbf{A}, F \subseteq G\}.$$

A model $\langle \mathbf{A}, F \rangle$ of a logic $\mathcal{L}$ is ***Suszko-reduced*** when its Suszko congruence $\widetilde{\Omega}^{\mathbf{A}}_{\mathcal{L}} F$ relative to $\mathcal{L}$ is the identity relation. The class of Suszko-reduced models of $\mathcal{L}$ is denoted by $\mathsf{Mod}^{\mathrm{Su}}\mathcal{L}$. The class of algebraic reducts of the matrices in $\mathsf{Mod}^{\mathrm{Su}}\mathcal{L}$ turns out to be the class $\mathsf{Alg}\mathcal{L}$; this fact reinforces the relevance of this class as a universal algebraic counterpart of a logic.

The map given by $F \longmapsto \widetilde{\Omega}^{\mathbf{A}}_{\mathcal{L}} F$ defined on $\mathcal{F}i_{\mathcal{L}}\mathbf{A}$ is called ***the Suszko operator*** (relative to $\mathcal{L}$) on $\mathbf{A}$. A logic is protoalgebraic if and only if the Suszko operator (relative to it) and the Leibniz operator, both on the formula algebra, coincide on its theories (Czelakowski, 2001, Theorem 1.5.4); or, equivalently, if and only if the two operators coincide on the filters of the logic on arbitrary algebras (Czelakowski, 2003, Theorem 1.10). Thus, it seems that the specific properties of the Suszko operator should be particularly relevant for algebraic studies of logics where protoalgebraicity is not assumed; for instance, it is one of the key tools in Raftery's study of truth-equational logics (Raftery, 2006). The paper by Albuquerque et al. (2016) studies a common framework that encompasses both the Leibniz and the Suszko operators, and obtains characterizations of several classes in the Leibniz hierarchy in terms of properties of the Suszko operator.

Each closure system $\mathcal{C}$ on a set A has an associated closure operator C over A, defined as $CX := \bigcap\{F \in \mathcal{C} : X \subseteq F\}$ for all $X \subseteq A$. Using it we can define ***the Frege relation*** of a closure system $\mathcal{C}$ as $\varLambda\mathcal{C} := \{\langle a, b \rangle \in A \times A : C\{a\} = C\{b\}\}$; notice that $\langle a, b \rangle \in \varLambda\mathcal{C}$ if and only if a and b belong to the same members of $\mathcal{C}$. This defines ***the Frege operator*** (relative to $\mathcal{C}$) as the map $F \longmapsto \varLambda_{\mathcal{C}} F := \varLambda\mathcal{C}^F$, for $F \in \mathcal{C}$. The relations $\varLambda\mathcal{C}$ and $\varLambda_{\mathcal{C}} F$ are equivalence relations, but not necessarily congruences; it turns out that the

---

[3] Czelakowski attributes its invention and first characterization to Suszko, in unpublished lectures.

largest congruence of $\mathbf{A}$ below the Frege relation ΛC is the Tarski congruence $\widetilde{\Omega}^{\mathbf{A}}C$, and the largest congruence of $\mathbf{A}$ below $\Lambda_C F$ is the Suszko congruence $\widetilde{\Omega}^{\mathbf{A}}_C F$.

The set $Th\mathcal{L}$ of theories of a logic $\mathcal{L}$ is a closure system, and so we can always view a logic as the g-matrix $\langle \mathbf{Fm}, Th\mathcal{L} \rangle$; the associated closure operator will be denoted by $C_{\mathcal{L}}$. Moreover, $\mathcal{F}i_{\mathcal{L}}\mathbf{Fm} = Th\mathcal{L}$. Thus, all the above definitions and constructions given for arbitrary g-matrices can in particular be given for a logic. In this case, the superscript that would correspond to the formula algebra will be omitted; thus, on the set of theories of $\mathcal{L}$ we have the Leibniz operator Ω and the Suszko operator $\widetilde{\Omega}_{\mathcal{L}}$, and we can also consider the Tarski operator $\widetilde{\Omega}$ on closure systems of theories. The Frege relation and operator relative to $Th\mathcal{L}$ are denoted by $\Lambda\mathcal{L}$ and $\Lambda_{\mathcal{L}}$ instead of $\Lambda Th\mathcal{L}$ and $\Lambda_{Th\mathcal{L}}$, respectively. The relation $\Lambda\mathcal{L}$ is also denoted by $\dashv\vdash_{\mathcal{L}}$, as it is simply the relation of interderivability with respect to the logic $\mathcal{L}$. As established in general, note that $\widetilde{\Omega}\mathcal{L}$ is the largest congruence below $\Lambda\mathcal{L}$.

So far we have recalled the definition of two classes of algebras associated with each logic $\mathcal{L}$, namely $\mathsf{Alg}^*\mathcal{L}$ and $\mathsf{Alg}\mathcal{L}$. A third class it is useful to consider, called the ***intrinsic variety*** of $\mathcal{L}$, is defined as $\mathbb{V}\mathcal{L} := \mathbb{V}(\mathbf{Fm}/\widetilde{\Omega}\mathcal{L})$, where for a class K of algebras the variety it generates is denoted by $\mathbb{V}(\mathsf{K})$. Since the congruence $\widetilde{\Omega}\mathcal{L}$ is fully invariant, it follows that $\mathbb{V}\mathcal{L} \vDash \alpha \approx \beta$ if and only if $\langle \alpha, \beta \rangle \in \widetilde{\Omega}\mathcal{L}$. The following facts about the three classes will be relevant to the paper:

$$\mathsf{Alg}^*\mathcal{L} \subseteq \mathsf{Alg}\mathcal{L} \subseteq \mathbb{V}\mathcal{L} \qquad \mathbb{V}(\mathsf{Alg}^*\mathcal{L}) = \mathbb{V}(\mathsf{Alg}\mathcal{L}) = \mathbb{V}\mathcal{L}.$$

An interesting fact we will need is the following.

Lemma 1. *Let $\mathcal{L}$ be a logic complete with respect to a class of (g-)matrices with the class K of algebras as algebraic reducts. For all $\alpha, \beta \in Fm$, if $\mathsf{K} \vDash \alpha \approx \beta$, then $\alpha \dashv\vdash_{\mathcal{L}} \beta$. As a consequence, $\mathbb{V}\mathcal{L} \subseteq \mathbb{V}(\mathsf{K})$, and hence both $\mathsf{Alg}^*\mathcal{L}$ and $\mathsf{Alg}\mathcal{L}$ are included in the variety generated by K.*

Proof. Assume that $\mathsf{K} \vDash \alpha \approx \beta$; this means that for any $\mathbf{A} \in \mathsf{K}$ and any $h \in \mathrm{Hom}(\mathbf{Fm}, \mathbf{A})$, $h\alpha = h\beta$. In particular, for any matrix $\langle \mathbf{A}, F \rangle$ in the class, $h\alpha \in F$ if and only if $h\beta \in F$. The completeness of $\mathcal{L}$ with respect to the class of matrices implies that $\alpha \dashv\vdash_{\mathcal{L}} \beta$. The case of g-matrices is proved similarly. That is, $\{ \langle \alpha, \beta \rangle \in Fm \times Fm : \mathsf{K} \vDash \alpha \approx \beta \} \subseteq \Lambda\mathcal{L}$. But since the set is clearly a congruence of the formula algebra, this fact implies that $\{ \langle \alpha, \beta \rangle \in Fm \times Fm : \mathsf{K} \vDash \alpha \approx \beta \} \subseteq \widetilde{\Omega}\mathcal{L}$. This shows that $\mathbb{V}\mathcal{L} \subseteq \mathbb{V}(\mathsf{K})$, and the fact that $\mathsf{Alg}^*\mathcal{L}$ and $\mathsf{Alg}\mathcal{L}$ generate the variety $\mathbb{V}\mathcal{L}$ proves the final assertion. $\square$

In practice, this may give interesting and workable information for a logic that is *defined* from a single (g-)matrix, or a small set of (g-)matrices: the equations that hold in the algebraic reducts of the defining (g-)matrices also hold in the three classes of algebras associated with the logic (see Example 23 for an application).

3 Assertional logics and truth-equational logics

Several classes of logics, of different strength, can be defined by considering how *the truth filter* in their matrix models (i.e., the set F of the matrices $\langle \mathbf{A}, F \rangle$ that are models of $\mathcal{L}$) is determined, and by their relation to the relative equational consequences of classes of algebras. To introduce them we need some further notation and terminology.

Equations are identified with pairs of formulas, which are conventionally denoted by $\alpha \approx \beta$ instead of $\langle \alpha, \beta \rangle$. Any set $\boldsymbol{\tau}(x)$ of equations in at most one variable x induces a map, denoted also by $\boldsymbol{\tau}$, that transforms (sets of) formulas into sets of equations; it is defined by putting $\boldsymbol{\tau}\varphi := \boldsymbol{\tau}(\varphi)$ for any $\varphi \in Fm$, and $\boldsymbol{\tau}\Gamma := \bigcup\{\boldsymbol{\tau}\varphi : \varphi \in \Gamma\}$ for any $\Gamma \subseteq Fm$. Then, for any algebra $\mathbf{A}$ we consider the set of "solutions" of the equations in $\boldsymbol{\tau}(x)$,

$$\boldsymbol{\tau}\mathbf{A} := \big\{a \in A : \mathbf{A} \vDash \boldsymbol{\tau}(x) \,[\![a]\!]\big\}$$
$$= \big\{a \in A : \delta^{\mathbf{A}}(a) = \varepsilon^{\mathbf{A}}(a) \ \text{ for all } \delta \approx \varepsilon \in \boldsymbol{\tau}(x)\big\},$$

and for each $a \in A$ we put

$$\boldsymbol{\tau}^{\mathbf{A}}(a) := \big\{\langle \delta^{\mathbf{A}}(a), \varepsilon^{\mathbf{A}}(a) \rangle : \delta \approx \varepsilon \in \boldsymbol{\tau}(x)\big\} \subseteq A \times A.$$

It is interesting to notice that $a \in \boldsymbol{\tau}\mathbf{A}$ if and only if $\boldsymbol{\tau}^{\mathbf{A}}(a) \subseteq \mathrm{Id}_A$, the identity relation on A.

Definition 2. Let $\langle \mathbf{A}, F \rangle$ be a matrix, M a class of matrices, and $\boldsymbol{\tau}(x)$ a set of equations.

- $\boldsymbol{\tau}$ ***defines the set*** F, or ***defines truth in*** $\langle \mathbf{A}, F \rangle$, when $F = \boldsymbol{\tau}\mathbf{A}$; i.e., when for any $a \in A$, $a \in F$ if and only if $\mathbf{A} \vDash \boldsymbol{\tau}(x) \,[\![a]\!]$, i.e., if and only if $\boldsymbol{\tau}^{\mathbf{A}}(a) \subseteq \mathrm{Id}_A$.
- $\boldsymbol{\tau}$ ***defines truth in*** M when it defines truth in all the matrices in M.
- ***Truth is equationally definable in*** M when there is a set of equations $\boldsymbol{\tau}(x)$ that defines truth in M.

In all these cases, the equations in the set $\boldsymbol{\tau}(x)$ are called the ***defining equations***.

Note that when this happens, for each algebra $\mathbf{A}$ there can be at most one subset F of A such that $\langle \mathbf{A}, F \rangle \in$ M; this (in general weaker) property is called in the literature the ***implicit definability*** of truth in M.

We are particularly interested in logics that have a complete matrix semantics where truth is equationally definable. These logics can be alternatively (and more intuitively) described with the help of the ***equational consequence relative to a class of algebras*** K. This is a closure relation $\vDash_{\mathsf{K}}$ on the set of equations, defined as follows. For any set Θ of equations and any equation $\delta \approx \varepsilon$,

$$\Theta \vDash_{\mathsf{K}} \delta \approx \varepsilon \iff \text{for every } \mathbf{A} \in \mathsf{K} \text{ and every } h \in \mathrm{Hom}(\mathbf{Fm}, \mathbf{A}),$$
$$\text{if } \mathbf{A} \vDash \alpha \approx \beta \ [\![h]\!] \text{ for all } \alpha \approx \beta \in \Theta, \text{ then } \mathbf{A} \vDash \delta \approx \varepsilon \ [\![h]\!].$$

With this definition, the following fact is easy to prove.

Lemma 3. *A logic $\mathcal{L}$ is complete with respect to some class M of matrices where truth is equationally definable if and only if there is a class K of algebras and a set of equations $\boldsymbol{\tau}(x)$ such that for all $\Gamma \cup \{\varphi\} \subseteq Fm$,*

$$\Gamma \vdash_{\mathcal{L}} \varphi \iff \boldsymbol{\tau}\Gamma \vDash_{\mathsf{K}} \boldsymbol{\tau}\varphi.$$

Proof. For one direction, take K as the class of algebraic reducts of M; for the other, take $\mathsf{M} := \{\langle \mathbf{A}, \boldsymbol{\tau}\mathbf{A} \rangle : \mathbf{A} \in \mathsf{K}\}$. $\qquad\square$

When the situation is as in the lemma, the class K is called an ***algebraic semantics*** for the logic $\mathcal{L}$.

A special kind of logics having an algebraic semantics correspond to those where there is a single defining equation with a particular and simple form, which we proceed to describe. A class of algebras is ***pointed*** when there is a term that is constant in the class. This term, usually denoted by $\top$, can be a primitive constant of the language, or be made up from primitive constants, or be a term with variables such that in the algebras of the class, all interpretations give it the same value; in this second case, it can safely be assumed that the term has only the variable x. So, we assume that $\top$ is a term with at most the variable x, and will occasionally write $\top(x)$ to emphasize this fact.

Now we can introduce the first main concept studied in the paper.

Definition 4. Let K be a pointed class of algebras, with $\top$ as the corresponding constant term. The ***assertional logic of*** K is the logic $\mathcal{L}$ determined by the following condition: for all $\Gamma \cup \{\varphi\} \subseteq Fm$,

$$\Gamma \vdash_{\mathcal{L}} \varphi \iff \{\gamma \approx \top : \gamma \in \Gamma\} \vDash_{\mathsf{K}} \varphi \approx \top.$$

A logic $\mathcal{L}$ is an ***assertional logic*** when it is the assertional logic of some pointed class of algebras.

In other words, $\mathcal{L}$ is the assertional logic of K if and only if $\mathcal{L}$ has K as an algebraic semantics with $x \approx \top$ as defining equation, and if and only if $\mathcal{L}$ is complete with respect to the class of matrices $\{\langle \mathbf{A}, \{\top^{\mathbf{A}}\} \rangle : \mathbf{A} \in \mathsf{K}\}$. We will see that the simplicity of the equation entails strong properties, not shared by logics having algebraic semantics with arbitrary defining equations.[4] Note that the constant term $\top$ must be a theorem of any assertional logic, because $\top \approx \top$ obviously holds in K. Note also that if a class of algebras is pointed,

---

[4] A logic having K as algebraic semantics with defining equations $\boldsymbol{\tau}$ is also called "the $\boldsymbol{\tau}$-assertional logic of K" in the literature; in such a case, the term "1-assertional" is used for our "assertional". In the present paper we will not need this more general terminology.

then the variety it generates is pointed as well. Using Lemma 1, we can obtain the following fact, which is also of interest.

Lemma 5. *If $\mathcal{L}$ is the assertional logic of a pointed class of algebras K, then $\mathbb{V}\mathcal{L} \subseteq \mathbb{V}(\mathsf{K})$. As a consequence, the classes $\mathsf{Alg}^*\mathcal{L}$ and $\mathsf{Alg}\mathcal{L}$ are also pointed, with the same constant term as K.* $\square$

Now we can introduce the second main concept studied in the paper.

Definition 6. A logic $\mathcal{L}$ is ***truth-equational*** when truth is equationally definable in $\mathsf{Mod}^*\mathcal{L}$.

The notion of truth-equational logic was studied by Raftery (2006); the above definition, which is more convenient for the present paper, is actually an equivalent characterization, which follows from Theorem 25 of Raftery (2006). Raftery proved that truth-equational logics need not be protoalgebraic, but nevertheless they can be characterized by properties of the Leibniz operator (see Theorem 18 below), and hence they belong in the Leibniz hierarchy.

It is clear from the definition, by the completeness of $\mathcal{L}$ with respect to the class $\mathsf{Mod}^*\mathcal{L}$, that if $\mathcal{L}$ is truth-equational, then it has an algebraic semantics, namely the class $\mathsf{Alg}^*\mathcal{L}$. Note that the converse is not true, as witnessed by Example 1 of Raftery (2006), but it is so when τ has the form $x \approx \top$, with $\top$ a constant term of $\mathsf{Alg}^*\mathcal{L}$; this fact is contained in the following characterization, essentially due to Raftery, of the assertional logics as a subclass of truth-equational logics.

Theorem 7. *For any logic $\mathcal{L}$ the following conditions are equivalent:*

(i) *$\mathcal{L}$ is an assertional logic.*
(ii) *$\mathcal{L}$ is truth-equational, with a truth definition of the form $x \approx \top$, where $\top$ is a constant term of $\mathsf{Alg}^*\mathcal{L}$ or, equivalently, of $\mathsf{Alg}\mathcal{L}$.*
(iii) *$\mathcal{L}$ has $\mathsf{Alg}^*\mathcal{L}$ as an algebraic semantics with $x \approx \top$ as defining equation, where $\top$ is a constant term of $\mathsf{Alg}^*\mathcal{L}$.*
(iv) *$\mathcal{L}$ has $\mathsf{Alg}\mathcal{L}$ as an algebraic semantics with $x \approx \top$ as defining equation, where $\top$ is a constant term of $\mathsf{Alg}\mathcal{L}$.*

Proof. Assertional logics satisfiy the conditions established in Corollary 40 of Raftery (2006) for a logic to be truth-equational, in this case with a truth definition of the form $x \approx \top$ where $\top$ is a theorem of the logic. Moreover, by Lemma 5, we know that $\top$ will be a constant of $\mathsf{Alg}^*\mathcal{L}$ and of $\mathsf{Alg}\mathcal{L}$. This shows that (i) implies (ii). It has already been observed as a general property that (ii) implies (iii). Similarly, (ii) implies (iv), because a logic is truth-equational if and only if truth is equationally definable in $\mathsf{Mod}^{\mathsf{Su}}\mathcal{L}$ (Raftery, 2006, Theorem 28), and the class of algebraic reducts of $\mathsf{Mod}^{\mathsf{Su}}\mathcal{L}$ is $\mathsf{Alg}\mathcal{L}$. Finally, each of (iii) and (iv) implies (i), simply by the involved definitions. $\square$

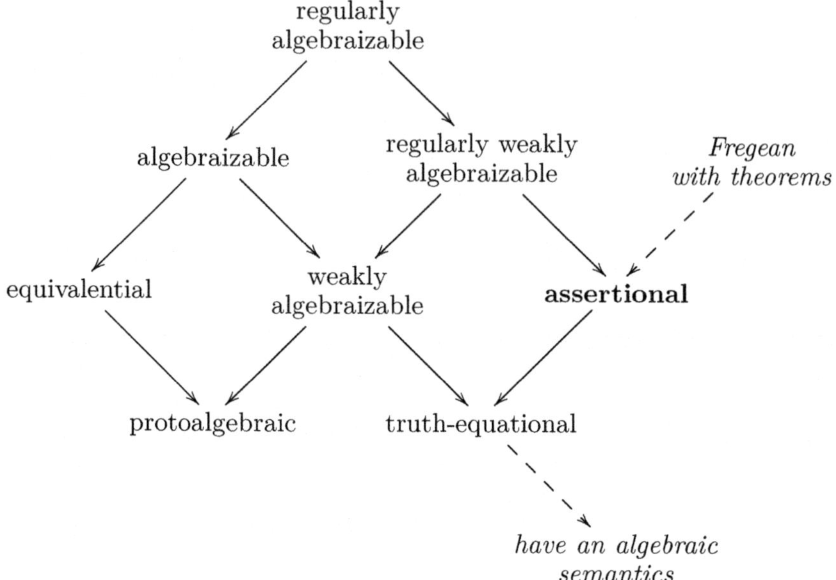

Fig. 1 The classes of logics in the fragment of the Leibniz hierarchy relevant to this paper, including the newly added class (in boldface), and showing (in italics) two related classes not belonging to it. Arrows indicate class inclusion.

Thus, all assertional logics are truth-equational; the class of the latter lies between the class of the former and that of the logics having an algebraic semantics (see Figure 1). This brings back into the Leibniz hierarchy many non-protoalgebraic logics that previously had seemed excluded from it. For instance, the $\langle \wedge, \vee, \top, \bot \rangle$-fragment of classical logic, which is the assertional logic of the variety of bounded distributive lattices; Visser's "basic logic" $BPL^\star$, shown to be non-protoalgebraic by Suzuki et al. (1998, Theorem 14); the implication-less fragment IPC^* of intuitionistic logic, proven to be non-protoalgebraic by Blok and Pigozzi (1989, § 5.2.5); and its denumerably many axiomatic extensions considered by Rebagliato and Verdú (1993). IPC^* and its extensions are examples where the constant term is not made up from primitive constants of the language; indeed, there $\top := \neg(x \wedge \neg x)$.

Observe that by Theorem 7, if $\mathcal{L}$ is the assertional logic of some class K of algebras, then it is the assertional logic of the class $\mathsf{Alg}^*\mathcal{L}$, and also of the class $\mathsf{Alg}\mathcal{L}$.

Assertional logics can be characterized in an independent way through the notion of a ***unital*** class of matrices, i.e., a class of matrices where all the filters are one-element sets:

Theorem 8. *For any logic,* $\mathcal{L}$ *the following conditions are equivalent:*

(i) $\mathcal{L}$ *is an assertional logic.*
(ii) *The class of matrices* $\mathsf{Mod}^*\mathcal{L}$ *is unital.*
(iii) *The class of matrices* $\mathsf{Mod}^{\mathrm{Su}}\mathcal{L}$ *is unital.*
(iv) $\mathcal{L}$ *has theorems and is complete with respect to a unital class of matrices.*

Proof. To show that (i) implies (ii) we use the characterizations of being assertional in Theorem 7. Thus, the assumption implies that $\mathsf{Mod}^*\mathcal{L} = \{\langle \mathbf{A}, \{\top^{\mathbf{A}}\}\rangle : \mathbf{A} \in \mathsf{Alg}^*\mathcal{L}\}$ and that this is a unital class. (i) implies (iii) for the same reason, applied to the class $\mathsf{Mod}^{\mathrm{Su}}\mathcal{L} = \{\langle \mathbf{A}, \{\top^{\mathbf{A}}\}\rangle : \mathbf{A} \in \mathsf{Alg}\mathcal{L}\}$. Trivially, each of (ii) and (iii) implies, separately, the second assertion of (iv), as a consequence of the completeness of $\mathcal{L}$ with respect to $\mathsf{Mod}^*\mathcal{L}$ and $\mathsf{Mod}^{\mathrm{Su}}\mathcal{L}$, respectively. Now observe that if $\mathcal{L}$ has no theorems, then for any algebra $\mathbf{A}$, the matrix $\langle \mathbf{A}, \emptyset\rangle$ is a model of $\mathcal{L}$. But, in particular, for a trivial (i.e., one-element) algebra $\mathbf{A}$, the matrix $\langle \mathbf{A}, \emptyset\rangle$ is always reduced and Suszko-reduced, because then $\varOmega^{\mathbf{A}}\{\emptyset\} = \widetilde{\varOmega}^{\mathbf{A}}_{\mathcal{L}}\{\emptyset\} = A \times A = \mathrm{Id}_A$. Thus, we would have that $\langle \mathbf{A}, \emptyset\rangle \in \mathsf{Mod}^*\mathcal{L}$ and $\langle \mathbf{A}, \emptyset\rangle \in \mathsf{Mod}^{\mathrm{Su}}\mathcal{L}$, respectively, against the assumption that the respective class is unital. This shows that $\mathcal{L}$ has theorems and completes the proof of (iv). Finally, in order to show that (iv) implies (i), let M be the unital class of matrices with respect to which $\mathcal{L}$ is complete, and let K be the class of their algebraic reducts. Observe that since $\mathcal{L}$ has theorems, all $\mathcal{L}$-filters are non-empty. Therefore, since the intersection of two $\mathcal{L}$-filters is always an $\mathcal{L}$-filter, and it cannot be empty, there can be at most one one-element $\mathcal{L}$-filter in each (arbitrary) algebra. The assumption that M is unital means that algebras in K have indeed one such $\mathcal{L}$-filter, and it is the only one on the algebra making the matrix reduced. Let $\top$ be a theorem of $\mathcal{L}$ in at most the variable x (which exists by the first assumption), and let $\mathbf{A} \in \mathsf{K}$. Since $\top$ is a theorem, for every $a \in A$ the point $\top^{\mathbf{A}}(a)$ must belong to the mentioned $\mathcal{L}$-filter, therefore this $\mathcal{L}$-filter must be exactly the set $\{\top^{\mathbf{A}}(a)\}$, for any $a \in A$. This also implies that $\top^{\mathbf{A}}(a) = \top^{\mathbf{A}}(b)$ for all $a, b \in A$. Therefore, $\top$ is a constant term of K, that is, the class K is pointed, and $\mathsf{M} = \{\langle \mathbf{A}, \{\top^{\mathbf{A}}\} : \mathbf{A} \in \mathsf{K}\rangle\}$. After this, the completeness of $\mathcal{L}$ with respect to M means that $\mathcal{L}$ is the assertional logic of K. $\qquad\square$

The fact that assertional logics have a unital class of reduced models has the following, seldom noticed consequence:

Corollary 9. *If* $\mathcal{L}$ *is an assertional logic, then the class of algebras* $\mathsf{Alg}^*\mathcal{L}$ *is relatively point-regular.*

Proof. Let $\top$ be the constant term of $\mathsf{Alg}^*\mathcal{L}$ witnessing that $\mathcal{L}$ is assertional, as in the previous results. Let $\mathbf{A} \in \mathsf{Alg}^*\mathcal{L}$ and let $\theta, \theta' \in \mathrm{Co}_{\mathsf{Alg}^*\mathcal{L}}\mathbf{A}$ such that $\top^{\mathbf{A}}/\theta = \top^{\mathbf{A}}/\theta'$. Since $\mathbf{A}/\theta \in \mathsf{Alg}^*\mathcal{L}$, by Theorem 8, $\langle \mathbf{A}/\theta, \{\top^{\mathbf{A}/\theta}\}\rangle \in \mathsf{Mod}^*\mathcal{L}$. Now, if $\pi\colon \mathbf{A} \to \mathbf{A}/\theta$ is the canonical projection, we have that

$$\theta = \pi^{-1}\mathrm{Id}_{A/\theta} = \pi^{-1}\varOmega^{\mathbf{A}/\theta}\{\top^{\mathbf{A}/\theta}\} = \varOmega^{\mathbf{A}}\pi^{-1}\{\top^{\mathbf{A}/\theta}\} = \varOmega^{\mathbf{A}}\big(\top^{\mathbf{A}}/\theta\big).$$

The same argument for θ' shows that $\theta' = \Omega^{\mathbf{A}}\left(\top^{\mathbf{A}}/\theta'\right)$. The assumption that $\top^{\mathbf{A}}/\theta = \top^{\mathbf{A}}/\theta'$ implies that $\theta = \theta'$. $\square$

The following characterization of assertional logics (if defined as in Theorem 8) is essentially due to Suszko (in unpublished lectures), according to Czelakowski (1981); the name "Suszko rules" was coined by Rautenberg (1993).

Theorem 10. *For any logic $\mathcal{L}$, the following conditions are equivalent:*

(i) $\mathcal{L}$ *is an assertional logic.*

(ii) $\mathcal{L}$ *has theorems and satisfies the so-called "Suszko rules":*

$$(2) \qquad\qquad x, y, \varphi(x, \vec{z}) \vdash_{\mathcal{L}} \varphi(y, \vec{z}),$$

for all $\varphi(x, \vec{z}) \in Fm$.

(iii) $\mathcal{L}$ *has theorems and satisfies that $\langle x, y \rangle \in \widetilde{\Omega}_{\mathcal{L}} C_{\mathcal{L}}\{x, y\}$.*

(iv) $\mathcal{L}$ *has theorems and satisfies that for every algebra $\mathbf{A}$ and every $a, b \in A$, $\langle a, b \rangle \in \widetilde{\Omega}_{\mathcal{L}}^{\mathbf{A}} Fi_{\mathcal{L}}^{\mathbf{A}}\{a, b\}$*

Proof. (i)$\Rightarrow$(ii) We know all assertional logics have theorems. Completeness of $\mathcal{L}$ with respect to some unital class of matrices, which Theorem 8 guarantees, directly implies the Suszko rules.

(ii)$\Rightarrow$(iii) and (iv) Let $\Gamma \in Th\mathcal{L}$ be such that $C_{\mathcal{L}}\{x, y\} \subseteq \Gamma$, that is, $x, y \in \Gamma$. Then by the Suszko rules, $\varphi(x, \vec{z}) \in \Gamma$ if and only if $\varphi(y, \vec{z}) \in \Gamma$, for all $\varphi(x, \vec{z}) \in Fm$. This means that $\langle x, y \rangle \in \Omega\Gamma$. Therefore, $\langle x, y \rangle \in \widetilde{\Omega}_{\mathcal{L}} C_{\mathcal{L}}\{x, y\}$, which proves (iii). Point (iv) is proved in the same way, but working on the $\mathcal{L}$-filters of an arbitrary algebra.

(iii)$\Rightarrow$(ii) follows by the same argument as the preceding implication; as a matter of fact, that the Suszko rules hold is equivalent to the condition that $\langle x, y \rangle \in \widetilde{\Omega}_{\mathcal{L}} C_{\mathcal{L}}\{x, y\}$.

(ii)$\Rightarrow$(i) Let $\langle \mathbf{A}, F \rangle \in \mathsf{Mod}^*\mathcal{L}$. Since $\mathcal{L}$ has theorems, $F \neq \emptyset$. Then the Suszko rules imply that F is a one-element set: If $a, b \in F$, then for every $\vec{c} \in A^n$, $\varphi^{\mathbf{A}}(a, \vec{c}) \in F$ if and only if $\varphi^{\mathbf{A}}(b, \vec{c}) \in F$, that is, $\langle a, b \rangle \in \Omega^{\mathbf{A}} F$; since the matrix is reduced, this implies that $a = b$. Thus, all the reduced models of $\mathcal{L}$ are unital, and by Theorem 8 this fact implies that $\mathcal{L}$ is an assertional logic.

(iv)$\Rightarrow$(iii) because the latter is a particular case of the former. $\square$

After the preceding results, we think it becomes clear that the class of assertional logics should be counted among those in the Leibniz hierarchy, as it can be defined by conditions on the Leibniz congruence: notice that the second conditions in points (iii) and (iv) of Theorem 10 can be paraphrased as "$\langle x, y \rangle \in \Omega\Gamma$ for all $\Gamma \in Th\mathcal{L}$ such that $x, y \in \Gamma$" and "$\langle a, b \rangle \in \Omega^{\mathbf{A}} F$ for all $F \in \mathcal{F}i_{\mathcal{L}}\mathbf{A}$ such that $a, b \in F$", respectively. For protoalgebraic logics, these can be simplified to "$\langle x, y \rangle \in \Omega C_{\mathcal{L}}\{x, y\}$" and "$\langle a, b \rangle \in \Omega^{\mathbf{A}} Fi_{\mathcal{L}}^{\mathbf{A}}\{a, b\}$", respectively.

It is well known that inside protoalgebraic logics, any of these conditions, or the equivalent ones found in Theorems 7 and 8, determine the classes of *regularly weakly algebraizable* logics; and inside equivalential logics, they produce the *regularly algebraizable* logics. These classes are usually considered as belonging to the Leibniz hierarchy, and by the same reason should the class of assertional logics be considered in it; its location in the hierarchy is parallel to the former ones, as Figure 1 on page 62 shows.

This new member of the hierarchy is different from the existing ones, and its location is really as shown in Figure 1, as the following examples confirm.

- *There are truth-equational logics that are not assertional.* Examples of this are all algebraizable logics that are not regularly algebraizable, such as all substructural logics associated with a variety of non-integral residuated lattices, described by Galatos et al. (2007); among the best known members of this class we find Relevance Logic (with and without the "Mingle" axiom; i.e., R and RM) and the multiplicative-additive fragment of Linear Logic $MALL$. Raftery (2006, Example 9) provides a non-protoalgebraic example: The logic in the language $\langle \rightarrow, \neg \rangle$ defined by taking as algebraic semantics the variety generated by the Sobociński three-element algebra, with $x \approx x \rightarrow x$ as defining equation. This logic, which has the same theorems as (but does not coincide with) the implication-negation fragment of RM, is neither protoalgebraic nor assertional, but is truth-equational. Notice that, although the term $x \rightarrow x$ is a theorem of the logic, it is not a constant term of the class of algebras.
- *There are assertional logics that are not (regularly) weakly algebraizable.* Examples of this will be all assertional logics that are not protoalgebraic, some of which are mentioned after Theorem 7.

From our Theorem 8, using Theorem 5.6.3 of Czelakowski (2001), it follows that the class of regularly weakly algebraizable logics is the intersection of the class of protoalgebraic logics and of assertional logics, and hence also the intersection of the classes of weakly algebraizable logics and of assertional logics; Figure 1 on page 62 shows these facts. In particular, a logic $\mathcal{L}$ is regularly weakly algebraizable if and only if it is protoalgebraic and $\mathsf{Mod}^*\mathcal{L}$ is unital; it is interesting to notice that the two just mentioned conditions can be formulated by making reference only to the class $\mathsf{Alg}^*\mathcal{L}$:

Corollary 11. *A logic $\mathcal{L}$ is regularly weakly algebraizable if and only if it has $\mathsf{Alg}^*\mathcal{L}$ as an algebraic semantics with $x \approx \top$ as defining equation, where $\top$ is a constant term of $\mathsf{Alg}^*\mathcal{L}$, and $\mathsf{Alg}^*\mathcal{L}$ is closed under subdirect products.*

Proof. Putting Theorems 7 and 8 together, we see that the condition that $\mathsf{Mod}^*\mathcal{L}$ is unital can be equivalently formulated in terms of $\mathsf{Alg}^*\mathcal{L}$ as stated. On the other hand, it is well-known (Czelakowski, 2001, Thm. 1.3.7) that a logic is protoalgebraic if and only if the class of matrices $\mathsf{Mod}^*\mathcal{L}$ is closed under subdirect products; but by the same theorems, the condition that $\mathsf{Mod}^*\mathcal{L}$ is unital implies that the logic is truth-equational, which implies that the truth filter of the matrices in $\mathsf{Mod}^*\mathcal{L}$ is unique, and therefore in our situation $\mathsf{Mod}^*\mathcal{L}$ is closed under subdirect products if and only if the class of algebras $\mathsf{Alg}^*\mathcal{L}$ is closed under subdirect products. $\qquad\qquad\square$

4 Full generalized models, and the Frege hierarchy

The reduction construction allows to introduce a special class of g-models of a logic. A ***basic full g-model*** of a logic $\mathcal{L}$ is one of the form $\langle \mathbf{A}, \mathcal{F}i_\mathcal{L}\mathbf{A}\rangle$, for some algebra $\mathbf{A}$. A ***full g-model*** of $\mathcal{L}$ is one whose reduction is a basic full g-model; that is, a g-matrix $\langle \mathbf{A}, \mathcal{C}\rangle$ such that $\mathcal{C}/\widetilde{\Omega}^\mathbf{A}\mathcal{C} = \mathcal{F}i_\mathcal{L}(\mathbf{A}/\widetilde{\Omega}^\mathbf{A}\mathcal{C})$. Note that a logic, viewed as a g-matrix, is a full g-model of itself, and indeed the largest one on the formula algebra. It turns out that $\mathsf{Alg}\mathcal{L}$ is also the class of algebraic reducts of the reduced full g-models of $\mathcal{L}$; in fact, the reduced full g-models of $\mathcal{L}$ are exacly those of the form $\langle \mathbf{A}, \mathcal{F}i_\mathcal{L}\mathbf{A}\rangle$ with $\mathbf{A} \in \mathsf{Alg}\mathcal{L}$. The notion of a full g-model of a logic, introduced by Font and Jansana (1996) and further studied in Font et al. (2006) and other papers, has allowed to develop a very general approach to the algebraic study of sentential logics, and in particular is instrumental in the following definitions.

The Frege hierarchy is a classification of logics according to what kind of replacement properties they (and their full g-models) satisfy. In abstract terms, replacement properties are defined algebraically as concerning congruences. A g-matrix $\langle \mathbf{A}, \mathcal{C}\rangle$ has ***the property of congruence*** when its Frege relation is a congruence of $\mathbf{A}$, i.e., when $\Lambda\mathcal{C} = \widetilde{\Omega}^\mathbf{A}\mathcal{C}$. A g-matrix $\langle \mathbf{A}, \mathcal{C}\rangle$ has ***the strong property of congruence*** when for any $F \in \mathcal{C}$, the g-matrix $\langle \mathbf{A}, \mathcal{C}^F\rangle$ has the property of congruence, i.e., when $\Lambda_\mathcal{C} F = \widetilde{\Omega}^\mathbf{A}_\mathcal{C} F$ for all $F \in \mathcal{C}$; note that this means that the Frege and the Suszko operators relative to $\mathcal{C}$ coincide. These two properties of congruence are preserved by reductions. Since the relation $\Lambda Th\mathcal{L}$ for a sentential logic $\mathcal{L}$ is its interderivability relation $\dashv\vdash_\mathcal{L}$, these two properties when formulated for a sentential logic amount to natural *replacement properties* of the interderivability relation.

These two properties originate the four classes of logics in the Frege hierarchy.

Definition 12. Let $\mathcal{L}$ be a logic.

- $\mathcal{L}$ is ***selfextensional*** when, viewed as the g-matrix $\langle \mathbf{Fm}, Th\mathcal{L}\rangle$, it has the property of congruence; i.e., when the interderivability relation $\dashv\vdash_\mathcal{L}$ is a congruence of $\mathbf{Fm}$.
- $\mathcal{L}$ is ***Fregean*** when, viewed as the g-matrix $\langle \mathbf{Fm}, Th\mathcal{L}\rangle$, it has the strong property of congruence; i.e., when for each $\Gamma \in Th\mathcal{L}$, the interderivability relation modulo Γ (i.e., the relation $\Lambda_\mathcal{L}\Gamma$) is a congruence of $\mathbf{Fm}$.
- $\mathcal{L}$ is ***fully selfextensional*** when all its full g-models have the property of congruence.

- $\mathcal{L}$ is ***fully Fregean*** when all its full g-models have the strong property of congruence.

The notion of a selfextensional logic is due to Wójcicki (1979); see also Wójcicki (1988, Chapter 5). The notion of a Fregean logic was introduced, in a slightly restricted form, by Pigozzi (1991) and Czelakowski (1992), and independently and as given here, by Font (1993). The other two classes of logics were introduced by Font and Jansana (1996); the hierarchy as such was first considered by Font (2003), and named after Frege in Font (2006). Observe that a logic is Fregean if and only if the Suszko and the Frege operators (relative to it) coincide on the theories of the logic. In Font and Jansana (1996, Proposition 2.40) it is shown that $\mathcal{L}$ is fully selfextensional if and only if for any algebra $\mathbf{A}$, the basic full g-model $\langle \mathbf{A}, \mathcal{F}i_{\mathcal{L}}\mathbf{A} \rangle$ has the property of congruence, and if and only if for every $\mathbf{A} \in \mathsf{Alg}\mathcal{L}$ the relation $\Lambda\mathcal{F}i_{\mathcal{L}}\mathbf{A}$ is the identity relation; that is, if and only if in the algebras in $\mathsf{Alg}\mathcal{L}$, different points generate different $\mathcal{L}$-filters ("$\mathcal{L}$-filters separate points"). This characterization is the clue to some of the interesting applications of fully selfextensional logics to the development of an abstract duality theory (Gehrke et al., 2010); it will be used in Theorem 15.

Some obvious relations hold between the four classes (taking into account that a logic is always a full g-model of itself), and are depicted in Figure 2. Two questions this graph naturally rises is whether the top (smallest) class of fully Fregean logics is the intersection of the two middle classes of Fregean logics and of fully selfextensional logics, and whether the lowest (largest) class of selfextensional logics is their union. These questions were posed as open problems in Font (2003, §6.2) and Font (2006, p. 202); the first one is answered affirmatively in the present paper for logics with theorems (Theorem 26), and the second one is answered negatively, even for logics with very strong properties (Example 23).

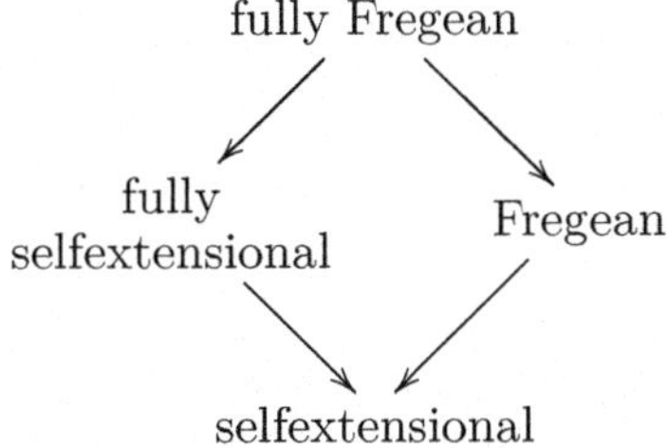

Fig. 2 The classes of logics in the Frege hierarchy. Arrows indicate class inclusion.

In order to find relations between the class of truth-equational logics and the classes in the Frege hierarchy, we start from the following observation.

Lemma 13. *Let $\mathcal{L}$ be a Fregean logic.*

1. *$\mathcal{L}$ satisfies the "Suszko rules" (2) displayed in Theorem 10.*
2. *If $\langle \mathbf{A}, F \rangle \in \mathsf{Mod}^{*}\mathcal{L}$, then F is either empty or a one-element set.*

Proof. 1. Trivially, for any logic $\mathcal{L}$ it holds that $\langle x, y \rangle \in \Lambda_{\mathcal{L}} C_{\mathcal{L}}\{x, y\}$. If $\mathcal{L}$ is Fregean, the relation $\Lambda_{\mathcal{L}} C_{\mathcal{L}}\{x, y\}$ will be a congruence, which implies that for all $\varphi(x, \vec{z}) \in Fm$, $\langle \varphi(x, \vec{z}), \varphi(y, \vec{z}) \rangle \in \Lambda_{\mathcal{L}} C_{\mathcal{L}}\{x, y\}$. That is, $C_{\mathcal{L}}\{x, y, \varphi(x, \vec{z})\} = C_{\mathcal{L}}\{x, y, \varphi(y, \vec{z})\}$, which amounts to the rules (2).
2. Assume that F is non-empty and take any $a, b \in F$. Since F is an $\mathcal{L}$-filter, from rules (2) it follows that for any $\varphi(x, \vec{z}) \in Fm$ and any $\vec{c} \in A^n$, $\varphi^{\mathbf{A}}(a, \vec{c}) \in F$ if and only if $\varphi^{\mathbf{A}}(b, \vec{c}) \in F$. By the classical characterization of Czelakowski (2001, Theorem 0.5.3), this says that $\langle a, b \rangle \in \Omega^{\mathbf{A}} F$, and since the matrix is reduced this implies that $a = b$. Thus, F is a one-element set. $\square$

Since, in general, the filter of a reduced matrix can be empty only when the algebra is trivial, we see that reduced models of Fregean logics on non-trivial algebras must be unital. This may be a practical criterion to disprove that a certain logic, (some of) whose reduced models are known, is Fregean. For instance, this shows that Belnap-Dunn's well-known four-valued logic is not Fregean, because it has reduced models on a nontrivial algebra with two-element designated sets, for instance those given by the two prime filters of the four-element De Morgan lattice (usually called $FOUR$) that defines the logic, which is a simple algebra; see Font (1997, p. 427).

One consequence of Lemma 13 (together with a result to be reviewed in the next section) is the following characterization:

Theorem 14. *Let $\mathcal{L}$ be a Fregean logic. The following conditions are equivalent:*

  (i) *$\mathcal{L}$ has theorems.*
  (ii) *$\mathcal{L}$ is assertional.*
 (iii) *$\mathcal{L}$ is truth-equational.*
 (iv) *The Leibniz operator is injective over the $\mathcal{L}$-filters of arbitrary algebras.*
  (v) *The Leibniz operator is injective over the theories of $\mathcal{L}$.*

Proof. By Lemma 13, if a Fregean logic has theorems, then its class of reduced models is unital, and therefore by Theorem 8, the logic is assertional; this proves that (i) implies (ii). That (ii) implies (iii) is contained in Theorem 7. Now, that (iii) implies (iv) follows from Theorem 28 of Raftery (2006), a result that you will find here as Theorem 18, because being completely order reflecting implies being injective. Clearly, (v) follows from (iv) as a particular case. Finally, (v) implies (i) because, if a logic has no theorems, then $\emptyset$ and Fm are both theories of the logic, and always $\Omega\emptyset = \Omega Fm = Fm \times Fm$, thus breaking injectivity of the Leibniz operator on the theories of the logic. $\square$

We thus see that Fregean logics with theorems are assertional, and hence truth-equational; the situation is that depicted in Figure 1 on page 62. The following examples confirm that there are no other relations.

- *There are assertional logics that are not Fregean.* We find many examples of this situation even among the regularly algebraizable logics, such as the global consequences of the usual normal modal logics ($K, T, S4, S5$, etc.), or Łukasiewicz's many-valued logics, or, more generally, the logics associated in Galatos et al. (2007) with any variety of integral residuated lattices that is not a variety of generalized Heyting algebras. In all these examples, the defining equation of the algebraization is of the form $x \approx \top$ for a constant $\top$, so that each is the assertional logic of the corresponding algebraic counterpart (a variety of normal modal algebras, or the corresponding variety of residuated lattices, respectively). But they are not selfextensional (the modal cases are easily shown by using Kripke models, and the second group is shown by Bou et al. (2009, Theorem 4.12)), hence *a fortiori* they are not Fregean.

- *There are Fregean logics with theorems that are not regularly weakly algebraizable.* Examples are the already mentioned logic IPC^* and its axiomatic extensions, which are not protoalgebraic, hence in particular not regularly weakly algebraizable. That IPC^* is Fregean is proved by Font and Jansana (1996, § 5.1.4), and all axiomatic extensions of a Fregean logic are Fregean as well; and all these logics have theorems (indeed, they are assertional).

Thus, the class of assertional logics is the smallest class in the Leibniz hierarchy containing the Fregean logics with theorems, as shown in Figure 1.

As a final application of Theorem 8, we obtain a (weakened) version of Theorem 14 for fully selfextensional logics.

Theorem 15. *A fully selfextensional logic is assertional if and only if it is truth-equational.*

Proof. By Theorem 7, all assertional logics are truth-equational. So let $\mathcal{L}$ be a fully selfextensional and truth-equational logic, and let $\langle \mathbf{A}, F \rangle \in \mathsf{Mod}^{\mathrm{Su}}\mathcal{L}$. Since $\widetilde{\varOmega}^{\mathbf{A}}_{\mathcal{L}}(\bigcap \mathcal{F}i_{\mathcal{L}}\mathbf{A}) \subseteq \widetilde{\varOmega}^{\mathbf{A}}_{\mathcal{L}}F = \mathrm{Id}_A$, it follows that $\widetilde{\varOmega}^{\mathbf{A}}_{\mathcal{L}}(\bigcap \mathcal{F}i_{\mathcal{L}}\mathbf{A}) = \widetilde{\varOmega}^{\mathbf{A}}_{\mathcal{L}}F = \mathrm{Id}_A$. One of the basic characterizations of truth-equational logics (Raftery, 2006, Theorem 28) is that the Suszko operator is injective on their filters, therefore $F = \bigcap \mathcal{F}i_{\mathcal{L}}\mathbf{A}$, that is, F is the smallest $\mathcal{L}$-filter of A. Thus, if $a, b \in F$, a and b belong to the same $\mathcal{L}$-filters (namely: all). But $\mathbf{A} \in \mathsf{Alg}\mathcal{L}$ and for these algebras, $\mathcal{L}$-filters separate points, because $\mathcal{L}$ is fully selfextensional (as commented on page 67 after Definition 12). Therefore, $a = b$. We have shown that $\mathsf{Mod}^{\mathrm{Su}}\mathcal{L}$ is unital, and by Theorem 8 this implies that $\mathcal{L}$ is assertional. $\qquad\square$

Some of the non-protoalgebraic examples of assertional logics mentioned before are fully selfextensional; for instance, Visser's logic BPL^*, as shown in Bou (2001), or the $\langle \wedge, \vee, \top, \bot \rangle$-fragment of classical logic, as follows from Theorem 4.28 of Font and Jansana (1996).

Notice that, unlike in the Fregean case (Theorem 14), the condition that $\mathcal{L}$ has theorems cannot be added as an equivalent one to those in Theorem 15; the logics that preserve degrees of truth with respect to certain varieties of commutative, integral residuated lattices (those that are not varieties of generalized Heyting algebras) provide an infinity of counterexamples: all these logics are fully selfextensional and have theorems but are not assertional; these properties are shown in, or follow from, Corollary 4.2, Lemma 2.6, Corollary 3.6 and Theorem 4.12 of Bou et al. (2009).

5 The full generalized models of truth-equational logics

The key characterization of truth-equational logics uses the following property.

Definition 16. Let $\mathcal{L}$ be a logic, and let $\mathbf{A}$ be an algebra. The Leibniz operator $\Omega^{\mathbf{A}}$ is ***completely order-reflecting*** over $\mathcal{F}i_{\mathcal{L}}\mathbf{A}$ when for all $\mathcal{F} \cup \{G\} \subseteq \mathcal{F}i_{\mathcal{L}}\mathbf{A}$, if $\bigcap_{F \in \mathcal{F}} \Omega^{\mathbf{A}}F \subseteq \Omega^{\mathbf{A}}G$ then $\bigcap \mathcal{F} \subseteq G$.

The following reformulation in terms of the Suszko operator, whose proof is an easy exercise, is very convenient:

Lemma 17. *Let $\mathcal{L}$ be a logic, and let $\mathbf{A}$ be an algebra. The Leibniz operator $\Omega^{\mathbf{A}}$ is completely order-reflecting over $\mathcal{F}i_{\mathcal{L}}\mathbf{A}$ if and only if for all $F, G \in \mathcal{F}i_{\mathcal{L}}\mathbf{A}$, if $\widetilde{\Omega}^{\mathbf{A}}_{\mathcal{L}}F \subseteq \Omega^{\mathbf{A}}G$, then $F \subseteq G$.* $\square$

The main result placing the class of truth-equational logics in the Leibniz hierarchy, due to Raftery (2006, Theorem 28), is the following.

Theorem 18. *For any logic $\mathcal{L}$, the following conditions are equivalent:*

  (i) *$\mathcal{L}$ is truth-equational.*
  (ii) *The Leibniz operator is completely order-reflecting over the $\mathcal{L}$-filters of arbitrary algebras.*
 (iii) *The Leibniz operator is completely order-reflecting over the theories of $\mathcal{L}$.* $\square$

In particular this implies that the Leibniz operator is order-reflecting, and hence injective, on the theories of $\mathcal{L}$ (and on the $\mathcal{L}$-filters of any algebra).

This characterization can be used to obtain an alternative proof of the truth-equationality of Fregean logics with theorems (Theorem 14), which needs not use assertional logics. To this end we show that the Leibniz operator is completely order-reflecting on the theories of $\mathcal{L}$, by using Lemma 17 over the formula algebra. Let $\Gamma, \Gamma' \in Th\mathcal{L}$ be such that $\widetilde{\Omega}_{\mathcal{L}}\Gamma \subseteq \Omega\Gamma'$. We have to show that $\Gamma \subseteq \Gamma'$, so let $\varphi \in \Gamma$. Take now any theorem ψ of $\mathcal{L}$, which exists by assumption; then in particular $\psi \in \Gamma$, and this implies that $C_{\mathcal{L}}(\Gamma, \varphi) = \Gamma = C_{\mathcal{L}}(\Gamma, \psi)$, that is, $\langle \varphi, \psi \rangle \in \Lambda_{\mathcal{L}}\Gamma$. But $\mathcal{L}$ is Fregean, which means that $\Lambda_{\mathcal{L}}\Gamma = \widetilde{\Omega}_{\mathcal{L}}\Gamma$. Therefore, by the assumption, $\langle \varphi, \psi \rangle \in \Omega\Gamma'$. Since also $\psi \in \Gamma'$, by compatibility it follows that $\varphi \in \Gamma'$, as desired.

The following technical but important property will allow us to obtain some characterizations of truth-equationality in terms of the full g-models of the logic.

Lemma 19. *Let $\mathcal{L}$ be any logic, $\mathbf{A}$ any algebra, and $F \in \mathcal{F}i_{\mathcal{L}}\mathbf{A}$. The following conditions are equivalent:*

    (i) *The g-matrix $\langle \mathbf{A}, (\mathcal{F}i_{\mathcal{L}}\mathbf{A})^F \rangle$ is a full g-model of $\mathcal{L}$.*
    (ii) *For all $G \in \mathcal{F}i_{\mathcal{L}}\mathbf{A}$, if $\widetilde{\Omega}^{\mathbf{A}}_{\mathcal{L}}F \subseteq \Omega^{\mathbf{A}}G$, then $F \subseteq G$.*

Proof. One of the central characterizations of the notion of full g-model of a logic (Font and Jansana, 1996, Theorem 2.14) is that a g-matrix $\langle \mathbf{A}, \mathcal{C} \rangle$ is a full g-model of $\mathcal{L}$ if and only if $\mathcal{C} = \{G \in \mathcal{F}i_{\mathcal{L}}\mathbf{A} : \widetilde{\Omega}^{\mathbf{A}}\mathcal{C} \subseteq \Omega^{\mathbf{A}}G\}$. Since by (1) on page 57, $\widetilde{\Omega}^{\mathbf{A}}(\mathcal{F}i_{\mathcal{L}}\mathbf{A})^F = \widetilde{\Omega}^{\mathbf{A}}_{\mathcal{L}}F$, in particular a g-matrix of the form $\langle \mathbf{A}, (\mathcal{F}i_{\mathcal{L}}\mathbf{A})^F \rangle$, for some $F \in \mathcal{F}i_{\mathcal{L}}\mathbf{A}$, is a full g-model of $\mathcal{L}$ if and only if $(\mathcal{F}i_{\mathcal{L}}\mathbf{A})^F = \{G \in \mathcal{F}i_{\mathcal{L}}\mathbf{A} : \widetilde{\Omega}^{\mathbf{A}}_{\mathcal{L}}F \subseteq \Omega^{\mathbf{A}}G\}$. But the direct inclusion holds by (1), and the reverse inclusion is exactly condition (ii). $\qquad\square$

This allows us to obtain our first characterization of truth-equational logics in terms of the form of their full g-models: a logic is truth-equational if and only if each of its filters determines a full g-model of the logic; we see that the same property, limited to the theories of the logic, is also sufficient to characterize truth-equationality.

Theorem 20. *For any logic $\mathcal{L}$, the following conditions are equivalent:*

    (i) *$\mathcal{L}$ is truth-equational.*
    (ii) *For every $\mathbf{A}$ and every $F \in \mathcal{F}i_{\mathcal{L}}\mathbf{A}$, the g-matrix $\langle \mathbf{A}, (\mathcal{F}i_{\mathcal{L}}\mathbf{A})^F \rangle$ is a full g-model of $\mathcal{L}$.*
    (iii) *For every $\Gamma \in Th\mathcal{L}$, the g-matrix $\langle \mathbf{Fm}, (Th\mathcal{L})^{\Gamma} \rangle$ is a full g-model of $\mathcal{L}$.*

Proof. In order to prove that (i) implies (ii), assume that $\mathcal{L}$ is truth-equational. Then, by Theorem 18 and Lemma 17, we see that for any $\mathbf{A}$, any $F \in \mathcal{F}i_{\mathcal{L}}\mathbf{A}$ satisfies condition (ii) in Lemma 19, therefore its condition (i) yields the present condition (ii). Clearly, (iii) is a particular case of (ii). And finally from (iii) we can prove (i): By applying Lemma 19 to the formula algebra, we see that (iii) amounts to saying that for every $\Gamma, \Gamma' \in Th\mathcal{L}$, if $\widetilde{\Omega}_{\mathcal{L}}\Gamma \subseteq \Omega\Gamma'$, then $\Gamma \subseteq \Gamma'$. But by Lemma 17 applied also to the formula algebra, this is to say that the Leibniz operator is completely order-reflecting over the theories of $\mathcal{L}$, and by Theorem 18, this implies that $\mathcal{L}$ is truth-equational. $\qquad\square$

The equivalence between (i) and (ii) is obtained in Albuquerque et al. (2016) as a by-product of a more general study of compatibility operators in abstract algebraic logic, of which the Suszko operator is a paradigmatic example; here we have given a direct proof. It is interesting to relate this characterization to that of protoalgebraic logics found in Font and Jansana (1996, Theorem 3.4): A logic $\mathcal{L}$ is protoalgebraic if and only if every full g-model of $\mathcal{L}$ is of the form $\langle \mathbf{A}, (\mathcal{F}i_{\mathcal{L}}\mathbf{A})^F \rangle$ for some algebra $\mathbf{A}$ and some $F \in \mathcal{F}i_{\mathcal{L}}\mathbf{A}$. This is in some sense "dual" to the characterization of truth-equational logics in Theorem 20. As a consequence, a logic is weakly algebraizable if and only if its full g-models are exactly the g-matrices of the form $\langle \mathbf{A}, (\mathcal{F}i_{\mathcal{L}}\mathbf{A})^F \rangle$ for some $F \in \mathcal{F}i_{\mathcal{L}}\mathbf{A}$; this was already obtained in Font and Jansana (1996, Theorem 3.8), and in fact this characterization of weakly algebraizable logics lies at the roots of the very definition of this class of logics in Font and Jansana (1996).

We are also interested in the following extension of the previous characterization: A logic is truth-equational if and only if the class of its full g-models is so-to-speak closed under the operation $\mathcal{C} \mapsto \mathcal{C}^F$; and again it is enough to require this property for the full g-models over the formula algebra.

Theorem 21. *For any logic $\mathcal{L}$, the following conditions are equivalent:*

  (i) *$\mathcal{L}$ is truth-equational.*
  (ii) *For every full g-model $\langle \mathbf{A}, \mathcal{C} \rangle$ of $\mathcal{L}$ and every $F \in \mathcal{C}$, the g-matrix $\langle \mathbf{A}, \mathcal{C}^F \rangle$ is a full g-model of $\mathcal{L}$.*
  (iii) *For every full g-model $\langle \mathbf{Fm}, \mathcal{C} \rangle$ of $\mathcal{L}$ over the formula algebra and every $\Gamma \in \mathcal{C}$, the g-matrix $\langle \mathbf{Fm}, \mathcal{C}^{\Gamma} \rangle$ is also a full g-model of $\mathcal{L}$.*

Proof. (i)$\Rightarrow$(ii) It is a general property of the theory of full g-models that the intersection of (the closure systems of) two full g-models of a logic produces another full g-model of the same logic; this is commented just before Theorem 1.20 of Font et al. (2006), and is also proved in Proposition 5.96 of Font (2016). If $\mathcal{L}$ is truth-equational, by Theorem 20, for every $F \in \mathcal{F}i_{\mathcal{L}}\mathbf{A}$, the g-matrix $\langle \mathbf{A}, (\mathcal{F}i_{\mathcal{L}}\mathbf{A})^F \rangle$ is a full g-model of $\mathcal{L}$. Now, if $\langle \mathbf{A}, \mathcal{C} \rangle$ is a full g-model of $\mathcal{L}$, $\mathcal{C} \subseteq \mathcal{F}i_{\mathcal{L}}\mathbf{A}$, and hence clearly $\mathcal{C} \cap (\mathcal{F}i_{\mathcal{L}}\mathbf{A})^F = \mathcal{C}^F$. Therefore, by the mentioned general property, the g-matrix $\langle \mathbf{A}, \mathcal{C}^F \rangle$ is a full g-model of $\mathcal{L}$, as desired.

(iii) is a particular case of (ii), and the implication (iii)$\Rightarrow$(i) is trivial because the g-matrix $\langle \mathbf{Fm}, Th\mathcal{L} \rangle$ is always full (indeed, it is a basic full g-model, by definition), therefore our (iii) implies the condition in Theorem 20(iii) as a particular case, and hence implies that $\mathcal{L}$ is truth-equational. $\square$

6 Applications to the hierarchies

The preceding characterization of the full g-models of truth-equational logics allows us to refine the Frege hierarchy inside this class.

Theorem 22. *A truth-equational logic is fully selfextensional if and only if it is fully Fregean.*

Proof. Every fully Fregean logic is fully selfextensional, so in one direction there is nothing to prove. Now assume that $\mathcal{L}$ is a truth-equational and fully selfextensional logic, and let $\langle \mathbf{A}, \mathcal{C} \rangle$ be any full g-model of $\mathcal{L}$ and $F \in \mathcal{C}$. Since $\mathcal{L}$ is truth-equational, by Theorem 21 the g-matrix $\langle \mathbf{A}, \mathcal{C}^F \rangle$ is also a full g-model of $\mathcal{L}$. Then, since $\mathcal{L}$ is fully selfextensional, the g-matrix has the property of congruence. This shows that all the full g-models of $\mathcal{L}$ have the strong property of congruence, that is, that $\mathcal{L}$ is fully Fregean. $\square$

Thus, for truth-equational logics (hence, in a large part of the Leibniz hierarchy) the Frege hierarchy reduces to three classes:

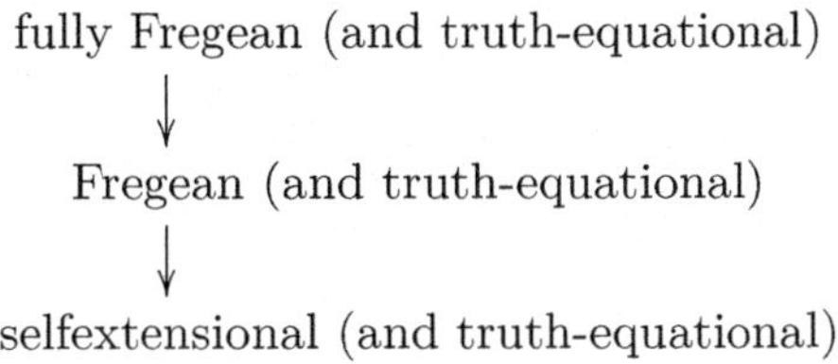

These three classes are different, and the lowest one is still a proper subclass of that of all truth-equational logics, as the following considerations show.

- The last mentioned fact (i.e., that not all truth-equational logics are self-extensional) is witnessed by the many algebraizable logics that are not selfextensional, as already mentioned after Theorem 14.
- Babenyshev (2003) has constructed a Fregean logic that is not fully Fregean. This logic, which has a proof-theoretic definition, has theorems, therefore by Theorem 14 it is assertional, and hence truth-equational.
- Next we construct a truth-equational and non-protoalgebraic logic that is selfextensional but not Fregean:

Example 23. Consider the algebra $\mathbf{A} = \langle \{0,1,2\}, \square, \neg \rangle$, where $\square$ and $\neg$ are two 1-ary operations, defined as follows:

$$\neg 1 = \square 1 = \square 0 = 0 \qquad \neg 0 = \square 2 = 1 \qquad \neg 2 = 2,$$

and consider the logic $\mathcal{L}$ in the language $\langle \square, \neg \rangle$ determined by the matrix $\langle \mathbf{A}, \{1\} \rangle$.

FACT 1. $\mathcal{L}$ is assertional, and hence truth-equational: To see this, note that in $\mathbf{A}$, $\neg\square\square a = 1$ for all $a \in A$, so that $\neg\square\square x$ is a constant term of $\mathbf{A}$, and $\mathcal{L}$ is the assertional logic of the class $\{\mathbf{A}\}$ with $\top := \neg\square\square x$. Therefore, by Theorem 7, $\mathcal{L}$ is truth-equational, with $\{x \approx \neg\square\square x\}$ as defining equation.

FACT 2. $\mathcal{L}$ is not Fregean: To see this, it is enough to check, from the definition, that

$$\square x, \neg\square x \dashv\vdash_{\mathcal{L}} \square x, \neg x$$

(because no evaluation makes the two premises on either side simultaneously equal to 1), and that

$$\Box x, \neg\neg\Box x \nvdash_{\mathcal{L}} \neg\neg x$$

(just evaluate x to 2). Therefore $\varLambda_{\mathcal{L}} C_{\mathcal{L}} \{\Box x\}$ is not a congruence with respect to the operation $\neg$.

FACT 3. If $\varphi \dashv\vdash_{\mathcal{L}} \psi$, then $\mathbf{A} \vDash \varphi \approx \psi$: Since all connectives are unary, there are two variables x and y such that $Var\{\varphi, \psi\} \subseteq \{x, y\}$.

We first construct all the terms in the variables x and y up to equivalence in $\mathbf{A}$. Observe that this is equivalent to ask for a set of representatives of the congruence classes that form the universe of the free algebra $\mathbf{Fm_A}\{x, y\}$ over the variety generated by $\mathbf{A}$ with two free generators. We reason as follows. The set A^{A^2} of binary functions on A can be given naturally the structure of an algebra $\mathbf{A}^{A^2}$. Then let $\pi_i \colon A^2 \to A$ be the projection map on the i-th component, for $i \in \{1, 2\}$, and let $\mathbf{C}$ be the subalgebra of $\mathbf{A}^{A^2}$ generated by $\{\pi_1, \pi_2\}$. We claim that

$$C = \big\{ \pi_1, \pi_2, \Box\pi_1, \Box\pi_2, \neg\pi_1, \neg\pi_2, \Box\Box\pi_1, \neg\Box\pi_1, \neg\Box\pi_2, \neg\Box\Box\pi_1 \big\}.$$

The inclusion from right to left follows from the fact that $\mathbf{C}$ is a subalgebra of $\mathbf{A}^{A^2}$, whereas the other one is a consequence of the fact that the identities $\neg\neg x \approx x$, $\Box\Box x \approx \Box\Box y$, $\Box\Box\Box x \approx \Box\Box x$ and $\Box\neg x \approx \Box x$ hold in $\mathbf{A}$.

Now, recall that the free algebra $\mathbf{Fm_A}\{x, y\}$ is isomorphic to $\mathbf{C}$ via the map sending the equivalence classes of x and y to π_1 and π_2 respectively; see for instance (Bergman, 2011, Theorem 4.9). Applying this fact to our claim, we conclude that

$$T(x, y) := \big\{ x, y, \Box x, \Box y, \neg x, \neg y, \Box\Box x, \neg\Box x, \neg\Box y, \neg\Box\Box x \big\}$$

is the set of terms in two variables up to equivalence in $\mathbf{A}$.

Since $Var\{\varphi, \psi\} \subseteq \{x, y\}$, there are $\varphi', \psi' \in T(x, y)$ such that $\mathbf{A} \vDash \varphi \approx \varphi'$ and $\mathbf{A} \vDash \psi \approx \psi'$. By Lemma 1, $\varphi \dashv\vdash_{\mathcal{L}} \varphi'$ and $\psi \dashv\vdash_{\mathcal{L}} \psi'$, and since by assumption $\varphi \dashv\vdash_{\mathcal{L}} \psi$, it follows that $\varphi' \dashv\vdash_{\mathcal{L}} \psi'$. But it is easy to check, working case by case, that no two distinct terms in $T(x, y)$ are interderivable in $\mathcal{L}$. Therefore, we conclude that $\varphi' = \psi'$. This implies that $\mathbf{A} \vDash \varphi \approx \psi$, as required.

FACT 4. $\mathcal{L}$ is selfextensional: As a consequence of Fact 3 and Lemma 1, $\alpha \dashv\vdash_{\mathcal{L}} \beta$ if and only if $\mathbf{A} \vDash \alpha \approx \beta$. But this last relation is clearly a congruence. Therefore, $\varLambda\mathcal{L} = \dashv\vdash_{\mathcal{L}}$ is a congruence, that is, $\mathcal{L}$ is selfextensional.

FACT 5. $\mathcal{L}$ is not protoalgebraic: This is because its language contains only unary connectives. As a direct consequence of the characterization in Theorem 1.1.3 of Czelakowski (2001), a protoalgebraic logic in such a language should be trivial, which $\mathcal{L}$ is not. $\hfill\Box$

This example also answers, in the negative, an open problem on the structure of the Frege hierarchy, posed in Font (2003, p. 78) and Font (2006, p. 202): that of whether the class of selfextensional logics is the union of the class of Fregean logics and the class of fully selfextensional logics. The logic constructed in Example 23 is selfextensional but not Fregean, hence not fully Fregean, and this in turn implies (by Theorem 22, since it is truth-equational) that it is not fully selfextensional either.

For *finitary* logics the result in Theorem 22 produces another refinement of the Frege hierarchy.

Corollary 24. *A finitary and weakly algebraizable logic is fully selfextensional if and only if it is Fregean, and if and only if it is fully Fregean.*

Proof. By Corollary 80 of Czelakowski and Pigozzi (2004), a finitary protoalgebraic logic is Fregean if and only if it is fully Fregean. Since weakly algebraizable logics are protoalgebraic, this applies to them, and since they are also truth-equational, merging this with Theorem 22 we obtain the stated result. $\qquad\square$

Thus, for finitary weakly algebraizable logics (hence, *a fortiori*, for finitary algebraizable logics), the Frege hierarchy reduces to only two classes, the selfextensional and the Fregean. That in this case these two classes are indeed different is shown by the following construction.

Example 25. Consider the language $\langle \rightarrow, \square, \mathbf{a}, \mathbf{b}, \mathbf{c}, \top \rangle$ of type $\langle 2, 1, 0, 0, 0, 0 \rangle$, and the set $A := \{a, b, c, 1\}$ with the order structure given by the following graph:

$$
\begin{array}{c}
\bullet\,1 \\
a\,\bullet\quad b\,\bullet\quad\bullet\,c
\end{array}
$$

We equip it with the structure of an algebra $\mathbf{A} = \langle A, \rightarrow, \square, a, b, c, 1 \rangle$ of the above similarity type, where the four constants are interpreted in the obvious way, and for every $x, y \in A$,

$$
x \rightarrow y := \begin{cases} 1 & \text{if } x \leqslant y, \\ y & \text{otherwise,} \end{cases} \qquad\qquad \square x := \begin{cases} b & \text{if } x \in \{1, a, c\}, \\ 1 & \text{otherwise.} \end{cases}
$$

Observe that the implicative reduct of $\mathbf{A}$ is a Hilbert algebra.

Let $\mathcal{L}$ be the logic determined by the g-matrix $\langle \mathbf{A}, \mathcal{C} \rangle$, where

$$
\mathcal{C} := \big\{ \{1\}, \{a, 1\}, \{c, 1\}, A \big\}.
$$

Observe that all the members of $\mathcal{C}$ are implicative filters.

FACT 1. $\mathcal{L}$ is finitary: It is well known that any logic defined by a finite set of finite matrices (hence, in particular, by a finite g-matrix) is finitary.

FACT 2. $\mathcal{L}$ is a finitely regularly algebraizable logic: The implicative fragment of $\mathcal{L}$ is a logic defined by a family of implicative filters of a Hilbert algebra, and is therefore an implicative logic in the sense of Rasiowa (1974). Moreover, it is easy to check that

$$x \to y, y \to x \vdash_{\mathcal{L}} \Box x \to \Box y.$$

As a consequence, $\mathcal{L}$ itself is an implicative logic, hence a finitely regularly algebraizable logic (Blok and Pigozzi, 1989, § 5.2).

FACT 3. $\mathcal{L}$ is selfextensional: Observe that the closure system $\mathcal{C}$ separates points in A, therefore $\Lambda\mathcal{C} = \mathrm{Id}_A$, and hence $\widetilde{\Omega}^{\mathbf{A}}\mathcal{C} = \mathrm{Id}_A$, that is, the g-matrix $\langle \mathbf{A}, \mathcal{C} \rangle$ has the property of congruence (and is reduced). This easily implies (Czelakowski and Pigozzi, 2004, Theorem 82) that $\langle \mathbf{Fm}, Th\mathcal{L} \rangle$ has the property of congruence, that is, the logic $\mathcal{L}$ is selfextensional.

FACT 4. $\mathcal{L}$ is not fully Fregean: It is easy to see that the following deductions hold

$$\emptyset \vdash_{\mathcal{L}} \top \qquad \mathbf{a}, \mathbf{c} \vdash_{\mathcal{L}} x \qquad \mathbf{b} \vdash_{\mathcal{L}} x,$$

and that this implies that $\mathcal{F}i_{\mathcal{L}}\mathbf{A} = \mathcal{C}$. Therefore, $\langle \mathbf{A}, \mathcal{C} \rangle$ is a full g-model of $\mathcal{L}$. Now, consider the closure system $\mathcal{C}^{\{a,1\}} = \{\{a,1\}, A\}$. It is clear that $\langle c, b \rangle \in \Lambda\mathcal{C}^{\{a,1\}}$, because c and b belong to the same members of $\mathcal{C}^{\{a,1\}}$, and that $\langle \Box c, \Box b \rangle \notin \Lambda\mathcal{C}^{\{a,1\}}$, because $\Box c = b \notin \{a,1\}$ while $\Box b = 1 \in \{a,1\}$. Hence the g-model $\langle \mathbf{A}, \mathcal{C}^{\{a,1\}} \rangle$ does not have the property of congruence, which is to say that the full g-model $\langle \mathbf{A}, \mathcal{C} \rangle$ has not the strong property of congruence. We conclude that $\mathcal{L}$ is not fully Fregean.

FACT 5. $\mathcal{L}$ is neither Fregean nor fully selfextensional: This follows from Fact 4 and Corollary 24, taking into account that $\mathcal{L}$ is finitary (Fact 1) and weakly algebraizable (Fact 2). $\qquad\qquad\Box$

This example also solves an old open problem in abstract algebraic logic: that of whether, for protoalgebraic logics, to be selfextensional implies to be fully selfextensional; the general case was solved by Babenyshev (2003). Example 25 shows that this is not the case, even for logics with much stronger properties, namely for finitely regularly algebraizable logics.

The reader may have noticed that Example 25 solves the issues addressed by Example 23 as well. However, the latter has the additional interest, over the former, of being in some sense "minimal" as an example of a matrix-determined non-Fregean logic, because it is defined by a 3-element matrix, and, trivially, all the logics determined by 2-element (g-)matrices are Fregean.

Finally, by combining several of the previous results, we obtain a result that clarifies the structure of the Frege hierarchy alone (although our proof goes through a class in the Leibniz hierarchy): for logics with theorems the top (smallest) class of the Frege hierarchy is actually the intersection of its two middle classes.

Theorem 26. *A logic with theorems is fully Fregean if and only if it is both Fregean and fully selfextensional.*

Proof. Trivially, if a logic is fully Fregean, then it is both Fregean and fully selfextensional. For the converse, suppose that a logic has these two properties. By Theorem 14, the logic will be truth-equational, and then we can apply Theorem 22, which tells us that, since it is assumed to be fully selfextensional, it is in fact fully Fregean. $\square$

This gives a partial, positive answer (for logics with theorems) to another of the open problems formulated in Font (2003, § 6.2) and Font (2006, p. 202). Now, it becomes an OPEN PROBLEM whether the assumption that the logic has theorems can be dispensed with in this result.

Acknowledgements

The authors acknowledge with thanks the following support: grant SFRH/ BD/79581/2011 from FCT of the government of Portugal, research grant 2014SGR-788 from the government of Catalonia, research project MTM2011-25747 from the government of Spain, which includes FEDER funds from the European Union, project GA13-14654S of the Czech Science Foundation, and an APIF grant from the University of Barcelona.

References

Albuquerque, H., Font, J. M. and Jansana, R. (2016). Compatibility operators in abstract algebraic logic, *The Journal of Symbolic Logic*, 81, 417–462.

Babenyshev, S. (2003). Fully Fregean logics, *Reports on Mathematical Logic*, 37, 59–78.

Bergman, C. (2011). *Universal Algebra. Fundamentals and Selected Topics*, CRC Press.

Blok, W. and Pigozzi, D. (1986). Protoalgebraic logics, *Studia Logica*, 45, 337–369.

Blok, W. and Pigozzi, D. (1989). *Algebraizable logics*, volume 396 of *Mem. Amer. Math. Soc.*, A.M.S., Providence, January, Out of print. Reprinted in the *Classic Reprints* series, Avanced Reasoning Forum, 2014.

Blok, W. and Pigozzi, D. (1992). Algebraic semantics for universal Horn logic without equality, in A. Romanowska and J. D. H. Smith, editors, *Universal Algebra and Quasigroup Theory*, pages 1–56. Heldermann, Berlin.

Bou, F. (2001). *Strict implication and subintuitionistic logics*, Master Thesis, University of Barcelona, In Spanish.

Bou, F., Esteva, F., Font, J. M., Gil, A. J., Godo, Ll., Torrens, A. and Verdú, V. (2009). Logics preserving degrees of truth from varieties of residuated lattices, *Journal of Logic and Computation*, 19, 1031–1069.

Czelakowski, J. (1981). Equivalential logics, I, II, *Studia Logica*, 40, 227–236 and 355–372.

Czelakowski, J. (1992). *Consequence operations: Foundational studies*. Reports of the research project *Theories, models, cognitive schemata*, Institute of Philosophy and Sociology, Polish Academy of Sciences, Warszawa.

Czelakowski, J. (2001). *Protoalgebraic logics*, volume 10 of *Trends in Logic - Studia Logica Library*, Kluwer Academic Publishers, Dordrecht.

Czelakowski, J. (2003). The Suszko operator. Part I, *Studia Logica (Special issue on Abstract Algebraic Logic, Part II)*, 74, 181–231.

Czelakowski, J. and Pigozzi, D. (2004). Fregean logics, *Annals of Pure and Applied Logic*, 127, 17–76.

Font, J. M. (1993). On the Leibniz congruences, In C. Rauszer, editor, *Algebraic Methods in Logic and in Computer Science*, volume 28 of *Banach Center Publications*, pages 17–36. Polish Academy of Sciences, Warszawa.

Font, J. M. (1997). Belnap's four-valued logic and De Morgan lattices, *Logic Journal of the IGPL*, 5, 413–440.

Font, J. M. (2003). Generalized matrices in abstract algebraic logic, in V. F. Hendriks and J. Malinowski, editors, *Trends in Logic. 50 years of Studia Logica*, volume 21 of *Trends in Logic - Studia Logica Library*, pages 57–86. Kluwer Academic Publishers, Dordrecht.

Font, J. M. (2006). Beyond Rasiowa's algebraic approach to non-classical logics, *Studia Logica*, 82, 172–209.

Font, J. M. (2015). Abstract algebraic logic – an introductory chapter, in N. Galatos and K. Terui, editors, *Hiroakira Ono on Residuated Lattices and Substructural Logics*, Outstanding Contributions to Logic, Springer, 68 p, to appear.

Font, J. M. (2016). *Abstract Algebraic Logic. An Introductory Textbook*, volume 60 of *Studies in Logic – Mathematical Logic and Foundations*, College Publications, London.

Font, J. M. and Jansana, R. (1996). *A general algebraic semantics for sentential logics*, volume 7 of *Lecture Notes in Logic*, Springer-Verlag, Out of print, *Second revised edition* published in 2009 by the Association for Symbolic Logic, and reprinted by Cambridge University Press (2017).

Font, J. M., Jansana, R. and Pigozzi, D. (2003). A survey of abstract algebraic logic, *Studia Logica (Special issue on Abstract Algebraic Logic, Part II)*, 74, 13–97, with an update in 91, 125–130, 2009.

Font, J. M., Jansana, R. and Pigozzi, D. (2006). On the closure properties of the class of full g-models of a deductive system, *Studia Logica (Special issue in memory of Willem Blok)*, 83, 215–278.

Galatos, N., Jipsen, P., Kowalski, T. and Ono, H. (2007). *Residuated lattices: an algebraic glimpse at substructural logics*, volume 151 of *Studies in Logic and the Foundations of Mathematics*, Elsevier, Amsterdam.

Gehrke, M., Jansana, R. and Palmigiano, A. (2010). Canonical extensions for congruential logics with the deduction theorem, *Annals of Pure and Applied Logic*, 161, 1502–1519.

Henkin, L., Monk, J. D. and Tarski, A. (1985). *Cylindric algebras, part II*, North-Holland, Amsterdam.

Pigozzi, D. (1991). Fregean algebraic logic. In H. Andréka, J. D. Monk, and I. Németi, editors, *Algebraic Logic*, volume 54 of *Colloquia mathematica Societatis János Bolyai*, pages 473–502, North-Holland, Amsterdam.

Raftery, J. (2006). The equational definability of truth predicates, *Reports on Mathematical Logic (Special issue in memory of Willem Blok)*, 41, 95–149.

Rasiowa, H. (1974). *An algebraic approach to non-classical logics*, volume 78 of *Studies in Logic and the Foundations of Mathematics*, North-Holland, Amsterdam.

Rautenberg, W. (1993). On reduced matrices, *Studia Logica*, 52, 63–72.

Rebagliato, J. and Verdú, V. (1993). On the algebraization of some Gentzen systems, *Fundamenta Informaticae (Special issue on Algebraic Logic and its Applications)*, 18, 319–338.

Suzuki, Y., Wolter, F. and Zacharyaschev, M. (1998). Speaking about transitive frames in propositional languages, *Journal of Logic, Language and Information*, 7, 317–339.

Wójcicki, R. (1979). Referential matrix semantics for propositional calculi, *Bulletin of the Section of Logic*, 8, 170–176.

Wójcicki, R. (1988). *Theory of logical calculi. Basic theory of consequence operations*, volume 199 of *Synthèse Library*, Reidel, Dordrecht.

Deduction-Detachment Theorem and Gentzen-Style Deductive Systems

Sergey Babenyshev

*The author is greatly indebted to Don Pigozzi,
under whose supervision were obtained
the results of this paper, for his caring
guidance and generous sharing of ideas.*

Abstract Logical implication is an attempt to catch the essence of cause-effect relationships of the real world in the context of formal deductive systems. The Deduction-Detachment Theorem (DDT) being, in its turn, a statement about essential logical properties of classical implication, was therefore of great interest for logicians. Although a statement about a Hilbert-style deductive system, DDT can be formulated by means of Gentzen-style rules, and as such seems to be a statement about the metatheoretical properties of Hilbert-style deductive systems. As is often the case with metatheoretical properties, DDT leaves the question about its meaning and scope a bit vague or at least requires a higher order abstraction layer to formalize them. Closure relations, discussed in this paper, present a convenient context to give a precise formal statement of the DDT and its connection with Hilbert- and pertinent Gentzen-style deductive systems.

2010 Mathematics Subject Classification: 03G27, 03B22

Key words: Deduction-Detachment theorem, Abstract Algebraic Logic, Gentzen systems, Closure relations, Protoalgebraic logics

---

Sergey Babenyshev
Siberian Fire-Rescue Academy, Severnaya Str 1, Zheleznogorsk, 662971, Russia, e-mail: sergey.babenyshev@gmail.com

© Springer International Publishing AG 2018
J. Czelakowski (eds.), *Don Pigozzi on Abstract Algebraic Logic,
Universal Algebra, and Computer Science*, Outstanding Contributions
to Logic 16, https://doi.org/10.1007/978-3-319-74772-9_3

1 Introduction

The Deduction-Detachment Theorem (DDT), independently discovered by
Tarski and Herbrand, is a metalogical property of deductive systems with
implication:

$$\Gamma, \alpha \vdash_S \beta \iff \Gamma \vdash_S \alpha \to \beta.$$

Being as it is a statement about proofs, DDT allows us in particular to encode
the proof theory of the deductive system into formulas, which leads to numerous and subtle consequences pertinent to the semantics of such deductive
systems. DDT can be formulated using a pair of Gentzen-style rules:

$$\frac{\Gamma, \alpha \vdash \beta}{\Gamma \vdash \alpha \to \beta}, \quad \frac{\Gamma \vdash \alpha \to \beta}{\Gamma, \alpha \vdash \beta}.$$

or through modus ponens ($\to$-detachment) and $\to$-deduction:

$$\alpha, \alpha \to \beta \vdash \beta, \quad \frac{\Gamma, \alpha \vdash \beta}{\Gamma \vdash \alpha \to \beta}.$$

Some of the goals of this paper are: 1) to establish the exact meaning of the
above-mentioned rules; 2) to show in what Gentzen-style system, associated
with a Hilbert-style system in question, those rules can formally present the
DDT.

To study the phenomenon of DDT in greater generality, we choose the
context of abstract Hilbert-style deductive systems:

In Abstract Algebraic Logic, an abstract Hilbert-style deductive system
$\mathcal{H}$ is identified with a family of sets, called $\mathcal{H}$-*theories*, of formulas of a given
propositional language, such that this family, usually denoted by $\mathrm{Th}\,\mathcal{H}$, is
closed under

1) arbitrary intersections, i.e., $\mathrm{Th}\,\mathcal{H}$ is a *closure system*,

2) unions of upward-directed families, i.e., $\mathrm{Th}\,\mathcal{H}$ is *algebraic*,

3) inverse substitutions, so that a preimage of any $\mathcal{H}$-theory under an
arbitrary substitution is an $\mathcal{H}$-theory again, i.e., $\mathrm{Th}\,\mathcal{H}$ is *invariant*.

So defined, $\mathcal{H}$ is indeed "abstract" because its definition does not refer to
any particular axiomatization. It is easy to see that the closure operator

$$(.)^{\mathrm{Th}\,\mathcal{H}} : X \mapsto \bigcap \{T \in \mathrm{Th}\,\mathcal{H} \mid X \subset T\},$$

associated with such a closure system $\mathrm{Th}\,\mathcal{H}$, defines a finitary, structural
consequence relation as follows

$$X \vdash_{\mathcal{H}} \alpha \iff \alpha \in X^{\mathrm{Th}\,\mathcal{H}}.$$

It was suggested by researchers of the Barcelona group to treat the
Gentzen-style deductive systems similarly and identify an abstract Gentzen-
style deductive system with the set of its theories with two distinctive features

1) a theory is a set of *sequents*, i.e., finite sequences of formulas,

2) a substitution acts on sequents componentwise.

The importance of Gentzen-style systems and related axiomatizations by Gentzen-style rules is largely due to the fact that various metatheoretical properties of Hilbert-style deductive systems can be formulated in terms of Gentzen-style rules, and the deduction-detachment theorem manifests one of these properties.

In this paper, we consider the Deduction-Detachment Theorem in a more general formulation (c.f. (Czelakowski, 1985, 2001; Blok and Pigozzi, 1997)):

A Hilbert-style deductive system admits a multiterm deduction-detachment theorem if there is a finite set of formulas $\Delta = \{\delta_i(x,y)\}_{i \in I}$ (may be empty) such that for all formulas α, β and every set of formulas Γ

$$\Gamma, \alpha \vdash_{\mathcal{H}} \beta \iff (\forall \delta \in \Delta)\ \Gamma \vdash_{\mathcal{H}} \delta(\alpha, \beta).$$

Even though the deduction-detachment theorem can and usually is formulated by the Gentzen-style rules, there is a tendency not to specify precisely the kind of Gentzen system in which those rules capture the DDT. To make this precise, we consider *axiomatic closure relations*:

Let $\mathcal{H}$ be an abstract Hilbert-style deductive system and $T \in \mathrm{Th}\,\mathcal{H}$. Then the set of sequents

$$\{\langle \alpha_1, \ldots, \alpha_n, \beta \rangle \mid T, \alpha_1, \ldots, \alpha_n \vdash_{\mathcal{H}} \beta\}$$

is called an axiomatic closure relation *for $\mathcal{H}$.*

In other words, an axiomatic closure relation lists all consequences that are possible in $\mathcal{H}$ if we add all formulas from some $\mathcal{H}$-theory as axioms (not axiom schemes). In this paper we will show that

An abstract Hilbert-style deductive system $\mathcal{H}$ admits the multiterm deduction-detachment theorem if and only if the set $\mathbf{Acr}\,\mathcal{H}$ of all axiomatic closure relations for $\mathcal{H}$ forms an abstract Gentzen-style deductive system, i.e., $\mathbf{Acr}\,\mathcal{H}$ is closed under

1) arbitrary intersections,
2) unions of upward-directed families,
3) inverse substitutions.

The paper is structured as follows: after the Introduction, in Section 2 we state basic definitions and notational conventions; in Section 3 we introduce general and axiomatic closure relations that can be associated with a given Hilbert-style deductive system and prove necessary facts about them; in Section 4 we recall the notion of the multiterm deduction-detachment theorem and prove the main result of this paper — Theorem 4.3; in Section 5 we discuss some connections between the obtained results and previous research, using yet another type of closure relation associated with Hilbert-style deductive systems — full closure relations.

The results of the current paper were first announced at the 9th Asian Logic Conference, Novosibirsk (Babenyshev, 2006).

2 Definitions and Preliminaries

Throughout this paper we will employ the following terminology and notational conventions:

Suppose A is a set. Then $\mathcal{P}(A) := \{X \mid X \subseteq A\}$ is the *power-set* of A. We write $X \subseteq_\omega A$ if X is a finite subset of A, furthermore $\mathcal{P}_\omega(A) := \{X \mid X \subseteq_\omega A\}$. The *n-th cartesian power* of A is denoted by A^n. A^+ denotes $\bigcup_{n=1}^\infty A^n$ — the set of all non-empty finite sequences of elements of A. For an arbitrary element $\bar{a} = \langle a_1, \ldots, a_n \rangle \in A^+$ we might write $\{\bar{a}\}$ instead of $\{a_1, \ldots, a_n\}$.

For a mapping $h : A \to A$ the operator-style notation ha will be routinely used instead of the function-style notation $h(a)$. Also any mapping h defined on A can be uniquely extended to a mapping on A^+ by the following definition:

$$h\langle a_1, \ldots, a_n \rangle = \langle ha_1, \ldots, ha_n \rangle, \text{ for all } \langle a_1, \ldots, a_n \rangle \in A^+.$$

The latter also uniquely extends to the *complex* (defined on sets) mapping on $\mathcal{P}(A^+)$ as follows:

$$hX = \{h\langle \bar{a} \rangle \mid \langle \bar{a} \rangle \in X\}, \text{ for all } X \subseteq A^+.$$

Note that the same symbol h will be routinely used for all these mappings.

A *propositional language type* is any set $\mathcal{L}$. The elements of $\mathcal{L}$ are called *functional symbols* in the algebraic context or *logical connectives* in the logical context. With $\mathcal{L}$ is associated a function $\rho : \mathcal{L} \to \omega$, where ρf is called the *arity* or *rank* of the functional symbol $f \in \mathcal{L}$. For each $n \in \omega$, $\mathcal{L}_n := \{f \in \mathcal{L} \mid \rho f = n\}$. An *algebra* $\mathbf{A}$ of type $\mathcal{L}$ is a pair $\langle A, \mathcal{L}^{\mathbf{A}} \rangle$, where A is a non-empty set called the *universe* of $\mathbf{A}$ and $\mathcal{L}^{\mathbf{A}} = \{f^{\mathbf{A}} \mid f \in \mathcal{L}\}$ is a list of operations over the set A such that for every $f \in \mathcal{L}_n$, $f^{\mathbf{A}} : A^n \to A$. Members of $\mathcal{L}^{\mathbf{A}}$ are called *basic operations* of $\mathbf{A}$. The set of all congruence relations on the algebra $\mathbf{A}$ is denoted by $\mathrm{Con}\,\mathbf{A}$ and forms a complete lattice ordered by inclusion with $0_{\mathbf{A}}$ as the smallest and $1_{\mathbf{A}}$ as the largest element.

Let $X = \{x_i\}_{i \in I}$ be a non-empty set. The set $\mathrm{Fm}_{\mathcal{L}} X$ of *formulas (or terms) of type $\mathcal{L}$ over the set of generators X* is defined recursively as follows
1) $X \subseteq \mathrm{Fm}_{\mathcal{L}} X$,
2) if $f \in \mathcal{L}_n$ and $\alpha_1, \ldots, \alpha_n \in \mathrm{Fm}_{\mathcal{L}} X$, then $f(\alpha_1, \ldots, \alpha_n) \in \mathrm{Fm}_{\mathcal{L}} X$,
3) $\mathrm{Fm}_{\mathcal{L}} X$ is the smallest set satisfying 1) and 2).
Formulas will be denoted usually by lowercase Greek letters. We write $\alpha(x_1, \ldots, x_n)$ or $\mathrm{Var}(\alpha) \subseteq \{x_1, \ldots, x_n\}$, if $\alpha \in \mathrm{Fm}_{\mathcal{L}}\{x_1, \ldots, x_n\}$. A *sequent* is a tuple $\langle \alpha_1, \ldots, \alpha_k \rangle$ of $\mathrm{Fm}_{\mathcal{L}}^+$. A sequent can also be written in the form $\alpha_1, \ldots, \alpha_{k-1} \triangleright \alpha_k$.

We can induce the structure of an algebra on $\mathrm{Fm}_{\mathcal{L}} X$ by associating with each $f \in \mathcal{L}_n$ an n-ary operation $f^{\mathbf{Fm}_{\mathcal{L}} X}$ on the set $\mathrm{Fm}_{\mathcal{L}} X$ defined as $f^{\mathbf{Fm}_{\mathcal{L}} X}\langle \bar{\alpha} \rangle = f(\bar{\alpha})$. The superscript in this case is usually omitted. This algebra $\mathbf{Fm}_{\mathcal{L}} X$ is called the *algebra of formulas (terms) of type $\mathcal{L}$ over the set of variables X*. We fix a countable set $\mathrm{Var} = \{x_0, x_1, x_2, \ldots\}$ of *propositional variables*. Then $\mathbf{Fm}_{\mathcal{L}} \mathrm{Var}$ is called *the formula algebra over the language of type $\mathcal{L}$* and will be denoted $\mathbf{Fm}_{\mathcal{L}}$. The universe of $\mathbf{Fm}_{\mathcal{L}}$ is denoted as $\mathrm{Fm}_{\mathcal{L}}$.

The algebra $\mathbf{Fm}_{\mathcal{L}} X$ is the *absolutely free algebra over the set X* in the class of all algebras of type $\mathcal{L}$. This means that, for every algebra $\mathbf{A}$ of type $\mathcal{L}$, an arbitrary mapping $h : X \to A$ can be uniquely extended to a homomorphism $h : \mathbf{Fm}_{\mathcal{L}} X \to A$. In particular any homomorphism $h : \mathbf{Fm}_{\mathcal{L}} X \to \mathbf{A}$ is determined by the mapping $h : X \to A$. A homomorphism $h : \mathbf{Fm}_{\mathcal{L}} \to \mathbf{A}$ is called an *evaluation*; a homomorphism $h : \mathbf{Fm}_{\mathcal{L}} \to \mathbf{Fm}_{\mathcal{L}}$ is called a *substitution*. A set $X \subseteq \mathrm{Fm}_{\mathcal{L}}$ $[\mathrm{Fm}_{\mathcal{L}}^+]$ is *invariant* if for every substitution $\sigma \colon \sigma X \subseteq X$.

A family $\mathcal{C} \subseteq \mathcal{P}(A)$ is *upward-directed* if for every pair $X, Y \in \mathcal{C}$ there is $Z \in \mathcal{C}$ such that $X, Y \subseteq Z$. A subset $\mathcal{C} \subseteq \mathcal{P}(A)$ is *algebraic* if $\bigcup \mathcal{D} \in \mathcal{C}$ for every upward-directed subfamily $\mathcal{D} \subseteq \mathcal{C}$. A family $\mathcal{C} \subseteq \mathcal{P}(A)$ is called a *closure system over A* if $A \in \mathcal{C}$ and $\bigcap \mathcal{D} \in \mathcal{C}$ for every non-empty subfamily $\mathcal{D} \subseteq \mathcal{C}$. A closure system $\mathcal{C}$ over $\mathrm{Fm}_{\mathcal{L}}$ is *[surjectively] invariant* if for every [surjective] substitution σ and every $T \in \mathcal{C}$, $\sigma^{-1} T := \{\alpha \mid \sigma \alpha \in T\} \in \mathcal{C}$, or, in other words, if $\sigma^{-1} \mathcal{C} \subseteq \mathcal{C}$ for all [surjective] substitutions $\sigma : \mathbf{Fm}_{\mathcal{L}} \to \mathbf{Fm}_{\mathcal{L}}$. Similarly, a closure system $\mathcal{C}$ over $\mathrm{Fm}_{\mathcal{L}}^+$ is *[surjectively] invariant* if for every [surjective] substitution σ and every $T \in \mathcal{C}$, $\sigma^{-1} T = \{\bar{\alpha} \triangleright \alpha \mid \sigma(\bar{\alpha} \triangleright \alpha) \in T\} \in \mathcal{C}$.

A *closure operator on A* is a mapping $\mathbf{C} : \mathcal{P}(A) \to \mathcal{P}(A)$ such that for any $X, Y \subseteq A$, $X \subseteq \mathbf{C}(X) = \mathbf{C}(\mathbf{C}(X)) \subseteq \mathbf{C}(X \cup Y)$. A set $X \in \mathcal{P}(A)$ such that $\mathbf{C}(X) = X$ is called a *closed set of $\mathbf{C}$*. A closure operator $\mathbf{C}$ is *finitary* if for any $X \subseteq A$, $\mathbf{C}(X) = \bigcup \{\mathbf{C}(Y) \mid Y \subseteq_\omega X\}$. The following relations between closure systems and closure operators are well known: 1) if $\mathbf{C}$ is a closure operator on A, then the family of its closed sets is a closure system over A; 2) if $\mathcal{C}$ is a closure system over A, then the mapping $\mathbf{C}_{\mathcal{C}} : \mathcal{P}(A) \to \mathcal{P}(A)$ defined for each $X \subseteq A$ as $\mathbf{C}_{\mathcal{C}} X := \bigcap \{Y \in \mathcal{C} \mid X \subseteq Y\}$ is a closure operator on A; 3) $\mathcal{C}$ is algebraic iff $\mathbf{C}_{\mathcal{C}}$ is finitary. We use interchangeably the exponential and prefix notations for closure operators, thus $X^{\mathcal{C}} = (X)^{\mathcal{C}} = \mathbf{C}_{\mathcal{C}}(X) = \mathbf{C}_{\mathcal{C}} X$.

Every closure system $\mathcal{C}$ over a set A, as a family of subsets ordered under set-inclusion, is a complete lattice. The infimum of a family $\{X_i\}_{i \in I} \subseteq \mathcal{C}$ is its intersection $\bigcap_{i \in I} X_i$, and its supremum is $\bigvee_{i \in I}^{\mathcal{C}} X_i := \mathbf{C}_{\mathcal{C}}(\bigcup_{i \in I} X_i)$; its largest element is A, and its smallest element is $\mathbf{C}_{\mathcal{C}}(\emptyset) = \bigcap \mathcal{C}$.

A *Hilbert-style system of type $\mathcal{L}$ ($HSS_{\mathcal{L}}$)* is a pair $\mathcal{H} = \langle \mathbf{Fm}_{\mathcal{L}}, \mathrm{Th}\,\mathcal{H} \rangle$ such that $\mathrm{Th}\,\mathcal{H} \subseteq \mathcal{P}(\mathrm{Fm}_{\mathcal{L}})$ is an algebraic invariant closure system over $\mathrm{Fm}_{\mathcal{L}}$. A *Gentzen-type system of type $\mathcal{L}$ ($GSS_{\mathcal{L}}$)* is a pair $\mathcal{G} = \langle \mathbf{Fm}_{\mathcal{L}}, \mathrm{Th}\,\mathcal{G} \rangle$ such that $\mathrm{Th}\,\mathcal{G} \subseteq \mathcal{P}(\mathrm{Fm}_{\mathcal{L}}^+)$ is an algebraic invariant closure system over $\mathrm{Fm}_{\mathcal{L}}^+$. For an $HSS_{\mathcal{L}}$ $\mathcal{H}$ and all $T \in \mathrm{Th}\,\mathcal{H}$, $[T)_{\mathrm{Th}\,\mathcal{H}} := \{U \in \mathrm{Th}\,\mathcal{H} \mid T \subseteq U\}$ denotes the principal filter of the lattice $\mathrm{Th}\,\mathcal{H}$ generated by T. If $\mathcal{R}$ is an $HSS_{\mathcal{L}}$ or $GSS_{\mathcal{L}}$, we denote $\mathrm{Thm}\,\mathcal{R} := \bigcap \mathrm{Th}\,\mathcal{R}$.

We take a Cantor-style approach towards Gentzen-style rules: we view a rule not as a "rule"— description of an action, but as a list of all its applications.

A *Gentzen-style sequent* is a sequence $\bar{s} \triangleright s$ of sequents. A *Gentzen-style rule* $\bar{s} \vdash s$ is the set of all substitution instances of the Gentzen-style sequent $\bar{s} \triangleright s$, i.e.,

$$\bar{s} \vdash s := \{\sigma(\bar{s} \triangleright s) \mid \sigma : \mathbf{Fm}_{\mathcal{L}} \to \mathbf{Fm}_{\mathcal{L}}\}.$$

In this context the sequent $\bar{s} \triangleright s$ is called the *scheme* for the rule $\bar{s} \vdash s$.

A Gentzen-style rule $s_1, \dots, s_n \vdash s$ can also be written as $\dfrac{s_1, \dots, s_n}{s}$.

Let x, y, z be variables. *Standard* Gentzen-style rules (sometimes called *structural*) are rules defined by the schemes

$$
\begin{array}{lll}
(\text{Ax}): & \vdash \Gamma, x, \Sigma \triangleright x & \text{Axioms} \\
(\text{Ex}): & \Gamma, x, y, \Sigma \triangleright z \vdash \Gamma, y, x, \Sigma \triangleright z & \text{Exchange} \\
(\text{W}): & \Gamma, \Sigma \triangleright y \vdash \Gamma, x, \Sigma \triangleright y & \text{Weakening} \\
(\text{Con}): & \Gamma, x, x, \Sigma \triangleright y \vdash \Gamma, x, \Sigma \triangleright y & \text{Contraction} \\
(\text{Cut}): & \Gamma, x, \Sigma \triangleright y; \Theta \triangleright x \vdash \Gamma, \Theta, \Sigma \triangleright y & \text{Cut}
\end{array}
$$

where Γ, Σ, Θ range over the set of finite, possibly empty, sequences of variables of $\mathrm{Fm}_{\mathcal{L}}$.

We denote the collection of the standard rules by (SR), i.e.,

$$(\text{SR}) = (\text{Ax}) \cup (\text{Ex}) \cup (\text{W}) \cup (\text{Con}) \cup (\text{Cut}).$$

Suppose $\mathcal{G} = \langle \mathbf{Fm}_{\mathcal{L}}, \mathrm{Th}\,\mathcal{G} \rangle$ is a $GSS_{\mathcal{L}}$. We say that a Gentzen-style rule $\bar{s} \vdash s$ *holds in* $\mathcal{G}$ (we write it as $\bar{s} \vdash_{\mathcal{G}} s$) if for every substitution σ and every $\mathcal{G}$-theory T

$$\sigma\{\bar{s}\} \subseteq T \implies \sigma s \in T.$$

3 Closure relations

Let us say that a sequent $\bar{\alpha} \triangleright \alpha \in \mathrm{Fm}_{\mathcal{L}}^{+}$ is *compatible* with the set of formulas $X \subseteq \mathrm{Fm}_{\mathcal{L}}$ (and, reciprocally, that the set X is *compatible* with the sequent $\bar{\alpha} \triangleright \alpha$) if the following implication holds

$$\{\bar{\alpha}\} \subseteq X \implies \alpha \in X.$$

Definition 3.1. *If $\mathcal{C} \subseteq \mathcal{P}(\mathrm{Fm}_{\mathcal{L}})$ is a family of sets of formulas, the* (finite) *closure relation $\mathbf{R}_{\mathcal{L}}\mathcal{C}$ for $\mathcal{C}$ consists of all sequents that are compatible with all sets of formulas in $\mathcal{C}$ (note that a sequent by definition is a finite sequence of formulas):*

$$\mathbf{R}_{\mathcal{L}}\mathcal{C} := \{\bar{\alpha} \triangleright \alpha \in \mathrm{Fm}_{\mathcal{L}}^{+} \mid (\forall X \in \mathcal{C})\, \{\bar{\alpha}\} \subseteq X \implies \alpha \in X\}. \qquad \square$$

If $\mathcal{C} \subseteq \mathcal{P}(\mathrm{Fm}_{\mathcal{L}})$ is a closure system then the definition above can be rewritten as

$$\mathbf{R}_{\mathcal{L}}\mathcal{C} := \{\bar{\alpha} \triangleright \alpha \in \mathrm{Fm}_{\mathcal{L}}^{+} \mid \alpha \in \{\bar{\alpha}\}^{\mathcal{C}}\}.$$

So defined, compatibility relation gives rise to a Galois connection between families of sets of formulas and sets of sequents. Finite closure relations represent the fixed-points for this connection on one side, while algebraic closure systems are the fixed-points for the other.

It is straightforward to show that the finite closure relations are exactly the sets of sequents that are closed under all standard rules (SR). In other words they are theories of the Gentzen-style system axiomatized purely by (SR).

Definition 3.2. *Let $\mathcal{H}$ be an $HSS_{\mathcal{L}}$. If $\mathcal{C} \subseteq \mathrm{Th}\,\mathcal{H}$ is an algebraic closure system over $\mathrm{Fm}_{\mathcal{L}}$, then $\mathbf{R}_{\mathcal{L}}\mathcal{C}$ is called a general (finite) closure relation for $\mathcal{H}$. The set of all general closure relations for $\mathcal{H}$ will be denoted by $\mathbf{Gcr}\,\mathcal{H}$.*

For every $HSS_{\mathcal{L}}$ $\mathcal{H}$ there is the distinguished general closure relation $\mathbf{R}_{\mathcal{L}}\mathrm{Th}\,\mathcal{H}$, which in its turn defines a Gentzen-style axiomatization (modulo (SR)) for a Gentzen-style system of the same type $\mathcal{L}$:

$$\vdash \mathbf{R}_{\mathcal{L}}\mathrm{Th}\,\mathcal{H} := \bigcup\{\vdash \bar{\alpha} \triangleright \alpha \mid \bar{\alpha} \triangleright \alpha \in \mathbf{R}_{\mathcal{L}}\mathrm{Th}\,\mathcal{H}\}.$$

Proposition 3.3. (Babenyshev, 2004, Theorem 2.2.10) *For every $HSS_{\mathcal{L}}$ $\mathcal{H}$*

1. $\mathbf{R}_{\mathcal{L}}\mathrm{Th}\,\mathcal{H} = \{\bar{\alpha} \triangleright \alpha \mid \bar{\alpha} \vdash_{\mathcal{H}} \alpha\}$.
2. $\mathbf{R}_{\mathcal{L}}\mathrm{Th}\,\mathcal{H}$ is invariant.
3. $\mathbf{Gcr}\,\mathcal{H}$ can be axiomatized by the standard rules (SR) and $\vdash \mathbf{R}_{\mathcal{L}}\mathrm{Th}\,\mathcal{H}$.
4. $\mathbf{Gcr}\,\mathcal{H}$ is a $GSS_{\mathcal{L}}$.
5. $\mathbf{R}_{\mathcal{L}}\mathrm{Th}\,\mathcal{H} = \mathrm{Thm}\,(\mathbf{Gcr}\,\mathcal{H})$.

Proof.

(1) $\mathbf{R}_{\mathcal{L}}\mathrm{Th}\,\mathcal{H} \overset{\mathrm{def}}{=} \{\bar{\alpha} \triangleright \alpha \mid \alpha \in \{\bar{\alpha}\}^{\mathrm{Th}\,\mathcal{H}}\} \overset{\mathrm{def}}{=} \{\bar{\alpha} \triangleright \alpha \mid \bar{\alpha} \vdash_{\mathcal{H}} \alpha\}$.

(2) For any substitution $\sigma : \mathbf{Fm}_{\mathcal{L}} \to \mathbf{Fm}_{\mathcal{L}}$

$$\bar{\alpha} \triangleright \alpha \in \mathbf{R}_{\mathcal{L}}\mathrm{Th}\,\mathcal{H} \overset{(1)}{\Longrightarrow} \bar{\alpha} \vdash_{\mathcal{H}} \alpha \implies \sigma\bar{\alpha} \vdash_{\mathcal{H}} \sigma\alpha \overset{(1)}{\Longrightarrow} \sigma\bar{\alpha} \triangleright \sigma\alpha \in \mathbf{R}_{\mathcal{L}}\mathrm{Th}\,\mathcal{H}.$$

(3) Suppose $\mathcal{A} \in \mathbf{Gcr}\,\mathcal{H}$. As a closure relation, $\mathcal{A}$ is a theory of (SR). By definition of $\mathbf{Gcr}\,\mathcal{H}$, $\mathcal{A} = \mathbf{R}_{\mathcal{L}}\mathcal{C}$ for some algebraic closure system $\mathcal{C} \subseteq \mathrm{Th}\,\mathcal{H}$, hence $\mathbf{R}_{\mathcal{L}}\mathrm{Th}\,\mathcal{H} \subseteq \mathbf{R}_{\mathcal{L}}\mathcal{C}$. Therefore for every rule $\vdash \bar{\alpha} \triangleright \alpha$ and every substitution σ

$$\vdash \bar{\alpha} \triangleright \alpha \subseteq \vdash \mathbf{R}_{\mathcal{L}}\mathrm{Th}\,\mathcal{H} \overset{\mathrm{def}}{\Longrightarrow} \bar{\alpha} \triangleright \alpha \in \mathbf{R}_{\mathcal{L}}\mathrm{Th}\,\mathcal{H}$$

$$\overset{(2)}{\Longrightarrow} \bar{\sigma}\alpha \triangleright \sigma\alpha \in \mathbf{R}_{\mathcal{L}}\mathrm{Th}\,\mathcal{H} \subseteq \mathbf{R}_{\mathcal{L}}\mathcal{C}.$$

Now let $\mathcal{A} \in \mathrm{Th}\,((\mathrm{SR}) \cup \vdash \mathbf{R}_{\mathcal{L}}\mathrm{Th}\,\mathcal{H})$. Then $\mathcal{A}$ is a finite closure relation over $\mathrm{Fm}_{\mathcal{L}}^{+}$, since it is a theory of (SR). We have that $\mathcal{A} = \mathbf{R}_{\mathcal{L}}\mathcal{C}$ for some algebraic closure system $\mathcal{C} \subseteq \mathcal{P}(\mathrm{Fm}_{\mathcal{L}})$. Since $\mathcal{A} \in \mathrm{Th}\,(\vdash \mathbf{R}_{\mathcal{L}}\mathrm{Th}\,\mathcal{H})$, then, for every $X \in \mathcal{C}$ and every $\bar{\alpha} \rhd \alpha \in \mathbf{R}_{\mathcal{L}}\mathrm{Th}\,\mathcal{H} \subseteq \mathcal{A}$:

$$\{\bar{\alpha}\}^{\mathcal{C}} \subseteq X \implies \alpha \in \{\bar{\alpha}\}^{\mathcal{C}} \subseteq X^{\mathcal{C}} = X \implies X \in \mathrm{Th}\,\mathcal{H}.$$

Thus $\mathcal{C} \subseteq \mathrm{Th}\,\mathcal{H}$, therefore $\mathcal{A} = \mathbf{R}_{\mathcal{L}}\,\mathcal{C} \in \mathbf{Gcr}\,\mathcal{H}$.

(4) It follows directly from (3).

(5) By (3), every $\mathcal{A} \in \mathbf{Gcr}\,\mathcal{H}$ is closed under the rules of $\vdash \mathbf{R}_{\mathcal{L}}\mathrm{Th}\,\mathcal{H}$. Thus, for every $\mathcal{A} \in \mathbf{Gcr}\,\mathcal{H}$, we have $\mathbf{R}_{\mathcal{L}}\mathrm{Th}\,\mathcal{H} \subseteq \mathcal{A}$ and hence $\mathbf{R}_{\mathcal{L}}\mathrm{Th}\,\mathcal{H} \subseteq \mathrm{Thm}\,\mathbf{Gcr}\,\mathcal{H}$. On the other hand, since $\mathbf{R}_{\mathcal{L}}\mathrm{Th}\,\mathcal{H} \in \mathbf{Gcr}\,\mathcal{H}$, then $\mathrm{Thm}\,\mathbf{Gcr}\,\mathcal{H} \subseteq \mathbf{R}_{\mathcal{L}}\mathrm{Th}\,\mathcal{H}$. $\square$

Closure relations were introduced in (Font et al., 2001) as a framework for studying metatheoretical properties of Hilbert-style systems. The fact that $\mathbf{Gcr}\,\mathcal{H}$ forms a Gentzen-style system was first observed also in (Font et al., 2001). The Gentzen-style system $\mathbf{Gcr}\,\mathcal{H}$ formalizes some metalogic of the Hilbert-style system $\mathcal{H}$. This metalogic is quite weak and equivalent in expressive power to strict universal Horn logic without equality (Bloom, 1975). Although $\mathbf{Gcr}\,\mathcal{H}$ is almost trivial, since it can be axiomatized, according to Proposition 3.3(3), by taking sequents representing the derivations in $\mathcal{H}$ and the standard Gentzen-style rules (SR), it has proved to be useful as a framework for working with other kinds of closure relations like *full* or *axiomatic* ones (Babenyshev, 2004).

Let us define for every $X \subseteq \mathrm{Fm}_{\mathcal{L}}$ and every $\mathcal{A} \subseteq \mathrm{Fm}_{\mathcal{L}}^{+}$

$$\rhd X := \{\rhd \alpha \mid \alpha \in X\}, \quad \mathrm{Ax}\,\mathcal{A} := \{\alpha \in \mathrm{Fm}_{\mathcal{L}} \mid \rhd \alpha \in \mathcal{A}\}, \quad \Theta\mathcal{A} := \mathcal{A} \cap \mathrm{Fm}_{\mathcal{L}}^{1},$$

where $\mathrm{Fm}_{\mathcal{L}}^{1}$ is by definition $\{\langle \alpha \rangle \mid \alpha \in \mathrm{Fm}_{\mathcal{L}}\}$. Thus we obtain the complex operators:

$$(\rhd) : \mathcal{P}(\mathrm{Fm}_{\mathcal{L}}) \to \mathcal{P}(\mathrm{Fm}_{\mathcal{L}}^{1}),$$
$$\mathrm{Ax} : \mathcal{P}(\mathrm{Fm}_{\mathcal{L}}^{+}) \to \mathcal{P}(\mathrm{Fm}_{\mathcal{L}}),$$
$$\Theta : \mathcal{P}(\mathrm{Fm}_{\mathcal{L}}^{+}) \to \mathcal{P}(\mathrm{Fm}_{\mathcal{L}}^{1}).$$

In the following proofs we mainly use the "exponential" notation for closures of sets. Namely, if $\mathcal{C}$ is a closure system over some set X, then for all $Y \subseteq X$:

$$Y^{\mathcal{C}} = (Y)^{\mathcal{C}} := \bigcap\nolimits_{Y \subseteq F \in \mathcal{C}} F.$$

For the sake of brevity and readability we will make no distinction between a formula α and an 1-tuple $\langle \alpha \rangle$ with this formula as its only component, so, for instance, for $X \subseteq \mathrm{Fm}_{\mathcal{L}}$: $(X)^{\mathbf{Gcr}\,\mathcal{H}}$ will be notationally equivalent to $(\rhd X)^{\mathbf{Gcr}\,\mathcal{H}}$.

Definition 3.4. *Let $\mathcal{H}$ be an $HSS_{\mathcal{L}}$. The set $\mathbf{Acr}\,\mathcal{H}$ of axiomatic closure relations for $\mathcal{H}$ is defined as follows:*

$$\mathbf{Acr}\,\mathcal{H} := \{(T)^{\mathbf{Gcr}\,\mathcal{H}} \mid T \in \mathrm{Th}\,\mathcal{H}\}.$$

An element of $\mathbf{Acr}\,\mathcal{H}$ *is called* an axiomatic closure relation for $\mathcal{H}$.   □

Note that in the definition above, in the expression $(T)^{\mathbf{Gcr}\,\mathcal{H}} = (\rhd\,T)^{\mathbf{Gcr}\,\mathcal{H}}$, the set $\rhd\,T$ contains sequents of the form $\rhd\,\alpha$, where $\alpha \in T \subseteq \mathrm{Fm}_{\mathcal{L}}$, and the closure of $\rhd\,T$ is taken in the family of Gentzen theories, where each of the theories is a set of sequents itself.

Proposition 3.5. *For every* $HSS_{\mathcal{L}}$ $\mathcal{H}$, $\Gamma \subseteq \mathrm{Fm}_{\mathcal{L}}$ *and* $\mathcal{A} \in \mathbf{Acr}\,\mathcal{H}$

1. $\mathbf{Acr}\,\mathcal{H} \subseteq \mathbf{Gcr}\,\mathcal{H}$.
2. $\mathrm{Ax}\,\mathcal{A} \in \mathrm{Th}\,\mathcal{H}$.
3. $\mathcal{A} = (\Theta\,\mathcal{A})^{\mathbf{Gcr}\,\mathcal{H}} = (\mathrm{Ax}\,\mathcal{A})^{\mathbf{Gcr}\,\mathcal{H}}$.
4. $\mathbf{Acr}\,\mathcal{H} = \{(X)^{\mathbf{Gcr}\,\mathcal{H}} \mid X \subseteq \mathrm{Fm}_{\mathcal{L}}\}$.
5. $\mathbf{Acr}\,\mathcal{H} = \{\mathbf{R}_{\mathcal{L}}[T)_{\mathrm{Th}\,\mathcal{H}} \mid T \in \mathrm{Th}\,\mathcal{H}\}$.
6. $(\Gamma)^{\mathbf{Gcr}\,\mathcal{H}} = (\Gamma)^{\mathbf{Acr}\,\mathcal{H}}$, *where* $(\Gamma)^{\mathbf{Acr}\,\mathcal{H}}$ *denotes here the smallest axiomatic closure relation containing the set of sequents* $\rhd\,\Gamma$. *It coincides with the usual closure* $(\Gamma)^{\mathbf{Acr}\,\mathcal{H}}$ *when* $\mathbf{Acr}\,\mathcal{H}$ *is a closure system.*
7. $\bar{\alpha} \rhd \alpha \in (\Gamma)^{\mathbf{Gcr}\,\mathcal{H}} \iff \alpha \in \{\bar{\alpha}\}^{\mathrm{Th}\,\mathcal{H}} \vee \Gamma^{\mathrm{Th}\,\mathcal{H}} \iff \Gamma, \bar{\alpha} \vdash_{\mathcal{H}} \alpha$.

Proof. (1) By definition.

(2) Suppose $\mathcal{A} \in \mathbf{Gcr}\,\mathcal{H}$, then, by definition, $\mathcal{A} = \mathbf{R}_{\mathcal{L}}\mathcal{C}$, for some algebraic closure system $\mathcal{C} \subseteq \mathrm{Th}\,\mathcal{H}$. We have $\mathrm{Ax}\,\mathcal{A} = \bigcap\mathcal{C}$, because

$$\alpha \in \bigcap\mathcal{C} = (\emptyset)^{\mathcal{C}} \iff \rhd\,\alpha \in \mathbf{R}_{\mathcal{L}}\mathcal{C} = \mathcal{A} \iff \alpha \in \mathrm{Ax}\,\mathcal{A}.$$

Therefore, since $\mathcal{C} \subseteq \mathrm{Th}\,\mathcal{H}$: $\mathrm{Ax}\,\mathcal{A} = \bigcap\mathcal{C} \in \mathrm{Th}\,\mathcal{H}$.

(3) If $\mathcal{A} \in \mathbf{Acr}\,\mathcal{H}$, then, by definition, $\mathcal{A} = (T)^{\mathbf{Gcr}\,\mathcal{H}}$ for some $T \in \mathrm{Th}\,\mathcal{H}$. Therefore

$$\rhd\,T \subseteq \Theta\,\mathcal{A} \subseteq \mathcal{A} \implies \mathcal{A} = (T)^{\mathbf{Gcr}\,\mathcal{H}} \subseteq (\Theta\,\mathcal{A})^{\mathbf{Gcr}\,\mathcal{H}} \subseteq \mathcal{A}^{\mathbf{Gcr}\,\mathcal{H}} = \mathcal{A}$$

$$\implies \mathcal{A} = (\Theta\,\mathcal{A})^{\mathbf{Gcr}\,\mathcal{H}}.$$

(4) If $\mathcal{A} \in \mathbf{Acr}\,\mathcal{H}$, then $\mathcal{A} = (\Theta\,\mathcal{A})^{\mathbf{Gcr}\,\mathcal{H}} = (\mathrm{Ax}\,\mathcal{A})^{\mathbf{Gcr}\,\mathcal{H}}$. For the other direction, suppose $\mathcal{A} = (X)^{\mathbf{Gcr}\,\mathcal{H}}$, for some $X \subseteq \mathrm{Fm}_{\mathcal{L}}$. Then $\mathcal{A} = (\Theta\,\mathcal{A})^{\mathbf{Gcr}\,\mathcal{H}} = (\mathrm{Ax}\,\mathcal{A})^{\mathbf{Gcr}\,\mathcal{H}}$, because

$$(\supseteq)\ \Theta\,\mathcal{A} \subseteq \mathcal{A} \implies (\Theta\,\mathcal{A})^{\mathbf{Gcr}\,\mathcal{H}} \subseteq \mathcal{A}^{\mathbf{Gcr}\,\mathcal{H}} = \mathcal{A},$$

$$(\subseteq)\ \mathcal{A} = (X)^{\mathbf{Gcr}\,\mathcal{H}} \implies \rhd\,X \subseteq \mathcal{A} \implies \rhd\,X \subseteq \Theta\,\mathcal{A}$$

$$\implies \mathcal{A} = (X)^{\mathbf{Gcr}\,\mathcal{H}} \subseteq (\Theta\,\mathcal{A})^{\mathbf{Gcr}\,\mathcal{H}}.$$

(5) Suppose $\mathcal{A} \in \mathbf{Acr}\,\mathcal{H}$. Then, by (3) and (2), $\mathcal{A} = (T)^{\mathbf{Gcr}\,\mathcal{H}}$, where $T = \mathrm{Ax}\,\mathcal{A} \in \mathrm{Th}\,\mathcal{H}$. Let $\mathcal{C} = [T)_{\mathrm{Th}\,\mathcal{H}}$. Being a general closure relation for $\mathcal{H}$, $\mathcal{A} = \mathbf{R}_{\mathcal{L}}\,\mathcal{D}$, for some algebraic closure system $\mathcal{D} \subseteq \mathrm{Th}\,\mathcal{H}$. Then $\mathcal{A} = \mathbf{R}_{\mathcal{L}}\mathcal{C}$, because

$$(\supseteq)\ T = \bigcap \mathcal{D} \implies \mathcal{D} \subseteq [T)_{\mathrm{Th}\,\mathcal{H}} = \mathcal{C} \implies \mathbf{R}_{\mathcal{L}}\mathcal{C} \subseteq \mathbf{R}_{\mathcal{L}}\mathcal{D} = \mathcal{A},$$

$$(\subseteq)\ \ominus\mathbf{R}_{\mathcal{L}}\mathcal{D} = \triangleright(\bigcap\mathcal{D}) = \triangleright T = \triangleright(\bigcap\mathcal{C}) = \ominus\mathbf{R}_{\mathcal{L}}\mathcal{C} \implies \triangleright T \subseteq \mathbf{R}_{\mathcal{L}}\mathcal{C}$$

$$\implies \mathcal{A} = (T)^{\mathbf{Gcr}\,\mathcal{H}} \subseteq \mathbf{R}_{\mathcal{L}}\mathcal{C}.$$

(6) Let $\Gamma \subseteq \mathrm{Fm}_{\mathcal{L}}$. Then, by (4), $(\Gamma)^{\mathbf{Gcr}\,\mathcal{H}} \in \mathbf{Acr}\,\mathcal{H}$. $(\Gamma)^{\mathbf{Gcr}\,\mathcal{H}}$ is the smallest axiomatic closure relation containing $\triangleright \Gamma$, because

$$\triangleright\Gamma \subseteq \mathcal{A} \in \mathbf{Acr}\,\mathcal{H} \implies (\Gamma)^{\mathbf{Gcr}\,\mathcal{H}} \subseteq \mathcal{A}^{\mathbf{Gcr}\,\mathcal{H}} = \mathcal{A}.$$

$$(7)\ \bar{\alpha}\triangleright\alpha \in (T)^{\mathbf{Gcr}\,\mathcal{H}} \overset{(6)}{=} (T)^{\mathbf{Acr}\,\mathcal{H}} \overset{(5)}{=} \mathbf{R}_{\mathcal{L}}[T)_{\mathrm{Th}\,\mathcal{H}}$$

$$\iff \alpha \in \{\bar{\alpha}\}^{[T)_{\mathrm{Th}\,\mathcal{H}}} = (T\cup\{\bar{\alpha}\})^{\mathrm{Th}\,\mathcal{H}} = T \vee \{\bar{\alpha}\}^{\mathrm{Th}\,\mathcal{H}} \iff T, \bar{\alpha} \vdash_{\mathcal{H}} \alpha.\ \Box$$

The following lemma, although simple, plays nevertheless a crucial role in the proof of the deduction-detachment theorem characterization.

Lemma 3.6. $\mathbf{Acr}\,\mathcal{H}$ *is a closure system iff for all families* $\{\mathcal{A}_i\}_{i\in I} \subseteq \mathbf{Acr}\,\mathcal{H}$

$$\bigcap_{i\in I}\mathcal{A}_i = (\bigcap_{i\in I}\ominus\mathcal{A}_i)^{\mathbf{Gcr}\,\mathcal{H}}.$$

Proof. It follows directly from the implications

$$(\Rightarrow)\ \ominus(\bigcap_{i\in I}\mathcal{A}_i) = \bigcap_{i\in I}\ominus\mathcal{A}_i$$

$$\implies \bigcap_{i\in I}\mathcal{A}_i \overset{3.5(3)}{=} (\ominus(\bigcap_{i\in I}\mathcal{A}_i))^{\mathbf{Gcr}\,\mathcal{H}} = (\bigcap_{i\in I}\ominus\mathcal{A}_i)^{\mathbf{Gcr}\,\mathcal{H}}.$$

$$(\Leftarrow)\ \bigcap_{i\in I}\mathcal{A}_i = (\bigcap_{i\in I}\ominus\mathcal{A}_i)^{\mathbf{Gcr}\,\mathcal{H}} \overset{3.5(4)}{\in} \mathbf{Acr}\,\mathcal{H}. \qquad \Box$$

4 Characterization of the Deduction-Detachment Theorem

Definition 4.1. *An* $HSS_{\mathcal{L}}$ $\mathcal{H}$ *admits a* multiterm deduction-detachment *theorem* (DDT$_{\Delta}$) *with respect to a finite (may be empty) set* $\Delta(x,y) \subseteq \mathrm{Fm}_{\mathcal{L}}\{x,y\}$ *of formulas in two variables if the following holds*

(1) $x, \Delta(x,y) \vdash_{\mathcal{H}} y,$ *(Δ-detachment)*

(2) *for all* $\alpha, \beta \in \mathrm{Fm}_{\mathcal{L}}$, *if* $\Gamma, \alpha \vdash_{\mathcal{H}} \beta$, *then* $\Gamma \vdash_{\mathcal{H}} \Delta(\alpha,\beta)$.   *($\Delta$-deduction)* $\Box$

Lemma 4.2. *Suppose that for some* $HSS_{\mathcal{L}}$ $\mathcal{H}$, $\mathbf{Acr}\,\mathcal{H}$ *is an invariant closure system, and let* $\Delta(x,y)$ *be a nonempty set of formulas in two variables. Then* $\mathcal{H}$ *admits* DDT$_{\Delta}$ *iff*

$$\{x\triangleright y\}^{\mathbf{Acr}\,\mathcal{H}} = (\Delta(x,y))^{\mathbf{Acr}\,\mathcal{H}}. \tag{Eq.1}$$

Proof. ($\Rightarrow$) If $\Gamma = \mathrm{Ax}((x \triangleright y)^{\mathbf{Acr}\,\mathcal{H}})$, then $(x \triangleright y)^{\mathbf{Acr}\,\mathcal{H}} \overset{3.5(3,6)}{=} (\Gamma)^{\mathbf{Acr}\,\mathcal{H}}$. Therefore

$$(\subseteq)\ x, \Delta(x,y) \vdash_{\mathcal{H}} y \overset{3.5(7)}{\Longrightarrow} x \triangleright y \in (\Delta(x,y))^{\mathbf{Gcr}\,\mathcal{H}} \overset{3.5(6)}{=} (\Delta(x,y))^{\mathbf{Acr}\,\mathcal{H}}$$

$$\Longrightarrow (x \triangleright y)^{\mathbf{Acr}\,\mathcal{H}} \subseteq (\Delta(x,y))^{\mathbf{Acr}\,\mathcal{H}}.$$

$$(\supseteq)\ x \triangleright y \in \{x \triangleright y\}^{\mathbf{Acr}\,\mathcal{H}} = (\Gamma)^{\mathbf{Acr}\,\mathcal{H}} \overset{3.5(6)}{=} (\Gamma)^{\mathbf{Gcr}\,\mathcal{H}}$$

$$\overset{3.5(7)}{\Longrightarrow} \Gamma, x \vdash_{\mathcal{H}} y \overset{\Delta-\mathrm{ded}}{\Longrightarrow} \Gamma \vdash_{\mathcal{H}} \Delta(x,y)$$

$$\Longrightarrow \Delta(x,y) \subseteq (\Gamma)^{\mathrm{Th}\,\mathcal{H}} \overset{3.5(2)}{=} \Gamma$$

$$\Longrightarrow \triangleright \Delta(x,y) \subseteq (\Gamma)^{\mathbf{Gcr}\,\mathcal{H}} \overset{3.5(6)}{=} (\Gamma)^{\mathbf{Acr}\,\mathcal{H}} = \{x \triangleright y\}^{\mathbf{Acr}\,\mathcal{H}}$$

$$\Longrightarrow (\Delta(x,y))^{\mathbf{Acr}\,\mathcal{H}} \subseteq \{x \triangleright y\}^{\mathbf{Acr}\,\mathcal{H}}.$$

($\Leftarrow$) The statement follows from the implications:

$$x \triangleright y \in (\Delta(x,y))^{\mathbf{Acr}\,\mathcal{H}} \overset{3.5(6)}{=} (\Delta(x,y))^{\mathbf{Gcr}\,\mathcal{H}}$$

$$\overset{3.5(7)}{\Longrightarrow} x, \Delta(x,y) \vdash_{\mathcal{H}} y. \qquad\qquad (\Delta\text{-detachment})$$

$$\Gamma, \alpha \vdash_{\mathcal{H}} \beta \overset{3.5(7)}{\Longrightarrow} \alpha \triangleright \beta \in (\Gamma)^{\mathbf{Gcr}\,\mathcal{H}} = (\Gamma)^{\mathbf{Acr}\,\mathcal{H}}$$

$$\Longrightarrow (\Delta(\alpha,\beta))^{\mathbf{Gcr}\,\mathcal{H}} \overset{3.5(6)}{=} (\Delta(\alpha,\beta))^{\mathbf{Acr}\,\mathcal{H}} \overset{(\mathrm{Eq}.1)}{=} \{\alpha \triangleright \beta\}^{\mathbf{Acr}\,\mathcal{H}} \subseteq (\Gamma)^{\mathbf{Gcr}\,\mathcal{H}}$$

$$\Longrightarrow \Delta(\alpha,\beta) \subseteq (\Gamma)^{\mathbf{Gcr}\,\mathcal{H}} \overset{3.5(7)}{\Longrightarrow} \Gamma \vdash_{\mathcal{H}} \Delta(\alpha,\beta). \qquad (\Delta\text{-deduction}) \quad \square$$

Examples. 1) Let us consider the normal modal logic $S4$, which we associate with a set of modal formulas — the set of theorems of $S4$. Let $\mathrm{Th} \vdash_{S4}^{g}$ be the family of all sets of modal formulas that are simultaneously closed under rules *modus ponens* $x, x \to y \vdash y$ and *necessitation rule* $x/\square x$ and each contain $S4$. Then $\mathrm{Th} \vdash_{S4}^{g}$ is the set of theories of the abstract Hilbert system for *global consequence relation* $\vdash_{S4}^{g}$ associated with the normal modal logic $S4$, while $S4$ is the set of its theorems: $S4 = \mathrm{Thm}\,(\mathrm{Th} \vdash_{S4}^{g})$. Let $\vdash_{S4}^{g}$ also denote the aforementioned abstract Hilbert system. Then, by deduction theorem for $\vdash_{S4}^{g}$,

$$\{x \triangleright y\}^{\mathbf{Acr} \vdash_{S4}^{g}} = \{\square x \to y\}^{\mathbf{Gcr} \vdash_{S4}^{g}} = \{\square x \to y\}^{\mathbf{Acr} \vdash_{S4}^{g}}.$$

2) The *inconsistent* Hilbert-style system $\mathcal{H} = \langle \mathbf{Fm}_{\mathcal{L}}, \{\mathrm{Fm}_{\mathcal{L}}\} \rangle$ over a language $\mathcal{L}$ admits DDT_{Δ} with respect to any finite set $\Delta \subseteq \mathrm{Fm}_{\mathcal{L}}$ of formulas, because

$$(\Delta)^{\mathbf{Acr}\,\mathcal{H}} = (\Delta)^{\mathbf{Gcr}\,\mathcal{H}} = \mathrm{Fm}_{\mathcal{L}}^{+} = \{x \triangleright y\}^{\mathbf{Acr}\,\mathcal{H}}.$$

3) Let us also note that the *almost inconsistent* $HSS_{\mathcal{L}}\mathcal{H} = \langle \mathbf{Fm}_{\mathcal{L}}, \{\emptyset, \mathrm{Fm}_{\mathcal{L}}\} \rangle$ over a language $\mathcal{L}$ admits $\mathrm{DDT}_{\emptyset}$, because

$$\{x \triangleright y\}^{\mathbf{Acr}\,\mathcal{H}} = \mathbf{R}_{\mathcal{L}}\{\emptyset, \mathrm{Fm}_{\mathcal{L}}\} = \mathrm{Fm}_{\mathcal{L}}^{+} \setminus \mathrm{Fm}_{\mathcal{L}}^{1} = (\emptyset)^{\mathbf{Gcr}\,\mathcal{H}} = (\emptyset)^{\mathbf{Acr}\,\mathcal{H}}.$$

Theorem 4.3. *Let $\mathcal{H}$ be a Hilbert-style system with theorems. Then $\mathbf{Acr}\,\mathcal{H}$ forms a Gentzen-style system iff $\mathcal{H}$ admits a multiterm deduction-detachment theorem.*

Proof. In view of the remark 2) above, it suffices to prove the theorem for $\mathcal{H}$ that is not inconsistent.

($\Rightarrow$) Since $\mathbf{Acr}\,\mathcal{H}$ is a closure system, there is a closure of the set $\{x \triangleright y\}$ in $\mathbf{Acr}\,\mathcal{H}$. If $\{x \triangleright y\}^{\mathbf{Acr}\,\mathcal{H}} = (\emptyset)^{\mathbf{Gcr}\,\mathcal{H}}$, then

$$\{x \triangleright y\}^{\mathbf{Acr}\,\mathcal{H}} = (\emptyset)^{\mathbf{Gcr}\,\mathcal{H}} \implies x \triangleright y \in (\emptyset)^{\mathbf{Gcr}\,\mathcal{H}} \overset{3.5(7)}{\implies} x \vdash_{\mathcal{H}} y,$$

so $\mathcal{H}$ is either inconsistent or almost inconsistent, a contradiction with the assumption. Thus $\{x \triangleright y\}^{\mathbf{Acr}\,\mathcal{H}} = (T)^{\mathbf{Gcr}\,\mathcal{H}}$, for some $T \in \mathrm{Th}\,\mathcal{H}$, such that $T \neq \mathrm{Thm}\,\mathcal{H}$, because $(\emptyset)^{\mathbf{Gcr}\,\mathcal{H}} = (\mathrm{Thm}\,\mathcal{H})^{\mathbf{Gcr}\,\mathcal{H}}$. Since $\{x \triangleright y\}^{\mathbf{Acr}\,\mathcal{H}}$ is compact in $\mathbf{Acr}\,\mathcal{H}$, there is a finite subset $\mathcal{O} \subseteq T$, such that $\{x \triangleright y\}^{\mathbf{Acr}\,\mathcal{H}} = \mathcal{O}^{\mathbf{Gcr}\,\mathcal{H}}$. Suppose σ is any substitution such that $\sigma\{x, y\} = \{x, y\}$ and $\sigma(\mathrm{Var} \setminus \{x, y\}) \subseteq \{x, y\}$ and let $\Delta(x, y) = \sigma\mathcal{O}$. Since $\mathbf{Acr}\,\mathcal{H}$ forms a Gentzen system, it is invariant under inverse substitutions, therefore

$$\{x \triangleright y\}^{\mathbf{Acr}\,\mathcal{H}} = \{\sigma x \triangleright \sigma y\}^{\mathbf{Acr}\,\mathcal{H}} = (\sigma\mathcal{O})^{\mathbf{Acr}\,\mathcal{H}} = (\Delta(x, y))^{\mathbf{Acr}\,\mathcal{H}}.$$

So, by Lemma 4.2, $\mathcal{H}$ admits DDT_{Δ}.

($\Leftarrow$) Suppose $\mathcal{H}$ admits DDT_{Δ}, where $\Delta \neq \emptyset$. Δ can be viewed as a function $\Delta : \mathrm{Fm}_{\mathcal{L}}^{2} \to \mathcal{P}(\mathrm{Fm}_{\mathcal{L}})$. Furthermore it can be extended to a function from $\mathrm{Fm}_{\mathcal{L}}^{+}$ to $\mathcal{P}(\mathrm{Fm}_{\mathcal{L}})$ inductively as follows

$$\Delta(\triangleright \alpha) := \alpha,$$
$$\Delta(\alpha_0, \ldots, \alpha_n \triangleright \alpha) := \Delta(\alpha_0, \ldots, \alpha_{n-1} \triangleright \Delta(\alpha_n, \alpha))$$
$$:= \bigcup_{\delta \in \Delta} \{\Delta(\alpha_0, \ldots, \alpha_{n-1} \triangleright \delta(\alpha_n, \alpha))\},$$

and further, in the usual way, to a complex function $\Delta : \mathcal{P}(\mathrm{Fm}_{\mathcal{L}}^{+}) \to \mathcal{P}(\mathrm{Fm}_{\mathcal{L}})$. Thus, for every $\mathcal{A} \in \mathbf{Acr}\,\mathcal{H}$, the following holds

$$(1) \quad \triangleright \alpha \in \mathcal{A} \iff \Delta(\triangleright \alpha) \overset{\mathrm{def}}{=} \alpha \in \mathrm{Ax}\,\mathcal{A}$$

$$(2) \quad \bar{\alpha}, \alpha_{|\bar{\alpha}|} \triangleright \alpha \in \mathcal{A} \overset{3.5(7)}{\iff} \mathrm{Ax}\,\mathcal{A}, \bar{\alpha}, \alpha_{|\bar{\alpha}|} \vdash_{\mathcal{H}} \alpha$$
$$\overset{4.1}{\iff} \mathrm{Ax}\,\mathcal{A}, \bar{\alpha} \vdash_{\mathcal{H}} \Delta(\alpha_{|\bar{\alpha}|}, \alpha)$$
$$\overset{3.5(7)}{\iff} \bar{\alpha} \triangleright \Delta(\alpha_{|\bar{\alpha}|}, \alpha) \subseteq \mathcal{A} \iff \ldots \iff \triangleright \Delta(\bar{\alpha}, \alpha_{|\bar{\alpha}|} \triangleright \alpha) \subseteq \mathcal{A}$$
$$\iff \Delta(\bar{\alpha}, \alpha_{|\bar{\alpha}|} \triangleright \alpha) \subseteq \mathrm{Ax}\,\mathcal{A}.$$

In other words:
$$\bar{\alpha} \triangleright \alpha \in \mathcal{A} \iff \Delta(\bar{\alpha} \triangleright \alpha) \subseteq \mathrm{Ax}\,\mathcal{A}. \qquad\qquad (\mathrm{Eq.2})$$

Therefore, for every family $\{\mathcal{A}_i\}_{i \in I} \subseteq \mathbf{Acr}\,\mathcal{H}$,

$$\bar{\alpha} \triangleright \alpha \in \bigcap_{i \in I} \mathcal{A}_i \iff (\forall i \in I)\ \bar{\alpha} \triangleright \alpha \in \mathcal{A}_i$$

$$\overset{(\mathrm{Eq.2})}{\iff} (\forall i \in I)\ \Delta(\bar{\alpha} \triangleright \alpha) \subseteq \mathrm{Ax}\,\mathcal{A}_i \iff \Delta(\bar{\alpha} \triangleright \alpha) \subseteq \bigcap_{i \in I} \mathrm{Ax}\,\mathcal{A}_i$$

$$\overset{(\mathrm{Eq.2})}{\iff} \bar{\alpha} \triangleright \alpha \in (\triangleright(\bigcap_{i \in I} \mathrm{Ax}\,\mathcal{A}_i))^{\mathbf{Gcr}\,\mathcal{H}} = (\bigcap_{i \in I} \Theta \mathcal{A}_i)^{\mathbf{Gcr}\,\mathcal{H}} \in \mathbf{Acr}\,\mathcal{H}.$$

Thus, by Lemma 3.6, $\mathbf{Acr}\,\mathcal{H}$ is closed under arbitrary intersections, hence it is a closure system.

Now suppose $\mathcal{A} \in \mathbf{Acr}\,\mathcal{H}$ and σ is any substitution. Then

$$\bar{\alpha} \triangleright \alpha \in \sigma^{-1}\mathcal{A} \iff \sigma(\bar{\alpha} \triangleright \alpha) \in \mathcal{A}$$

$$\overset{(\mathrm{Eq.2})}{\iff} \Delta(\sigma\bar{\alpha} \triangleright \sigma\alpha) = \sigma\Delta(\bar{\alpha} \triangleright \alpha) \subseteq \mathrm{Ax}\,\mathcal{A}$$

$$\iff \Delta(\bar{\alpha} \triangleright \alpha) \subseteq \sigma^{-1}(\mathrm{Ax}\,\mathcal{A}) \overset{(\mathrm{Eq.2})}{\iff} \bar{\alpha} \triangleright \alpha \in (\sigma^{-1}\Theta\mathcal{A})^{\mathbf{Gcr}\,\mathcal{H}} \in \mathbf{Acr}\,\mathcal{H}.$$

Thus, in addition to being a closure system, $\mathbf{Acr}\,\mathcal{H}$ is invariant.

Finally, suppose $\{\mathcal{A}_i\}_{i \in I}$ is an upward-directed family of axiomatic closure relations. To prove that $\mathbf{Acr}\,\mathcal{H}$ is algebraic, by Proposition 3.5(4), it suffices to show that

$$\bigcup_{i \in I} \mathcal{A}_i = (\Theta(\bigcup_{i \in I} \mathcal{A}_i))^{\mathbf{Gcr}\,\mathcal{H}}.$$

Indeed

$$(\subseteq)\ \mathcal{A}_i \subseteq \bigcup_{i \in I} \mathcal{A}_i \implies \Theta \mathcal{A}_i \subseteq \Theta(\bigcup_{i \in I} \mathcal{A}_i)$$

$$\implies (\Theta \mathcal{A}_i)^{\mathbf{Gcr}\,\mathcal{H}} \subseteq (\Theta(\bigcup_{i \in I} \mathcal{A}_i))^{\mathbf{Gcr}\,\mathcal{H}}$$

$$\implies \mathcal{A}_i \overset{3.5(3)}{=} (\Theta \mathcal{A}_i)^{\mathbf{Gcr}\,\mathcal{H}} \subseteq (\Theta(\bigcup_{i \in I} \mathcal{A}_i))^{\mathbf{Gcr}\,\mathcal{H}}$$

$$\implies \bigcup_{i \in I} \mathcal{A}_i \subseteq (\Theta(\bigcup_{i \in I} \mathcal{A}_i))^{\mathbf{Gcr}\,\mathcal{H}}.$$

$(\supseteq)$ Since $\mathbf{Gcr}\,\mathcal{H}$ is algebraic and $\mathbf{Acr}\,\mathcal{H} \subseteq \mathbf{Gcr}\,\mathcal{H}$, $\bigcup_{i \in I} \mathcal{A}_i \in \mathbf{Gcr}\,\mathcal{H}$. Then

$$\Theta(\bigcup_{i \in I} \mathcal{A}_i) \subseteq \bigcup_{i \in I} \mathcal{A}_i \implies (\Theta(\bigcup_{i \in I} \mathcal{A}_i))^{\mathbf{Gcr}\,\mathcal{H}} \subseteq (\bigcup_{i \in I} \mathcal{A}_i)^{\mathbf{Gcr}\,\mathcal{H}} = \bigcup_{i \in I} \mathcal{A}_i.$$

$\square$

The characterization provided by Theorem 4.3 is closely related to some known partial characterizations of the multiterm deduction-detachment theorem (see Remark 2 later in the paper for a short historical overview). For instance, a protoalgebraic deductive system $\mathcal{H}$ admits a multiterm deduction-detachment theorem if and only if the lattice $\mathrm{Th}\,\mathcal{H}$ is infinitely meet-distributive over its compact elements (Czelakowski, 2001, Theorem 2.6.8), where

Definition 4.4. *A complete lattice L is* infinitely meet-distributive over its compact elements *if for every compact element $c \in L$ and every family $\{a_i\}_{i \in I} \subseteq L$*

$$c \vee \left(\bigwedge_{i \in I} a_i \right) = \bigwedge_{i \in I} (c \vee a_i). \qquad \square$$

The relation that the condition given by Definition 4.4 has with Theorem 4.3 is established in the following theorem.

Theorem 4.5. $\mathbf{Acr}\,\mathcal{H}$ *is a closure system iff the lattice* $\mathrm{Th}\,\mathcal{H}$ *is infinitely meet-distributive over its compact elements.*

Proof. In this theorem $\vee$ denotes the join in the complete lattice $\mathrm{Th}\,\mathcal{H}$.
($\Rightarrow$) Suppose $\mathbf{Acr}\,\mathcal{H}$ is closed under arbitrary intersections. Let $\{T_i\}_{i \in I} \subseteq \mathrm{Th}\,\mathcal{H}$ and $\{\bar\alpha\} \subseteq_\omega \mathrm{Fm}_\mathcal{L}$. Then for every $\alpha \in \mathrm{Fm}_\mathcal{L}$

$$\alpha \in \bigcap_{i \in I}(\{\bar\alpha\}^{\mathrm{Th}\,\mathcal{H}} \vee T_i) \iff (\forall i \in I)\ \alpha \in \{\bar\alpha\}^{\mathrm{Th}\,\mathcal{H}} \vee T_i$$

$$\overset{3.5(7)}{\iff} (\forall i \in I)\ \bar\alpha \rhd \alpha \in (\rhd T_i)^{\mathbf{Gcr}\,\mathcal{H}} \iff \bar\alpha \rhd \alpha \in \bigcap_{i \in I}(\rhd T_i)^{\mathbf{Gcr}\,\mathcal{H}}$$

$$\overset{3.6}{\iff} \bar\alpha \rhd \alpha \in \left(\bigcap_{i \in I}(\rhd T_i) \right)^{\mathbf{Gcr}\,\mathcal{H}} = \left(\rhd \left(\bigcap_{i \in I} T_i \right) \right)^{\mathbf{Gcr}\,\mathcal{H}}$$

$$\overset{3.5(7)}{\iff} \alpha \in \{\bar\alpha\}^{\mathrm{Th}\,\mathcal{H}} \vee \left(\bigcap_{i \in I} T_i \right).$$

Thus $\bigcap_{i \in I}(\{\bar\alpha\}^{\mathrm{Th}\,\mathcal{H}} \vee T_i) = \{\bar\alpha\}^{\mathrm{Th}\,\mathcal{H}} \vee (\bigcap_{i \in I} T_i)$.

($\Leftarrow$) By contradiction. Suppose $\mathrm{Th}\,\mathcal{H}$ is infinitely meet-distributive over compact elements, but there is a family $\{\mathcal{A}_i\}_{i \in I} \subseteq \mathbf{Acr}\,\mathcal{H}$ and a sequent $\bar\alpha \rhd \alpha$ such that $\bar\alpha \rhd \alpha \in \bigcap_{i \in I}\mathcal{A}_i$ and $\bar\alpha \rhd \alpha \notin (\bigcap_{i \in I} \Theta\,\mathcal{A}_i)^{\mathbf{Gcr}\,\mathcal{H}}$. Then

1) $\bar\alpha \rhd \alpha \notin (\bigcap_{i \in I} \Theta\,\mathcal{A}_i)^{\mathbf{Gcr}\,\mathcal{H}} \overset{3.5(7)}{\iff} \alpha \notin \{\bar\alpha\}^{\mathrm{Th}\,\mathcal{H}} \vee (\bigcap_{i \in I} \mathrm{Ax}\,\mathcal{A}_i)$,

2) $\bar\alpha \rhd \alpha \in \bigcap_{i \in I}\mathcal{A}_i \iff (\forall i \in I)\ \bar\alpha \rhd \alpha \in \mathcal{A}_i$

$\iff (\forall i \in I)\ \alpha \in \{\bar\alpha\}^{\mathrm{Th}\,\mathcal{H}} \vee \mathrm{Ax}\,\mathcal{A}_i \iff \alpha \in \bigcap_{i \in I}(\{\bar\alpha\}^{\mathrm{Th}\,\mathcal{H}} \vee \mathrm{Ax}\,\mathcal{A}_i)$.

But, by assumption, $\{\bar\alpha\}^{\mathrm{Th}\,\mathcal{H}} \vee (\bigcap_{i \in I} \mathrm{Ax}\,\mathcal{A}_i) = \bigcap_{i \in I}(\{\bar\alpha\}^{\mathrm{Th}\,\mathcal{H}} \vee \mathrm{Ax}\,\mathcal{A}_i)$, a contradiction. $\qquad \square$

Remarks:

1. The fact that the set of axiomatic closure relations is closed under finite intersections if and only if the lattice of all $\mathcal{H}$-theories is distributive, was first observed in (Font et al., 2006).

2.[1] The equivalence between equational definability of principal congruences in a variety (EDPC) and the fact that the join-semilattice of compact congruences of every algebra in the variety is dually Brouwerian was established by P. Köhler and D. Pigozzi in 1980 (Köhler and Pigozzi, 1980). The latter paper marks the starting point in the research of algebraic aspects of DDT. The paper (Czelakowski, 1985) by J. Czelakowski was a positive response on the Köhler-Pigozzi work and was published in the abstracted form

---

[1] The author would like to thank Janusz Czelakowski and the anonymous reviewer for clarifying the historical aspects of the issue.

under the same title "Algebraic aspects of Deduction Theorems" in Bulletin of the Section of Logic, Vol. 12, No. 3 (1983), 111-116. The main result of (Czelakowski, 1985) is that DDT for any non-pathological finitary logic is equivalent to the fact that the join-semilattice of its finitely axiomatizable closed theories is dually Brouwerian (non-pathological logical systems turned out to be equivalent to protoalgebraic logics defined later by Blok and Pigozzi (Blok and Pigozzi, 1986)). This equivalence carries over to the associated lattices of deductive filters. The general fact that in any algebraic lattice L, the join-semilattice of compact elements of L is dually Brouwerian if and only if the lattice is infinitely meet-distributive over its compact elements is noted in (Czelakowski, 1985, Theorem 2.12) (see also (Czelakowski, 2001, Proposition 2.6.7)). This fact, in the context of congruence lattices for varieties, was independently proved by E. Kiss in 1983 (Kiss, 1983). Wim Blok and Don Pigozzi had similar ideas at that time. They were aware of the connection between the deduction theorem and EDPC about the time when Peter Köhler and Don Pigozzi were writing their paper (Köhler and Pigozzi, 1980) during the winter of 1978–79 (see the remarks placed in (Czelakowski, 1985)). Wim Blok and Don Pigozzi wrote a paper on the deduction theorem in algebraic logic, viz. (Blok and Pigozzi, 1997), which was not published when Wim Blok lived. (He passed away in 2002.) The full text of (Blok and Pigozzi, 1997) appears only now in this volume. The results on the deduction theorem they included in (Blok and Pigozzi, 1997) overlap to some extent the ones presented in (Czelakowski, 1985), but the focus of their paper is different, being as it is on the equivalence between the deduction theorem and EDPC, e.g. the basic fact that for strongly algebraizable Hilbert-style deductive systems the DDT amounts to EDPC for the equivalent varieties was established in (Blok and Pigozzi, 1997).

3. Note that $\mathrm{Th}\,\mathcal{H}$ is always infinitely join-distributive over its compact elements, i.e.,

$$\{\bar{\alpha}\}^{\mathrm{Th}\,\mathcal{H}} \cap \left(\bigvee_{i \in I} \mathrm{Ax}\,\mathcal{A}_i \right) = \bigvee_{i \in I} \left(\{\bar{\alpha}\}^{\mathrm{Th}\,\mathcal{H}} \cap \mathrm{Ax}\,\mathcal{A}_i \right),$$

by (Czelakowski, 2001, Proposition 2.5.1), since $\mathrm{Th}\,\mathcal{H}$ is algebraic.

A number of prominent Hilbert-style systems fail to admit DDT.

Examples. 4) The Hilbert-style deductive system corresponding to global consequence relation $\vdash^g_K$ associated with the smallest modal propositional logic K does not admit DDT, because the lattice $\mathrm{Th} \vdash^g_K$ is not infinitely meet-distributive over its compact elements, even though it is distributive. Thus $\mathbf{Acr} \vdash^g_K$ is closed under finite intersections, but not under arbitrary intersections.

5) On the other hand, let us consider a Hilbert-style deductive system $\vdash_R$ with $x, x \to y \vdash y$ and $x, y \vdash x \wedge y$ as inference rules schemes for the Anderson and Belnap's logic of *relevant implication* R (without constants, see (Czelakowski, 2001, Example 2.1.3) for axiomatization details). Then $\vdash_R$ does not admit DDT because $\mathbf{Acr} \vdash_R$ is not invariant. It does admit a weaker form

of DDT though ((Maximova, 1966) and Dziobiak, unpublished):

$$X, \alpha \vdash_R \beta \iff (\exists n)(\exists \gamma_0, \ldots, \gamma_n)\ X \vdash_R \left[\left(\bigwedge_{i \leq n} (\gamma_i \to \gamma_i) \right) \wedge \alpha \right] \to \beta.$$

5 Deduction-Detachment Theorem and protoalgebraic systems

The deduction-detachment theorem was originally introduced to investigate the algebraizability phenomenon of the classical logic. Therefore it is particularly interesting to investigate DDT in the context of protoalgebraic Hilbert-style deductive systems, which were specifically introduced to capture weak forms of the algebraizability phenomena (Blok and Pigozzi, 1986).

Definition 5.1. *An $HSS_{\mathcal{L}}$ $\mathcal{H}$ is protoalgebraic if there exists a finite (may be empty) set $\Delta(x,y)$ of formulas in two variables, such that*

$$(1) \quad \vdash_{\mathcal{H}} \Delta(x,x),$$

$$(2) \quad x, \Delta(x,y) \vdash_{\mathcal{H}} y. \qquad\qquad (\Delta\text{-detachment})$$

The class of protoalgebraic Hilbert-style systems contains many well-known propositional deductive systems, such as deductive systems of propositional intuitionistic, modal and relevant logics and, of course, that of the classical logic. It is easy to see that DDT_{Δ} implies protoalgebraicity with the same set of formulas Δ. But protoalgebraicity is strictly weaker than DDT, as the above-mentioned examples of deductive systems for K and R show.

For protoalgebraic systems, the conditions for DDT can be weakened, as was shown by J. Czelakowski (Czelakowski, 2001). Further we will show how some of J. Czelakowski's results can be derived using the approach of this paper.

Lemma 5.2. *If $\mathcal{H}$ is protoalgebraic, then $\mathbf{Acr}\,\mathcal{H}$ is surjectively invariant.*

Proof. Suppose $\mathcal{A} \in \mathbf{Acr}\mathcal{H}$ and σ is a surjective substitution. By Lemma 3.5(5), $\mathcal{A} = \mathbf{R}_{\mathcal{L}}[T)_{\mathrm{Th}\,\mathcal{H}}$ for some $T \in \mathrm{Th}\,\mathcal{H}$. Then, by a characteristic property of protoalgebraic deductive systems (Font et al., 2001, Theorem 2.6):

$$\sigma^{-1}[T)_{\mathrm{Th}\,\mathcal{H}} = [\sigma^{-1}T)_{\mathrm{Th}\,\mathcal{H}}, \qquad\qquad (\mathrm{Eq}.3)$$

therefore $\sigma^{-1}\mathcal{A} = \mathbf{R}_{\mathcal{L}}[\sigma^{-1}T)_{\mathrm{Th}\,\mathcal{H}} \in \mathbf{Acr}\,\mathcal{H}$. $\square$

The following lemma is a Gentzen-style analogue of the Lemma 2.3 (Czelakowski and Jansana, 2000) for Hilbert-style deductive systems.

Lemma 5.3. *Suppose $\mathcal{R}$ is an algebraic closure system over $\mathrm{Fm}_{\mathcal{L}}^{+}$. Then $\mathcal{R}$ is invariant, if it is surjectively invariant.*

Proof. Based on correspondence between closure systems and consequence relations, it suffices to show that for all $X, Y \subseteq \mathrm{Fm}_{\mathcal{L}}^{+}$ and every substitution σ

$$X^{\mathcal{R}} \subseteq Y^{\mathcal{R}} \implies (\sigma X)^{\mathcal{R}} \subseteq (\sigma Y)^{\mathcal{R}}.$$

Since $\mathcal{R}$ is an algebraic closure system, for every sequent $s \in X$, there is a finite subset $Y_s \subseteq_\omega Y$, such that $\{s\}^{\mathcal{R}} \subseteq Y_s^{\mathcal{R}}$. Since $Y_s \cup \{s\}$ is finite, then $\mathrm{Var}(Y_s \cup \{s\})$ is finite, therefore there is a surjective substitution σ_s, such that $\sigma_s s = \sigma s$, $\sigma_s Y_s = \sigma Y_s$. Thus

$$(\sigma X)^{\mathcal{R}} = (\textstyle\bigcup_{s \in X} \sigma_s s)^{\mathcal{R}} \subseteq (\textstyle\bigcup_{s \in X} \sigma_s Y_s)^{\mathcal{R}} = (\textstyle\bigcup_{s \in X} \sigma Y_s)^{\mathcal{R}} \subseteq (\sigma Y)^{\mathcal{R}}. \qquad \square$$

Corollary 5.4. (Czelakowski, 2001, Theorem 2.6.8(i,iii)) *Let $\mathcal{H}$ be a protoalgebraic Hilbert-style deductive system. If the lattice* $\mathrm{Th}\,\mathcal{H}$ *is infinitely meet-distributive over its compact elements, then $\mathcal{H}$ admits a multiterm deduction-detachment theorem.*

Proof. By Lemma 5.2, since $\mathcal{H}$ is protoalgebraic, $\mathbf{Acr}\,\mathcal{H}$ is surjectively invariant. On the other hand, since $\mathrm{Th}\,\mathcal{H}$ is infinitely meet-distributive over its compact elements, then $\mathbf{Acr}\,\mathcal{H}$ is a closure system over $\mathrm{Fm}_{\mathcal{L}}^{+}$, by Theorem 4.5. Thus, according to Lemma 5.3, $\mathbf{Acr}\,\mathcal{H}$ must be invariant. Altogether, we have that $\mathbf{Acr}\,\mathcal{H}$ is an algebraic, invariant, closure system over $\mathrm{Fm}_{\mathcal{L}}^{+}$, hence it is a Gentzen-style deductive system. Therefore, by Theorem 4.3, $\mathcal{H}$ admits a multiterm deduction-detachment theorem. $\qquad \square$

Another case, where the characterization of DDT given by Theorem 4.3 can be enlightening, is related to the question of existence of fully adequate Gentzen-style systems for weakly algebraizable Hilbert-style systems.

Definition 5.5. *Let $\mathcal{H}$ be an $HSS_{\mathcal{L}}$. The set of* full closure relations *of $\mathcal{H}$ is defined as follows:*

$$\mathbf{Fcr}\,\mathcal{H} := \{(\theta)^{\mathbf{Gcr}\,\mathcal{H}} \mid \theta \in \mathrm{Con}\,\mathbf{Fm}_{\mathcal{L}}\}.$$

An element of $\mathbf{Fcr}\,\mathcal{H}$ *is called a* full closure relation *for $\mathcal{H}$.* $\qquad \square$

This notion is well-defined. Indeed $\theta \in \mathrm{Con}\,\mathbf{Fm}_{\mathcal{L}}$ contains sequents of the form $\alpha \rhd \beta \overset{\mathrm{def}}{=} \langle \alpha, \beta \rangle$, for every $\langle \alpha, \beta \rangle \in \theta$, and the closure of θ is taken in the closure system of theories of sequents.

It is more traditional to define full closure relations via closure systems of compatible $\mathcal{H}$-theories:

An $\mathcal{H}$-theory T is *compatible* with a congruence $\theta \in \mathrm{Con}\,\mathbf{Fm}_{\mathcal{L}}$, if $\langle \alpha, \beta \rangle \in \theta$ and $\alpha \in T$ imply $\beta \in T$. Abusing our notation, we denote the set of all theories compatible with θ by $[\theta)_{\mathrm{Th}\,\mathcal{H}}$.

The largest congruence compatible with an $\mathcal{H}$-theory T (which always exists) is called the *Leibniz congruence for T* and denoted by ΩT. If $\mathcal{C}$ is a family of $\mathcal{H}$-theories, then the intersection $\bigcap_{T \in \mathcal{C}} \Omega T$ is called the *Tarski* *congruence for* $\mathcal{C}$ and denoted by $\widetilde{\Omega}\mathcal{C}$. Leibniz congruences are also defined for closure relations, namely if $\mathcal{A}$ is a closure relation then

$$\Omega\mathcal{A} = \sup\{\theta \in \mathrm{Con}\,\mathbf{Fm}_{\mathcal{L}} \mid \theta \subseteq \mathcal{A}\}.$$

There is a direct connection between Leibniz congruences for closure relations and Tarski congruences for families of theories, namely

$$\Omega\mathbf{R}_{\mathcal{L}}\mathcal{C} = \widetilde{\Omega}\mathcal{C}.$$

Proposition 5.6. *For every* $HSS_{\mathcal{H}}$ $\mathcal{H}$ *and* $\mathcal{F} \in \mathbf{Fcr}\,\mathcal{H}$:

1. $\mathbf{Fcr}\,\mathcal{H} \subseteq \mathbf{Gcr}\,\mathcal{H}$.
2. $\mathcal{F} = (\Omega\mathcal{F})^{\mathbf{Gcr}\,\mathcal{H}}$.
3. $\mathcal{F} = \mathbf{R}_{\mathcal{L}}\{T \in \mathrm{Th}\,\mathcal{S} \mid \Omega\mathcal{F} \subseteq \Omega T\} = \mathbf{R}_{\mathcal{L}}[\Omega\mathcal{F})_{\mathrm{Th}\,\mathcal{H}}$.
4. $\mathbf{Fcr}\,\mathcal{H}$ *is a closure system iff for every non-empty family* $\{\mathcal{F}_i\}_{i \in I} \subseteq \mathbf{Fcr}\,\mathcal{H}$

$$\bigcap_{i \in I}\mathcal{F}_i = \left(\bigcap_{i \in I}\Omega\mathcal{F}_i\right)^{\mathbf{Gcr}\,\mathcal{H}}.$$

Proof. (1) By definition of $\mathbf{Fcr}\,\mathcal{H}$.

(2) ($\subseteq$) It holds by definition of $\mathbf{Fcr}\,\mathcal{H}$, since $\Omega\mathcal{A} \in \mathrm{Con}\,\mathbf{Fm}_{\mathcal{L}}$.

($\supseteq$) Suppose $\mathcal{A} = (\theta)^{\mathbf{Gcr}\,\mathcal{H}}$, for some $\theta \in \mathrm{Con}\,\mathbf{Fm}_{\mathcal{L}}$. Then

$$\theta \subseteq \Omega\mathcal{A} \subseteq \mathcal{A} \implies \mathcal{A} = (\theta)^{\mathbf{Gcr}\,\mathcal{H}} \subseteq (\Omega\mathcal{A})^{\mathbf{Gcr}\,\mathcal{H}} \subseteq \mathcal{A}^{\mathbf{Gcr}\,\mathcal{H}} = \mathcal{A}$$

$$\implies \mathcal{A} = (\Omega\mathcal{A})^{\mathbf{Gcr}\,\mathcal{H}}.$$

(3) Let $\mathcal{F} \in \mathbf{Fcr}\,\mathcal{H}$. Since $\mathcal{F}$ is a general closure relation for $\mathcal{H}$, then $\mathcal{F} = \mathbf{R}_{\mathcal{L}}\mathcal{C}$, for some algebraic closure system $\mathcal{C} \subseteq \mathrm{Th}\,\mathcal{H}$. Let

$$\mathcal{D} = \{T \in \mathrm{Th}\,\mathcal{H} \mid \Omega\mathcal{F} \subseteq \Omega T\}.$$

Then it is straightforward to show that $\mathcal{D}$ is an algebraic closure system over $\mathrm{Fm}_{\mathcal{L}}$. Also $\mathcal{C} \subseteq \mathcal{D}$, because for all $T \in \mathcal{C}$

$$\Omega\mathcal{F} = \Omega\mathbf{R}_{\mathcal{L}}\mathcal{C} = \widetilde{\Omega}\mathcal{C} \overset{\mathrm{def}}{=} \bigcap_{S \in \mathcal{C}} \Omega S \subseteq \Omega T.$$

Thus $\mathcal{F} = \mathbf{R}_{\mathcal{L}}\mathcal{D}$, because

($\supseteq$) $\mathcal{C} \subseteq \mathcal{D} \implies \mathbf{R}_{\mathcal{L}}\mathcal{D} \subseteq \mathbf{R}_{\mathcal{L}}\mathcal{C} = \mathcal{F}$,

($\subseteq$) $\Omega\mathcal{F} \subseteq \bigcap\{\Omega T \mid T \in \mathcal{D}\} \overset{\mathrm{def}}{=} \widetilde{\Omega}\mathcal{D} = \Omega\mathbf{R}_{\mathcal{L}}\mathcal{D} \subseteq \mathbf{R}_{\mathcal{L}}\mathcal{D}$

$$\implies \mathcal{F} \overset{5.6(2)}{=} (\Omega\mathcal{F})^{\mathbf{Gcr}\,\mathcal{H}} \subseteq (\mathbf{R}_{\mathcal{L}}\mathcal{D})^{\mathbf{Gcr}\,\mathcal{H}} = \mathbf{R}_{\mathcal{L}}\mathcal{D}.$$

(4) ($\Leftarrow$) Since $\bigcap_{i \in I}\Omega\mathcal{F}_i \in \mathrm{Con}\,\mathbf{Fm}_{\mathcal{L}}$, then $\bigcap_{i \in I}\mathcal{F}_i = (\bigcap_{i \in I}\Omega\mathcal{F}_i)^{\mathbf{Gcr}\,\mathcal{H}} \in \mathbf{Fcr}\,\mathcal{H}$, by Definition 5.5.

($\Rightarrow$) Let $\{\mathcal{F}_i\}_{i \in I} \subseteq \mathbf{Fcr}\,\mathcal{H}$ be a non-empty family of full closure relations, then, since $\mathbf{Fcr}\,\mathcal{H}$ is a closure system, there is an $\mathcal{F} = \bigcap_{i \in I}\mathcal{F}_i \in \mathbf{Fcr}\,\mathcal{H}$.

Therefore

$$\mathcal{F} = \bigcap_{i \in I} \mathcal{F}_i$$
$$\implies \Omega \mathcal{F} = \Omega(\bigcap_{i \in I} \mathcal{F}_i) = \bigcap_{i \in I} \Omega \mathcal{F}_i$$
$$\implies \bigcap_{i \in I} \mathcal{F}_i = \mathcal{F} \stackrel{(2)}{=} (\Omega \mathcal{F})^{\mathbf{Gcr}\,\mathcal{H}} = (\bigcap_{i \in I} \Omega \mathcal{F}_i)^{\mathbf{Gcr}\,\mathcal{H}}.$$

For the case of the empty family of full closure relations, the intersection is the largest full closure relation, which, by Definition 5.5, always exists and is equal to $(1_{\mathbf{Fm}_{\mathcal{L}}})^{\mathbf{Gcr}\,\mathcal{S}}$. More precisely it is equal to $\mathrm{Fm}_{\mathcal{L}}^{+}$ if $\mathcal{H}$ has theorems, and to $\mathrm{Fm}_{\mathcal{L}}^{+} \setminus \mathrm{Fm}_{\mathcal{L}}$ if $\mathcal{H}$ does not. $\square$

Note that the full closure relations for $\mathcal{H}$ arise from the Galois connection between $\mathrm{Con}\,\mathbf{Fm}_{\mathcal{L}} \subseteq \mathcal{P}(\mathrm{Fm}_{\mathcal{L}}^{+})$ and the closure system $\mathbf{Gcr}\,\mathcal{H} \subseteq \mathcal{P}(\mathrm{Fm}_{\mathcal{L}}^{+})$ (for more general categorical treatment of full closure relations see (Raftery, 2006)).

Full closure relations had previously got attention mainly in connection with algebraizability phenomena beyond the class of protoalgebraic systems (see (Font and Jansana, 1996) for details). If, for a Hilbert-style system $\mathcal{H}$ with theorems, $\mathbf{Fcr}\,\mathcal{H}$ forms a Gentzen-style system, it is called a *fully adequate Gentzen-style system for* $\mathcal{H}$ (Font and Jansana, 1996; Font et al., 2001). For weakly algebraizable Hilbert-style systems there is a criterion for the existence of a fully adequate Gentzen-style system:

Corollary 5.7. (Font et al., 2001, Corollary 5.7) *Let $\mathcal{H}$ be a weakly algebraizable Hilbert-style system. Then $\mathcal{H}$ has a fully adequate Gentzen-style system iff it admits a multiterm deduction-detachment theorem.*

Proof. By Theorem 3.6(i,iv) from (Czelakowski and Jansana, 2000), for a weakly algebraizable Hilbert-style system $\mathcal{H}$, there is an one-to-one correspondence between families of closure systems $\{[\theta)_{\mathrm{Th}\,\mathcal{H}} \mid \theta \in \mathrm{Con}\,\mathbf{Fm}_{\mathcal{L}}\}$ and $\{[T)_{\mathrm{Th}\,\mathcal{H}} \mid T \in \mathrm{Th}\,\mathcal{H}\}$, therefore

$$\mathbf{Fcr}\,\mathcal{H} = \{\mathbf{R}_{\mathcal{L}}[\theta)_{\mathrm{Th}\,\mathcal{H}} \mid \theta \in \mathrm{Con}\,\mathbf{Fm}_{\mathcal{L}}\} = \{\mathbf{R}_{\mathcal{L}}[T)_{\mathrm{Th}\,\mathcal{H}} \mid T \in \mathrm{Th}\,\mathcal{H}\} = \mathbf{Acr}\,\mathcal{H}.$$

The rest follows directly from Theorem 4.3. $\square$

Suppose $\mathcal{H}$ is a Hilbert-style deductive system. If $\mathcal{F} \in \mathbf{Fcr}\,\mathcal{H}$, then $\mathrm{Ax}\,\mathcal{F} \in \mathrm{Th}\,\mathcal{H}$ is called a *Leibniz theory of* $\mathcal{H}$. Let us denote the set of all Leibniz theories of $\mathcal{H}$ by $\mathrm{Th}^{L}\mathcal{H}$, i.e,

$$\mathrm{Th}^{L}\mathcal{H} = \{\mathrm{Ax}\,\mathcal{F} \mid \mathcal{F} \in \mathbf{Fcr}\,\mathcal{H}\} = \mathrm{Ax}\,\mathbf{Fcr}\,\mathcal{H}.$$

$\mathrm{Th}^{L}\mathcal{H}$, also denoted as $\mathrm{Th}^{S}\mathcal{H}$, in the protoalgebraic case, may represent the *strong* version of the original deductive system $\mathcal{H}$ (Font and Jansana, 2001).

Every Leibniz theory for $\mathcal{H}$ is always the least among theories with the same Leibniz congruence. In the case of protoalgebraic deductive systems,

$\mathbf{Fcr}\,\mathcal{H} \subseteq \mathbf{Acr}\,\mathcal{H}$ and Leibniz theories are exactly $\mathcal{H}$-theories that generate full closure relations as axiomatic ones:

$$\mathbf{Fcr}\,\mathcal{H} = \{(T)^{\mathbf{Gcr}\,\mathcal{H}} \mid T \in \mathrm{Th}^{L}\mathcal{H}\}.$$

Corollary 5.8. *Suppose $HSS_{\mathcal{L}}\ \mathcal{H}$ admits a multiterm deduction-detachment theorem. Then $\mathcal{H}$ has a fully adequate Gentzen-style system, whenever $\mathrm{Th}^{L}\mathcal{H}$ is closed under non-empty intersections.*

Proof. Suppose $\{\mathcal{F}_i\}_{i \in I} \subseteq \mathbf{Fcr}\,\mathcal{H}$, then $\{\mathrm{Ax}\,\mathcal{F}_i\}_{i \in I} \subseteq \mathrm{Th}^{L}\,\mathcal{H}$. Since $\mathcal{H}$ is protoalgebraic, then $\mathbf{Fcr}\,\mathcal{H} \subseteq \mathbf{Acr}\,\mathcal{H}$, so $\{\mathcal{F}_i\}_{i \in I} \subseteq \mathbf{Acr}\,\mathcal{H}$. By Theorem 4.5

$$\bigcap_{i \in I}\mathcal{F}_i = \Big(\bigcap_{i \in I} \Theta\,\mathcal{F}_i\Big)^{\mathbf{Gcr}\,\mathcal{H}} = \Big(\bigcap_{i \in I} \mathrm{Ax}\,\mathcal{F}_i\Big)^{\mathbf{Gcr}\,\mathcal{H}} \in \mathbf{Fcr}\,\mathcal{H},$$

since, by assumption, $\bigcap_{i \in I} \mathrm{Ax}\,\mathcal{F}_i \in \mathrm{Th}^{L}\,\mathcal{H}$.

Furthermore, $\mathbf{Fcr}\,\mathcal{H}$ is always algebraic and surjectively invariant, therefore, being a closure system, it is invariant, by Lemma 5.3. $\qquad\square$

References

[*]Babenyshev, S. (2006). Another characterization of the Deduction-Detachment theorem, *Mathematical Logic in Asia*, Proceedings of the 9th Asian Logic Conference, pp. 1–16.

Babenyshev, S. (2004). Metatheories of deductive systems. *Ph.D. Thesis*, Iowa State University. (available at `http://lib.dr.iastate.edu/rtd/928/`)

Blok, W. and Pigozzi, D. (1986). Protoalgebraic logics, *Studia Logica*, 45, 337–369.

Blok, W. and Pigozzi, D. (1997). Abstract algebraic logic and the deduction theorem, *Manuscript*. (See hhtp://orion.math.iastate.edu/dpigozzi/ for updated version, 2001.). The final version published in this volume.

Bloom, S.L. (1975). Some theorems on structural consequence operations, *Studia Logica*, 34, 1–9.

Czelakowski, J. (1985). Algebraic aspects of deduction theorems, *Studia Logica*, 44, 369–387.

Czelakowski, J. (2001). *Protoalgebraic logics*, Kluwer Academic Publishers, Dordrecht.

Czelakowski, J. and Jansana, R. (2000). Weakly algebraizable logics. *The Journal of Symbolic Logic*, 65, 2, 641–668.

Font, J.M. and Jansana, R. (1996). *A general algebraic semantics for sentential logics*, Springer-Verlag, Berlin. Number 7 in *Lecture Notes in Logic*.

Font, J.M. and Jansana, R. (2001). Leibniz filters and the strong version of a protoalgebraic logic, *Archive for Mathematical Logic*, 40, 6, 437–465.

---

[*]In his first passport the name of the author was spelled as in French orthography — "(Saccording diplomatic tradition of that period)". The current, already English, spelling is "Sergey Babenyshev". (*The Editor of the volume*)

Font, J.M., Jansana, R. and Pigozzi, D. (2006). On the closure properties of the class of full g-models of a deductive system, *Studia Logica*, 83, 215–278.

Font, J.M., Jansana, R. and Pigozzi, D. (2001). Fully adequate Gentzen systems and the deduction theorem, *Reports on Mathematical Logic*, 35, 115–165.

Köhler, P. and Pigozzi, D. (1980). Varieties with equationally definable principal congruences, *Algebra Universalis*, 11, 213–219.

Kiss, E.W. (1983). Complemented and skew congruences, *Annali dell' Università di Ferrara*, Sez. VII 29, 111–127.

Maximova, L.L. (1966). Formal proofs in the calculus of strong implication (in Russian), *Algebra i Logika*, 5, No. 6, 33–38.

Raftery, J.G. (2006). Correspondences between Gentzen and Hilbert systems, *Journal of Symbolic Logic*, 71, 903–957.

Introducing Boolean Semilattices

Clifford Bergman*

Abstract We present and discuss a variety of Boolean algebras with operators that is closely related to the variety generated by all complex algebras of semilattices. We consider the problem of finding a generating set for the variety, representation questions, and axiomatizability. Several interesting subvarieties are presented. We contrast our results with those obtained for a number of other varieties generated by complex algebras of groupoids.

Key words: Boolean algebra, BAO, semilattice, Boolean semilattice, Boolean groupoid, canonical extension, equationally definable principal congruence

2010 Mathematics Subject Classification: 03G25, 06E25, 06A12

The study of Boolean algebras with operators (BAOs) has been a consistent theme in algebraic logic throughout its history. It provides a unifying framework for several branches of logic including relation algebras, cylindric algebras, and modal algebras. From a purely algebraic standpoint, a class of BAOs provides a rich field of study, combining the strength of Boolean algebras with whatever structure is imposed on the operators.

In fact, with all this structure, one might expect that analyzing a variety of BAOs would border on the trivial. The variety of Boolean algebras, after all, is generated by a primal algebra. As such, it is congruence-distributive, congruence-permutable, semisimple, equationally complete, has EDPC,..., the list goes on. And yet it turns out that for all but the most degenerate operators, the analysis is anything but simple. The explanation is in the

---

Clifford Bergman
Department of Mathematics, Iowa State University, Ames, Iowa, 50011, USA, e-mail:
`cbergman@iastate.edu`

* This research was partially supported by the National Science Foundation under Grant No. 1500218 and the Barbara J. Janson Professorship

J. Czelakowski (eds.), *Don Pigozzi on Abstract Algebraic Logic,
Universal Algebra, and Computer Science*, Outstanding Contributions
to Logic 16, https://doi.org/10.1007/978-3-319-74772-9_4

intricate and unexpected interplay between the Boolean operations and the additional operators that arise from standard constructions.

In this paper we consider Boolean algebras with one very simple operator, namely an (almost) semilattice operation, that is, a binary operation that is associative, commutative, and (almost) idempotent. The qualification on idempotence will be explained below. We shall develop some of the arithmetic of these algebras, discuss some representation questions, and pose some problems. While there are no deep results in this work, we hope that it will stimulate further research in this interesting class of algebras.

Peter Jipsen is responsible, for better or worse, for introducing me to representation questions for BAOs. Several of the results presented here are due to him, or jointly to the two of us. Other theorems described here are the result of joint work with Wim Blok in the early 1990s. My interest in algebraic logic in general stems from my long working relationship with Don Pigozzi. Don was my first mentor as a professional mathematician. He has played a large role in my subsequent development.

Our universal algebraic terminology and notation follows the book (Bergman, 2012). That reference should be consulted for any notions not defined here. Jipsen's thesis (Jipsen, 1992), and Jónsson's survey article (Jónsson, 1993) provide a good introduction to the subject of Boolean algebras with operators. Goldblatt's paper (Goldblatt, 1989) is a detailed study of the complex algebra construction.

1 Complex algebras

We begin with a motivating construction. Let $\mathbf{G} = \langle G, \cdot \rangle$ be an algebra with a single binary operation (a groupoid, in common parlance). We form a new structure, the *complex algebra of* $\mathbf{G}$ by $\mathbf{G}^+ = \langle \mathrm{Sb}(G), \cap, \cup, \sim, \odot, \varnothing, G \rangle$. Here, $\mathrm{Sb}(G)$ is the family of all subsets of G, "$\cap$" and "$\cup$" are the usual operations of intersection and union, $\sim X = G - X$ is the complement of the subset X, and $X \odot Y = \{ x \cdot y : x \in X, y \in Y \}$.

The operation "$\odot$" is called a complex operation. In practice, it seems unnecessary to use different notation for an operation and its induced complex operation, so we will generally write $X \cdot Y$ in place of $X \odot Y$. We want to stress that there is nothing special about one binary operation. The complex algebra construction makes sense for any number of operations of any rank. We restrict our attention to groupoids because it already captures the intricacies of the situation. Generalization to arbitrary algebraic structures is straightforward.

The complex algebra is, of course, an expansion of a Boolean algebra. The new operation satisfies several additional identities, namely

$$X \cdot \varnothing = \varnothing, \qquad\qquad \varnothing \cdot X = \varnothing$$
$$X \cdot (Y \cup Z) = (X \cdot Y) \cup (X \cdot Z), \quad (Y \cup Z) \cdot X = (Y \cdot X) \cup (Z \cdot X) \,.$$

The first pair of identities assert that the complex operation is *normal*, the latter pair that it is *additive*. We can actually say a bit more, although not in a first-order manner. The Boolean algebra is complete and atomic and the complex operation distributes over arbitrary union, not just finite union.

This is our "ur-example" of a BAO. We formalize it as follows.

Definition 1.1. A *Boolean groupoid* is an algebra $\mathbf{B} = \langle B, \wedge, \vee, ', \cdot, 0, 1 \rangle$ such that $\langle B, \wedge, \vee, ', 0, 1 \rangle$ is a Boolean algebra and "$\cdot$" is an additional binary operation satisfying the identities

$$
\begin{aligned}
x \cdot 0 &\approx 0 \cdot x \approx 0 \\
x \cdot (y \vee z) &\approx (x \cdot y) \vee (x \cdot z) \\
(y \vee z) \cdot x &\approx (y \cdot x) \vee (z \cdot x).
\end{aligned}
$$

(1)

We shall often use the notation $\mathbf{B}_0$ to denote the Boolean algebra reduct of the Boolean groupoid $\mathbf{B}$ and write $\mathbf{B} = \langle \mathbf{B}_0, \cdot \rangle$.

Since it is defined equationally, the class of Boolean groupoids forms a variety, (that is, a class of algebras closed under subalgebra, homomorphic image, and product) which we denote BG. From our observations above, the complex algebra of every groupoid lies in BG. It is natural to wonder whether the converse could be true: is every Boolean groupoid a complex algebra? A moment's reflection shows that this is impossible on cardinality grounds. There is no complex algebra of cardinality $\aleph_0$, but it is easy to see that there are indeed Boolean groupoids that are countably infinite.

More generally, we can ask whether the complex algebras generate BG as a variety. The answer turns out to be "yes" as we discuss in Sect. 3. In order to demonstrate this, we must develop a technique to extend an arbitrary Boolean groupoid to one that is complete and atomic. We do this in Sect. 2.

Before continuing, we introduce some terminology that we use in the sequel. In the language of Boolean algebras, we write $x - y$ in place of $x \wedge y'$ and

$$x \oplus y = (x - y) \vee (y - x).$$

With this definition we obtain a ring $\langle B, \oplus, \wedge, 0, 1 \rangle$ of characteristic 2 from the Boolean algebra $\mathbf{B}_0$.

One important consequence of additivity in a Boolean groupoid is *monotonicity*: if $x_1 \le x_2$ then $x_1 \cdot y \le x_2 \cdot y$ and $y \cdot x_1 \le y \cdot x_2$.

Definition 1.2. Let $p(x_1, x_2, \ldots, x_n)$ and $q(x_1, x_2, \ldots, x_n)$ be terms in the language of groupoids.

1. The identity $p(x_1, \ldots, x_n) \approx q(x_1, \ldots, x_n)$ is *linear* if each variable occurs exactly once in each of p and q.

2. The identity $p(x_1, \ldots, x_n) \approx q(x_1, \ldots, x_n)$ is *semilinear* if p has no repeated variables and every variable of q occurs in p. (But q can have repeated variables.)

3. The identity $p(x_1, \ldots, x_n) \approx q(x_1, \ldots, x_n)$ is *regular* if exactly the same variables appear in p and q. (But each variable may occur any number of times.)

Note that semilinearity is nonsymmetric, that is, $p \approx q$ semilinear does not imply $q \approx p$ semilinear, unless the identity is actually linear. The significance of linear and semilinear identities is delineated in the following proposition whose proof is a simple verification. Regular identities will be addressed in Sect. 2. Figure 1 shows a few familiar identities and their relationship to these properties.

	linear	semilinear	regular
$x(yz) \approx (xy)z$	✓	✓	✓
$xy \approx yx$	✓	✓	✓
$x \approx x^2$		✓	✓
$xy \approx x$		✓	
$(xy)y \approx (yx)y$			✓

Fig. 1 Some linear, semilinear, and regular identities

Proposition 1.3 ((Shafaat, 1974), (Grätzer and Whitney, 1984)). *Let* **G** *be a groupoid and* $p \approx q$ *an identity.*

1. If $p \approx q$ is linear then $\mathbf{G} \vDash p \approx q \iff \mathbf{G}^+ \vDash p \approx q$.

2. If $p \approx q$ is semilinear then $\mathbf{G} \vDash p \approx q \iff \mathbf{G}^+ \vDash p \leq q$.

By a *partial groupoid* we mean a set with a partially defined binary operation. For example, every subset of a groupoid inherits a partial groupoid structure. We shall say that a partial groupoid, **P**, satisfies an identity $p(x_1, \ldots, x_n) \approx q(x_1, \ldots, x_n)$ if, for every $a_1 \ldots, a_n \in P$, we have $p(a_1, \ldots, a_n)$ is defined in **P** if and only if $q(a_1, \ldots, a_n)$ is defined in **P**, and in that case, the two quantities coincide.

2 Duality

In this section we explore the passage from an object to its complex algebra. In particular, we are interested in reversing the process. In order to do this we must, on the groupoid side, expand our attention to ternary relational structures, and on the complex side, extend a Boolean groupoid to one that is complete and atomic.

To begin with, observe that it is quite easy to recover the structure of a groupoid from its complex algebra. For a groupoid $\mathbf{G}$ and $a, b \in G$, we have $a \cdot b = c$ in $\mathbf{G}$ if and only if $\{a\} \odot \{b\} = \{c\}$ in $\mathbf{G}^+$. Speaking abstractly, the singletons $\{a\}$, $\{b\}$, and $\{c\}$ are atoms of the Boolean algebra $\mathbf{G}_0^+$. In the complex algebra, the product of two atoms is always an atom.

Unfortunately, an arbitrary Boolean groupoid may not have any atoms, and even when it does, the product of two atoms need not be an atom. Thus, when we attempt to generalize the passage from complex algebra to groupoid, we obtain, not an algebra, but a ternary relational structure. By a *ternary relational structure* we simply mean a pair $\langle H, \theta \rangle$ in which H is a set and θ is a subset of H^3.

Definition 2.1. Let $\mathbf{B} = \langle \mathbf{B}_0, \cdot \rangle$ be a Boolean groupoid. The *atom structure of* $\mathbf{B}$ is the ternary relational structure $\mathbf{B}_+ = \langle A, \theta \rangle$ in which A is the set of atoms of $\mathbf{B}_0$ and $\theta = \{ (x, y, z) \in A^3 : z \leq x \cdot y \}$.

To each groupoid $\mathbf{G} = \langle G, \cdot \rangle$ we can associate the ternary relational structure $\mathbf{G}^\square = \langle G, \theta \rangle$ in which $\theta = \{ (x, y, x \cdot y) : x, y \in G \}$. It is easy to verify that $\mathbf{G}^\square \cong (\mathbf{G}^+)_+$. In fact, we can extend this association to any partial groupoid $\mathbf{P}$. Notice that in this case, if $x, y \in P$ with $x \cdot y$ undefined, then there will be no triple in $\mathbf{P}^\square$ of the form (x, y, z) for any z. Put another way, in the complex algebra $\mathbf{P}^+$ we will have $\{x\} \odot \{y\} = \varnothing$.

To proceed further, we must extend the complex algebra construction to ternary relational structures.

Definition 2.2. Let $\mathbf{H} = \langle H, \psi \rangle$ be a ternary relational structure. The *complex algebra of* $\mathbf{H}$ is the Boolean groupoid $\mathbf{H}^+ = \langle \mathrm{Sb}(H), \cap, \cup, \sim, \odot, \varnothing, H \rangle$ in which
$$X \odot Y = \{ z \in H : (\exists x \in X)(\exists y \in Y) \ (x, y, z) \in \psi \} \ .$$

It is not difficult to verify that for a ternary relational structure $\mathbf{H}$, $(\mathbf{H}^+)_+ \cong \mathbf{H}$, and dually, for a complete and atomic Boolean groupoid $\mathbf{B}$, $(\mathbf{B}_+)^+ \cong \mathbf{B}$.

We still have the problem of a lack of atoms in an arbitrary Boolean groupoid. This was addressed in 1951 by Jónsson and Tarski (Jónsson and Tarski, 1951) as an extension of the Stone representation theorem. We do this in two steps. Start with a Boolean groupoid, $\mathbf{B}$. Let $\mathbf{B}^*$ denote the set of ultrafilters (i.e., maximal filters) of $\mathbf{B}_0$. We impose a ternary relational structure on $\mathbf{B}^*$ by defining

$$\theta = \{ (U, V, W) \in (B^*)^3 : W \supseteq \{ u \cdot v : u \in U, \ v \in V \} \} \ .$$

Finally we define $\mathbf{B}^\sigma$ to be $\langle B^*, \theta \rangle^+$.

In his exposition (Jónsson, 1993), Jónsson summarizes the relationship between $\mathbf{B}$ and $\mathbf{B}^\sigma$ as follows (specialized to the case of Boolean groupoids).

Theorem 2.3. *Let* $\mathbf{B}$ *be a Boolean groupoid. There is a unique Boolean groupoid* $\mathbf{B}^\sigma$, *called the* canonical extension of $\mathbf{B}$ *such that*

1. $\mathbf{B}_0^\sigma$ *is a complete and atomic extension of* $\mathbf{B}_0$*;*
2. *For all distinct atoms p and q of* $(\mathbf{B}^\sigma)$*, there exists $a \in B$ such that $p \le a$ and $q \le a'$;*
3. *Every subset of B that joins to 1 in* $\mathbf{B}^\sigma$ *has a finite subset that also joins to 1;*
4. *For atoms p,q of* $\mathbf{B}^\sigma$*, $p \cdot q = \bigwedge \{a \cdot b : a,b \in B,\ a \ge p,\ b \ge q\}$. The product is extended completely additively to the remainder of* $\mathbf{B}^\sigma$*.*

As an example computation, let $\mathbf{B}$ be a Boolean groupoid, A denote the set of atoms of $\mathbf{B}_0^\sigma$, and let $p \in A$. Then using Theorem 2.3(4) we compute

$$(2) \quad p \cdot 1 = \bigvee_{q \in A} p \cdot q = \bigvee_{q \in A} \bigwedge_{a \ge p} \bigwedge_{b \ge q} a \cdot b =$$

$$\bigwedge_{a \ge p} \bigvee_{q \in A} \bigwedge_{b \ge q} a \cdot b = \bigwedge_{a \ge p} \bigvee_{q \in A} a \cdot q = \bigwedge_{a \ge p} a \cdot 1.$$

In practice, it is unnecessary to make reference to B^*. We start from an arbitrary Boolean groupoid $\mathbf{B}$, move first to the canonical extension, $\mathbf{B}^\sigma$, and then to the atom structure $\mathbf{B}_+^\sigma$. This ternary relational structure must serve as an approximation to a groupoid induced by $\mathbf{B}$. The utility of this approximation varies depending upon the particular situation at hand.

A class, or property, preserved by canonical extensions, is called *canonical*. A Boolean groupoid term is called *strictly positive* if it does not involve complementation. One of the deep theorems on the subject is the following.

Theorem 2.4 (Jónsson and Tarski, 1951). *Let s, t, and u be strictly positive terms. Then each of the following is canonical.*

$$s \approx t$$

$$s \approx 0 \to t \approx u$$

$$s \not\approx 0 \to t \approx u.$$

Consider now two ternary relational structures $\langle G, \theta \rangle$ and $\langle H, \psi \rangle$. A function $h \colon G \to H$ induces a complete Boolean algebra homomorphism $\overleftarrow{h} \colon \mathbf{H}_0^+ \to \mathbf{G}_0^+$ given by $\overleftarrow{h}(X) = \{g \in G : h(g) \in X\}$. A necessary and sufficient condition for $\overleftarrow{h}$ to be a Boolean groupoid homomorphism is that h be a bounded morphism, as in the following definition.

Definition 2.5. A function $h \colon G \to H$ is a *bounded morphism* between the ternary relational structures $\langle G, \theta \rangle$ and $\langle H, \psi \rangle$ if

(i) $(\forall \mathbf{x} \in G^3)\ \mathbf{x} \in \theta \implies h(\mathbf{x}) \in \psi$ and

(ii) $(\forall z \in G)(\forall y_1, y_2 \in H)\ (y_1, y_2, h(z)) \in \psi \implies$
$(\exists x_1, x_2 \in G)\ h(x_1) = y_1,\ h(x_2) = y_2,\ (x_1, x_2, z) \in \theta.$

It is straightforward to verify that h is an injective bounded morphism if and only if $\overleftarrow{h}$ is a surjective Boolean groupoid homomorphism, and h is surjective iff $\overleftarrow{h}$ is injective. Let us study those two special situations a little more closely.

Suppose first that $\langle G, \theta \rangle$ and $\langle H, \psi \rangle$ are ternary relational structures, with $G \subseteq H$. If the inclusion map is a bounded morphism, we call $\langle G, \theta \rangle$ an *inner substructure* of $\langle H, \psi \rangle$. Unwinding Definition 2.5, we have the following characterization.

Lemma 2.6. *$\langle G, \theta \rangle$ is an inner substructure of $\langle H, \psi \rangle$ if $G \subseteq H$ and*

$$(i) \quad \theta = \psi \cap G^3 \text{ and}$$
$$(ii) \quad (\forall z \in G)(\forall y_1, y_2 \in H)\, (y_1, y_2, z) \in \psi \implies y_1, y_2 \in G\,.$$

When these conditions hold, $\langle G, \theta \rangle^+$ is a homomorphic image of $\langle H, \psi \rangle^+$.

Based on the first of the two conditions in the lemma, we often refer to G as an inner substructure of $\langle H, \psi \rangle$ without explicitly mentioning θ.

Now suppose that $\mathbf{G}$ is a groupoid. A subset K is called a *sink* if

$$(x \in K \ \& \ y \in G) \implies (x \cdot y \in K \ \& \ y \cdot x \in K)\,.$$

(It is common to call K an ideal of $\mathbf{G}$, but we wish to avoid conflict with the use of "ideal" in the Boolean algebra context.) Consider $\mathbf{G}$ as a ternary relational structure $\mathbf{G}^{\square} = \langle G, \theta \rangle$. It follows immediately from Lemma 2.6 that a subset H will be an inner substructure of $\mathbf{G}^{\square}$ if and only if $G - H$ is a sink.

We now turn to bounded morphic images of a partial groupoid. These correspond to certain quotient structures. Let $\mathbf{P}$ be a partial groupoid, and α an equivalence relation on P. For an element $a \in P$ we write a/α for the equivalence class of a modulo α. We call α a *bounded equivalence* if for all $a, b \in P$ the image of the partial map $p \colon a/\alpha \times b/\alpha \to P$ given by $p(x, y) = x \cdot y$ is a union of α-classes. The bounded equivalence α induces a ternary relational structure $\langle P/\alpha, \psi \rangle$ in which $\psi = \{\, (a/\alpha, b/\alpha, c/\alpha) : c = a \cdot b \,\}$.

Lemma 2.7. *Let α be a bounded equivalence on the partial groupoid $\mathbf{P}$. Then the natural map $q \colon \mathbf{P}^{\square} \to \mathbf{P}/\alpha$ is a surjective bounded morphism.*

Proof. We must check the two conditions in Definition 2.5. The first condition is simply the definition of the relation ψ on P/α. For the second, let $z \in P$, y_1/α, $y_2/\alpha \in P/\alpha$ and suppose that $(y_1/\alpha, y_2/\alpha, z/\alpha) \in \psi$. Then there are x_1, x_2, w such that $x_1 \, \alpha \, y_1$, $x_2 \, \alpha \, y_2$, $w \, \alpha \, z$, and $x_1 \cdot x_2 = w$. By assumption, the image of the partial map p on $y_1/\alpha \times y_2/\alpha$ is a union of α-classes. Since w lies in that image and $z \, \alpha \, w$, we must have $z = p(u_1, u_2) = u_1 \cdot u_2$. Hence $q(u_1) = y_1/\alpha$, $q(u_2) = y_2/\alpha$, and $(u_1, u_2, z) \in \theta$. $\qquad\square$

The converse of lemma 2.7 is true as well: the kernel of a surjective bounded morphism is a bounded equivalence.

The correspondence $\mathbf{G} \mapsto \mathbf{G}^+$ and $\mathbf{B} \mapsto \mathbf{B}_+$, together with the bounded morphisms and homomorphisms, form the basis of a dual equivalence between the categories of ternary relational structures and of complete and atomic Boolean groupoids. This duality is explored in great detail in (Goldblatt, 1989). We need only one additional aspect of the duality, which is quite easy to verify.

The coproduct of a family $\langle \langle G_i, \theta_i \rangle : i \in I \rangle$ of ternary relational structures is simply the disjoint union $\langle \biguplus_i G_i, \biguplus_i \theta_i \rangle$. The complex algebra of a disjoint union is isomorphic to the direct product of the complex algebras of the components:

$$(3) \qquad \left(\biguplus_{i \in I} \mathbf{G}_i \right)^+ \cong \prod_{i \in I} \mathbf{G}_i^+ .$$

The isomorphism maps the complex, X, of the disjoint union, to the I-tuple, $\langle X \cap G_i : i \in I \rangle$, in the product. We leave the details to the reader.

Let $\mathbf{P}$ be a partial groupoid. We can extend $\mathbf{P}$ to a total groupoid $\overline{\mathbf{P}}$ by adjoining a new element, ∞, to P, and defining $x \cdot y = \infty$ whenever $x, y \in \overline{P}$ and their product is undefined in $\mathbf{P}$. This construction is surprisingly robust. It preserves associativity, commutativity, idempotence; in fact, any *regular identity*, as in Definition 1.2. See (Romanowska, 1986) for the importance of these identities.

It is immediate from Lemma 2.6 that $\mathbf{P}^\square$ is an inner substructure of $\overline{\mathbf{P}}^\square$. Thus, every partial groupoid is an inner substructure of a groupoid. And conversely, every inner substructure of a groupoid is itself a partial groupoid.

Now suppose that $\langle \mathbf{G}_i : i \in I \rangle$ is a family of groupoids (or even partial groupoids). Then the disjoint union is a partial groupoid, $\mathbf{G}$, which can be extended to a total groupoid, $\overline{\mathbf{G}}$. Taking complex algebras, and applying the duality principles that we have developed, we have a surjective Boolean groupoid homomorphism, h:

$$\overline{\mathbf{G}}^+ \xrightarrow{\ h\ } \left(\biguplus_{i \in I} \mathbf{G}_i \right)^+ \cong \prod_{i \in I} \mathbf{G}_i^+ .$$

We summarize these observations as a theorem.

Theorem 2.8. *Let Σ be a set of regular identities and $\mathcal{K}$ be the variety of groupoids defined by Σ.*

1. *Every partial groupoid that satisfies Σ can be embedded as an inner substructure of a groupoid in $\mathcal{K}$.*
2. $\mathbf{P}(\mathcal{K}^+) \subseteq \mathbf{H}(\mathcal{K}^+)$.

3 Representation of Boolean Groupoids

We now return to our examination of the relationship between the finitely based variety, BG, of Boolean groupoids, and the class of complex algebras of groupoids. Our discussion is lifted almost verbatim from (Jipsen, 1992, Theorem 3.20).

Lemma 3.1. *Every Boolean groupoid,* $\mathbf{B}$*, can be embedded into* $\mathbf{P}^+$ *for some partial groupoid,* $\mathbf{P}$*.*

Proof. In light of Theorem 2.3, we can assume that $\mathbf{B}$ is complete and atomic. Let $\langle A, \psi \rangle = \mathbf{B}_+$. Thus A is the set of atoms of the Boolean algebra $\mathbf{B}_0$ and

$$\psi = \left\{ (u, v, w) \in A^3 : w \leq u \cdot v \right\} .$$

Let $P = A \times A$. Fix a surjection $g \colon P \to P$, and let $p_1 \colon P \to A$ denote the first projection, i.e., $p_1(x, y) = x$. We define a partial binary operation on P by

$$(a, b) \cdot (c, d) = g(b, d) \quad \text{if } p_1 g(b, d) \leq a \cdot c .$$

In this definition, both the computations of $a \cdot c$ and $p_1 g(b, d) \leq a \cdot c$ take place in $\mathbf{B}$.

We claim that $p_1 \colon \mathbf{P} \to \mathbf{A}$ is a surjective, bounded morphism. If this is so, then by our observations following Definition 2.5, $\breve{p}_1$ embeds $\mathbf{B} = \mathbf{A}^+$ into $\mathbf{P}^+$, proving the lemma.

Clearly p_1 is surjective. To verify the two conditions in Definition 2.5, observe that if $(a, b) \cdot (c, d) = (u, v)$ then $u = p_1 g(b, d) \leq a \cdot c$. Consequently, $(a, c, u) \in \psi$. Thus the first requirement holds.

For the second, let $(u, v) \in P$, $a, c \in A$, and $(a, c, u) \in \psi$. By the definition of ψ, $u \leq a \cdot c$. By the surjectiveness of g, there is a pair $(b, d) \in P$ such that $g(b, d) = (u, v)$. Then $p_1(a, b) = a$, $p_1(c, d) = c$, and $(a, b) \cdot (c, d) = (u, v)$ as desired. $\qquad\square$

Theorem 3.2. *Every Boolean groupoid,* $\mathbf{B}$*, lies in* $\mathbf{SH}(\mathbf{G}^+)$ *for some groupoid* $\mathbf{G}$*. If B is finite, then G can be taken to be finite as well.*

Proof. By Lemma 3.1, $\mathbf{B}$ can be embedded into $\mathbf{P}^+$ for a partial groupoid, $\mathbf{P}$. By Theorem 2.8, $\mathbf{P}$ is an inner substructure of a total groupoid $\mathbf{G}$. Therefore $\mathbf{P}^+$ is a homomorphic image of $\mathbf{G}^+$. Thus $\mathbf{B} \in \mathbf{SH}(\mathbf{G}^+)$. If B is finite, then, in the proof of Lemma 3.1, A is finite, so P, hence G, is finite as well. $\qquad\square$

As a result, we see that the variety generated by all complex algebras of groupoids is axiomatized by the identities of Boolean algebras, together with those of (1.1). In particular, it is a finitely based variety. If we write G for the variety of groupoids, and G^+ for the class of complex algebras of groupoids, we can state this relationship succinctly as follows.

Corollary 3.3. $\mathsf{BG} = \mathbf{SH}(\mathsf{G}^+) = \mathbf{V}(\mathsf{G}^+)$.

A Boolean groupoid is *commutative* if the binary operator is commutative. We have analogous statements to the results above for the commutative case.

Theorem 3.4. *Every commutative Boolean groupoid lies in* $\mathbf{SH}(\mathbf{G}^+)$ *for some commutative groupoid,* $\mathbf{G}$*. Consequently, the variety of commutative Boolean groupoids is generated by the complex algebras of all commutative groupoids.*

Proof. The construction of the partial groupoid $\mathbf{P}$ in Lemma 3.1 must be modified to make it commutative. Let A be the set of atoms as before. Choose a set W of cardinality $2|A|$. (If A is infinite, we can simply take $W = A$.) Let $P = A \times W$. Fix a surjective function $g \colon W \times W \to P$ such that $g(x,y) = g(y,x)$. (This is possible because $|P| \le \frac{1}{2}|W \times W|$.) Now we define the partial binary operation on P just as before

$$(a,b) \cdot (c,d) = g(b,d) \quad \text{if } p_1 g(b,d) \le a \cdot c$$

but note that now, a and c lie in A, while b and d lie in W. Thus $p_1 g(b,d) \in A$. The remainder of the argument now proceeds as before. $\qquad\square$

These last two results can be looked at in a couple of different ways. On the one hand, two fairly natural varieties of BAOs (Boolean groupoids and commutative Boolean groupoids) are shown to be generated by an easy-to-characterize class of complex algebras. Following Jónsson, (Jónsson, 1993), we might call the complex algebras of groupoids the *primary models* of the system defined in Definition 1.1. Viewed this way, Theorems 3.2 and 3.4 are *generation theorems:* the variety of (commutative) Boolean groupoids can be generated by the complex algebras of all (commutative) groupoids.

On the other hand, we can consider the two theorems of this section to be providing a finite axiomatization for two naturally occurring varieties of algebras, namely the varieties generated by complex algebras of groupoids and of commutative groupoids. And not just any axiomatization. The axiom sets consist of only the identities that "must" be included: the axioms for Boolean algebras, additivitiy, normality, and (in the commutative case), the commutative law. Note that the commutative law is linear. According to Proposition 1.3 it is preserved by the passage to complex algebras so it must be present in the axiomatization.

The simplicity of this axiomatization tells us that in the passage to the complex algebra of a groupoid, there are no "unexpected" interactions between the complex operation and the Boolean operations. Peter Jipsen first presented this topic to the author in a seminar in 1991, in the context of semigroups, rather than groupoids. Note that the associative law is also linear. Thus, "of course," this author thought, "there will be no unexpected interactions between the semigroup operation and the Boolean ones, besides associativity."[2] How wrong that was!

---

[2] This author also recalls Don Pigozzi rolling his eyes and proclaiming "You have no idea what you are getting yourself into."

Theorem 3.5 (Jipsen, 2004). *The variety generated by all complex algebras of semigroups is not finitely based.*

Jipsen's theorem is in striking contrast to Corollary 3.3. Not only do "unexpected" interactions exist, but there are infinitely many. In fact, at the time this paper is being written it is unknown whether the variety generated by all complex algebras of semigroups even has a decidable equational base.

Somewhat stronger than a generation theorem is a *representation* theorem. Let $\mathcal{V}$ be a variety of Boolean groupoids, and $\mathcal{K}$ a finitely axiomatizable class of ternary relational structures. We say $\mathcal{V}$ is *representable by* $\mathcal{K}$ if $\mathcal{V} = \mathbf{SP}(\mathcal{K}^{+})$. At this time it is not known whether Corollary 3.3 or Theorem 3.4 can be strengthened to representations.

Thus we are presented with a wealth of possible problems that we can pose in the following general framework. Let $\mathcal{V}$ denote a finitely based variety of Boolean groupoids, and $\mathcal{K}$ a finitely axiomatizable class of ternary relational structures (preferably groupoids).

A generation problem. Given $\mathcal{V}$, find a class $\mathcal{K}$ so that $\mathcal{V} = \mathbf{V}(\mathcal{K}^{+})$.

A representation problem. Given $\mathcal{V}$, does there exist a class $\mathcal{K}$ so that $\mathcal{V} = \mathbf{SP}(\mathcal{K}^{+})$?

A finite basis/decidability problem. Given $\mathcal{K}$, is $\mathbf{V}(\mathcal{K}^{+})$ finitely based/decidable?

Problem 3.6. Is BG represented by the class of all groupoids?

We have positive answers to these questions in a couple of other interesting cases.

Theorem 3.7. 1. (Bergman) *Let* Lz *denote the variety of left-zero semigroups. Then* $\mathbf{V}(\mathsf{Lz}^{+})$ *is finitely based. This variety is representable by left-zero semigroups.*

2. (Jipsen) *Let* Rb *denote the variety of rectangular bands. Then* $\mathbf{V}(\mathsf{Rb}^{+})$ *is finitely based. This variety is representable by rectangular bands.*

We were surprised to discover that the two varieties of complex algebras in the above theorem are term-equivalent to the varieties of diagonal-free cylindric algebras of dimensions 1 and 2, respectively. Note also that both Lz and Rb satisfy the associative law. So it is not the associative law *per se* that is responsible for destroying the finite axiomatizability of the complex algebras in Theorem 3.5. The situation is apparently more subtle. Recently, Peter Jipsen announced the following theorem.

Theorem 3.8. *Let* IG *(respectively* CIG*) denote the variety of idempotent (respectively commutative and idempotent) groupoids. Then* $\mathbf{V}(\mathsf{IG}^{+})$ *coincides with the variety of Boolean groupoids satisfying the additional identity* $x \leq x^2$. *Similarly,* $\mathbf{V}(\mathsf{CIG}^{+})$ *is equal to the variety of commutative Boolean groupoids satisfying* $x \leq x^2$.

Motivated by all of this, a natural next class to investigate is that of complex algebras of semilattices. This turns out to be a rich field of study in and of itself, and constitutes the remainder of this paper. The doctoral dissertation (Reich, 1996) contains a similar analysis of the variety generated by complex algebras of semigroups.

4 Boolean semilattices

We now turn to our primary object of study, namely complex algebras of semilattices. Let Sl denote the variety of semilattices, that is, groupoids satisfying

$$x \cdot (y \cdot z) \approx (x \cdot y) \cdot z$$
$$x \cdot y \approx y \cdot x$$
$$x \cdot x \approx x \,.$$

These are the identities of *associativity, commutativity, and idempotence* respectively. As before, we can form the complex algebra of any semilattice, and consider the variety generated by all such complex algebras: $\mathsf{HSP}(\mathsf{Sl}^+)$. Once again, we are faced with fascinating questions about this variety: Can we find an axiomatization? Is it finitely axiomatizable? Is the equational theory even decidable?

Unfortunately, we don't know the answers to any of these questions. The evidence suggests that they are all negative. As an approximation to the theory, we assemble a short list of identities, all of which are easily seen to hold in Sl^+, and derive some interesting algebraic properties.

To begin with, we have the axioms for Boolean groupoids listed in (1). Guided by Proposition 1.3 we add both the associative and commutative laws. They are linear, so are inherited by the complex algebras. Idempotence is semilinear. Thus we add the identity $x \leq x \cdot x$, which is called the *square-increasing law*.

Definition 4.1. A *Boolean semilattice* is a Boolean groupoid (Definition 1.1) satisfying the additional axioms

$$
\begin{aligned}
\mathrm{bsl}_1 \quad & x \cdot (y \cdot z) \approx (x \cdot y) \cdot z \\
\mathrm{bsl}_2 \quad & x \cdot y \approx y \cdot x \\
\mathrm{bsl}_3 \quad & x \leq x \cdot x
\end{aligned}
$$

The variety of Boolean semilattices will be denoted BSl.

We introduce the term "Boolean semilattice" with no small amount of trepidation. Is this the right definition for such a natural piece of terminology?

Only time will tell. Our axiomatization has the merit of being short, natural (in light of Proposition 1.3), and equational. As we shall demonstrate in the next few pages, a number of interesting consequences of these axioms can be derived that demonstrate the strength and interest of this system. However, it is certainly possible that further research will suggest additional identities that should be added to the above set.

Since every semilattice is idempotent, it is reasonable to expect that the term "Boolean semilattice" should imply idempotence as well, that is, that bsl_3 should be replaced by the stronger identity $x \approx x \cdot x$. However it is not hard to see that the complex algebra of a semilattice, $\mathbf{S}$, satisfies this stronger identity if and only if $\mathbf{S}$ is linearly ordered. In fact, as we show in Sect. 5, the variety defined by that stronger identity is generated by the complex algebras of all linear semilattices. For this reason, we chose to define Boolean semilattice using the square-increasing law.

As we already noted, the complex algebra of every semilattice is a Boolean semilattice. Thus we have $\mathbf{V}(\mathsf{Sl}^{+}) \subseteq \mathsf{BSl}$. Conversely, if $\mathbf{G}$ is a Boolean groupoid and $\mathbf{G}^{+} \in \mathsf{BSl}$, then $\mathbf{G}$ must be a semilattice. To see this, note that in $\mathbf{G}^{+}$, the product of two atoms is an atom. Thus, by bsl_1 and bsl_2, $\mathbf{G}$ is associative and commutative. Further, if $a, b \in G$ and $a \cdot a = b$, then, in $\mathbf{G}^{+}$ we have $\{a\} \subseteq \{b\}$ by bsl_3, so $a = b$.

It is easy to see that each of the 3 identities are independent from the others by considering the complex algebra of a groupoid that is either associative or not, commutative or not, etc.

We list next several additional identities and other formulae that are consequences of the definition of Boolean semilattice. These are useful in practice.

Proposition 4.2. *Every Boolean semilattice satisfies the following formulae.*

$$(4) \qquad 1 \cdot 1 \approx 1$$

$$(5) \qquad x \wedge y \leq x \cdot y$$

$$(6) \qquad x \cdot y \cdot 1 \approx (x \cdot 1) \wedge (y \cdot 1)$$

$$(7) \qquad x \cdot ((x \cdot 1) - x) \leq x^2 \vee ((x \cdot 1 - x)^2 \,.$$

In fact, bsl_3 *can be replaced by* (5).

Proof. By the square-increasing law, $1 \leq 1 \cdot 1 \leq 1$, proving (4). In any Boolean semilattice, $x \wedge y \leq (x \wedge y) \cdot (x \wedge y) \leq x \cdot y$ by monotonicity. Thus (5) holds. Conversely, bsl_3 can be derived from (5) by taking $x = y$.

For (6), first observe that $x \cdot y \cdot 1 \leq x \cdot 1 \cdot 1 = x \cdot 1$ and similarly $x \cdot y \cdot 1 \leq y \cdot 1$. Thus $x \cdot y \cdot 1 \leq (x \cdot 1) \wedge (y \cdot 1)$. On the other hand by (5), $(x \cdot 1) \wedge (y \cdot 1) \leq x \cdot 1 \cdot y \cdot 1 = x \cdot y \cdot 1$ by bsl_1–bsl_3.

Let us derive (7). First, by monotonicity, $x \cdot ((x \cdot 1) - x) \leq x \cdot 1$. Second, $x \cdot 1 - x = (x \cdot 1) \wedge x'$ by definition. Note that

$$x \vee ((x \cdot 1) \wedge x') = (x \vee (x \cdot 1)) \wedge (x \vee x') = x \cdot 1 \,.$$

Hence $x \cdot ((x \cdot 1) - x) \le x \vee ((x \cdot 1) - x) \le x^2 \vee ((x \cdot 1) - x)^2$. $\square$

As we have already noted, $\mathbf{V}(\mathsf{SI}^+) \subseteq \mathsf{BSI}$. It was, of course, our hope that these two varieties would coincide. Alas, that is not the case. We present two examples. Consider first the identity

$$\tag{8} x \wedge (y \cdot 1) \le x \cdot y \,.$$

This identity is easily seen to hold in $\mathbf{S}^+$, for any semilattice, $\mathbf{S}$. However, let $\mathbf{H}$ denote the ternary relational structure $\langle \{a,b\}, \theta \rangle$ in which

$$\theta = \{(a,a,a), (a,b,b), (b,a,b), (b,b,a), (b,b,b)\} \,.$$

One can conveniently represent this relation with the multiplication table

$\cdot$	a	b
a	a	b
b	b	1

This table can be thought of as a subset of the multiplication table for $\mathbf{H}^+$. Since this particular complex algebra has two atoms, $a \vee b = 1$. The remainder of the table can be deduced from normality and additivity. $\mathbf{H}^+$ is easily checked to be associative, commutative, and, square-increasing. We see that $\mathbf{H}^+$ fails to satisfy (8) with $x = a$ and $y = b$.

As a second example, let $\mathbf{A}$ be the 8-element Boolean groupoid, with atoms $\{a,b,c\}$ that multiply as follows:

	a	b	c
a	a	a	a
b	a	$a \vee b$	$b \vee c$
c	a	$b \vee c$	1

The algebra $\mathbf{A}$ satisfies bsl_1–bsl_3, so $\mathbf{A} \in \mathsf{BSI}$. In fact it also satisfies (8). However, $\mathbf{A}$ fails to satisfy the identity

$$\tag{9} x \cdot \tau \le (x \cdot z \wedge v) \cdot y \vee (x \cdot z - v) \cdot \tau$$

with τ shorthand for $u \wedge (y \cdot z)$. It is a simple computation to verify that the complex algebra of any semilattice satisfies equation (9). Thus $\mathbf{A} \notin \mathbf{V}(\mathsf{SI}^+)$.

These examples were relatively easy to find, involving algebras with 2 or 3 atoms. It certainly suggests to us that it will be possible to find longer and longer identities that fail in larger and larger finite algebras. Based on this, we conjecture that the answer to the following finite basis problem is 'no'.

Problem 4.3. Is $\mathbf{V}(\mathsf{SI}^+)$ finitely based? Is the equational theory decidable?

A useful source of tools for attacking Problem 4.3 might be (Hodkinson et al., 2001). Perhaps there is more hope for a positive answer to one of the following problems. (But see Theorem 7.2.)

Problem 4.4. Is either BSl or $\mathbf{V}(\mathsf{Sl}^+)$ generated by its finite members?

Problem 4.5. Is there a finitely axiomatizable class, $\mathcal{K}$, of ternary relational structures, such that $\mathsf{BSl} = \mathbf{V}(\mathcal{K}^+)$?

Algebraic theory of Boolean semilattices

Let $\downarrow x$ denote the term $x \cdot 1$. Notice that for a semilattice $\mathbf{S}$ and $X \subseteq S$, the complex $\downarrow X$ is the downset (i.e., the ideal) generated by X. (We view the semilattice operation to be the greatest lower bound.) This operator plays a key role in the structure theory of Boolean semilattices.

Proposition 4.6. *In any Boolean semilattice, '$\downarrow$' yields a closure operator, that is, for $\mathbf{B} \in \mathsf{BSl}$ and $x, y \in B$, $x \leq \downarrow x = \downarrow\downarrow x$, and $x \leq y \implies \downarrow x \leq \downarrow y$.*

Proof. $x \leq x \cdot x \leq x \cdot 1 = \downarrow x$ by bsl_3 and monotonicity. Also

$$\downarrow\downarrow x = (x \cdot 1) \cdot 1 = x \cdot (1 \cdot 1) = x \cdot 1 = \downarrow x$$

by associativity and (4). Finally, if $x \leq y$ then $\downarrow x = x \cdot 1 \leq y \cdot 1 = \downarrow y$, again, by monotonicity. $\qquad\square$

An element x of a Boolean semilattice is called *closed* if $x = \downarrow x$. By normality, we always have $\downarrow 0 = 0$ and by identity (4), $\downarrow 1 = 1$. Thus 0 and 1 are always closed elements.

It is well-known that if θ is a congruence relation on a Boolean algebra $\mathbf{B}_0$, then $I = 0/\theta$ is an ideal of $\mathbf{B}_0$. Conversely, every ideal, I, gives rise to a congruence by defining $\theta_I = \big\{ (x, y) \in B_0^2 : x \oplus y \in I \big\}$. This correspondence provides a lattice isomorphism between the congruences and ideals of $\mathbf{B}_0$. It can be extended to Boolean groupoids, indeed, to BAOs in general, as follows.

Definition 4.7. Let $\mathbf{B}$ be a Boolean groupoid, and I an ideal of $\mathbf{B}_0$. Then I is a *congruence ideal of* $\mathbf{B}$ if, for some $\theta \in \mathrm{Con}(\mathbf{B})$, we have $I = 0/\theta$.

Proposition 4.8 (Jipsen, 1992). *Let $\mathbf{B}$ be a Boolean groupoid, and I and ideal of $\mathbf{B}_0$. Then I is a congruence ideal of $\mathbf{B}$ if and only if $x \in I$ implies $x \cdot 1 \in I$ and $1 \cdot x \in I$. There is a lattice isomorphism between the congruences and the congruence ideals of $\mathbf{B}$.*

Corollary 4.9. *Let $\mathbf{B}$ be a Boolean semilattice.*

1. *Let I be an ideal of $\mathbf{B}_0$. Then I is a congruence ideal of $\mathbf{B}$ if and only if $x \in I \implies \downarrow x \in I$.*
2. *Let $a \in B$. Then the smallest congruence ideal of $\mathbf{B}$ containing a is*

$$(\downarrow a] = \big\{ x \in B : x \leq \downarrow a \big\} .$$

An element a such that $(a]$ is a congruence ideal is called a *congruence element*. It follows from the above corollary, that on a Boolean semilattice, the congruence elements are precisely the closed elements. If $\mathbf{S}$ is a semilattice, then the congruence elements of $\mathbf{S}^{+}$ are the downsets of $\mathbf{S}$.

It is easy to see that if x and y are congruence elements in any Boolean semilattice, then so are $x \vee y$ and $x \cdot y$. In fact, in the lattice of congruence ideals, $(x] \vee (y] = (x \vee y]$ and $(x] \wedge (y] = (x \cdot y]$, for congruence elements x and y. Thus the principal congruence ideals form a sublattice of the lattice of all congruence ideals.

Recall from the discussion following Lemma 2.6 the definition of a sink in a groupoid. We noted there that the inner substructures of a groupoid coincide with the complements of the sinks. In the case of a semilattice, the sinks are precisely the downsets, and the complements of the downsets are the upsets. We state this formally.

Lemma 4.10. *Let* $\mathbf{S}$ *be a semilattice. The inner substructures of* $\mathbf{S}^{\square}$ *are the upsets of* $\mathbf{S}$.

We reiterate that every partial semilattice, $\mathbf{P}$, is an upset, hence an inner substructure, of a semilattice, simply by adjoining a smallest element to P. Lemma 4.10 can be generalized somewhat.

Proposition 4.11. *Let* $\mathbf{B}$ *be a complete and atomic Boolean semilattice, and let* c *be a closed element of* $\mathbf{B}$. *Then* $U = \{\, z \in B_{+} : z \leq c' \,\}$ *is an inner substructure of* $\mathbf{B}_{+}$.

Proof. Write $\mathbf{B}_{+} = \langle A, \psi \rangle$. We need to check the condition in Lemma 2.6. Let $z \in U$ and $y_1, y_2 \in A$. The condition $(y_1, y_2, z) \in \psi$ is equivalent to $z \leq y_1 \cdot y_2$. Suppose that $y_1 \notin U$. Then, since y_1 is an atom, $y_1 \leq c$, hence $z \leq y_1 \cdot y_2 \leq c \cdot 1 = c$, since c is closed. But this implies $z \leq c \wedge c' = 0$, which is false. Similarly, $y_2 \in U$. $\hfill\square$

In a landmark series of papers, (Köhler and Pigozzi, 1980; Blok and Pigozzi, 1982; Blok et al., 1984; Blok and Pigozzi, 1994a,b), Don Pigozzi, together with Wilem Blok and Peter Köhler, developed the notion of equationally definable principal congruences (EDPC). Varieties with EDPC exhibit remarkable properties. The variety of Boolean semilattices has EDPC, and provides a very interesting case study in its application.

Definition 4.12. A variety, $\mathcal{V}$, has EDPC if there are 4-variable terms $p_i(x, y, z, w)$, and $q_i(x, y, z, w)$, for $i = 1, \ldots, n$, such that for every $\mathbf{A} \in \mathcal{V}$ and every $a, b, c, d \in A$

$$(c, d) \in \mathrm{Cg}^{\mathbf{A}}(a, b) \iff \mathbf{A} \models p_i(a, b, c, d) = q_i(a, b, c, d), \text{ for } i = 1, \ldots, n.$$

Theorem 4.13. *The variety* BSl *has EDPC.*

Proof. Let $\mathbf{B}$ be a Boolean semilattice, $a, b, c, d \in B$. Then from the theory of Boolean algebras we know that $(c, d) \in \mathrm{Cg}^{\mathbf{B}}(a, b)$ iff $(c \oplus d, 0) \in \mathrm{Cg}^{\mathbf{B}}(a \oplus b, 0)$. From our observations above, in a Boolean semilattice, this latter condition is equivalent to $c \oplus d \leq \downarrow(a \oplus b)$. Thus, in the definition of EDPC, we can take $n = 1$, $p_1(x, y, z, w) = (z \oplus w) \wedge ((x \oplus y) \cdot 1)$ and $q_1(x, y, z, w) = z \oplus w$. $\qquad\square$

Every variety with EDPC is congruence distributive and has the congruence extension property. Of course the first of these holds in any variety of BAOs. But the second is significant.

Corollary 4.14. *The variety* BSl *has the* congruence extension property (CEP). *That is, for every* $\mathbf{C} \leq \mathbf{B} \in$ BSl *and* $\theta \in \mathrm{Con}(\mathbf{C})$, *there is* $\bar{\theta} \in \mathrm{Con}\,\mathbf{B}$ *such that* $\bar{\theta} \cap C^2 = \theta$.

It is actually quite easy to see from Corollary 4.9 that BSl has the congruence extension property. Suppose that $\mathbf{C} \leq \mathbf{B}$. For a congruence ideal, I, on $\mathbf{C}$, let $J = \{x \in B : (\exists y \in I)\, x \leq y\}$. It is easy to see that J is an ideal of $\mathbf{B}_0$ and that $J \cap C = I$. To apply Corollary 4.9, let $x \in J$. By definition, there is $y \in I$ with $x \leq y$. Then $\downarrow x \leq \downarrow y \in I$ since I is assumed to be a congruence ideal.

An important application of the congruence extension property is the following relationship which is useful in understanding the generation of varieties. The proof is a straightforward verification.

Corollary 4.15. *Let* $\mathcal{K}$ *be a class of algebras with the congruence extension property. Then* $\mathbf{HS}(\mathcal{K}) = \mathbf{SH}(\mathcal{K})$.

Let us turn now to a consideration of subdirect irreducibility. Recall that an algebra is *subdirectly irreducible* if it is nontrivial and has a smallest nontrivial congruence, called the *monolith*. Subdirectly irreducible algebras form the basic building blocks for analyzing varieties. The notion tends to disappear from view in the study of Boolean algebras because the only subdirect irreducible is the 2-element algebra. However the situation for Boolean semilattices is radically different.

Lemma 4.16. *A Boolean semilattice is subdirectly irreducible if and only if it has a smallest nonzero closed element.*

Proof. Let $\mathbf{B}$ be a subdirectly irreducible Boolean semilattice and let I be the congruence ideal associated with the monolith. Choose any $a \in I$, $a \neq 0$ and let $c = \downarrow a$. Note that c is a nonzero closed element. Since I is a congruence ideal, $c \in I$, so $(c] \subseteq I$. But by the minimality of I, $(c] = I$. Now, if b is any nonzero closed element, then $(b]$ is a congruence ideal, so $(c] \subseteq (b]$, which is to say, $c \leq b$. $\qquad\square$

Proposition 4.17. *Let* $\mathbf{S}$ *be a semilattice. Then* $\mathbf{S}^+$ *is subdirectly irreducible if and only if* $\mathbf{S}$ *has a lower bound. In particular, every finite semilattice has a subdirectly irreducible complex algebra.*

Proof. Recall that the closed elements of $\mathbf{S}^+$ are the downsets of $\mathbf{S}$. The smallest nonempty downset of a semilattice (if it exists) will always be of the form $\{a\}$, where a is the lower bound. $\qquad\square$

It is usually easier to work with congruence ideals rather than congruences. We will frequently consider the monolith to the the smallest nonzero congruence ideal on a subdirectly irreducible Boolean semilattice.

Theorem 4.18. *Let* $\mathbf{B}$ *be a subdirectly irreducible Boolean semilattice. Then* $\mathbf{B}^\sigma$ *is subdirectly irreducible.*

Proof. By Lemma 4.16, $\mathbf{B}$ has a smallest nonzero closed element, a. Thus, for every $x \in B - \{0\}$, $x \cdot 1 \geq a$. Let y be an atom of $\mathbf{B}^\sigma$. Since $\mathbf{B}$ is a subalgebra of $\mathbf{B}^\sigma$, the condition $a = a \cdot 1$ continues to hold in $\mathbf{B}^\sigma$. By Equation (2)

$$y \cdot 1 = \bigwedge \{x \cdot 1 : y \leq x \in B\} \geq a .$$

Therefore a generates the monolith of $\mathbf{B}^\sigma$. $\qquad\square$

Two concepts related to subdirect irreducibility are simplicity and finite subdirect irreducibility. A nontrivial algebra $\mathbf{A}$ is *simple* if $\mathrm{Con}(\mathbf{A})$ has exactly 2 elements. $\mathbf{A}$ is *finitely subdirectly irreducible* if, for any two congruences θ and ψ on $\mathbf{A}$, $\theta > 0$ & $\psi > 0 \implies \theta \wedge \psi > 0$. Finally, we call a Boolean groupoid *integral* if $x > 0$ & $y > 0 \implies x \cdot y > 0$.

Proposition 4.19. *Let* $\mathbf{B}$ *be a Boolean semilattice.*

1. $\mathbf{B}$ *is finitely subdirectly irreducible if and only if it is integral.*
2. $\mathbf{B}$ *is simple if and only if* $x \neq 0 \implies \downarrow x = 1$.
3. $\mathbf{B}$ *simple implies* $\mathbf{B}^\sigma$ *simple.*

Proof. Suppose that $\mathbf{B}$ is finitely subdirectly irreducible and that $x \cdot y = 0$. Then $0 = x \cdot y \cdot 1 = (x \cdot 1) \cdot (y \cdot 1) = \downarrow x \cdot \downarrow y$. Consequently $(\downarrow x] \wedge (\downarrow y] = (0]$. Then by our assumption, either $\downarrow x = 0$, which implies $x = 0$, or $\downarrow y = 0$, so $y = 0$. Thus $\mathbf{B}$ is integral.

Conversely, suppose that $\mathbf{B}$ is integral and that I and J are nonzero congruence ideals of $\mathbf{B}$. Then there are nonzero closed elements $x \in I$ and $y \in J$. We have $x \cdot y \leq x \cdot 1 = x \in I$ and similarly, $x \cdot y \leq y \cdot 1 \in J$. By integrality, $0 \neq x \cdot y \in I \cap J$. This is enough to show that $\mathbf{B}$ is finitely subdirectly irreducible.

Part (2) follows easily from Corollary 4.9. Part (3) follows from Theorem 2.4 since the terms in part (2) are all strictly positive. $\qquad\square$

Corollary 4.20. *Let* $\mathbf{S}$ *be a semilattice. Then* $\mathbf{S}^+$ *is finitely subdirectly irreducible.* $\mathbf{S}^+$ *is simple iff* $|S| = 1$.

Proof. The complex algebra of a groupoid is always integral. If $\mathbf{S}^+$ is simple, then the only downset of $\mathbf{S}$ is S itself, so $\mathbf{S}$ must be trivial. $\qquad\square$

In fact, this corollary, as well as Proposition 4.17 holds whenever $\mathbf{S}$ is a partial semilattice.

Discriminator algebras

The *discriminator* on a set A is the ternary operation

$$d_A(x,y,z) = \begin{cases} z & \text{if } x = y \\ x & \text{if } x \neq y. \end{cases}$$

A nontrivial algebra, $\mathbf{A}$, is a *discriminator algebra* if d_A is a term operation of $\mathbf{A}$. Discriminator algebras have powerful structure. They are simple, every nontrivial subalgebra is again a discriminator algebra, and they generate an arithmetical variety. As an example, every finite field is a discriminator algebra.

A variety is called a *discriminator variety* if there is a single term that induces the discriminator on every subdirectly irreducible member. The varieties of Boolean algebras, relation algebras, and cylindric algebras (of a fixed dimension) are examples of discriminator varieties.

The discriminator is a kind of "if-then-else" operation on a set. Because of its connection to propositional logic, it is perhaps not surprising that on a Boolean algebra with operators, there is a convenient shortcut to building a discriminator term. We define the *unary discriminator* on a Boolean algebra $\mathbf{B}_0$ to be the function

$$c(x) = \begin{cases} 0 & \text{if } x = 0 \\ 1 & \text{if } x \neq 0. \end{cases}$$

The ternary and unary discriminators are interdefinable by

$$c(x) = d(0,x,1)' \text{ and } d(x,y,z) = \left(x \wedge c(x \oplus y)\right) \vee \left(z \wedge c(x \oplus y)'\right).$$

Thus, a Boolean algebra with operators has a term defining the (ternary) discriminator if and only if it has a term defining the unary discriminator.

Proposition 4.19 tells us that every simple Boolean semilattice is a discriminator algebra, with unary discriminator $c(x) = \downarrow x$. The variety BSl, of all Boolean semilattices, is not a discriminator variety since there are many subdirectly irreducible algebras that are not simple. However, BSl has a largest discriminator subvariety, which is easily described (see also (Jipsen, 1993; McKenzie, 1975)).

Theorem 4.21. *Let* $\mathsf{BSl_D}$ *be the subvariety of* BSl *defined by the identity*

$$(10) \qquad\qquad (x \cdot 1)' \cdot 1 \approx (x \cdot 1)'.$$

$\mathsf{BSl_D}$ *is a discriminator variety, in fact, it is the largest discriminator subvariety of* BSl. *$\mathsf{BSl_D}$ is generated by the class of all simple Boolean semilattices.*

Proof. Let $\mathbf{B}$ be a subdirectly irreducible member of $\mathsf{BSl_D}$, with minimal nonzero congruence ideal, M. Let a be a nonzero element of M. From equation (10), the element $b = (a \cdot 1)'$ is closed, consequently $I = (b]$ is a congruence ideal. If $b \neq 0$, then by the minimality of M, we must have $M \subseteq I$, so $a \leq b$. But then $a \cdot 1 \leq b \cdot 1 = b = (a \cdot 1)'$ which is impossible as $a > 0$. Consequently, we must have $b = 0$, which is to say, $\downarrow a = 1$. Thus by Proposition 4.21, $\mathbf{B}$ is simple. As we have already argued that every simple algebra is a discriminator, we conclude that $\mathsf{BSI_D}$ is a discriminator variety.

On the other hand, let D be any discriminator subvariety of BSI. Then each of its subdirectly irreducible algebras is simple. It is easy to see that every simple Boolean semilattice satisfies equation (10). Consequently, $\mathsf{D} \subseteq \mathsf{BSI_D}$.

□

Equation (10) says that the complement of a closed element is closed. From this we obtain another property that is characteristic of discriminator varieties—numerous direct decompositions. If $\mathbf{B} \in \mathsf{BSI_D}$ and a is any closed element of $\mathbf{B}$, then we have the decomposition $\mathbf{B} \cong \mathbf{B}/(a] \times \mathbf{B}/(a']$.

Equation (10) also implies that if $\mathbf{S}$ is a nontrivial semilattice, then $\mathbf{S}^+ \notin \mathsf{BSI_D}$. For, the closed elements of $\mathbf{S}^+$ are the downsets of $\mathbf{S}$. And the complement of a downset is never a downset.

Thus we have to look harder for primary models for $\mathsf{BSI_D}$. Here is one interesting class of such structures. Let Tot denote the class of all ternary relational structures $\langle H, H^3 \rangle$ (i.e., total relations) for any set H.

Theorem 4.22. *Every member of* Tot^+ *is a simple Boolean semilattice.* $\mathbf{V}(\mathsf{Tot}^+)$ *is the subvariety of* $\mathsf{BSI_D}$ *defined by the identity* $x \cdot y \cdot 1 \approx x \cdot y$. *This subvariety is represented by* Tot.

Proof. It is straightforward to verify that any member of Tot^+ satisfies bsl_1–bsl_3. Let $\mathbf{H} \in \mathsf{Tot}$. Then for any $a, b, c \in H$, we have $(a, b, c) \in H^3$, hence $c \leq a \cdot b$ in $\mathbf{H}^+$. Since c represents an arbitrary atom of the complete and atomic $\mathbf{H}^+$, we conclude that $a \cdot b = 1$. Since a and b are themselves arbitrary atoms we deduce that for any $x > 0$ and $y > 0$ in $\mathbf{H}^+$, $x \cdot y = 1$. In particular, $\downarrow x = 1$, so by Proposition 4.19, $\mathbf{H}^+$ is simple. Furthermore, we conclude that $\mathbf{H}^+ \vDash x \cdot y \cdot 1 \approx x \cdot y$ since if either x or y is 0 then both sides of the identity are 0.

Let $\mathcal{W}$ be the subvariety of $\mathsf{BSI_D}$ defined by the identity $x \cdot y \cdot 1 \approx x \cdot y$. By the previous paragraph and Theorem 4.21, $\mathbf{V}(\mathsf{Tot}^+) \subseteq \mathcal{W}$. We shall show that, conversely, $\mathcal{W} \subseteq \mathbf{SP}(\mathsf{Tot}^+)$. suppose that $\mathbf{A}$ is a subdirectly irreducible member of $\mathcal{W}$. It is enough to show that $\mathbf{A} \in \mathbf{S}(\mathsf{Tot}^+)$. Since $\mathsf{BSI_D}$ is a discriminator variety containing $\mathbf{A}$, we must have $\mathbf{A}$ simple. Therefore, by Theorem 4.18, $\mathbf{A}^\sigma$ is simple, hence $\mathbf{A}^\sigma \in \mathsf{BSI_D}$. Since $x \cdot y \cdot 1 \approx x \cdot y$ is a strictly positive identity satisfied by $\mathbf{A}$, by Theorem 2.4 we get $\mathbf{A}^\sigma \vDash x \cdot y \cdot 1 \approx x \cdot y$. Hence $\mathbf{A}^\sigma \in \mathcal{W}$.

Now, for any three atoms a, b, c of $\mathbf{A}^\sigma$, we have $a \cdot b > 0$ (since simple algebras are integral), so $a \cdot b = \downarrow(a \cdot b) = 1$. Thus $c \leq a \cdot b$. This means that the ternary relational structure $(\mathbf{A}^\sigma)_+$ is a total relation. Therefore $\mathbf{A}^\sigma \in \mathsf{Tot}^+$. Since $\mathbf{A}$ is a subalgebra of $\mathbf{A}^\sigma$, we get $\mathbf{A} \in \mathbf{S}(\mathsf{Tot}^+)$.

□

The class $\mathbf{V}(\mathsf{Tot}^+)$ is a proper subvariety of $\mathsf{BSI_D}$. The algebras $\mathbf{B}_2$ and $\mathbf{B}_3$ in Figure 2 of Sect. 7 are both simple (so they lie in $\mathsf{BSI_D}$) but fail to satisfy the identity $x \cdot y \cdot 1 \approx x \cdot y$ with $x = y = a$. Thus the question of a nice class of generators for $\mathsf{BSI_D}$ remains open.

Problem 4.23. Is $\mathsf{BSI_D} = \mathbf{V}(\mathcal{K}^+)$ for some finitely axiomatizable class, $\mathcal{K}$, of ternary relational structures?

Finally, notice that equation (10) is not strictly positive. Thus we can not apply Theorem 2.4 to conclude that $\mathsf{BSI_D}$ is closed under canonical extension. However, let $\mathbf{B}$ be a subdirectly irreducible member of $\mathsf{BSI_D}$. Then $\mathbf{B}$ is simple, hence $\mathbf{B}^\sigma$ is simple, so $\mathbf{B}^\sigma \in \mathsf{BSI_D}$. This suggests the following question.

Problem 4.24. Is $\mathsf{BSI_D}$ canonical?

5 Linear Semilattices

It would seem, based on a rational naming convention, that a "Boolean semilattice" should always satisfy the identity $x \cdot x \approx x$. However, as we explain in this section, this identity is too strong to be of much use.

In fact, for a semilattice $\mathbf{S}$, $\mathbf{S}^+ \vDash x^2 \approx x$ precisely when $\mathbf{S}$ is linearly ordered. To see this, observe that for $X \subseteq S$, the condition $X \cdot X = X$ is equivalent to X being a subsemilattice of $\mathbf{S}$. Thus $\mathbf{S}^+ \vDash x^2 \approx x$ says that every subset is a subsemilattice, and this in turn holds exactly when $\mathbf{S}$ is linearly ordered.

We shall call a Boolean semilattice *idempotent* if it satisfies the identity $x^2 \approx x$. Let LS denote the class of linearly ordered semilattices. We have just argued that every member of LS^+ is idempotent. Thus every member of $\mathbf{V}(\mathsf{LS}^+)$ is idempotent. In this section, we shall establish the converse. Let us write IBSI for the variety of idempotent Boolean semilattices.

Lemma 5.1 (Bergman-Jipsen). *The following identities hold in* IBSI.

1. $x \wedge y \le x \cdot y \le x \vee y$;
2. $x \wedge (y \cdot 1) \le x \cdot y$;
3. $x \cdot y \approx (x \wedge (y \cdot 1)) \vee (y \wedge (x \cdot 1))$.

Proof. $x \wedge y \le x \cdot y$ holds in any Boolean semilattice, by Proposition 4.2(5). By idempotence and additivity,

$$x \vee y = (x \vee y)^2 = x^2 \vee (x \cdot y) \vee y^2 \ge x \cdot y$$

proving (1).

For (2),

$$x \wedge (y \cdot 1) = x \wedge y \cdot (x \vee x') = x \wedge ((x \cdot y) \vee (x' \cdot y)) \le x \wedge ((x \cdot y) \vee (x' \vee y))$$
$$= (x \wedge (x \cdot y)) \vee (x \wedge x') \vee (x \wedge y) \le (x \cdot y) \vee (x \wedge y) = x \cdot y$$

where (1) is used in the first inequality and the last equality.

Finally, $(x \wedge (y \cdot 1)) \vee (y \wedge (x \cdot 1)) \leq x \cdot y$ follows from (2). Conversely, by (1), monotonicity, and distributivity

$$x \cdot y \leq (x \cdot 1) \wedge (y \cdot 1) \wedge (x \vee y) =$$
$$((x \cdot 1) \wedge (y \cdot 1) \wedge x) \vee ((x \cdot 1) \wedge (y \cdot 1) \wedge y) = (x \wedge (y \cdot 1)) \vee (y \wedge (x \cdot 1)).$$

$\square$

The third identity in the above lemma can be written

$$(11) \qquad\qquad x \cdot y \approx (x \wedge {\downarrow} y) \vee (y \wedge {\downarrow} x).$$

Thus an idempotent Boolean semilattice is term-equivalent to its closure-reduct.

Lemma 5.2. *let* **B** *be an idempotent Boolean semilattice. Then for atoms* a, b,

$$a \cdot b = \begin{cases} a & \text{if } {\downarrow}a < {\downarrow}b \\ a \vee b & \text{if } {\downarrow}a = {\downarrow}b \\ b & \text{if } {\downarrow}a > {\downarrow}b \\ 0 & \text{otherwise.} \end{cases}$$

Proof. Suppose that ${\downarrow}a < {\downarrow}b$. Then $a < {\downarrow}b$ and $b \not\leq {\downarrow}a$ so $b \wedge {\downarrow}a = 0$, since b is an atom. Consequently $a \cdot b = a$ by (11). The second and third alternatives are argued similarly. Finally, if ${\downarrow}a$ and ${\downarrow}b$ are incomparable then $a \not\leq {\downarrow}b$ and $b \not\leq {\downarrow}a$. Then from (11), $a \cdot b = 0$. $\square$

Lemma 5.3. *Let* **B** *be an atomic, subdirectly irreducible, idempotent Boolean semilattice. Then* $\mathrm{Con}(\mathbf{B})$ *is linearly ordered.*

Proof. We shall first show that the closed elements of **B** are linearly ordered. Suppose that b and c are incomparable, closed elements. By atomicity, there are atoms b_0 and c_0 such that $b_0 \leq b$, $b_0 \not\leq c$, $c_0 \leq c$, and $c_0 \not\leq b$. If ${\downarrow}b_0 \leq {\downarrow}c_0$ then (since c is closed)

$$b_0 \leq {\downarrow}b_0 \leq {\downarrow}c_0 \leq c$$

which is a contradiction. Similarly ${\downarrow}c_0 \not\leq {\downarrow}b_0$, i.e., ${\downarrow}b_0$ and ${\downarrow}c_0$ are incomparable. Therefore by Lemma 5.2, $b_0 \cdot c_0 = 0$. But **B** is subdirectly irreducible, hence integral (by Proposition 4.19), which is a contradiction. Thus our original elements b and c must be comparable.

Now we address the statement in the lemma. Because of the correspondence between congruences and congruence ideals, it is enough to show that for any two congruence ideals I and J, either $I \subseteq J$ or $J \subseteq I$. So assume instead that there are elements $b \in I - J$ and $c \in J - I$. By Lemma 4.9, ${\downarrow}b \in I$ and, since $b \leq {\downarrow}b$, we have ${\downarrow}b \notin J$. Similarly ${\downarrow}c \in J - I$. By our deductions above, either ${\downarrow}b \leq {\downarrow}c$ or ${\downarrow}c \leq {\downarrow}b$. But then either ${\downarrow}b \in J$ or ${\downarrow}c \in I$, which is a contradiction. $\square$

Theorem 5.4. *Let $\mathbf{B}$ be a complete, atomic, idempotent Boolean semilattice, and suppose that $\mathrm{Con}(\mathbf{B})$ is linearly ordered. Then $\mathbf{B} \in \mathsf{S}(\mathsf{LS}^+)$.*

Proof. Let A be the set of atoms of $\mathbf{B}_0$. Fix a linear ordering, $\unlhd$, on A. Let $S = \{\,(\downarrow a, n, a) : a \in A, n \in \mathbb{N}\,\}$ ordered lexicographically. That is

$$(\downarrow a, n, a) < (\downarrow b, m, b) \text{ if } \quad \downarrow a < \downarrow b \text{ computed in } \mathbf{B}, \text{ or}$$
$$\downarrow a = \downarrow b \ \& \ n < m \text{ or}$$
$$\downarrow a = \downarrow b \ \& \ n = m \ \& \ a \lhd b\,.$$

Because of our assumption on the congruence lattice of $\mathbf{B}$, the closed elements are linearly ordered. So with this definition, $\mathbf{S}$ becomes a linearly ordered meet-semilattice.

Write $\mathbf{S}^{\square} = \langle S, \theta \rangle$ and $\mathbf{B}_+ = \langle A, \psi \rangle$. Recall that

$$\theta = \{\,(u, v, u \cdot v) : u, v \in S\,\}$$
$$\psi = \{\,(a, b, c) : c \leq a \cdot b\,\}\,.$$

Define $h\colon \mathbf{S}^{\square} \to \mathbf{B}_+$ by $h(\downarrow a, n, a) = a$. Clearly h is surjective. We shall show that h is a bounded morphism. From our comments in Sect. 2 it will follow that $\overleftarrow{h}$ embeds $\mathbf{B} = (\mathbf{B}_+)^+$ into $\mathbf{S}^+$, thereby proving the theorem.

We apply Definition 2.5. To verify the first condition, let $(u, v, u \cdot v) \in \theta$, say, $u = (\downarrow a, n, a)$ and $v = (\downarrow b, m, b)$. Since $\mathbf{S}$ is linear, we can assume that $u \leq v$, so $u \cdot v = u$. Then $(h(u), h(v), h(u \cdot v)) = (a, b, a)$. The condition $u \leq v$ implies $\downarrow a \leq \downarrow b$. By Lemma 5.2 we must have $a \leq a \cdot b$, so $(a, b, a) \in \psi$.

For the second condition in the definition of bounded morphism, let $a, b \in A$, $u = (\downarrow c, n, c) \in S$, and assume that $(a, b, h(u)) \in \psi$. This implies that $c \leq a \cdot b$. By Lemma 5.2 we must have $c = a$ or $c = b$. If $c = a$ then $\downarrow a \leq \downarrow b$ and $u = (\downarrow a, n, a)$. Take $v = u$ and $w = (\downarrow b, n+1, b)$. Then $v \leq w$ in $\mathbf{S}$, so $(v, w, u) \in \theta$, satisfies the condition. On the other hand, if $c = b$, take $v = (\downarrow a, n+1, a)$, and $w = u = (\downarrow b, n, b)$. Then $w \leq v$, so $(v, w, u) \in \theta$ again satisfies the condition. $\square$

Corollary 5.5 (Bergman-Blok). *The variety of idempotent Boolean semilattices is equal to $\mathsf{SP}(\mathsf{LS}^+)$.*

Proof. At the beginning of the section we verified that LS^+ is contained in IBSI, from which one inclusion of the theorem follows. We must verify that every idempotent Boolean semilattice lies in $\mathsf{V}(\mathsf{LS}^+)$. For this, it suffices to show that every subdirectly irreducible member of IBSI lies in $\mathsf{S}(\mathsf{LS}^+)$.

So let $\mathbf{A}$ be a subdirectly irreducible, idempotent Boolean semilattice, and let $\mathbf{B} = \mathbf{A}^{\sigma}$. Since the identities defining IBSI are strictly positive, $\mathbf{B}$ is itself an idempotent Boolean semilattice. By Theorem 4.18, $\mathbf{B}$ is subdirectly irreducible as well. And of course $\mathbf{B}$ is complete and atomic.

Then by Lemma 5.3, $\mathrm{Con}(\mathbf{B})$ is linearly ordered, and therefore by Theorem 5.4, $\mathbf{B} \in \mathsf{S}(\mathsf{LS}^+)$. Since $\mathbf{A}$ is a subalgebra of $\mathbf{B}$, the result follows. $\square$

Thus we have a satisfactory resolution to the representation problem: the finitely based variety IBSI is represented by the (finitely axiomatizable) class of linearly ordered semilattices. In fact, the variety IBSI is term-equivalent to the variety $\mathsf{S4.3}$ of modal algebras via the interpretations $\Diamond x = x \cdot 1$ and $x \cdot y = (\Diamond x \wedge y) \vee (x \wedge \Diamond y)$. From this equivalence it follows from known results that IBSI has only countably many subvarieties, each of which is finitely axiomatizable and generated by its finite members.

6 Semilattice Representability

Let us return to the relationship between the members of BSI and the complex algebras of semilattices. An integral Boolean semilattice is called *semilattice representable* if it can be embedded into $\mathbf{S}^+$ for some semilattice $\mathbf{S}$. In this section we shall simply say "representable" instead of "semilattice representable." It may also be of interest to determine whether a finite Boolean semilattice can be embedded into the complex algebra of a finite semilattice. When this occurs we say that the Boolean semilattice is *finitely representable*.

Lemma 6.1. *Let $\mathbf{B}$ be a Boolean semilattice, and $r \in B$. Suppose that $\downarrow r = 1$. Then for any homomorphism $h\colon \mathbf{B} \to \mathbf{S}^+$ for a semilattice, $\mathbf{S}$, the complex $h(r)$ must contain all maximal elements of $\mathbf{S}$.*

Proof. Let $R = h(r) \subseteq S$. Then $\downarrow r = 1$ implies that the downset generated by R is all of S. Thus if u is a maximal element of $\mathbf{S}$, then for some $x \in R$, $u \leq x$. By maximality, $u = x \in R$. $\qquad\square$

Corollary 6.2. *Let $\mathbf{B}$ be a Boolean semilattice, $r \in B$. Suppose that $\downarrow r = \downarrow(r') = 1$. Then there is no homomorphism from $\mathbf{B}$ to $\mathbf{S}^+$ for any semilattice with a maximal element. In particular $\mathbf{B}$ is not finitely representable.*

Proof. Let $h\colon \mathbf{B} \to \mathbf{S}^+$ be a homomorphism. By Lemma 6.1, both $h(r)$ and $h(r') = h(r)'$ must contain all maximal elements. Since these sets are disjoint, $\mathbf{S}$ has no maximal elements. $\qquad\square$

Corollary 6.3. *No simple Boolean semilattice is finitely representable.*

Proof. Follows from Proposition 4.19 and Corollary 6.2. $\qquad\square$

Recall that every partial semilattice is an inner substructure (i.e., an upset) of a semilattice. It is easy to see that the proofs of Lemma 6.1 and Corollary 6.2 remain valid when $\mathbf{S}$ is only a partial semilattice. Thus no simple Boolean semilattice can be embedded into the complex algebra of an upset of a semilattice.

Finally, we make one observation that may be useful in addressing Problems 4.3 and 4.4. Since the identities defining semilattices are regular, we can apply Theorem 2.8 to obtain $\mathbf{P}(\mathsf{SI}^+) \subseteq \mathbf{H}(\mathsf{SI}^+)$ and then Corollary 4.15 yields

$$\mathbf{V}(\mathsf{SI}^+) = \mathbf{HSP}(\mathsf{SI}^+) = \mathbf{HS}(\mathsf{SI}^+) = \mathbf{SH}(\mathsf{SI}^+).$$

7 Varieties of Boolean semilattices

The lattice of subvarieties of BSl is itself a rich and complex structure. At this time, we content ourselves with a few simple observations.

Because of normality and square-increasingness, $\{0,1\}$ forms a subalgebra of any nontrivial Boolean semilattice, in which $0 \cdot 0 = 0 \cdot 1 = 1 \cdot 0 = 0$ and $1 \cdot 1 = 1$. This algebra can be represented as $\mathbf{1}^+$, in which $\mathbf{1}$ represents a 1-element semilattice. Consequently, this algebra generates the smallest nontrivial subvariety of BSl. This subvariety is defined, relative to BSl, by the identity $x \cdot y \approx x \wedge y$. Thus, this subvariety is term-equivalent to the variety of Boolean algebras.

There are seven 4-element Boolean semilattices. Two of them are $\mathbf{1}^+ \times \mathbf{1}^+$ and $\mathbf{2}^+$, where $\mathbf{2}$ represents the 2-element semilattice. Figure 2 describes the product of the two atoms, a and b on each of the 7 algebras.

	$a \cdot a$	$b \cdot b$	$a \cdot b$
$\mathbf{1}^+ \times \mathbf{1}^+$	a	b	0
$\mathbf{2}^+$	a	b	a
$\mathbf{A}$	a	1	b
$\mathbf{B}_1$	a	1	a
$\mathbf{B}_2$	a	b	1
$\mathbf{B}_3$	a	1	1
$\mathbf{B}_4$	1	1	1

Fig. 2 The 4-element Boolean semilattices, with atoms a and b

The algebra $\mathbf{A}$ in the figure is identical to the complex algebra $\mathbf{H}^+$ discussed in conjunction with equation (8). As we demonstrated at that time, $\mathbf{A}$ is not semilattice representable. $\mathbf{B}_1$ can be embedded into $\mathbf{S}^+$, where $\mathbf{S}$ is the 3-element, nonlinear semilattice. The remaining three algebras are not finitely representable, by Corollary 6.2. However it is not hard to show that each can be represented on an infinite semilattice.

$\mathbf{1}^+ \times \mathbf{1}^+$ of course lies in the variety generated by $\mathbf{1}^+$. Both $\mathbf{2}^+$ and $\mathbf{B}_2$ are idempotent, so they lie in IBSl. $\mathbf{A}, \mathbf{B}_2, \mathbf{B}_3$, and $\mathbf{B}_4$ are simple, so they lie in BSl$_\mathsf{D}$. (In fact, $\mathbf{B}_4 \in \mathsf{Tot}^+$, see Theorem 4.22.) Finally, since all six (except for $\mathbf{1}^+ \times \mathbf{1}^+$) are subdirectly irreducible and have the same finite size, by Jónsson's lemma (see for example (Bergman, 2012, Cor. 5.13)) they must generate pairwise incomparable varieties. All six generate varieties that cover $\mathbf{V}(\mathbf{1}^+)$.

Problem 7.1. Determine all covers of $\mathbf{V}(\mathbf{1}^+)$ in the lattice of subvarieties of BSl. Does every subdirectly irreducible, 8-element Boolean semilattice contain a 4-element subalgebra?

We have already observed that the variety of Boolean semilattices has EDPC. In (Blok and Pigozzi, 1982), Blok and Pigozzi discuss the significance of quotients via compact congruences. For a class $\mathcal{K}$ of algebras we write

$$\mathbf{H}_\omega(\mathcal{K}) = \{\, \mathbf{B}/\theta : \mathbf{B} \in \mathcal{K},\ \theta \text{ a compact congruence of } \mathbf{B}\,\}\ .$$

In a Boolean semilattice, compact congruences correspond precisely to closed elements. In a semilattice, $\mathbf{S}$, a closed element of $\mathbf{S}^+$ is precisely a downset, D, of $\mathbf{S}$. The complex $S - D$ is an upset, which is to say, an inner substructure of $\mathbf{S}$. The resulting quotient, $\mathbf{S}^+/[D)$ is isomorphic to the complex algebra $(\mathbf{S} - D)^+$.

Let $\mathbf{A}$ be a member of a fixed variety, $\mathcal{V}$. The algebra $\mathbf{A}$ is called a *splitting algebra* (relative to $\mathcal{V}$), if $\mathcal{V}$ has a largest subvariety excluding $\mathbf{A}$. This variety, if it exists, is denoted $\mathcal{V}/\mathbf{A}$, and is called the *conjugate variety to* $\mathbf{A}$. The conjugate variety is defined by a single equation (relative to $\mathcal{V}$) called the conjugate equation. Blok and Pigozzi prove that if $\mathcal{V}$ has EDPC, then every finitely presented, subdirectly irreducible algebra in $\mathcal{V}$ is a splitting algebra, with conjugate variety

$$(12) \qquad\qquad \mathcal{V}/\mathbf{A} = \{\, \mathbf{B} \in \mathcal{V} : \mathbf{A} \notin \mathbf{SH}_\omega(\mathbf{B})\,\}\ .$$

In particular, if $\mathcal{V}$ has finite similarity type, which is the case for Boolean semilattices, then every finite subdirectly irreducible algebra is splitting.

As an application of this idea, we offer the following. Let $\mathsf{SI}_{\mathrm{fin}}$ denote the class of finite semilattices.

Theorem 7.2. $\mathbf{V}(\mathsf{SI}_{\mathrm{fin}}{}^+) \neq \mathbf{V}(\mathsf{SI}^+)$.

Proof. Let $\mathbf{B}_2$ be the 4-element algebra in Figure 2. We have already observed that $\mathbf{B}_2$ is finite and simple, hence splitting. Suppose $\mathbf{S}$ is a semilattice and $\mathbf{B}_2 \in \mathbf{SH}_\omega(\mathbf{S}^+)$. Then $\mathbf{B}_2$ is a subalgebra of $\mathbf{C}^+$ in which $\mathbf{C}$ is an inner substructure, i.e., an upset, of $\mathbf{S}$. By the remark following Corollary 6.3, $\mathbf{C}$, hence $\mathbf{S}$, must be infinite.

Therefore, by Equation (12), $\mathsf{SI}_{\mathrm{fin}}^+ \subseteq \mathbf{V}(\mathsf{SI}^+)/\mathbf{B}_2$. Since the latter class is a variety, $\mathbf{V}(\mathsf{SI}_{\mathrm{fin}}{}^+) \subseteq \mathbf{V}(\mathsf{SI}^+)/\mathbf{B}_2$. Since $\mathbf{V}(\mathsf{SI}^+)/\mathbf{B}_2$ obviously omits $\mathbf{B}_2$ itself, it must be a proper subvariety of $\mathbf{V}(\mathsf{SI}^+)$. $\square$

We close with a construction of $2^{\aleph_0}$ distinct subvarieties of $\mathbf{V}(\mathsf{SI}^+)$. Several other constructions are known. For example, it is known that there are uncountably many varieties of closure algebras, and this can be transformed into a construction for Boolean semilattices.

For any positive integer n, let A_n denote an antichain of size n, and let $\mathbf{Y}_n$ be the semilattice obtained from A_n by adjoining a new least element, z. It is easy to see that the only upsets of $\mathbf{Y}_n$ are Y_n itself and sets of the form A_k for some $k \leq n$.

Clearly, a bounded morphic image of $\mathbf{A}_k$ is of the form $\mathbf{A}_l$ for $l \leq k$. Also, no proper bounded morphic image of $\mathbf{Y}_n$ is a semilattice. To see this, we use Lemma 2.7. Suppose that α is a proper, nontrivial, bounded equivalence on $\mathbf{Y}_n$. There must be distinct elements a, b, c with $(a, b) \in \alpha$, $(a, c) \notin \alpha$ and

$a \in A_n$. If $b = z$ then the set $a/\alpha \cdot c/\alpha$ is not a union of α-classes, since it contains b but not a. This contradicts Lemma 2.7. Hence $b \neq z$, so the ternary relation on $\mathbf{Y}_n/\alpha$ contains $(a/\alpha, a/\alpha, z/\alpha)$ which is impossible in a semilattice.

Since $\mathbf{Y}_n$ is a lower-bounded semilattice, $\mathbf{Y}_n^+$ is subdirectly irreducible (Proposition 4.17). Applying duality to the previous two paragraphs, we deduce that

$$(13) \qquad\qquad n \neq m \implies \mathbf{Y}_m^+ \notin \mathsf{SH}(\mathbf{Y}_n^+).$$

From this and the Blok-Pigozzi relationship (12), we obtain the following.

Proposition 7.3. *Let S be a set of natural numbers and define $\mathcal{V}_S = \mathbf{V}\{\mathbf{Y}_n^+ : n \in S\}$. Then $\mathbf{Y}_m^+ \in \mathcal{V}_S$ if and only if $m \in S$. Consequently, $\{\mathcal{V}_S : S \subseteq \mathbb{N}\}$ forms an uncountable family of subvarieties of $\mathbf{V}(\mathsf{SI}^+)$.*

Proof. If $m \notin S$ then by (12) and (13), $\mathcal{V}_S \subseteq \mathbf{V}(\mathsf{SI}^+)/\mathbf{Y}_m^+$. Since $\mathbf{Y}_m^+$ is finite and subdirectly irreducible, it is a splitting algebra, so this latter class is a variety. $\qquad\qquad\qquad\square$

The proof of 7.3 actually shows something stronger. The variety $\mathbf{V}(\mathsf{SI}_{\mathrm{fin}}^+)$ has uncountably many subvarieties.

References

Bergman, C. (2012). Universal algebra. Fundamentals and selected topics, *Pure and Applied Mathematics (Boca Raton)*, vol. 301, CRC Press, Boca Raton, FL.

Blok, W. and Pigozzi, D. (1982). On the structure of varieties with equationally definable principal congruences I, *Algebra Universalis* **15**, 195–227.

Blok, W.J., Köhler, P. and Pigozzi, D. (1984). On the structure of varieties with equationally definable principal congruences. II, *Algebra Universalis* **18**(3), 334–379. DOI 10.1007/BF01203370. URL http://dx.doi.org/10.1007/BF01203370

Blok, W.J. and Pigozzi, D. (1984). On the structure of varieties with equationally definable principal congruences. III, *Algebra Universalis* **32**(4), 545–608. DOI 10.1007/BF01195727. URL http://dx.doi.org/10.1007/BF01195727

Blok, W.J. and Pigozzi, D. (1994). On the structure of varieties with equationally definable principal congruences. IV, *Algebra Universalis* **31**(1), 1–35. DOI 10.1007/BF01188178. URL http://dx.doi.org/10.1007/BF01188178

Goldblatt, R. (1989). Varieties of complex algebras, *Annals of Pure and Applied Logic* **44**, 173–242.

Grätzer, G. and Whitney, S. (1984). Infinitary varieties of structures closed under the formation of complex structures, *Colloquium Mathematicum* **48**, 1–5.

Hodkinson, I., Mikulas, S. and Venema, Y. (2001). Axiomatizing complex algebras by games. *Algebra Universalis* **46**, 455–478.

Jipsen, P. (1992). Computer aided investigations of relation algebras, *Ph.D. thesis*, Vanderbilt University.

Jipsen, P. (1993). Discriminator varieties of Boolean algebras with residuated operators, In: *Algebraic methods in logic and in computer science* (Warsaw, 1991), *Banach Center Publications*, vol. 28, pp. 239–252, Polish Academy of Sciences, Warsaw.

Jipsen, P. (2004). A note on complex algebras of semigroups, In: R. Berghammer (ed.) *Relational and Kleene-Algebraic Methods in Computer Science*, *Lecture Notes in Computer Science*, vol. 3051, pp. 171–177, Springer-Verlag, Berlin Heidelberg.

Jónsson, B. (1993). A survey of Boolean algebras with operators, In: *Algebras and orders* (Montreal, PQ, 1991), *NATO Adv. Sci. Inst. Ser. C Math. Phys. Sci.*, vol. 389, pp. 239–286, Kluwer Acad. Publ., Dordrecht.

Jónsson, B. and Tarski, A. (1951). Boolean algebras with operators. I, *The American Journal of Mathematics* **73**, 891–939.

Köhler, P. and Pigozzi, D. (1980). Varieties with equationally definable principal congruences, *Algebra Universalis* **11**(2), 213–219. URL `http://dx.doi.org/10.1007/BF02483100`

McKenzie, R. (1975). On spectra, and the negative solution of the decision problem for identities having a finite nontrivial model, *The Journal of Symbolic Logic* **40**, 186–195.

Reich, P. (1996). *Complex algebras of semigroups*, Retrospective theses and dissertations, paper 11765, Iowa State University, Ames, Iowa, USA.

Romanowska, A. (1986). On regular and regularized varieties, *Algebra Universalis* **23**, 215–241.

Shafaat, A. (1974). On varieties closed under the construction of power algebras, *Bulletin of the Australian Mathematical Society* **11**, 213–218.

The Equationally-Defined Commutator in Quasivarieties Generated by Two-Element Algebras

Janusz Czelakowski

Dedicated to Professor Don Pigozzi
on the occasion of his 80th birthday

Abstract The notion of the equationally-defined commutator was introduced and thoroughly investigated in (Czelakowski, 2015). In this work the properties of the equationally-defined commutator in quasivarieties generated by two-element algebras are examined. It is proved: *If a quasivariety $\mathbf{Q}$ is generated by a finite set of two-element algebras, then the equationally-defined commutator of $\mathbf{Q}$ is additive* (Theorem 3.1) *Moreover it satisfies the associativity law* (Theorem 3.6). The second result is strengthened if the quasivariety is generated by a single two-element algebra $\mathbf{2}$: *If $\mathbf{Q} = \boldsymbol{SP}(\mathbf{2})$, then the equationally-defined commutator of $\mathbf{Q}$ universally validates one of the following laws: $[x,y] = x \wedge y$ or $[x,y] = 0$* (Theorem 3.9). In other words, any quasivariety generated by a single two-element algebra is either relatively congruence-distributive or Abelian. A syntactical characterization of all quasivarieties generated by finite sets of two-element algebras is also presented (Theorems 2.2–2.3).

Key words: quasivariety, congruence, commutator equation, consequence operation, the equationally-defined commutator.

2010 Mathematics Subject Classification: 03C05, 03G27, 08C15

Janusz Czelakowski
Insitute of Mathematics and Informatics, Opole University, Oleska 48, 45-052 Opole, Poland, e-mail: `jczel@math.uni.opole.pl`

© Springer International Publishing AG 2018
J. Czelakowski (eds.), *Don Pigozzi on Abstract Algebraic Logic,*
Universal Algebra, and Computer Science, Outstanding Contributions
to Logic 16, https://doi.org/10.1007/978-3-319-74772-9_5

1 Preliminary Remarks

The paper is a companion piece to the monograph Czelakowski (2015). While this monograph provides general definitions and results concerning the equationally-defined commutator, the present paper is focused on the properties of the commutator for quasivarieties generated by finite sets of two-element algebras. Some partial observations about the equationally-defined commutator in selected quasivarieties generated by two-elements algebras were placed in the above monograph (e.g. in Section 5.4). Theorem 3.1 of this paper generalizes them. It is proved that the equationally-defined commutator of each quasivariety generated by a finite set of two-element algebras is additive and associative.

As to the narrative structure of the paper, the first two sections contain indispensable definitions and general facts about quasivarieties and the equationally-defined commutator. These results are mostly taken from Czelakowski (2015). Section 3 is central; it contains new theorems.

Quasivarieties

The standard algebraic terminology and notation is applied. Let τ be a fixed algebraic signature and $\boldsymbol{L}$ the corresponding first-order language with equality $\approx$. $Var = \{\nu_n : n \in \omega\}$ is the set of individual variables of $\boldsymbol{L}$. $\boldsymbol{Te}$ is the algebra of *terms* of $\boldsymbol{L}$ and $Eq(\tau)$ is the set of *equations* of $\boldsymbol{L}$.

An algebra $\boldsymbol{A}$ of signature τ is often referred to as a τ-*algebra*.

If $t = t(x_1,\ldots,x_n)$ is a term in at most n individual variables $\underline{x} = x_1,\ldots,x_n$, and $\underline{a} = a_1,\ldots,a_n$ is a sequence of elements of a τ-algebra $\boldsymbol{A}$, then $t^{\boldsymbol{A}}(a_1,\ldots,a_n)$ is the *value* of the term t for $\underline{a} = a_1,\ldots,a_n$ in $\boldsymbol{A}$. $t^{\boldsymbol{A}}(a_1,\ldots,a_n)$ is defined in the standard way by induction on complexity of terms. We shall also use the abbreviation $t^{\boldsymbol{A}}(\underline{a})$ or $t(\underline{a})$ for $t^{\boldsymbol{A}}(a_1,\ldots,a_n)$, often omitting the superscript "$\boldsymbol{A}$".

A *quasivariety* is a class of algebras closed under the formation of subalgebras $\boldsymbol{S}$, direct products $\boldsymbol{P}$ and ultraproducts $\boldsymbol{P_u}$. The operation $\boldsymbol{S}, \boldsymbol{P}$ and $\boldsymbol{P_u}$ are viewed here in the inclusive sense—they also include isomorphic copies of subalgebras, direct products and ultraproducts, respectively. The trivial one element algebra is treated as a direct product of the void family. Every quasivariety contains (all isomorphic copies of) the trivial algebra.

A well-known result due to Maltsev states that a class of algebras closed under isomorphisms is a quasivariety iff it is axiomatized by means of a set of quasi-identities, i.e., a set of the universal closures of implications $\alpha_1 \approx \beta_1 \wedge \ldots \wedge \alpha_n \approx \beta_n \rightarrow \alpha \approx \beta$.

Let $\boldsymbol{Q}$ be a quasivariety of τ-algebras and $\boldsymbol{A}$ a τ-algebra, not necessarily in $\boldsymbol{Q}$. A congruence Φ on $\boldsymbol{A}$ is called a $\boldsymbol{Q}$-*congruence* if $\boldsymbol{A}/\Phi \in \boldsymbol{Q}$. $Con_{\boldsymbol{Q}}(\boldsymbol{A})$ is the set of $\boldsymbol{Q}$-congruences on $\boldsymbol{A}$. Thus $Con_{\boldsymbol{Q}}(\boldsymbol{A}) = \{\Phi \in Con(\boldsymbol{A}) : \boldsymbol{A}/\Phi \in \boldsymbol{Q}\}$.

$Con_{\mathbf{Q}}(\boldsymbol{A})$ contains the universal congruence $\mathbf{1}_{\boldsymbol{A}} := A^2$ and it contains the smallest $\mathbf{Q}$-congruence being the intersection of all $\mathbf{Q}$-congruences of $\boldsymbol{A}$. This smallest $\mathbf{Q}$-congruence is the identity congruence $\mathbf{0}_{\boldsymbol{A}}$ ($=$ diagonal relation on A) if and only if $\boldsymbol{A} \in \mathbf{Q}$.

$Con_{\mathbf{Q}}(\boldsymbol{A})$ is a finitary closure system on A^2 and therefore forms the universe of an algebraic lattice $\boldsymbol{Con}_{\mathbf{Q}}(\boldsymbol{A})$ called the *lattice of* $\mathbf{Q}$-*congruences*.

If $\mathbf{V}$ is a variety, and $\boldsymbol{A}$ is an algebra of type τ, then $Con_{\mathbf{V}}(\boldsymbol{A})$ forms a principal filter in the lattice $\boldsymbol{Con}(\boldsymbol{A})$ of all congruences of $\boldsymbol{A}$. But if $\boldsymbol{A}$ is in $\mathbf{V}$, then $Con_{\mathbf{V}}(\boldsymbol{A})$ coincides with $Con(\boldsymbol{A})$.

For any $X \subseteq A^2$, $\Theta^{\boldsymbol{A}}_{\mathbf{Q}}(X)$ denotes the least $\mathbf{Q}$-congruence of $\boldsymbol{A}$ that contains X. Thus

$$\Theta^{\boldsymbol{A}}_{\mathbf{Q}}(X) = \bigcap \{\Phi \in Con_{\mathbf{Q}}(\boldsymbol{A}) : X \subseteq \Phi\}.$$

Given a class $\mathbf{K}$ of τ-algebras, we let $\mathbf{K}^{\vDash}$ denote the consequence *operation* (on the set of τ-equations) determined by $\mathbf{K}$. Thus, for $\{\alpha_i \approx \beta_i : i \in I\} \cup \{\alpha \approx \beta\} \subseteq Eq(\tau)$, $\alpha \approx \beta \in \mathbf{K}^{\vDash}(\{\alpha_i \approx \beta_i : i \in I\}$ if and only if, for every algebra $\boldsymbol{A} \in \mathbf{K}$ and every $h \in Hom(\boldsymbol{Te}, \boldsymbol{A})$, $h(\alpha) = h(\beta)$ whenever $h(\alpha_i) = h(\beta_i)$ for all $i \in I$.

The consequence $\mathbf{K}^{\vDash}$ is structural, i.e., $\alpha \approx \beta \in \mathbf{K}^{\vDash}(\{\alpha_i \approx \beta_i : i \in I\})$ implies that $e\alpha \approx e\beta \in \mathbf{K}^{\vDash}(\{e\alpha_i \approx e\beta_i : i \in I\})$ for all endomorphisms e of the term algebra $\boldsymbol{Te}$. The consequence $\mathbf{K}^{\vDash}$ validates the well known Birkhoff's rules. Furthermore, if $\mathbf{K}$ is closed under the formation of ultraproducts, the consequence $\mathbf{K}^{\vDash}$ is finitary. $\alpha \approx \beta \in \mathbf{K}^{\vDash}(\emptyset)$ means that the equation $\alpha \approx \beta$ is valid in the class $\mathbf{K}$.

Following common practice we suppress parentheses as much as possible and in case of finite set of equations we usually write $\alpha \approx \beta \in \mathbf{K}^{\vDash}(\alpha_1 \approx \beta_1, \ldots, \alpha_n \approx \beta_n)$ instead of $\alpha \approx \beta \in \mathbf{K}^{\vDash}(\{\alpha_1 \approx \beta_1, \ldots, \alpha_n \approx \beta_n\})$.

There is an obvious translation of $\mathbf{K}^{\vDash}$ into the language of quasi-identities over $\boldsymbol{Te}$:

$\alpha \approx \beta \in \mathbf{K}^{\vDash}(\alpha_1 \approx \beta_1, \ldots, \alpha_n \approx \beta_n)$ iff the implication

$$\alpha_1 \approx \beta_1 \wedge \ldots \wedge \alpha_n \approx \beta_n \to \alpha \approx \beta \text{ is valid in } \mathbf{K}.$$

If $\mathbf{K}$ is a finite set of finite algebras, then $\mathbf{Q} := \boldsymbol{SP}(\mathbf{K})$ is the quasi-variety generated by $\mathbf{K}$. It follows that the (finitary) consequence operations $\mathbf{K}^{\vDash}$ and $\mathbf{Q}^{\vDash}$ coincide.

Commutator Equations

Let m and n be positive integers and let $\underline{x} = x_1, \ldots, x_m$, $\underline{y} = y_1, \ldots, y_m$, $\underline{z} = z_1, \ldots, z_n$, $\underline{w} = w_1, \ldots, w_n$, and $\underline{u} = u_1, \ldots, u_k$ be sequences of *pairwise distinct* individual variables. The lengths of the strings $\underline{x}$ and $\underline{y}$ are equal, $|\underline{x}| = |\underline{y}| = m$ and, similarly, $|\underline{z}| = |\underline{w}| = n$.

Let τ be an algebraic signature.

$$p(\underline{x}, \underline{y}, \underline{z}, \underline{w}, \underline{u}) := p(x_1, \ldots, x_m, y_1, \ldots, y_m, z_1, \ldots, z_n, w_1, \ldots, w_n, u_1, \ldots, u_k)$$

marks a term in Te_τ built up with at most the variables $\underline{x} = x_1, \ldots, x_m$, $\underline{y} = y_1, \ldots, y_m$, $\underline{z} = z_1, \ldots, z_n$, $\underline{w} = w_1, \ldots, w_n$, and $\underline{u} = u_1, \ldots, u_k$.

Let $\mathbf{K}$ be a class of algebras. $\alpha(\underline{x}, \underline{y}, \underline{z}, \underline{w}, \underline{u}) \approx \beta(\underline{x}, \underline{y}, \underline{z}, \underline{w}, \underline{u})$ is a *commutator equation of* $\mathbf{K}$ *in the variables* $\underline{x}, \underline{y}$ and $\underline{z}, \underline{w}$ if $\mathbf{K}$ validates the equations

$$\alpha(\underline{x}, \underline{x}, \underline{z}, \underline{w}, \underline{u}) \approx \beta(\underline{x}, \underline{x}, \underline{z}, \underline{w}, \underline{u}) \text{ and } \alpha(\underline{x}, \underline{y}, \underline{z}, \underline{z}, \underline{u}) \approx \beta(\underline{x}, \underline{y}, \underline{z}, \underline{z}, \underline{u}).$$

$CoEq(\mathbf{K})$ is the set of all commutator equations of $\mathbf{K}$.

A *quaternary commutator equation* of $\mathbf{K}$ (with parameters) is any commutator equation $\alpha(x, y, z, w, \underline{u}) \approx \beta(x, y, z, w, \underline{u})$ for $\mathbf{K}$ in the variables x, y and z, w.

It follows from the above definition that $\alpha(\underline{x}, \underline{y}, \underline{z}, \underline{w}, \underline{u}) \approx \beta(\underline{x}, \underline{y}, \underline{z}, \underline{w}, \underline{u})$ is a commutator equation of $\mathbf{K}$ (in the variables $\underline{x}, \underline{y}$ and $\underline{z}, \underline{w}$) if and only if it is a commutator equation (in the variables $\underline{x}, \underline{y}$ and $\underline{z}, \underline{w}$) of the variety $\boldsymbol{Va}(\mathbf{K})$ generated by $\mathbf{K}$. Consequently, the classes $\overline{\mathbf{K}}$, $\boldsymbol{Qv}(\mathbf{K})$ and $\boldsymbol{Va}(\mathbf{K})$ possess the same commutator equations.

The above definition is reformulated in terms of the consequence operation $\mathbf{K}^{\models}$ as follows: for fixed $m, n \geqslant 1$,

$$\mathbf{K}^{\models}(x_1 \approx y_1, \ldots, x_m \approx y_m) \cap \mathbf{K}^{\models}(z_1 \approx w_1, \ldots, z_n \approx w_n)$$

is the set of all commutator equations of $\mathbf{K}$ in the variables $\underline{x} = x_1, \ldots, x_m$, $\underline{y} = y_1, \ldots, y_m$, $\underline{z} = z_1, \ldots, z_n$, $\underline{w} = w_1, \ldots, w_n$. In particular

$$\mathbf{K}^{\models}(x \approx y) \cap \mathbf{K}^{\models}(z \approx w)$$

is the set of all quaternary commutator equations of $\mathbf{K}$ (with parameters) in the variables x, y and z and w.

Definition 1.1. Let $\mathbf{Q}$ be a quasivariety of algebras of signature τ. Let $\boldsymbol{A}$ be a τ-algebra, and let Φ and Ψ be $\mathbf{Q}$-congruences on $\boldsymbol{A}$. The *equationally-defined commutator of* Φ *and* Ψ *on* $\boldsymbol{A}$ *relative to* $\mathbf{Q}$, in symbols

$$[\Phi, \Psi]$$

is the least $\mathbf{Q}$-congruence on $\boldsymbol{A}$ which contains the following set of pairs:

$$\{\langle \alpha(\underline{a}, \underline{b}, \underline{c}, \underline{d}, \underline{e}), \beta(\underline{a}, \underline{b}, \underline{c}, \underline{d}, \underline{e}) \rangle : \alpha(\underline{x}, \underline{y}, \underline{z}, \underline{w}, \underline{u}) \approx \beta(\underline{x}, \underline{y}, \underline{z}, \underline{w}, \underline{u}) \in CoEq(\mathbf{Q}),$$
$$\underline{a} \equiv \underline{b}\,(\Phi),\ \underline{c} \equiv \underline{d}\,(\Psi),\ \text{and}\ \underline{e} \in A^{<\omega}\}.$$

The equationally-defined commutator of $\mathbf{Q}$ is *additive on* $\boldsymbol{A}$ if for any Φ_1, $\Phi_2, \Psi \in Con_{\mathbf{Q}}(\boldsymbol{A})$:

(C1) $[\Phi_1 +_{\mathbf{Q}} \Phi_2, \Psi] = [\Phi_1, \Psi] +_{\mathbf{Q}} [\Phi_2, \Psi].$

The equationally-defined commutator of $\mathbf{Q}$ is *additive on* $\mathbf{Q}$ if it is additive on every algebra $\mathbf{A} \in \mathbf{Q}$.

If the equationally-defined commutator of $\mathbf{Q}$ is additive, then it has the following property (the Correspondence Property):

(C2)   If $h : \mathbf{A} \to \mathbf{B}$ is a surjective homomorphism between $\mathbf{Q}$-algebras and $\Phi, \Psi \in Con_{\mathbf{Q}}(\mathbf{A})$, then

$$\ker_{\mathbf{Q}}(h) +_{\mathbf{Q}} [\Phi, \Psi]^{\mathbf{A}} = h^{-1}([\Theta_{\mathbf{Q}}^{\mathbf{B}}(h\Phi), \Theta_{\mathbf{Q}}^{\mathbf{B}}(h\Psi)]^{\mathbf{B}})$$

(see (Czelakowski, 2015), Theorem 5.1.1).

The additivity property is extensively applied in the commutator theory—see (Freese, McKenzie, 1987) and (Kearnes, McKenzie, 1992). The monograph (Czelakowski, 2015) provides various criteria of additivity of the equationally-defined commutator. We mention here one:

Theorem 1.2. *Let $\mathbf{Q}$ be a quasivariety. The following conditions are equivalent:*

(1) *The equationally-defined commutator of $\mathbf{Q}$ is additive;*
(2) *There exists a set $\Delta(x, y, z, w, \underline{u})$ of quaternary commutator equations for $\mathbf{Q}$ in x, y and z, w (possibly with k parameters $\underline{u} = u_1, u_2, \ldots, k \leqslant \omega$) such that for every τ-algebra $\mathbf{A}$ and for every pair of sets $X, Y \subseteq A^2$,*

$$[\Theta_{\mathbf{Q}}^{\mathbf{A}}(X), \Theta_{\mathbf{Q}}^{\mathbf{A}}(Y)] = \Theta_{\mathbf{Q}}^{\mathbf{A}}(\bigcup\{(\forall \underline{e})\, \Delta^{\mathbf{A}}(a, b, c, d, \underline{e}) : \langle a, b \rangle \in X, \langle c, d \rangle \in Y\}). \square$$

(Here $(\forall \underline{e})\, \Delta^{\mathbf{A}}(a, b, c, d, \underline{e})$ is the set of all pairs $\{\langle \alpha(a, b, c, d, \underline{e}), \beta(a, b, c, d, \underline{e}) \rangle :$ $\alpha(x, y, z, w, \underline{u}) \approx \beta(x, y, z, w, \underline{u}) \in \Delta$, and $\underline{e} \in A^{<\omega}\}$.)

It should be underlined that in any relatively congruence-modular (RCM) quasivariety $\mathbf{Q}$, the equationally-defined commutator of $\mathbf{Q}$ coincides with the one introduced and studied in (Kearnes, McKenzie, 1992) and therefore it is additive. But there are quasivarieties which are *not* RCM and whose equationally-defined commutator retains the property of additivity. In this paper further examples of such quasivarieties are provided.

The definition of the equationally-defined commutator also makes sense for the closed theories of the equational consequence $\mathbf{Q}^{\vDash}$. Thus if $X, Y \in Th(\mathbf{Q}^{\vDash})$, then

$$[X, Y] := \mathbf{Q}^{\vDash}(\{\langle \alpha(\underline{p}, \underline{q}, \underline{r}, \underline{s}, \underline{t}), \beta(\underline{p}, \underline{q}, \underline{r}, \underline{s}, \underline{t}) \rangle : \alpha(\underline{x}, \underline{y}, \underline{z}, \underline{w}, \underline{u}) \approx \beta(\underline{x}, \underline{y}, \underline{z}, \underline{w}, \underline{u})$$
$$\in CoEq(\mathbf{Q}), \underline{p} \approx \underline{q} \subseteq X, \underline{r} \approx \underline{s}(Y), \text{ and } \underline{t} \in Te^{<\omega}\}.$$

The commutator is additive on $\mathbf{Q}$ iff it is additive in the lattice of theories of $\mathbf{Q}^{\vDash}$ ((Czelakowski, 2015), Theorem 5.2.1). ($\underline{p} \approx \underline{q}$ marks a finite sequence of equations $p_1 \approx q_1, \ldots, p_n \approx q_n$, and $\underline{p} \approx \underline{q} \subseteq X$ means that $p_i \approx q_i \in X$ for $i = 1, \ldots, n$.)

2 Quasivarieties Generated by Two-Element Algebras

Generally, there may be several two-element algebras in a given signature τ. For instance, if one takes a two-element Boolean algebra **2** with the carrier $\{0, 1\}$, then it can be extended to a modal algebra in two ways: either by declaring that $\Box 0 = 0$ and $\Box 1 = 1$ or by putting: $\Box 0 = 1$ and $\Box 1 = 1$. Yet another example is the variety **HSI** of *the high school identities*. **HSI** contains five two-element algebras (see e.g. (Burris, Lee, 1993), (Burris, Yeats, 2004)). Generally, if the signature τ is finite, there are only finitely many non-isomorphic two-element algebras of type τ.

In what follows, τ is a fixed algebraic signature and **2** denotes an arbitrary but fixed two-element algebra of type τ with the carrier $\{0, 1\}$. The quasivariety **Q** generated by **2** is equal to the class $SP(\mathbf{2})$.

It is well-known that all relatively finitely subdirectly-irreducible algebras in $SP(\mathbf{2})$ belong to $S(\mathbf{2})$. It follows that there is only one (up to isomorphism) non-trivial algebra in $S(\mathbf{2})$, viz. **2**. Therefore $SP(\mathbf{2})_{\mathrm{RFSI}}$ consists of isomorphic copies of **2**.

The quasivariety $SP(\mathbf{2})$ need not be a variety. Let $'$ be a unary operation symbol and let **2** be the truth-table of negation. Let the three-element algebra **3** result from **2** by augmenting it with a fixed-point a for $'$. (Thus $\mathbf{0}' = \mathbf{1}$, $\mathbf{1}' = \mathbf{0}$ and $a' = a$). **3** is a homomorphic image of $\mathbf{2} \times \mathbf{2}$. Moreover **2** and **3** are the only subdirectly irreducible algebras in $HSP(\mathbf{2})$. ($\Theta(0, 1)$ is the smallest non-zero congruence in **3**.) The algebra **3** is not a member of $SP(\mathbf{2})$, because no algebra in $SP(\mathbf{2})$ admits a fixed point. Therefore the class $HSP(\mathbf{2})$ is larger than $SP(\mathbf{2})$.

Berman (1980) proved the following general fact: *Let **2** be any two-element algebra of arbitrary similarity type. The variety $HSP(\mathbf{2})$ has at most three subdirectly irreducible algebras, and any subdirectly irreducible member of $HSP(\mathbf{2})$ has at most three elements.*

Thus $HSP(\mathbf{2})$ is always residually less than 4.

In the paper we are concerned with the properties of the equationally-defined commutator in the quasivarieties generated by finite sets of two-element algebras. Before passing to the commutator we shall first give some general facts concerning such quasivarieties.

Let τ be a fixed signature. Given a τ-equation $p(x, y, z, \underline{u}) \approx q(x, y, z, \underline{u})$ in the variables x, y, z and possibly some other variables $\underline{u}$, we define (by making appropriate substitutions of variables in $p \approx q$) the following three-premiss rule

$$r_{p \approx q}: \quad p(x, x, z, \underline{u}) \approx q(x, x, z, \underline{u}),\ p(x, y, x, \underline{u}) \approx q(x, y, x, \underline{u}),$$

$$p(x, y, y, \underline{u}) \approx q(x, y, y, \underline{u}) / p \approx q$$

Lemma 2.1. *Let* $\mathbf{Q}$ *be a quasivariety generated by a finite set* $\mathbf{K}$ *of two-element* τ*-algebras, that is,* $\mathbf{Q} = \boldsymbol{SP}(\mathbf{K})$. *Then* $r_{p\approx q}$ *is a rule of* $\mathbf{Q}^{\vDash}$ *for each equation* $p(x,y,z,\underline{u}) \approx q(x,y,z,\underline{u})$.

Proof (of the lemma). Fix terms p and q. It suffices to show that every algebra $\mathbf{2} \in \mathbf{K}$ validates $r_{p\approx q}$. Let $h : \boldsymbol{Te} \to \mathbf{2}$ be a homomorphism such that

$$p(hx, hx, hz, h\underline{u}) = q(hx, hx, hz, h\underline{u}),$$
$$p(hx, hy, hx, h\underline{u}) = q(hx, hy, hx, h\underline{u}),$$
$$p(hx, hy, hy, h\underline{u}) = q(hx, hy, hy, h\underline{u}).$$

As $\mathbf{2}$ has two elements, $\mathbf{0}$ and $\mathbf{1}$, there are two variables u and v among x, y, z such that $hu = hv$. Suppose e.g. that $hx = hy$. It then follows that $p(hx, hx, hz, h\underline{u}) = p(hx, hy, hz, h\underline{u})$ and $q(hx, hx, hz, h\underline{u}) = q(hx, hy, hz, h\underline{u})$. Hence $p(hx, hy, hz, h\underline{u}) = q(hx, hy, hz, hu)$. So h validates $p \approx q$. The other cases are similarly handled. $\qquad\square$

The following two theorems provide a characterization of quasivarieties generated by finite sets of two-element algebras:

Theorem 2.2. *For any quasivariety* $\mathbf{Q}$ *the following conditions are equivalent:*

(i) $\mathbf{Q}$ *is generated by a finite set* $\mathbf{K}$ *of two-element algebras.*

(ii)   *For any set of equations* X *and any three distinct variables* x, y, z,

$$\mathbf{Q}^{\vDash}(X) = \mathbf{Q}^{\vDash}(X, x \approx y) \cap \mathbf{Q}^{\vDash}(X, x \approx z) \cap \mathbf{Q}^{\vDash}(X, y \approx z).$$

(iii) *There are three distinct variables* x, y, z *such that for any set of equations* X,

$$\mathbf{Q}^{\vDash}(X) = \mathbf{Q}^{\vDash}(X, x \approx y) \cap \mathbf{Q}^{\vDash}(X, x \approx z) \cap \mathbf{Q}^{\vDash}(X, y \approx z).$$

(iv)   *There are three distinct variables* x, y, z *such that for any equation* $p(x, y, z, \underline{u}) \approx q(x, y, z, \underline{u})$, $r_{p\approx q}$ *is a rule of* $\mathbf{Q}^{\vDash}$.

Proof. The equivalence of (i), (ii) and (iii) is a particular instance of Theorem 8.3.8 of (Czelakowski, 2015). The implication (i) $\Rightarrow$ (iv) is the content of the above lemma. We shall prove that (iv) implies (ii). Note however that due to the structurality of $\mathbf{Q}^{\vDash}$, the quantification "*There are three distinct variables x, y, z*" in (iv) can be replaced by the universal quantification "*For any three distinct variables x, y, z*". It suffices to show that (ii) holds for any finite set X and any three variables x, y, z that *do not* occur in the equations of X. Assume that an equation $p(x, y, z, \underline{u}) \approx q(x, y, z, \underline{u})$ belongs to $\mathbf{Q}^{\vDash}(X, x \approx y) \cap \mathbf{Q}^{\vDash}(X, x \approx z) \cap \mathbf{Q}^{\vDash}(X, y \approx z)$. Then by structurality we get that

$$p(x,x,z,\underline{u}) \approx q(x,x,z,\underline{u}) \in \mathbf{Q}^{\vDash}(X),$$

$$p(x,y,x,\underline{u}) \approx q(x,y,x,\underline{u}) \in \mathbf{Q}^{\vDash}(X),$$

$$p(x,y,y,\underline{u}) \approx q(x,y,y,\underline{u}) \in \mathbf{Q}^{\vDash}(X).$$

Applying the rule $r_{p \approx q}$ we obtain that $p \approx q \in \mathbf{Q}^{\vDash}(X)$. So (ii) holds.

It follows that conditions (i)–(iv) are mutually equivalent. $\qquad\square$

Note. Conditions (ii) and (iii) can be equivalently reformulated in terms of $\mathbf{Q}$-congruences on the free algebra $\boldsymbol{F}_{\mathbf{Q}}(\omega)$. Similarly, (iv) can be equivalently expressed as the $\mathbf{Q}$-validity of the implication

$$p(x,x,z,\underline{u}) \approx q(x,x,z,\underline{u}) \wedge p(x,y,x,\underline{u}) \approx q(x,y,x,\underline{u}) \wedge$$
$$p(x,y,y,\underline{u}) \approx q(x,y,y,\underline{u}) \to p \approx q$$

for any equation $p(x,y,z,\underline{u}) \approx q(x,y,z,\underline{u})$. $\qquad\square$

Given a quasivariety $\mathbf{Q}$, we define the consequence operation $\mathbf{Q}^{\vDash(2)}$ on $Eq(\tau)$ as follows. For any set $X \subseteq Eq(\tau)$ and any equation $p \approx q \in Eq(\tau)$ we put:

$$p \approx q \in \mathbf{Q}^{\vDash(2)}(X) \Leftrightarrow_{df} ep \approx eq \in \mathbf{Q}^{\vDash}(eX)$$

$$\text{for every endomorphism } e : Var \to \{x,y\}.$$

It follows that $\mathbf{Q}^{\vDash} \leqslant \mathbf{Q}^{\vDash(2)}$ and therefore $\mathbf{Q}^{\vDash(2)}$ validates the rules of Birkhoff logic.

The following observation supplements Theorem 2.2. It is an adaptation to equational logic of a theorem from propositional logic ((Wójcicki, 1988), Theorem 4.1.4):

Theorem 2.3. *For any quasivariety* $\mathbf{Q}$ *the following conditions are equivalent:*

(1) $\mathbf{Q} = \boldsymbol{SP}(\mathbf{K})$ *for some finite set* $\mathbf{K}$ *of two-element algebras.*

(2) $\mathbf{Q}^{\vDash(2)} = \mathbf{Q}^{\vDash}$ *and for every ternary equation* $p(x,y,z) \approx q(x,y,z)$, $r_{p \approx q}$ *is a rule of* $\mathbf{Q}^{\vDash}$.

Note that the equations $p \approx q$ of (2) do not contain parametric variables.

Proof. (1) $\Rightarrow$ (2). Assume (1). We must prove that for any set $X \subseteq Eq(\tau)$ and any equation $p \approx q \in Eq(\tau)$:

(a) $p \approx q \in \mathbf{Q}^{\vDash}(X) \Leftrightarrow ep \approx eq \in \mathbf{Q}^{\vDash}(eX)$ for every every endomorphism e of $\boldsymbol{Te}$ such that $e : Var \to \{x,y\}$.

We have $\mathbf{Q} = \boldsymbol{SP}(\mathbf{K})$, where $\mathbf{K}$ is a finite set $\{\boldsymbol{2}_1,\dots,\boldsymbol{2}_n\}$ of two-element algebras. Since $\mathbf{Q}^{\vDash}$ is structural, to establish (a) it suffices to show that

(b) $p \approx q \notin \mathbf{Q}^{\vDash}(X)$

implies

(c) $\qquad ep \approx eq \notin \mathbf{Q}^{\vDash}(eX)$

for some endomorphism e such that $e : Var \to \{x, y\}$.

(b) implies that $p \approx q \notin \mathbf{2}_i^{\vDash}(X)$ for some i, $1 \leqslant i \leqslant n$, where $\mathbf{2}_i = \{\mathbf{0}, \mathbf{1}\}$. Let $h : \mathbf{Te} \to \mathbf{2}_i$ be a homomorphism which validates the equations of X and refutes $p \approx q$, that is, $hp \neq hq$. Let l $(1 \leqslant l \leqslant 2)$ be the number of values h takes on the variables. Suppose for simplicity that $l = 2$. (In case when $l = 1$, the reasoning is similar.) Let then z, w be variables such that $hz \neq hw$. We define the substitution e_h in $\mathbf{Te}$ by $e_h u = x$ iff $hu = hz$ and $e_h u = y$ iff $hu = hw$. Thus e_h maps Var onto $\{x, y\}$. Then (c) holds for this e_h, because taking any homomorphism $h' : \mathbf{Te} \to \mathbf{2}_i$ such that $h'eu = hu$ for all u, we see that that h' validates eX and refutes $ep \approx eq$.

The fact that for every ternary equation $p(x, y, z) \approx q(x, y, z)$, $r_{p \approx q}$ is a rule of $\mathbf{Q}^{\vDash}$ is immediate. So (2) holds.

Note. Let $n \geqslant 2$ be a positive integer and $x_1, \ldots, x_n$ different variables. $\mathbf{Q}^{\vDash(n)}$ is the consequence operation in $Eq(\tau)$ defined as follows:

$$p \approx q \in \mathbf{Q}^{\vDash(n)}(X) \Leftrightarrow_{df} ep \approx eq \in \mathbf{Q}^{\vDash}(eX) \text{ for every every endomorphism of } \mathbf{Te} \text{ such that } e : Var \to \{x_1, \ldots, x_n\}.$$

The above proof shows that the equality $\mathbf{Q}^{\vDash(2)} = \mathbf{Q}^{\vDash}$ occurring in condition (2) of the above theorem is equivalent to $\mathbf{Q}^{\vDash(n)} = \mathbf{Q}^{\vDash}$ for all $n \geqslant 2$. $\qquad \square$

(2) $\Rightarrow$ (1). We fix three variables x, y, z and assume (2). It follows that

(∗) $\quad \mathbf{Q}^{\vDash(2)} = \mathbf{Q}^{\vDash}$ *and for any set of equations* $X \subseteq Eq(x, y, z)$,

$$Eq(x, y, z) \cap \mathbf{Q}^{\vDash}(X, x \approx y) \cap \mathbf{Q}^{\vDash}(X, x \approx z) \cap \mathbf{Q}^{\vDash}(X, y \approx z) \subseteq \mathbf{Q}^{\vDash}(X).$$

In view of the above note we also have that $\mathbf{Q}^{\vDash(3)} = \mathbf{Q}^{\vDash}$. In virtue of Theorem 2.2, to prove (1) it suffices to show that

$$\mathbf{Q}^{\vDash}(X, x \approx y) \cap \mathbf{Q}^{\vDash}(X, x \approx z) \cap \mathbf{Q}^{\vDash}(X, y \approx z) \subseteq \mathbf{Q}^{\vDash}(X),$$

for any set $X \subseteq Eq(\tau)$.

Assume $p \approx q \in \mathbf{Q}^{\vDash}(X, x \approx y) \cap \mathbf{Q}^{\vDash}(X, x \approx z) \cap \mathbf{Q}^{\vDash}(X, y \approx z)$. It then follows that $ep \approx eq \in \mathbf{Q}^{\vDash}(eX, x \approx y) \cap \mathbf{Q}^{\vDash}(eX, x \approx z) \cap \mathbf{Q}^{\vDash}(eX, y \approx z)$ for every endomorphism e of $\mathbf{Te}$ such that $e : Var \to \{x, y, z\}$ and which is the identity map on $\{x, y, z\}$. The second conjunct of (∗) then yields that $ep \approx eq \in \mathbf{Q}^{\vDash}(eX)$ for every such an e. $\mathbf{Q}^{\vDash(3)} = \mathbf{Q}^{\vDash}$ then gives that $p \approx q \in \mathbf{Q}^{\vDash}(X)$. $\qquad \square$

3 The Equationally-Defined Commutator

The following observation is central in this work:

Theorem 3.1. *Let* **Q** *be any quasivariety generated by a finite set of two-element algebras. Then the equationally-defined commutator for* **Q** *is additive.*

Particular instances of the above theorem are known in the literature. In any relatively congruence-distributive quasivariety, the (equationally-defined) commutator of two relative congrunces coincides with their meet. Therefore the commutator is additive. In (Czelakowski, 2015) it was proved that the equationally-defined commutator in the variety of semilattices is Abelian. (This fact does *not* hold for other commutators defined for semilattices—see e.g. (McKenzie, McNulty, Taylor, 1987), Exercise 5, p. 258.)

The methods presented in this paper do not refer to the classifications of clones of two-element algebras due to Post. Instead the techniques based on the notion of a commutator equation, as well as on simple combinatorial arguments will be applied.

Before proving the theorem, some additional remarks are appropriate. Let x, y, z, w be different variables. $\mathbf{Q}^{\vDash}(x \approx y) \cap \mathbf{Q}^{\vDash}(z \approx w)$ is the set of all quaternary commutator equations of **Q** in the variables x, y and z, w. Any set of equations $\Delta = \Delta(x, y, z, w, \underline{u})$ such that $\mathbf{Q}^{\vDash}(\Delta(x, y, z, w, \underline{u})) = \mathbf{Q}^{\vDash}(x \approx y) \cap \mathbf{Q}^{\vDash}(z \approx w)$ is called a *generating* set for the equationally-defined commutator of **Q**. Since **Q** is finitely generated, there is a finite generating set Δ (Theorem 8.1.1 of (Czelakowski, 2015)). Accordingly, we may assume Δ is a fixed finite generating set.

Lemma 3.2. *Let* **Q** *be generated by a finite set of two-element algebras. Then* **Q** *possesses a finite generating set* $\Delta(x, y, z, w)$ *in four variables only (without parameters).*

Proof. Let $\mathbf{Q} = \boldsymbol{SP}(\mathbf{K})$, where **K** is a finite family of two-element algebras. **Q** has a finite generating set $\Delta' = \Delta'(x, y, z, w, u_1, \ldots, u_k)$, possibly with parameters. We then define new set of equations $\Delta'' = \Delta''(x, y, z, w)$ in the four variables x, y, z, w only. The idea is to substitute binary terms t in x, y for the parameters $u_1, \ldots, u_k$ occurring in the equations of Δ'. Accordingly, we define:

$$\Delta''(x, y, z, w) := \bigcup \{ \Delta'(x, y, z, w, u_1/t_1, \ldots, u_k/t_k) :$$

$$t_1, \ldots, t_k \text{ are arbitrary terms in at most the variables } x, y \}.$$

Δ'' is a set of quaternary commutator equations of **K**. As Δ' is a generating set and $\mathbf{Q}^{\vDash} = \mathbf{K}^{\vDash}$, we obtain that $\Delta'' \subseteq \mathbf{K}^{\vDash}(\Delta')$. But we also have that $\Delta' \subseteq \mathbf{K}^{\vDash}(\Delta'')$. (One directly works with homomorphisms h in the algebras of **K** and uses the fact that for every parameter u, the value hu is equal to ht for some term t involving at most x, y.) It follows that $\mathbf{K}^{\vDash}(\Delta'') = \mathbf{K}^{\vDash}(\Delta')$.

Since Δ' is finite and $\mathbf{K}^{\models}$ finitary, there is a finite set $\Delta \subset \Delta''$ such that $\mathbf{K}^{\models}(\Delta) = \mathbf{K}^{\models}(\Delta')$. The lemma follows. $\square$

Note. A slight modification of the above proof shows that if $\mathbf{Q}$ is a quasivariety generated by a finite family of finite, at most four-generated algebras, then the equationally-defined commutator of $\mathbf{Q}$ possesses a finite, parameter-free generating set $\Delta(x,y,z,w)$. $\square$

In what follows we shall assume that $\Delta(x,y,z,w)$ is a finite parameter-free generating set for the equationally-defined commutator for $\mathbf{Q}$. (In the reasonings we shall carry out, the assumption that a generating set does not contain parameters is unnecessary but it will simplify notational issues.)

The structurality of the consequence operation $\mathbf{Q}^{\models}$ implies that

$$(1) \qquad \mathbf{Q}^{\models}(\Delta(x,y,z,w)) = \mathbf{Q}^{\models}(\Delta(x,y,w,z)) = \mathbf{Q}^{\models}(\Delta(y,x,z,w)) =$$
$$\mathbf{Q}^{\models}(\Delta(y,x,w,z)) = \mathbf{Q}^{\models}(\Delta(z,w,x,y)).$$

Let $p(x,y,z,w) \approx q(x,y,z,w)$ be an equation and $\alpha,\beta,\gamma,\delta$ arbitrary terms. We form the equation $p(\alpha,\beta,\gamma,\delta) \approx q(\alpha,\beta,\gamma,\delta)$ obtained from $p \approx q$ by means of the substitution x/α, y/β, z/γ, w/δ.

Let $h : \boldsymbol{Te} \to \boldsymbol{A}$ be a homomorphism. The satisfiability of $p(\alpha,\beta,\gamma,\delta) \approx q(\alpha,\beta,\gamma,\delta)$ in the algebra $\boldsymbol{A}$ under h entirely depends on the values $h\alpha$, $h\beta$, $h\gamma$, $h\delta$. It follows that the satisfiability of $p(\alpha,\beta,\gamma,\delta) \approx q(\alpha,\beta,\gamma,\delta)$ in $\boldsymbol{A}$ under h is equivalent to the satisfiablity of the equation $p(x,y,z,w) \approx q(x,y,z,w)$ under an appropriate assignment of the elements of $\boldsymbol{A}$ to the variables x,y,z,w, viz. under the mapping which assigns $h\alpha$ to x, $h\beta$ to y, $h\gamma$ to z, and $h\delta$ to w.

We define the set of equations $\Delta(\alpha,\beta,\gamma,\delta)$ by making the above substitution in all equations $p \approx q$ of Δ. It follows from the above remarks that a homomorphism $h : \boldsymbol{Te} \to \boldsymbol{2}$ validates the equations of $\Delta(\alpha,\beta,\gamma,\delta)$ in $\boldsymbol{2}$ if and only if the mapping which assigns $h\alpha$ to x, $h\beta$ to y, $h\gamma$ to z and $h\delta$ to w validates the set of equations $\Delta(x,y,z,w)$ in $\boldsymbol{2}$. This trivial observation is crucial in the further reasoning.

To simplify the notation, we shall mark by

$$\Delta(a,b,c,d)$$

the sentence stating that a homomorphism $h : \boldsymbol{Te} \to \boldsymbol{2}$ such that $a = hx$, $b = hy$, $c = hz$, $d = hw$ validates the equations of $\Delta(x,y,z,w)$ in $\boldsymbol{2}$, when $\boldsymbol{2}$ is clear from context. $\Delta(a,b,c,d)$ is thus either true or false. In view of (1), there then hold in $\boldsymbol{2}$ the following equivalences:

$$(2) \qquad \Delta(a,b,c,d) \Leftrightarrow \Delta(a,b,d,c) \Leftrightarrow \Delta(b,a,c,d) \Leftrightarrow \Delta(b,a,d,c) \Leftrightarrow$$
$$\Delta(b,a,d,c) \Leftrightarrow \Delta(c,d,a,b),$$

for all a,b,c,d.

Let $\mathbf{Q} = \boldsymbol{SP}(\mathbf{K})$, where $\mathbf{K}$ is a finite family of two-element algebras. To simplify notation, we put $C := \mathbf{Q}^{\models}$. We shall first prove the following lemma, being a necessary condition for the equationally-defined commutator to be additive:

Lemma 3.3. *The consequence operation C validates, for arbitrary positive integers m and n, the following equation:*

> *For any disjoint finite sequences $\underline{x}_m, \underline{y}_m, \underline{z}_n, \underline{w}_n$ of pairwise different individual variables, where $\underline{x}_m = x_1, \ldots, x_m$, $\underline{y}_m = y_1, \ldots, y_m$, $\underline{z}_n = z_1, \ldots, z_n$, $\underline{w}_n = w_1, \ldots, w_n$,*

$$(\text{EqDistr})_{m,n} \quad C(\underline{x}_m \approx \underline{y}_m) \cap C(\underline{z}_n \approx \underline{w}_n) =$$
$$C\left(\bigcup_{1 \leqslant i \leqslant m, 1 \leqslant j \leqslant n} C(x_i \approx y_i) \cap C(z_j \approx w_j) \right).$$

Proof. $(\text{EqDistr})_{m,n}$ for $m \geqslant 1$ and $n \geqslant 1$, are certain restricted laws of distributivity tailored for simplest atomic equations. Conditions $(\text{EqDistr})_{m,n}$ do not continue to hold (with the exception of the trivial case $m = 1$ and $n = 1$) if the individual variables occurring in these laws are uniformly replaced by arbitrary terms.

We shall first consider the case when $m = 2$ and $n = 1$. We must prove (omitting some subscripts):

$$(\text{EqDistr})_{2,1} \quad C(x_1 \approx y_1, x_2 \approx y_2) \cap C(z \approx w) =$$
$$C(C(x_1 \approx y_1) \cap C(z \approx w) \cup C(x_2 \approx y_2) \cap C(z \approx w)).$$

The proof of $(\text{EqDistr})_{2,1}$ will give clues about the proof in the general case.

Let $\Delta_2 = \Delta_2(x_1, y_1, x_2, y_2; z, w)$ be a set of equations (in the variables x_1, y_1, x_2, y_2, z, w) such that

$$C(\Delta_2(x_1, y_1, x_2, y_2; z, w)) = C(x_1 \approx y_1, x_2 \approx y_2) \cap C(z \approx w).$$

Such a set (without parameters) exists. (To show this, suitably modify the proof of Lemma 3.2.) Thus parametric variables will be discarded in further reasonings.) Structurality of C yields equations for Δ_2 analogous to (1), that is.

$$(3) \qquad C(\Delta_2(x_1, y_1, x_2, y_2; z, w)) = C(\Delta_2(x_2, y_2, x_1, y_1; z, w)) =$$
$$C(\Delta_2(x_1, y_1, y_2, x_2; z, w)) = C(\Delta_2(y_1, x_1, x_2, y_2; z, w)) =$$
$$C(\Delta_2(y_1, x_1, y_2, x_2; z, w)) = \text{ etc.}$$

But structurality also gives that

$$(3a) \qquad C(\Delta_2(x_1, y_1, x_1, y_1; z, w)) = C(x_1 \approx y_1) \cap C(z \approx w)$$

and

(3b) $\qquad C(\Delta_2(x_2,y_2,x_2,y_2;z,w)) = C(x_2 \approx y_2) \cap C(z \approx w).$

As to (3a), $\Delta_2(x_1,y_1,x_1,y_1;z,w)$ is a set of commutator equations in x_1,y_1 and z, w. Hence $C(\Delta_2(x_1,y_1,x_1,y_1;z,w)) \subseteq C(x_1 \approx y_1) \cap C(z \approx w)$. Let $\Delta(x_1,y_1;z,w)$ be a set of equations such that $C(x_1 \approx y_1) \cap C(z \approx w) = C(\Delta(x_1,y_1;z,w))$. Evidently, $\Delta(x_1,y_1;z,w) \subseteq C(x_1 \approx y_1, x_2 \approx y_2) \cap C(z \approx w) = C(\Delta_2(x_1,y_1,x_2,y_2;z,w))$. Hence, by structurality, $\Delta(x_1,y_1;z,w) \subseteq C(\Delta_2(x_1,y_1,x_1,y_1;z,w))$. This gives that $C(x_1 \approx y_1) \cap C(z \approx w) \subseteq C(\Delta_2(x_1, y_1,x_1,y_1;z,w))$. (3a) follows. (3b) is similarly checked.

We must therefore prove that

(4) $\quad \Delta_2(x_1,y_1,x_2,y_2;z,w) \subseteq C(\Delta_2(x_1,y_1,x_1,y_1;z,w) \cup \Delta_2(x_2,y_2,x_2,y_2;z,w)),$

because the reverse inclusion is immediate. Suppose (4) does not hold. Hence there is a homomorphism $h : \boldsymbol{Te} \to \mathbf{2}$ in some algebra $\mathbf{2} \in \mathbf{K}$ which *validates* the equations of $\Delta_2(x_1,y_1,x_1,y_1;z,w) \cup \Delta_2(x_2,y_2,x_2,y_2;z,w)$ and *falsifies* an equation belonging to $\Delta_2(x_1,y_1,x_2,y_2;z,w)$. We assume that $\{\mathbf{0},\mathbf{1}\}$ is the carrier of $\mathbf{2}$.

As h does *not* validate $\Delta_2(x_1,y_1,x_2,y_2;z,w)$, it follows that

(∗) $\qquad (hx_1 \neq hy_1 \text{ or } hx_2 \neq hy_2) \text{ and } hz \neq hw,$

because otherwise h would validate Δ_2 due to the definition of Δ_2.

We shall first consider the case when $hz = \mathbf{1}$ and $hw = \mathbf{0}$. Following (∗), we shall consider the following three cases.

Case A. $hx_1 \neq hy_1$ and $hx_2 \neq hy_2$.

Then $\Delta_2(hx_1,hy_1,hx_2,hy_2;hz,hw)$ reduces to one of the following conditions:

(5)(a) $\Delta_2(1,0,1,0;1,0), \Delta_2(1,0,0,1;1,0), \Delta_2(0,1,1,0;1,0), \Delta_2(0,1,0,1;1,0).$

Case B. $hx_1 = hy_1$ and $hx_2 \neq hy_2$.

Then $\Delta_2(hx_1,hy_1,hx_2,hy_2;hz,hw)$ reduces to one of the following conditions:

(5)(b) $\Delta_2(1,1,1,0;1,0), \Delta_2(0,0,1,0;1,0), \Delta_2(1,1,0,1;1,0), \Delta_2(0,0,0,1;1,0).$

Case C. $hx_1 \neq hy_1$ and $hx_2 = hy_2$.

Then $\Delta_2(hx_1,hy_1,hx_2,hy_2;hz,hw)$ is one of the conditions:

(5)(c) $\Delta_2(1,0,1,1;1,0), \Delta_2(1,0,0,0;1,0), \Delta_2(0,1,1,1;1,0), \Delta_2(0,1,0,0;1,0).$

Thus at least one condition *among* the above conditions listed in (5)(a) + (5)(b) + (5)(c) is false.

On the other hand, h *validates* $\Delta_2(x_1, y_1, x_1, y_1; z, w) \cup \Delta_2(x_2, y_2, x_2, y_2; z, w)$, i.e., h validates $\Delta_2(x_1, y_1, x_1, y_1; z, w)$ *and* h validates $\Delta_2(x_2, y_2, x_2, y_2; z, w)$. The condition $\Delta_2(hx_1, hy_1, hx_1, hy_1; hz, hw) \wedge \Delta_2(hx_2, hy_2, hx_2, hy_2; hz, hw)$ cannot take the form in which $hx_1 = hy_1$ and $hx_2 = hy_2$ or $hz = hw$, because h would then validate $\Delta_2(x_1, y_1, x_2, y_2; z, w)$, which is assumed not to hold. Consequently, the *true* condition $\Delta_2(hx_1, hy_1, hx_1, hy_1; hz, hw) \wedge \Delta_2(hx_2, hy_2, hx_2, hy_2; hz, hw)$ takes one of the following three forms that are parallel to the above cases.

Case A. $hx_1 \neq hy_1$ and $hx_2 \neq hy_2$.

Then at least one of the following conjunctions is true:

$$(6)(a) \qquad \Delta_2(1,0,1,0;1,0) \wedge \Delta_2(1,0,1,0;1,0),$$
$$\Delta_2(1,0,1,0;1,0) \wedge \Delta_2(0,1,0,1;1,0),$$
$$\Delta_2(0,1,0,1;1,0) \wedge \Delta_2(1,0,1,0;1,0),$$
$$\Delta_2(0,1,0,1;1,0) \wedge \Delta_2(0,1,0,1;1,0).$$

We may simplify these conditions (because e.g. in some conditions conjuncts are repeated). But for our purposes it is unnecessary.

If at least one of the conditions listed in (6)(a) is true, they *all* are true, because they are equivalent due to the properties of $\Delta_2(x_1, y_1, x_2, y_2; z, w)$ exhibited in (3).

Case B. $hx_1 = hy_1$ and $hx_2 \neq hy_2$.

Then at least one of the following conjunctions is true:

$$(6)(b) \qquad \Delta_2(1,1,1,1;1,0) \wedge \Delta_2(1,0,1,0;1,0),$$
$$\Delta_2(1,1,1,1;1,0) \wedge \Delta_2(0,1,0,1;1,0),$$
$$\Delta_2(0,0,0,0;1,0) \wedge \Delta_2(1,0,1,0;1,0),$$
$$\Delta_2(0,0,0,0;1,0) \wedge \Delta_2(0,1,0,1;1,0).$$

Again, if at least one of the conditions listed in (6)(b) is true, they are *all* true. Indeed, the first conjuncts of these conditions are all true and the second conjuncts are equivalent in virtue of (3).

Case C. $hx_1 \neq hy_1$ and $hx_2 = hy_2$.

Then at least one of the following is true:

$$(6)(c) \qquad \Delta_2(1,0,1,0;1,0) \wedge \Delta_2(1,1,1,1;1,0),$$
$$\Delta_2(1,0,1,0;1,0) \wedge \Delta_2(0,0,0,0;1,0),$$
$$\Delta_2(0,1,0,1;1,0) \wedge \Delta_2(1,1,1,1;1,0),$$
$$\Delta_2(0,1,0,1;1,0) \wedge \Delta_2(0,0,0,0;1,0).$$

Again, if at least one of the conditions of (6)(c) is true, they are *all* true, because the second conjuncts of these conditions are all true and the first conjuncts are equivalent.

If Case A holds, holds, then at least one of the conditions of (5)(a) is false. Then notice that *all* conditions of (5)(a) are *false*, because, in virtue of (3), they are equivalent. It then follows that the conjuctions listed in (6)(a) are all false as well. This contradicts the fact that the conditions of (6)(a) are all true.

To handle Cases B and C, some other facts are needed.

The structurality of C implies that

$$(7) \qquad \Delta_2(x_1, x_1, x_2, y_2; z, w) \subseteq C(\Delta_2(x_2, y_2, x_2, y_2; z, w)).$$

Indeed, we have that $C(\Delta_2(x_2, y_2, x_2, y_2; z, w)) = C(x_2 \approx y_2) \cap C(z \approx w)$, by (3a). But $\Delta_2(x_1, x_1, x_2, y_2; z, w)$ is a set of quaternary commutator equations in $x_2, y_2; z, w$. Therefore $\Delta_2(x_1, x_1, x_2, y_2; z, w) \subseteq C(x_2 \approx y_2) \cap C(z \approx w)$. (7) then follows.

By a similar argument we also get:

$$(8) \qquad \Delta_2(x_1, y_1, x_2, x_2; z, w) \subseteq C(\Delta_2(x_1, y_1, x_1, y_1; z, w)).$$

If Case B holds, one of the conditions listed in (5)(b) is false and the conditions of (6)(b) are all true.

(7) entails the following true implications:

$$\Delta_2(1, 0, 1, 0; 1, 0) \text{ implies } \Delta_2(1, 1, 1, 0; 1, 0)$$
$$\Delta_2(1, 0, 1, 0; 1, 0) \text{ implies } \Delta_2(0, 0, 1, 0; 1, 0)$$
$$\Delta_2(0, 1, 0, 1; 1, 0) \text{ implies } \Delta_2(1, 1, 0, 1; 1, 0)$$
$$\Delta_2(0, 1, 0, 1; 1, 0) \text{ implies } \Delta_2(0, 0, 0, 1; 1, 0)$$

Since in virtue of (6)(b), the antecedents of the above implications are true, it follows that their succeedents are true as well. But this means that all conditions listed in (5)(b) are true. This contradicts the assumption that some condition of (5)(b) is false.

If Case C holds, one of the conditions listed in (5)(c) is false and the conditions of (6)(c) are all true. To get a contradiction, a similar argument is applied. (8) yields the following true implications:

$$\Delta_2(1, 0, 1, 0; 1, 0) \text{ implies } \Delta_2(1, 0, 1, 1; 1, 0)$$
$$\Delta_2(1, 0, 1, 0; 1, 0) \text{ implies } \Delta_2(1, 0, 0, 0; 1, 0)$$
$$\Delta_2(0, 1, 0, 1; 1, 0) \text{ implies } \Delta_2(0, 1, 1, 1; 1, 0)$$
$$\Delta_2(0, 1, 0, 1; 1, 0) \text{ implies } \Delta_2(0, 1, 0, 0; 1, 0)$$

Since in virtue of (6)(c), the antecedents of the above implications are true, it follows that their succedents are true as well. But this means that all conditions listed in (5)(c) are true. This contradicts the assumption that some condition of (5)(c) is false.

Thus in the above three cases we arrive at a contradiction.

In the above reasoning it was assumed that $hz = \mathbf{1}$ and $hw = \mathbf{0}$. But the situation when $hz = \mathbf{0}$ and $hw = \mathbf{1}$ is fully symmetric. In this case we also get a contradiction.

We now pass to the proof of $(\mathrm{EqDistr})_{m,n}$ for all $m \geqslant 1$ and $n \geqslant 1$.

Let $m \geqslant 1$. We assume that $(\mathrm{EqDistr})_{m+1,1}$ holds. We then show that $(\mathrm{EqDistr})_{m+2,1}$. Let $x_1 \approx y_1,\ x_2 \approx y_2,\ x_3 \approx y_3, \ldots, x_{m+2} \approx y_{m+2},\ z \approx w$ be equations of different variables. Put $\underline{x} := x_3, \ldots, x_{m+2}$, $\underline{y} := y_3, \ldots, y_{m+2}$. Let $\Delta(\underline{x}, \underline{y}, x_1, y_1, x_2, y_2; z, w)$ be a set of equations in the variables $\underline{x}, x_1, x_2, \underline{y}, y_1, y_2; z, w$ such that

$$(1)_{m+2,1} \qquad C(\Delta(\underline{x}, \underline{y}, x_1, y_1, x_2, y_2; z, w)) =$$
$$C(\underline{x} \approx \underline{y}, x_1 \approx y_1, x_2 \approx y_2) \cap C(z \approx w).$$

Structurality of C gives that

$$(2\mathrm{a})_{m+1,1} \qquad C(\Delta(\underline{x}, \underline{y}, x_1, y_1, x_1, y_1; z, w)) = C(\underline{x} \approx \underline{y}, x_1 \approx y_1) \cap C(z \approx w),$$
$$(2\mathrm{b})_{m+1,1} \qquad C(\Delta(\underline{x}, \underline{y}, x_2, y_2, x_2, y_2; z, w)) = C(\underline{x} \approx \underline{y}, x_2 \approx y_2) \cap C(z \approx w).$$

Suitably modifying the above proof of $(\mathrm{EqDistr})_{2,1}$ one proves that

$$(3)_{m+2,1} \qquad C(\Delta(\underline{x}, \underline{y}, x_1, y_1, x_2, y_2; z, w)) =$$
$$C(\Delta(\underline{x}, \underline{y}, x_1, y_1, x_1, y_1; z, w) \cup \Delta(\underline{x}, \underline{y}, x_2, y_2, x_2, y_2; z, w)).$$

From $(3)_{m+2,1}$ and $(2\mathrm{a})_{m+1,1}$, $(2\mathrm{b})_{m+1,1}$ it follows that

$$(4)_{m+2,1} \qquad C(\underline{x} \approx \underline{y}, x_1 \approx y_1, x_2 \approx y_2) \cap C(z \approx w) =$$
$$C(C(\underline{x} \approx \underline{y}, x_1 \approx y_1) \cap C(z \approx w) \cup C(\underline{x} \approx \underline{y}, x_2 \approx y_2) \cap C(z \approx w)).$$

But as $(\mathrm{EqDistr})_{m+1,1}$ holds by IH, we have that

$$C(\underline{x} \approx \underline{y}, x_1 \approx y_1) \cap C(z \approx w) =$$
$$C\left(\bigcup_{3 \leqslant i \leqslant m} C(x_i \approx y_i) \cap C(z \approx w) \cup C(x_1 \approx y_1) \cap C(z \approx w) \right)$$

and

$$C(\underline{x} \approx \underline{y}, x_2 \approx y_2) \cap C(z \approx w) =$$
$$C\left(\bigcup_{3 \leqslant i \leqslant m} C(x_i \approx y_i) \cap C(z \approx w) \cup C(x_2 \approx y_2) \cap C(z \approx w) \right).$$

Combining the last two equalities with $(4)_{m+2,1}$ we obtain that

$$C(\underline{x} \approx \underline{y}, x_1 \approx y_1, x_2 \approx y_2) \cap C(z \approx w) =$$
$$C(\bigcup_{1 \leqslant i \leqslant m} C(x_i \approx y_i) \cap C(z \approx w) \cup C(x_1 \approx y_1) \cap C(z \approx w)).$$

So $(\mathrm{EqDistr})_{m+2,1}$ also holds. This shows $(\mathrm{EqDistr})_{m,1}$ for all positive m.

Fix m. We shall prove $(\mathrm{EqDistr})_{m,n}$ for all positive n. The case $(\mathrm{EqDistr})_{m,1}$ is already established. We assume $(\mathrm{EqDistr})_{m,n}$. We show $(\mathrm{EqDistr})_{m,n+1}$ for $n \geqslant 1$.

We consider sets of equations $x_1 \approx y_1, \ldots, x_m \approx y_m$ and $z_1 \approx w_1, z_2 \approx w_2, z_3 \approx w_3, \ldots, z_n \approx y_n, z_{n+1} \approx y_{n+1}$. We mark the set $x_1 \approx y_1, \ldots, x_m \approx y_m$ by $\underline{x} \approx \underline{y}$ and the set $z_3 \approx w_3, \ldots, z_n \approx y_n, z_{n+1} \approx y_{n+1}$ by $\underline{z} \approx \underline{w}$. (This set is empty when $n = 2$.)

Let $\Delta(\underline{x}, \underline{y}; z_1, w_1, z_2, w_2, \underline{z}, \underline{w})$ be a set of equations of terms in the above variables such that

$$(5)_{m,n+1} \qquad C(\Delta(\underline{x}, \underline{y}; \underline{z}, \underline{w}, z_1, w_1, z_2, w_2)) =$$
$$C(\underline{x} \approx \underline{y}) \cap C(\underline{z} \approx \underline{w}, z_1 \approx w_1, z_2 \approx w_2).$$

Structurality of C gives that

$$(6a)_{m,n} \qquad C(\Delta(\underline{x}, \underline{y}; \underline{z}, \underline{w}, z_1, w_1, z_1, w_1)) = C(\underline{x} \approx \underline{y}) \cap C(\underline{z} \approx \underline{w}, z_1 \approx w_1),$$
$$(6b)_{m,n} \qquad C(\Delta(\underline{x}, \underline{y}; \underline{z}, \underline{w}, z_2, w_2, z_2, w_2)) = C(\underline{x} \approx \underline{y}) \cap C(\underline{z} \approx \underline{w}, z_2 \approx w_2).$$

By way of an appropriate modification of the proof of $(\mathrm{EqDistr})_{2,1}$ one proves that

$$(7)_{m,n+1} \qquad C(\Delta(\underline{x}, \underline{y}; \underline{z}, \underline{w}, z_1, w_1, z_2, w_2)) =$$
$$C(\Delta(\underline{x}, \underline{y}; \underline{z}, \underline{w}, z_1, w_1, z_1, w_1) \cup \Delta(\underline{x}, \underline{y}; \underline{z}, \underline{w}, z_2, w_2, z_2, w_2)).$$

From $(7)_{m,n+1}$ and $(6a)_{m,n}$, $(6b)_{m,n}$ it follows that

$$(8)_{m,n+1} \quad C(\underline{x} \approx \underline{y}) \approx C(\underline{z} \approx \underline{w}, z_1 \approx w_1, z_2 \approx w_2) =$$
$$C(C(\underline{x} \approx \underline{y}) \cap C(\underline{z} \approx \underline{w}, z_1 \approx w_1) \cup C(\underline{x} \approx \underline{y}) \cap C(\underline{z} \approx \underline{w}, z_2 \approx w_2)).$$

But as $(\mathrm{EqDistr})_{m,n}$ holds by IH, we have that

$$C(\underline{x} \approx \underline{y}) \cap C(\underline{z} \approx \underline{w}, z_1 \approx w_1) =$$
$$C(\bigcup_{1 \leqslant i \leqslant m, 3 \leqslant j \leqslant n+1} C(x_i \approx y_i) \cap C(z_j \approx w_j) \cup \bigcup_{1 \leqslant i \leqslant m} C(x_i \approx y_i) \cap C(z_1 \approx w_1))$$

and

$$C(\underline{x} \approx \underline{y}) \cap C(\underline{z} \approx \underline{w}, z_2 \approx w_2) =$$

$$C\left(\bigcup_{1 \leqslant i \leqslant m, 3 \leqslant j \leqslant n+1} C(x_i \approx y_i) \cap C(z_j \approx w_j) \cup \bigcup_{1 \leqslant i \leqslant m} C(x_i \approx y_i) \cap C(z_2 \approx w_2) \right).$$

Combining the last two equalities with $(8)_{m,n+1}$ we obtain that

$$C(\underline{x} \approx \underline{y}) \cap C(\underline{z} \approx \underline{w}, z_1 \approx w_1, z_2 \approx w_2) =$$

$$C\left(\bigcup_{1 \leqslant i \leqslant m, 1 \leqslant j \leqslant n+1} C(x_i \approx y_i) \cap C(z_j \approx w_j) \right).$$

So $(\mathrm{EqDistr})_{m,n+1}$ holds. This concludes the proof of the lemma. $\qquad\square$

The proof of Lemma 3.3 does not refer to intrinsic properties of two-element algebras. The lemma is purely combinatorial, because its proof counts the numbers of pertinent possibilities directly resulting from the fact that the carrier of the algebra has two-elements. One may pose a natural question whether the reasoning presented in Lemma 3.3 carries over to finite algebras of cardinalities greater than 2. We leave this combinatorial question open here.

The above lemma and Proposition 5.2.13 of (Czelakowski, 2015) imply:

Corollary 3.4. *Assume that* $\mathbf{Q} = \boldsymbol{SP}(\mathbf{K})$ *for some finite family* $\mathbf{K}$ *of two-element algebras. Then for every algebra* $\boldsymbol{A} \in \mathbf{Q}$ *and any* $\Phi, \Psi \in Con_{\mathbf{Q}}(\boldsymbol{A})$,

$$[\Phi, \Psi] = \sup\{[\Theta_{\mathbf{Q}}(a,b), \Theta_{\mathbf{Q}}(c,d)] : a \equiv b\,(\Phi), c \equiv d\,(\Psi)\}$$

in the algebra $\boldsymbol{A}$. $\qquad\square$

(Here the supremum is taken in the lattice $\boldsymbol{Con_{\mathbf{Q}}}(\boldsymbol{A})$.)

The property of the equationally-defined commutator shown in Corollary 3.4 is weaker than additivity. To prove additivity we need the following fact.

Lemma 3.5. *Assume* $\mathbf{Q} = \boldsymbol{SP}(\mathbf{K})$ *for some finite family* $\mathbf{K}$ *of two-element algebras. Let* $C := \mathbf{Q}^{\models}$ *and* $\Delta(x,y,z,w)$ *be a finite generating set. Assume that*

$$(r) \qquad\qquad \alpha \approx \beta \in C(\alpha_1 \approx \beta_1, \ldots, \alpha_n \approx \beta_n).$$

Then for any variables z, w *not occurring in the equations* $\alpha_1 \approx \beta_1, \approx, \alpha_n \approx \beta_n,$ $\alpha \approx \beta$ *it holds:*

$$(r_\Delta) \qquad\qquad \Delta(\alpha, \beta, z, w) \subseteq C(\Delta(\alpha_1, \beta_1, z, w) \cup \ldots \cup \Delta(\alpha_n, \beta_n, z, w)).$$

Proof (of the lemma). Suppose that (r_Δ) does *not* hold. There is an algebra $\mathbf{2} \in \mathbf{K}$ and a homomorphism $h : \boldsymbol{Te} \to \mathbf{2}$ which *validates* the equations of $\Delta(\alpha_1, \beta_1, z, w) \cup \ldots \cup \Delta(\alpha_n, \beta_n, z, w)$ and *falsifies* some equation belonging to

$\Delta(\alpha, \beta, z, w)$. As above, we assume that $\{\mathbf{0}, \mathbf{1}\}$ is the carrier of $\mathbf{2}$. As h does not validate $\Delta(\alpha, \beta, z, w)$, the properties of generating sets (they were discussed in the introduction) imply that $h\alpha \neq h\beta$ and $hz \neq hw$. It follows that the situation when some equation of $\Delta(\alpha, \beta, z, w)$ is falsified by h occurs exactly when the false condition $\Delta(h\alpha, h\beta, hz, hw)$ takes one of the forms:

$$(*) \qquad \Delta(\mathbf{1}, \mathbf{0}, \mathbf{1}, \mathbf{0}), \Delta(\mathbf{1}, \mathbf{0}, \mathbf{0}, \mathbf{1}), \Delta(\mathbf{0}, \mathbf{1}, \mathbf{1}, \mathbf{0}), \Delta(\mathbf{0}, \mathbf{1}, \mathbf{0}, \mathbf{1}).$$

In virtue of (2), these four conditions are equivalent and therefore they *all* are *false*.

On the other hand, (r) and $h\alpha \neq h\beta$ imply that h *falsifies* at least one equation $\alpha_i \approx \beta_i$ for some i ($1 \leqslant i \leqslant n$), i.e., $h\alpha_i \neq h\beta_i$. As h validates the set $\Delta(\alpha_1, \beta_1, z, w) \cup \ldots \cup (\alpha_n, \beta_n, z, w)$ and $hz \neq hw$, therefore each of the *true* conditions $\Delta(h\alpha_i, h\beta_i, hz, hw)$ takes one of the forms $\Delta(\mathbf{1}, \mathbf{0}, \mathbf{1}, \mathbf{0}), \Delta(\mathbf{1}, \mathbf{0}, \mathbf{0}, \mathbf{1})$, $\Delta(\mathbf{0}, \mathbf{1}, \mathbf{1}, \mathbf{0}), \Delta(\mathbf{0}, \mathbf{1}, \mathbf{0}, \mathbf{1})$. According to (2), all of them are equivalent and therefore *true*. But these conditions are already listed in $(*)$ and all are *false*. We thus arrive at a contradiction. $\qquad\square$

Applying Theorem 5.2.3 of (Czelakowski, 2015) to Lemma 3.5 and Corollary 3.4 we obtain that the equationally-defined commutator for $\mathbf{Q}$ is additive. The theorem has been proved. $\qquad\square$

Theorem 3.6. *Let $\mathbf{Q}$ be a quasivariety generated by finite family of two-element algebras. Then equationally-defined commutator for $\mathbf{Q}$ satisfies the associativity law*

$$[[\boldsymbol{x_1}, \boldsymbol{x_2}], \boldsymbol{x_3}] \approx [\boldsymbol{x_1}, [\boldsymbol{x_2}, \boldsymbol{x_3}]].$$

Comments. $[[\boldsymbol{x_1}, \boldsymbol{x_2}], \boldsymbol{x_3}] \approx [\boldsymbol{x_1}, [\boldsymbol{x_2}, \boldsymbol{x_3}]]$ means that

$$[[\Phi_1, \Phi_2], \Phi_3] = [\Phi_1, [\Phi_2, \Phi_3]]$$

holds in any algebra $\boldsymbol{A} \in \mathbf{Q}$ and for any congruences $\Phi_1, \Phi_2, \Phi_3 \in Con_{\mathbf{Q}}(\boldsymbol{A})$.

Let $x_1, y_1, x_2, y_2, x_3, y_3$ be different individual variables from the first-order language $\boldsymbol{L}_\tau$ corresponding to τ. In view of additivity of the commutator, the equation

$$[\mathbf{Q}^\models (x_1 \approx y_1), [\mathbf{Q}^\models (x_2 \approx y_2), \mathbf{Q}^\models (x_3 \approx y_3)]] =$$
$$[[\mathbf{Q}^\models (x_1 \approx y_1), \mathbf{Q}^\models (x_2 \approx y_2)], \mathbf{Q}^\models (x_3 \approx y_3)]$$

(holding in the lattice of the equational theories of $\mathbf{Q}^\models$) already implies the commutator identity $[[\boldsymbol{x_1}, \boldsymbol{x_2}], \boldsymbol{x_3}] \approx [\boldsymbol{x_1}, [\boldsymbol{x_2}, \boldsymbol{x_3}]]$ in $\mathbf{Q}$. This follows from the remarks placed in Sections 3.2 and 5.2 of (Czelakowski, 2015). (The reverse implication also holds.)

Proof (of the theorem). As in the proof of the previous theorem, it is assumed that $\Delta(x, y, z, w)$ is a generating set of quarternary commutator equations for $\mathbf{Q}$, that is, it satisfies the condition: $\mathbf{Q}^{\models}(\Delta(x,y,z,w)) = \mathbf{Q}^{\models}(x \approx y) \cap \mathbf{Q}^{\models}(z \approx w)$. According to Theorem 9.2.1 and 9.2.2 of (Czelakowski, 2015), we have that

$$[\mathbf{Q}^{\models}(x_1 \approx y_1), [\mathbf{Q}^{\models}(x_2 \approx y_2), \mathbf{Q}^{\models}(x_3 \approx y_3)]] = \mathbf{Q}^{\models}(\Delta(x_1, y_1, \Delta(x_2, y_2, x_3, y_3)))$$

and

$$[[\mathbf{Q}^{\models}(x_1 \approx y_1), \mathbf{Q}^{\models}(x_2 \approx y_2)], \mathbf{Q}^{\models}(x_3 \approx y_3)] = \mathbf{Q}^{\models}(\Delta(\Delta(x_1, y_1, x_2, y_2), x_3, y_3)).$$

In view of the above remarks, it suffices to show that

$$(*) \qquad \mathbf{Q}^{\models}(\Delta(x_1, y_1, \Delta(x_2, y_2, x_3, y_3))) = \mathbf{Q}^{\models}(\Delta(\Delta(x_1, y_1, x_2, y_2), x_3, y_3)).$$

The proof of $(*)$ is based on two lemmas—Lemmas 3.7 and 3.8. We shall first prove that the theory on the RHS of $(*)$ is included in the theory on the LHS. After proving Lemma 3.7, we shall show the reverse inclusion (Lemma 3.8).

Lemma 3.7. $\Delta(\Delta(x_1, y_1, x_2, y_2), x_3, y_3) \subseteq \mathbf{Q}^{\models}(\Delta(x_1, y_1, \Delta(x_2, y_2, x_3, y_3)))$.

Proof (of the lemma). We have that $\mathbf{Q} = \mathbf{SP}(\mathbf{K})$ for some finite set $\mathbf{K}$ of two-element algebras. Suppose that the lemma does not hold. Hence there is an algebra $\mathbf{2} \in \mathbf{K}$ with the carrier $\{0, 1\}$ such that for some homomorphism $h : \mathbf{Te} \to \mathbf{2}$ it is the case that h validates the equations of $\Delta(x_1, y_1, \Delta(x_2, y_2, x_3, y_3))$ and it falsifies an equation

$$p(r(x_1, y_1, x_2, y_2), s(x_1, y_1, x_2, y_2), x_3, y_3) \approx$$
$$q(r(x_1, y_1, x_2, y_2), s(x_1, y_1, x_2, y_2), x_3, y_3)$$

belonging to $\Delta(\Delta(x_1, y_1, x_2, y_2), x_3, y_3)$. Hence

$$(1) \qquad p(r(hx_1, hy_1, hx_2, hy_2), s(hx_1, hy_1, hx_2, hy_2), hx_3, hy_3)$$
$$\neq q(r(hx_1, hy_1, hx_2, hy_2), s(hx_1, hy_1, hx_2, hy_2), hx_3, hy_3)$$

and as $p \approx q$ and $r \approx s$ are quaternary commutator equations, it follows that

$$(2) \qquad r(hx_1, hy_1, hx_2, hy_2) \neq s(hx_1, hy_1, hx_2, hy_2)$$

and

$$(3) \qquad hx_1 \neq hy_1, \; hx_2 \neq hy_2, \; hx_3 \neq hy_3.$$

We shall consider several cases and subcases implied by (1)–(3).

Case 1. $p(r(hx_1, hy_1, hx_2, hy_2), s(hx_1, hy_1, hx_2, hy_2), hx_3, hy_3) = \mathbf{1}$ and $q(r(hx_1, hy_1, hx_2, hy_2), s(hx_1, hy_1, hx_2, hy_2), hx_3, hy_3) = \mathbf{0}$.

Subcase 1.1. $r(hx_1, hy_1, hx_2, hy_2) = \mathbf{1}$ and $s(hx_1, hy_1, hx_2, hy_2) = \mathbf{0}$.

According to Subcase 1.1 and (3) we isolate further 8 (sub)subcases:

Subcase 1.1.1. $hx_1 = 1$, $hy_1 = 0$, $hx_2 = 1$, $hy_2 = 0$, $hx_3 = 1$, $hy_3 = 0$.

Subcase 1.1.2. $hx_1 = 1$, $hy_1 = 0$, $hx_2 = 1$, $hy_2 = 0$, $hx_3 = 0$, $hy_3 = 1$.

Subcase 1.1.3. $hx_1 = 1$, $hy_1 = 0$, $hx_2 = 0$, $hy_2 = 1$, $hx_3 = 1$, $hy_3 = 0$.

Subcase 1.1.4. $hx_1 = 1$, $hy_1 = 0$, $hx_2 = 0$, $hy_2 = 1$, $hx_3 = 0$, $hy_3 = 1$.

Subcase 1.1.5. $hx_1 = 0$, $hy_1 = 1$, $hx_2 = 1$, $hy_2 = 0$, $hx_3 = 1$, $hy_3 = 0$.

Subcase 1.1.6. $hx_1 = 0$, $hy_1 = 1$, $hx_2 = 1$, $hy_2 = 0$, $hx_3 = 0$, $hy_3 = 1$.

Subcase 1.1.7. $hx_1 = 0$, $hy_1 = 1$, $hx_2 = 0$, $hy_2 = 1$, $hx_3 = 1$, $hy_3 = 0$.

Subcase 1.1.8. $hx_1 = 0$, $hy_1 = 1$, $hx_2 = 0$, $hy_2 = 1$, $hx_3 = 0$, $hy_3 = 1$.

We shall separately analyse them. The general idea of the proof is to define equations of the form

$$(*) \qquad p'(x_1, y_1, r'(x_2, y_2, x_3, y_3), s'(x_2, y_2, x_3, y_3)) \approx$$
$$q'(x_1, y_1, r'(x_2, y_2, x_3, y_3), s'(x_2, y_2, x_3, y_3)),$$

a separate equation for each subcase 1.1.1—1.1.8, and to show that the above homomorphism h, pertinent to each such subcase, falsifies $(*)$, that is, it assigns **1** to the left hand side of $(*)$ and **0** to the right hand side.

Subcase 1.1.1. $hx_1 = 1$, $hy_1 = 0$, $hx_2 = 1$, $hy_2 = 0$, $hx_3 = 1$, $hy_3 = 0$.

We define:

$$r'(x_2, y_2, x_3, y_3) := r(x_2, y_2, x_3, y_3), \quad s'(x_2, y_2, x_3, y_3) := s(x_2, y_2, x_3, y_3),$$
$$p'(x, y, z, w) := p(z, w, x, y), \quad q'(x, y, z, w) := q(z, w, x, y),$$

and form the equation

$$(1.1.1) \qquad p'(x_1, y_1, r'(x_2, y_2, x_3, y_3), s'(x_2, y_2, x_3, y_3)) \approx$$
$$q'(x_1, y_1, r'(x_2, y_2, x_3, y_3), s'(x_2, y_2, x_3, y_3))$$

This equation belongs to $\mathbf{Q}^{\models}(\Delta(x_1, y_1, \Delta(x_2, y_2, x_3, y_3)))$. We then compute the value of (1.1.1) under h. We have:

$(1.1.1)_{\mathrm{L}} \quad h(p'(x_1, y_1, r'(x_2, y_2, x_3, y_3), s'(x_2, y_2, x_3, y_3))) =$

$p'(hx_1, hy_1, r'(hx_2, hy_2, hx_3, hy_3), s'(hx_2, hy_2, hx_3, hy_3)) =$

$p(r'(hx_2, hy_2, hx_3, hy_3), s'(hx_2, hy_2, hx_3, hy_3), hx_1, hy_1) =$

$p(r(hx_2, hy_2, hx_3, hy_3), s(hx_2, hy_2, hx_3, hy_3), hx_1, hy_1) =$ (by Subcase 1.1.1)

$p(r(hx_1, hy_1, hx_2, hy_2), s(hx_1, hy_1, hx_2, hy_2), hx_3, hy_3) = 1$ (by Case 1)

Analogously,

$(1.1.1)_{\mathrm{R}}$ $h(q'(x_1, y_1, r'(x_2, y_2, x_3, y_3), s'(x_2, y_2, x_3, y_3))) =$

$\quad q'(hx_1, hy_1, r'(hx_2, hy_2, hx_3, hy_3), s'(hx_2, hy_2, hx_3, hy_3)) =$

$\quad q(r'(hx_2, hy_2, hx_3, hy_3), s'(hx_2, hy_2, hx_3, hy_3), hx_1, hy_1) =$

$\quad q(r(hx_2, hy_2, hx_3, hy_3), s(hx_2, hy_2, hx_3, hy_3), hx_1, hy_1) =$(by Subcase 1.1.1)

$\quad q(r(hx_1, hy_1, hx_2, hy_2), s(hx_1, hy_1, hx_2, hy_2), hx_3, hy_3) = \mathbf{0}$ (by Case 1).

As h validates the equations of $\Delta(x_1, y_1, \Delta(x_2, y_2, x_3, y_3))$, it also validates the equation (1.1.1). But in virtue of $(1.1.1)_{\mathrm{L}}$ and $(1.1.1)_{\mathrm{R}}$, h falsifies (1.1.1). A contradiction.

The other subcases are similarly handled. The details are omitted.

Subcase 1.1.2. $hx_1 = 1$, $hy_1 = 0$, $hx_2 = 1$, $hy_2 = 0$, $hx_3 = 0$, $hy_3 = 1$.

We define:

$$r'(x_2, y_2, x_3, y_3) := r(x_2, y_2, y_3, x_3), \quad s'(x_2, y_2, x_3, y_3) := s(x_2, y_2, y_3, x_3),$$
$$p'(x, y, z, w) := p(z, w, y, x), \quad\quad\quad q'(x, y, z, w) := q(z, w, y, x),$$

and form the equation

$$(1.1.2) \quad\quad p'(x_1, y_1, r'(x_2, y_2, x_3, y_3), s'(x_2, y_2, x_3, y_3)) \approx$$
$$q'(x_1, y_1, r'(x_2, y_2, x_3, y_3), s'(x_2, y_2, x_3, y_3))$$

This equation belongs to $\mathbf{Q}^{\vDash}(\Delta(x_1, y_1, \Delta(x_2, y_2, x_3, y_3)))$. Then proceeding as in the above case we compute the value of (1.1.2) under h. We have:

$(1.1.2)_{\mathrm{L}}$ $h(p'(x_1, y_1, r'(x_2, y_2, x_3, y_3), s'(x_2, y_2, x_3, y_3))) =$

$\quad p(r(hx_2, hy_2, hy_3, hx_3), s(hx_2, hy_2, hy_3, hx_3), hy_1, hx_1) =$ (by Subcase 1.1.2)

$\quad p(r(hx_1, hy_1, hx_2, hy_2), s(hx_1, hy_1, hx_2, hy_2), hx_3, hy_3) = \mathbf{1}$ (by Case 1).

Analogously,

$(1.1.2)_{\mathrm{R}}$ $h(q'(x_1, y_1, r'(x_2, y_2, x_3, y_3), s'(x_2, y_2, x_3, y_3))) =$

$\quad q(r(hx_2, hy_2, hy_3, hx_3), s(hx_2, hy_2, hy_3, hx_3), hy_1, hx_1) =$ (by Subcase 1.1.2)

$\quad q(r(hx_1, hy_1, hx_2, hy_2), s(hx_1, hy_1, hx_2, hy_2), hx_3, hy_3) = \mathbf{0}$ (by Case 1).

As h validates the equations of $\Delta(x_1, y_1, \Delta(x_2, y_2, x_3, y_3))$, it also validates the equation (1.1.2). But in virtue of $(1.1.2)_{\mathrm{L}}$ and $(1.1.2)_{\mathrm{R}}$, h falsifies (1.1.2). A contradiction.

Subcase 1.1.3. $hx_1 = 1$, $hy_1 = 0$, $hx_2 = 0$, $hy_2 = 1$, $hx_3 = 1$, $hy_3 = 0$.

We define:

$$r'(x_2, y_2, x_3, y_3) := r(y_2, x_2, x_3, y_3), \quad s'(x_2, y_2, x_3, y_3) := s(y_2, x_2, x_3, y_3),$$
$$p'(x, y, z, w) := p(z, w, y, x), \quad\quad\quad q'(x, y, z, w) := q(z, w, y, x),$$

and form the equation

$$(1.1.3) \qquad p'(x_1, y_1, r'(x_2, y_2, x_3, y_3), s'(x_2, y_2, x_3, y_3)) \approx$$
$$q'(x_1, y_1, r'(x_2, y_2, x_3, y_3), s'(x_2, y_2, x_3, y_3))$$

This equation belongs to $\mathbf{Q}^{\vDash}(\Delta(x_1, y_1, \Delta(x_2, y_2, x_3, y_3)))$. We then compute the value of (1.1.3) under h. We have:

$$(1.1.3)_{\mathrm{L}} \qquad h(p'(x_1, y_1, r'(x_2, y_2, x_3, y_3), s'(x_2, y_2, x_3, y_3))) = \mathbf{1}.$$

Analogously,

$$(1.1.3)_{\mathrm{R}} \qquad h(q'(x_1, y_1, r'(x_2, y_2, x_3, y_3), s'(x_2, y_2, x_3, y_3))) = \mathbf{0}.$$

As h validates the equations of $\Delta(x_1, y_1, \Delta(x_2, y_2, x_3, y_3))$, it also validates the equation (1.1.3). But in virtue of $(1.1.3)_{\mathrm{L}}$ and $(1.1.3)_{\mathrm{R}}$, h falsifies (1.1.3). A contradiction.

Subcase 1.1.4. $hx_1 = \mathbf{1}$, $hy_1 = \mathbf{0}$, $hx_2 = \mathbf{0}$, $hy_2 = \mathbf{1}$, $hx_3 = \mathbf{0}$, $hy_3 = \mathbf{1}$.

We define:

$$r'(x_2, y_2, x_3, y_3) := r(y_2, x_2, x_3, y_3), \quad s'(x_2, y_2, x_3, y_3) := s(y_2, x_2, x_3, y_3),$$
$$p'(x, y, z, w) := p(z, w, y, x), \qquad q'(x, y, z, w) := q(z, w, y, x),$$

and form the equation

$$(1.1.4) \qquad p'(x_1, y_1, r'(x_2, y_2, x_3, y_3), s'(x_2, y_2, x_3, y_3)) \approx$$
$$q'(x_1, y_1, r'(x_2, y_2, x_3, y_3), s'(x_2, y_2, x_3, y_3))$$

This equation belongs to $\mathbf{Q}^{\vDash}(\Delta(x_1, y_1, \Delta(x_2, y_2, x_3, y_3)))$. We then compute the value of (1.1.4) under h. We have:

$$(1.1.4)_{\mathrm{L}} \qquad h(p'(x_1, y_1, r'(x_2, y_2, x_3, y_3), s'(x_2, y_2, x_3, y_3))) = \mathbf{1}.$$

Analogously,

$$(1.1.4)_{\mathrm{R}} \qquad h(q'(x_1, y_1, r'(x_2, y_2, x_3, y_3), s'(x_2, y_2, x_3, y_3))) = \mathbf{0}.$$

As h validates the equations of $\Delta(x_1, y_1, \Delta(x_2, y_2, x_3, y_3))$, it also validates the equation (1.1.4). But in virtue of $(1.1.4)_{\mathrm{L}}$ and $(1.1.4)_{\mathrm{R}}$, h falsifies (1.1.4). A contradiction.

Subcase 1.1.5. $hx_1 = \mathbf{0}$, $hy_1 = \mathbf{1}$, $hx_2 = \mathbf{1}$, $hy_2 = \mathbf{0}$, $hx_3 = \mathbf{1}$, $hy_3 = \mathbf{0}$.

We define:

$$r'(x_2, y_2, x_3, y_3) := r(y_2, x_2, x_3, y_3), \quad s'(x_2, y_2, x_3, y_3) := s(y_2, x_2, x_3, y_3),$$
$$p'(x, y, z, w) := p(z, w, y, x), \qquad q'(x, y, z, w) := q(z, w, y, x),$$

and form the equation

$$(1.1.5) \qquad p'(x_1, y_1, r'(x_2, y_2, x_3, y_3), s'(x_2, y_2, x_3, y_3)) \approx$$
$$q'(x_1, y_1, r'(x_2, y_2, x_3, y_3), s'(x_2, y_2, x_3, y_3))$$

We then compute the value of (1.1.5) under h. We have:

$$(1.1.5)_{\mathrm{L}} \qquad h(p'(x_1, y_1, r'(x_2, y_2, x_3, y_3), s'(x_2, y_2, x_3, y_3))) = \mathbf{1}.$$

Analogously,

$$(1.1.5)_{\mathrm{R}} \qquad h(q'(x_1, y_1, r'(x_2, y_2, x_3, y_3), s'(x_2, y_2, x_3, y_3))) = \mathbf{0}.$$

As h validates the equations of $\Delta(x_1, y_1, \Delta(x_2, y_2, x_3, y_3))$, it also validates the equation (1.1.5). But in virtue of $(1.1.5)_{\mathrm{L}}$ and $(1.1.5)_{\mathrm{R}}$, h falsifies (1.1.5). A contradiction.

Subcase 1.1.6. $hx_1 = \mathbf{0}$, $hy_1 = \mathbf{1}$, $hx_2 = \mathbf{1}$, $hy_2 = \mathbf{0}$, $hx_3 = \mathbf{0}$, $hy_3 = \mathbf{1}$.

We define:

$$r'(x_2, y_2, x_3, y_3) := r(y_2, x_2, y_3, x_3), \quad s'(x_2, y_2, x_3, y_3) := s(y_2, x_2, y_3, x_3),$$
$$p'(x, y, z, w) := p(z, w, x, y), \qquad\qquad q'(x, y, z, w) := q(z, w, x, y),$$

and form the equation

$$(1.1.6) \qquad p'(x_1, y_1, r'(x_2, y_2, x_3, y_3), s'(x_2, y_2, x_3, y_3)) \approx$$
$$q'(x_1, y_1, r'(x_2, y_2, x_3, y_3), s'(x_2, y_2, x_3, y_3))$$

We then compute the value of (1.1.6) under h. We have:

$$(1.1.6)_{\mathrm{L}} \qquad h(p'(x_1, y_1, r'(x_2, y_2, x_3, y_3), s'(x_2, y_2, x_3, y_3))) = \mathbf{1}.$$

Analogously,

$$(1.1.6)_{\mathrm{R}} \qquad h(q'(x_1, y_1, r'(x_2, y_2, x_3, y_3), s'(x_2, y_2, x_3, y_3))) = \mathbf{0}.$$

As h validates the equations of $\Delta(x_1, y_1, \Delta(x_2, y_2, x_3, y_3))$, it also validates the equation (1.1.6). But in virtue of $(1.1.6)_{\mathrm{L}}$ and $(1.1.6)_{\mathrm{R}}$, h falsifies (1.1.6). A contradiction.

Subcase 1.1.7. $hx_1 = \mathbf{0}$, $hy_1 = \mathbf{1}$, $hx_2 = \mathbf{0}$, $hy_2 = \mathbf{1}$, $hx_3 = \mathbf{1}$, $hy_3 = \mathbf{0}$.

We define:

$$r'(x_2, y_2, x_3, y_3) := r(x_2, y_2, x_3, y_3), \quad s'(x_2, y_2, x_3, y_3) := s(x_2, y_2, x_3, y_3),$$
$$p'(x, y, z, w) := p(z, w, y, x), \qquad\qquad q'(x, y, z, w) := q(z, w, y, x),$$

and form the equation

$$(1.1.7) \qquad p'(x_1, y_1, r'(x_2, y_2, x_3, y_3), s'(x_2, y_2, x_3, y_3)) \approx$$
$$q'(x_1, y_1, r'(x_2, y_2, x_3, y_3), s'(x_2, y_2, x_3, y_3))$$

We then compute the value of (1.1.7) under h. We have:

$$(1.1.7)_{\mathrm{L}} \qquad h(p'(x_1, y_1, r'(x_2, y_2, x_3, y_3), s'(x_2, y_2, x_3, y_3))) = \mathbf{1}.$$

Analogously,

$$(1.1.7)_{\mathrm{R}} \qquad h(q'(x_1, y_1, r'(x_2, y_2, x_3, y_3), s'(x_2, y_2, x_3, y_3))) = \mathbf{0}.$$

As h validates the equations of $\Delta(x_1, y_1, \Delta(x_2, y_2, x_3, y_3))$, it also validates the equation (1.1.7). But in virtue of $(1.1.7)_{\mathrm{L}}$ and $(1.1.7)_{\mathrm{R}}$, h falsifies (1.1.7). A contradiction.

Subcase 1.1.8. $hx_1 = \mathbf{0}$, $hy_1 = \mathbf{1}$, $hx_2 = \mathbf{0}$, $hy_2 = \mathbf{1}$, $hx_3 = \mathbf{0}$, $hy_3 = \mathbf{1}$.

We define:

$$r'(x_2, y_2, x_3, y_3) := r(x_2, y_2, x_3, y_3), \quad s'(x_2, y_2, x_3, y_3) := s(x_2, y_2, x_3, y_3),$$
$$p'(x, y, z, w) := p(z, w, x, y), \qquad q'(x, y, z, w) := q(z, w, x, y),$$

and form the equation

$$(1.1.8) \qquad p'(x_1, y_1, r'(x_2, y_2, x_3, y_3), s'(x_2, y_2, x_3, y_3)) \approx$$
$$q'(x_1, y_1, r'(x_2, y_2, x_3, y_3), s'(x_2, y_2, x_3, y_3))$$

We then compute the value of (1.1.8) under h. We have:

$$(1.1.8)_{\mathrm{L}} \qquad h(p'(x_1, y_1, r'(x_2, y_2, x_3, y_3), s'(x_2, y_2, x_3, y_3))) = \mathbf{1}.$$

Analogously,

$$(1.1.8)_{\mathrm{R}} \qquad h(q'(x_1, y_1, r'(x_2, y_2, x_3, y_3), s'(x_2, y_2, x_3, y_3))) = \mathbf{0}.$$

As h validates the equations of $\Delta(x_1, y_1, \Delta(x_2, y_2, x_3, y_3))$, it also validates the equation (1.1.8). But in virtue of $(1.1.8)_{\mathrm{L}}$ and $(1.1.8)_{\mathrm{R}}$, h falsifies (1.1.8). A contradiction.

The above remarks show that each (sub)subcase of Subcase 1.2 yields a contradiction.

The other subcase

Subcase 1.2. $r(hx_1, hy_1, hx_2, hy_2) = \mathbf{0}$ and $s(hx_1, hy_1, hx_2, hy_2) = \mathbf{1}$

is handled similarly. Subcase 1.2 yields 8 further (sub)subcases, identical with the subcses (1.1.1)–(1.1.8) analyzed as above. In each such a (sub)subcase the identical equation of the form $(*)$ is defined. The only difference in comparison with Subcase 1.1 is that the equation $(*)$ corresponding to each subcase among (1.1.1)–(1.1.8) is now falsified by h in such a way that its LHS now receives value $\mathbf{0}$ and RHS is equal to $\mathbf{1}$.

It remains to consider

Case 2. $p(r(hx_1, hy_1, hx_2, hy_2), s(hx_1, hy_1, hx_2, hy_2), hx_3, hy_3) = \mathbf{0}$ and $q(r(hx_1, hy_1, hx_2, hy_2), s(hx_1, hy_1, hx_2, hy_2), hx_3, hy_3) = \mathbf{1}$.

The situation is now symmetric to Case 1. Case 2 gives rise to the two subcases:

Subcase 2.1. $r(hx_1, hy_1, hx_2, hy_2) = \mathbf{1}$ and $s(hx_1, hy_1, hx_2, hy_2) = \mathbf{0}$

and

Subcase 2.2. $r(hx_1, hy_1, hx_2, hy_2) = \mathbf{1}$ and $s(hx_1, hy_1, hx_2, hy_2) = \mathbf{0}$.

They are both handled in the identical way as in the above Subcases 1.1 and 1.2.

Summing up, in each resulting case the supposition that the above lemma does not hold leads to a contradiction. $\qquad\qquad\qquad\qquad\qquad\qquad\qquad\square$

We now prove the other implication pertinent to $(*)$.

Lemma 3.8. $\Delta(x_1, y_1, \Delta(x_2, y_2, x_3, y_3))) \subseteq \mathbf{Q}^{\models}(\Delta(\Delta(x_1, y_1, x_2, y_2), x_3, y_3)))$.

To prove the lemma, we suitably modify the above proof of Lemma 3.7.

Suppose that the lemma does not hold. Hence for some algebra $\mathbf{2} \in \mathbf{K}$ and some homomorphism $h : \boldsymbol{Te} \to \mathbf{2}$ it is the case that h validates the equations of $\Delta(\Delta(x_1, y_1, x_2, y_2), x_3, y_3)$ and falsifies an equation

$$(**) \qquad p(x_1, y_1, r(x_2, y_2, x_3, y_3), s(x_2, y_2, x_3, y_3)) \approx$$
$$q(x_1, y_1, r(x_2, y_2, x_3, y_3), s(x_2, y_2, x_3, y_3))$$

belonging to $\Delta(x_1, y_1, \Delta(x_2, y_2, x_3, y_3))$. Hence

$$(4) \qquad p(hx_1, hy_1, r(hx_2, hy_2, hx_3, hy_3), s(hx_2, hy_2, hx_3, hy_3)) \neq$$
$$q(hx_1, hy_1, r(hx_2, hy_2, hx_3, hy_3), s(hx_2, hy_2, hx_3, hy_3))$$

and as $p \approx q$ and $r \approx s$ are quaternary commutator equations, it follows that

$$(5) \qquad r(hx_2, hy_2, hx_3, hy_3) \neq s(hx_2, hy_2, hx_3, hy_3)$$

and

$$(6) \qquad hx_1 \neq hy_1, \ hx_2 \neq hy_2, \ hx_3 \neq hy_3.$$

We work with $(**)$ and consider several cases and subcases implied by (4)–(6).

Case 1. $p(hx_1, hy_1, r(hx_2, hy_2, hx_3, hy_3), s(hx_2, hy_2, hx_3, hy_3)) = \mathbf{1}$ and $q(hx_1, hy_1, r(hx_2, hy_2, hx_3, hy_3), s(hx_2, hy_2, hx_3, hy_3)) = \mathbf{0}$.

Subcase 1.1. $r(hx_2, hy_2, hx_3, hy_3) = 1$ and $s(hx_2, hy_2, hx_3, hy_3) = 0$.

According to Subcase 1.1 and (6) we isolate further 8 (sub)subcases, the same as in the proof of Lemma 3.7:

Subcase 1.1.1. $hx_1 = 1$, $hy_1 = 0$, $hx_2 = 1$, $hy_2 = 0$, $hx_3 = 1$, $hy_3 = 0$.

We define

$$r'(x_1, y_1, x_2, y_2) := r(x_1, y_1, x_2, y_2), \quad s'(x_1, y_1, x_2, y_2) := s(x_1, y_1, x_2, y_2)$$

and

$$p'(x, y, z, w) := p(z, w, x, y), \qquad q'(x, y, z, w) := q(z, w, x, y)$$

and form the equation

$$(2.1.1) \qquad p'(r'(x_1, y_1, x_2, y_2), s'(x_1, y_1, x_2, y_2), x_3, y_3) \approx$$
$$q'(r'(x_1, y_1, x_2, y_2), s'(x_1, y_1, x_2, y_2), x_3, y_3).$$

This equation belongs to $\mathbf{Q}^\vDash(\Delta(\Delta(x_1, y_1, x_2, y_2), x_3, y_3))$. We then compute the value of (2.1.1) under h. We have:

$(2.1.1)_{\mathrm{L}} \quad hp'(r'(x_1, y_1, x_2, y_2), s'(x_1, y_1, x_2, y_2), x_3, y_3) =$
$p'(r'(hx_1, hy_1, hx_2, hy_2), s'(hx_1, hy_1, hx_2, hy_2), hx_3, hy_3) =$
$p(hx_3, hy_3, r'(hx_1, hy_1, hx_2, hy_2), s'(hx_1, hy_1, hx_2, hy_2)) =$
$p(hx_3, hy_3, r(hx_1, hy_1, hx_2, hy_2), s(hx_1, hy_1, hx_2, hy_2)) =$ (by Subcase 1.1.1)
$p(hx_1, hy_1, r(hx_2, hy_2, hx_3, hy_3), s(hx_2, hy_2, hx_3, hy_3)) = 1$ (by Case 1).

Analogously

$(2.1.1)_{\mathrm{R}} \quad hq'(r'(x_1, y_1, x_2, y_2), s'(x_1, y_1, x_2, y_2), x_3, y_3) =$
$q'(r'(hx_1, hy_1, hx_2, hy_2), s'(hx_1, hy_1, hx_2, hy_2), hx_3, hy_3) =$
$q(hx_3, hy_3, r'(hx_1, hy_1, hx_2, hy_2), s'(hx_1, hy_1, hx_2, hy_2)) =$
$q(hx_3, hy_3, r(hx_1, hy_1, hx_2, hy_2), s(hx_1, hy_1, hx_2, hy_2)) =$ (by Subcase 1.1.1)
$q(hx_1, hy_1, r(hx_2, hy_2, hx_3, hy_3), s(hx_2, hy_2, hx_3, hy_3)) = 0$ (by Case 1).

As h validates the equations of $\Delta(\Delta(x_1, y_1, x_2, y_2), x_3, y_3)$, it also validates the equation (2.1.1). But in virtue of $(2.1.1)_{\mathrm{L}}$ and $(2.1.1)_{\mathrm{R}}$, h falsifies (2.1.1). A contradiction.

The other subcases are similarly handled. (Details are omitted.)

Subcase 1.1.2. $hx_1 = 1$, $hy_1 = 0$, $hx_2 = 1$, $hy_2 = 0$, $hx_3 = 0$, $hy_3 = 1$.

We define:

$$r'(x_1, y_1, x_2, y_2) := r(x_1, y_1, y_2, x_2), \quad s'(x_1, y_1, x_2, y_2) := s(x_1, y_1, y_2, x_2),$$
$$p'(x, y, z, w) := p(w, z, x, y), \qquad q'(x, y, z, w) := q(w, z, x, y),$$

and form the equation

$$(2.1.2) \qquad p'(r'(x_1, y_1, x_2, y_2), s'(x_1, y_1, x_2, y_2), x_3, y_3) \approx$$
$$q'(r'(x_1, y_1, x_2, y_2), s'(x_1, y_1, x_2, y_2), x_3, y_3).$$

This equation belongs to $\mathbf{Q}^{\models}(\Delta(\Delta(x_1, y_1, x_2, y_2), x_3, y_3))$. We then compute the value of (2.1.2) under h. We have:

$(2.1.2)_\mathrm{L}$ $hp'(r'(x_1, y_1, x_2, y_2), s'(x_1, y_1, x_2, y_2), x_3, y_3) =$
   $p(hy_3, hx_3, r(hx_1, hy_1, hy_2, hx_2), s(hx_1, hy_1, hy_2, hx_2)) = $ (by Subcase 1.1.2)
   $p(hx_1, hy_1, r(hx_2, hy_2, hx_3, hy_3), s(hx_2, hy_2, hx_3, hy_3)) = \mathbf{1}$ (by Case 1).

Analogously

$(2.1.2)_\mathrm{R}$ $hq'(r'(x_1, y_1, x_2, y_2), s'(x_1, y_1, x_2, y_2), x_3, y_3) =$
   $q(hy_3, hx_3, r(hx_1, hy_1, hy_2, hx_2), s(hx_1, hy_1, hy_2, hx_2)) = $ (by Subcase 1.1.2)
   $q(hx_1, hy_1, r(hx_2, hy_2, hx_3, hy_3), s(hx_2, hy_2, hx_3, hy_3)) = \mathbf{0}$ (by Case 1).

As h validates the equations of $\Delta(\Delta(x_1, y_1, x_2, y_2), x_3, y_3)$, it also validates the equation (2.1.2). But in virtue of $(2.1.2)_\mathrm{L}$ and $(2.1.2)_\mathrm{R}$, h falsifies (2.1.2). A contradiction.

Subcase 1.1.3. $hx_1 = \mathbf{1}, \ hy_1 = \mathbf{0}, \ hx_2 = \mathbf{0}, \ hy_2 = \mathbf{1}, \ hx_3 = \mathbf{1}, \ hy_3 = \mathbf{0}$.
   We define:

$$r'(x_1, y_1, x_2, y_2) := r(x_2, y_2, x_1, y_1), \quad s'(x_1, y_1, x_2, y_2) := s(x_2, y_2, x_1, y_1),$$
$$p'(x, y, z, w) := p(z, w, xy), \qquad\qquad q'(x, y, z, w) := q(z, w, x, y),$$

and form the equation

$$(2.1.3) \qquad p'(r'(x_1, y_1, x_2, y_2), s'(x_1, y_1, x_2, y_2), x_3, y_3) \approx$$
$$q'(r'(x_1, y_1, x_2, y_2), s'(x_1, y_1, x_2, y_2), x_3, y_3).$$

This equation belongs to $\mathbf{Q}^{\models}(\Delta(\Delta(x_1, y_1, x_2, y_2), x_3, y_3))$. We then compute the value of (2.1.3) under h. We have:

$(2.1.3)_\mathrm{L}$ $hp'(r'(x_1, y_1, x_2, y_2), s'(x_1, y_1, x_2, y_2), x_3, y_3) = \mathbf{1}$.

Analogously

$(2.1.3)_\mathrm{R}$ $hq'(r'(x_1, y_1, x_2, y_2), s'(x_1, y_1, x_2, y_2), x_3, y_3) = \mathbf{0}$ (by Case 1).

As h validates the equations of $\Delta(\Delta(x_1, y_1, x_2, y_2), x_3, y_3)$, it also validates the equation (2.1.3). But in virtue of $(2.1.3)_\mathrm{L}$ and $(2.1.3)_\mathrm{R}$, h falsifies (2.1.3). A contradiction.

Subcase 1.1.4. $hx_1 = 1$, $hy_1 = 0$, $hx_2 = 0$, $hy_2 = 1$, $hx_3 = 0$, $hy_3 = 1$.
   We define:

$$r'(x_1, y_1, x_2, y_2) := r(x_2, y_2, y_1, x_1), \quad s'(x_1, y_1, x_2, y_2) := s(x_2, y_2, y_1, x_1),$$
$$p'(x, y, z, w) := p(z, w, x, y), \qquad\qquad q'(x, y, z, w) := q(z, w, x, y),$$

and form the equation

$$(2.1.4) \qquad p'(r'(x_1, y_1, x_2, y_2), s'(x_1, y_1, x_2, y_2), x_3, y_3) \approx$$
$$q'(r'(x_1, y_1, x_2, y_2), s'(x_1, y_1, x_2, y_2), x_3, y_3).$$

This equation belongs to $\mathbf{Q}^{\vDash}(\Delta(\Delta(x_1, y_1, x_2, y_2), x_3, y_3))$. We then compute
the value of (2.1.4) under h. We have:

$$(2.1.4)_{\mathrm{L}} \qquad hp'(r'(x_1, y_1, x_2, y_2), s'(x_1, y_1, x_2, y_2), x_3, y_3) = 1.$$

Analogously

$$(2.1.4)_{\mathrm{R}} \qquad hq'(r'(x_1, y_1, x_2, y_2), s'(x_1, y_1, x_2, y_2), x_3, y_3) = 0.$$

As h validates the equations of $\Delta(\Delta(x_1, y_1, x_2, y_2), x_3, y_3)$, it also validates
the equation (2.1.4). But in virtue of $(2.1.4)_{\mathrm{L}}$ and $(2.1.4)_{\mathrm{R}}$, h falsifies (2.1.4).
A contradiction.

Subcase 1.1.5. $hx_1 = 0$, $hy_1 = 1$, $hx_2 = 1$, $hy_2 = 0$, $hx_3 = 1$, $hy_3 = 0$.
   We define:

$$r'(x_1, y_1, x_2, y_2) := r(x_2, y_2, y_1, x_1), \quad s'(x_1, y_1, x_2, y_2) := s(x_2, y_2, y_1, x_1),$$
$$p'(x, y, z, w) := p(z, w, x, y), \qquad\qquad q'(x, y, z, w) := q(z, w, x, y),$$

and form the equation

$$(2.1.5) \qquad p'(r'(x_1, y_1, x_2, y_2), s'(x_1, y_1, x_2, y_2), x_3, y_3) \approx$$
$$q'(r'(x_1, y_1, x_2, y_2), s'(x_1, y_1, x_2, y_2), x_3, y_3).$$

This equation belongs to $\mathbf{Q}^{\vDash}(\Delta(\Delta(x_1, y_1, x_2, y_2), x_3, y_3))$. We then compute
the value of (2.1.5) under h. We have:

$$(2.1.5)_{\mathrm{L}} \qquad hp'(r'(x_1, y_1, x_2, y_2), s'(x_1, y_1, x_2, y_2), x_3, y_3) = 1.$$
$$(2.1.5)_{\mathrm{R}} \qquad hq'(r'(x_1, y_1, x_2, y_2), s'(x_1, y_1, x_2, y_2), x_3, y_3) = 0.$$

As h validates the equations of $\Delta(\Delta(x_1, y_1, x_2, y_2), x_3, y_3)$, it also validates
the equation (2.1.5). But in virtue of $(2.1.5)_{\mathrm{L}}$ and $(2.1.5)_{\mathrm{R}}$, h falsifies (2.1.5).
A contradiction.

Subcase 1.1.6. $hx_1 = 0$, $hy_1 = 1$, $hx_2 = 1$, $hy_2 = 0$, $hx_3 = 0$, $hy_3 = 1$.

We define:

$$r'(x_1,y_1,x_2,y_2) := r(x_2,y_2,x_1,y_1), \quad s'(x_1,y_1,x_2,y_2) := s(x_2,y_2,x_1,y_1),$$
$$p'(x,y,z,w) := p(z,w,x,y), \qquad\qquad q'(x,y,z,w) := q(z,w,x,y),$$

and form the equation

$$(2.1.6) \qquad p'(r'(x_1,y_1,x_2,y_2),s'(x_1,y_1,x_2,y_2),x_3,y_3) \approx$$
$$q'(r'(x_1,y_1,x_2,y_2),s'(x_1,y_1,x_2,y_2),x_3,y_3).$$

This equation belongs to $\mathbf{Q}^{\vDash}(\Delta(\Delta(x_1,y_1,x_2,y_2),x_3,y_3))$. We then compute the value of (2.1.6) under h. We have:

$$(2.1.6)_{\mathrm{L}} \qquad hp'(r'(x_1,y_1,x_2,y_2),s'(x_1,y_1,x_2,y_2),x_3,y_3) = \mathbf{1}.$$

and

$$(2.1.6)_{\mathrm{R}} \qquad hq'(r'(x_1,y_1,x_2,y_2),s'(x_1,y_1,x_2,y_2),x_3,y_3) = \mathbf{0}.$$

As h validates the equations of $\Delta(\Delta(x_1,y_1,x_2,y_2),x_3,y_3)$, it also validates the equation (2.1.6). But in virtue of $(2.1.6)_{\mathrm{L}}$ and $(2.1.6)_{\mathrm{R}}$, h falsifies (2.1.6). A contradiction.

Subcase 1.1.7. $hx_1 = \mathbf{0}$, $hy_1 = \mathbf{1}$, $hx_2 = \mathbf{0}$, $hy_2 = \mathbf{1}$, $hx_3 = \mathbf{1}$, $hy_3 = \mathbf{0}$.

We define:

$$r'(x_1,y_1,x_2,y_2) := r(x_2,y_2,y_1,x_1), \quad s'(x_1,y_1,x_2,y_2) := s(x_2,y_2,y_1,x_1),$$
$$p'(x,y,z,w) := p(z,w,x,y), \qquad\qquad q'(x,y,z,w) := q(z,w,x,y),$$

and form the equation

$$(2.1.7) \qquad p'(r'(x_1,y_1,x_2,y_2),s'(x_1,y_1,x_2,y_2),x_3,y_3) \approx$$
$$q'(r'(x_1,y_1,x_2,y_2),s'(x_1,y_1,x_2,y_2),x_3,y_3).$$

This equation belongs to $\mathbf{Q}^{\vDash}(\Delta(\Delta(x_1,y_1,x_2,y_2),x_3,y_3))$. We then compute the value of (2.1.7) under h. We have:

$$(2.1.7)_{\mathrm{L}} \qquad hp'(r'(x_1,y_1,x_2,y_2),s'(x_1,y_1,x_2,y_2),x_3,y_3) = \mathbf{1}.$$

and

$$(2.1.7)_{\mathrm{R}} \qquad hq'(r'(x_1,y_1,x_2,y_2),s'(x_1,y_1,x_2,y_2),x_3,y_3) = \mathbf{0}.$$

As h validates the equations of $\Delta(\Delta(x_1,y_1,x_2,y_2),x_3,y_3)$, it also validates the equation (2.1.7). But in virtue of $(2.1.7)_{\mathrm{L}}$ and $(2.1.7)_{\mathrm{R}}$, h falsifies (2.1.7). A contradiction.

Subcase 1.1.8. $hx_1 = 0,\ hy_1 = 1,\ hx_2 = 0,\ hy_2 = 1,\ hx_3 = 0,\ hy_3 = 1.$

We define:

$$r'(x_1,y_1,x_2,y_2) := r(x_2,y_2,x_1,y_1), \quad s'(x_1,y_1,x_2,y_2) := s(x_2,y_2,x_1,y_1),$$
$$p'(x,y,z,w) := p(z,w,x,y), \qquad\qquad q'(x,y,z,w) := q(z,w,x,y),$$

and form the equation

$$(2.1.8) \qquad p'(r'(x_1,y_1,x_2,y_2),s'(x_1,y_1,x_2,y_2),x_3,y_3) \approx$$
$$q'(r'(x_1,y_1,x_2,y_2),s'(x_1,y_1,x_2,y_2),x_3,y_3).$$

This equation belongs to $\mathbf{Q}^{\models}(\Delta(\Delta(x_1,y_1,x_2,y_2),x_3,y_3))$. We then compute the value of (2.1.8) under h. We have:

$$(2.1.8)_{\mathrm{L}} \qquad hp'(r'(x_1,y_1,x_2,y_2),s'(x_1,y_1,x_2,y_2),x_3,y_3) = 1$$

and

$$(2.1.8)_{\mathrm{R}} \qquad hq'(r'(x_1,y_1,x_2,y_2),s'(x_1,y_1,x_2,y_2),x_3,y_3) = 0.$$

As h validates the equations of $\Delta(\Delta(x_1,y_1,x_2,y_2),x_3,y_3)$, it also validates the equation (2.1.8). But in virtue of $(2.1.8)_{\mathrm{L}}$ and $(2.1.8)_{\mathrm{R}}$, h falsifies (2.1.8). A contradiction. $\qquad\square$

From Lemmas 3.7–3.8 the theorem follows. $\qquad\square$

Theorem 3.6 can be strengthened if $\mathbf{Q}$ is generated by a *single* two-element algebra.

If $\mathbf{Q}$ is a quasivariety generated by a two-element algebra, it may happen that the commutator is Abelian (or nullary), which means that for any algebra $\boldsymbol{A} \in \mathbf{Q}$ and any $a,b,c,d \in A$ it is the case that $[\Theta_{\mathbf{Q}}(a,b),[\Theta_{\mathbf{Q}}(c,d)] = 0_A$. This situation occurs e.g. in the variety of semilattices. On the other hand, in the variety of Boolean algebras we have that $[\Theta(a,b),[\Theta(c,d)] = \Theta(a,b) \cap \Theta(c,d)]$ for any Boolean algebra $\boldsymbol{A}$ and any $a,b,c,d \in A$. (The last equation characterizes the commutator in relatively congruence-distributive quasivarieties—see e.g. (Kearnes, McKenzie, 1992).) In turn, the variety $\mathbf{CEA}$ of classical equivalence algebras, that is, the variety generated by the algebra represented by the two-element truth-table $\mathbf{2}$ for the classical equivalence connective, is congruence-modular and it is not congruence distributive. ($\mathbf{CEA}$ is generated as a quasivariety by $\mathbf{2}$, i.e., $\mathbf{CEA} = SP(\mathbf{2})$.) The equationally-defined commutator for $\mathbf{CEA}$ coincides with the standard commutator (as defined for any CM variety) and hence it is also additive. $\mathbf{CEA}$ is term equivalent to the variety of Boolean groups, i.e., the variety of Abelian groups in which every element is an idempotent and therefore it is its own inverse. It can be proved that any two term equivalent varieties have the same equationally-defined commutators. It then follows that the equationally-defined commutator for $\mathbf{CEA}$ is nullary.

The next theorem is a refinement of Theorem 3.6:

Theorem 3.9. *Let* $\mathbf{Q}$ *be a quasivariety generated by a two-element algebra. Then the equationally-defined commutator of* $\mathbf{Q}$ *universally validates one of the following laws:* $[\boldsymbol{x}, \boldsymbol{y}] = \boldsymbol{x} \wedge \boldsymbol{y}$ *or* $[\boldsymbol{x}, \boldsymbol{y}] = \boldsymbol{0}$.

In other words, the above theorem states that any quasivariety generated by a single two-element algebra is either relatively congruence-distributive or Abelian.

Before proving the theorem, we recall some definitions and facts from (Czelakowski, 2015).

Let $\mathbf{Q}$ be a quasivariety and $\boldsymbol{A}$ an algebra in $\mathbf{Q}$. Let $\varPhi$ be a $\mathbf{Q}$-congruence on $\boldsymbol{A}$. $\varPhi$ is said to be *prime* (in the lattice $\boldsymbol{Con}_{\mathbf{Q}}(\boldsymbol{A})$) if, for any congruences $\varPhi_1, \varPhi_2 \in Con_{\mathbf{Q}}(\boldsymbol{A})$, $[\varPhi_1, \varPhi_2] \subseteq \varPhi$ implies that $\varPhi_1 \subseteq \varPhi$ or $\varPhi_2 \subseteq \varPhi$. (Here $[\varPhi_1, \varPhi_2]$ is the equationally-defined commutator of the congruences $\varPhi_1, \varPhi_2$ in $\boldsymbol{A}$ in the sense of $\mathbf{Q}$.)

$\boldsymbol{A} \in \mathbf{Q}$ is said to be *prime* (in $\mathbf{Q}$) if the identity congruence $\boldsymbol{0}_{\boldsymbol{A}}$ is prime in $\boldsymbol{Con}_{\mathbf{Q}}(\boldsymbol{A})$. Thus $\boldsymbol{A}$ is prime in $\mathbf{Q}$ if and only if $[\varPhi_1, \varPhi_2] = \boldsymbol{0}_{\boldsymbol{A}}$ holds for no pair of nonzero congruences $\varPhi_1, \varPhi_2 \in Con_{\mathbf{Q}}(\boldsymbol{A})$.

$\mathbf{Q}_{\mathrm{PRIME}}$ denotes the class of all prime algebras in $\mathbf{Q}$. It is easy to see that $\mathbf{Q}_{\mathrm{PRIME}} \subseteq \mathbf{Q}_{\mathrm{RFSI}}$ (Proposition 7.2.1 of (Czelakowski, 2015).)

Proposition 3.10. *Let* $\mathbf{Q}$ *be quasivariety whose equationally-defined commutator is additive. Let* $\Delta(x, y, z, w, \underline{u})$ *be a generating set. Let* $\boldsymbol{A}$ *be an algebra in* $\mathbf{Q}$ *and* $\varPhi \in Con_{\mathbf{Q}}(\boldsymbol{A})$. *The following conditions are equivalent:*

(i) $\varPhi$ *is prime in* $\boldsymbol{Con}_{\mathbf{Q}}(\boldsymbol{A})$;

(ii)   *For all* $a, b, c, d \in A$, $[\Theta_{\mathbf{Q}}^{\boldsymbol{A}}(a, b), \Theta_{\mathbf{Q}}^{\boldsymbol{A}}(c, d)]^{\boldsymbol{A}} \subseteq \varPhi$ *implies* $\langle a, b \rangle \in \varPhi$ *or* $\langle c, d \rangle \in \varPhi$;

(iii)   *For all* $a, b, c, d \in A$, $\Theta_{\mathbf{Q}}^{\boldsymbol{A}}((\forall \underline{e})\, \Delta(a, b, c, d, \underline{e})) \subseteq \varPhi$ *implies* $\langle a, b \rangle \in \varPhi$ *or* $\langle c, d \rangle \in \varPhi$. $\square$

Note. In the above proposition, the generating set Δ may contains parametric variables, because in the general case Lemma 3.2 does not apply. $\square$

In view of Theorem 7.2.4 of (Czelakowski, 2015) we have:

Proposition 3.11. *Let* $\mathbf{Q}$ *be a quasivariety whose equationally-defined commutator is additive. The class* $\boldsymbol{SP}(\mathbf{Q}_{\mathrm{PRIME}})$ *is the largest RCD quasivariety included in* $\mathbf{Q}$. $\square$

Note. It follows from Proposition 3.10 that if $\Delta(x, y, z, w, \underline{u})$ is finite, the class $\mathbf{Q}_{\mathrm{PRIME}}$ is axiomatized relative to $\mathbf{Q}$ by the universal-existential first-order sentence $(\forall xyzw)((\forall \underline{u}) \wedge \Delta \rightarrow (x \approx y \vee z \wedge w))$. If moreover $\mathbf{Q}$ is generated by a finite family of finite at most four-generated algebras, then Δ can be assumed to do not contain parametric variables (see the note following Lemma 3.2). Then $\mathbf{Q}_{\mathrm{PRIME}}$ is a universal class. In fact, as $\mathbf{Q}_{\mathrm{PRIME}}$ is a finite set of finite algebras, it is a finitely axiomatized universal class (irrespective of the fact the signature of $\mathbf{Q}$ is finite or not). The RCD quasivariety

$SP(\mathbf{Q}_{\mathrm{PRIME}})$ is therefore finitely axiomatized. This follows from Pigozzi's Theorem (see (Pigozzi, 1988)) or from Theorem 3.4 of (Czelakowski, Dziobiak, 1990). $\qquad\square$

In purely syntactical contexts it is sometimes convenient to operate with prime equational theories rather than prime congruences.

Let $\mathbf{Q}$ be as in Proposition 3.10. A theory T of $\mathbf{Q}^{\vDash}$ is *prime* if for any terms p, q, r, s

$$\Delta(p,q,r,s,\underline{t}) \subseteq T \text{ for all sequences } \underline{t} \text{ of terms implies } p \approx q \in T \text{ or } r \approx s \in T.$$

If T is a proper prime theory of $\mathbf{Q}^{\vDash}$, then by factoring the term algebra $\boldsymbol{Te}$ by means of the congruence ΩT, where $\alpha \equiv \beta \,(\mathrm{mod}\,\Omega T)$ means that $\alpha \approx \beta \in T$, we obtain a non-trivial countable algebra belonging to $\mathbf{Q}_{\mathrm{PRIME}}$.

We pass to the proof of Theorem 3.9. We first prove:

Lemma 3.12. *Assume $\mathbf{Q}$ is generated by a finite family $\mathbf{K}$ of two-element algebras. Then for every equation $p \approx q \in \mathbf{Q}^{\vDash}(x \approx y) \cap \mathbf{Q}^{\vDash}(z \approx w) \setminus \mathbf{Q}^{\vDash}(\emptyset)$, there exists a prime theory T of $\mathbf{Q}^{\vDash}$ such that $p \approx q \notin T$.*

Proof (of the lemma). We assume $\mathbf{Q} = \boldsymbol{SP}(\mathbf{K})$ for a finite class $\mathbf{K}$ of two-element algebras. In view of Theorem 3.1 the equationally-defined commutator of $\mathbf{Q}$ is additive. Let $\Delta = \Delta(x,y,z,w)$ be a generating set (without parameters), that is, $\mathbf{Q}^{\vDash}(x \approx y) \cap \mathbf{Q}^{\vDash}(z \approx w) = \mathbf{Q}^{\vDash}(\Delta(x,y,z,w))$. We recall that the structurality of $\mathbf{Q}^{\vDash}$ gives that

$$(1) \qquad \mathbf{Q}^{\vDash}(\Delta(x,y,z,w)) = \mathbf{Q}^{\vDash}(\Delta(x,y,w,z)) =$$
$$\mathbf{Q}^{\vDash}(\Delta(y,x,z,w)) = \mathbf{Q}^{\vDash}(\Delta(y,x,w,z)).$$

Let $\boldsymbol{A}$ be a fixed non-trivial algebra in $\mathbf{Q}$. The fact that the equations of Δ hold in a $\boldsymbol{A}$ under a homomorphism $h : \boldsymbol{Te} \to \boldsymbol{A}$ such that $a = hx$, $b = hy$, $c = hz$ and $d = hw$ is marked as

$$\Delta(a,b,c,d).$$

In virtue of (1), there hold the following equivalences:

$$(2) \qquad \Delta(a,b,c,d) \Leftrightarrow \Delta(a,b,d,c) \Leftrightarrow \Delta(b,a,c,d) \Leftrightarrow \Delta(b,a,d,c).$$

Assume $p \approx q \in \mathbf{Q}^{\vDash}(x \approx y) \cap \mathbf{Q}^{\vDash}(z \approx w) \setminus \mathbf{Q}^{\vDash}(\emptyset)$. We may assume $p \approx q \in \Delta$. There is an algebra $\mathbf{2} \in \mathbf{K}$ whose carrier is $\{\mathbf{0}, \mathbf{1}\}$ and a homomorphism $h : \boldsymbol{Te} \to \mathbf{2}$ such that $hp \neq hq$. As $p \approx q \in \Delta$, it follows that h does not validate $\Delta(x,y,z,w)$.

Putting $a = hx$, $b = hy$, $c = hz$ and $d = hw$ and taking into account the fact that $p \approx q$ is a quaternary commutator equation, we then obtain that $a \neq b$, $c \neq d$ and $\Delta(a,b,c,d)$ does not hold. Hence each of the constituent conditions of (2) is also excluded.

As $\{a,b,c,d\} = \{\mathbf{0},\mathbf{1}\}$, we obtain from (2) that each of the equivalent conditions

$$(3) \qquad \Delta(\mathbf{1},\mathbf{0},\mathbf{1},\mathbf{0}), \Delta(\mathbf{1},\mathbf{0},\mathbf{0},\mathbf{1}), \Delta(\mathbf{0},\mathbf{1},\mathbf{0},\mathbf{1}), \Delta(\mathbf{0},\mathbf{1},\mathbf{1},\mathbf{0})$$

is excluded.

Let $T := \{\alpha \approx \beta \in Eq(\tau) : h\alpha = h\beta \text{ in } \mathbf{A}\}$. T is a finitely meet-irreducible theory of $\mathbf{Q}^{\vDash}$ and $p \approx q \notin T$.

Claim. T *is a prime theory of* $\mathbf{Q}^{\vDash}$.

Proof (of the claim). We must prove that for any equations $\alpha \approx \beta$ and $\gamma \approx \delta$, if $\Delta(\alpha,\beta,\gamma,\delta) \subseteq T$, then $\alpha \approx \beta \in T$ or $\gamma \approx \delta \in T$. Assume that $\Delta(\alpha,\beta,\gamma,\delta) \subseteq T$. Thus $\Delta(h\alpha,h\beta,h\gamma,h\delta)$ holds. As the values of h belong to $\{\mathbf{0},\mathbf{1}\}$, (3) implies that either $h\alpha = h\beta$ or $h\gamma = h\delta$. Hence $\alpha \approx \beta \in T$ or $\gamma \approx \delta \in T$. $\square$

This proves the lemma. $\square$

To conclude the proof of Theorem 3.9, assume $\mathbf{Q} = \mathbf{SP}(\mathbf{2})$. The class $\mathbf{SP}(\mathbf{Q}_{\mathrm{PRIME}})$ is the largest RCD quasivariety included in $\mathbf{Q}$. As $\mathbf{Q}$ is a minimal quasivariety, it follows that either $\mathbf{SP}(\mathbf{Q}_{\mathrm{PRIME}}) = \mathbf{Q}$ or $\mathbf{SP}(\mathbf{Q}_{\mathrm{PRIME}})$ is a trivial quasivariety.

Suppose that the equationally-defined commutator of $\mathbf{Q}$ does not validate $[\boldsymbol{x},\boldsymbol{y}] = \boldsymbol{x} \wedge \boldsymbol{y}$. Then the equality $\mathbf{SP}(\mathbf{Q}_{\mathrm{PRIME}}) = \mathbf{Q}$ is excluded (because otherwise $\mathbf{Q}$ would be RCD and therefore it would validate $[\boldsymbol{x},\boldsymbol{y}] = \boldsymbol{x} \wedge \boldsymbol{y}$). It then follows that $\mathbf{SP}(\mathbf{Q}_{\mathrm{PRIME}})$ is a trivial quasivariety. This means that $\mathbf{Q}$ contains only trivial prime algebras.

To prove that $\mathbf{Q}$ is Abelian, it suffices to show that $\mathbf{Q}^{\vDash}(x \approx y) \cap \mathbf{Q}^{\vDash}(z \approx w) = \mathbf{Q}^{\vDash}(\emptyset)$ (see (Czelakowski, 2015), Theorem 9.3.3). Suppose otherwise that $\mathbf{Q}^{\vDash}(x \approx y) \cap \mathbf{Q}^{\vDash}(z \approx w) \setminus \mathbf{Q}^{\vDash}(\emptyset) \neq \emptyset$. In view of the above lemma, $\mathbf{Q}^{\vDash}$ possesses a proper prime theory T. By taking the quotient algebra $\boldsymbol{Te}/\Omega T$, we obtain a non-trivial prime algebra belonging to $\mathbf{Q}$. This contradicts the fact that $\mathbf{Q}$ contains only trivial prime algebras.

This completes the proof of Theorem 3.9. $\square$

References

Berman, J. (1980). A proof of Lyndon's Finite Basis Theorem, *Discrete Mathematics* **29**, 229–233.

Burris, S. and Lee, S. (1993). Tarski's high school identities, *American Mathematical Monthly* **100**, No. 3, 231–236.

Burris, S. and Yeats, K.A. (2004). The saga of the high school identities, *Algebra Universalis* **52**, Nos. 2–3, 325–342.

Czelakowski, J. (2015). *The Equationally-Defined Commutator. A Study in Equational Logic and Algebra*, Birkhäuser, Basel.

Czelakowski, J. and Dziobiak, W. (1990). Congruence distributive quasivarieties whose subdirectly irreducible members form a universal class, *Algebra Universalis* **27**, 128–149.

Freese, R. and McKenzie, R.N. (1987). *Commutator Theory for Congruence Modular Varieties*, London Mathematical Society Lecture Note Series **125**, Cambridge University Press, Cambridge-New York. (The second edition is available online.)

Kearnes, K. and McKenzie, R.N. (1992). Commutator theory for relatively modular quasivarieties, *Transactions of the American Mathematical Society* **331**, No. 2, 465–502.

Lyndon, R. (1951). Identities in two-valued calculi, *Transactions of the American Mathematical Society* **71**, 457–465.

McKenzie, R.N., McNulty, G.F. and Taylor, W. (1987). *Algebras, Lattices, Varieties. Volume I*, Wadsworth & Brooks/Cole. Monterey, CA.

Pigozzi, D. (1988). Finite basis theorems for relatively congruence-distributive quasivarieties, *Transactions of the American Mathematical Society* **310**, No. 2, 499–533.

Taylor, W. (1976). Pure compactifications in quasi-primal varieties, *Canadian Journal of Mathematics* **28**, 50–62.

Wójcicki, R. (1988). *Theory of Logical Calculi. Basic Theory of Consequence Operations*, Kluwer, Dordrecht.

A short overview of Hidden Logic

Isabel Ferreirim and Manuel A. Martins*

Abstract In this paper we review a hidden (sorted) generalization of k-deductive systems—*hidden k-logics*. They encompass deductive systems as well as hidden equational logics and inequational logics. The special case of hidden equational logics has been used to specify and to verify properties in program development of behavioral systems within the dichotomy visible vs. hidden data. We recall one of the main applications of this work—the study of behavioral equivalence. Related results are obtained through combinatorial properties of the Leibniz congruence relation.

In addition we obtain a few new developments concerning hidden equational logic, namely we present a new characterization of the behavioral consequences of a theory.

2010 Mathematics Subject Classification: 03G27, 03B22

Isabel Ferreirim
Department of Mathematics, University of Lisbon, Portugal, e-mail: `imferreirim@ciencias.ulisboa.pt`

Manuel A. Martins
Department of Mathematics and CIDMA, University of Aveiro, Portugal, e-mail: `martins@ua.pt`

* This work was funded by ERDF - European Regional Development Fund through the COMPETE Programme (operational programme for competitiveness) and by National Funds through the FCT (Portuguese Foundation for Science and Technology) within project UID/MAT/04106/2013 at CIDMA. The second author also acknowledges the financial assistance by the EU FP7 Marie Curie PIRSES-GA-2012-318986 project GeT-Fun: Generalizing Truth-Functionality.

1 Introduction

This paper is intended in part as a survey of results which have been developed in a series of papers in a broader context - the hidden k-logics. These logics are a generalization of k-deductive systems and their study originated in a series of lectures on Abstract Algebraic Logic - Application to Computer Science, during Don Pigozzi's visit to CAUL, Lisbon, in 1999 (Pigozzi, 1999).

When we refer to a deductive system we usually mean a 1-dimensional deductive system such as the deductive system of the classical, the intuitionistic and modal propositional calculi. The notion of deductive system, as an abstract consequence operator, is due to Tarski (see (Tarski, 1930)). In their groundbreaking work (Blok and Pigozzi, 1986), Blok and Pigozzi introduced and studied the notion of Leibniz congruence in the context of 1-deductive systems. These systems have been studied for years by several logicians, for example Blok and Pigozzi (Blok and Pigozzi, 1989), Czelakowski (Czelakowski, 1981) and (Czelakowski, 2001), the Barcelona group led by Font and Jansana (Font and Jansana, 1996), and also Hermann (Herrmann, 1996), Pigozzi (Pigozzi, 2001) and Wójcicki (Wójcicki, 1988).

The higher dimensional systems, called *k-deductive systems*, constitute a natural generalization including many logical systems, e.g., equational and inequational logics. k-deductive systems were introduced by Blok and Pigozzi in (Blok and Pigozzi, 1992); other important references are (Czelakowski and Pigozzi, 1999) and (Pałasińska, 1994). For these higher dimension deductive systems Blok and Pigozzi developed a theory similar to the theory of 1-deductive systems. The notion of Leibniz congruence still plays the central role in the study of k-deductive systems. There is another generalization of k-deductive systems, called K-deductive systems, that includes Gentzen systems. This theory was developed by Pałasińska and presented in her PhD thesis (Pałasińska, 1994).

We also should call attention to the work by Voutsadakis on the study of deductive systems within category theory. Voutsadakis has done extensive work on categorical abstract algebraic logic of π-institutions (Voutsadakis, 2004).

Abstract algebraic logic (AAL) is an area of algebraic logic that focuses on the study of the relationship between logical equivalence and logical truth. Moreover, AAL is centered on the process of associating a class of algebras to a logical system. This approach contrasts with the usual treatment given in algebraic logic where the emphasis is on the study of the class of algebras obtained by this process. A logical system, a *deductive system* as it has been called, is taken to be a pair formed by a signature Σ and a substitution-invariant closure relation on the set of terms over Σ in a countably infinite fixed set of variables X, $\mathrm{Te}_\Sigma(X)$ (we will use the word 'formula' as a synonym for 'term'). By a *closure relation* on $\mathrm{Te}_\Sigma(X)$ we mean a binary relation $\vdash$, where $\vdash \subseteq \mathcal{P}(\mathrm{Te}_\Sigma(X)) \times \mathrm{Te}_\Sigma(X)$, between subsets of terms and individual terms satisfying the following conditions: (1) $\Gamma \vdash \gamma$ for each $\gamma \in \Gamma$ and (2)

$\Gamma \vdash \varphi$ and $\Delta \vdash \gamma$ for each $\gamma \in \Gamma$ imply $\Delta \vdash \varphi$. The relation $\vdash$ is said to be *substitution-invariant* if $\Gamma \vdash \varphi$ implies $\sigma(\Gamma) \vdash \sigma(\varphi)$ for every substitution $\sigma : X \to \mathrm{Te}_\Sigma(X)$. Moreover, $\vdash$ is said to be *finitary* if $\Gamma \vdash \varphi$ implies $\Delta \vdash \varphi$ for some finite subset Δ of Γ.

The main paradigm in AAL is the representation of the classical propositional calculus in the equational theory of Boolean algebras by means of the *Lindenbaum-Tarski process*. In its traditional form, the Lindenbaum-Tarski process relies on the fact that the classical propositional calculus has a biconditional "$\leftrightarrow$" that defines logical equivalence. The set of all formulas is partitioned into logical equivalence classes and then abstracted by the familiar algebraic process of forming the quotient algebra. This algebra is called the *Lindenbaum-Tarski algebra*. There are many deductive systems that do not have a biconditional, and hence the Lindenbaum-Tarski process cannot be applied directly. However, there is an abstract notion of logical equivalence in every deductive system called the *Leibniz congruence*. In this way the Lindenbaum-Tarski process can be generalized so as to apply to many deductive systems.

The Leibniz congruence $\Omega(T)$ on the term algebra over a theory T is characterized in the following way: for any pair of terms α, β, $\alpha \equiv \beta \, (\Omega(T))$ if for every term φ and any variable p occurring in φ, $\varphi(p/\alpha) \in T$ if and only if $\varphi(p/\beta) \in T$. The Leibniz congruence is extended in a natural way to the power set of an arbitrary algebra. Given a Σ-algebra A and a designated subset F of A, the pair $\langle \mathbf{A}, F \rangle$ is called a *matrix*. The relation $\Omega(F)$ identifies any two elements which cannot be distinguished by any property defined by a formula. More precisely, for any pair of elements a, b of $\mathbf{A}$, $a \equiv b \, (\Omega(F))$ if for each formula $\varphi(x, u_0, \cdots, u_{k-1})$, and all parameters $\bar{c} \in A^k$, $\varphi^{\mathbf{A}}(a, \bar{c}) \in F$ if and only if $\varphi^{\mathbf{A}}(b, \bar{c}) \in F$. Moreover, $\Omega(F)$ is a congruence on $\mathbf{A}$. A matrix $\langle \mathbf{A}, F \rangle$ is said to be *reduced* if $\Omega(F)$ is the identity relation. The congruence $\Omega(F)$ is called the Leibniz congruence since it may be seen as the sentential version of the second order definition of equality given by Leibniz. He defined two objects to be equal if they have exactly the same properties. In the model of a world given by the matrix $\langle \mathbf{A}, F \rangle$, a property is determined by a formula $\varphi(x, \bar{u})$ and parameters $\bar{c} \in A^k$. Thus two elements are equal, in the Leibniz sense, if the condition above holds.

Equational logic serves as the underlying logic in many formal approaches to program specification. The algebraic data types specified in this formal way may be viewed as abstract machines on which the programs are to be run. This is one way of giving a precise algebraic semantics for programs, against which the correctness of a program can be tested. Equational logic can be seen as a 2-deductive system and then the tools and results of AAL can be applied to it.

However, object oriented (OO) programs present a special challenge for equational methods. This is due to properties inherent to the OO programs. A more appropriate model for the abstract machine in the case of an OO program is, arguably, a state transition system: as in the case of a state of such a system, a state of an OO program can be viewed as encapsulating all pertinent information about the abstract machine when it reaches the state during execution of the program. As a way of meeting the aforementioned challenge the standard equality predicate can be augmented by *behavioral equivalence*; in this way many of the characteristic properties of state transition systems can be grafted onto equational logic.

Two terms are said to be behaviorally equivalent if and only if they cannot be distinguished by any visible context. This is the primitive notion of behavioral equivalence due to Reichel (Reichel, 1984). The idea of looking at the satisfaction relation of hidden terms as behavioral equivalence was also introduced by Reichel in the 80's (Reichel, 1984) and it seems to be the correct way of interpreting equality between hidden terms. Since then, it has been adopted and generalized by many people. The most significant contributions have been given by Goguen and Malcolm (Goguen and Malcolm, 2000), Bidoit and Hennicker (Bidoit and Hennicker, 1996) and their coworkers.

Generalizations of the notion of behavioral equivalence have been considered in the literature. Goguen et al. consider Γ-*behavioral equivalence*, where Γ is a subset of the set of all operation symbols in the signature (see, e.g., (Goguen and Roşu, 1999)). Γ-behavioral equivalence is defined in a manner analogous to ordinary behavioral equivalence, but making use only of the contexts built from the operation symbols in Γ. It can be proved that the Γ-behavioral equivalence is the largest Γ-congruence with the identity as the visible part. Thus, coinduction methods, based on this fact, may still be formulated for this more general notion. Other interesting questions concerning Γ-behavioral equivalence may arise, such as the study of the compatibility of some operation symbols outside of Γ with respect to Γ-behavioral equivalence. This problem has been studied by Diaconescu and Futatsugui (Diaconescu and Futatsugi, 2000) and Bidoit and Hennicker (Bidoit and Hennicker, 1999).

On the other hand, Bidoit and Hennicker (Bidoit and Hennicker, 1996) generalize this notion by endowing the hidden algebras with a binary relation that may be partial. As a particular case we can apply their algebraic approach to the behavioral setting by considering algebras endowed with the Γ-behavioral equivalence.

One important feature of behavioral equivalence in computer science is that it is the largest congruence that is the identity on the visible part. This is, in some way, similar to the property of the Leibniz congruence being the largest congruence compatible with the filter. To apply AAL to the theory of the specification of abstract data types, we have to view specification logic as a deductive system (i.e., as a substitution-invariant closure relation on an appropriate set of formulas) and behavioral equivalence as a generalized notion of Leibniz congruence. The class of deductive systems has to be expanded so as to include multisorted as well as one-sorted systems. The notion of Leibniz congruence has to be considered in the context of the dichotomy of visible vs. hidden; namely, the formulas used in the characterization of the Leibniz congruence also have to be restricted to an appropriate proper subset of all formulas, namely the visible formulas, which are called *contexts*. Therefore, the notion of k-deductive systems has to be generalized by considering the data to be heterogeneous in the sense that the data elements may be of different kinds. Specifically, there are the basic data, like integers, reals and Boolean, whose properties are well-known and for which well-defined and easily manipulated representations are available; and there are the auxiliary data such as arrays, lists, stacks, whose properties are specified by their behavior under the programs with visible output, and hence ultimately in terms of the basic data. Thus, we use distinct representations for each kind of data elements.

This leads to the notion of hidden k-logics. They are a natural generalization of k-deductive systems. They encompass deductive systems as well as equational logics and inequational logics and their respective hidden versions. Hidden k-logics are used to specify systems whose data may be heterogeneous, i.e., split in different kinds, usually called sorts. Moreover, in hidden k-logics we are also able to distinguish internal data (*hidden data*) and the real data (*visible data*). This advantage is central in the specification of OO systems.

Hidden k-logics, as a natural generalization of k-deductive systems, were introduced by Martins and Pigozzi in (Martins and Pigozzi, 2007). Preliminary work on applications of AAL to the specification of abstract data types had been discussed in Lisbon, CAUL, in a series of lectures given by Don Pigozzi in 1999. The theory was then developed in (Martins, 2004), where improvements concerning specification and verification of programs were established using tools from *abstract algebraic logic* (AAL). A generalization of the AAL theory to the hidden setting has been successfully explored and several applications to the OO paradigm have been developed using AAL methods (cf. (Martins, 2004, 2006, 2007, 2008; Martins and Pigozzi, 2007)). This generalization is not straightforward. The multisort aspect is present for example in the following: in the one-sorted case one can show that a hidden k-logic is *protoalgebraic* (an important semantic property) if and only if it admits a protoequivalence system without parameters; however, in the broader context of multi-sorted logics, a generic protoequivalence system contains parameters (cf. (Martins, 2007)). This new bridge between AAL and the specification and verification theory of software systems has yet to be further developed. On the other hand, it must be mentioned that behavioral specification theory has also influenced the development of AAL, namely the recent theory of behavioral algebraization of logics (cf. (Caleiro et al., 2009) and (Voutsadakis, 2014)).

For the purposes of this paper, it is useful to define a hidden k-logic as an abstract closure relation on the set of k-formulas. That is, a *hidden k-logic* is a pair $\mathcal{L} = \langle \Sigma, \vdash_{\mathcal{L}} \rangle$, where Σ is a hidden signature and $\vdash_{\mathcal{L}}$ is a substitution-invariant closure relation on the set of visible k-formulas, called *the consequence relation of* $\mathcal{L}$. This consequence relation may be finitary or not. It is finitary just in case it admits a presentation by axioms and inference rules, in the usual Hilbert style. In this case, $\vdash_{\mathcal{L}}$ is said to be *specifiable*. An $\mathcal{L}$-*theory* of a hidden k-logic $\mathcal{L}$ is a set of visible k-formulas that are closed under the consequence relation $\vdash_{\mathcal{L}}$. The set of all $\mathcal{L}$-theories is denoted by $\mathrm{Th}(\mathcal{L})$.

Hidden k-logics are useful mainly because they encompass not only the 2-dimensional hidden and standard equational logics, but also *Boolean logics*; these are 1-dimensional multisorted logics with Boolean as the only visible sort, and with equality-test operations for some of the hidden sorts in place of equality predicates. They also include all assertional logics, the purview of AAL. In this way we obtain a unified theory for a variety of logical systems. We give special attention to a special hidden 2-logic, the *hidden equational logic*. In the hidden equational case we only consider a primitive notion of equality between visible data. It is defined as a sorted equational logic, using reflexivity, symmetry, transitivity and congruence rules, but only on the visible part. The expression "hidden equational logic" comes from the fact that the equality predicate is restricted so as to apply only to visible data elements. There is no primitive notion of equality for hidden data elements in the logic.

There is an important assumption about the syntax of hidden k-logics, as we define them, that arises from the fact that they are intended to serve as the underlying logic in the specification of object oriented systems. The assumption is that the specification can use only visible axioms since we only have access to the internal behavior by programs with visible output, i.e., the equality between two hidden data elements of the same sort is not specifiable by abstract equality axioms as in the standard equational logic. This assumption follows the work of Leavens and Pigozzi (Leavens and Pigozzi, 2000, 2002). The restriction to axioms of visible type is natural from the perspective of operational semantics. That is, in operational terms, one views the axioms as specifying the output of programs, which indirectly determine the behavior of the hidden data objects the programs manipulate. Hence, only the visible part of the system is specified. This does not follow some approaches in the area (see, e.g., (Goguen and Roşu, 1999) and (Roşu, 2000)) but it does not restrict the power of specifications in practice. On the contrary, it endows the underlying theory with even richer modes of specification. Indeed, we may also specify internal properties of the system after checking that they do not produce unexpected behavioral changes on the system, i.e., that by adding those properties to the specification we do not obtain a different set of behavioral consequences. In the hidden equational case, we show that if a conditional equation is behaviorally valid, then it may be added as a new axiom without any undesired consequences (Martins and Pigozzi, 2007).

The semantics of hidden k-logics reflects all the special features of these logics that have been discussed. A k-*data structure* is a collection of data items of different sorts, such as lists, Booleans, numbers, and operations involving them, together with a set of k-tuples of elements, called a *filter*, that serves as a set of generalized truth values (the term filter comes from the one-sorted case in AAL). Moreover, the universe, which is a sorted algebra, is split into two disjoint sets, namely the *hidden* part, which corresponds to the states of a state transition system, and the *visible* part. There are also two different interpretations of the operation symbols: the *attributes* return visible data and are used to observe the state of the system while the *methods* may change the state. As in AAL, the main object is to understand and clarify the relationship between logical truth and logical equivalence, which for hidden k-logics correspond, respectively, to the visible properties of states, as specified by the axioms of the logic, and their behavioral equivalence.

Since the consequence relation $\vdash$ is a closure relation on the set of visible k-formulas, the filter consists exclusively of k-tuples of visible elements of the k-data structure. The designated filter F of a k-data structure $\mathcal{A} = \langle \mathbf{A}, F \rangle$ is considered as the set of "truth values" in $\mathcal{A}$. Thus, we say that $\mathcal{A} = \langle \mathbf{A}, F \rangle$ is a *model of* a hidden k-logic $\mathcal{L}$ if every consequence $\Gamma \vdash \varphi$ of $\mathcal{L}$ is a semantic consequence of $\mathcal{A}$, in the sense that for every assignment $h : X \to A$, $h(\varphi) \in F$ whenever $h(\Gamma) \subseteq F$. In this case, we say that F is an $\mathcal{L}$-*filter*. The $\mathcal{L}$-filters on the term algebra are the theories of $\mathcal{L}$ and consequently $\langle \mathrm{Te}_\Sigma(X), T \rangle$, with T a theory of $\mathcal{L}$, is always a model of $\mathcal{L}$.

1.1 Related work

Many computer scientists have studied behavioral equivalence for the last 20 years. Here we present some approaches which are important to contextualize our work.

1.1.1 Hidden algebras

Hidden algebras were introduced by Goguen in (Goguen, 1989) and further developed in (Goguen and Malcolm, 2000; Goguen and Roşu, 1999; Roşu, 2000), in order to generalize many-sorted algebras and give an algebraic semantics for the object oriented paradigm.

When they first appeared, hidden algebras were considered over restricted signatures. These were assumed to have the visible part fixed, in the sense that all sorted algebras over it have the same visible part. Usually, this visible part was a standard algebra such as the natural numbers or the two-element Boolean algebra. This is called *fixed-data semantics*. Another restriction which is sometimes assumed in order to apply coalgebraic methods and results to the study of behavioral equivalence is the requirement that the methods and the attributes must have exactly one hidden argument. In this case it is called *monadic semantics*.

The behavioral aspects of modern software make hidden algebras more suitable than standard algebras for abstract machine implementations. Consequently, there has been an increasing development in this field. In the last fifteen years the theory on hidden algebras has been further developed and applied to more general settings, first by Goguen and Malcolm (Goguen and Malcolm, 2000) and more recently by his former collaborators (Goriac et al., 2010; Moore and Roşu, 2015). Currently, almost all of the results may be established for *polyadic loose-data semantics*. Polyadic loose-data semantics allows any kind of operation symbols. Furthermore, in order to have more freedom to choose an adequate implementation, the visible part of the algebras is no longer fixed: it may be any sorted algebra in which the requirements (axioms) of the given specification are valid. Moreover, some authors are interested in applying coalgebraic methods, and then they have to restrict their signatures to the monadic fixed-data semantics. Malcolm (Malcolm, 1996) has shown that behavioral equivalence may be formulated in the context of coalgebra (see also (Reichel, 1995)).

1.1.2 Behavioral equivalence and behavioral validity

Two terms are said to be behaviorally equivalent if and only if they cannot be distinguished by any visible context. This is the primitive notion of behavioral equivalence due to Reichel (Reichel, 1984).

The idea of looking at the satisfaction relation of hidden terms as behavioral equivalence was also introduced by Reichel in the 80's (Reichel, 1984) and it seems to be the correct way of interpreting equality between hidden terms. Since then, it has been adopted and generalized by many authors. The most significant contributions have been given by Goguen, Bidoit, Bouhoula and their co-authors (e.g., (Goguen and Malcolm, 1999; Bidoit and Hennicker, 1996; Bouhoula and Rusinowitch, 2002)).

Generalizations of the notion of behavioral equivalence have also been considered in the literature. Goguen and Roşu (Goguen and Roşu, 1999) introduced and studied Γ-behavioral equivalence, where Γ is a subset of the set of all operation symbols in the signature. Γ-behavioral equivalence is defined analogously to ordinary behavioral equivalence, but making use only of the contexts built from the operation symbols in Γ. It can be proved that the Γ-behavioral equivalence is the largest Γ-congruence with the identity as the visible part. Thus, coinduction methods, based on this fact, may still be formulated for this more general notion. Other interesting questions concerning Γ-behavioral equivalence may arise, such as the study of the compatibility of some operation symbols outside of Γ with respect to Γ-behavioral equivalence. This problem has been studied by Diaconescu and Futatsugi (Diaconescu and Futatsugi, 2000) and Bidoit and Hennicker (Bidoit and Hennicker, 1999).

On the other hand, Bidoit et al. (Bidoit et al., 1995) generalize this notion by endowing the hidden algebras with a binary relation. As a particular case we can apply their algebraic approach to the behavioral setting by considering the algebras together with the Γ-behavioral equivalence.

1.1.3 Hidden logics

Many behavioral logics have been defined and studied in the literature. The most relevant versions are *hidden logic* (Goguen and Malcolm, 1999, 2000) and *observational logic* (Hennicker and Bidoit, 1999; Bidoit and Hennicker, 1996). There is also another observational logic due to Padawitz (Padawitz, 2000), called *swinging types logic*, but it is similar to the observational logic of Bidoit and his coworkers (see `http://ls5-www.cs.uni-dort mund.de/~peter/ Swinging.html` for more details).

Hidden logic is a variant of the equational logic in which some part of the specification is visible and another is hidden. The formulas are just equations and the satisfaction relation is taken behaviorally.

Observational logic is different from hidden logic but both are based on behavioral equivalence, which means indistinguishability under contexts. Observational logic was introduced by Bidoit and Hennicker (see (Bidoit and Hennicker, 1996), (Hennicker and Bidoit, 1999) and (Hennicker, 1997)) to formalize behavioral validity (correctness). Tarski's satisfaction relation of first-order formulas (with equality) is considered as a "behavioral satisfaction relation" which is determined, in a natural way, by the family of congruence relations (possibly partial) with which each algebra is endowed. This relation is called *behavioral equality*. The behavioral satisfaction relation is just defined by considering the equality symbol interpreted as the behavioral equality. First-order theories are generalized to the so-called *behavioral theories* where the equality symbol is interpreted as the behavioral equality. In (Bidoit and Hennicker, 1996) Bidoit and Hennicker develop a method for proving behavioral theorems whenever an axiomatization of the behavioral equality is provided. This is based on reducing behavioral satisfaction to ordinary satisfaction. Consequently any proof system for first-order logic can be used to prove the behavioral validity, with respect to a given behavioral equality, of first-order formulas.

1.1.4 Automatic methods for behavioral reasoning in hidden logics

As far as we know, the languages that support automated behavioral reasoning are Spike (Berreged et al., 1998), CafeOBJ (Diaconescu and Futatsugi, 1998) and BOBJ (Goguen et al., 2000).

In (Bouhoula and Rusinowitch, 2002), Bouhoula and Rusinowitch set forth an automatic method for proving behavioral validity of conditional equations in conditional specifications. They use the fact that there are specifications for which a smaller set of contexts is enough to know what the outputs of the remaining ones are. They call them *critical contexts*. The work of Bouhoula and Rusinowitch was the genesis of the SPIKE language which is based on context induction.

The CafeOBJ language was developed by Diaconescu and Futatsugi (Diaconescu and Futatsugi, 1998). It implements behavioral rewriting to make behaviorally sound reductions of terms. It is based on a behavioral version of the well known efficient method of rewriting for automated theorem proving (see `http://www.ldl.jaist.ac.jp/Research/CafeOBJ/`).

Goguen et al. have been developing algorithms for automating behavioral reasoning based on their techniques of coinduction and have been making use of cobases. Coinduction in its pure form requires human intervention in the choice of the cobasis. A cobasis is just a set of operation symbols that generates a relation on the set of terms which is a subset of the behavioral equivalence. A good choice of a cobasis can simplify the proof enormously. Those algorithms have been improved in order to be applied to more general situations and have been implemented in the BOBJ language. In (Goguen et al., 2000) Goguen et al. presented a new technique which combines behavioral rewriting and coinduction. The most recent version is CCCRW, called *conditional circular coinductive rewriting with case analysis*. The authors claim that it is in fact the most powerful automated proof technique available at present (Goguen et al., 2002) (see also (Goriac et al., 2010; Moore and Roşu, 2015)). Besides the fact that this new algorithm uses conditional circular coinductive rewriting to prove behavioral validity, it also allows for case analysis (see `http://www-cse.ucsd.edu/groups/tatami/bobj`). This theoretical result supports the automated behavioral prover **Circ** based on the circularity principle (`http://fsl.cs.illinois.edu/index.php/Circ`), which generalizes both circular coinduction and structural induction.

1.1.5 AAL approach to behavioral equivalence

As mentioned earlier, Pigozzi gave a series of lectures on the application of AAL to Computer Science. These lectures marked the starting point of Martins' investigations on the algebraic theory of hidden k-logics, which led to his PhD thesis (Martins, 2004). In an introductory paper with Pigozzi (Martins and Pigozzi, 2007) the instantiation to hidden equational logic was studied in depth. Closure properties of the class of behavioral models (reduced models) are studied in (Martins, 2007). Refinement and institutions for behavioral logics in the context of our approach were discussed in (Martins, 2006). A natural generalization of the Nerode equivalence of finite automata to k-data structures concerning this general notion of behavioral equivalence of a k-data structure can be found in (Martins, 2008). Recently, a deduction-detachment theorem for hidden k-logics was presented in (Babenyshev and Martins, 2014) and the behavioral equivalence between hidden k-logics is investigated in (Babenyshev and Martins, 2016) by Babenyshev and Martins.

Outline of the paper

This paper is organized in two main sections. In section 2 we give an overview of basic concepts and recent results pertaining to hidden k-logic. The Leibniz congruence is one of the tools developed in this context. As in standard AAL, it plays a crucial role in the theory. Section 3 is devoted to behavioral equivalence. Theorem 3.3 shows the adequacy of this approach: the behavioral equivalence is, in fact, the (generalized) Leibniz congruence. In Sections 3.1 and 3.2 we present new characterizations for an equation to be a behavioral consequence of a theory of a HEL. In Section 3.2 we get a simpler characterization for the case of strict equational HEL's.

2 Hidden k-logic

A *hidden (sorted) signature* is a triple $\Sigma = \langle \text{SORT}, \text{VIS}, \langle \text{OP}_\tau \mid \tau \in \text{TYPE} \rangle \rangle$, where: SORT is a nonempty, countable set whose elements are called *sorts*; VIS is a subset of SORT, called the set of *visible sorts*; TYPE is a set of nonempty sequences $S_0, \ldots, S_n$ of sorts, called *types* and usually written as $S_0, \ldots, S_{n-1} \to S_n$; and, for each $\tau \in \text{TYPE}$, OP_τ is a countable set; the elements of OP_τ are called *operation symbols* of type τ. Operation symbols of type $\to S$ are said to be *constants*. We will denote $\langle \text{OP}_\tau \mid \tau \in \text{TYPE} \rangle$ by OP.

The sorts in $\text{SORT} \setminus \text{VIS}$, that are not visible, are called *hidden sorts*. The set of hidden sorts is denoted by HID. For simplicity, we require the sets of operation symbols to be pairwise disjoint in order to avoid overloading of names (i.e., for any distinct $\tau, \tau' \in \text{TYPE}$, $\text{OP}_\tau \cap \text{OP}_{\tau'} = \varnothing$).

From each hidden signature Σ we obtain the associated *un-hidden* signature Σ^{UH} by making all sorts of Σ visible.

A Σ-*algebra* is a pair $\langle A, \langle O^A \mid \tau \in \text{TYPE}, O \in \text{OP}_\tau \rangle \rangle$, where A is a SORT-sorted set, such that $A_S \neq \varnothing$, for all $S \in \text{SORT}$, and for any $\tau \in \text{TYPE}$ and $O \in \text{OP}_\tau$, O^A is an operation on A of type τ (i.e., if $\tau = S_0, \ldots, S_{n-1} \to S_n$ then $O^A : A_{S_0} \times \cdots \times A_{S_{n-1}} \to A_{S_n}$). As usual, we use the same symbol to denote an algebra and the carrier of the algebra.

We assume for carrier sets A of data structures that $A_S \neq \varnothing$ for all $S \in \text{SORT}$, a condition similar to one used to define regular universal algebras. With this assumption we exclude some data structures of practical interest. However, the mathematics is simpler in this case and most results of universal algebra hold in their usual form.

A (*sorted*) *congruence on* a Σ-algebra A is a sorted binary relation $\Theta \subseteq A^2$ such that: (i) for each $S \in \text{SORT}$, Θ_S is an equivalence relation on A_S and (ii) Θ satisfies the congruence condition: for every operation symbol $O \in \text{OP}_\tau$ with $\tau = S_0, \ldots, S_{n-1} \to S_n$, and all $a_0, a_0' \in A_{S_0}, \ldots, a_{n-1}, a_{n-1}' \in A_{S_{n-1}}$

such that $a_i \Theta_{S_i} a_i'$, $O^A(a_0,\ldots,a_{n-1}) \Theta_{S_n} O^A(a_0',\ldots,a_{n-1}')$ holds. The set of all congruences over A is denoted by $\mathrm{Con}(A)$.

The sorted notions of subalgebra, homomorphism, isomorphism, etc. are defined in a natural way (see (Meinke and Tucker, 1992) for the formal definitions).

For each set of sorts SORT we fix a locally countably infinite sorted set $X = \langle X_S : S \in \mathrm{SORT} \rangle$ of *(propositional sorted) variables*. We assume the components of the sorted set of variables are pairwise disjoint. The elements in X_S are called *S-variables*. To denote that a variable x is of sort S (i.e., that $x \in X_S$) we write $x\!:\!S$.

We say that a term $O(t_0,\ldots,t_{n-1})$, where $O \in \mathrm{OP}_\tau$ with $\tau = S_0,\ldots,S_{n-1} \to S_n$, has type S_n. Given a signature Σ we define the SORT-sorted set $\mathrm{Te}_\Sigma(X)$ of terms over the signature Σ with variables in X as usual. Note that, since the components of the family $\mathrm{Te}_\Sigma(X)$ are pairwise disjoint, a SORT-sorted subset Γ of $\mathrm{Te}_\Sigma(X)$ can be identified with the unsorted set $\bigcup_{S \in \mathrm{SORT}} \Gamma_S$. A hidden signature Σ is said to be *standard* if there is a ground term (i.e., a term without variables) of every sort. We use the lower case Greek letters $\varphi, \psi, \vartheta, \ldots$ to represent terms, possibly with annotations to indicate sorts of terms and variables. Specifically, writing φ in the form

$$(1) \qquad\qquad \varphi(x_0\!:\!S_0,\ldots,x_{n-1}\!:\!S_{n-1})\!:\!S$$

indicates that φ is of sort S and that the variables that actually occur in φ are included in the list $x_0,\ldots,x_{n-1}$ of variables of sorts $S_0,\ldots,S_{n-1}$, respectively.

We define, in the usual way, operations over $\mathrm{Te}_\Sigma(X)$ to obtain the *term algebra* over the signature Σ. It is well known that $\mathrm{Te}_\Sigma(X)$ has the *universal mapping property over X* in the sense that, for every Σ-algebra A and every sorted map $h : X \to A$, called an *assignment*, there is a unique sorted homomorphism $h^* : \mathrm{Te}_\Sigma(X) \to A$ that extends h. In the sequel, we will not distinguish between these two maps. If φ is the term (1), and $a_i \in A_{S_i}$, $i < n$, we write $\varphi^A(a_0,\ldots,a_{n-1})$ for the image $h(\varphi)$ under any homomorphism $h : \mathrm{Te}_\Sigma(X) \to A$ such that $h(x_i) = a_i$ for all $i < n$. A map from X to the set of terms, and its unique extension to an endomorphism of $\mathrm{Te}_\Sigma(X)$, is called a *substitution*.

To provide a context that allows us to deal simultaneously with specification logics that are assertional (for example the ones with a Boolean sort but no equality) and equational, we introduce the notion of a *k-term* for any nonzero natural number k. In the sequel k denotes a fixed nonzero natural number. A *k-variable of sort S* is a sequence of k variables all of the same sort S. A *k-term (k-formula in logical context) of sort S over Σ* is a sequence of k Σ-terms all of the same sort S. We indicate k-terms by overlining, so $\bar\varphi(x_1,\cdots,x_n)\!:\!S = \langle \varphi_0(x_1,\cdots,x_n)\!:\!S,\ldots,\varphi_{k-1}(x_1,\cdots,x_n)\!:\!S \rangle$. When we do not need to make the common sort S of each term of $\bar\varphi\!:\!S$ explicit, we simply write it as $\bar\varphi$. $\mathrm{Te}_\Sigma^k(X)$ is the sorted set of all k-terms over

Σ. Thus $\mathrm{Te}_\Sigma^k(X) = \langle (\mathrm{Te}_\Sigma(X))_S^k : S \in \mathrm{SORT} \rangle$. The set of all *visible k-terms* $(\mathrm{Te}_\Sigma^k(X))_{\mathrm{VIS}}$ is the VIS-sorted set $\langle (\mathrm{Te}_\Sigma(X))_V^k : V \in \mathrm{VIS} \rangle$.

2.1 Data structures and Leibniz congruence

Let Σ be a hidden signature. A *visible k-data structure* (a *k-data structure* for short) *over* Σ is a pair $\mathcal{A} = \langle A, F \rangle$, where A is a Σ-algebra and $F \subseteq A_{\mathrm{VIS}}^k$; A is called the *underlying algebra* and F the *designated filter of* $\mathcal{A}$ (see (Martins and Pigozzi, 2007) for examples in the hidden equational case).

Let $\mathcal{A} = \langle A, F \rangle$ be a k-data structure. A congruence relation Θ on A is VIS-*compatible* (or simply *compatible*) with F if for all $V \in \mathrm{VIS}$ and for all $\bar{a}, \bar{a}' \in A_V^k$ the following condition holds.

$$\text{if } a_i \equiv a_i'(\Theta_V) \text{ for all } i \leq k \text{ then, } \bar{a} \in F_V \text{ iff } \bar{a}' \in F_V;$$

that is, each F_V is the union of a cartesian product of Θ_V-classes i.e.,

$$F_V = \bigcup_{\bar{a} \in F_V} (a_1/\Theta_V) \times (a_2/\Theta_V) \times \cdots \times (a_k/\Theta_V).$$

Lemma 2.1. *Let $\mathcal{A} = \langle A, F \rangle$ be a k-data structure. There is a largest congruence relation on A compatible with F.*

Proof. Let Φ and Ψ be two congruences on A compatible with F. The relational product $\Phi \circ \Psi$, defined for each $S \in \mathrm{SORT}$ by

$$(\Phi \circ \Psi)_S := \big\{ \langle a, b \rangle \in A_S^2 : \exists c \in A_S \left(\langle a, c \rangle \in \Phi_S \text{ and } \langle c, b \rangle \in \Psi_S \right) \big\},$$

is also compatible with F. Since the join $\Phi \vee \Psi$, in the lattice of congruences, is given by $\bigcup_{i<\omega} \Phi \circ^i \Psi$, where $\Phi \circ^0 \Psi = \Delta_A$ and $\Phi \circ^{i+1} \Psi = (\Phi \circ^i \Psi) \circ (\Phi \circ \Psi)$, we have that $\Phi \vee \Psi$ is also compatible with F. Hence, the set of all congruence relations on A compatible with F is directed in the sense that, for any pair of congruences compatible with F, there is a third congruence with the same property that includes both of them. We can conclude from this that the union of all compatible congruences is again a compatible congruence. Therefore, the largest congruence compatible with F always exists. $\square$

Definition 2.2. Let $\mathcal{A} = \langle A, F \rangle$ be a k-data structure. The largest congruence relation on A compatible with F is called the *Leibniz congruence of F on A* and is denoted by $\Omega_A(F)$.

The Leibniz congruence plays a central role in abstract algebraic logic when restricted to single-sorted, k-data structures; see for example (Pigozzi, 2001) and (Font et al., 2003). The term was introduced in (Blok and Pigozzi, 1989), but the concept appeared much earlier. The motivation behind the choice of the term *Leibniz* will become clear after Theorem 3.3. A systematic study of the Leibniz congruence in hidden k-logics can be found in (Martins, 2004), in particular a proof of its characterization is given in Theorem 3.3. In the case of single-sorted 1-data structures, this result was well known in the literature of sentential logic; see for example (Blok and Pigozzi, 1989).

An interesting property of the Leibniz congruence is its preservation under inverse images of surjective homomorphisms, i.e., given a k-data structure $\mathcal{A} = \langle A, F \rangle$ over Σ, a Σ-algebra B and a surjective homomorphism $h : B \to A$, we have that $h^{-1}(\Omega_A(F)) = \Omega_B(h^{-1}(F))$.

2.2 Hidden k-logic

For each nonzero natural number k, a hidden k-logic is considered to be a consequence relation on the set of visible k-terms of some hidden signature, independently of any specific choice of axioms and rules of inference. More precisely, it is defined as a substitution invariant consequence relation on the set of visible k-terms.

Definition 2.3. A *hidden k-logical system* (*hidden k-logic* for short) is a pair $\mathcal{L} = \langle \Sigma, \vdash_{\mathcal{L}} \rangle$, where Σ is a hidden signature with VIS as its set of visible sorts, and $\vdash_{\mathcal{L}} \subseteq \mathcal{P}((\mathrm{Te}_{\Sigma}^{k}(X))_{\mathrm{VIS}}) \times (\mathrm{Te}_{\Sigma}^{k}(X))_{\mathrm{VIS}}$ is an (unsorted) relation that satisfies for all $\Gamma \cup \Delta \cup \{\bar{\gamma}, \bar{\varphi}\} \subseteq (\mathrm{Te}_{\Sigma}^{k}(X))_{\mathrm{VIS}}$ the following conditions:

(i) $\Gamma \vdash_{\mathcal{L}} \bar{\gamma}$ for each $\bar{\gamma} \in \Gamma$;
(ii) if $\Gamma \vdash_{\mathcal{L}} \bar{\varphi}$, and $\Delta \vdash_{\mathcal{L}} \bar{\gamma}$ for each $\bar{\gamma} \in \Gamma$, then $\Delta \vdash_{\mathcal{L}} \bar{\varphi}$;
(iii) if $\Gamma \vdash_{\mathcal{L}} \bar{\varphi}$, then $\sigma(\Gamma) \vdash_{\mathcal{L}} \sigma(\bar{\varphi})$ for every substitution σ.

Note, that being unsorted, $\vdash_{\mathcal{L}}$ can relate premises and consequences of different visible sorts.

A hidden k-logic is *specifiable* if $\vdash_{\mathcal{L}}$ is finitary (or compact), i.e., if $\Gamma \vdash_{\mathcal{L}} \bar{\varphi}$ implies $\Delta \vdash_{\mathcal{L}} \bar{\varphi}$ for some globally finite subset Δ of Γ (recall that a set Γ is said to be globally finite if for every $S \in \mathrm{SORT}$ A_S is a finite set and A_S is empty except for a finite number of sorts). The relation $\vdash_{\mathcal{L}}$ is called *the consequence relation of $\mathcal{L}$*; when $\mathcal{L}$ is clear from the context we simply write $\vdash$. A hidden k-logic with $\mathrm{VIS} = \mathrm{SORT}$ will be called a *visible k-logic*, or simply a k-logic.

As it is usual in a sentential logic framework, we treat formulas (k-formulas) as synonymous to terms (k-terms, respectively). Moreover, for a given hidden k-logic $\mathcal{L} = \langle \Sigma, \vdash_{\mathcal{L}} \rangle$ we denote $\mathrm{Te}_{\Sigma}^{k}(X)$ and $(\mathrm{Te}_{\Sigma}^{k}(X))_{\mathrm{VIS}}$ by $\mathrm{Fm}(\mathcal{L})$ and $\mathrm{Fm}_{\mathrm{VIS}}(\mathcal{L})$, respectively.

Hidden k-logics were introduced by Martins and Pigozzi (cf. (Martins and Pigozzi, 2007)) in the context of the algebraic specification and verification of software systems. The basic theory of hidden k-logics was presented in (Martins, 2004). The class of hidden k-logics includes such well-known logical systems as the 2-dimensional hidden and standard equational logics, as well as the Boolean logic (for more examples see (Martins, 2004)).

Every consequence relation $\vdash$ has a natural extension to a relation, also denoted by $\vdash$, between sets of visible k-terms; it is defined by $\Gamma \vdash \Delta$ if $\Gamma \vdash \bar{\varphi}$ for each $\bar{\varphi} \in \Delta$. We define the relation of *interderivability* between sorted sets in the following way: $\Gamma \dashv\vdash \Delta$ if, $\Gamma \vdash \Delta$ and $\Delta \vdash \Gamma$. We will abbreviate $\{\bar{\psi}\} \vdash \bar{\varphi}$, $\Gamma \cup \{\bar{\varphi}_0, \ldots, \bar{\varphi}_{n-1}\} \vdash \bar{\varphi}$ and $\Gamma_0 \cup \cdots \cup \Gamma_{n-1} \vdash \bar{\varphi}$ by $\bar{\psi} \vdash \bar{\varphi}$, $\Gamma, \bar{\varphi}_0, \ldots, \bar{\varphi}_{n-1} \vdash \bar{\varphi}$ and $\Gamma_0, \ldots, \Gamma_{n-1} \vdash \bar{\varphi}$, respectively.

Let $\mathcal{L}$ be a (not necessarily specifiable) hidden k-logic. By a *theorem* of $\mathcal{L}$ we mean a visible k-term $\bar{\varphi}$ such that $\vdash_{\mathcal{L}} \bar{\varphi}$, i.e., $\varnothing \vdash_{\mathcal{L}} \bar{\varphi}$. The set of all theorems is denoted by $\mathrm{Thm}(\mathcal{L})$. A rule such as

$$(2) \qquad \frac{\bar{\varphi}_0 : V_0, \ldots, \bar{\varphi}_{n-1} : V_{n-1}}{\bar{\varphi}_n : V_n},$$

where $\bar{\varphi}_0, \ldots, \bar{\varphi}_n$ are all visible k-terms, is said to be a *derivable rule* of $\mathcal{L}$ if $\{\bar{\varphi}_0, \ldots, \bar{\varphi}_{n-1}\} \vdash_{\mathcal{L}} \bar{\varphi}_n$. A set of visible k-terms T closed under the consequence relation, i.e., $T \vdash_{\mathcal{L}} \bar{\varphi}$ implies $\bar{\varphi} \in T$, is called a *theory* of $\mathcal{L}$ or $\mathcal{L}$-*theory*. The set of all theories is denoted by $\mathrm{Th}(\mathcal{L})$; it forms a complete lattice under set-theoretic inclusion, which is algebraic if $\mathcal{L}$ is specifiable. Let $T_i \in \mathrm{Th}(\mathcal{L})$, for $i \in I$. Their meet is $\bigcap_{i \in I} T_i$ and their joint is the intersection of all theories that contain each T_i, i.e., $\bigvee_{i \in I}^{\mathcal{L}} T_i = \bigcap \{T \in \mathrm{Th}(\mathcal{L}) : T_i \subseteq T \text{ for all } i \in I\}$. Given any set Γ of visible k-terms, the set $\mathrm{Con}_{\mathcal{L}}(\Gamma)$ is the smallest $\mathcal{L}$-theory containing Γ. It is easy to see that $\mathrm{Con}_{\mathcal{L}}(\Gamma) = \{\bar{\varphi} \in (\mathrm{Te}_{\Sigma}^k(X))_{\mathrm{VIS}} : \Gamma \vdash_{\mathcal{L}} \bar{\varphi}\}$, i.e., the set of all consequences of Γ.

Very often, a specifiable hidden k-logic has a Hilbert style presentation, i.e., it is given by a set of axioms (visible k-terms) and inference rules of the general form (2). We say that a visible k-term $\bar{\psi}$ is *directly derivable* from a set Γ of visible k-terms by a rule such as (2) if there is a substitution $h : X \to \mathrm{Te}_{\Sigma}(X)$ such that $h(\bar{\varphi}_n) = \bar{\psi}$ and $h(\bar{\varphi}_0), \ldots, h(\bar{\varphi}_{n-1}) \in \Gamma$.

Given a set AX of visible k-terms and a set IR of inference rules, we say that $\bar{\psi}$ is *derivable* from Γ by the set AX and the set IR, in symbols $\Gamma \vdash_{\mathrm{AX,IR}} \bar{\psi}$, if there is a finite sequence of k-terms, $\bar{\psi}_0, \ldots, \bar{\psi}_{n-1}$ such that $\bar{\psi}_{n-1} = \bar{\psi}$, and for each $i < n$ one of the following conditions hold:

(a) $\bar{\psi}_i \in \Gamma$,
(b) $\bar{\psi}_i$ is a substitution instance of a k-term in AX
(c) $\bar{\psi}_i$ is directly derivable from $\{\bar{\psi}_j\}_{j<i}$ by one of the inference rules in IR.

It is clear that $\langle \Sigma, \vdash_{\mathrm{AX,IR}} \rangle$ is a specifiable hidden k-logic. Moreover, a hidden k-logic $\mathcal{L}$ is specifiable iff there exist (possibly infinite) sets AX and IR, of axioms and inference rules, respectively, such that, for any visible k-terms $\bar{\psi}$ and any set Γ of visible k-terms, $\Gamma \vdash_{\mathcal{L}} \bar{\psi}$ iff $\Gamma \vdash_{\mathrm{AX,IR}} \bar{\psi}$. The pair $\langle \mathrm{AX}, \mathrm{IR} \rangle$ is called a *presentation* of $\mathcal{L}$ by axioms and inference rules. Hence we can present our examples of specifiable logics by exhibiting their set of axioms and of inference rules. If $\mathcal{L} = \langle \Sigma, \vdash_{\mathrm{AX},\mathrm{IR}} \rangle$, for some AX and IR with $|\mathrm{AX} \cup \mathrm{IR}| < \omega$, we say that $\mathcal{L}$ is *finitely axiomatizable*.

2.2.1 Semantics.

Let $\mathcal{A} = \langle A, F \rangle$ be a k-data structure. A visible k-term $\bar{\varphi} : V$ is said to be a *semantic consequence* of a set of visible k-terms Γ in $\mathcal{A}$, in symbols $\Gamma \models_{\mathcal{A}} \bar{\varphi}$, if, for every assignment $h : X \to A$, $h(\bar{\varphi}) \in F_V$ whenever $h(\bar{\psi}) \in F_W$ for every $\bar{\psi} : W \in \Gamma$. A visible k-term $\bar{\varphi}$ is a *validity* of $\mathcal{A}$, and conversely $\mathcal{A}$ is a *model* of $\bar{\varphi}$, if $\varnothing \models_{\mathcal{A}} \bar{\varphi}$. A rule such as (2) is a *valid rule* of $\mathcal{A}$, and conversely $\mathcal{A}$ is a *model* of the rule, if $\{\bar{\varphi}_0, \ldots, \bar{\varphi}_{n-1}\} \models_{\mathcal{A}} \bar{\varphi}_n$. A visible formula $\bar{\varphi}$ is a *semantic consequence* of a set of visible k-terms Γ for an arbitrary class $\mathcal{M}$ of k-data structures over Σ, in symbols $\Gamma \models_{\mathcal{M}} \bar{\varphi}$, if $\Gamma \models_{\mathcal{A}} \bar{\varphi}$ for each $\mathcal{A} \in \mathcal{M}$. It can be proved that $\models_{\mathcal{M}}$ is always a k-logic, however it may not be specifiable. A visible k-term or rule such as (2) is a *valid* rule of $\mathcal{M}$ if it is a validity of each member of $\mathcal{M}$.

A k-data structure $\mathcal{A}$ is a *model* of a hidden k-logic $\mathcal{L}$ if $\Gamma \vdash_{\mathcal{L}} \bar{\varphi}$ implies $\Gamma \models_{\mathcal{A}} \bar{\varphi}$, for every $\Gamma \cup \{\bar{\varphi}\} \subseteq (\mathrm{Te}_{\Sigma}^k(X))_{\mathrm{VIS}}$. When $\mathcal{A}$ is a model of $\mathcal{L}$ the designated filter of $\mathcal{A}$ is called an $\mathcal{L}$-*filter over* A. The set of all $\mathcal{L}$-filters over an algebra A is denoted by $\mathrm{Fi}_{\mathcal{L}}(A)$. The special models whose underlying algebra is the formula algebra, i.e., of the form $\langle \mathrm{Te}_{\Sigma}(X), T \rangle$, with $T \in \mathrm{Th}(\mathcal{L})$ are called *Lindenbaum-Tarski models*. The class of all models of $\mathcal{L}$ is denoted by $\mathrm{Mod}(\mathcal{L})$. If $\mathcal{L}$ is a specifiable hidden k-logic, then $\mathcal{A}$ is a model of $\mathcal{L}$ iff every axiom and rule of inference is a validity of $\mathcal{A}$. The class of all *reduced models of* $\mathcal{L}$, i.e., all models $\langle A, F \rangle$ such that $\Omega_A(F) = \mathrm{id}_A$, is denoted by $\mathrm{Mod}^*(\mathcal{L})$. A class of k-data structures $\mathcal{M}$ is a *data structure semantics* for $\mathcal{L}$ if $\vdash_{\mathcal{L}} = \models_{\mathcal{M}}$. The Completeness Theorem holds for hidden k-logics (cf. (Martins and Pigozzi, 2007)), i.e., for every $\Gamma \cup \{\bar{\varphi}\} \subseteq (\mathrm{Te}_{\Sigma}^k(X))_{\mathrm{VIS}}$,

$$\Gamma \vdash_{\mathcal{L}} \bar{\varphi} \quad \text{iff} \quad \Gamma \models_{\mathrm{Mod}(\mathcal{L})} \bar{\varphi} \quad \text{iff} \quad \Gamma \models_{\mathrm{Mod}^*(\mathcal{L})} \bar{\varphi}.$$

An important class of hidden 2-logics is the class of *hidden equational logics*, where the notion of equality is only considered for visible data. It is defined (using the reflexivity, symmetry, transitivity and congruence rules) as a sorted equational logic, restricted to the visible part (cf. (Martins and Pigozzi, 2007)).

In an equational logic framework, a pair of terms of the same sort $\langle s, t \rangle$ is called *an equation* and it is denoted by $s \approx t$.

Definition 2.4 (Free hidden equational logic, cf. (Martins, 2004)). Let Σ be a hidden signature and VIS its set of visible sorts.

1. The *free hidden equational logic over* Σ (*free* HEL_{Σ} for short) is the specifiable hidden 2-logic presented as follows:

   Axioms: for all $V \in \mathrm{VIS}$

$$x\!:\!V \approx x\!:\!V$$

Inference rules: for each $V, W \in \mathrm{VIS}$,

(IR_1) $\qquad \dfrac{x\!:\!V \approx y\!:\!V}{y\!:\!V \approx x\!:\!V};$

(IR_2) $\qquad \dfrac{x\!:\!V \approx y\!:\!V,\, y\!:\!V \approx z\!:\!V}{x\!:\!V \approx z\!:\!V};$

(IR_3) $\qquad \dfrac{\varphi\!:\!V \approx \psi\!:\!V}{\vartheta(x/\varphi)\!:\!W \approx \vartheta(x/\psi)\!:\!W}$ for each $\vartheta \in \mathrm{Te}_W$ and each $x \in X_V$.

2. The *free un-hidden equational logic over* Σ (*free* UHEL_Σ, for short) contains an equality predicate for each sort, visible and hidden. The axioms and inference rules are the same as those of the free HEL_Σ, except that V and W are now allowed to range over all sorts. Thus $\mathrm{UHEL}_\Sigma = \mathrm{HEL}_{\Sigma^{\mathrm{UH}}}$.

An *applied hidden equational logic over* Σ (or simply a HEL_Σ) is any hidden 2-logic $\mathcal{L}$ over Σ that satisfies all axioms and rules of inference of the free HEL_Σ. An *applied un-hidden equational logic over* Σ (UHEL_Σ) is defined similarly; the subscript Σ may be omitted if it is clear from the context. We say that a specified applied hidden or unhidden equational logic is *strict equational* if it does not have extra-logical inference rules.

Definition 2.5. Let $\mathcal{L}$ be a HEL_Σ and E a set of equations of arbitrary, possibly un-hidden, sort. We define $\mathcal{L}^{\mathrm{UH}}[E]$ as the natural extension of $\mathcal{L}$ by E to a UHEL over the same signature (when E is empty we just write $\mathcal{L}^{\mathrm{UH}}$).

The standard model of the free HEL_Σ is of the form $\langle A, \mathrm{id}_{A_{\mathrm{VIS}}} \rangle$, where A is a Σ-algebra and $\mathrm{id}_{A_{\mathrm{VIS}}}$ is the identity relation on the visible part of A, but one gets more general 2-data structures as models by taking any congruence relation on the visible part of A in place of $\mathrm{id}_{A_{\mathrm{VIS}}}$. By a *congruence relation on the visible part of* A, or simply a VIS-*congruence*, we mean a VIS-sorted set $\langle F_V : V \in \mathrm{VIS} \rangle$ such that, for every $V \in \mathrm{VIS}$, F_V is an equivalence relation on A_V, and for every term $\varphi(x_0\!:\!V_0, \ldots, x_{n-1}\!:\!V_{n-1}, y_0\!:\!H_0, \ldots, y_{m-1}\!:\!H_{m-1})\!:\!V$ with $V_0, \ldots, V_{n-1}, V \in \mathrm{VIS}$ and $H_0, \ldots, H_{m-1}, \in \mathrm{HID}$, if $\langle a_i, b_i \rangle \in F_{V_i}$ for all $i < n$, then for all $c_j \in A_{H_j}$ $j < m$,

$$\langle \varphi^A(a_0, \ldots, a_{n-1}, c_0, \ldots, c_{m-1}), \varphi^A(b_0, \ldots, b_{n-1}, c_0, \ldots, c_{m-1}) \rangle \in F_V.$$

The basic notions and results about hidden k-logics, as well as many examples of HELs may be found in (Martins, 2004) and (Martins and Pigozzi, 2007). An interesting fact about HELs is that the visible consequences of any set of visible equations are the same either for $\mathcal{L}^{\mathrm{UH}}$ or for $\mathcal{L}$, i.e., for any $\Gamma \cup \{t \approx t'\} \subseteq \mathrm{Te}_\Sigma^2(X)_{\mathrm{VIS}}$, $\Gamma \vdash_{\mathcal{L}^{\mathrm{UH}}} t \approx t'$ iff $\Gamma \vdash_{\mathcal{L}} t \approx t'$.

2.3 Concrete examples

We give several examples of specifiable hidden logics. We have purposely chosen simple, well-known ones that allow us to illustrate the basic ideas without burdening the reader with irrelevant detail. The first two illustrate how the logic of a particular data structure can be alternatively formalized as a Boolean 1-logic and as an equational 2-logic, a HEL. The flag logics provide two different ways of specifying semaphores, which are commonly used in scheduling resources (Goguen and Malcolm, 1999).

Example 2.6. (Flags as a Boolean 1-logic)

Consider the hidden signature Σ_{flag}:

SORT $= \{\text{flag}, \text{bool}\}$, with bool the unique visible sort, and the following operation symbols:

$$\text{up} : \text{flag} \to \text{flag}; \qquad \text{rev} : \text{flag} \to \text{flag};$$
$$\text{dn} : \text{flag} \to \text{flag}; \qquad \text{up?} : \text{flag} \to \text{bool},$$

and the operation symbols for the Boolean part: $\neg, \wedge, \vee, \text{true}$ and false. The Boolean biconditional $\varphi \leftrightarrow \psi$ is an abbreviation for the compound operation $(\neg \varphi \vee \psi) \wedge (\neg \psi \vee \varphi)$.

The Boolean logic of flags, $\mathcal{L}_{\text{bflag}}$, is the 1-logic with the following extra-logical axioms:

$$\text{up?}(\text{up}(F)) \qquad\qquad \text{up?}(\text{rev}(F)) \leftrightarrow \neg(\text{up?}(F))$$
$$\neg\text{up?}(\text{dn}(F))$$

and including usual logical axioms for the classical propositional logic. There are no extra-logical rules of inference. $\qquad\qquad\qquad\qquad\qquad\qquad\diamond$

Example 2.7. (Flags as a HEL) The signature is the same as above.

The equational logic of flags, $\mathcal{L}_{\text{eflag}}$, is the $\text{HEL}_{\Sigma_{\text{flag}}}$ with the following extra-logical axioms:

$$\text{up?}(\text{up}(F)) \approx \text{true} \qquad\qquad \text{up?}(\text{rev}(F)) \approx \neg(\text{up?}(F))$$
$$\text{up?}(\text{dn}(F)) \approx \text{false}$$

and including the usual logical axioms for Boolean algebra. There are no extra-logical rules of inference. $\qquad\qquad\qquad\qquad\qquad\qquad\qquad\qquad\diamond$

As expected, $\mathcal{L}_{\text{bflag}}$ and $\mathcal{L}_{\text{eflag}}$ are equivalent. Precisely, $\dfrac{\varphi_1 \leftrightarrow \varphi_1', \ldots, \varphi_n \leftrightarrow \varphi_n'}{\psi \leftrightarrow \psi'}$ is a derivable rule of $\mathcal{L}_{\text{bflag}}$ iff $\dfrac{\varphi_1 \approx \varphi_1', \ldots, \varphi_n \approx \varphi_n'}{\psi \approx \psi'}$ is a derivable rule of $\mathcal{L}_{\text{eflag}}$.

Example 2.8. (Stacks of Natural Numbers as a HEL) As in the standard specification of the logic of stacks, only the natural numbers are visible. Consequently, the axioms and rules of inference can only reference "numerical behavior" of stacks rather than the stacks themselves. In particular there can be no axiom or rule involving equality between stacks. Because of this we get an infinite number of axioms, while in the standard formalizations, where assertions about the equality of stacks are allowed, the axiomatization is finite and conceptually simpler.

The specification differs from the usual one in another regard. The top of the empty stack is zero and pushing zero on the empty stack gives the empty stack. This is done to simplify the specification logic and agrees with what is done in (Goguen and Malcolm, 2000).

Consider the hidden signature Σ_{stacks}:

SORT = {nat, stack}, with nat the unique visible sort and the following operation symbols:

empty : $\to$ stack	top : stack $\to$ nat
zero : $\to$ nat	pop : stack $\to$ stack
push : nat, stack $\to$ stack	s : nat $\to$ nat

The specification logic of stacks, $\mathcal{L}_{\text{stacks}}$, is the logic with hidden signature Σ_{stacks} and the following axioms and inference rules:

Extra-logical axioms:

$\text{top}(\text{pop}^n(\text{empty})) \approx \text{zero}$, for all n;

$\text{top}(\text{push}(x,y)) \approx x$;

$\text{top}(\text{pop}^{n+1}(\text{push}(x,y))) \approx \text{top}(\text{pop}^n(y))$, for all n.

Extra-logical inference rule:

$s(x) \approx s(y) \to x \approx y$. $\Diamond$

2.3.1 Other hidden k-logics

Example 2.9 (Free inequational logic). Let Σ be any one-sorted signature. The *free inequational logic* is the one-sorted 2-logic over Σ defined by the axioms and inference rules in Fig. 1. As in the equational case, we use a special symbol to denote the 2-formula $\langle \varphi, \psi \rangle$; namely we write $\varphi \preceq \psi$ for $\langle \varphi, \psi \rangle$. This logic is relevant in the context of ordered (universal) algebra (see

$$
\begin{array}{l}
\text{Axioms:} \\
x \preceq x; \\
\text{Inference rules:} \\
\dfrac{x \preceq y, y \preceq z}{x \preceq z}; \\
\dfrac{x_0 \preceq y_0, \ldots, x_{n-1} \preceq y_{n-1}}{O(x_0, \ldots, x_{n-1}) \preceq O(y_0, \ldots, y_{n-1})}, \\
\text{for every operation symbol } O.
\end{array}
$$

Fig. 1 Free inequational logic.

(Wechler, 1992)) and abstract algebra. We can generalize the inequational logic to the sorted case and, more generally, to the hidden sorted case in the same way we generalized the equational logic to the hidden equational logic. A more general notion of inequational logic can be found in (Babenyshev and Martins, 2016).   ◇

Example 2.10 (Stacks of natural numbers with Booleans). The signature is obtained from the signature of stacks of natural numbers by adjoining a new sort bool, for the Boolean operation symbols, and one new attribute $eq : \mathrm{nat}, \mathrm{nat} \to \mathrm{bool}$, the equality test for natural numbers. The sort bool is the only visible sort. Informally, we can say that the axioms and inference rules are obtained by applying *eq* to each of the axioms and inference rules of the specification of stacks (see Fig. 2). The operation symbol *eq* is called an *equational test function* and the models are called **generalized equality test models**. These models have been studied in (Pigozzi, 1991).

Axioms:

$eq(x,x)$

$eq(\mathrm{top}(\mathrm{pop}^n(empty)), zero)$, for all n;

$eq(\mathrm{top}(\mathrm{push}(x,y)), x)$;

$eq(\mathrm{top}(\mathrm{pop}^{n+1}(\mathrm{push}(x,y))), \mathrm{top}(\mathrm{pop}^n(y)))$, for all n;

Inference rules:

$$\frac{eq(x,y)}{eq(y,x)} \qquad\qquad \frac{eq(x,y)}{eq(s(x),s(y))}$$

$$\frac{eq(x,y), eq(y,z)}{eq(x,z)} \qquad\qquad \frac{eq(s(x),s(y))}{eq(x,y)}$$

Fig. 2 Stacks of natural numbers with Booleans.

◇

3 Behavioral equivalence

In hidden equational logic, we may say that two hidden data elements of the same sort are *behaviorally equivalent* if any visible procedure returns the same value when executed with either of the two objects as input. The notion arises from the alternative view of a data structure as a transition system in which the hidden data elements represent states of the system and the operations (i.e., the *methods*) that return hidden, as opposed to visible, elements induce transitions between states.

In the formalism of HEL, the concept of procedure takes the form of a context. Formally, a *S-context* over a hidden signature Σ is a term

$$(3) \qquad\qquad \varphi(z{:}S, u_1{:}T_1, \ldots, u_m{:}T_m){:}U$$

with a distinguished variable z of sort S and parametric variables $u_1, \ldots, u_m$ of arbitrary (visible or hidden) sort. It is a *visible context* if the sort U of φ is visible.

Definition 3.1. Let A be a Σ-algebra and let S be a arbitrary sort. Then, $a, a' \in A_S$ are *behaviorally equivalent in* A, in symbols $a \equiv^{\mathrm{beh}}_A a'$, if for every visible S-context $\varphi(z\!:\!S, u_1\!:\!T_1, \ldots, u_m\!:\!T_m)$ and for all $b_1 \in A_{T_1}, \ldots, b_m \in A_{T_m}$,

$$\varphi^A(a, b_1, \ldots, b_m) = \varphi^A(a', b_1, \ldots, b_m).$$

Variants of this notion of behavioral equivalence have occurred in the literature. For example, Goguen and Malcolm (Goguen and Malcolm, 2000) restrict the set of contexts to the ones built from a predefined set of observational operational symbols.

In order to generalize the notion of behavioral equivalence to hidden k-logics we first generalize the notion of context. A (k, S)-*context* over a hidden signature Σ is a k-term

$$(4) \qquad \bar{\varphi}(z\!:\!S, u_1\!:\!T_1, \ldots, u_m\!:\!T_m)\!:\!U$$
$$= \langle \varphi_1(z\!:\!S, u_1\!:\!T_1, \ldots, u_m\!:\!T_m), \ldots, \varphi_k(z\!:\!S, u_1\!:\!T_1, \ldots, u_m\!:\!T_m) \rangle\!:\!U$$

with a distinguished variable z of sort S and parametric variables $u_1, \ldots, u_m$. It is a *visible context* if the sort U of $\bar{\varphi}$ is visible.

Definition 3.2. Let $\mathcal{A} = \langle A, F \rangle$ be a k-data structure over a hidden signature Σ. Two elements a, a' of A of arbitrary sort S are said to be *behaviorally equivalent* in $\mathcal{A}$, in symbols $a \equiv^{\mathrm{beh}}_{\mathcal{A}} a'$, if for every visible (k, S)-context $\bar{\varphi}(z\!:\!S, u_1\!:\!T_1, \ldots, u_m\!:\!T_m)\!:\!V$ and for all $b_1 \in A_{T_1}, \ldots, b_m \in A_{T_m}$,

$$(5) \qquad \bar{\varphi}^A(a, b_1, \ldots, b_m) \in F_V \quad \text{iff} \quad \bar{\varphi}^A(a', b_1, \ldots, b_m) \in F_V.$$

This notion does indeed generalize behavioral equivalence in equational logic, since, as a consequence of Theorem 3.4 below, we have that a and a' are behaviorally equivalent in a Σ-algebra A iff they are behaviorally equivalent in the 2-dimensional equality data structure $\langle A, \mathrm{id}_{A_{\mathrm{VIS}}} \rangle$ in the sense of Definition 3.2.

Behavioral equivalence over a k-data structure turns out to be a congruence relation on the underlying algebra of the data structure with special properties. In the 1-sorted, 1-data structures (called *matrices*) which constitute the natural models of sentential logic, the detailed combinatorial analysis of this congruence constitutes the basis of a branch of mathematical logic called *abstract algebraic logic* (cf. (Pigozzi, 2001)). Our intention here is to extend this analysis to the behavioral congruences of arbitrary multi-sorted k-data structures and in particular to the models of hidden equational logic. The following two theorems constitute the basis of this approach. They are due to Manuel Martins and Don Pigozzi in (Martins and Pigozzi, 2007). We include their proofs here since this paper is intended also as a survey.

Theorem 3.3. *Let Σ be a hidden signature and let al$A = \langle A, F \rangle$ be a k-data structure over Σ. Then, $\equiv^{\mathrm{beh}}_{\mathcal{A}} = \Omega_A(F)$, i.e., for every $S \in \mathrm{SORT}$ and for all $a, a' \in A_S$, $a \equiv^{\mathrm{beh}}_{\mathcal{A}} a'$ iff $a \equiv a' (\Omega_A(F)_S)$.*

Proof. It is easy to see that $\equiv^{\mathrm{beh}}_{\mathcal{A}}$ is an equivalence relation on A. To see that it is a congruence relation, let O be an operation symbol of type $T_1, \ldots, T_n \to S$ and suppose $a_i \equiv^{\mathrm{beh}}_{\mathcal{A}} a'_i$, $1 \le i \le n$. We must show that, for any visible (k, S)-context $\bar{\varphi}(z{:}S, \bar{u}{:}\bar{Q}){:}V$, with the designated variable $z{:}S$, and for all parameters $\bar{b} \in A_{\bar{Q}}$, we have

$$(6) \qquad \bar{\varphi}^A\big(O^A(\bar{a}), \bar{b}\big) \in F_V \quad \text{iff} \quad \bar{\varphi}^A\big(O^A(\bar{a}'), \bar{b}\big) \in F_V.$$

Consider any $i \le n$. Using the assumption $a_i \equiv^{\mathrm{beh}}_{\mathcal{A}} a'_i$, and taking x_i as the designated variable, $x_1, \ldots, x_{i-1}, x_{i+1}, \ldots, x_n, u_1, \ldots, u_n$ as parametric variables, and $a_1, \ldots, a_{i-1}, a'_{i+1}, \ldots, a'_n, b_1, \ldots, b_m$ as parameters we have

$$\bar{\varphi}^A\big(O^A(a_1, \ldots, a_{i-1}, a_i, a'_{i+1}, \ldots, a'_n), \bar{b}\big) \in F_V$$
$$\text{iff} \quad \bar{\varphi}^A\big(O^A(a_1, \ldots, a_{i-1}, a'_i, a'_{i+1}, \ldots, a'_n), \bar{b}\big) \in F_V.$$

Since this equivalence holds for all $i \le n$, (6) holds, and hence $\equiv^{\mathrm{beh}}_{\mathcal{A}}$ is a congruence on A.

To see that $\equiv^{\mathrm{beh}}_{\mathcal{A}}$ is compatible with F, consider $\bar{a}, \bar{a}' \in A^k_V$ such that $\bar{a} \big(\equiv^{\mathrm{beh}}_{\mathcal{A}}\big)^k_V \bar{a}'$. Consider the k-sequence of pairwise distinct variables $\bar{x} = \langle x_1{:}V, \ldots, x_k{:}V \rangle$ (called a *k-variable*, a special *k*-formula). For each i, $1 \le i \le k$, view $x_1, \cdots, x_n$ as a (k, V)-context with designated variable x_i and treat $a_1, \ldots, a_{i-1}, a'_{i+1}, \ldots, a'_k$ as parameters. Then from the assumption $a_i \big(\equiv^{\mathrm{beh}}_{\mathcal{A}}\big)_V a'_i$ we conclude that

$$\langle a_1, \ldots, a_{i-1}, a_i, a'_{i+1}, \ldots, a'_n \rangle \in F_V \quad \text{iff} \quad \langle a_1, \ldots, a_{i-1}, a'_i, a'_{i+1}, \ldots, a'_n \rangle \in F_V.$$

So $\bar{a} \in F_V$ iff $\bar{a}' \in F_V$. Thus $\equiv^{\mathrm{beh}}_{\mathcal{A}}$ is compatible with F.

Finally, we must show that $\equiv^{\mathrm{beh}}_{\mathcal{A}}$ is the largest congruence on A compatible with F. Let Θ be any congruence on A that is compatible with F. Assume $a \equiv a' (\Theta_S)$. Let $\bar{\varphi}(z{:}S, \bar{u}{:}\bar{Q}){:}V$ be a visible (k, S)-formula with designated variable $z{:}S$, and let $\bar{b} \in A_{\bar{Q}}$ be a system of parameters. By the congruence property, $\bar{\varphi}^A(a, \bar{b}) \equiv \bar{\varphi}^A(a', \bar{b}) (\Theta^k_V)$. So by the compatibility of Θ with F we have $\bar{\varphi}^A(a, \bar{b}) \in F_V$ iff $\bar{\varphi}^A(a', \bar{b}) \in F_V$. Thus $\Theta \subseteq \equiv^{\mathrm{beh}}_{\mathcal{A}}$. $\qquad\square$

This theorem shows the adequacy of using the Leibniz congruence to study behavioral equivalence. Moreover, when applied to hidden equational logics, Theorem 3.3 takes a more natural form in terms of 1-dimensional contexts as we now see.

Theorem 3.4. *Let Σ be a hidden signature and let $\mathcal{A} = \langle A, F \rangle$ be a model of the free HEL_Σ, i.e., F is a VIS-congruence on A. Then, for every*

$S \in \mathrm{SORT}$ *and all* $a, a' \in A_S$, $a \equiv_{\Omega(F)_S} a'$ *iff, for every visible* S-*context* $\varphi(z\!:\!S, u_1\!:\!Q_1, \ldots, u_m\!:\!Q_m)\!:\!V$ *and for all* $b_1 \in A_{Q_1}, \ldots, b_m \in A_{Q_m}$,

$$(7) \qquad \varphi^A(a, b_1, \ldots, b_m) \equiv \varphi^A(a', b_1, \ldots, b_m)\,(F_V).$$

Proof. By Theorem 3.3, $a \equiv_{\Omega(F)_S} a'$ iff, for every $(2, S)$-context $\langle \varphi(z\!:\!S, \bar{u}\!:\!\bar{Q}),$ $\psi(z\!:\!S, \bar{u}\!:\!\bar{Q})\rangle$ of sort V, and every $\bar{b} \in A_{\bar{Q}}$,

$$(8) \qquad \varphi^A(a, \bar{b}) \equiv \psi^A(a, \bar{b})\,(F_V) \quad \text{iff} \quad \varphi^A(a', \bar{b}) \equiv \psi^A(a', \bar{b})\,(F_V).$$

Suppose (7) holds for every S-context $\varphi(z, \bar{u})$ and every $\bar{b} \in A_{\bar{Q}}$. If $\varphi^A(a, \bar{b})$ $\equiv_{F_V} \psi^A(a, \bar{b})$, then

$$\varphi^A(a', \bar{b}) \equiv \varphi^A(a, \bar{b}) \equiv \psi^A(a, \bar{b}) \equiv \psi^A(a', \bar{b})\,(F_V)$$

(the first and third equivalences hold because F is a VIS-congruence). Thus (8) holds for every pair of S-contexts and every sequence of parameters $\bar{b}$, hence, $a \equiv_{\Omega(F)_V} a'$.

Conversely, assume $a \equiv_{\Omega(F)_V} a'$. Let $\varphi(z\!:\!S, \bar{u}\!:\!\bar{Q})\!:\!V$ be an arbitrary visible S-context, where $\bar{u}\!:\!\bar{Q} = \langle u_1\!:\!Q_1, \ldots, u_n\!:\!Q_n \rangle$. Let u_{n+1} be a new parametric variable of sort V; the single term u_{n+1} can be viewed as a visible S-context with designated variable z (which does not actually occur) and parametric variables $\bar{u}^+ := \langle u_1, \ldots, u_n, u_{n+1} \rangle$. φ can also be viewed as an S-context with the same parametric variables. Let $\langle b_1, \ldots, b_n \rangle$ be any system of parameters of sort $\bar{Q}$, and extend it to a system $\bar{b}^+ := \langle b_1, \ldots, b_n, b_{n+1} \rangle$, where $b_{n+1} = \varphi^A(a, \bar{b})$. Thus $\varphi^A(a, \bar{b}^+) = b_{n+1} = u_{n+1}^A(a, \bar{b}^+)$. So by (8), $\varphi^A(a', \bar{b}^+) \equiv_{F_V} u_{n+1}^A(a', \bar{b}^+)$. But $u_{n+1}^A(a', \bar{b}^+)$ also equals b_{n+1}. So $\varphi^A(a, \bar{b}) \equiv_{F_V} \varphi^A(a', \bar{b})$. Thus (7) holds for every S context $\varphi(z, \bar{u})$ and every $\bar{b} \in A_{\bar{Q}}$. $\qquad\square$

Applying this result to equality models (i.e., models whose filter is the identity), we get that a and a' are behaviorally equivalent, in the sense of Definition 3.1, iff $a \equiv a'\,\big(\Omega_A(\mathrm{id}_{A_{\mathrm{VIS}}})\big)$; hence behavioral equivalence over k-data structures does indeed generalize the familiar notion of behavioral equivalence over a sorted algebra. This result was obtained independently by Goguen and Malcolm (Goguen and Malcolm, 2000).

For hidden equational logics the Leibniz relation has the following useful property; this can also be found in (Goguen and Malcolm, 1999, 2000) for the case of equality models.

Corollary 3.5. *Let* $\mathcal{A} = \langle A, F \rangle$ *be a model of the free* HEL_{Σ}. *Then* $\Omega_A(F)$ *is the largest congruence on* A *whose visible part is* F.

Proof. Suppose $a \equiv a'\,\big(\Omega_A(F)_V\big)$ with $V \in \mathrm{VIS}$. Let z be a variable of sort V. Then z is a visible V-context and hence $a = z^A(a) \equiv z^A(a') = a' \mod F_V$. Thus $\Omega_A(F)_{\mathrm{VIS}} \subseteq F$. Conversely, assume $a \equiv a' \mod F_V$. Then for every V-context $\varphi(z, \bar{u})$ and every choice of parameters $\bar{b} \in A_{\bar{Q}}$, we have $\varphi^A(a, \bar{b}) \equiv$

$\varphi^A(a', \bar{b}) \mod F_V$. Thus $a \equiv a' (\Omega_A(F)_V)$ and hence $\Omega_A(F)_{\mathrm{VIS}} = F$. If Θ is any congruence on A such that $\Theta_{\mathrm{VIS}} = F$, then Θ is compatible with F, and hence $\Theta \subseteq \Omega_A(F)$. $\qquad\square$

As a special case (for equality models) we have that $\Omega_A(\mathrm{id}_{A_{\mathrm{VIS}}})_{\mathrm{VIS}} = \mathrm{id}_{A_{\mathrm{VIS}}}$, i.e., two visible elements of a Σ-algebra are behaviorally equivalent iff they are equal. However, in computer science it is important to establish procedures to check if two elements are behaviorally equivalent. The last result allows the following method of coinduction.

Given a data structure $\langle A, F \rangle$ we want to know if a pair $\langle a, a' \rangle \in A_S^2$ is in $\Omega(F)_S$. Our method consists of the following three steps:

1 - Define a suitable relation R on A, such that the visible part is F;
2 - Show that this relation is a congruence on A;
3 - Finally show that a and a' are equivalent modulo R.

At first glance, step 1 of this method seems to be a very hard task, however it works very well in many concrete examples. In Example 2.7, we can use this method to prove that $\mathrm{rev}(\mathrm{rev}(F)) \approx \mathrm{rev}(F)$ is behaviorally equivalent in any equality model A of $\mathcal{L}_{\mathrm{eflag}}$. It is enough to consider $R := \big\{(a, a') \in A_{\mathrm{flag}}^2 : \mathrm{up?}^A(a) = \mathrm{up?}^A(a')\big\}$.

When applied to Lindenbaum models, Corollary 3.5 gives rise to the following results.

Corollary 3.6. *Let $\mathcal{L}$ be a HEL and G an $\mathcal{L}^{\mathrm{UH}}$-theory. Then $G \subseteq \Omega(G_{\mathrm{VIS}})$ and $G_{\mathrm{VIS}} = \Omega(G_{\mathrm{VIS}})_{\mathrm{VIS}}$.*

We can also conclude from Theorem 3.4 that for every sorted algebra A, the operator $\Omega_A : \mathrm{Fi}_{\mathcal{L}}(A) \to \mathrm{Con}(A)$ defined by mapping each $F \in \mathrm{Fi}_{\mathcal{L}}(A)$ into $\Omega_A(F)$ is injective and monotonic.

Corollary 3.7. *Let $\mathcal{L}$ be a HEL. Then for every sorted algebra A, Ω_A is injective and monotonic.*

Proof. The proof of injectivity is obvious.

Let $F, G \in \mathrm{Fi}_{\mathcal{L}}(A)$ such that $F \subseteq G$. Suppose that $a \equiv a' (\Omega(F)_S)$. Then for every S-context $\varphi(z{:}S, \bar{x}){:}V$ and for all $\bar{b} \in A^k$, $\varphi^A(a, \bar{b}) \equiv \varphi^A(a', \bar{b}) (F_V)$. Hence, for every S-context $\varphi(z{:}S, \bar{x}){:}V$ and for all $\bar{b} \in A^k$, $\varphi^A(a, \bar{b}) \equiv \varphi^A(a', \bar{b}) (G_V)$. Therefore, $a \equiv a' (\Omega(G)_S)$. $\qquad\square$

For arbitrary hidden logics this result is false, even in the one-sorted case. The class of logics for which Ω is injective and monotonic is called *Weakly Algebraizable Logics*. (This class was investigated by Czelakowski and Jansana in (Czelakowski, 2000)). Injectivity and monotonicity are independent properties in sense that neither one of them implies the other (see (Descalço and Martins, 2005) for an example of a non monotonic injective logic).

For some specifications, certain intuitive properties are not satisfied in the usual sense. This is the case, for instance, of the usual specification of Flags (see (Goguen and Malcolm, 1999)), where equation $\text{rev}(\text{rev}(F)) \approx F$ should be a property of the specification of Flags but is not a theorem of the $\mathcal{L}_{\text{eflag}}^{\text{UH}}$. Therefore, we consider, next, a weaker notion of satisfaction, called *behavioral satisfaction*. The perspective according to which satisfaction is to be considered behaviorally is called *behavioral approach* (see, e.g., (Roşu, 2000)).

Definition 3.8. Let $t \approx t'$ be an equation of arbitrary sort, and $\mathcal{A} = \langle A, F \rangle$ a k-data structure. We say that the equation $t \approx t'$ is *behaviorally satisfied in* $\mathcal{A}$, in symbols $\models_{\mathcal{A}}^{beh} t \approx t'$, if for all $h : X \to A$, $h(t) \equiv_{\langle A, F \rangle}^{beh} h(t')$. Let $\mathcal{L}$ be a HEL and let $\text{Mod}(\mathcal{L})^=$ denote $\{A : \langle A, id_{A_{\text{VIS}}} \rangle \in \text{Mod}(\mathcal{L})\}$. We say that $t \approx t'$ is *behaviorally valid over* $\mathcal{L}$, in symbols $\models_{\mathcal{L}}^{beh} t \approx t'$, if for every $A \in \text{Mod}(\mathcal{L})^=$, $\models_{\langle A, id_{A_{\text{VIS}}} \rangle}^{beh} t \approx t'$. If $\mathcal{L}$ is clear from the context we simply write, $\models^{beh} t \approx t'$.

In the example of $\mathcal{L}_{\text{eflag}}$, we have just sketched the proof that equation $\text{rev}(\text{rev}(F)) \approx F$ is behaviorally satisfied in each algebra in $\text{Mod}(\mathcal{L}_{\text{eflag}})^=$.

Lemma 3.9. *Let* $\mathcal{A} = \langle A, F \rangle$ *be a data structure over a hidden signature* Σ *and* S *be an arbitrary sort. Then,* $\models_{\mathcal{A}}^{beh} t \approx t'$ *iff for every* $h : X \to A$ *and every visible* S-*context* $\varphi(z{:}S, x_1, \ldots, x_n) {:} V$,

$$h(\varphi(t, x_1, \ldots, x_n)) \equiv h(\varphi(t', x_1, \ldots, x_n)) \; (F_V)$$

We are going to introduce some notation.

Definition 3.10. Let $t \approx t'$ be an equation of type S. We define $\Delta[t \approx t']$ to be the set $\{t \approx t'\}$, if $S \in \text{VIS}$; and $\Delta[t \approx t']$ to be the set $\{\varphi(t, x_1, \ldots, x_n) \approx \varphi(t', x_1, \ldots, x_n) | \varphi(z{:}S, x_1, \ldots, x_n) \in (\text{Te}_\Sigma(X))_{\text{VIS}}\}$, if $S \notin \text{VIS}$.

The following lemma is a useful characterization for an equation to be behaviorally satisfied in an algebra A (the version for equality models was proved by Roşu in (Roşu, 2000)).

Lemma 3.11. *Let* $\mathcal{A} = \langle A, F \rangle$ *be a data structure over a hidden signature* Σ *and* S *be an arbitrary sort. Then the following conditions are equivalent:*

(i) $\models_{\mathcal{A}}^{beh} t \approx t'$;
(ii) $\models_{\mathcal{A}} \Delta[t \approx t']$.

In the special case of data structures of form $\langle \text{Te}_\Sigma(X), T \rangle$, with T a theory, we call the equations in $\Omega(T)$, *behavioral consequences* of T. Behavioral consequences of a substitution invariant theory T may be characterized in the following way.

Corollary 3.12. *Let* $\mathcal{L}$ *be a HEL,* T *a substitution invariant theory of* $\mathcal{L}$ *and* $t, t' \in (\text{Te}_\Sigma(X))_S$. *Then the following are equivalent:*

(i) $t \equiv t'(\ \Omega(T)_S)$;

(ii) *for every visible S-context $\varphi(z\!:\!S, x_1, \ldots, x_n)$,*
$T \vdash_{\mathcal{L}} \varphi(t, x_1, \ldots, x_n) \approx \varphi(t', x_1, \ldots, x_n)$.

A detailed study of the properties of $\Omega(T)$ can be found in (Martins and Pigozzi, 2007).

3.1 Formal Behavioral Consequence Relation

In this section we present a characterization of the Leibniz congruence in terms of the consequence relation of $\mathcal{L}^{\mathrm{UH}}$. This characterization justifies the name we have been using for the elements in $\Omega(T)$ - behavioral consequences of T.

This result generalizes, in two directions, the work of Leavens and Pigozzi in (Leavens and Pigozzi, 2002). On the one hand, we allow conditional equations as axioms (instead of equations only); on the other hand, we characterize $\Omega(T)$ for all theories (instead of just $\Omega(\mathrm{Thm}(\mathcal{L}))$).

In the second part of this section, we present a simpler characterization for the special case of strict HEL, by dropping the condition that G has to range over all theories.

Now we consider the following generalization of the definition of formal behavioral consequence (introduced in (Leavens and Pigozzi, 2002)).

Definition 3.13. Let $\mathcal{L}$ be a HEL, $T \in \mathrm{Th}(\mathcal{L})$ and F a set of equations over Σ. Then we say that F is a *global formal behavioral consequence* of T, in symbols $T \vdash_{\mathcal{L}}^{GFB} F$, if for every $G \in \mathrm{Th}(\mathcal{L}^{\mathrm{UH}})$, such that $T \subseteq G$, and every visible equation $s \approx s'$, $G \cup F \vdash_{\mathcal{L}^{\mathrm{UH}}} s \approx s'$ implies that $G \vdash_{\mathcal{L}^{\mathrm{UH}}} s \approx s'$. If $F = \{t \approx t'\}$, then we say that $t \approx t'$ is a *global formal behavioral consequence* of T and we write $T \vdash_{\mathcal{L}}^{GFB} t \approx t'$.

Let $\mathcal{L}$ be a HEL and $T \in \mathrm{Th}(\mathcal{L})$. We define the following relation on the term algebra. For each sort S, $\mathcal{GFB}(T)_S$ is the set of all pairs (t, t') of formulas of type S, such that $T \vdash_{\mathcal{L}}^{GFB} t \approx t'$. Thus, $\mathcal{GFB}(T) = \langle \mathcal{GFB}(T)_S : S \in \mathrm{SORT} \rangle$.

The global formal behavioral consequence relation, as a property of a set, can actually be reduced to a property of its individual members.

Theorem 3.14. *Let $\mathcal{L}$ be a specifiable HEL and F be a set of equations. Then for every $T \in \mathrm{Th}(\mathcal{L})$,*

$$(9) \qquad T \vdash_{\mathcal{L}}^{GFB} F \Leftrightarrow (T \vdash_{\mathcal{L}}^{GFB} t \approx t', \ \text{for all } t \approx t' \in F)$$

Proof. The implication from left to right is obvious. To prove the converse we first show that it holds for any finite F, using induction on the number of elements in F.

Let $G \in \mathrm{Th}(\mathcal{L}^{\mathrm{UH}})$ such that $T \subseteq G$ and $s, s' \in (\mathrm{Te}_\Sigma(X))_{\mathrm{VIS}}$. Suppose that $G \cup F \vdash_{\mathcal{L}^{\mathrm{UH}}} s \approx s'$.

If F has only one equation, we have, by hypothesis, that $G \vdash_{\mathcal{L}^{\mathrm{UH}}} s \approx s'$.

Let now F be the union of F' with $\{t \approx t'\}$. Suppose that $G \cup F' \cup \{t \approx t'\} \vdash_{\mathcal{L}^{\mathrm{UH}}} s \approx s'$. Hence, $\mathrm{Cn}_{\mathcal{L}^{\mathrm{UH}}}(G \cup F') \cup \{t \approx t'\} \vdash_{\mathcal{L}^{\mathrm{UH}}} s \approx s'$. Since, $T \vdash_{\mathcal{L}}^{GFB} t \approx t'$, we have that $\mathrm{Cn}_{\mathcal{L}^{\mathrm{UH}}}(G \cup F') \vdash_{\mathcal{L}^{\mathrm{UH}}} s \approx s'$. Hence, $G \cup F' \vdash_{\mathcal{L}^{\mathrm{UH}}} s \approx s'$. Finally, by the induction hypothesis, $G \vdash_{\mathcal{L}^{\mathrm{UH}}} s \approx s'$. So, we have just proved that $T \vdash_{\mathcal{L}}^{GFB} F$.

Let now F be any set of equations. Let $G \in \mathrm{Th}(\mathcal{L}^{\mathrm{UH}})$ such that $T \subseteq G$ and $s, s' \in (\mathrm{Te}_\Sigma(X))_{\mathrm{VIS}}$. Suppose that $G \cup F \vdash_{\mathcal{L}^{\mathrm{UH}}} s \approx s'$. Then, there is a finite subset F_0 of F such that $G \cup F_0 \vdash_{\mathcal{L}^{\mathrm{UH}}} s \approx s'$. From the discussion above, $G \vdash_{\mathcal{L}^{\mathrm{UH}}} s \approx s'$. $\qquad\square$

The following lemma shows that the global formal behavioral consequence is closed under ordinary equational deduction.

Lemma 3.15. *Let $\mathcal{L}$ be a specifiable HEL, $T \in \mathrm{Th}(\mathcal{L})$ and F a set of equations. Then $T \vdash_{\mathcal{L}}^{GFB} F$ and $F \vdash_{\mathcal{L}^{\mathrm{UH}}} t \approx t'$ implies that $T \vdash_{\mathcal{L}}^{GFB} t \approx t'$.*

Proof. Let $G \in \mathrm{Th}(\mathcal{L}^{\mathrm{UH}})$, such that $T \subseteq G$, and $s \approx s'$ be a visible equation. Suppose that $G \cup \{t \approx t'\} \vdash_{\mathcal{L}^{\mathrm{UH}}} s \approx s'$. Then, $G \cup F \vdash_{\mathcal{L}^{\mathrm{UH}}} s \approx s'$. Since $T \vdash_{\mathcal{L}}^{GFB} F$, we have that $G \vdash_{\mathcal{L}^{\mathrm{UH}}} s \approx s'$. Therefore, $T \vdash_{\mathcal{L}}^{GFB} t \approx t'$. $\qquad\square$

Corollary 3.16. *Let $\mathcal{L}$ be a specifiable HEL and $T \in \mathrm{Th}(\mathcal{L})$. Then $\mathcal{GFB}(T)$ is a theory of $\mathcal{L}^{\mathrm{UH}}$.*

Theorem 3.17. *Let $\mathcal{L}$ be a specifiable HEL and $T \in \mathrm{Th}(\mathcal{L})$. Then*

1. *$\mathcal{GFB}(T)_V = T_V$, for all $V \in \mathrm{VIS}$.*
2. *$\mathcal{GFB}(T) \subseteq \Omega(T)$.*

Proof. Obviously, $T \subseteq \mathcal{GFB}(T)_{\mathrm{VIS}}$. To prove the other inclusion, suppose that $(t, t') \in \mathcal{GFB}(T) \cap \mathrm{Te}_\Sigma^2(X)_{\mathrm{VIS}}$. Since $T \cup \{t \approx t'\} \vdash_{\mathcal{L}^{\mathrm{UH}}} t \approx t'$, then by definition of $\mathcal{GFB}(T)$, $T \vdash_{\mathcal{L}^{\mathrm{UH}}} t \approx t'$. Therefore, since $T \vdash_{\mathcal{L}} t \approx t'$ iff $T \vdash_{\mathcal{L}^{\mathrm{UH}}} t \approx t'$, for any $t, t' \in \mathrm{Te}_\Sigma(X)_{\mathrm{VIS}}$, we have $T \vdash_{\mathcal{L}} t \approx t'$, i.e. $t \approx t' \in T$.

Since $\mathcal{GFB}(T)$ is a congruence which coincides with T in the visible part and $\Omega(T)$ is the largest congruence with this property, we have that $\mathcal{GFB}(T) \subseteq \Omega(T)$. $\qquad\square$

The following theorem is one of the main results of this section.

Theorem 3.18. *Let $\mathcal{L}$ be a specifiable HEL and $T \in \mathrm{Th}(\mathcal{L})$. Then*

$$\Omega(T) = \mathcal{GFB}(T)$$

Proof. Suppose that $t \equiv t'\ (\Omega(T))$. Let $s, s' \in \mathrm{Te}_\Sigma(X)_{\mathrm{VIS}}$ and $G \in \mathrm{Th}(\mathcal{L}^{\mathrm{UH}})$, such that $T \subseteq G$. Suppose that

$$(10) \qquad\qquad G \cup \{t \approx t'\} \vdash_{\mathcal{L}^{\mathrm{UH}}} s \approx s'$$

Since $T \subseteq G_{\mathrm{VIS}}$ and the HEL logics are monotonic (see Corollary 3.7), $\Omega(T) \subseteq \Omega(G_{\mathrm{VIS}})$. Hence, $t \equiv t'\,(\Omega(G_{\mathrm{VIS}}))$. Since $G \subseteq \Omega(G_{\mathrm{VIS}})$ (see Corollary 3.6), (10) implies $s \equiv s'\,(\Omega(G_{\mathrm{VIS}}))$. Moreover, as $s \approx s'$ is visible, $s \approx s' \in G_{\mathrm{VIS}}$. So, $G_{\mathrm{VIS}} \vdash_{\mathcal{L}^{\mathrm{UH}}} s \approx s'$. Since the visible $\mathcal{L}$-consequences and the $\mathcal{L}^{\mathrm{UH}}$-consequences of a set of visible equations are the same, we have $G \vdash_{\mathcal{L}} s \approx s'$. That is, $T \vdash_{\mathcal{L}}^{GFB} t \approx t'$.

The other inclusion is Part 2 of Theorem 3.17. $\qquad\qquad\qquad\qquad\qquad$ $\square$

3.2 *Strict equational* HEL.

In this section we define a simpler notion of behavioral consequence of a theory -*formal behavioral consequence of T*. It considers only the theory T instead of considering all theories of $\mathcal{L}^{\mathrm{UH}}$ that contain T. For strict equational HEL's, we obtain simpler results.

First we define a relation on the term algebra that will play an important role in the sequel. A similar relation was already considered by Leavens and Pigozzi in the context of equational reasoning with subtyping (cf. (Leavens and Pigozzi, 2002)). The difference is that here we do not restrict the type of the arguments of the term r and we do not consider the substitution instances of the axioms.

Definition 3.19. Let E be a set of equations and $\widetilde{E} = \{t' \approx t : t \approx t' \in E\}$. We define the SORT-sorted relation $\equiv_E$ in the following way:

for each sort S, $t\,(\equiv_E)_S\,t'$ iff there is a term $r(z\!:\!S, y_1\!:\!T_1, \ldots, y_m\!:\!T_m)$ and an equation $e(x_1\!:\!S_1, \ldots, x_n\!:\!S_n) \approx e'(x_1\!:\!S_1, \ldots, x_n\!:\!S_n) \in E \cup \widetilde{E}$ of type S such that $t = r(e(x_1, \cdots, x_n), y_1, \ldots, y_m)$ and $t' = r(e'(x_1, \ldots, x_n), y_1, \ldots, y_m)$.

Finally, we define $\equiv_E^*$ as the reflexive, transitive closure of $\equiv_E$.

In the sequel we will use the following immediate consequence of the definition of $\equiv_E^*$:

Lemma 3.20. $t \equiv_E^* t'$ *iff there are terms* $s_1, \ldots, s_n$ *such that* $t = s_1 \equiv_E \cdots \equiv_E s_i \equiv_E s_{i+1} \equiv_E \cdots \equiv_E s_n = t'$.

By reformulating Lemma 2.21 of (Leavens and Pigozzi, 2002) we obtain the following characterization of equational consequence:

Lemma 3.21. *Let $\mathcal{L}$ be a strict equational HEL with set of equations E. Let F be a set of equations and $t \approx t'$ an equation. Then*

$$(11) \qquad\qquad F \vdash_{\mathcal{L}^{\mathrm{UH}}} t \approx t' \Leftrightarrow t\,(\equiv_{F \cup E^{\mathcal{L}}})^*\,t',$$

where $E^{\mathcal{L}}$ is the set of all substitution instances of equations in E.

Proof. $[\Leftarrow]$ Suppose that $t(\equiv_{F\cup E^{\mathcal{L}}})^*t'$. Then, there are terms $s_0,\ldots s_n$ such that $t = s_0 \equiv_{F\cup E^{\mathcal{L}}} s_1 \equiv_{F\cup E^{\mathcal{L}}} \cdots s_i \equiv_{F\cup E^{\mathcal{L}}} s_{i+1}\cdots \equiv_{F\cup E^{\mathcal{L}}} s_n = t'$. So, for each $i \leq n$, there is a S-context $r(z{:}S,y_1,\ldots,y_m)$ and an equation $e(x_1,\ldots,x_n) \approx e'(x_1,\ldots,x_n) \in (F\cup E^{\mathcal{L}})\cup(\widetilde{F}\cup\widetilde{E^{\mathcal{L}}})$ such that:

- $s_i = r(e(x_1,\ldots,x_n),y_1,\ldots,y_m)$ and
- $s_{i+1} = r(e'(x_1,\cdots,x_n),y_1,\ldots,y_m)$.

If $e \approx e' \in E^{\mathcal{L}}\cup\widetilde{E^{\mathcal{L}}}$ then obviously $F\vdash_{\mathcal{L}\text{UH}} s_i \approx s_{i+1}$. Otherwise, by the congruence inference rule, $F\vdash_{\mathcal{L}\text{UH}} s_i \approx s_{i+1}$. Therefore, by transitivity, we obtain $F\vdash_{\mathcal{L}\text{UH}} t \approx t'$.

$[\Rightarrow]$ We are going to prove that all equations in F and all substitution instances of the extralogical axioms are in $(\equiv_{F\cup E^{\mathcal{L}}})^*$ and $(\equiv_{F\cup E^{\mathcal{L}}})^*$ is closed under equation deduction. It is clear that all substitution instances of the equations in E and all equations in F, considered as ordered pairs, are in $(\equiv_{F\cup E^{\mathcal{L}}})^*$ (by considering the context $r = z{:}S$). By definition, $(\equiv_{F\cup E^{\mathcal{L}}})^*$ is obviously closed under the rules of reflexivity, symmetry and transitivity. To show that $\equiv_{F\cup E^{\mathcal{L}}}$ is closed under the congruence rule, let g be an operation symbol of type $S_1,\ldots,S_n \to S$ and $t_1,\ldots,t_n$ and $t'_1,\ldots,t'_n$ of type $S_1,\ldots,S_n$ respectively. Suppose that $t_i \equiv_{F\cup E^{\mathcal{L}}} t'_i$, for all $i \leq n$. To prove that $g(t'_1,\cdots,t'_{i-1},t'_i,\ldots,t'_n) \equiv_{F\cup E^{\mathcal{L}}} g(t_1,\ldots,t_i,t_{i+1},\ldots,t_n)$ it is enough to prove that

$$g(t'_1,\ldots,t'_{i-1},t_i,\ldots,t_n) \equiv_{F\cup E^{\mathcal{L}}} g(t'_1,\ldots,t'_i,t_{i+1},\ldots,t_n), \text{ for } i \leq n.$$

We know that $t_i = r(e(x_1,\ldots,x_n),y_1,\ldots,y_m)$ and $t'_i = r(e'(x_1,\ldots,x_n),y_1,\ldots,y_m)$, for some S_i-context r and some $e(x_1,\ldots,x_n) \approx e'(x_1,\ldots,x_n) \in (F\cup E^{\mathcal{L}})\cup(\widetilde{F}\cup\widetilde{E^{\mathcal{L}}})$.

So,
$$g(t'_1,\ldots,t'_{i-1},t_i,\ldots,t_n) = g(t'_1,\ldots,t'_{i-1},r(e(x_1,\ldots,x_n),y_1,\ldots,y_m),\,t_{i+1},\ldots,t_n)$$
and
$$g(t'_1,\ldots,t'_i,t_{i+1},\ldots,t_n) = g(t'_1,\ldots,t'_{i-1},r(e'(x_1,\ldots,x_n),y_1,\ldots,y_m),t_{i+1},\ldots,t_n).$$
Hence,
$$g(t'_1,\ldots,t'_{i-1},r(e(x_1,\ldots,x_n),y_1,\ldots,y_m),t_{i+1},\ldots,t_n) =$$
$$\varphi(r(e(x_1,\ldots,x_n),y_1,\ldots,y_m),y_1,\ldots,y_m)$$
and
$$g(t'_1,\ldots,t'_{i-1},r(e'(x_1,\ldots,x_n),y_1,\ldots,y_m),t_{i+1},\ldots,t_n) =$$
$$\varphi(r(e'(x_1,\ldots,x_n),y_1,\ldots,y_m),y_1,\ldots,y_m),$$
where $\varphi(z,y_1,\ldots,y_m) = g(t'_1,\ldots,t'_{i-1},z{:}T_i,t_{i+1},\ldots,t_n)$.

Therefore, by the definition of $\equiv_{F\cup E^{\mathcal{L}}}$,

$$g(t'_1,\ldots,t'_{i-1},t_i,\ldots,t_n) \equiv_{F\cup E^{\mathcal{L}}} g(t'_1,\ldots,t'_i,t_{i+1},\ldots,t_n).$$

Now, we show that $(\equiv_{F\cup E^{\mathcal{L}}})^*$ is also closed under the congruence rule. Suppose now that $t_i(\equiv_{F\cup E^{\mathcal{L}}})^*t'_i$, for all $i \leq n$. By definition of $(\equiv_{F\cup E^{\mathcal{L}}})^*$ and reflexivity, for all $i \leq n$, there are terms $s_i^1,\ldots,s_i^m$ such that

$$t_i = s_i^1 \equiv_{F\cup E^{\mathcal{L}}} \cdots \equiv_{F\cup E^{\mathcal{L}}} s_i^m = t'_i.$$

Hence, by the previous discussion,

$$g(s_1^1,\ldots,s_n^1) \equiv_{F \cup E^{\mathcal{L}}} \cdots \equiv_{F \cup E^{\mathcal{L}}} g(s_1^m,\ldots,s_n^m).$$

Finally, by transitivity we get $g(t_1,\ldots,t_n)\,(\equiv_{F \cup E^{\mathcal{L}}})^* g(t_1',\ldots,t_n')$. $\qquad\square$

Definition 3.22. Let $\mathcal{L}$ be a HEL, $T \in \mathrm{Th}(\mathcal{L})$ and F a set of equations. We say that F is a *formal behavioral consequence* of T, in symbols $T \vdash_{\mathcal{L}}^{FBE} F$, if for every visible equation $s \approx s'$, $T \cup F \vdash_{\mathcal{L}\mathrm{UH}} s \approx s'$ implies that $T \vdash_{\mathcal{L}\mathrm{UH}} s \approx s'$. We say that an equation $t \approx t'$ is a *formal behavioral consequence* of T, if $\{t \approx t'\}$ is and we write $T \vdash_{\mathcal{L}}^{FBE} t \approx t'$.

Similarly to the global behavioral consequence relation, the formal behavioral consequence relation, as a property of a set, is actually a local property.

Theorem 3.23. *Let $\mathcal{L}$ be a strict equational HEL and F be a set of equations. Then for every $T \in \mathrm{Th}(\mathcal{L})$,*

$$(12)\qquad\qquad T \vdash_{\mathcal{L}}^{FBE} F \Leftrightarrow (T \vdash_{\mathcal{L}}^{FBE} t \approx t', \text{ for all } t \approx t' \in F)$$

Proof. The implication from right to left is straightforward. To prove the other implication, let $s,s' \in (\mathrm{Te}_{\Sigma}(X))_{\mathrm{VIS}}$ and assume that $T \cup F \vdash_{\mathcal{L}\mathrm{UH}} s \approx s'$. Since $E^{\mathcal{L}} \subseteq T$, by Lemma 3.21, $s \equiv_{T \cup F}^* s'$. Hence, there are $s_1,\cdots,s_n \in \mathrm{Te}_{\Sigma}(X)$ such that $s = s_1 \equiv_{T \cup F} \cdots \equiv_{T \cup F} s_m = s'$. Then, for every $i \leq n$, $s_i = r(e(x_1,\ldots,x_n),y_1,\ldots,y_m)$ and $s_{i+1} = r(e'(x_1,\ldots,x_n),y_1,\ldots,y_m)$ for some $r(z,y_1,\ldots,y_m)$ and $e(x_1,\ldots,x_n) \approx e'(x_1,\ldots,x_n) \in T \cup F \cup \widetilde{F}$.

If $e(x_1,\ldots,x_n) \approx e'(x_1,\ldots,x_n) \in T$ then, by the congruence rule, $T \vdash_{\mathcal{L}\mathrm{UH}} r(e(x_1,\ldots,x_n),y_1,\ldots,y_m) \approx r(e'(x_1,\ldots,x_n),y_1,\ldots,y_m)$. I.e., $T \vdash_{\mathcal{L}\mathrm{UH}} s_i \approx s_{i+1}$. Otherwise, $e(x_1,\ldots,x_n) \approx e'(x_1,\ldots,x_n) \in F \cup \widetilde{F}$. Then, by the congruence rule,
$\{e \approx e'\} \vdash_{\mathcal{L}\mathrm{UH}} s_i \approx s_{i+1}$. So, $T \cup \{e \approx e'\} \vdash_{\mathcal{L}\mathrm{UH}} s_i \approx s_{i+1}$. Since $T \vdash_{\mathcal{L}}^{FBE} e \approx e'$, we get $T \vdash_{\mathcal{L}\mathrm{UH}} s_i \approx s_{i+1}$.

So, for every $i \leq n$, $T \vdash_{\mathcal{L}\mathrm{UH}} s_i \approx s_{i+1}$. Hence, by transitivity rule, $T \vdash_{\mathcal{L}\mathrm{UH}} s \approx s'$. Therefore $T \vdash_{\mathcal{L}}^{FBE} F$ $\qquad\square$

Let $\mathcal{L}$ be a HEL and $T \in \mathrm{Th}(\mathcal{L})$. We define a relation $\mathcal{FB}(T)$ on the term algebra in the following way: for each sort S, $\mathcal{FB}(T)_S$ is the set of all pairs (t,t') of formulas of type S, such that $T \vdash_{\mathcal{L}}^{FBE} t \approx t'$. Thus, $\mathcal{FB}(T) = \langle \mathcal{FB}(T)_S : S \in \mathrm{SORT}\rangle$.

Lemma 3.24. *Let $\mathcal{L}$ be a strict HEL. Then for every $T \in \mathrm{Th}(\mathcal{L})$, $\mathcal{GFB}(T) \subseteq \mathcal{FB}(T)$.*

Proof. Let $T \in \mathrm{Th}(\mathcal{L})$. Suppose that $t \approx t' \in \mathcal{GFB}(T)$. Let $s \approx s'$ be a visible equation. Suppose that $T \cup \{t \approx t'\} \vdash_{\mathcal{L}\mathrm{UH}} s \approx s'$. Then, $\mathrm{Cn}_{\mathcal{L}\mathrm{UH}}(T) \cup \{t \approx t'\} \vdash_{\mathcal{L}\mathrm{UH}} s \approx s'$. Hence, by hypothesis, $\mathrm{Cn}_{\mathcal{L}\mathrm{UH}}(T) \vdash_{\mathcal{L}\mathrm{UH}} s \approx s'$. So, $T \vdash_{\mathcal{L}\mathrm{UH}} s \approx s'$. I.e., $(t,t') \in \mathcal{FB}(T)$. $\qquad\square$

The following corollary is a consequence of the previous Lemma and Theorem 3.18. It provides a necessary condition for an equation to be a behavioral consequence of a theory.

Corollary 3.25. *Let $\mathcal{L}$ be a strict equational HEL and $t \approx t'$ an equation. Then, for every $T \in \mathrm{Th}(\mathcal{L})$*

$$(13) \qquad t \equiv t'\,(\Omega(T)) \Rightarrow T \vdash_{\mathcal{L}}^{FBE} t \approx t'.$$

An important result of this paper is the converse of (13). First we show that, similarly to the global behavioral consequence, the formal behavioral consequence relation is closed under $\mathcal{L}^{\mathrm{UH}}$-consequences.

Lemma 3.26. *Let $\mathcal{L}$ be a strict equational HEL, $T \in \mathrm{Th}(\mathcal{L})$ and F a set of equations. Then $T \vdash_{\mathcal{L}}^{FBE} F$ and $F \vdash_{\mathcal{L}^{\mathrm{UH}}} t \approx t'$ implies that $T \vdash_{\mathcal{L}}^{FBE} t \approx t'$.*

Proof. Let $s \approx s'$ be a visible equation such that $T \cup \{t \approx t'\} \vdash_{\mathcal{L}^{\mathrm{UH}}} s \approx s'$. Then, $T \cup F \vdash_{\mathcal{L}^{\mathrm{UH}}} s \approx s'$. Since $T \vdash_{\mathcal{L}}^{FBE} F$, we have that $T \vdash_{\mathcal{L}^{\mathrm{UH}}} s \approx s'$. Therefore, $T \vdash_{\mathcal{L}}^{FBE} t \approx t'$. $\qquad \square$

Therefore,

Corollary 3.27. *Let $\mathcal{L}$ be a strict equational HEL and $T \in \mathrm{Th}(\mathcal{L})$. Then $\mathcal{FB}(T)$ is a theory of $\mathcal{L}^{\mathrm{UH}}$.*

Theorem 3.28. *Let $\mathcal{L}$ be a strict equational HEL and $T \in \mathrm{Th}(\mathcal{L})$. Then*

1. *$\mathcal{FB}(T)_V = T_V$, for all $V \in \mathrm{VIS}$.*
2. *$\mathcal{FB}(T) \subseteq \Omega(T)$.*

Proof. Clearly $T_V \subseteq \mathcal{FB}(T)_V$ for all $V \in \mathrm{VIS}$. Let $t \approx t' \in \mathcal{FB}(T)_V$. Then, by definition, for every visible equation $s \approx s'$, $T \cup \{t \approx t'\} \vdash_{\mathcal{L}^{\mathrm{UH}}} s \approx s'$ implies $T \vdash_{\mathcal{L}^{\mathrm{UH}}} s \approx s'$. In particular, $T \vdash_{\mathcal{L}^{\mathrm{UH}}} t \approx t'$, i.e., $T \vdash_{\mathcal{L}} t \approx t'$. Since T is a theory, $t \approx t' \in T$.

$\mathcal{FB}(T) \subseteq \Omega(T)$ is a consequence of Corollary 3.27 and the fact that $\Omega(T)$ is the largest congruence equal to T on the visible part. $\qquad \square$

We now have gathered all the necessary results to prove the following characterization of the behavioral consequences of a theory.

Theorem 3.29. *Let $\mathcal{L}$ be a strict equational HEL and $T \in \mathrm{Th}(\mathcal{L})$. Then, for every $t, t' \in \mathrm{Te}_\Sigma(X)$,*

$$(14) \qquad t \equiv t'\,(\Omega(T)) \Leftrightarrow T \vdash_{\mathcal{L}}^{FBE^*} t \approx t'.$$

Proof. The direct implication is just Corollary 3.25, and the converse follows from Part (2) of Theorem 3.28. $\qquad \square$

4 Conclusion

The discussion of AAL in the many sorted case started in the early nineties. In (Blok and Pigozzi, 1992, Section 15) Blok and Pigozzi presented some results for equality-test algebras, where notions such as reduced matrix are dealt with. However, one can credit Don Pigozzi with the seminal ideas presented in 1999 in a course at University of Lisbon, that opened this new area of research. This application of AAL to computer science, namely to behavioral equivalence, produced several results, based on the theory of hidden k-logic. We believe that there is space for further developments. Actually, we are currently trying to use tools from AAL to deal with behavioral transitions, another topic in computer science. Preordered algebra is the natural algebraic framework to specify and reason about transitions. In (Diaconescu, 2011) Diaconescu studies a combination of preordered algebra and hidden algebra, which he calls *hidden preordered algebra*. A new concept appears in this context - *behavioral transition*. Behavioral transitions are already presented in CafeOBJ, however there are still several aspects that need more attention. For instance its methodological aspects remain unexplored. As shown in (Diaconescu, 2011), the coinduction proof method for behavioral equivalence can be extended to proving behavioral transitions. This method is based on the fact that the behavioral preordered algebra congruence on an ordered algebra $(A, \leq)$ is the largest hidden preordered algebra congruence on $(A, \leq)$. This is similar to the case of hidden congruence (behavioral equivalence) vs. Leibniz congruence. Hence, we believe that a model with two filters, one for equations and another for transitions might prove a good fit as a semantics for this computer science paradigm. Raftery's work (Raftery, 2013) on ordered algebraizable logics will probably play an important role in our intended application.

References

Babenyshev, S. and Martins, M. A. (2014). Deduction detachment theorem in hidden k-logics, *Journal of Logic and Computation*, 24(1), 233-255.

Babenyshev, S. and Martins, M. A. (2016). Behavioral equivalence of hidden logics: an abstract algebraic approach, accepted in *Journal of Applied Logic*.

Berreged, N., Bouhoula, A. and Rusinowitch, M. (1998). Observational proofs with critical contexts, in *Fundamental Approaches to Software Engineering, volume 1382 of LNCS*, Springer, Berlin, pages 38–53.

Bidoit, M. and Hennicker, R. (1996). Behavioural theories and the proof of behavioural properties, *Theoretical Computer Science*, 165(1), 3–55.

Bidoit, M. and Hennicker, R. (1999). Observer complete definitions are behaviourally coherent, preliminary long version available as Report LSV 99-4.

Bidoit, M., Hennicker, R. and Wirsing, M. (1995). Behavioural and abstractor specifications, *Science of Computer Programming*, 25(2-3), 149–186.

Blok, W. J. and Pigozzi, D. (1986). Protoalgebraic logics, *Studia Logica*, 45, 337–369.

Blok, W. J. and Pigozzi, D. (1989). *Algebraizable logics*, Memoirs of the American Mathematical Society, vol. 77, 396.

Blok, W. J and Pigozzi, D. (1992). Algebraic semantics for universal Horn logic without equality, 19. In: A. Romanowska and J.D.H. Smith (eds.), *Universal Algebra and Quasigroup Theory*, Lectures of the Conference in Jadwisin 1989, Research Exposition in Mathematics, Heldermann, Berlin, pages 1–56.

Bouhoula, A. and Rusinowitch, M. (2002). Observational proofs by rewriting, *Theoretical Computer Science*, 275, no. 1-2, 675–698.

Caleiro, C., Gonçalves, R. and Martins, M. A. (2009). Behavioral Algebraization of Logics, *Studia Logica*, 91(1), 63–111.

Czelakowski, J. (1981). Equivalential logics. (I, II), *Studia Logica*, 40, 227–236 and 355–372.

Czelakowski, J. (2001). *Protoalgebraic logics*, Trends in Logic–Studia Logica Library, 10, Dordrecht: Kluwer Academic Publishers.

Czelakowski, J. and Jansana, R. (2000). Weakly algebraizable logics, *The Journal of Symbolic Logic*, 65, no. 2, 641–668.

Czelakowski, J. and Pigozzi, D. (1999). Amalgamation and interpolation in abstract algebraic logic, in Caicedo, Xavier et al. (ed.), *Models, algebras, and proofs. Selected papers of the X Latin American symposium on mathematical logic, Bogotá, Colombia, June 24–29, 1995*, New York, NY: Marcel Dekker. Lect. Notes Pure Appl. Math. 203, pages 187–265.

Descalço, L. and Martins, M. A. (2005). On the injectivity of the Leibniz operator, *Bulletin of the Section of Logic*, 34(4), 203–211.

Diaconescu, R. (2011). Coinduction for preordered algebra, *Information and Computation*, 209(2), 108–117.

Diaconescu, R. and Futatsugi, K. (1998). CafeOBJ report: The language, proof techniques, and methodologies for object-oriented algebraic specification, in *A. series in Computing*, editor, *World Scientific*, volume 6.

Diaconescu, R. and Futatsugi, K. (2000). Behavioural coherence in object-oriented algebraic specification. *Journal of Universal Computer Science*, 6(1), 74–96.

Font, J. and Jansana, R. (1996). *A general algebraic semantics for sentential logics*, Lecture Notes in Logic, 7. Springer, Berlin.

Font, J., Jansana, R. and Pigozzi, D. (2003). A survey of abstract algebraic logic, *Studia Logica*, 74, 13–97.

Goguen, J. (1989). Types as theories, in *Topology and category theory in computer science*, Oxford Sci. Publ., pages 357–390. Oxford University Press, New York.

Goguen, J., Lin, K. and Roşu, G. (2002). Conditional circular coinductive rewriting with case analysis, *Recent Trends in Algebraic Development Tech-*

niques, 16th International Workshop, WADT 2002, Revised Selected Papers, 216–232.

Goguen, J. and Malcolm, G. (1999). Hidden coinduction: Behavioural correctness proofs for objects, *Mathematical Structures in Computer Science*, 9, no. 3, 287–319.

Goguen, J. and Malcolm, G. (2000). A hidden agenda, *Theoretical Computer Science*, 245, no. 1, 55–101.

Goguen, J. and Roşu, G. (1999). Hiding more of hidden algebra., Wing, Jeannette M. (ed.) et al., FM '99. Formal methods, *World congress on Formal methods in the development of computing systems, Toulouse, France, September 20-24, 1999*, Proceedings, in 2 vols. Springer, Berlin; 3-540-66588-9 (vol. 2)). Lect. Notes Comput. Sci. 1708; 1709, 1704-1719.

Goriac, E., Lucanu, D. and Roşu, G. (2010). Automating coinduction with case analysis, in *Formal Methods and Software Engineering - 12th International Conference on Formal Engineering Methods, ICFEM 2010, Shanghai, China, November 17-19, 2010. Proceedings*, pages 220–236.

Hennicker, R. (1997). *Structural specifications with behavioural operators: semantics, proof methods and applications*, Habilitationsschrift.

Hennicker, R. and Bidoit, M. (1999). Observational logic, in *Proceedings of AMAST '98, 7th International Conference on Algebraic Methodology and Software Technology*, Lecture Notes in Computer Science, Springer, Berlin, pag. 263–277.

Herrmann, B. (1996). Equivalential and algebraizable logic, *Studia Logica*, 57(2-3), 419–436.

Goguen, J., Lin, K. and Roşu, G. (2000). Circular coinductive rewriting, in *Proceedings, Automated Software Engineering 2000 (Grenoble France)*, IEEE Press, pages 123–131, September.

Leavens, G. T. and Pigozzi, D. (2000). A complete algebraic characterization of behavioral subtyping, *Acta Informatica*, 36(8), 617–663.

Leavens, G. T. and Pigozzi, D. (2002). Equational reasoning with subtypes, Technical Report TR #02-07, Iowa State University, July, Available at `ftp://ftp.cs.iastate.edu/pub/techreports/TR02-07/TR.pdf`.

Malcolm, G. (1996). Behavioural equivalence, bisimulation, and minimal realisation, in Magne Haveraaen and Olaf Owe and Ole-Johan Dahl (eds.), *Recent Trends in Data Type Specifications, 11th Workshop on Specification of Abstract Data Types*, Springer-Verlag Lecture Notes in Computer Science.

Martins, M. A. (2004). *Behavioral reasoning in generalized hidden logics*, PhD thesis, University of Lisbon.

Martins, M. A. (2006). Behavioral institutions and refinements in generalized hidden logics, *Journal of Universal Computer Science*, 12(8), 1020–1049.

Martins, M. A. (2007). Closure properties for the class of behavioral models, *Theoretical Computer Science*, 379(1–2), 53–83.

Martins, M. A. (2008). On the behavioral equivalence between k-data structures, *Computer Journal*, 51(2), 181–191.

Martins, M. A. and Pigozzi, D. (2007). Behavioural reasoning for conditonal equations, *Mathematical Structures in Computer Science*, 17(5), 1075–1113.

Meinke, K. and Tucker, J. V. (1992). *Universal algebra, Handbook of Logic in Computer Science*, vol. 1, Oxford Univ. Press, New York, pp. 189–411.

Moore, B. and Roşu, G. (2015). Program verification by coinduction, Technical Report, available at `http://hdl.handle.net/2142/73177`, University of Illinois, February.

Padawitz, P. (2000). Swinging types=functions + relations + transition systems, *Theoretical Computer Science*, 243, no. 1-2, 93–165.

Pałasińska, K. M.(1994). *Deductive systems and finite axiomatization properties*, PhD thesis, Iowa State University.

Pigozzi, D. (1991). Equality-test and if-then-else algebras: Axiomatization and specification, *SIAM Journal on Computing*, 20(4), 766–805.

Pigozzi, D. (1999). *Abstract algebraic logic and the specification of abstract data types*, Preprint, June.

Pigozzi, D. (2001). *Abstract algebraic logic*, Encyclopedia of Mathematics, Supplement III (M. Hazewinkel, ed.), Kluwer Academic Publishers, Dordrecht, December, pp. 2–13.

Raftery, J.G. (2013). Order algebraizable logics, *Annals of Pure and Applied Logic*, 164 (3), 251–283.

Reichel, H. (1984). Behavioural validity of conditional equations in abstract data types, in *Proceedings of the Vienna Conference*.

Reichel, H. (1995). An approach to object semantics based on terminal coalgebras, *Mathematical Structures in Computer Science*, 5(2), 129–152.

Roşu, G. (2000). *Hidden Logic*, PhD thesis, University of California, San Diego.

Tarski, A. (1930). Fundamentale begriffe der methodologie der deduktiven wissenschaften. i, *Monatshefte für Mathematik and Physik*, 37, 361–404, English translaction as Fundamental concepts of the methodology of the deductive systems.

Voutsadakis, G. (2004). Categorical Abstract Algebraic Logic: Categorical Algebraization of Equational Logic, *Logic Journal of the IGPL*, 12(4), 313–333.

Voutsadakis, G. (2014). Categorical Abstract Algebraic Logic: Behavioral π-Institutions, *Studia Logica*, 102 (3), 617-646.

Wechler, W. (1992). *Universal algebra for computer scientists*, EATCS Monographs on Theoretical Computer Science. 25. Berlin: Springer-Verlag.

Wójcicki, R. (1988). *Theory of logical caculi. Basic theory of consequence operations.* Synthese Library, 199. Dordrecht: Kluwer Academic Publishers.

Absorption and directed Jónsson terms

Alexandr Kazda, Marcin Kozik, Ralph McKenzie, and Matthew Moore*

Abstract We prove that every congruence distributive variety has directed Jónsson terms, and every congruence modular variety has directed Gumm terms. The directed terms we construct witness every case of absorption witnessed by the original Jónsson or Gumm terms. This result is equivalent to a pair of claims about absorption for admissible preorders in congruence distributive and congruence modular varieties, respectively. For finite algebras, these absorption theorems have already seen significant applications, but until now, it was not clear if the theorems hold for general algebras as well. Our method also yields a novel proof of a result by P. Lipparini about the existence of a chain of terms (which we call Pixley terms) in varieties that are at the same time congruence distributive and k-permutable for some k.

2010 Mathematics Subject Classification: 08B05, 08B10

Alexandr Kazda
IST Austria, Am Campus 1, 3400 Klosterneuburg, Austria, e-mail: `alex.kazda@gmail.com`

Marcin Kozik
Theoretical Computer Science, Faculty of Mathematics and Computer Science, Jagiellonian University, Cracow, Poland, e-mail: `kozik@tcs.uj.edu.pl`

Ralph McKenzie
Department of Mathematics, Vanderbilt University, Nashville, TN 37240, USA, e-mail: `ralph.n.mckenzie@vanderbilt.edu`

Matthew Moore
Department of Mathematics and Statistics, McMaster University, Hamilton, Ontario, Canada L8S 4K1 e-mail: `matthew.moore@math.mcmaster.ca`

* The second author was supported by National Science Center grant DEC-2011-/01/B/ST6/01006.

1 Introduction

In 1967, Bjarni Jónsson (Jónsson, 1967) proved that a variety $\mathcal{V}$ is congruence distributive (CD) if and only if it has, for some n, a sequence of terms $J_0(x,y,z)$, ..., $J_n(x,y,z)$ satisfying a certain system of equations, namely, $J_0(x,y,z) = x$, $J_n(x,y,z) = z$, $J_i(x,y,x) = x$ for each $0 \leq i \leq n$, and for each $0 \leq i < n$, either the equation $J_i(x,x,y) = J_{i+1}(x,x,y)$ or the equation $J_i(x,y,y) = J_{i+1}(x,y,y)$. This Maltsev condition can be formulated more specifically in several equivalent ways. The following formulation is convenient for our purposes: for some $n \geq 0$ and terms $J_0(x,y,z),\ldots,J_{2n+1}(x,y,z)$, consider the system of equations

$$(\mathrm{J}(n)) \qquad \begin{aligned} &J_1(x,x,y) = x, \qquad J_{2n+1}(x,y,y) = y, \\ &J_i(x,y,x) = x, &&\text{for } 0 \leq i \leq 2n+1, \\ &J_{2i+1}(x,y,y) = J_{2i+2}(x,y,y) &&\text{for } 0 \leq i \leq n-1, \\ &J_{2i}(x,x,y) = J_{2i+1}(x,x,y) &&\text{for } 1 \leq i \leq n, \end{aligned}$$

and call this package of equations $\mathrm{J}(n)$. By a chain of *Jónsson terms* for a variety $\mathcal{V}$, we mean a sequence of terms satisfying over $\mathcal{V}$ the equations $\mathrm{J}(n)$ for some n. Jónsson proved that an algebra $\mathbf{A}$ has terms obeying the equations $\mathrm{J}(n)$, for some n, if and only if the congruence lattice of every algebra in the variety generated by $\mathbf{A}$ is distributive. A system of *directed Jónsson terms* for $\mathcal{V}$ consists, for some $n \geq 1$, of terms $D_1(x,y,z),\ldots,D_n(x,y,z)$ satisfying over $\mathcal{V}$ the equations $\mathrm{DJ}(n)$:

$$(\mathrm{DJ}(n)) \qquad \begin{aligned} &D_1(x,x,y) = x, \qquad D_n(x,y,y) = y, \\ &D_i(x,y,x) = x &&\text{for } 1 \leq i \leq n, \\ &D_i(x,y,y) = D_{i+1}(x,x,y) &&\text{for } 1 \leq i < n. \end{aligned}$$

Our chief purpose is to show that a variety has Jónsson terms if and only if it has directed Jónsson terms. Moreover, in such a case, one can find a sequence of terms which satisfy $\mathrm{J}(n)$ and $\mathrm{DJ}(2n+1)$ for some n at the same time. These two results are contained in Corollary 4.1 and Observation 1.2.

H.P. Gumm (Gumm, 1981) proved that a variety $\mathcal{V}$ is congruence modular (CM) if and only if it has, for some $n \geq 0$, a sequence of terms $J_1(x,y,z),\ldots,J_{2n+1}(x,y,z)$, and $P(x,y,z)$ satisfying the equations $\mathrm{G}(n)$:

$$(\mathrm{G}(n)) \qquad \begin{aligned} &J_1(x,x,y) = x, \qquad J_{2n+1}(x,y,y) = P(x,y,y), \qquad P(x,x,y) = y, \\ &J_i(x,y,x) = x &&\text{for } 0 \leq i \leq 2n+1, \\ &J_{2i+1}(x,y,y) = J_{2i+2}(x,y,y) &&\text{for } 0 \leq i \leq n-1, \\ &J_{2i}(x,x,y) = J_{2i+1}(x,x,y) &&\text{for } 1 \leq i \leq n. \end{aligned}$$

Directed Gumm terms are terms $D_1(x,y,z), \ldots, D_n(x,y,z)$, and $Q(x,y,z)$ satisfying $\mathrm{DG}(n)$ for some $n \geq 1$:

$$
\begin{array}{lll}
D_1(x,x,y) = x, & D_n(x,y,y) = Q(x,y,y), & Q(x,x,y) = y, \\
(\mathrm{DG}(n)) \quad D_i(x,y,x) = x & \text{for } 1 \leq i \leq n, \\
D_i(x,y,y) = D_{i+1}(x,x,y) & \text{for } 1 \leq i < n.
\end{array}
$$

Similarly to the congruence distributive case, we show that a variety has Gumm terms if and only if it has directed Gumm terms, and that given Gumm terms we can find terms satisfying $\mathrm{G}(n)$ and $\mathrm{DG}(2n+1)$ for some n at the same time. These two results are contained in Theorem 6.1 and Observation 1.2.

Our context makes it natural to introduce another Maltsev condition that looks similar to directed Jónsson terms but is actually much stronger. The condition is that for some $n \geq 1$ there are terms $P_1(x,y,z), \ldots, P_n(x,y,z)$ satisfying $\mathrm{P}(n)$:

$$
\begin{array}{lll}
P_1(x,y,y) = x, & P_n(x,x,y) = y, \\
(\mathrm{P}(n)) \quad P_i(x,y,x) = x & \text{for } 1 \leq i \leq n, \\
P_i(x,x,y) = P_{i+1}(x,y,y) & \text{for } 1 \leq i < n.
\end{array}
$$

This condition, which we call *Pixley terms*, first appeared in P. Lipparini (Lipparini, 1995).

Observe that if we remove the equations "$J_i(x,y,x) = x$" from $\mathrm{DJ}(n)$, we obtain a Maltsev condition that is always trivially satisfied by taking $D_1(x,y,z) = y$ and $D_i(x,y,z) = z$ for all $1 < i \leq n$. For contrast, removing the equations $P_i(x,y,x) = x$ from $\mathrm{P}(n)$ produces the classical Hagemann-Mitschke terms (Hagemann and Mitschke, 1973), and these have highly nontrivial consequences. A variety has a chain of n Hagemann-Mitschke terms if and only if it has $(n+1)$-permuting congruences. The variety of lattices, for example, satisfies $\mathrm{J}(1)$ but does not have Hagemann-Mitschke terms.

A. Pixley (Pixley, 1963) proved that a variety is congruence distributive and all its congruences permute if and only if it satisfies $\mathrm{P}(1)$. A term $P_1(x,y,z)$ for which

$$
P_1(x,y,x) = P_1(x,y,y) = P_1(y,y,x) = x
$$

holds has long been called a *Pixley term*. In this connection, note that the term $J_1(x,y,z)$ with the equations $J_1(x,y,x) = J_1(x,x,y) = J_1(y,x,x) = x$ constituting Jónsson terms $\mathrm{J}(0)$ is familiarly known as a *majority* term; and both $\mathrm{J}(0)$ and $\mathrm{DJ}(1)$ are just asserting that we have a majority term.

Here is our principal result about these Maltsev conditions.

Theorem 1.1. *Let $\mathcal{V}$ be any variety of algebras.*

1. *$\mathcal{V}$ is congruence distributive if and only if it has directed Jónsson terms. In such a case there is a sequence of terms satisfying $DJ(2n+1)$ and $J(n)$ at the same time (for some $n \geq 1$). See Corollary 4.1 and Observation 1.2.*
2. *For any integer $k \geq 1$, a variety $\mathcal{V}$ is congruence distributive and has $(k+1)$-permuting congruences if and only if it satisfies $P(k)$. See Theorem 5.1 for the "$\Rightarrow$" implication.*
3. *$\mathcal{V}$ is congruence modular if and only if it has directed Gumm terms. In such a case there is a sequence of terms satisfying $G(n)$ and $DG(2n+1)$ at the same time (for some $n \geq 1$). See Theorem 6.1 and Observation 1.2.*

Statement (2) is Proposition 5 in P. Lipparini (Lipparini, 1995). However, our proof, given in Section 5, is new, and shows more.

Observation 1.2. *Let $\mathcal{V}$ be a variety that admits a chain of terms satisfying $DJ(n)$. Then $\mathcal{V}$ admits a chain of terms that satisfy $J(n-1)$ and $DJ(2n-1)$ at the same time. Similarly, $DG(n)$ implies the existence of a chain of terms that simultaneously satisfies $G(n-1)$ and $DG(2n-1)$.*

Proof. Given directed Jónsson terms $D_1, \ldots, D_n$, we produce the new terms by letting

$$J_1(x,y,z) = D_1(x,y,z), \qquad J_{2i}(x,y,z) = D_{i+1}(x,x,z),$$
$$J_{2i+1}(x,y,z) = D_{i+1}(x,y,z) \qquad \text{for } 1 \leq i \leq n-1.$$

We leave to the reader the easy proof that $DG(n)$ implies appropriate terms for congruence modular varieties.

Similarly, $P(n)$ implies $J(n)$: given some Pixley terms $P_1, \ldots, P_n$, take

$$J_1(x,y,z) = x, \qquad J_{2n+1}(x,y,z) = z,$$
$$J_{2i}(x,y,z) = P_i(x,y,z) \qquad \qquad \text{for } 0 \leq i < n,$$
$$J_{2i+1}(x,y,z) = P_{i+1}(x,z,z) \qquad \text{for } 0 \leq i < n.$$

It is an easy exercise to show that $P(k)$ implies $(k+1)$-permuting congruences.

Our proof of the converse implications, that is, $J(n)$ implies $DJ(k)$ for some k, and $G(n)$ implies $DG(k)$ for some k, will take some work and will be concluded in Sections 4 and 6. The fact that a $(k+1)$-permutable variety with Jónsson terms satisfies $P(k)$ is demonstrated in Section 5.

2 Absorption

The notion of *absorption* was introduced by L. Barto and M. Kozik (Barto and Kozik, 2012), who proved deep results about absorption in finite algebras and used this theory as a powerful tool for applying universal algebraic methods in the study of constraint satisfaction problems (this area where universal algebra and theoretical computer science meet has blossomed over the past decade).

If $\mathbf{C}$ and $\mathbf{D}$ are subalgebras of an algebra $\mathbf{A}$ we say that $\mathbf{C}$ *absorbs* $\mathbf{D}$ if $\emptyset \neq C \subseteq D$ and there is a term operation $s(x_1,\ldots,x_n)$ of the algebra $\mathbf{A}$ such that $\mathbf{A} \models s(x,\ldots,x) = x$ (i.e. s is idempotent) and whenever $\bar{d} \in D^n$ with $d_i \in D \setminus C$ for at most one $i \in \{1,\ldots,n\}$, then $s(\bar{d}) \in C$. We denote the fact that $\mathbf{C}$ absorbs $\mathbf{D}$ in this sense by $\mathbf{C} \lhd \mathbf{D}$, or $\mathbf{C} \lhd_s \mathbf{D}$ where s is the term operation that witnesses the absorption.

In this paper, however, a different variant of absorption is needed. We will say that a sequence $J_1,\ldots,J_{2n+1}$ of terms is a chain of *weak Jónsson terms* if $J_1,\ldots,J_{2n+1}$ satisfy all of the equations $J(n)$ except perhaps $J_i(x,y,x) = x$. We define *weak directed Jónsson chains*, *weak Gumm chains*, and *weak directed Gumm chains* similarly, always dropping the requirement that $J_i(x,y,x) = x$.

If $\mathbf{C}$ and $\mathbf{D}$ are subalgebras of $\mathbf{A}$, $\emptyset \neq C \subseteq D$, and $t(x,y,z)$ is a ternary idempotent term operation of $\mathbf{A}$, then we write $\mathbf{C} \lhd_t^m \mathbf{D}$ and say that $\mathbf{C}$ *middle absorbs* $\mathbf{D}$ *with respect to* t if $t(a,b,c) \in C$ whenever $a,c \in C$ and $b \in D$. If $\mathcal{T}$ is a set of ternary idempotent term operations of $\mathbf{A}$, we say that $\mathbf{C}$ middle absorbs $\mathbf{D}$ with respect to $\mathcal{T}$, written $\mathbf{C} \lhd_{\mathcal{T}}^m \mathbf{D}$, provided that $\mathbf{C} \lhd_t^m \mathbf{D}$ for every $t \in \mathcal{T}$.

We are interested in four special cases of middle absorption: Jónsson absorption, Gumm absorption, and directed versions thereof. We save Gumm absorption for the end of this paper and concentrate on Jónsson absorption for now.

We say that $\mathbf{C}$ *Jónsson absorbs* $\mathbf{D}$ if $\mathbf{C} \lhd_{\mathcal{J}}^m \mathbf{D}$, where $\mathcal{J}$ is a sequence of weak Jónsson terms. Directed Jónsson absorption is defined analogously, with weak directed Jónsson terms. We shall write $\mathbf{C} \lhd_J \mathbf{D}$ (in words, C Jónsson absorbs D) to indicate either that $\mathbf{C} \lhd_{\mathcal{J}}^m \mathbf{D}$ for some chain $\mathcal{J}$ of weak Jónsson terms, or that $\mathbf{C} \lhd_{\mathcal{J}}^m \mathbf{D}$ for a specific system of terms that is being held fixed. The context will make clear which is meant. Our use of the notation $\mathbf{C} \lhd_{DJ} \mathbf{D}$ (directed Jónsson absorption) is analogous.

One can show that if $\mathbf{A}$ is a finite idempotent algebra—equivalently, every one-element subset of A is a subuniverse—then $\mathbf{A}$ admits a chain of Jónsson terms (respectively, directed Jónsson terms) if and only if for every $a \in A$ we have $\{a\} \lhd_J \mathbf{A}$ (respectively, $\{a\} \lhd_{DJ} \mathbf{A}$). Moreover, it is immediate that standard absorption, $\mathbf{C} \lhd \mathbf{D}$, implies $\mathbf{C} \lhd_{DJ} \mathbf{D}$, which in turn implies $\mathbf{C} \lhd_J \mathbf{D}$. Indeed, suppose that $\mathbf{C} \lhd_t \mathbf{D}$ for $t = t(x_1,\ldots,x_n)$. Take $Q_1(x,y,z) = t(x,\ldots,x,y)$,

$$Q_j(x,y,z) = t(x,\ldots,x,y,z,\ldots,z) \qquad \text{with } y \text{ in the } (n-j+1)\text{-th place,}$$

for $1 < j < n$, and $Q_n(x,y,z) = t(y,z,\ldots,z)$. This is a system of directed Jónsson operations with respect to which $\mathbf{C}$ middle absorbs $\mathbf{D}$. The proof that $\mathbf{C} \lhd_{DJ} \mathbf{D}$ implies $\mathbf{C} \lhd_J \mathbf{D}$ is similar to the argument that if $\mathcal{V}$ is a variety with a chain of terms that satisfy $DJ(n)$, then $\mathcal{V}$ has a chain of terms satisfying $J(n-1)$.

The second principal result of our paper is included in Theorem 2.2. Before introducing it we present a proof of the same result for finite algebras. The result was motivated by Barto (Barto, 2013) and the proof essentially follows the argument presented there.

Theorem 2.1. *Suppose that E and F are admissible preorders on $\mathbf{A}$ (that is, they are subalgebras of $\mathbf{A}^2$ that are reflexive and transitive over A). If $E \lhd_J F$, then $E = F$.*

Proof (Proof (assuming $\mathbf{A}$ is finite)). Suppose that E and F are admissible preorders of the finite algebra $\mathbf{A}$ and $E \lhd_J F$. Let $J_1,\ldots,J_{2n+1}$ be the terms that witness the Jónsson absorption, and let $(a,b) \in F$. We must show that $(a,b) \in E$. For ease of notation, we will write $x \dashrightarrow y$ for $(x,y) \in F$ and $x \to y$ for $(x,y) \in E$ (so we want to show $(a \dashrightarrow b) \Rightarrow (a \to b)$).

Without loss of generality, we can assume that $\mathbf{A}$ is idempotent, and is generated by $\{a,b\}$. Thus, b is a top element in the order $\dashrightarrow$, since if $c \in A$ then we can write $c = t(a,b)$ for some term t, and then $c = t(a,b) \dashrightarrow t(b,b) = b$ because $\dashrightarrow$ respects all term operations. Since $\mathbf{A}$ is finite, we can also assume that a is $\to$-maximal in $\mathbf{A}$. (If there was a c strictly $\to$-larger than a in the algebra generated by $\{a,b\}$, we could replace a by c.) Using $J_1(\to,\dashrightarrow,\to) \subset \to$ and a Jónsson equation, we have

$$a = J_1(a,a,b) \to J_1(a,b,b).$$

Now we prove by induction on i that $a \to J_{2i+1}(a,b,b)$ for all $0 \le i \le n$. Suppose that $a \to J_{2i+1}(a,b,b) = J_{2i+2}(a,b,b) = q$. Let $p = J_{2i+2}(a,a,b)$. Absorption gives that $p \to q$, and that $p = J_{2i+3}(a,a,b) \to J_{2i+3}(a,b,b)$, so all we need to show is that $a \to p$.

The maximality of a yields $q \to a$. Since p lies in the subalgebra generated by $\{a,b\}$, we have $a \dashrightarrow p$. Putting it together, we have $q \to a \dashrightarrow p \to q$.

We have obtained $q \dashrightarrow p \dashrightarrow q$. Absorption now allows us to prove that $q \to p$:

$$q = J_1(q,q,p) \to J_1(q,p,p) = J_2(q,p,p) \to J_2(q,q,p)$$
$$= J_3(q,q,p) \to \cdots \to J_{2n+1}(q,q,p) \to J_{2n+1}(q,p,p) = p.$$

Therefore, $a \to q \to p \to J_{2i+3}(a,b,b)$ (see Figure 1 as a reference to what we did) and we have $a \to J_{2i+1}(a,b,b)$ for all i. In particular, $a \to J_{2n+1}(a,b,b) = b$, and we are done.

Note that there is a straightforward proof of the conclusion of the above Theorem if we assume that $E \lhd_{DJ} F$ instead of $E \lhd_J F$.

Using Theorem 2.1, we will now prove part 1 of Theorem 1.1 in the finite case. Let $\mathcal{V}$ be an idempotent CD variety, and let $\mathbf{F}_2(x,z)$ and $\mathbf{F}_3(x,y,z)$ be the free two and three generated algebras in $\mathcal{V}$. Let

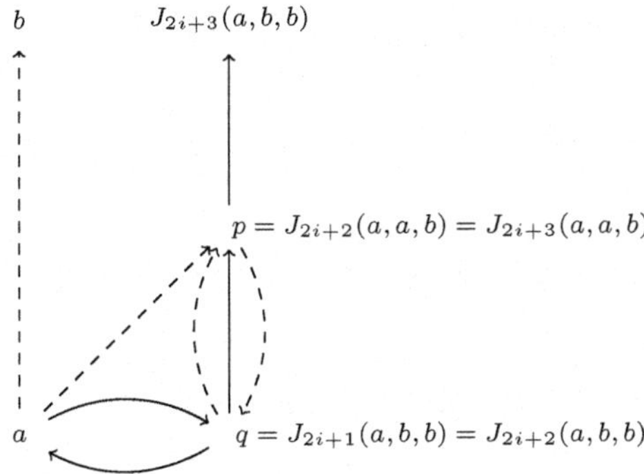

Fig. 1 The elements a,b,p,q in the finite case of Theorem 2.1.

$$\mathcal{G} = \{t(x,y,z) \in \mathbf{F}_3 \colon t(x,y,x) = x \text{ holds in } \mathcal{V}\},$$
$$F = \{(t(x,x,z),t(x,z,z)) \colon t \in \mathbf{F}_3\}, \quad \text{and} \quad E = \{(t(x,x,z),t(x,z,z)) \colon t \in \mathcal{G}\}.$$

Denote by $\to$ (resp. $\dashrightarrow$) the transitive closures of E (resp. F). It is straightforward to show that E, F, $\to$, and $\dashrightarrow$ are admissible relations on $\mathbf{F}_2$ (using reflexivity). Since E, F are reflexive, the relations $\to$ and $\dashrightarrow$ are preorders on $\mathbf{F}_2$.

Observe that $(x,z) \in F$ (we can choose t to be the projection to the second coordinate). Let $\mathcal{J}$ be a chain of Jónsson terms in $\mathcal{V}$. One can easily verify that then $E \vartriangleleft_{\mathcal{J}}^m F$, from which it follows that $\to \vartriangleleft_{\mathcal{J}}^m \dashrightarrow$. Using Theorem 2.1, we then have that $\to$ and $\dashrightarrow$ are the same. In particular, $x \to z$, and there is a sequence of terms $D_1,\ldots,D_m \in \mathcal{G}$ witnessing this fact. Examining the terms $D_1,\ldots,D_m$, we get the following system of equalities in $\mathcal{V}$:

$$D_1(x,x,z) = x, \qquad D_m(x,z,z) = z,$$
$$D_i(x,y,x) = x \qquad \text{for } 1 \le i \le m,$$
$$D_i(x,z,z) = D_{i+1}(x,x,z) \qquad \text{for } 1 \le i < m,$$

which means that $D_1,\ldots,D_m$ are directed Jónsson terms.

Of course, the sequence of proofs presented so far only works when $\mathbf{F}_2$ is finite, but we will improve that. In fact, we will show that one can always make Jónsson absorption into directed Jónsson absorption.

Theorem 2.2. *Let $\mathcal{V}$ be a variety, and $\mathcal{J}$ be a chain of weak Jónsson terms of $\mathcal{V}$. Then there exists a chain $\mathcal{D}$ of weak directed Jónsson terms of $\mathcal{V}$ such that for all $\mathbf{A}, \mathbf{B} \in \mathcal{V}$ we have $\mathbf{B} \vartriangleleft_{\mathcal{J}}^m \mathbf{A} \Rightarrow \mathbf{B} \vartriangleleft_{\mathcal{D}}^m \mathbf{A}$.*

The proof of Theorem 2.2 will have to wait until Section 4, after we have constructed suitable tools.

3 Paths in the free algebra

This section contains the core of this paper – a proof of a somewhat technical result from which Theorem 2.2 follows.

We choose and fix a variety $\mathcal{W}$ whose only basic operations are $J_1 \ldots, J_{2k+1}$, which satisfy the equations

$$
\begin{aligned}
&J_1(x,x,y) = x, \\
&J_{2i+1}(x,y,y) = J_{2i+2}(x,y,y) && \text{for } 0 \le i \le k-1, \\
&J_{2i}(x,x,y) = J_{2i+1}(x,x,y) && \text{for } 1 \le i \le k.
\end{aligned}
\tag{1}
$$

By adding more equations and operations, we could make $\mathcal{W}$ congruence distributive or congruence modular. Our aim is to turn the chain $J_1, \ldots, J_{2k+1}$ into a longer chain of directed terms that ends at something like $J_{2k+1}(x,z,z)$.

Notice that the operations of $\mathcal{W}$ are idempotent. Let $\mathbf{F}_3$ be the free algebra on three generators in $\mathcal{W}$, freely generated (relative to $\mathcal{W}$) by the elements x, y, z. Let $\mathbf{F}_2 \le \mathbf{F}_3$ be the subalgebra of $\mathbf{F}_3$ freely generated by x and z.

We shall be working with two binary relations E, F on $\mathbf{F}_2$. Define $\mathbf{F}$ to be the subalgebra of $\mathbf{F}_2^2$ generated by the pairs (x,x), (x,z) and (z,z), that is

$$
F = \big\{ (t(x,x,z), t(x,z,z)) : t \text{ is a ternary term of } \mathcal{W} \big\}.
$$

Let $\mathcal{J} = \{J_1, \ldots, J_{2k+1}\}$ and define $\mathcal{G}$ to be the set of all $\mathcal{W}$-terms $t(x,y,z)$ such that whenever $\mathbf{A}, \mathbf{B} \in \mathcal{W}$ are algebras such that $\mathbf{B} \vartriangleleft_{\mathcal{J}}^m \mathbf{A}$, then $\mathbf{B} \vartriangleleft_{t}^m \mathbf{A}$. While the set $\mathcal{G}$ is hard to describe explicitly, one can easily see that $\mathcal{J} \subseteq \mathcal{G}$ and that $\mathcal{G}$ is a subalgebra of $\mathbf{F}_3$.

From this it immediately follows that

$$
E = \big\{ (t(x,x,z), t(x,z,z)) : t(x,y,z) \in \mathcal{G} \big\}
$$

is an admissible relation over $\mathbf{F}_2$. Moreover, it is straightforward to verify from the definition of absorption that $\mathcal{G} \vartriangleleft_{\mathcal{J}}^m \mathbf{F}_3$, from which it follows that $E \vartriangleleft_{\mathcal{J}}^m F$. We will view the pair $E \vartriangleleft_{\mathcal{J}}^m F$ as a generic instance of absorption in $\mathcal{W}$. Notice that $(x,x), (z,z) \in E$ since the projections x, z belong to $\mathcal{G}$. Thus, since all operations are idempotent, we have that the relations E and F are reflexive over F_2. That is, $(a,a) \in E$ for all $a \in F_2$.

It is important to notice that for every $a \in F_2$ we have $(x,a), (a,z) \in F$. To see this, write $a = t(x,z)$ for a term t, and apply the term operation $t^{\mathbf{F}_2}$ of $\mathbf{F}_2$ to the pairs (x,x) and (x,z) and use that $t(x,x) = x$, yielding $(x,a) \in F$; and for $(a,z) \in F$ apply $t^{\mathbf{F}_2}$ to (x,z) and (z,z).

We shall write $p \dashrightarrow q$ to indicate that the pair (p,q) belongs to the transitive closure of F and $p \rightarrow q$ to indicate that (p,q) belongs to the transitive closure of E. Both relations $\dashrightarrow$ and $\rightarrow$ are admissible preorders of $\mathbf{F}_2$ (i.e. they are transitive and reflexive). We leave it to the reader to verify that $\rightarrow \vartriangleleft_{\mathcal{J}}^m \dashrightarrow$.

We now introduce *left powers of elements* of $\mathbf{F}_2$: for any $a = a(x,z) \in \mathbf{F}_2$ define $a^0 = z$ and, inductively,

$$a^{k+1}(x,z) = a(x,a^k).$$

In more complicated expressions, we evaluate powers first, so for example $a^2(b,c)$ means "take $a^2(x,z)$ and substitute $x = b, z = c$", giving us $a(b,a(b,c))$. Observe that thus defined, exponentiation satisfies the equalities $(a^k)^\ell = a^{k\ell}$ and $z^k = z$ for any $a \in \mathbf{F}_2$ and any k, ℓ nonnegative integers.

Letting $J = J(x,z) = J_{2k+1}(x,z,z)$, we can state the core result of this paper, whose proof takes up the remainder of this section.

Theorem 3.1. *There exists $b \in \mathbf{F}_2$ such that $x \to J^{2^k}(b, J^{2^k - 1})$.*

The next lemma is essential for our proof of Theorem 3.1. Every endomorphism of $\mathbf{F}_2$ is uniquely determined by the elements to which it sends x and z, and, conversely, for any pair $a,b \in \mathbf{F}_2$ there is an endomorphism σ of $\mathbf{F}_2$ that sends each $c(x,z) \in \mathbf{F}_2$ to $c(a,b) = c(a(x,z),b(x,z))$ (in particular $\sigma(x) = a$ and $\sigma(z) = b$). An endomorphism σ of $\mathbf{F}_2$ will be called *special* if $\sigma(x) \dashrightarrow \sigma(z)$.

Lemma 3.2. *Every special endomorphism of $\mathbf{F}_2$ respects $\dashrightarrow$ and $\to$. That is, given $a \dashrightarrow b$,*

- *if $c = c(x,z) \dashrightarrow d(x,z) = d$ then $c(a,b) \dashrightarrow d(a,b)$; and*
- *if $c = c(x,z) \to d(x,z) = d$ then $c(a,b) \to d(a,b)$.*

Proof. To show that σ, moving x to a and z to b with $a \dashrightarrow b$, respects $\dashrightarrow$, it suffices to show that $c\,F\,d$ implies $c(a,b) \dashrightarrow d(a,b)$. Let $c(x,z)\,F\,d(x,z)$. Thus there is a term $s(u,v,w)$ so that

$$c(x,z) = s(x,x,z) \qquad \text{and} \qquad d(x,z) = s(x,z,z).$$

Applying σ to these equations, we have that

$$c(a,b) = s(a,a,b) \qquad \text{and} \qquad d(a,b) = s(a,b,b),$$

or in a more suggestive matrix form:

$$\begin{pmatrix} c(a,b) \\ d(a,b) \end{pmatrix} = s \begin{pmatrix} a & a & b \\ a & b & b \end{pmatrix}.$$

Now observe that in each of the three columns of the matrix on the right hand side, the rows are related by $\dashrightarrow$. Since s preserves $\dashrightarrow$, we have $c(a,b) \dashrightarrow d(a,b)$, as required.

To show that σ respects $\to$, it again suffices to show that $c\,E\,d$ implies $c(a,b) \to d(a,b)$. Let $c(x,z)\,E\,d(x,z)$. As before, there is a term $s(u,v,w)$ such that

$$c(x,z) = s(x,x,z) \qquad \text{and} \qquad d(x,z) = s(x,z,z),$$

but this time we also know that $s(x,y,z) \in \mathcal{G}$. We again apply σ and write the result in a matrix form:

$$\begin{pmatrix} c(a,b) \\ d(a,b) \end{pmatrix} = s \begin{pmatrix} a & a & b \\ a & b & b \end{pmatrix}.$$

Observe that in the first and third columns on the right hand side, the rows are $\rightarrow$-related, while the middle column is $\dashrightarrow$-related. Since $s \in \mathcal{G}$ it follows that $\rightarrow \lhd_s^m \dashrightarrow$ and hence the pair on the left hand side must be $\rightarrow$-related. Therefore $c(a,b) \rightarrow d(a,b)$, as required.

Using Lemma 3.2, it is an easy exercise to show that if $a \rightarrow b$, then $a^n \rightarrow b^n$ for any positive integer n. Figures 2 and 3 illustrate the next definition.

Definition 3.3. Let n be a nonnegative integer. An *n-fence* from c to d, denoted by $F(c,d)$, is a sequence of elements of $\mathbf{F}_2$ satisfying

$$c = a_0 \rightarrow b_1 \leftarrow a_1 \rightarrow b_2 \leftarrow a_2 \rightarrow \cdots \leftarrow a_n \rightarrow b_{n+1} = d.$$

Let n be a positive integer. An *n-box* B is a sequence $q_1 \dashrightarrow p_1 \dashrightarrow q_2 \dashrightarrow p_2 \dashrightarrow q_3 \dashrightarrow \cdots \dashrightarrow q_n \dashrightarrow p_n$ such that

$$p_1 \rightarrow p_2 \rightarrow \cdots \rightarrow p_n \qquad \text{and} \qquad q_1 \rightarrow q_2 \rightarrow \cdots \rightarrow q_n.$$

An *n-box* from c to b and d, denoted by $B(c;b,d)$, is an n-box with $c = q_1$, $q_n \rightarrow b$, and $p_n \rightarrow d$. Note that a 0-fence from c to d is simply $c \rightarrow d$.

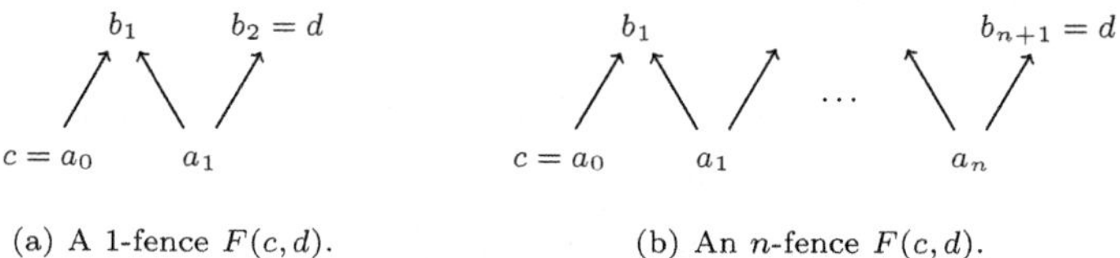

(a) A 1-fence $F(c,d)$.   (b) An n-fence $F(c,d)$.

Fig. 2 Pictures of fences.

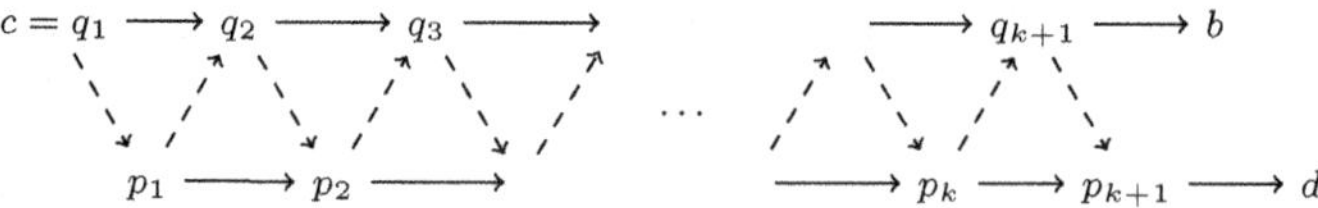

Fig. 3 A $(k+1)$-box $B(c;b,d)$.

The next three lemmas contain the heart of the proof of Theorem 3.1.

Lemma 3.4. *Suppose that $B(c;b,d)$ is a $(k+1)$-box. Then $c \rightarrow J_{2k+1}(b,d,d)$.*

Proof. Label the vertices of the box from left to right according to Figure 3 as $q_1, p_1, q_2, p_2, \ldots, q_{k+1}, p_{k+1}, b, d$.

Observe that since $q_1 \to q_2$, $p_1 \to p_2$, and $q_1 \dashrightarrow p_1 \dashrightarrow q_2$ (and $\to \vartriangleleft_{\mathcal{J}}^m \dashrightarrow$), we have the sequence:

$$c = J_1(q_1, q_1, p_1) \to J_1(q_2, p_1, p_1) = J_2(q_2, p_1, p_1)$$
$$\to J_2(q_2, q_2, p_2) = J_3(q_2, q_2, p_2).$$

Continuing in this vein, we obtain for i ranging from 1 to k the sequence:

$$c \to J_{2i-1}(q_i, q_i, p_i) \to J_{2i-1}(q_{i+1}, p_i, p_i) = J_{2i}(q_{i+1}, p_i, p_i)$$
$$\to J_{2i}(q_{i+1}, q_{i+1}, p_{i+1}) = J_{2i+1}(q_{i+1}, q_{i+1}, p_{i+1}).$$

Letting $i = k$ (and thus $2i + 1 = n$), we conclude that

$$c \to J_{2k+1}(q_{k+1}, q_{k+1}, p_{k+1}) \to J_{2k+1}(q_{k+1}, p_{k+1}, p_{k+1}).$$

Finally, using $q_{k+1} \to b$ and $p_{k+1} \to d$, we get $c \to J_{2k+1}(b, d, d)$.

Lemma 3.5. *Assume that there is a 1-fence $x \to b \leftarrow a \to d$. Then for every $\ell > 1$ there is an ℓ-box $B(x; b, d(b, d))$.*

Proof. We put $q_1 = x$ and $p_1 = a(x, a)$. For $2 \le i \le \ell$, let

$$q_i = b(q_{i-1}, a) \qquad \text{and} \qquad p_i = a(q_i, a).$$

We claim that the result is an ℓ-box $B(x; b, d(b, d))$. The rest of the proof consists of verifying the various $\dashrightarrow$ and $\to$ relations involved. We invite the reader to use Figure 4 for a reference (note that some diagonal edges are solid where the definition of a box required only dashed edges – this is all right since $\to$ is a subset of $\dashrightarrow$).

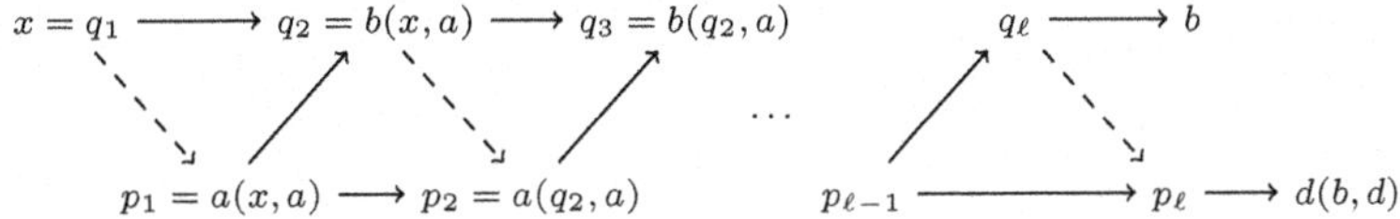

Fig. 4 The ℓ-box $B(x; b, d(b, d))$.

Observe that $x \dashrightarrow a$, so the endomorphism σ sending x to x and z to a is special. It is easy to see that $\sigma(b) = b(x, a) = q_2$ and $\sigma(a) = a(x, a) = p_1$. Since $x \to b \leftarrow a$, it follows by Lemma 3.2 that $x \to q_2 \leftarrow p_1$.

We now proceed by induction to prove that $q_i \to q_{i+1}$ and $p_i \to p_{i+1}$ for all $i = 1, \ldots, \ell - 1$. We already know the arrows for $i = 1$, and from $q_{i-1} \to q_i$, we easily get both $q_i = b(q_{i-1}, a) \to b(q_i, a) = q_{i+1}$ and $p_i \to p_{i+1}$ for all applicable values of i.

Observe that $q_1 = x \dashrightarrow a$. Since $q_i = b(q_{i-1}, a)$, induction gives us that $q_i \dashrightarrow a$ for all i. Repeated use of this set of dashed arrows allows us to prove that $p_i \to q_{i+1}$ and $q_i \dashrightarrow p_i$ for all i in the following way. Consider first the endomorphism σ sending x to q_i and z to a. Since $q_i \dashrightarrow a$, this is a special endomorphism. Since $a \to b$, we have $p_i = \sigma(a) \to \sigma(b) = q_{i+1}$ for all i. To see $q_i \dashrightarrow p_i$, observe that $q_i = a(q_i, q_i) \dashrightarrow a(q_i, a) = p_i$.

All that remains now is to get the two arrows at the rightmost end of the box. Similarly to the previous paragraph, it is easy to prove by induction on i that $q_i \to b$ for all i, so in particular $q_\ell \to b$. To obtain $p_\ell \to d(b, d)$, observe that $p_\ell = a(q_\ell, a) \to d(q_\ell, a) \to d(b, a) \to d(b, d)$ (we have used first Lemma 3.2, then $q_\ell \to b$, and finally $a \to d$).

Lemma 3.6. *For each $0 \leq i < k$, there exists a $(k-i)$-fence from x to $J^{2^{i+1}-1}$. (Recall that $J = J(x, z) = J_{2k+1}(x, z, z)$.)*

Proof. We proceed by induction on i. For $i = 0$, we get a k-fence from x to J by putting $b_\ell = J_{2\ell-1}(x, z, z)$ and $a_\ell = J_{2\ell}(x, x, z)$, for $1 \leq \ell \leq k$.

Suppose now that $1 \leq i < k$ and we have a $(k-i+1)$-fence

$$(2) \quad x \to b_1 \leftarrow a_1 \to b_2 \leftarrow a_2 \to \cdots \leftarrow a_{k-i} \to b_{k-i+1} \leftarrow a_{k-i+1} \to J^{2^i-1}.$$

We proceed to construct a $(k-i)$-fence from x to $J^{2^{i+1}-1}$.

Applying first Lemma 3.5 and then Lemma 3.4 to the 1-fence with vertices x, b_1, a_1, b_2 above, we get

$$x \to J_{2k+1}(b_1, b_2(b_1, b_2), b_2(b_1, b_2)) = J(b_1, b_2(b_1, b_2)).$$

Denote the term on the right hand side of the above arrow by b_1'. Using $b_1 \leftarrow x$, we get

$$b_1' = J(b_1, b_2(b_1, b_2)) \leftarrow J(x, b_2(x, b_2)) = J(x, b_2^2).$$

Since $b_2^2 \leftarrow a_2^2$, we obtain $b_1' \leftarrow J(x, a_2^2)$. Consider the sequence $b_1', a_1' = J(x, a_2^2)$, and

$$a_\ell' = J(x, a_{\ell+1}^2) \qquad \text{and} \qquad b_\ell' = J(x, b_{\ell+1}^2)$$

for $2 \leq \ell \leq k - i$. It is easy to verify that

$$x \to b_1' \leftarrow a_1' \to b_2' \leftarrow a_3' \to \cdots \leftarrow a_{k-i}'.$$

Let us look at the element a_{k-i}' in this fence. We have

$$a_{k-i}' = J(x, a_{k-i+1}^2) \to J\left(x, \left(J^{2^i-1}\right)^2\right) = J(x, J^{2^{i+1}-2}) = J^{2^{i+1}-1}$$

(we use $a_{k-i+1} \to J^{2^i-1}$ from (2) above). We have therefore found a $(k-i)$-fence from x to $J^{2^{i+1}-1}$, as was needed.

We are now ready to prove Theorem 3.1.

Theorem (Theorem 3.1). *There exists $b \in F_2$ such that $x \to J^{2^k}(b, J^{2^k-1})$.*

Proof. By taking $i = k-1$ in Lemma 3.6, we obtain a 1-fence $x \to b \leftarrow a \to J^{2^k-1}$. Applying Lemmas 3.5 and 3.4, and observing that

$$J(b, J^{2^k-1}(b, J^{2^k-1})) = J^{2^k}(b, J^{2^k-1}),$$

we get $x \to J^{2^k}(b, J^{2^k-1})$.

4 Directed Jónsson terms

Theorem (Theorem 2.2). *Let $\mathcal{V}$ be a variety, and $\mathcal{J}$ be a chain of weak Jónsson terms of $\mathcal{V}$. Then there exists a chain $\mathcal{D}$ of weak directed Jónsson terms of $\mathcal{V}$ such that for all $\boldsymbol{A}, \boldsymbol{B} \in \mathcal{V}$ we have $\boldsymbol{B} \lhd_{\mathcal{J}}^m \boldsymbol{A}$ implies $\boldsymbol{B} \lhd_{\mathcal{D}}^m \boldsymbol{A}$.*

Proof. Let $J_1, \ldots, J_{2k+1}$ be a chain of weak Jónsson terms in $\mathcal{V}$. By taking an inessential expansion of $\mathcal{V}$, we can assume that J_i are basic operations of $\mathcal{V}$. Consider the variety $\mathcal{W}$ from the previous chapter. Since the equational basis of $\mathcal{W}$ is a subset of the identities true in $\mathcal{V}$, the variety $\mathcal{W}$ interprets into $\mathcal{V}$.

Theorem 3.1 gives us that there is a chain $\mathcal{D} = \{D_1, \ldots, D_m\} \subseteq \mathcal{G}$ such that the system of equalities

$$\begin{aligned}
D_1(x,x,z) &= x, \\
D_i(x,z,z) &= D_{i+1}(x,x,z) \quad \text{for each } i = 1, \ldots, m-1, \\
D_m(x,z,z) &= J^{2^k}(b, J^{2^k-1})
\end{aligned}$$

holds in $\mathcal{W}$. Since $\mathcal{W}$ interprets into $\mathcal{V}$, these equalities must also hold in $\mathcal{V}$. Moreover, in $\mathcal{V}$ we have the equality $J(x,z) = J_{2k+1}(x,z,z) = z$, so $J^{2^k}(b, J^{2^k-1}) = z$.

Finally, let $\boldsymbol{B} \leq \boldsymbol{A}$ be algebras in $\mathcal{V}$. By removing all of the basic operations except $J_1, \ldots, J_{2k+1}$, we obtain a pair of reducts $\boldsymbol{B}^\star \leq \boldsymbol{A}^\star$ which both lie in $\mathcal{W}$. If $\boldsymbol{B} \lhd_{\mathcal{J}}^m \boldsymbol{A}$, then trivially $\boldsymbol{B}^\star \lhd_{\mathcal{J}}^m \boldsymbol{A}^\star$, and $D_1, \ldots, D_m \in \mathcal{G}$ gives us $\boldsymbol{B}^\star \lhd_{\mathcal{D}}^m \boldsymbol{A}^\star$. Since $\boldsymbol{A}^\star$ is a reduct of $\boldsymbol{A}$, we immediately have $\boldsymbol{B} \lhd_{\mathcal{D}}^m \boldsymbol{A}$.

The chain $D_1, \ldots, D_m$ middle absorbs anything that $\mathcal{J}$ absorbs, and satisfies in $\mathcal{V}$ the system of equalities

$$D_1(x,x,z) = x,$$
$$D_i(x,z,z) = D_{i+1}(x,x,z) \quad \text{for each } i = 1,\ldots,k-1,$$
$$D_m(x,z,z) = z.$$

Therefore, $D_1,\ldots,D_m$ is the weak directed Jónsson chain $\mathcal{D}$ we were looking for.

Corollary 4.1. *Let $\mathcal{V}$ be a variety with a system of Jónsson terms $\mathcal{J}$. Then $\mathcal{V}$ has a system of directed Jónsson terms.*

Proof. Let $\mathbf{F}_3^{id}$ be idempotent reduct of the free three generated algebra in $\mathcal{V}$. Then $\mathbf{F}_3^{id}$ contains a chain of Jónsson terms $\mathcal{J}$ such that $\{x\} \lhd_{\mathcal{J}}^m \mathbf{F}_3^{id}$. Applying Theorem 2.2 with $\mathbf{B} = \{x\}$ and $\mathbf{A} = \mathbf{F}_3^{id}$ gives us that there is a chain of directed weak Jónsson terms $\mathcal{D}$ such that $\{x\} \lhd_{\mathcal{D}}^m F_3^{id}$. Every D_i in $\mathcal{D}$ satisfies $D_i(x,y,x) = x$, making $\mathcal{D}$ a chain of directed Jónsson terms for $\mathcal{V}$.

We are now ready to give a full proof of Theorem 2.1.

Theorem (Theorem 2.1). *Suppose that E and F are admissible preorders on $\mathbf{A}$ (that is, they are subalgebras of $\mathbf{A}^2$ that are reflexive and transitive). If $E \lhd_J F$ then $E = F$.*

Proof. Let $\mathbf{A} \in \mathcal{V}$, where $\mathcal{V}$ has a weak Jónsson system of terms $\mathcal{J}$ and suppose that E, F are admissible preorders of $\mathbf{A}$ with $E \lhd_{\mathcal{J}} F$. Let $\mathcal{D} = \{D_1,\ldots,D_m\}$ be the system of weak directed Jónsson terms for $\mathbf{A}$ supplied by Theorem 2.2. Then $E \lhd_{\mathcal{D}}^m F$ and so for every $(a,b) \in F$ we have:

$$a = D_1(a,a,b) \, E \, D_1(a,b,b) = D_2(a,a,b) \, E \, D_2(a,b,b) = \cdots$$
$$\cdots = D_m(a,a,b) \, E \, D_m(a,b,b) = b,$$

yielding $(a,b) \in E$.

5 Pixley terms

We now proceed to prove the statement (2) of Theorem 1.1 (Lipparini's Proposition 5 in (Lipparini, 1995)).

Theorem 5.1. *Let k be any positive integer and let $\mathcal{V}$ be a $(k+1)$-permutable variety with a system of Jónsson terms $\mathcal{J}$. Then $\mathcal{V}$ has a system of Pixley terms $\mathcal{P} = \{P_1,\ldots,P_k\}$ such that whenever $\mathbf{A}, \mathbf{B} \in \mathcal{V}$ and $\mathbf{B} \lhd_{\mathcal{J}}^m \mathbf{A}$, then $\mathbf{B} \lhd_{\mathcal{P}}^m \mathbf{A}$.*

Proof. The proof is a variant of the proof of Theorem 3.1. Choose and fix an arbitrary idempotent variety $\mathcal{V}$ that has a system $\mathcal{J}$ of Jónsson terms and a system $H_1, \ldots, H_k$ of Hagemann-Mitschke terms, i.e. terms that satisfy the equations

$$H_1(x, z, z) = x, \qquad H_k(x, x, z) = z,$$
$$H_i(x, x, z) = H_{i+1}(x, z, z) \qquad \text{for } 1 \le i < k.$$

Starting as in Section 3, we let $\mathbf{F}_2$ be the free algebra of rank two in $\mathcal{V}$ freely generated by x and z. Let F be the subalgebra of $\mathbf{F}_2^2$ generated by the pairs (x, x), (x, z), and (z, z), that is

$$F = \big\{(t(x, x, z), t(x, z, z)) : t \text{ a term of } \mathcal{V}\big\}.$$

As before, we define $\mathcal{G}$ to be the set of all $\mathcal{V}$-terms $t(x, y, z)$ such that whenever $\mathbf{A}, \mathbf{B} \in \mathcal{V}$ and $\mathbf{B} \triangleleft_{\mathcal{J}}^m \mathbf{A}$, then also $\mathbf{B} \triangleleft_t^m \mathbf{A}$, and let

$$E = \big\{(t(x, x, z), t(x, z, z)) : t(x, y, z) \in \mathcal{G}\big\}.$$

As before, E and F are idempotent admissible relations over $\mathbf{F}_2$ and we have $E \triangleleft_{\mathcal{J}}^m F$.

Using $p \to q$ to denote that (p, q) belongs to the transitive closure of E, we proved in Sections 3 and 4 that $x \to z$. Since the operations H_i respect E and $\to$, we have that $z \to x$. This is a classical observation, but the proof is easy and so we give it in the following paragraph.

Since $x \to x$, $z \to z$, and $x \to z$, we have

$$z = H_1(z, x, x) \to H_1(z, z, x) = H_2(z, x, x)$$
$$\to H_2(z, z, x) = H_3(z, x, x) \to \cdots \to H_k(z, z, x) = x.$$

Transitivity of $\to$ gives $z \to x$.

We now demonstrate the classical fact that $E^{k+1} = E^k$, which gives us that E^k is the transitive closure of E (and in particular $(z, x) \in E^k$). Since E is reflexive, we have $E^k \subseteq E^{k+1}$. Suppose that we have $(a, b) \in E^{k+1}$. Then there are a_i for $i \le k+1$ such that

$$a = a_0 \, E \, a_1 \, E \cdots E \, a_i \, E \, a_{i+1} \, E \cdots E \, a_{k+1} = b.$$

Letting $c_i = H_{i+1}(a_i, a_{i+1}, a_{i+1})$ for $0 \le i < k$, it is easy to verify that

$$a \, E \, c_1 \, E \, c_2 \, E \cdots E \, c_{k-1} \, E \, b,$$

so $(a, b) \in E^k$.

Continuing with the main proof, we have $(z, x) \in E^k$. This means that there are $\mathcal{V}$-terms
$$D_1(x, y, z), \ldots, D_k(x, y, z) \in \mathcal{G}$$

satisfying $z = D_1(x, x, z)$, $D_i(x, z, z) = D_{i+1}(x, x, z)$ for $1 \le i < k$, and $D_k(x, z, z) = x$. As before, whenever $\mathbf{A} \triangleleft_{\mathcal{J}}^m \mathbf{B}$, we also have $\mathbf{A} \triangleleft_{D_1, \ldots, D_k}^m \mathbf{B}$, so in particular $D_i(x,z,x) = x$. The terms $P_i = D_{k-i+1}$ for $1 \leq i \leq k$ then satisfy P(k) in $\mathcal{V}$.

6 Directed Gumm terms

To conclude the proof of Theorem 1.1, we focus on Gumm terms and introduce Gumm absorption. Gumm terms G(n), directed Gumm terms DG(n), and weak versions of both were defined in the introduction. (Recall that weak versions drop the conditions $J_i(x,y,x) = x$.)

When a variety $\mathcal{V}$ has a chain of weak Gumm terms (respectively, weak directed Gumm terms) $J_1,\ldots,J_n,P$, and $\mathbf{A},\mathbf{B} \in \mathcal{V}$ are such that $\mathbf{B} \leq \mathbf{A}$, we say that $\mathbf{B}$ *Gumm absorbs* $\mathbf{A}$ (respectively, *directed Gumm absorbs* $\mathbf{A}$) with respect to these chains if $\mathbf{B} \triangleleft^m_{J_1,\ldots,J_n} \mathbf{A}$. We next state and prove a variant of Theorem 2.2 for Gumm terms.

Theorem 6.1. *Let $\mathcal{V}$ be a variety, and $J_1,\ldots,J_{2k+1},P$ be a chain of weak Gumm terms of $\mathcal{V}$. Then there exists a chain $D_1,\ldots,D_m,Q$ of weak directed Gumm terms of $\mathcal{V}$ such that whenever $\mathbf{A},\mathbf{B} \in \mathcal{V}$ and $\mathbf{B} \triangleleft^m_{J_1,\ldots,J_{2k+1}} \mathbf{A}$, then $\mathbf{B} \triangleleft^m_{D_1,\ldots,D_m} \mathbf{A}$.*

In particular, if $J_1,\ldots,J_{2k+1},P$ is a chain of Gumm terms, then it follows that $D_1,\ldots,D_m,Q$ is a chain of directed Gumm terms.

Note that Theorem 6.1 immediately gives us the third assertion of Theorem 1.1.

Proof. The argument follows the same pattern as our proof of Theorem 2.2. We consider the variety $\mathcal{W}$ and use Theorem 3.1 to obtain terms $D_1,\ldots,D_m$ in $\mathcal{V}$ such that

$$D_1(x,x,z) = x,$$
$$D_i(x,z,z) = D_{i+1}(x,z,z) \quad \text{for each } i = 1,\ldots,m-1,$$
$$D_m(x,z,z) = J^{2^k}(b,J^{2^k-1}),$$

where b is some term composed from $J_1,\ldots,J_{2k+1}$, and

$$J(x,y) = J_{2k+1}(x,y,y) = P(x,y,y) \quad \text{in } \mathcal{V}.$$

The term $J^{2^k}(b,J^{2^k-1})$ can be expressed as

$$\underbrace{J(b,J(b,\ldots,J(b,}_{2^k\text{-many } J\text{'s}} \underbrace{J(x,J(x,\ldots,J}_{(2^k-1)\text{-many } J\text{'s}}(x,z))\ldots))\ldots)),$$

More formally, if we let $d_0(x,z) = z$ and

$$d_i(x,z) = J_{2k+1}(x, d_{i-1}(x,z), d_{i-1}(x,z)) \qquad \text{for } 1 \leq i < 2^k,$$
$$d_i(x,z) = J_{2k+1}(b(x,z), d_{i-1}(x,z), d_{i-1}(x,z)) \qquad \text{for } 2^k \leq i < 2^{k+1},$$

then we will have $d_{2^{k+1}-1}(x,z) = D_m(xzz)$.

Now we systematically rewrite $J^{2^k}(b, J^{2^k-1})$, replacing all but the right-most occurrence of z by y, and replacing all occurrences of J_{2k+1} by P, to obtain a term $Q(xyz)$.

More formally, we let $Q_0(xyz) = z$, $Q_1(x,y,z) = P(x,y,z)$, and

$$Q_i(x,y,z) = P(x, Q_{i-1}(x,y,y), Q_{i-1}(x,y,z))$$

for $2 \leq i < 2^k$, and

$$Q_i(x,y,z) = P(b(x,y), Q_{i-1}(x,y,y), Q_{i-1}(x,y,z))$$

for $2^k \leq i < 2^{k+1}$. Having done that, we let $Q(x,y,z) = Q_{2^{k+1}-1}(x,y,z)$.

Using the equality $J_{2k+1}(x,z,z) = P(x,z,z)$, one can easily prove that $Q(x,z,z) = J^{2^k}(b, J^{2^k-1})$ in $\mathbf{F}_2$. Idempotence of b together with $P(x,x,z) = z$ then gives us that $Q(x,x,z) = z$.

Thus we have a chain of weak directed Gumm terms $D_1, \ldots, D_m, Q$. Since $D_1, \ldots, D_m \in \mathcal{G}$, the middle absorption property follows as in Theorem 2.2. Showing that ordinary Gumm terms imply the existence of a chain of directed Gumm terms is analogous to the proof of Corollary 4.1.

We can now also state and prove a version of Theorem 2.1 for Gumm terms.

Theorem 6.2. *Suppose that $\mathbf{E}$ and $\mathbf{F}$ are reflexive subalgebras of $\mathbf{A}^2$ and that $\mathbf{E}$ Gumm absorbs $\mathbf{F}$. Whenever $(a,b) \in F$, there is $c \in A$ such that $(b,c) \in F$ and (a,c) belongs to the transitive closure of E.*

Proof. Apply Theorem 6.1 to get weak directed Gumm terms $D_1, \ldots, D_m, Q$ for the variety generated by $\mathbf{A}$ so that $\mathbf{E} \lhd_{D_1, \ldots, D_m} \mathbf{F}$. Then

$$a\, E\, D_1(a,b,b)\, E\, D_2(a,b,b)\, E \cdots E\, D_m(a,b,b) = Q(a,b,b) = c,$$

where $b = Q(a,a,b)\, F\, Q(a,b,b) = c$.

7 Final Remarks

We have worked through the various parts of the proof of Theorem 3.1, calculating the precise lengths of the E-chains produced. The final formula for the length of the E-chain connecting x to $J^{2^k}(b, J^{2^k-1})$ simplifies to

$$\frac{(2k+1)(k+1)((k+1)^{k-2}-1)}{k}.$$

Thus, to be precise, we have proved that $J(k)$ implies $DJ(m)$ with m equal to the displayed number. This is our best value for m. It would be interesting to know if a different approach, or the introduction of some new tricks, can lower this value of m substantially. We close the paper by posing a problem stemming from our work here.

Problem 7.1. Does there exist a sequence of algebras $\mathbf{A}_1, \mathbf{A}_2, \ldots$ such that each $\mathbf{A}_n$ is $J(n)$, but the least m such that $\mathbf{A}_n$ is $DJ(m)$ grows at least exponentially in n?

References

Barto, L. (2013). Finitely related algebras in congruence distributive varieties have near unanimity terms, *Canadian Journal of Mathematics* 65/1, 3–21.

Barto, L. and Kozik, M. (2012). Absorbing subalgebras, cyclic terms and the constraint satisfaction problem, *Logical Methods in Computer Science* 8/1:07, 1–26.

Gumm, H.-P. (1981). Congruence modularity is permutability composed with distributivity, *Archiv der Mathematik* (Basel) **36**, 569–576.

Hagemann, J. and Mitschke, A. (1973). On n-permutable congruences, *Algebra Universalis* **3**, 8–12.

Jónsson, B. (1967). Algebras whose congruence lattices are distributive, *Mathematica Scandinavica* **21**, 110–121.

Lipparini, P. (1995). n-Permutable varieties satisfy non trivial congruence identities, *Algebra Universalis* **33**, 159–168.

Pixley, A. (1963). Distributivity and permutability of congruences in equational classes of algebras, *Proceedings of the American Mathematical Society* **14**, 105–109.

Relatively congruence modular quasivarieties of modules

Keith A. Kearnes

Dedicated to Don Pigozzi

Abstract We show that the quasiequational theory of a relatively congruence modular quasivariety of left R-modules is determined by a two-sided ideal in R together with a filter of left ideals. The two-sided ideal encodes the identities that hold in the quasivariety, while the filter of left ideals encodes the quasiidentities. The filter of left ideals defines a generalized notion of torsion.

It follows from our result that if R is left Artinian, then any relatively congruence modular quasivariety of left R-modules is axiomatizable by a set of identities together with at most one proper quasiidentity, and if R is a commutative Artinian ring then any relatively congruence modular quasivariety of left R-modules is a variety.

Key words: Quasivariety, relatively congruence modular, relatively congruence distributive, finitely axiomatizable quasivariety, quasivarieties of modules

2010 Mathematics Subject Classification: 03C05, 08C15

1 Introduction

This paper is inspired by Problem 9.13 of Don Pigozzi's paper, *Finite basis theorems for relatively congruence-distributive quasivarieties*, (Pigozzi, 1988). The problem asks:

Keith A. Kearnes
Department of Mathematics, University of Colorado, Boulder, CO 80309-0395, USA,
e-mail: Keith.Kearnes@Colorado.EDU

© Springer International Publishing AG 2018
J. Czelakowski (eds.), *Don Pigozzi on Abstract Algebraic Logic,
Universal Algebra, and Computer Science*, Outstanding Contributions
to Logic 16, https://doi.org/10.1007/978-3-319-74772-9_8

Is it true that every finitely generated and relatively congruence modular quasivariety is finitely based?

This question is still open. Pigozzi's paper shows the answer to be affirmative if "modular" is strengthened to "distributive". Analogous problems for varieties were shown to have positive solutions in (Baker, 1977; McKenzie, 1987; Willard, 2000; Kearnes et al., to appear), and a related problem for quasivarieties was shown to have an affirmative solution in (Maróti and McKenzie, 2004). The best partial answer to Problem 9.13 that now exists is Theorem 8 of (Dziobiak et al., 2009), which says that if a quasivariety $\mathcal{K}$ *and the variety it generates* are finitely generated and relatively congruence modular, then $\mathcal{K}$ is finitely based.

A relatively congruence distributive quasivariety is nothing other than a relatively congruence modular quasivariety in which no member has a non-trivial abelian congruence, so Pigozzi's paper solves the part of Problem 9.13 that does not involve abelian congruences. Tools for dealing with abelian congruences in relatively congruence modular quasivarieties were developed in (Kearnes and McKenzie, 1992; Kearnes and Szendrei, 1998), but they have not yet yielded a full solution to the problem. What these tools show is that abelian congruences in such quasivarieties are quasiaffine, which means the blocks of an abelian congruence support a module-like structure. In this paper we study the purest relatively congruence modular quasivarieties which are not distributive, namely quasivarieties of modules. Our main result is a description of all relatively congruence modular quasivarieties of modules.

2 the classification theorem

For a unital ring R let R-Mod be the variety of left R-modules. If $\mathcal{K}$ is a subquasivariety of R-Mod and $M \in \mathcal{K}$, then a $\mathcal{K}$-*submodule* (or *relative submodule*) of M is an R-submodule $S \leq M$ such that $M/S \in \mathcal{K}$. $\mathcal{K}$ is *relatively congruence modular (RCM)* if every $M \in \mathcal{K}$ has a modular lattice of $\mathcal{K}$-submodules.

For the simplest example of these definitions take $R = \mathbb{Z}$, so that R-Mod is the variety of abelian groups. Let $\mathcal{K}$ be the subquasivariety of $\mathbb{Z}$-Mod consisting of torsion-free abelian groups. The only relative submodules on, say, $\mathbb{Z} \in \mathcal{K}$ are (0) and $\mathbb{Z}$. That is, $\mathbb{Z}$ is *relatively simple* (although $\mathbb{Z}$ is far from being a simple module). It can be shown that this quasivariety, the quasivariety of torsion-free abelian groups, is a minimal quasivariety which happens to be RCM.

Let's generalize the example above in an artificial way. Let $R = \mathbb{Z}[t]$ and let $\mathcal{K}$ be the subquasivariety of R-Mod consisting of all torsion-free abelian groups considered as R-modules by defining t to act as zero on any module in $\mathcal{K}$ Then $\mathcal{K}$ is axiomatized by the identity $tx = 0$ together with a family of quasiidentities of the form

$$nx = 0 \rightarrow x = 0.$$

This is essentially the same as the preceding example, so in particular it is RCM.

The point of this paper is to show that every RCM quasivariety of modules looks like the one from the previous paragraph. For any RCM quasivariety $\mathcal{K}$ of R-modules there is a set Σ of 1-variable identities along with a specific torsion notion which realizes $\mathcal{K}$ as the subquasivariety of R-Mod consisting of the torsion-free R-modules that satisfy Σ. Σ corresponds to a two-sided ideal in R while the torsion notion corresponds to a filter in the lattice of left ideals of R.

Let's begin by identifying the role played by Σ.

Lemma 2.1. *Let $\mathcal{V}$ be a subvariety of R-Mod and let $I = \{r \in R \mid \mathcal{V} \models rx = 0\}$ be its annihilator. Then*

(1) *I is a two-sided ideal in R,*
(2) *$\Sigma := \{rx = 0 \mid r \in I\}$ axiomatizes $\mathcal{V}$ relative to R-Mod, and*
(3) *$\mathcal{V}$ is definitionally equivalent to R/I-Mod.*

We imagine applying this in the situation where $\mathcal{K}$ is a subquasivariety of R-Mod and $\mathcal{V}$ is the variety generated by $\mathcal{K}$.

We do not prove Lemma 2.1, but do point out that a key idea in the proof is that a single module identity $r_1 x_1 + \cdots + r_k x_k = 0$ has the same strength as the set $\{r_1 x_1 = 0, \ldots, r_k x_k = 0\}$ of 1-variable module identities.

Lemma 2.1 allows us to pass from R to R/I and henceforth consider only the situation where $\mathcal{K}$ generates R-Mod. We shall make this assumption as we work out the main result of the paper.

Next we describe the torsion concept that plays a role in this paper.

Definition 2.2. Let $\mathcal{L}$ be the poset of finitely generated left ideals of R ordered by inclusion. A *torsion notion* for R is a subset $\mathcal{F} \subseteq \mathcal{L}$ satisfying the following conditions:

(1) $\mathcal{F}$ is a nonempty order filter in $\mathcal{L}$. ($A \in \mathcal{F}$, $B \in \mathcal{L}$, and $A \subseteq B$ implies $B \in \mathcal{F}$.)
(2) $\mathcal{F}$ is downward directed. ($A, B \in \mathcal{F}$ implies there is a $C \in \mathcal{F}$ such that $C \subseteq A$ and $C \subseteq B$.)
(3) If $X, Y \subseteq R$ are finite subsets such that the left ideals (X) and (Y) belong to $\mathcal{F}$, then the left ideal (XY) belongs to $\mathcal{F}$.
(4) For all $A \in \mathcal{F}$ and $r \in R$ there is $B \in \mathcal{F}$ such that $Br \subseteq A$.
(5) (Regularity of elements of $\mathcal{F}$) If $A \in \mathcal{F}$, $r \in R$, and $Ar = 0$, then $r = 0$.

Given a torsion notion $\mathcal{F}$ we say that an element m of an R-module M is an *$\mathcal{F}$-torsion element* if $Am = 0$ for some $A \in \mathcal{F}$. If M has no nonzero $\mathcal{F}$-torsion elements, then it is *$\mathcal{F}$-torsion-free*. This may also be expressed by saying that $M \models Ax = 0 \rightarrow x = 0$ for each $A \in \mathcal{F}$. It is easy to see that a statement of the form $Ax = 0 \rightarrow x = 0$ for a finitely generated left ideal A is equivalent to a quasiidentity. Namely if $A = (a_1, \ldots, a_m)$, then $Ax = 0 \to x = 0$ is satisfied if and only if

$$(q_A) \qquad\qquad (a_1 x = 0) \wedge \cdots \wedge (a_m x = 0) \to (x = 0)$$

is satisfied, so the class of $\mathcal{F}$-torsion-free R-modules is a quasivariety.

Any torsion notion $\mathcal{F}$ contains R, and the set $\{R\}$ is always a torsion notion. Every R-module is torsion-free with respect to this trivial torsion notion.

Items (1)–(4) of Definition 2.2 simplify quite a bit when R is commutative. Namely, (4) automatically holds when R is commutative, since we can choose $B = A$. Item (3) now asserts that $\mathcal{F}$ is closed under multiplication. When this holds, (2) will also hold, since for commutative rings the product of two ideals is contained in each of them. Thus (1)–(4) merely say that $\mathcal{F}$ is a multiplicatively closed order filter in the poset of finitely generated ideals. (For any ring R, item (5) of the definition asserts that the free R-module, R, is $\mathcal{F}$-torsion-free.)

Here is the statement of the main theorem of the paper.

Theorem 2.3. *Let $\mathcal{K}$ be a quasivariety of R-modules such that the variety generated by $\mathcal{K}$ is all of R-Mod. Then $\mathcal{K}$ is RCM iff there is a torsion notion $\mathcal{F}$ such that $\mathcal{K}$ is the quasivariety of $\mathcal{F}$-torsion-free R-modules.*

We prove Theorem 2.3 in the next two sections, but here we derive a corollary.

Corollary 2.4. *Let R be a left Artinian ring. If $\mathcal{K}$ is an RCM quasivariety of R-modules, then $\mathcal{K}$ may be axiomatized relative to R-Mod by a finite set of identities together with at most one proper quasiidentity. In particular, if R is a finitely presentable ring, then $\mathcal{K}$ is finitely axiomatizable.*

Proof. By the Hopkins-Levitski Theorem, a left Artinian ring is left Noetherian, so every left ideal of R is finitely generated. In particular, we can effect the passage from R-modules to R/I-modules (as indicated in Lemma 2.1) by imposing finitely many identities on the variety R-Mod. Thus we may assume henceforth that $\mathcal{K}$ generates R-Mod as a variety and our goal is now to prove that $\mathcal{K}$ can be axiomatized relative to R-Mod by at most one quasiidentity.

Let $\mathcal{F}$ be the torsion notion guaranteed by Theorem 2.3. Since R is Artinian, $\mathcal{F}$ is generated as an order filter by its minimal elements. By item (2) of Definition 2.2, $\mathcal{F}$ is a principal order filter in $\mathcal{L}$, say $\mathcal{F}$ is the order filter generated by $A \in \mathcal{L}$. Now the notion of '$\mathcal{F}$-torsion-free' is expressible by $Ax = 0 \to x = 0$, or equivalently by the single quasiidentity q_A. (What has been left unsaid so far is that if $A \subseteq B$, then $Ax = 0 \to x = 0$ is stronger than $Bx = 0 \to x = 0$.)

The last assertion of the corollary follows from the fact that R-Mod is finitely axiomatizable when R is finitely presentable.

The one proper quasiidentity mentioned in Corollary 2.4 can be eliminated when R is commutative.

Corollary 2.5. *Let R be a commutative Artinian ring. Any RCM quasivariety of R-modules is a variety.*

Proof. Let $\mathcal{K}$ be an RCM quasivariety of R-modules. We shall argue the proof for arbitrary R until it is necessary to appeal to commutativity.

Using Lemma 2.1 we may reduce to the case where the variety generated by $\mathcal{K}$ is all of R-Mod. In the proof of Corollary 2.4 we showed that the torsion notion $\mathcal{F}$ associated to $\mathcal{K}$ is a principal filter in the poset of finitely generated left ideals of R. Let A be the generator of this principal filter.

Item (4) of the definition of 'torsion notion' implies that A is a two-sided ideal. For if $r \in R$, then there must be a $B \in \mathcal{F}$ such that $Br \subseteq A$. Since $A \subseteq B$ this yields $Ar \subseteq Br \subseteq A$.

Another special property that A must satisfy is that $A^2 = A$. To see this, choose a finite set X that generates A as a left ideal. Then item (3) implies that $(X^2) \in \mathcal{F}$. But clearly $(X^2) \subseteq A^2$, so $A \subseteq (X^2) \subseteq A^2 \subseteq A$.

Now we invoke the commutativity hypothesis. A finitely generated idempotent ideal is generated by an idempotent element, so $A = (e)$ for some element e satisfying $e^2 = e$. For $r = 1 - e$ we have $Ar = 0$, so item (5) of the definition of 'torsion notion' yields that $1 - e = 0$, i.e. $e = 1$, or equivalently $A = R$. This forces $\mathcal{F} = \{R\}$. As noted after Definition 2.2, this implies that every R-module is $\mathcal{F}$-torsion-free, so $\mathcal{K} = R$-Mod.

Corollaries 2.4 and 2.5 are not true if you weaken 'Artinian' to 'Noetherian', since the quasivariety of torsion-free abelian groups is not finitely axiomatizable and is not a variety. Also, Corollary 2.5 is not true without the commutativity hypothesis. To see this, let R be the ring of upper triangular 2×2 matrices over some field. If the matrix units in R are e_{11}, e_{12}, e_{22}, then the quasivariety of R-modules axiomatized relative to R-Mod by $(e_{11}x = 0) \wedge (e_{12}x = 0) \to (x = 0)$ is RCM and is not a variety.

3 RCM $\Longrightarrow$ torsion notion

In this section we prove that if $\mathcal{K}$ is an RCM quasivariety of R-modules and the variety generated by $\mathcal{K}$ is all of R-Mod, then there is a torsion notion $\mathcal{F}$ such that $\mathcal{K}$ is the quasivariety of $\mathcal{F}$-torsion-free R-modules. This is one direction of the proof of Theorem 2.3.

To prove what is needed we make use of the fact, proved in (Kearnes and McKenzie, 1992), that an RCM quasivariety has an 'almost equational axiomatization', and that the $\mathcal{K}$-extension of a submodule can be computed easily with the aid of that axiomatization. Here the $\mathcal{K}$-*extension* of a submodule $S \leq M$ is the least $\mathcal{K}$-submodule $\overline{S} \leq M$ that contains S.

We recall the necessary concept from (Kearnes and McKenzie, 1992). A Δ-*axiom* is a first-order sentence, involving pairs of terms $(p_j(x,y,\bar{u},\bar{v},\bar{z})$, $q_j(x,y,\bar{u},\bar{v},\bar{z}))$, $j < n$, expressing that

(1) the identities
$$p_j(x,x,\bar{u},\bar{u},\bar{z}) = q_j(x,x,\bar{u},\bar{u},\bar{z})$$

hold for $j < n$, and
(2) the quasiidentity

$$\bigwedge(p_j(x,y,\bar{u},\bar{u},\bar{z}) = q_j(x,y,\bar{u},\bar{u},\bar{z}))) \to (x = y)$$

holds.

We label this Δ-axiom $\Delta(p,q)$.

The two theorems from (Kearnes and McKenzie, 1992) that we will use are:

Theorem 3.1. *(Theorem 5.1 of (Kearnes and McKenzie, 1992)) Let $\mathcal{K}$ be an RCM quasivariety. $\mathcal{K}$ is axiomatized by a set of Δ-axioms combined with a set of identities.*

For the following theorem, the $\mathcal{K}$-extension of a congruence θ is the least $\mathcal{K}$-congruence containing θ.

Theorem 3.2. *(Theorem 5.2 of (Kearnes and McKenzie, 1992)) Let $\mathcal{K}$ be an RCM quasivariety. Let $\mathbf{A} \in \mathcal{K}$, $\theta \in \mathrm{Con}(\mathbf{A})$, and $u,v \in A$. Then (u,v) belongs to the $\mathcal{K}$-extension of θ iff there is some Δ-axiom $\Delta(p,q)$ valid in $\mathcal{K}$, some pairs $(a_i,b_i) \in \theta$, and some elements $\bar{c}$ such that $p_i(u,v,\bar{a},\bar{b},\bar{c}) = q_i(u,v,\bar{a},\bar{b},\bar{c})$ for all i.*

When dealing with quasivarieties of modules it is possible to code a Δ-axiom $\Delta(p,q)$ as a left ideal in such a way that the following are true.

Theorem 3.3. *Let $\Delta(p,q)$ be a Δ-axiom and let A be its encoding as a left ideal.*

(1) *An R-module M satisfies $\Delta(p,q)$ iff it satisfies $Ax = 0 \to x = 0$.*
(2) *If $\mathcal{K}$ is an RCM quasivariety of R-modules, $M \in \mathcal{K}$, $S \leq M$ is a submodule, and $\overline{S}$ is its $\mathcal{K}$-extension, then $m \in M$ can be shown to lie in $\overline{S}$ using $\Delta(p,q)$ (in the way described in Theorem 3.2) iff $Am \subseteq S$.*

In order to prove the theorem we must first describe how to encode a Δ-axiom as a left ideal.

The first step of the construction uses the fact that equations of the form $p = q$ can be rewritten as equations of the form $(p - q) = 0$. So take a Δ-axiom for R-modules, $\Delta(p,q)$, and rewrite its pairs as differences

$$\begin{aligned}
D_j(x,y,\bar{u},\bar{v},\bar{z}) &:= p_j(x,y,\bar{u},\bar{v},\bar{z}) - q_j(x,y,\bar{u},\bar{v},\bar{z}) \\
&= a_j x + b_j y + \sum_i c_{ij} u_i + \sum_i d_{ij} v_i + \sum_i e_{ij} z_i,
\end{aligned}$$

where $a_j, b_j, c_{ij}, d_{ij}, e_{ij} \in R$. Item (1) from the definition of a Δ-axiom now reads

$$(1)' \qquad D_j(x,x,\bar{u},\bar{u},\bar{z}) = 0 = (a_j + b_j)x + \sum_i (c_{ij} + d_{ij})u_i + \sum_i e_{ij}z_i.$$

We will be working in the situation where $\mathcal{K}$ is a quasivariety of modules and the variety it generates is all of R-Mod. For $(1)'$ to hold in such a quasivariety the coefficients in the righthand expression must all be zero, i.e. $a_j + b_j = c_{ij} + d_{ij} = e_{ij} = 0$. Thus

$$D_j(x,y,\bar{u},\bar{v},\bar{z}) = a_j(x-y) + \sum_i c_{ij}(u_i - v_i)$$

(no dependence on the last block of variables). Introducing new variables X, U_i to represent $x - y, u_i - v_i$ we shall find that the module term operation

$$(1) \qquad\qquad E_j(X,\bar{U}) = a_j X + \sum_i c_{ij} U_i$$

can be used to replace the pair (p_j, q_j) in the definition of 'Δ-axiom'. That is, $\Delta(p,q)$ can be rewritten in a reduced form in an obvious way using the terms $E_j(X,\bar{U})$.

The left ideal associated to $\Delta(p,q)$ is defined to be $A = (a_0, \ldots, a_{n-1})$, the left ideal generated by the coefficients of X in the module terms $E_j(X,\bar{U})$, $j < n$.

Proof (Proof of Theorem 3.3). For part (1) of Theorem 3.3 consider a Δ-axiom $\Delta(p,q)$ and write it using the terms from (1). Condition $(1)'$ of the definition of a Δ-axiom now reads

$$(1)'' \qquad\qquad\qquad E_j(0,\bar{0}) = 0,$$

which must hold simply because E_j is a module term. Condition (2) of the definition of a Δ-axiom reads

$$\bigwedge_j (E_j(X,\bar{0}) = 0) \to (X = 0).$$

This is equivalent to $Ax = 0 \to x = 0$ for $x = X$. This establishes Theorem 3.3 (1).

Now we turn to Theorem 3.3 (2). Suppose that $M \in \mathcal{K}$, $S \leq M$ and $m \in \bar{S}$. Choose $\Delta(p,q)$ witnessing that $m \in \bar{S}$. With $\Delta(p,q)$ written in terms of the E_j's we have that there exist a tuple $\bar{s}$ with entries in S such that $E_j(m,\bar{s}) = a_j m + \sum_i c_{ij} s_{ij}$ belongs to S for all j. This means that $a_j m \in -\sum_i c_{ij} s_{ij} + S = S$ for all j, or $Am \subseteq S$ for A equal to the associated left ideal. Conversely, assume that $Am \subseteq S$. Then for $\bar{s} = \bar{0}$ we have that $E_j(m,\bar{0}) \in S$ for all j, so the form of $\Delta(p,q)$ that uses the terms $E_j(X,\bar{U})$ shows that $m \in \bar{S}$. This establishes Theorem 3.3 (2).

Now we state and prove the main theorem of this section.

Theorem 3.4. *Let $\mathcal{K}$ be an RCM quasivariety such that the variety generated by $\mathcal{K}$ is all of R-Mod. If $\mathcal{F}$ is the set of left ideals of R that code Δ-axioms true in $\mathcal{K}$, then $\mathcal{F}$ is a torsion notion for R-modules and $\mathcal{K}$ is the quasivariety of $\mathcal{F}$-torsion-free R-modules.*

Proof. According to Theorem 3.1, $\mathcal{K}$ is axiomatized relative to R-Mod by a set of Δ-axioms together with a set of identities. Since the variety generated by $\mathcal{K}$ is all of R-Mod, we will not use any identities other than those that hold in R-Mod. According to Theorem 3.3, the Δ-axioms true in $\mathcal{K}$ are equivalent to a family of statements of the form $Ax = 0 \to x = 0$ where A is a finitely generated left ideal. Let $\mathcal{F}$ be the set of finitely generated left ideals of R such that $Ax = 0 \to x = 0$ holds in $\mathcal{K}$. Since the subset of these left ideals that arise from Δ-axioms already serves to axiomatize $\mathcal{K}$ relative to R-Mod, the full set also serves to axiomatize $\mathcal{K}$ relative to R-Mod. It follows from this that, if we show that $\mathcal{F}$ is a torsion notion, then $\mathcal{K}$ must be the quasivariety of $\mathcal{F}$-torsion-free R-modules.

Item (1) from definition of a 'torsion notion' is the claim that $\mathcal{F}$ is an order filter in the poset of finitely generated left ideals of R. That is, if A and B are finitely generated left ideals of R, $A \subseteq B$, and $Ax = 0 \to x = 0$ holds in $\mathcal{K}$, then $Bx = 0 \to x = 0$ also holds in $\mathcal{K}$. This is true because $A \subseteq B$ implies that $Bx = 0 \to Ax = 0$.

Item (5) is the next easiest to verify. Since the variety generated by $\mathcal{K}$ is R-Mod, both $\mathcal{K}$ and R-Mod have the same free modules. Hence the 1-generated free module R belongs to $\mathcal{K}$. Hence R satisfies $Ax = 0 \to x = 0$ for each $A \in \mathcal{F}$, which is exactly what (5) claims.

Item (2) asserts $\mathcal{F}$ is down directed. Choose $A, B \in \mathcal{F}$. We shall apply Theorem 3.3 to the situation $M := R \oplus R \,(\in \mathcal{K})$ and $S := A \oplus B \leq M$. Note that the pair $(1,0) \in M$ belongs to the $\mathcal{K}$-extension of S, since $A \in \mathcal{F}$ and $A(1,0) \subseteq A \oplus B$. Similarly $(0,1)$ belongs to the $\mathcal{K}$-extension of S, since $B \in \mathcal{F}$ and $B(0,1) \subseteq A \oplus B$. Since $\bar{S}$ is a submodule, the element $(1,0) + (0,1) = (1,1)$ must belong to the $\mathcal{K}$-extension of $A \oplus B$. Hence there must exist $C \in \mathcal{F}$ such that $C(1,1) \subseteq A \oplus B$. Necessarily $C \subseteq A \cap B$.

Item (4) asserts that for all $A \in \mathcal{F}$ and $r \in R$ there is a $B \in \mathcal{F}$ such that $Br \subseteq A$. To prove this we again apply the second part of Theorem 3.3. Let $M = R \in \mathcal{K}$ and let $S = A$. The element $1 \in R(= M)$ belongs to the $\mathcal{K}$-extension of $A(= S)$, since $A \cdot 1 \subseteq S$. The $\mathcal{K}$-extension of S is a submodule, so for any $r \in R$ we have that $r \cdot 1 = r \in M$ also belongs to the $\mathcal{K}$ extension of $S = A$. Theorem 3.3 guarantees the existence of $B \in \mathcal{F}$ such that $B \cdot r \subseteq A$, which is what item (4) requires.

Item (3) asserts that if $X, Y \subseteq R$ are finite subsets such that the left ideals (X) and (Y) belong to $\mathcal{F}$, then the left ideal (XY) belongs to $\mathcal{F}$. To prove this, assume that $X = \{a_0, \ldots, a_{m-1}\}$ and $Y = \{b_0, \ldots, b_{n-1}\}$. The fact that $(X), (Y) \in \mathcal{F}$ implies that $\mathcal{K}$ satisfies the quasiidentities

$$\bigwedge_i (a_i x = 0) \to (x = 0) \quad \text{and} \quad \bigwedge_j (b_j x = 0) \to (x = 0).$$

But this means that $\mathcal{K}$ satisfies

$$(2) \qquad \bigwedge_j \left(\bigwedge_i (a_i(b_j x) = 0) \right) \to (x = 0).$$

For, if $\bigwedge_i (a_i(b_j x) = 0)$ holds for a fixed j, then the quasiidentity associated to X guarantees that $b_j x = 0$. But if this holds for all j, then the quasiidentity associated to Y guarantees that $x = 0$. Now (2) is just the quasiidentity associated to $XY = \{a_i b_j \mid i < m, j < n\}$. Since we have shown that it holds in $\mathcal{K}$ we conclude that $(XY) \in \mathcal{F}$.

4 torsion notion $\implies$ RCM

In this section we prove that if $\mathcal{F}$ is a torsion notion for R-modules, then the quasivariety of $\mathcal{F}$-torsion-free R-modules is RCM and the variety it generates is all of R-Mod. This is other direction of the proof of Theorem 2.3.

Lemma 4.1. *Assume that $\mathcal{F}$ is a torsion notion for R-modules, and that $\mathcal{K}$ is the quasivariety of $\mathcal{F}$-torsion-free R-modules. If $M \in \mathcal{K}$ and $S \leq M$ is a submodule of M, then the $\mathcal{K}$-extension of S is the set*

$$\overline{S} := \{m \in M \mid \exists A \in \mathcal{F}(Am \subseteq S)\}.$$

(In this lemma we are not assuming that $\mathcal{K}$ is RCM, so we cannot refer to Theorem 3.3.)

Proof. The set $\overline{S}$ defined in the statement contains S because S is a submodule. (One can take $A = R \in \mathcal{F}$ to prove any $m \in S$ belongs to $\overline{S}$.)

Let's prove that $\overline{S}$ is closed under addition. If $x, y \in \overline{S}$, then there exist $A, B \in \mathcal{F}$ such that $Ax, By \subseteq S$. By the down directedness of $\mathcal{F}$ there is a $C \subseteq A \cap B$ such that $C \in \mathcal{F}$. For this C we have

$$C(x + y) \subseteq Cx + Cy \subseteq Ax + By \subseteq S,$$

yielding $x + y \in \overline{S}$.

Now we argue that $\overline{S}$ is closed under scalar multiplication. Assume that $x \in \overline{S}$ and $r \in R$. Since $x \in \overline{S}$ there is some $A \in \mathcal{F}$ such that $Ax \subseteq S$. By item (4) of Definition 2.2 there exists $B \in \mathcal{F}$ such that $Br \subseteq A$. Thus

$$B(rx) \subseteq Ax \subseteq S,$$

yielding $rx \in \overline{S}$.

Next we argue that $\overline{S}$ is a $\mathcal{K}$-submodule of M. For this we must show that $M/\overline{S} \in \mathcal{K}$, or that $M/\overline{S}$ is $\mathcal{F}$-torsion-free. This can be established by showing that if $A \in \mathcal{F}$, $m \in M$, and $Am \subseteq \overline{S}$, then $m \in \overline{S}$. Suppose that $A = (a_0, \ldots, a_{m-1})$ as a left ideal. The statement $Am \subseteq \overline{S}$ now means $\{a_0 m, \ldots, a_{m-1} m\} \subseteq \overline{S}$. For each k there must exist $A_k \in \mathcal{F}$ such that $A_k(a_k m) \subseteq S$. By the down directedness of $\mathcal{F}$ there is a $B \subseteq \cap A_k$, and this B has the property that $Ba_k m \subseteq S$ for all k. Suppose that $B = (b_0, \ldots, b_{n-1})$ as a left ideal. By item (3) of Definition 2.2 the left ideal C generated by the set $\{b_j a_i \mid i < m, j < n\}$ belongs to $\mathcal{F}$. Cm is the submodule of M generated by all elements $b_j a_i m$, all of which belong to S. Thus $Cm \subseteq S$. This forces $m \in \overline{S}$, concluding the proof that $M/\overline{S}$ is $\mathcal{F}$-torsion-free.

We have shown that $\overline{S}$ is a $\mathcal{K}$-submodule extending S, but still must show that it is the least such. For this it suffices to observe that, from the definition of $\overline{S}$, if $m \in \overline{S}$, then for any submodule $T \leq M$ satisfying $S \leq T \leq \overline{S}$ we have that m/T is an $\mathcal{F}$-torsion element of M/T.

For the next theorem, which is the main result of the section, we need another fact from (Kearnes and McKenzie, 1992). In Theorem 4.1 of that paper it is shown that a quasivariety is RCM if it satisfies the 'extension principle' and the 'relative shifting lemma'. The second of these properties will hold for any subquasivariety of an RCM quasivariety. Thus, since R-Mod is RCM, any subquasivariety of R-Mod satisfies the 'relative shifting lemma'. The 'extension principle' is not typically inherited by subquasivarieties.

The *extension principle* for a quasivariety $\mathcal{K}$ of modules is the property that, for $M \in \mathcal{K}$, the function mapping a submodule $S \leq M$ to its $\mathcal{K}$-extension $\overline{S}$ is a lattice homomorphism from the lattice of submodules of M to the lattice of $\mathcal{K}$-submodules of M. In the presence of the 'relative shifting lemma', the extension principle is equivalent to the *weak extension principle*, which asserts that if $S \cap T = 0$ for submodules $S, T \leq M$, $M \in \mathcal{K}$, then $\overline{S} \cap \overline{T} = 0$. (The equivalence of the weak and full extension principles for quasivarieties satisfying the 'relative shifting lemma' is explained at the foot of page 482 of (Kearnes and McKenzie, 1992).)

Altogether, this means that a subquasivariety of R-Mod is RCM iff it satisfies the weak extension principle. We need this fact to prove the following theorem.

Theorem 4.2. *If $\mathcal{F}$ is a torsion notion for R-modules, then the quasivariety $\mathcal{K}$ of $\mathcal{F}$-torsion-free R-modules is RCM and the variety it generates is R-Mod.*

Proof. As discussed before the statement of the theorem, to prove that the quasivariety of $\mathcal{F}$-torsion-free modules is RCM it suffices to establish the weak extension principle. So choose an $\mathcal{F}$-torsion-free module $M \in \mathcal{K}$ and two submodules $S, T \leq M$ satisfying $S \cap T = 0$. Let's prove that their $\mathcal{K}$-extensions $\overline{S}$ and $\overline{T}$ are disjoint.

Choose $m \in \overline{S} \cap \overline{T}$. By Lemma 4.1 there exist $A, B \in \mathcal{F}$ such that $Am \subseteq S$ and $Bm \subseteq T$. By the down directedness of $\mathcal{F}$ there is a $C \subseteq A \cap B$ that belongs to $\mathcal{F}$, and for this C we have $Cm \subseteq Am \cap Bm \subseteq S \cap T = 0$, so m is an $\mathcal{F}$-torsion element. This forces $m = 0$, as desired. We conclude that the quasivariety of $\mathcal{F}$-torsion-free R-modules is RCM.

To show that the variety generated by the $\mathcal{F}$-torsion-free R-modules is all of R-Mod, it suffices to note that the 1-generated free R-module is $\mathcal{F}$-torsion-free. This is the content of item (5) of Definition 2.2. Thus $R \in \mathcal{K}$, so the variety generated by $\mathcal{K}$ is R-Mod.

5 final statement

Given a fixed ring R, we now know that a typical RCM quasivariety $\mathcal{K}$ of R-modules can be described by a pair $(I, \mathcal{F})$ where I is an ideal – the annihilator of $\mathcal{K}$ – and $\mathcal{F}$ is a torsion notion for R/I. This information can be expressed entirely in terms of the left ideal structure of R by replacing $\mathcal{F}$ with the set $\mathcal{G}$ defined to consist of all $\nu^{-1}(A)$ for $A \in \mathcal{F}$ and $\nu \colon R \to R/I$ the natural map. This yields the following statement.

Theorem 5.1. *Let R be a ring. A quasivariety $\mathcal{K}$ of R-modules is RCM iff there is a pair $(I, \mathcal{G})$ such that $\mathcal{K}$ is the collection of R-modules satisfying $Ix = 0$ and $Ax = 0 \to x = 0$ for all $A \in \mathcal{G}$. Here we require that I be a two-sided ideal of R and $\mathcal{G}$ be a family of left ideals of R, each containing I and finitely generated over I, such that items (1)–(4) of Definition 2.2 hold, along with*

(5)$'$ *(Regularity modulo I of elements of $\mathcal{G}$) If $A \in \mathcal{G}$, $r \in R$ and $Ar \subseteq I$, then $r \in I$.*

References

Baker, K. A. (1977). Finite equational bases for finite algebras in a congruence-distributive equational class, *Advances in Mathematics*, 24(3), 207–243.

Dziobiak, W., Maróti, M., McKenzie, R. and Nurakunov, A. (2009). The weak extension property and finite axiomatizability for quasivarieties, *Fundamenta Mathematicae*, 202(3), 199–223.

Kearnes, K. and McKenzie, R. (1992). Commutator theory for relatively modular quasivarieties, *Transactions of the American Mathematical Society*, 331(2), 465–502.

Kearnes, K. and Szendrei, Á. (1998). The relationship between two commutators, *International Journal of Algebra and Computation*, 8(4), 497–531.

Kearnes, K., Szendrei, Á and Willard, R. D. (2016). A finite basis theorem for difference-term varieties with a finite residual bound, *Transactions of the American Mathematical Society*, 368(3), 2115–2143.

Maróti, M. and McKenzie, R. (2004). Finite basis problems and results for quasivarieties, *Studia Logica*, 78(1-2), 293–320.

McKenzie, R. (1987). Finite equational bases for congruence modular varieties, *Algebra Universalis*, 24(3), 224–250.

Pigozzi, D. (1988). Finite basis theorems for relatively congruence-distributive quasivarieties, *Transactions of the American Mathematical Society*, 310(2), 499–533.

Willard, R. (2000). A finite basis theorem for residually finite, congruence meet-semidistributive varieties, *The Journal of Symbolic Logic*, 65(1), 187–200.

The computational complexity of deciding whether a finite algebra generates a minimal variety

George F. McNulty*

In celebration of the contributions of Don Pigozzi

Abstract We prove that the problem of deciding whether a finite algebra of finite signature generates a minimal variety is complete for deterministic 2EXPTIME.

Key words: minimal variety of algebras; NP-hard; finite algebra decision problems

1 Introduction

This paper addresses the complexity of the following computational problem:

THE MINIMAL VARIETY PROBLEM

Input: A finite algebra **A** of finite signature.

Problem: Decide if the variety generated by **A** is minimal.

We understand an *algebra* to be a nonempty set endowed with a system (here always a finite system) of operations, each of some finite rank. A *variety of algebras* is a class of algebras, all of the same signature, that is closed with respect to the formation of homomorphic images, subalgebras, and arbitrary direct products—or what is the same according to a classical theorem of Garrett Birkhoff, that is the class of all models of some set of equations. The

George F. McNulty
Department of Mathematics, University of South Carolina, Columbia, South Carolina 29208, United States of America e-mail: `mcnulty@math.sc.edu`

* The author was supported by NSF grant: 1500216

variety generated by **A** consists of all those algebras that are homomorphic images of subalgebras of direct powers of **A**. Varieties of a fixed signature are (lattice) ordered by the inclusion relation $\subseteq$. The variety of all algebras is the largest variety in this ordering while the class consisting of all the one-element algebras is the smallest—it is called the *trivial variety*. A variety is *minimal* if it is nontrivial but has no nontrivial proper subvarieties.

The Minimal Variety Problem is one of a series of similar problems that, roughly, take finite algebras as inputs and ask whether the variety they generate has some particular property. Perhaps the most famous problem of this kind is

TARSKI'S FINITE BASIS PROBLEM

Input: A finite algebra **A** of finite signature.

Problem: Decide if the variety generated by **A** is axiomatized by some finite set of equations.

Ralph McKenzie (1996c) settled this problem, outstanding for more than thirty years, by showing that there is no algorithm to make such decisions. Some techniques McKenzie developed in his solution find their way into our arguments below.

Despite the common wisdom that most everything is undecidable, it turns out that a number of these finite algebra problems are decidable. One of the key results was found early.

THE TWO FINITE ALGEBRA MEMBERSHIP PROBLEM

Input: Two finite algebra **A** and **B** of the same finite signature.

Problem: Decide if **A** belongs to the variety generated by **B**.

Jan Kalicki (1952) provided a brute force algorithm to settle this problem. Soon afterwards, Dana Scott (1956) used Kalicki's algorithm in devising another brute force algorithm to settle that Minimal Variety Problem: First build the algebra freely generate by a two-element set for the variety generated by **A** (this is a certain sublagebra of $\mathbf{A}^{A^2}$); second generate a list of all the nontrivial homomorphic images of this free algebra; last, check for each **B** on the list that **A** belongs to the variety generated by **B**. Now Kalicki's algorithm is itself very costly, but even if it weren't Scott's algorithm would be.

Our task here is to provide more information concerning the computational complexity of the Minimal Variety Problem. To obtain some lower bound on the complexity, we rely on another problem in this series. Fix a finite algebra **B** of finite signature.

THE FINITE ALGEBRA MEMBERSHIP PROBLEM FOR **B**.

Input: A finite algebra **A** of finite signature.

Problem: Decide if **A** belongs to the variety generated by **B**.

In 1998 Zoltán Székely in his dissertation (Székely, 1998, 2002) devised a 7-element algebra **S** with an NP-complete Finite Algebra Membership Problem. Building on this result, in 2006 Marcel Jackson and Ralph McKenzie invented a finite semigroup with an NP-hard membership problem. By 2009, Marcin Kozik (2009, 2007) created a finite algebra with a 2EXPTIME-complete Finite Algebra Membership Problem. In 2000, Clifford Bergman and Giora Slutzki proved that the Finite Algebra Membership Problem belongs to 2EXPTIME. Further results of Bergman and Slutzki on finite algebra decision problems can be found in (Bergman and Slutzki, 2002a,b).

Suppose that **B** generates a minimal variety. Then an algebra **A** belongs to the variety generated by **B** if and only if $\mathbf{A} \times \mathbf{B}$ generates a minimal variety. In this way we can conclude that the Minimal Variety Problem is at least as hard as the Finite Algebra Membership Problem for **B**. The algebras constructed by Székely and later by Kozik do not generated minimal varieties. Fortunately, Don Pigozzi (1979) discovered a way to convert a non-finitely based finite algebra into one that also generates a minimal variety. This is exactly the method we need to apply to the results of Székely and of Kozik to obtain the conclusion that Minimal Variety Problem is complete for deterministic 2EXPTIME.

It has long been known that the problems of deciding if a finite algebra generates a congruence distributive variety (or a variety with most other Mal'cev conditions) is decidable. Freese and Valeriote (2009) were able to show that the problems associated with congruence distributivity and congruence modularity are actually EXPTIME-complete. Kaarli and Pixley (2002) showed that the problem of deciding of a finite algebra whether it generates an affine complete variety is decidable. The complexity of this problem, seemingly, as yet to be investigated.

A remark on technical matters
There is a technical difficulty in the way these problems have been formulated. Not all finite algebras of finite signature are suitable as inputs to computational procedures. The individual elements of an algebra might themselves not be amenable as inputs. A way to remedy this is to insist that the members of these algebras must be finite strings of 0's and 1's. Then the universe of the algebra can be regarded as a list of those finite strings. Moreover, each basic operation of an algebra can be regarded as a list $r + 1$-tuples of elements of the algebra, where r is the rank of the operation. By reserving some short strings of 0's and 1's to use as delimiters, the whole finite algebra can be regarded as a finite string of 0's and 1's. In this way, each finite algebra (up to isomorphism) can be regarded as a suitable input for a Turing machine. There are many ways in which this reduction of a finite algebra to a string of 0's and 1's might be accomplished, some more efficient than others. Since our interest here is in procedures that require at least exponential time complexity, the details of how this reduction is done have small importance. So we assume that one of the more or less obvious methods of reduction has been adopted. For an algebra **A** we take the length of the resulting string of

0's and 1's as the **magnitude** of **A**. The only constraint we impose on the reduction is that there is a polynomial $p(x)$ with natural number coefficients so that there is an algorithm for evaluating the basic operations of **A** that uses time bounded above by $p(\alpha)$, where α is the magnitude of **A**. (Actually, even this constraint can be relaxed.)

2 An upper bound

Scott's algorithm produces a list containing an isomorphic copy of each 2-generated member for the variety generated by **A** and then invokes Kalicki's algorithm to determine whether each algebra on the list generates the same variety as **A**. Bergman and Slutzki (2000) have shown that Kalicki's algorithm is in 2EXPTIME and Marcin Kozik (2009) has given a 2EXPTIME-complete instance of the Finite Algebra Membership Problem—so Kalicki's algorithm cannot be replaced by anything substantially cheaper. It is conceivable that the list produced by Scott's algorithm is quite long and that Scott's algorithm is even more expensive than Kalicki's algorithm.

In 1997 Keith Kearnes and Ágnes Szendrei gave a deep characterization of minimal locally finite varieties, see (Kearnes and Szendrei, 1997). This leads to an algorithm that is like Scott's but much simpler. Consider a given finite nontrivial algebra **A** of finite signature. Among the homomorpic images of subalgebras of **A** one can find a nontrivial algebra **B** of smallest cardinality. This algebra **B** is *strictly simple* in the sense that it has no proper nontrivial subalgebras and no proper nontrivial congruences (and hence no proper nontrivial homomorphic images). Were the variety generated by **A** minimal, then **B** would have two further properties:

(a) The algebras **A** and **B** would generate the same variety, and
(b) The algebra **B** would generate a minimal variety.

On the other hand, the failure of either (a) or (b) ensures that **A** does not generate a minimal variety. Of course (a) can be tested by one use of Kalicki's algorithm. Kearnes and Szendrei prove that in a minimal locally finite variety the strictly simple generator is unique up to isomorphism and is embeddable into every nontrivial algebra in the variety. This gives us another modest simplification: we can take be **B** to be minimal among the nontrivial subalgebras of **A** rather than among their homomorphic images. So the broad outline of the Kearnes-Szendrei algorithm is:

(1) Obtain the nontrivial subalgebra **B** of **A** of smallest cardinality.
(2) Test whether **B** is simple.
(3) Test whether **A** and **B** generate the same variety.
(4) Test whether **B** generates a minimal variety.

Steps (1) and (2) can be accomplished in polynomial time. Step (3) can be handled be one use of Kalicki's algorithm. So we see that our problem reduces to

THE STRICTLY SIMPLE MINIMAL VARIETY PROBLEM

Input: A finite strictly simple algebra $\mathbf{B}$ of finite signature.

Problem: Decide if the variety generated by $\mathbf{B}$ is minimal.

Addressing this problem was principal point of the investigation of Kearnes and Szendrei.

A one-variable term function e of $\mathbf{A}$ is said to be *idempotent* provided $e \circ e = e$ and, further, is called a *minimal idempotent* provided the image $e(A)$ of A under e has more than one element and $f(A)$ is not properly included in $e(A)$ for any idempotent nonconstant term function f. The term consisting of just the variable x names the identity map on A, which is idempotent and not constant. Evidently, every nontrivial finite algebra has minimal idempotents. Here is the Kearnes-Szendrei characterization:

The Kearnes-Szendrei Characterization
Let $\mathbf{B}$ be a finite strictly simple algebra and let e be a minimal idempotent term operation on $\mathbf{B}$. The following are equivalent.

(a) $\mathbf{B}$ *generates a minimal variety.*
(b) $\mathbf{B}$ *is not Abelian or has a trivial subalgebra and for some positive natural number n, there exist binary terms f_i and unary terms g_i and h_i for $0 \leq i \leq n$ such that each of the following equations hold in $\mathbf{B}$.*

$$x \approx f_0(x, eg_0(x))$$
$$(\circledast) \quad f_i(x, eh_i(x)) \approx f_{i+1}(x, eg_{i+1}(x)) \qquad (0 \leq i \leq n-1)$$
$$f_n(x, eh_n(x)) \approx e(x)$$

To determine the complexity of the algorithm outlined above for the Minimal Variety Problem we must determine the complexity, in terms of the magnitude of a given finite algebra $\mathbf{A}$ of finite signature, of each of the following.

- Constructing a nontrivial subalgebra $\mathbf{B}$ of $\mathbf{A}$ of smallest cardinality.
- Determining whether $\mathbf{B}$ is simple.
- Determining whether $\mathbf{B}$ is Abelian or has a trivial subalgebra.
- Constructing a minimal idempotent term function e of $\mathbf{B}$.
- Determining whether there are finite systems of unary and binary terms so that the equations $(\circledast)$ hold in $\mathbf{B}$.
- Determining whether $\mathbf{A}$ and $\mathbf{B}$ generate the same variety.

Let α be the magnitude of $\mathbf{A}$.

To find the nontrivial sublagebra $\mathbf{B}$ of least cardinality, one need only examine the subalgebras generated by the two-elements subsets. As the number of two-element subsets is bounded above by $|A|^2 \leq \alpha^2$ and closing each such subset using the basic operations is also bounded by a polynomial in α, this can be accomplished in polynomial time.

Once $\mathbf{B}$ is in hand, to determine whether it is simple is just a matter of constructing all its congruences generated by collapsing pairs of distinct elements. With the help of Mal'cev's Congruence Generation Theorem, this can be accomplished in time polynomial in the magnitude of $\mathbf{B}$ and hence of $\mathbf{A}$.

Determining whether $\mathbf{B}$ is Abelian we need only determine if the diagonal $\{\langle b,b \rangle \mid b \in B\}$ of B is a congruence class of some congruence of $\mathbf{B}^2$. As above, we can do this in time polynomial in the magnitude of $\mathbf{B}^2$ and hence of $\mathbf{A}$.

To facilitate addressing the next two tasks, it pays to begin by listing term functions in the variables x and y on the algebra $\mathbf{B}$. Doing this in any reasonably parsimonious way will cost only an exponential amount of time—this amounts to listing the elements of the clone $\mathrm{Clo}_2\,\mathbf{B}$, which is the subuniverse of $\mathbf{B}^{B^2}$ generated be the two projection functions. The elements of $\mathrm{Clo}_2\,\mathbf{B}$ are recorded as certain lists of 3-tuples of elements of B. Finding a minimal idempotent e is now a matter of searching through this (perhaps exponentially) long list of terms for all the nonconstant idempotents and selecting one that gives the smallest image of B. Since $|B|^{|B|}$ is an upper bound on the number of term fuctions in x, this process can be completed in time exponential in the magnitode of $\mathbf{B}$ and hence also in time exponential in the magnitude of $\mathbf{A}$.

To discover if there is a way to fulfill ($\circledast$). We impose the structure of a directed graph on $\mathrm{Clo}_2\,\mathbf{B}$ as follows. The vertices will just be the elements of $\mathrm{Clo}_2\,\mathbf{B}$. Given such vertice f and f', we create an edge from f to f' provided there are term functions h and g in $\mathrm{Clo}_2\,\mathbf{B}$ (in the subclone generated by the first projection function) so that

$$f(x, eh(x)) = f'(x, eg(x))$$

holds in $\mathrm{Clo}_2\,\mathbf{B}$. The ($\circledast$) is just the contention that there is a directed path from the first projection function to e (considered as a member of $\mathrm{Clo}_2\,\mathbf{B}$). There at most $|B|^{|B|^2}$ vertices in this directed graph. So this reachability problem can be settled in time exponential in the magnitude of $\mathbf{B}$ and, hence also exponential in the magnitude of $\mathbf{A}$. The author would like to thank the referee for suggesting this reachability argument.

Finally, the last task can be accomplished by Kalicki's algorithm, which requires only doubly exponential time.

In this way, we find the following corollary of the work of Kearnes and Szendrei:

Corollary 1. *The Minimal Variety Problem belongs to the deterministic complexity class* 2EXPTIME.

3 The Minimal Variety Problem is NP-hard

In the next section, we will prove that the Minimal Variety Problem is complete for deterministic 2EXPTIME. So the content of this section will be superceded. The method is roughly the same in these two sections, but in the current section it is applied to algebras built from finite graphs that have only a few basic operations, whereas in the next section the algebras are built from arbitrary alternating Turing machines working in exponential space. Such algebras have many somewhat complicated operations. So the basic idea of the method is more transparent here in the case of graph algebras.

The graph $\mathbb{S}$ displayed in Figure 1 is one of a sequence of graphs devised by Zoltán Székely in 1998 (Székely, 1998, 2002) to characterize k-colorable graphs and to build a finite algebra with an NP-complete finite algebra membership problem. In 1979 William Wheeler, as part of his inves-

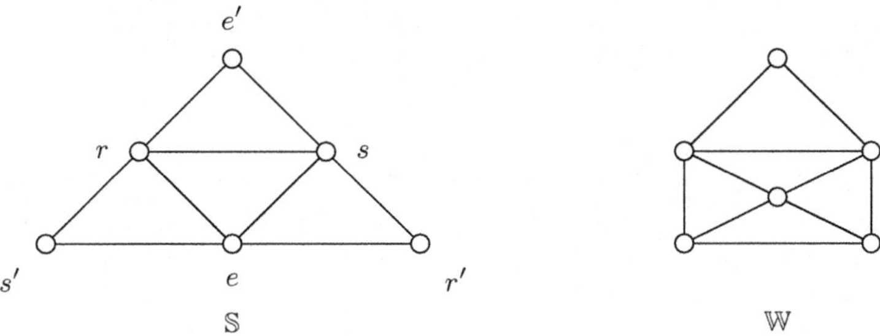

Fig. 1 Zoltán Székely's graph $\mathbb{S}$ and Willaim Wheeler's graph $\mathbb{W}$

tigation (Wheeler, 1979) of the first-order theory of k-colorable graphs, gave a similar sequence of graphs. Székely and Wheeler show that the two graphs displayed in Figure 1 capture 3-colorability in the sense that a graph is 3-colorable if and only if it is isomorphic to an induced subgraph of a direct power of $\mathbb{S}$ (in Székely's case) or of $\mathbb{W}$ (for Wheeler). Below we will use $\mathbb{S}$ since its symmetrical nature simplifies the discussion.

Let $\mathbb{H}$ be any graph so that $\mathbb{H}$ and $\mathbb{S}$ have no common vertices. The graph $\mathbb{S}_{\mathbb{H}}$ is the disjoint union $\mathbb{H}$ with $\mathbb{S}$. So there are no edges between $\mathbb{S}$ and $\mathbb{H}$. See Figure 2.

We think of the vertices of $\mathbb{S}$ as colors: r for ruby, s for sapphire, and e for emerald and three "primed" variants. By an *admissable* coloring of $\mathbb{H}$ we mean a homomorphism from $\mathbb{H}$ into $\mathbb{S}$, where a homomorphism is a map that preserves adjacency.

We associate with the graph $\mathbb{S}_{\mathbb{H}}$ an algebra $\mathbf{S}_{\mathbb{H}}^{*}$ as follows. The universe of $\mathbf{S}_{\mathbb{H}}^{*}$ is the set of vertices of $\mathbb{S}_{\mathbb{H}}$ extended by a single new element $\perp$, called the default element. The signature provides eight two-place operation symbols: $\cdot, \wedge, Q_r, Q_s, Q_e, Q_{r'}, Q_{s'}$, and $Q_{e'}$ as well as seven constant symbols: $\perp, c_r, c_s, c_e, c_{r'}, c_{s'}$, and $c_{e'}$ to name the default element and each vertex of $\mathbb{S}$. Here is how the operations of $\mathbf{S}_{\mathbb{H}}^{*}$ are defined:

$$a \cdot b = \begin{cases} a & \text{if } a \text{ and } b \text{ are adjacent vertices of } \mathbb{S}_{\mathbb{H}} \\ \perp & \text{otherwise} \end{cases}$$

$$a \wedge b = \begin{cases} a & \text{if } a = b \\ \perp & \text{otherwise} \end{cases}$$

$$Q_d(a,b) = \begin{cases} b & \text{if } a = d \\ \perp & \text{otherwise} \end{cases} \qquad \text{for } d \in \{r, s, e, r', s', e'\}$$

By *proper* elements of this algebra (and others below) we mean those different from $\perp$. The operation $\cdot$ is the graph algebra operation introduced by Caroline Shallon in her dissertation (Shallon, 1979), see also (McNulty and Shallon, 1983). The operation $\wedge$ is a semilattice operation making our algebra into a flat graph algebra, that is a graph algebra that is also a height 1 semilattice. Flat graph algebras where investigated by Dejan Delić in (Delić, 2001) and by William Lampe, the author, and Ross Willard in (Lampe et al., 2001). Here we use the semilattice operation to access a construction of Ralph McKenzie from (McKenzie, 1996c). Finally, the operations of the form Q_a where invented by Don Pigozzi (1979) to construct a finite nonfinitely based algebra

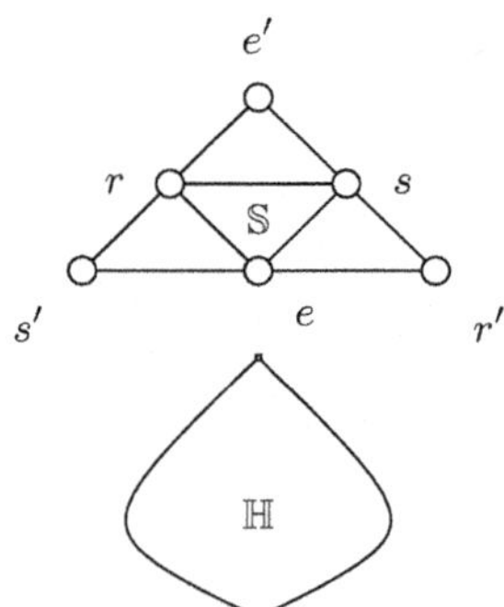

Fig. 2 The graph $\mathbb{S}_{\mathbb{H}}$: The disjoint union of the graph $\mathbb{H}$ with the graph $\mathbb{S}$

that generates a minimal variety—an alternative approach to Pigozzi's result was given by Ágnes Szendrei (1993).

We use $\mathbf{S}^*$ to denote $\mathbf{S}^*_{\mathbb{H}}$ when $\mathbb{H}$ is empty. The algebra $\mathbf{S}$ is the associated flat graph algebra—it is the reduct of $\mathbf{S}^*$ resulting from forgetting the constant symbols and the operations Q_a. Likewise, $\mathbf{H}$ is the flat graph algebra associated with the graph $\mathbb{H}$.

Theorem 2. *The problem of deciding whether a finite algebra of finite signature generates a minimal variety is at least NP-hard.*

Proof. Zoltan Székely proved, for any finite connected graph $\mathbb{H}$ that

$$\mathbf{H} \in \mathsf{HSPS} \text{ if and only if } \mathbb{H} \text{ is 3-colorable.}$$

Determining 3-colorability of finite connected graphs an NP-complete problem. So our theorem will be proven if we can show for every finite connected graph $\mathbb{H}$ that

$$\mathbf{S}^*_{\mathbb{H}} \text{ generates a minimal variety if and only if } \mathbf{H} \in \mathsf{HSPS}.$$

Our first task is to see that $\mathbf{S}^*$ generates a minimal variety. We duplicate here Pigozzi's reasoning that $\mathbf{S}^*$ is embeddable into any nontrivial algebra in the variety generated by $\mathbf{S}^*$. First, observe that the following equations are true in $\mathbf{S}^*$:

$$
(\star) \quad
\begin{aligned}
Q_a c_a y &\approx y && \text{for all proper elements } a \in S \\
Q_a c_b y &\approx \bot && \text{for all proper elements } a, b \in S \text{ with } a \neq b \\
Q_a \bot y &\approx \bot && \text{for all proper elements } a \in S
\end{aligned}
$$

Now let $\mathbf{B}$ be any nontrivial algebra belonging to the variety generated by $\mathbf{S}^*$. Let $h : S \to B$ be defined by

$$
h(a) = \begin{cases} c_a^{\mathbf{B}} & \text{if } a \text{ is a proper element of } S \\ \bot^{\mathbf{B}} & \text{if } a \text{ is } \bot^{\mathbf{S}} \end{cases}
$$

Because $\mathbf{B}$ is in the variety generated by $\mathbf{S}^*$ and because every proper element of S is named by a constant symbol, we find that h is a homomorphism. So we only have to argue that h is one-to-one. To this end suppose $a, b \in S$ with $a \neq b$. It does no harm to assume that a is proper. From the last two equations in $(\star)$ we have in $\mathbf{S}^*$

$$Q_a d y \approx \bot$$

where d is either the constant symbol c_b or the constant symbol $\bot$. So this equation must hold in $\mathbf{B}$ as well. We also see from the first equation in $(\star)$ that

$$Q_a c_a y \approx y$$

holds in $\mathbf{B}$.

Now let e be any proper element of B. So we have

$$Q_a^{\mathbf{B}}(d^{\mathbf{B}}, e) = \bot \neq e = Q_a(c_a^{\mathbf{B}}, e).$$

It follows that $h(b) = d^{\mathbf{B}} \neq c_a^{\mathbf{B}} = h(a)$. So h is one-to-one as desired. As a consequence, $\mathbf{S}^*$ generates a minimal variety.

Observe that $\mathbf{S}^*$ is a subalgebra of $\mathbf{S}_{\mathbb{H}}^*$. Consequently,

$$\mathbf{S}_{\mathbb{H}}^* \text{ generates a minimal variety if and only if } \mathbf{S}_{\mathbb{H}}^* \in \mathsf{HSPS}^*.$$

So what we need to prove our theorem is to establish

$$\mathbf{S}_{\mathbb{H}}^* \in \mathsf{HSPS}^* \text{ if and only if } \mathbf{H} \in \mathsf{HSPS}.$$

Now the left-to-right direction of this equivalence can be established by considering the reduct to the signature of flat graph algebras: notice that the set $H \cup \{\bot\} \subseteq S_{\mathbb{H}}^*$, where H is the set of vertices of $\mathbb{H}$, is closed under the flat graph operations.

This leaves the right-to-left implication. So suppose that $\mathbb{H}$ is a finite connected graph such that $\mathbf{H} \in \mathsf{HSPS}$. Following Székely, this means that $\mathbb{H}$ is 3-colorable. Székely shows that this entails the existence of a particular natural number t, a particular subalgebra $B_{\mathbb{H}}$ of $\mathbf{S}^t$, and a particular congruence θ of $\mathbf{B}_{\mathbb{H}}$ so that $\mathbf{H} \cong \mathbf{B}_{\mathbb{H}}/\theta$. Székely provides fuller details of this construction in (Székely, 1998). Below, you will find an outline of Székely's method that covers the points we will need to adapt.

Let n be the number of vertices of $\mathbb{H}$. Let

$$H = \{p_0, p_1, \ldots, p_{n-1}\}$$

The algebra $\mathbf{B}_{\mathbb{H}}$ is generated by n t-tuples $\pi_0, \ldots, \pi_{n-1}$. We construe each of these t-tuples as a column vector and the whole sequence of them as an array with t rows and n columns. This array is constructed with the help of admissable 3-colorings of $\mathbb{H}$, using the colors $e, r,$ and s. Such a coloring is construed as an n-tuple of colors called a *color row*. Székely's $t \times n$ array has the following properties:

(a) Each row is a copy of a color row modified so that one or two entries are primed.
(b) For any pair of distinct nonadjacent vertices of $\mathbb{H}$ there is a row in the array were the entries associated to the two vertices are primed.
(c) Every column has both primed and unprimed entries.
(d) No two distinct columns are the same.

Now let $\mathbf{B}_{\mathbb{H}}$ be the subalgebra of $\mathbf{S}^t$ generated by the columns of this array. Let $B_{\bot}$ consist of those elements of $B_{\mathbb{H}}$ that have at least one entry equal to $\bot$.

As Székely points out, $B_{\mathbb{H}} = \{\pi_0, \ldots, \pi_{n-1}\} \cup B_\perp$ and the equivalence relation that collapses $B_\perp$ and isolates each of the π_k's is a congruence relation θ such that $\mathbf{H} \cong \mathbf{B}_{\mathbb{H}}/\theta$. The fact that θ is a congruence is a consequence of the fact that $B_\perp$ is an *absorbing set* in the sense that the output of any of the basic operations of $\mathbf{B}_{\mathbb{H}}$ is in $B_\perp$ when at least one of the inputs is in $B_\perp$. The isomorphism holds because each of e, r, and s is adjacent to the other two in $\mathbb{S}$, e' is adjacent to r and s (and likewise for r' and s'), but no primed vertex is adjacent to a primed vertex in $\mathbb{S}$.

We will adapt this strategy of Székely to construct $\mathbf{B}_{\mathbb{H}}^*$. In our case the signature is expanded by constant symbols to name each element of S and by the Pigozzi operations. In any direct power of $\mathbf{S}^*$ the elements named by the constant symbol ε_d will be a tuple (here construed as a column vector) all of whose entries are d. We must have these tuples. Since there are no edges connecting vertices in H with vertices in S, we must arrange matters in $\mathbf{B}_{\mathbb{H}}^*$ so that $\perp$ is produced on at least one coordinate when the graph product is applied in these cases.

To complete this proof, we have to extend Székely's array, making each column longer and adding six new columns. Each new column will have all entries equal to one of e, r, s, e', r' and s'. So in the direct power of $\mathbf{S}^*$ they will be named by the new constant symbols. We add extra rows to capture those adjacencies and nonadjacencies involved in this extension.

We add $3n$ new rows to obtain a $(t+3n) \times n$ array with columns denoted by $\pi_0^\dagger, \ldots, \pi_{n-1}^\dagger$. What we require is the following constraints to hold:

(a†) Each row is an admissible color row modified so that one or two entries are primed.

(b†) For any pair of distinct nonadjacent vertices of $\mathbb{H}$ there is a row in the array were the entries associated to the two vertices are primed.

(c†) Every column has both primed and unprimed entries.

(d†) No two distinct columns are the same.

(e†) Each column $\pi_k^\dagger$ with $0 \le k < n$ has each of e', r', and s' as entries.

Provided the $3n$ new rows satisfy (a†), we only have to contend with the last constraint. For each vertex q among the n vertices $p, \ldots, p_{n-1}$ of $\mathbb{H}$ we select three admissible r, s, e-colorings of $\mathbb{H}$ as follows. The first assigns r to q. The second coloring assigns s to q, while the third coloring assigns e to q. Now modify each of these colorings just at the vertex q by priming the color. Each of these modified colorings is admissable, so constraint (a†) is not injured. So by using three entries we have ensured that constraint (e†) holds for q. We do this for each of the n vertices of $\mathbb{H}$.

Let $\mathbf{B}_{\mathbb{H}}^\dagger$ to the subalgebra of the appropriate direct power of $\mathbf{S}$ generated by $\{\pi_0^\dagger, \ldots, \pi_{n-1}^\dagger\}$ and let $B_\perp^\dagger$ be the subset of those elements of $B_{\mathbb{H}}^\dagger$ that have $\perp$ as an entry. The set $B_\perp^\dagger$ is absorbing. It is clear that

$$B_{\mathbb{H}}^\dagger = \{\pi_0^\dagger, \ldots, \pi_{n-1}^\dagger\} \cup B_\perp^\dagger.$$

Now for each $d \in \{e, r, s, e', r', s'\}$ let ε_d be the $t + 3n$-tuple all entries of which are d. Let $\mathbf{B}^*_{\mathbb{H}}$ be the subalgebra of the appropriate power of $\mathbf{S}^*$ generated by

$$\{\varepsilon_e, \varepsilon_r, \varepsilon_s, \varepsilon_{e'}, \varepsilon_{r'}, \varepsilon_{s'}\} \cup \{\pi_0^\dagger, \ldots, \pi_{n-1}^\dagger\}.$$

Let θ be the equivalence relation that collapses $B^*_\perp$ and isolates every other element. Since $B^*_\perp$ is an absorbing set, θ will a congruence relation. To see that the associated quotient map η is really a homomorphism from $\mathbf{B}^*_{\mathbb{H}}$ onto $\mathbf{S}^*_{\mathbb{H}}$, first observe that the set

$$\{\varepsilon_e, \varepsilon_r, \varepsilon_s, \varepsilon_{e'}, \varepsilon_{r'}, \varepsilon_{s'}\} \cup B^*_\perp$$

is a subuniverse of $\mathbf{B}^*_{\mathbb{H}}$ and the restriction of η to this subuniverse is a homomorphism. Second, observe that the set

$$\{\pi_0^\dagger, \ldots, \pi_{n-1}^\dagger\} \cup B^*_\perp$$

is closed with respect to the flat graph algebra operations and that η restricted to this set respects all the flat graph algebra operations. So to confirm that η is a homomorphism we need to consider the Pigozzi operations and the flat graph algebra operations when inputs come from both parts of our generating set. The Pigozzi operation $Q_d(\mu, \nu)$ produces an output in $B^*_\perp$ unless $\mu = \varepsilon_d$, in which case the output is ν. This means η always respects the Pigozzi operations. Now consider the graph algebra operation $\cdot$. Let us suppose, without loss of generality, that $\mu = \varepsilon_d$, where $d \in \{r, s, e, e', r', s'\}$, and $\nu = \pi_k^\dagger$ with $0 \leq k < n$. Since in $\mathbb{S}$ no vertex is adjacent to itself nor to its primed version, constraint (e†) ensures that $\mu \cdot \nu, \nu \cdot \mu \in B^*_\perp$. So η respects each instance of the graph algebra operation of this kind. Last we must consider the meet. But this is easy since $\mu \wedge \nu = \mu$ if $\mu = \nu$ and otherwise $\mu \wedge \nu \in B^*_\perp$. So η is a homomorphism mapping $\mathbf{B}^*_{\mathbb{H}}$ onto $\mathbf{S}^*_{\mathbb{H}}$.

So we reach the conclusion we desire: $\mathbf{S}^*_{\mathbb{H}} \in \mathsf{HSPS}^*$. This completes the proof of the theorem. $\qquad\square$

4 The Minimal Variety Problem is 2EXPTIME Complete

Theorem 2

The Minimal Variety Problem is 2EXPTIME *complete.*

Marcin Kozik (2009) proved that there is a finite algebra $\mathbf{E}$ that has a 2EXPTIME complete finite algebra membership problem. In this section we will show that Kozik's work can be modified, roughly in the way that Székely's work was modified in the previous section, to show that the Minimal Variety Problem is 2EXPTIME complete.

In 1981 it was shown by Chandra, Kozen, and Stockmeyer, see (Chandra et al., 1981), that 2EXPTIME=AEXPSPACE. That is, the class of languages recognized in doubly exponential time by deterministic Turing machines is the same as the class of languages recognized using in exponential space by alternating Turing machines. What Kozik does is associate with each alternating Turing machine $\mathbb{T}$ an algebra $\mathbf{E}(\mathbb{T})$ and with each input word w of $\mathbb{T}$ an algebra $\mathbf{S}_w$ of magnitude bounded polynomially in the length of w, so that

$$\mathbf{S}_w \in \mathsf{HSPE}_{\mathbb{T}} \text{ if and only if } w \notin \mathcal{L},$$

where $\mathcal{L}$ is the language recognized by $\mathbb{T}$ is exponential space. By way of Chandra, Kozen and Stockmeyer, this result about alternating Turing machines operating in exponential space is equivalent to a deterministic result for doubly expontential time. Because deterministic complexity classes are closed with respect to complementation, this establishes Kozik's theorem.

While Kozik's machine algebras take their inspiration from McKenzie's machine algebras (McKenzie, 1996a), those of Kozik are more involved: they must reflect the behavior of $\mathbb{T}$ on all inputs (where McKenzie needed only the blank tape as input) and they must take into account the limitation to exponential space.

Rather than repeating here the details of Kozik's method from (Kozik, 2009), we adopt almost all his notation and presume that the reader has a copy of (Kozik, 2009) in hand for reference. From this point we regard $\mathbb{T}$ as a fixed alternating Turing machine (with the minor restrictions imposed by Kozik) and we take w to be a word of length n on the tape alphabet of $\mathbb{T}$. We take $\mathbf{E}$ to be the algebra Kozik associates with the machine $\mathbb{T}$. Just as we did in the previous section we let $\mathbf{E}^*$ be the expansion of $\mathbf{E}$ by using new constant symbols to name each element of E and by the inclusion of a two-place Pigozzi operation for each proper element of E. So for each element $d \in E$ our signature has a constant symbol c_d and, in case d is proper, a two-place operation symbol Q_d.

Following Don Pigozzi (1979), as we did at the beginning of the proof of Theorem 1, we see that $\mathbf{E}^*$ generates a minimal variety. We will form $\mathbf{S}_w^*$ by an amalgamation of $\mathbf{E}^*$ with $\mathbf{S}_w$ by identifying $\perp^{\mathbf{E}}$ and $\perp^{\mathbf{S}_w}$ and letting $\perp^{\mathbf{E}}$ be the value of any operation when it has inputs from both parts of this amalgamation, with a few well-controlled exceptions. Evidently, $\mathbf{E}^*$ is a subalgebra of $\mathbf{S}_w^*$. This means

$$\mathbf{S}_w^* \text{ generates a minimal variety if and only if } \mathbf{S}_w^* \in \mathsf{HSPE}^*.$$

So the proof of Theorem 2 will be complete once we establish

Main Contention

$$\mathbf{S}_w^* \in \mathsf{HSPE}^* \textit{ if and only if } \mathbf{S}_w \in \mathsf{HSPE}.$$

Proof. The left-to-right implication is easy. We can simply forget the Pigozzi operations and ignore the constant symbols naming the elements of E. Once this is done $\mathbf{S}_w$ is a subalgebra of the reduct of $\mathbf{S}_w^*$. So $\mathbf{S}_w \in \mathsf{HSPE}$.

For the reverse implication, we suppose that $\mathbf{S}_w \in \mathsf{HSPE}$. In this case, Kozik constructs a subalgebra $\mathbf{B}(w)$ of $\mathbf{E}^{2^n}$ as well as a congruence θ of $\mathbf{B}(w)$ such that $\mathbf{S}_w \cong \mathbf{B}(w)/\theta$. Moreover, the θ-class of $\perp^{\mathbf{B}(w)}$ is the only congruence class that is not a singleton. We will follow the same strategy. The main difficulty is that our $\mathbf{B}^*(w)$ must have the constant tuples named by our new constant symbols in the direct power as well as all the elements generated with their help and the help of those elements in Kozik's construction.

Since the word w is fixed, to simplify notation, we suppress it from this point on. So we use $\mathbf{S}$ and $\mathbf{B}$ where Kozik uses $\mathbf{S}_w$ and $\mathbf{B}(w)$. We also dispense with the use of superscripts to indicate the operations denoted by the operation symbols is our various algebras, relying on context to resolve any ambiguities.

To proceed we must describe the algebras $\mathbf{E}, \mathbf{S}$, and $\mathbf{B}$, as well as their corresponding modified versions $\mathbf{E}^*, \mathbf{S}^*$, and $\mathbf{B}^*$.

The signature of the algebra $\mathbf{E}$ is complicated because it reflects the alternating Turing machine $\mathbb{T}$. This machine has a tape alphabet $\mathcal{A}$ and a set $\mathcal{S}$ of states. We use β to denote the blank tape symbol. Notice that $\beta \in \mathcal{A}$. There are two special states: 1 the initial state and 0 the accepting state. The universe E of the algebra $\mathbf{E}$ is the union of the following eleven disjoint sets:

$$\mathcal{Y} = \{\Delta^{L,0}, \Delta^{L,1}, \Delta^{H,0}, \Delta^{H,1}, \Delta^{R,0}, \Delta^{R,1}\}$$

$$\mathcal{X}_L = \{L^{Z,a}_{(i,b),c} \mid a,b,c \in \mathcal{A}, Z \in \{R,H,L\}, \text{ and } i \in S\}$$

$$\mathcal{X}_H = \{H^{Z,a}_{(i,b),c} \mid a,b,c \in \mathcal{A}, Z \in \{R,H,L\}, \text{ and } i \in S\}$$

$$\mathcal{X}_R = \{R^{Z,a}_{(i,b),c} \mid a,b,c \in \mathcal{A}, Z \in \{R,H,L\}, \text{ and } i \in S\}$$

$$\widehat{\mathcal{X}}_L = \{L^{Z,a}_{(\hat{i},b),c} \mid a,b,c \in \mathcal{A}, Z \in \{R,H,L\}, \text{ and } i \in S - \{0\}\}$$

$$\widehat{\mathcal{X}}_H = \{H^{Z,a}_{(\hat{i},b),c} \mid a,b,c \in \mathcal{A}, Z \in \{R,H,L\}, \text{ and } i \in S - \{0\}\}$$

$$\widehat{\mathcal{X}}_R = \{R^{Z,a}_{(\hat{i},b),c} \mid a,b,c \in \mathcal{A}, Z \in \{R,H,L\}, \text{ and } i \in S - \{0\}\}$$

$$\mathcal{Z}_L = \{L^{Z,a} \mid a \in \mathcal{A} \text{ and } Z \in \{R,H,L\}\}$$

$$\mathcal{Z}_H = \{H^{Z,a} \mid a \in \mathcal{A} \text{ and } Z \in \{R,H,L\}\}$$

$$\mathcal{Z}_R = \{R^{Z,a} \mid a \in \mathcal{A} \text{ and } Z \in \{R,H,L\}\}$$

$$\{\perp\}$$

Further, we let

$$\mathcal{X} = \mathcal{X}_L \cup \mathcal{X}_H \cup \mathcal{X}_R$$
$$\widehat{\mathcal{X}} = \widehat{\mathcal{X}}_L \cup \widehat{\mathcal{X}}_H \cup \widehat{\mathcal{X}}_R$$
$$\mathcal{Z} = \mathcal{Z}_L \cup \mathcal{Z}_H \cup \mathcal{Z}_L.$$

So $E = \mathcal{Y} \cup \mathcal{X} \cup \widehat{\mathcal{X}} \cup \mathcal{Z} \cup \{\bot\}$.

The three element set $\{L, H, R\}$ has a binary relation $\prec$ all of whose instances are displayed below.

$$L \prec L \prec H \prec R \prec R.$$

Finally, with each state i other than the accepting state 0 there is a distinct associated state $\hat{i}$.

Each element of E, apart from the default element $\bot$, is determined by specifying a number of parameters. For example, the elements of $\mathcal{X}$ have a main symbol (one of L, H, or R), two superscripts, and three subscripts. The basic operations of $\mathbf{E}$ are described by telling how to use the parameters of the inputs to obtain the parameters of the output. Kozik invented a convenient notational scheme for this purpose. Two examples will suffice (but the scheme is carefully described in (Kozik, 2009)).

$$\blacksquare^{\square,\blacksquare}_{(\blacksquare,\square),\blacksquare}(R^{H,a}_{(2,b),c}) = \blacksquare^{\square,\blacksquare}_{(\blacksquare,\square),\blacksquare}(R^{L,a}_{(2,b),c}).$$
$$\square^{\blacksquare,\square}_{(\square,\square),\square}(R^{H,a}_{(2,b),c}) = H \prec R = \square^{\blacksquare,\square}(\Delta^{R,1}).$$

The first displayed formula asserts (correctly) that the corresponding parameters of the inputs at the $\blacksquare$ positions are equal. The second asserts that the parameters at the $\blacksquare$ positions stand in the relation $\prec$.

The basic operations of $\mathbf{E}$ fall into three categories: space building operations, computation and checking operations, and a handful of auxiliary operations. The element $\bot$ is an absorbing element for each of the basic operations of $\mathbf{E}$. That is, if $\bot$ is among the inputs of the operation, then the output is $\bot$. By the **proper domain** of an operation Q of rank r we mean the set of all $\bar{t} \in E^r$ so that $Q(\bar{t}) \neq \bot$.

The signature of $\mathbf{E}$ is rather involved. More details are found below. Nevertheless we can describe here two important algebras. Let w be a word in the tape alphabet of the underlying alternating Turing machine and suppose the length of w is n. The signature of $\mathbf{E}$ has among its operation symbols a one-place operation symbol G, a two-place operation symbol Ω, a three-place operation symbol F^a for each letter a of the tape alphabet, and a zero-place operation symbol $\bot$. The algebra $\mathbf{S}_w = \mathbf{S}$ has $2n + 3$ distinct elements

$$S = \{s_0, \ldots, s_n\} \cup \{\ell_0, \ldots, \ell_n\} \cup \{\bot\}.$$

The basic operations of $\mathbf{S}$ return the output $\bot$ on all inputs with only the following exceptions:

$$G(s_0) = s_0$$

$$F^{w(k)}(\ell_k, \ell_{k+1}, s_k) = s_{k+1} \text{ for all } k < n$$

Here $w(k)$ is the kth letter of w. So the leftmost letter of w is $w(0)$ and the rightmost is $w(n-1)$. [We make a minor departure here from Kozik's notation: he uses Σ_k where we have s_k and Λ_k where we have ℓ_k.]

The algebra $\mathbf{S}^*$ has additional elements and operations.

$$S^* = E \cup \{s_0, \ldots, s_n\} \cup \{\ell_0, \ldots, \ell_n\} \cup \{\bot\} = E \cup S_w,$$

where E, the universe of $\mathbf{E}$, is disjoint from $\{s_0, \ldots, s_n\} \cup \{\ell_0, \ldots, \ell_n\}$. The basic operations of $\mathbf{S}^*$ are defined so that $\mathbf{E}$ is a subalgebra, so that $\mathbf{S}$ is a subalgebra of the reduct to the signature of $\mathbf{E}$, and with every basic operation applied to other sorts of inputs returns $\bot$ as the output with only the following exceptions:

$$\Omega(e,c) = \begin{cases} c & \text{if } \square^{\square,\square}_{(\blacksquare,\square),\square}(e) = \widehat{1} \\ \bot & \text{Otherwise.} \end{cases}$$

$$Q_d(e,c) = \begin{cases} c & \text{If } d = e \\ \bot & \text{Otherwise} \end{cases}$$

and where each constant symbol c_e names the element e of $E - \{\bot\}$.

The algebra $\mathbf{S}^*$ is, in essence, an amalgamation of the algebra $\mathbf{E}^*$ and the algebra $\mathbf{S}$ with very little interaction between these two parts. From another viewpoint, $\mathbf{S}^*$ is an extension of $\mathbf{E}^*$ where the interplay of the operations with the additional elements is sharply restricted.

4.1 Space Building Operations of E

For each $a \in \mathcal{A}$ there is a three-place operation F^a. A triple $\langle b, c, d \rangle$ belongs to the proper domain of F^a if and only if all of the following hold:

- $b, c \in \mathcal{Y}$,
- $d \in \mathcal{X} \cup \mathcal{Z}$,
- $d \in \mathcal{Z}$ or $\square^{\square,\square}_{(\blacksquare,\square),\square}(d) = 1$,
- $\square^{\blacksquare,\square}(b) = \square^{\blacksquare,\square}_{(\square,\square),\square}(d) \prec \square^{\blacksquare,\square}(c)$, and
- $\square^{\blacksquare,\square}(b) \neq H$ or $\square^{\square,\blacksquare}_{(\square,\square).\square}(d) = a$.

When $\langle b, c, d \rangle$ belongs to the proper domain of F^a and $d \in \mathcal{X}$ then $F^a(b, c, d)$ is determined as follows:

- $\square^{\square,\square}_{(\blacksquare,\blacksquare),\square}(F^a(b,c,d)) = (1, a)$,

- $\square^{\blacksquare,\square}_{(\square,\square),\square}(F^a(b,c,d)) = \square^{\blacksquare,\square}(c),$
- $\square^{\square,\blacksquare}_{(\square,\square),\square}(F^a(b,c,d)) = \square^{\square,\blacksquare}_{(\square,\square),\square}(d),$
-

$$\blacksquare^{\square,\square}_{(\square,\square),\square}(F^a(b,c,d)) = \begin{cases} R & \text{if } \blacksquare^{\square,\square}_{(\square,\square),\square}(d) = H \text{ and } \square^{\square,\blacksquare}(b) = 1 \\ \blacksquare^{\square,\square}_{(\square,\square),\square}(d) & \text{otherwise} \end{cases}$$

-

$$\square^{\square,\square}_{(\square,\square),\blacksquare}(F^a(b,c,d)) = \begin{cases} a & \text{if } \blacksquare^{\square,\square}_{(\square,\square),\square}(d) = H \text{ and } \square^{\square,\blacksquare}(b) = 0 \\ \square^{\square,\square}_{(\square,\square),\blacksquare}(d) & \text{otherwise.} \end{cases}$$

In this case, the outputs of F^a all belong to $\mathcal{X} \cup \{\bot\}$.

In case $\langle b,c,d \rangle$ belongs to the proper domain of F^a and $d \in \mathcal{Z}$ then $F^a(b,c,d)$ belongs to $\mathcal{Z}$ and is determined as follows:

- $\square^{\blacksquare,\square}(F^a(b,c,d)) = \square^{\blacksquare,\square}(c),$
- $\square^{\square,\blacksquare}(F^a(b,c,d)) = \square^{\square,\blacksquare}(d),$
-

$$\blacksquare^{\square,\square}(F^a(b,c,d)) = \begin{cases} R & \text{if } \blacksquare^{\square,\square}(d) = H \text{ and } \square^{\square,\blacksquare}(b) = 1 \\ \blacksquare^{\square,\square}(d) & \text{otherwise} \end{cases}$$

For each $a \in \mathcal{A}$ there is another three-place operation F^a_{compl}. This operation has the same proper domain as F^a and its proper outputs always belong to $\mathcal{Z}$. Here is how $F^a_{\text{compl}}(b,c,d)$ is determined, when $d \in \mathcal{X}$.

- $\square^{\blacksquare,\square}(F^a(b,c,d)) = \square^{\blacksquare,\square}(c),$
- $\square^{\square,\blacksquare}(F^a(b,c,d)) = \square^{\square,\blacksquare}_{(\square,\square),\square}(d),$
-

$$\blacksquare^{\square,\square}(F^a(b,c,d)) = \begin{cases} L & \text{if } \blacksquare^{\square,\square}_{(\square,\square),\square}(d) = H \text{ and } \square^{\square,\blacksquare}(b) = 0 \\ \blacksquare^{\square,\square}_{(\square,\square),\square}(d) & \text{otherwise} \end{cases}$$

Here is how $F^a_{\text{compl}}(b,c,d)$ is determined, when $d \in \mathcal{Z}$.

- $\square^{\blacksquare,\square}(F^a(b,c,d)) = \square^{\blacksquare,\square}(c),$
- $\square^{\square,\blacksquare}(F^a(b,c,d)) = \square^{\square,\blacksquare}(d),$
-

$$\blacksquare^{\square,\square}(F^a(b,c,d)) = \begin{cases} L & \text{if } \blacksquare^{\square,\square}(d) = H \text{ and } \square^{\square,\blacksquare}(b) = 0 \\ \blacksquare^{\square,\square}(d) & \text{otherwise} \end{cases}$$

250

4.2 Computing and Checking Operations of **E**

These operations have the most complex definitions, as they reflect the instructions of the Turing machine $\mathbb{T}$ and the involved manner in which acceptance/rejection is understood for alternating Turing machines. Fortunately, we only need to describe some features of the proper domains of these operations. For each instruction of $\mathbb{T}$ there is a three-place operation F. These are called computing operations by Kozik. If $F(a,b,c) \neq \perp$, then

- $a, b \in \mathcal{Z}$ and $c \in \mathcal{X} \cup \widehat{\mathcal{X}}$ and
- $\square^{\blacksquare,\blacksquare}(a) = \square^{\blacksquare,\blacksquare}(b) = \square^{\blacksquare,\blacksquare}_{(\square,\square),\square}(c)$.

The checking operations arise from certain sets of machine instruction, with the ranks of the operations depending on the size of these sets. All the tuples in the proper domains of such operations are tuples of elements of $\mathcal{X} \cup \widehat{\mathcal{X}}$ and, as with the computing operations, a tuple that belongs to the proper domain has the property that all its superscript agree.

4.3 Auxiliary Operations of **E**

There is a one-place operation G with proper domain those elements of $b \in \mathcal{X}$ so that $\blacksquare^{\square,\square}_{(\blacksquare,\blacksquare),\blacksquare}(b) = H_{(1,\beta),\beta}$. On its proper domain, G is the identity function.

There is a one-place operation Π and a belongs to the proper domain of Π if and only if

- $a \in \mathcal{X} \cup \mathcal{Z}$, and
- $a \in \mathcal{Z}$ or $\square^{\square,\square}_{(\blacksquare,\square),\square}(a) = 1$. When a belongs to the proper domain of Π we put

$$\Pi(a) = \blacksquare^{\blacksquare,\blacksquare}_{(\square,\square),\square}(a).$$

There is a two-place operation Ω. The tuple $\langle a,b \rangle$ belongs to the proper domain of Ω if and only if $b \neq \perp$ and $a \in \widehat{\mathcal{X}}$ with $\square^{\square,\square}_{(\blacksquare,\square),\square}(a) = \widehat{1}$. When $\langle a,b \rangle$ belongs to the proper domain of Ω, we put

$$\Omega(a,b) = b.$$

4.4 The Pigozzi Operations of **E***

For each $d \in E - \{\perp\}$ we had the two-place Pigozzi operation Q_d defined via

$$Q_d(a,b) = \begin{cases} b & \text{if } d = a \\ \bot & \text{otherwise.} \end{cases}$$

Finally, for each $d \in E$ we provide a zero-place operation c_d, that is a constant to name the element d.

4.5 The Algebras $\mathbf{B}$, $\mathbf{B}^\dagger$, and $\mathbf{B}^*$

For our fixed word w of length n, Marcin Kozik (2009) constructs the algebra $\mathbf{B}$ as a subalgebra of $\mathbf{E}^{2^n}$ by specifying a set

$$\{\sigma_0\} \cup \{\lambda_0, \ldots, \lambda_n\}$$

of $n+2$ generators. Here we need a slightly modification. We use one more co-ordinate, listing it first. We display these modified generators $\sigma_0^\dagger, \lambda_0^\dagger, \lambda_1^\dagger, \ldots, \lambda_n^\dagger$ as column vectors of length $1 + 2^n$.

$\sigma_0^\dagger$	$\lambda_0^\dagger$	$\lambda_1^\dagger$	$\lambda_2^\dagger$	$\lambda_3^\dagger$	$\cdots$	$\lambda_{n-1}^\dagger$	$\lambda_n^\dagger$
$H^{R,\beta}_{(1,\beta),\beta}$	$\Delta^{R,0}$	$\Delta^{R,0}$	$\Delta^{R,0}$	$\Delta^{R,0}$	$\cdots$	$\Delta^{R,0}$	$\Delta^{R,0}$
$H^{H,w(0)}_{(1,\beta),\beta}$	$\Delta^{H,0}$	$\Delta^{R,0}$	$\Delta^{R,0}$	$\Delta^{R,0}$	$\cdots$	$\Delta^{R,0}$	$\Delta^{R,1}$
$H^{L,w(1)}_{(1,\beta),\beta}$	$\Delta^{L,1}$	$\Delta^{H,0}$	$\Delta^{R,0}$	$\Delta^{R,0}$	$\cdots$	$\Delta^{R,0}$	$\Delta^{R,1}$
$H^{L,w(2)}_{(1,\beta),\beta}$	$\Delta^{L,0}$	$\Delta^{L,1}$	$\Delta^{H,0}$	$\Delta^{R,0}$	$\cdots$	$\Delta^{R,0}$	$\Delta^{R,1}$
$H^{L,w(3)}_{(1,\beta),\beta}$	$\Delta^{L,1}$	$\Delta^{L,1}$	$\Delta^{L,0}$	$\Delta^{H,0}$	$\cdots$	$\Delta^{R,0}$	$\Delta^{R,1}$
$H^{L,w(4)}_{(1,\beta),\beta}$	$\Delta^{L,0}$	$\Delta^{L,0}$	$\Delta^{L,1}$	$\Delta^{L,0}$	$\cdots$	$\Delta^{R,0}$	$\Delta^{R,1}$
$\vdots$	$\vdots$	$\vdots$	$\vdots$	$\vdots$	$\ddots$	$\vdots$	$\vdots$
$H^{L,w(n-1)}_{(1,\beta),\beta}$	$\Delta^{L,*}$	$\Delta^{L,*}$	Δ^{L*}	$\Delta^{L,*}$	$\cdots$	$\Delta^{H,0}$	$\Delta^{R,1}$
$H^{L,w(n-1)}_{(1,\beta),\beta}$	$\Delta^{L,*}$	$\Delta^{L,*}$	Δ^{L*}	$\Delta^{L,*}$	$\cdots$	$\Delta^{L,1}$	$\Delta^{H,1}$
$H^{L,w(n-1)}_{(1,\beta),\beta}$	$\Delta^{L,*}$	$\Delta^{L,*}$	Δ^{L*}	$\Delta^{L,*}$	$\cdots$	$\Delta^{L,1}$	$\Delta^{L,1}$
$\vdots$	$\vdots$	$\vdots$	$\vdots$	$\vdots$	$\ddots$	$\vdots$	$\vdots$

Observe that the word w is recorded in the second superscripts of $\sigma_0^\dagger$. The three rows partially displayed are row n, row $n+1$, and row $n+2$. The $*$ in the second superscript indicates that the pattern of 0's and 1's set out in the first few rows should continue—apart from the top row, it sweeps systematically through all 2^n strings of 0's and 1's of length n. For example, reading down column 0 these second superscripts simply alternate, while reading down column $n-1$ there are 2^{n-1} instances of 0 followed by 2^{n-1} instances of 1.

The second superscripts for λ_n are always 1. Our modifications of Kozik's generators is the insertion of a new first element, so σ_0 and each λ_k can be obtained by deleting the first entry of the variants marked with $\dagger$.

Now consider $F^{w(0)}(\lambda_0^\dagger, \lambda_1^\dagger, \sigma_0^\dagger)$.

$\lambda_0^\dagger$	$\lambda_1^\dagger$	$\sigma_0^\dagger$	$F^{w(0)}(\lambda_0^\dagger, \lambda_1^\dagger, \sigma_0^\dagger)$
$\Delta^{R,0}$	$\Delta^{R,0}$	$R^{R,\beta}_{(1,\beta),\beta}$	$H^{R,\beta}_{(1,w(0)),w(0)}$
$\Delta^{H,0}$	$\Delta^{R,0}$	$H^{H,w(0)}_{(1,\beta),\beta}$	$H^{R,w(0)}_{(1,w(0)),w(0)}$
$\Delta^{L,1}$	$\Delta^{H,0}$	$H^{L,w(1)}_{(1,\beta),\beta}$	$R^{H,w(1)}_{(1,w(0)),w(1)}$
$\Delta^{L,0}$	$\Delta^{L,1}$	$H^{L,w(2)}_{(1,\beta),\beta}$	$H^{L,w(2)}_{(1,w(0)),w(0)}$
$\Delta^{L,1}$	$\Delta^{L,1}$	$H^{L,w(3)}_{(1,\beta),\beta}$	$R^{L,w(3)}_{(1,w(0)),w(3)}$
$\Delta^{L,0}$	$\Delta^{L,0}$	$H^{L,w(4)}_{(1,\beta),\beta}$	$H^{L,w(4)}_{(1,w(0)),w(0)}$
$\vdots$	$\vdots$	$\vdots$	$\vdots$

Following Kozik, we put $\sigma_1^\dagger = F^{w(0)}(\lambda_0^\dagger, \lambda_1^\dagger, \sigma_0^\dagger)$. More generally, by recursion we put

$$\sigma_{k+1}^\dagger = F^{w(k)}(\lambda_k^\dagger, \lambda_{k+1}^\dagger, \sigma_k^\dagger)$$

for all $k < n$. Our modification of Kozik's generators introduces no significant changes in Kozik's analysis. This is a point that we will verify.

Marcin Kozik (2009) proved that the equivalence relation that isolates the $2n + 2$ elements (unmodified)

$$\sigma_0, \lambda_0, \sigma_1, \lambda_1, \ldots, \sigma_n . \lambda_n$$

and collapses all the other elements of B is a congruence of $\mathbf{B}$. As the quotient algebra is isomorphic with $\mathbf{S}$, this establishes that $\mathbf{S} \in \mathsf{HSPE}$.

We reserve $\rho : E^{1+2^n} \to E^{2^n}$ to denote the projection onto the last 2^n coordinates. So we see that $\rho(\sigma_k^\dagger) = \sigma_k$ and $\rho(\lambda_k^\dagger) = \lambda_k$, for all $k \leq n$.

We let $\mathbf{B}^\dagger$ be the subalgebra of $\mathbf{E}^{1+2^n}$ that is generated by our modified

$$\{\sigma_0^\dagger\} \cup \{\lambda_0^\dagger, \ldots, \lambda_n^\dagger\}.$$

We let $\mathbf{B}^*$ be the subalgebra of $(\mathbf{E}^*)^{1+2^n}$ generated by the modified

$$\{\sigma_0^\dagger\} \cup \{\lambda_0^\dagger, \ldots, \lambda_n^\dagger\}.$$

The main difference between $\mathbf{B}^*$ and $\mathbf{B}^\dagger$ is that B^* contains all the constant tuples that are named by the new constant symbols that were added to the expanded signature of $\mathbf{E}^*$. Of course, we also have to account for the behavior of the Pigozzi operations.

For each proper element $d \in E$ we use ε_d to denote that constantly d tuple in E^{1+2^n}.

In (Kozik, 2009), Kozik establishes for the subalgebra $\mathbf{B}$ of $\mathbf{E}^{2^n}$ the following assertions.

(A) $B - \left(\{\sigma_0, \ldots, \sigma_n\} \cup \{\lambda_0, \ldots, \lambda_n\}\right)$ is an absorbing set for $\mathbf{B}$.

(B) The only applications of operations in $\mathbf{B}$ that do not give outputs in this absorbing set are

$$G(\sigma_0) = \sigma_0$$
$$F^{w(k)}(\lambda_k, \lambda_{k+1}, \sigma_k) = \sigma_{k+1} \text{ for all } k < n.$$

(C) Except for σ_0 and λ_0 all the elements of B of full support exhibit each of R, H, and L in their first superscripts. In particular, the only constant 2^n-tuple belonging to B is the tuple that is constantly $\perp$.

(D) There is no element $\gamma \in B$ that also belongs to $\mathcal{X} \cup \widehat{\mathcal{X}}$ so that $\square_{(\blacksquare, \square), \square}^{\square, \square}(\gamma)$ is $\widehat{1}$ on all coordinates.

It follows from (B) that for any $\gamma \in B^\dagger$ and any $k \leq n$

$$\rho(\gamma) = \lambda_k \Rightarrow \gamma = \lambda_k^\dagger, \text{ and}$$
$$\rho(\gamma) = \sigma_k \Rightarrow \gamma = \sigma_k^\dagger.$$

This means that

$$B^\dagger - \left(\{\sigma_0^\dagger, \ldots, \sigma_n^\dagger\} \cup \{\lambda_0^\dagger, \ldots, \lambda_n^\dagger\}\right) = \rho^{-1}\left(B - \left(\{\sigma_0, \ldots, \sigma_n\} \cup \{\lambda_0, \ldots, \lambda_n\}\right)\right).$$

It is straightforward to check that the inverse image of an absorbing set with respect to a homomorphism is again an absorbing set.

This means that for the subalgebra $\mathbf{B}^\dagger$ of $\mathbf{E}^{1+2^n}$ we obtain

($\mathrm{A}^\dagger$) $B^\dagger - \left(\{\sigma_0^\dagger, \ldots, \sigma_n^\dagger\} \cup \{\lambda_0^\dagger, \ldots, \lambda_n^\dagger\}\right)$ is an absorbing set for $\mathbf{B}^\dagger$.

($\mathrm{B}^\dagger$) The only applications of operations in $\mathbf{B}^\dagger$ that do not give outputs in this absorbing set are

$$G(\sigma_0^\dagger) = \sigma_0^\dagger$$
$$F^{w(k)}(\lambda_k^\dagger, \lambda_{k+1}^\dagger, \sigma_k^\dagger) = \sigma_{k+1}^\dagger \text{ for all } k < n.$$

($\mathrm{C}^\dagger$) All the elements of $B^\dagger$ of full support exhibit each of R, H, and L in their first superscripts. In particular, the only constant $1+2^n$-tuple belonging to $B^\dagger$ is the tuple that is constantly $\perp$.

($\mathrm{D}^\dagger$) There is no element $\gamma \in B^\dagger$ that also belongs to $\mathcal{X} \cup \widehat{\mathcal{X}}$ so that $\square_{(\blacksquare, \square), \square}^{\square, \square}(\gamma)$ is $\widehat{1}$ on all coordinates.

Let $B_\perp^*$ consist of those elements of B^* that have $\perp$ on at least one position.

Claim
The set $B^\dagger \cup \{\varepsilon_d \mid d \in E - \{\perp\}\} \cup B_\perp^*$ is a subuniverse of $(\mathbf{E}^*)^{1+2^n}$.

Proof. The set $\{\varepsilon_d \mid d \in E\}$ is certainly a subuniverse and the set

$$B^\dagger \cup B^*_\perp$$

is closed with respect to all the operations of $\mathbf{E}^{1+2^n}$. So we need only consider what happens with operations of rank at least 2 with at least one input from $\{\varepsilon_d \mid d \in E\}$ and at least one input from $B^\dagger$. By considering the proper domains of the tape building operations, the computing operations, and the checking operations, in this setting the outputs always belong to $B^*_\perp$. In more detail, for the tape building operations the fact that each of R, H, and L occur in the first superscripts of any element of B^* of full support must disrupt the constraint dealing with $\prec$ as long as some input is a constant tulple and some other input is in $B^\dagger$ and has full support. [The absence of R has a first superscript in σ_0 and λ_0 was the reason for the introduction of a new coordinate.] The disruption in the proper domains for the computing and checking operations is even more straightforward.

This leaves the operation Ω and the Pigozzi operations. But these operations either produce a member of $B^*_\perp$ or else return their second input. $\qquad\square$

Now let θ be the equivalence relation that isolates each element of

$$\{\varepsilon_d \mid d \in E - \{\perp\}\} \cup \{\sigma^\dagger, \ldots, \sigma^\dagger_n\} \cup \{\lambda^\dagger_0, \ldots . \lambda^\dagger_n\}$$

and collapses all the other elements of B^* into a single θ-class. We claim that θ is a congruence of $\mathbf{B}^*$.

Let us first consider the Pigozzi operation Q_d. Suppose that $\gamma \, \theta \, \gamma'$ and $\delta \, \theta \, \delta'$. If these are both equalities, then $Q_d(\gamma, \delta) \, \theta \, Q_d(\gamma', \delta')$ is clear. If γ and γ' belong to the big block, then neither pair of inputs belongs to the proper domain of Q_d and our result follows. If δ and δ' belong to the big block, then both outputs belong to the big block.

Next consider the operation Ω. Observe that $\Omega(\gamma, \delta) = \delta$ when $\gamma = \varepsilon_d$ with $\Box^{\Box,\Box}_{(\blacksquare,\Box),\Box}(d) = \widehat{1}$ and otherwise the output belongs to the big block, by $\mathrm{D}^\dagger$. So θ respects Ω.

For the remaining operations, we are back in the signature of $\mathbf{E}$, but without Ω. Notice that mixed inputs (some from $B^\dagger$ and some not) always produce outputs in the big block. This means that such instances cannot prevent θ from being a congruence relation. On the other hand, the restriction of θ to the subuniverse $\{\varepsilon_d \mid d \in E\}$ is just the identity relation, while the restriction of θ to $B^\dagger$ isolates the elements $\sigma^\dagger_0, \ldots, \sigma^\dagger_n$ and $\lambda^\dagger_0, \ldots, \lambda^\dagger_n$ while identifying the remaining elements. This big block is an absorbing set. So this restriction of θ is also a congruence.

So θ is a congruence of $\mathbf{B}^*$. But $\mathbf{B}^*/\theta$ is evidently isomorphic to $\mathbf{S}^*$. Therefore $\mathbf{S}^* \in \mathsf{HSP}\,\mathbf{E}^*$, completing the proof of the Main Contention and so the proof of the theorem. $\qquad\square$

References

Bergman, C. and Slutzki, G. (2000). Complexity of some problems concerning varieties and quasi-varieties of algebras, *SIAM Journal on Computing,* 30(2), 359–382.

Bergman, C. and Slutzki, G. (2002a). Computational complexity of generators and nongenerators in algebra, *International Journal of Algebra and Computation,* 12(5), 719–735.

Bergman, C. and Slutzki, G. (2002b). Computational complexity of some problems involving congruences on algebras, *Theoretical Computer Science,* 270(1–2), 591–608.

Chandra, A. K., Kozen, D. C. and Stockmeyer, L. J. (1981). Alternation, *Journal of the Association for Computing Machinery,* 28(1), 114–133.

Delić, D. (2001). Finite bases for flat graph algebras, *Journal of Algebra,* 246(1), 453–469.

Freese, R. and Valeriote, M. A. (2009). On the complexity of some Maltsev conditions, *International Journal of Algebra and Computation,* 19(1), 41–77.

Jackson, M. and McKenzie, R. (2006). Interpreting graph colorability in finite semigroups, *International Journal of Algebra and Computation,* 16(1), 119–140.

Kaarli, K. and Pixley, A. (2002). The decidability of the affine completeness generation problem, *Algebra Universalis,* 48(1), 129–131.

Kalicki, J. (1952). On comparison of finite algebras, *Proceedings of the American Mathematical Society,* 3, 36–40.

Kearnes, K. A. and Szendrei, Á. (1997). A characterization of minimal locally finite varieties, *Proceedings of the American Mathematical Society,* 349(5), 1749–1768.

Kozik, M. (2007). Computationally and algebraically complex finite algebra membership problems, *Transactions of the American Mathematical Society,* 17(8), 1635–1666.

Kozik, M. (2009). A 2EXPTIME complete varietal membership problem, *SIAM Journal on Computing,* 38(6), 2443–2467.

Lampe, W. A., McNulty, G. F. and Willard, R. (2001). Full duality among graph algebras and flat graph algebras. *Algebra Universalis,* 45(2-3), 311–334, Conference on Lattices and Universal Algebra (Szeged, 1998).

McKenzie, R. (1996a). The residual bound of a finite algebra is not computable, *International Journal of Algebra and Computation,* 6(1), 29–48.

McKenzie, R. (1996b). The residual bounds of finite algebras. *International Journal of Algebra and Computation,* 6(1), 1–28.

McKenzie, R. (1996c). Tarski's finite basis problem is undecidable, *International Journal of Algebra and Computation,* 6(1), 49–104.

McNulty, G. F. and Shallon, C. R. (1983). Inherently nonfinitely based finite algebras, In *Universal algebra and lattice theory (Puebla, 1982),* Volume 1004 of *Lecture Notes in Math.,* pp. 206–231. Berlin: Springer.

Pigozzi, D. (1979). Minimal, locally-finite varieties that are not finitely axiomatizable, *Algebra Universalis,* 9(3), 374–390.

Scott, D. (1956). Equationally complete extensions of finite algebras, *Nederl. Akad. Wetensch. Proc. Ser. A.* **59** = *Indagationes Mathematicae,* 18, 35–38.

Shallon, C. R. (1979). *Non-Finitely based binary algebras derived from lattices,* Ph. D. thesis, University of California, Los Angeles.

Székely, Z. (1998). *Complexity of the finite algebra membership problem for varieties,* Ph. D. thesis, University of South Carolina.

Székely, Z. (2002). Computational complexity of the finite algebra membership problem for varieties, *International Journal of Algebra and Computation,* 12(6), 811–823.

Szendrei, Á. (1993). Non-finitely based finite groupoids generating minimal varieties, *Acta Scientiarum Mathematicarum (Szeged),* 57(1–4), 593–600.

Wheeler, W. H. (1979). The first order theory of N-colorable graphs, *Transactions of the American Mathematical Society,* 250, 289–310.

Characterization of protoalgebraic k-deductive systems

Katarzyna Pałasińska

*Dedicated to Professor Don Pigozzi
on his 80th Birthday.*

Abstract A sentential logic is protoalgebraic iff it has a finite system of equivalence formulas (Blok and Pigozzi, 1986). This can be generalized to the context of universal Horn logic without equality, (Blok and Pigozzi, 1992). In this paper we revise this characterization.

2010 Mathematics Subject Classification: 03G27, 03B22

Key words: deductive system, protoalgebraic system,, congruence, system of equivalence formulas, system of congruence formulas

1 Introduction

Among many contributions of Don Pigozzi to universal algebra and logic, the concepts of *protoalgebraic* and *algebraizable logics*, introduced in his joint research with Willem J. Blok (1986; 1989), are of particular importance. In their fundamental paper, (Blok and Pigozzi, 1992), the authors generalize these concepts to universal Horn logic with one predicate that need not be equality. If the arity of the predicate is k, then a strict universal Horn theory in such logic is called a k-*deductive system* (Blok and Pigozzi, 1992). If $k = 1$ and the predicate is interpreted as "truth", then a k-deductive system becomes a sentential logic, understood as a system of axioms and rules. On the other hand, the quasi-equational logic of a quasivariety of algebras is an

Katarzyna Pałasińska
Cracow University of Technology Faculty of Physics, Mathematics and Computer Science, Cracow, Poland, e-mail: `kpalasinska@gmail.com`

© Springer International Publishing AG 2018
J. Czelakowski (eds.), *Don Pigozzi on Abstract Algebraic Logic,
Universal Algebra, and Computer Science*, Outstanding Contributions
to Logic 16, https://doi.org/10.1007/978-3-319-74772-9_10

example of a 2-deductive system with the binary predicate interpreted as a congruence, or as equality in the quotient structure. In (Blok and Pigozzi, 1992), the semantics of k-deductive systems is studied along the lines of universal algebra. The concepts of protoalgebraic and algebraizable k-deductive systems lead to considering the whole hierarchy of properties between these two; the stronger the condition on a k-deductive system in this hierarchy, the more its semantics resembles that of a quasivariety.

The authors thoroughly explain the roots of their research, carefully acknowledging the work of others before them. In return, their work in (Blok and Pigozzi, 1986, 1989, 1992) has opened a new chapter in the area of algebraic logic, motivating a lot of subsequent research, see (Czelakowski, 2001; Font, 2016) for recent overviews.

The protoalgebraic property of a deductive system is a Malcev-style condition: a 1-deductive system is protoalgebraic if and only if it has a system of binary *equivalence formulas*, (Blok and Pigozzi, 1986). The aim of Theorem 13.2 in (Blok and Pigozzi, 1992) is to generalize this fact to k-deductive systems. This characterization requires a small modification. It is stated there that a k-deductive system is protoalgebraic iff there exists a finite system of binary equivalence k-formulas. It should be noticed, however, that in case of $k > 1$, the equivalence formulas should not be claimed to be binary, as justified by Example 3 below, see also (Pynko, 1999). We indicate the place in the original proof where a mistake has crept in and state a corrected version, Theorem 6. This has been noticed in (Pałasińska, 1994) (published as (Pałasińska, 2003, Theorem 5.12)), independently in (Elgueta and Jansana, 1999) and (Pynko, 1999), with (Elgueta and Jansana, 1999, Theorem 6) being more general than Theorem 6. A natural definition of a congruence system of k-formulas with $k - 1$ parameters is also proposed and the Leibniz operator in protoalgebraic k-deductive systems is characterized with their use. We give an example of a k-deductive systems with a finite system of congruence k-formulas with parameters that does not have such a system without parameters (Example 15) and we correct the proof of (Blok and Pigozzi, 1992, Theorem 13.10). The paper (Blok and Pigozzi, 1992) is a thorough exposition of an important topic with a lasting potential to motivate further research on k-deductive systems. We believe that it is worthwhile to discuss explicitly these fragments of (Blok and Pigozzi, 1992, section 13).

2 Preliminaries

Let Λ be a *propositional language* i.e., a finite set of symbols with an associated *arity* function ρ assigning a natural number to each symbol. A Λ-algebra is a pair $\mathbf{A} = \langle A, \Lambda^{\mathbf{A}} \rangle$, where A is a nonempty set and $\Lambda^{\mathbf{A}} := \{ \lambda^{\mathbf{A}} : \lambda \in \Lambda \}$ is a set of operations on A, one for each symbol in Λ, with $\lambda^{\mathbf{A}}$ being a $\rho(\lambda)$-ary operation on A. Let V be an infinite countable set of variables, disjoint with

Λ. Assume that $x, y, z, y_1, y_2, \ldots, x_1, x_2 \ldots, z_1, \ldots, z_k$ are in V. Notice that we have listed infinitely many variables x_i's and y_i's and finitely many z_i's. The symbols u, v and w, with indices, will be meta-variables ranging over V. By Te we mean the set of all terms in the language Λ over the variables V and by $\mathrm{Te}(x, y_1, \ldots, y_n)$ the set of those terms in which no other variable than possibly $x, y_1, \ldots, y_n$ occurs. Terms involving only two variables are called *binary*. The symbol **Te** denotes the algebra of terms defined on the set Te in the usual way. Terms are also called *formulas* and for a natural number k, a *k-term*, or a *k-formula*, is a k-tuple of terms. Boldface symbols $\mathbf{t}, \mathbf{s}, \mathbf{r}$ are used to denote k-terms, with the convention that $\mathbf{t} = \langle t_1, \ldots, t_k \rangle$, similarly for $\mathbf{s}$ and $\mathbf{r}$. We will often use a special sequence of $k - 1$ variables $\langle z_1, \ldots, z_{k-1} \rangle$ that will also be denoted by $\mathbf{z}$ (even though it is a $(k-1)$-term and not a k-term). As usual, an operator C on the power set of k-terms, $\mathrm{C} : \mathcal{P}(\mathrm{Te}^k) \longrightarrow \mathcal{P}(\mathrm{Te}^k)$ that is increasing, monotone and idempotent is called a closure operator. It is a *consequence operator* if it is in addition finitary and structural, i.e.,

$$\mathrm{C}(X) = \bigcup \{ \mathrm{C}(Y) : Y \subseteq X, Y \text{ is finite} \}$$

and

$$\sigma(\mathrm{C}(X)) \subseteq \mathrm{C}(\sigma(X)),$$

for every set $X \subseteq \mathrm{Te}^k$ and every endomorphism σ on the algebra **Te**, with $\sigma(X)$ defined as the set resulting from application of σ to each term in X.

Following (Blok and Pigozzi, 1992), by a *k-deductive system* we mean a pair $\mathcal{S} = \langle \Lambda, \mathrm{C} \rangle$ such that Λ is a propositional language and C is a consequence operator on the power set of k-terms for this language. We write $\mathrm{Cn}_{\mathcal{S}}$ for C in this case. With each k-deductive system there is associated a *consequence relation* $\vdash_{\mathcal{S}}$:

$$X \vdash_{\mathcal{S}} \mathbf{t} \qquad \text{iff} \qquad \mathbf{t} \in \mathrm{Cn}_{\mathcal{S}}(X),$$

where $X \cup \{\mathbf{t}\} \subseteq \mathrm{Te}^k$. The subscript $\mathcal{S}$ is omitted, if known from the context. Each set of the form $\mathrm{Cn}(X)$ is a *theory* of $\mathcal{S}$. Equivalently, theories are sets of k-terms closed under Cn. The members of $\mathrm{Cn}(\emptyset)$ are called *theorems of $\mathcal{S}$*. A rule of $\mathcal{S}$ is a pair $\langle X, \mathbf{t} \rangle$ such that $X \cup \{\mathbf{t}\}$ is a finite subset of Te^k and $X \vdash_{\mathcal{S}} \mathbf{t}$. Conversely, a given set of k-terms (*axioms*) and a set of rules, determine a consequence operator and therefore a k-deductive system.

A *k-matrix* is a pair $\mathfrak{M} = \langle \mathbf{M}, R \rangle$, where $\mathbf{M} = \langle M, \Lambda^{\mathbf{M}} \rangle$ is a Λ-algebra and R is a k-ary relation on M. A *model* of a k-deductive system $\mathcal{S}$ is a k-matrix $\mathfrak{M} = \langle \mathbf{M}, R \rangle$, where the relation R is closed under all axioms and rules of $\mathcal{S}$. In such case R is called an *$\mathcal{S}$-filter* on $\mathbf{M}$ and $\mathfrak{M}$ is called an *$\mathcal{S}$-matrix*.

A congruence θ on $\mathbf{M}$ is *compatible* with the k-ary relation R on M if for all $a_1, b_1, \ldots, a_k, b_k \in M$, the conditions $(a_1, b_1), \ldots, (a_k, b_k) \in \theta$ and $(a_1, \ldots, a_k) \in R$ imply $(b_1, \ldots, b_k) \in R$. The largest congruence on $\mathbf{M}$ compatible with R always exists; it is denoted by $\Omega_{\mathbf{M}}(R)$ and called the *Leibniz congruence* of R on $\mathbf{M}$; the subscript $\mathbf{M}$ is often omitted. The Leibniz congruence $\Omega_{\mathbf{M}}(R)$ can also be characterized as follows. A pair $(a, b) \in M^2$ is in

$\Omega(R)$ iff for every n, every $\mathbf{t} \in \mathrm{Te}(x, y_1, \ldots, y_n)$ and $\bar{c} \in M^n$,

$$\mathbf{t}^{\mathbf{M}}(a, \bar{c}) \in R \implies \mathbf{t}^{\mathbf{M}}(b, \bar{c}).$$

The operator $\Omega_{\mathbf{M}}$ assigning to each k-ary relation R on M its Leibniiz congruence, is called the *Leibniz operator* on $\mathbf{M}$. Let $\mathcal{S}$ be a k-deductive system and let $\mathbf{M}$ be a Λ-algebra. The restriction of $\Omega_{\mathbf{M}}$ to $\mathcal{S}$-filters is denoted by $\Omega_{\mathbf{M}}^{\mathcal{S}}$. Thus $\Omega_{\mathbf{M}}^{\mathcal{S}}$ assigns to each $\mathcal{S}$-filter R on $\mathbf{M}$ the Leibniz congruence of R on $\mathbf{M}$.

3 Protoalgebraic k-deductive systems

Definition 1. (Blok and Pigozzi, 1992, Definition 7.1.) A k-deductive system $\mathcal{S}$ is protoalgebraic if for each Λ-algebra $\mathbf{M}$ the operator $\Omega_{\mathbf{M}}^{\mathcal{S}}$ is monotone, i.e., for every two $\mathcal{S}$-filters R and S on $\mathbf{M}$,

$$R \subseteq S \implies \Omega_{\mathbf{M}}^{\mathcal{S}}(R) \subseteq \Omega_{\mathbf{M}}^{\mathcal{S}}(S).$$

The property of being "protoalgebraic" is a Malcev type condition. Specifically, a 1-deductive system is protoalgebraic iff it has a finite system of equivalence formulas (Blok and Pigozzi, 1986). A *finite system of equivalence k-formulas* was defined in (Blok and Pigozzi, 1992) as follows.

Definition 2. (Blok and Pigozzi, 1992, Definition 13.1.) By a finite system of equivalence k-formulas for $\mathcal{S}$ we mean a sequence $\Delta_1, \ldots, \Delta_n$ of binary k-formulas, with $n \in \mathbf{N}$, $\Delta_j = \Delta_j(x, y)$, such that

$$(1) \qquad \vdash_{\mathcal{S}} \Delta_j(x, x), \text{ for all } j = 1, \ldots, n \qquad \text{and}$$

$$(2) \qquad \begin{array}{c} \langle z_1, \ldots, z_{i-1}, x, z_{i+1}, \ldots, z_k \rangle, \ \Delta_1(x, y), \ldots, \Delta_n(x, y) \\ \vdash_{\mathcal{S}} \langle z_1, \ldots, z_{i-1}, y, z_{i+1}, \ldots, z_k \rangle, \end{array}$$

for all $i = 1, \ldots, k$.

Since in the deduction the order of premisses does not matter, we identify the system of equivalence k-formulas with the set $\mathbf{\Delta} = \{\Delta_1, \ldots, \Delta_n\}$. So $\mathbf{\Delta}(x, y)$ abbreviates $\{\Delta_1(x, y), \ldots, \Delta_n(x, y)\}$ and hence (1) and (2) can be written equivalently as

$$(3) \qquad \vdash_{\mathcal{S}} \mathbf{\Delta}(x, x) \qquad \text{and}$$

$$(4) \qquad \langle z_1, \ldots, z_{i-1}, x, z_{i+1}, \ldots, z_k \rangle, \ \mathbf{\Delta}(x, y) \vdash_{\mathcal{S}} \langle z_1, \ldots, z_{i-1}, y, z_{i+1}, \ldots, z_k \rangle,$$

for all $i = 1,\ldots,k$. It follows from Definition 2 that for $k = 1$, $\boldsymbol{\Delta}$ is a finite system of equivalence k-formulas for a 1-deductive system $\mathcal{S}$ iff

$$(5) \qquad \vdash_{\mathcal{S}} \boldsymbol{\Delta}(x,x) \qquad \text{and}$$

$$(6) \qquad x, \boldsymbol{\Delta}(x,y) \vdash_{\mathcal{S}} y$$

The characterization of protoalgebraic 1-deductive systems in the style of Malcev is a special case of the following statement (Blok and Pigozzi, 1992, Theorem 13.2).

A k-deductive system $\mathcal{S}$ is protoalgebraic if and only if it has a finite system of equivalence k-formulas.

Therefore a 1-deductive system $\mathcal{S}$ is protoalgebraic iff there is a finite system $\boldsymbol{\Delta}(x,y)$ of binary formulas such that

$$(7) \qquad \vdash_{\mathcal{S}} \boldsymbol{\Delta}(x,y) \qquad \text{and}$$

$$(8) \qquad x, \boldsymbol{\Delta}(x,y) \vdash_{\mathcal{S}} y.$$

This characterization of 1-deductive protoalgebraic systems has proved to be fundamental in abstract algebraic logic. As already mentioned, for k greater than 1, it requires a correction.

4 Example

Consider the following example of a 2-deductive system.

Example 3. (Pałasińska, 1994) *Let $\Lambda = \{\cdot\}$ with $\rho(\cdot) = 2$. The operation symbol $\cdot$ is omitted when writing terms. Let $\mathcal{S}$ be the 2-deductive system over Λ defined by the following axiom and rules.*

$$(9) \qquad\qquad\qquad \vdash \langle x,x \rangle$$
$$(10) \qquad \langle xz,yz \rangle, \langle x,z \rangle \vdash \langle y,z \rangle$$
$$(11) \qquad \langle xz,yz \rangle, \langle z,x \rangle \vdash \langle z,y \rangle.$$

We first show that this 2-deductive system does not have a finite system of binary equivalence formulas.

Proposition 4. *There is no finite set $\boldsymbol{\Delta}(x,y)$ of binary 2-formulas, such that (3) and (4) hold for the 2-deductive system $\mathcal{S}$ of Example 3.*

Proof. For the proof by contradiction assume that $\mathbf{\Delta} = \mathbf{\Delta}(x,y)$ is a finite system of equivalence 2-formulas for $\mathcal{S}$. Then

$$(12) \qquad \mathbf{\Delta} \cup \{\langle x,z \rangle\} \vdash_{\mathcal{S}} \langle y,z \rangle.$$

Notice that

$$(13) \qquad \mathbf{\Delta} \subseteq \mathrm{Cn}_{\mathcal{S}}\mathbf{\Delta} \cap \mathbf{Te}^2(x,y)$$

and that

$$(14) \qquad \{\langle t,t \rangle : t \in \mathbf{Te}\} \subseteq \mathrm{Cn}_{\mathcal{S}}(\emptyset).$$

Consider the $\mathcal{S}$-theory T generated by $\mathbf{\Delta} \cup \{\langle x,z \rangle\}$. By definition

$$(15) \qquad T = \mathrm{Cn}_{\mathcal{S}}\left(\mathbf{\Delta} \cup \{\langle x,z \rangle\}\right).$$

We claim that

$$(16) \qquad T = \left(\mathrm{Cn}_{\mathcal{S}}\mathbf{\Delta} \cap \mathbf{Te}^2(x,y)\right) \cup \{\langle x,z \rangle\} \cup \{\langle t,t \rangle : t \in \mathbf{Te}\}.$$

Let $\tilde{T}$ be the right hand side of (16). By (14) and (15) $\tilde{T} \subseteq T$.

Recall that by assumption, $\mathbf{\Delta} \subseteq \mathrm{Te}^2(x,y)$, so

$$\mathbf{\Delta} \cup \{\langle x,z \rangle\} \subseteq \left(\mathrm{Cn}_{\mathcal{S}}\mathbf{\Delta} \cap \mathbf{Te}^2(x,y)\right) \cup \{\langle x,z \rangle\} \subseteq \tilde{T}.$$

Therefore $T \subseteq \mathrm{Cn}_{\mathcal{S}}\tilde{T}$. To finish the proof of (16) it remains to show that $\tilde{T}$ is an $\mathcal{S}$-theory. By definition, $\tilde{T}$ is closed under the axiom (9). To show the closure of $\tilde{T}$ under (10), let t,r,s be arbitrary terms and assume that

$$(17) \qquad \langle ts, rs \rangle \in \tilde{T}$$

and

$$(18) \qquad \langle t,s \rangle \in \tilde{T}.$$

We want to show that

$$(19) \qquad \langle r,s \rangle \in \tilde{T}.$$

If $t = r$ there is nothing to prove. Also, $\langle ts, rs \rangle$ cannot be $\langle x,z \rangle$. So it is enough to assume that

$$(20) \qquad \langle ts, rs \rangle \in (\mathrm{Cn}_{\mathcal{S}}\mathbf{\Delta}) \cap \mathrm{Te}^2(x,y).$$

It follows that

$$(21) \qquad t,r,s \in \mathrm{Te}(x,y).$$

By (18) and(21)

$$(22) \qquad \langle t, s \rangle \in \tilde{T} \cap \mathrm{Te}^2(x, y).$$

Now by (14)

$$\tilde{T} \cap \mathrm{Te}^2(x,y) \subseteq (\mathrm{Cn}_{\mathcal{S}} \mathbf{\Delta} \cap \mathbf{Te}^2(x,y)) \cup (\mathrm{Cn}_{\mathcal{S}} \emptyset \cap \mathbf{Te}^2(x,y))$$
$$\subseteq (\mathrm{Cn}_{\mathcal{S}} \mathbf{\Delta} \cap \mathbf{Te}^2(x,y)).$$

So by (22)

$$\langle t, s \rangle \in \mathrm{Cn}_{\mathcal{S}}(\mathbf{\Delta}) \cap \mathbf{Te}^2(x,y).$$

We have by (20)

$$\langle ts, rs \rangle \in \mathrm{Cn}_{\mathcal{S}}(\mathbf{\Delta})$$

and

$$\langle t, s \rangle \in \mathrm{Cn}_{\mathcal{S}}(\mathbf{\Delta}).$$

It follows by (10) that

$$\langle r, s \rangle \in \mathrm{Cn}_{\mathcal{S}}(\mathbf{\Delta})$$

and by (21) that

$$\langle r, s \rangle \in \mathrm{Cn}_{\mathcal{S}}(\mathbf{\Delta}) \cap \mathbf{Te}^2(x,y) \subseteq \tilde{T}.$$

This finishes the proof of (19) and the proof of the closure of $\tilde{T}$ under (10). Closure under (11) is proved similarly. This establishes (16).

By (16), the 2-formula $\langle y, z \rangle$ is not a member of T. This contradicts (12).

The assumption that $\mathbf{\Delta}$ is finite does not play any role in the above argument. This is not accidental, because the existence of any set $\mathbf{\Delta}$ satisfying conditions (10) and (2) of Definition 2 implies the existence of a finite subset with these properties. In spite of not having a finite system of binary 2-formulas, the 2-deductive system $\mathcal{S}$ of Example 3 is protoalgebraic. This will be stated as Corollary 7 in the next section.

5 Characterization of protoalgebraic k-deductive systems

Theorem 6 below characterizes protoalgebraic k-deductive systems by the existence of a finite system of $k + 1$-ary equivalence k-formulas with $k - 1$ parameters. The proof we give is a modification of that of Theorem 13.2 in (Blok and Pigozzi, 1992); similar ones occur in the literature for 1-deductive systemss (e.g., (Herrmann, 1997; Czelakowski, 2001)) and for *atomic theories* of first order structures in (Elgueta and Jansana, 1999).

5.1 Equivalence k-formulas with parameters

Definition 5. Let $\mathcal{S}$ be a k-deductive system. By a system of $k+1$-ary equivalence k-formulas with $k-1$ parameters for $\mathcal{S}$ we mean a sequence $\boldsymbol{\Delta} = \langle \Delta_1, \ldots, \Delta_n \rangle$ of k-formulas, $\Delta_i = \Delta_i(x, y, z_1, \ldots, z_{k-1})$ such that

$$(23) \qquad \vdash_{\mathcal{S}} \Delta_i(x, x, z_1, \ldots, z_{k-1}), \ \text{ for all } i = 1, \ldots, n$$

$$
\langle z_1, \ldots, z_{i-1}, x, z_i, \ldots, z_{k-1} \rangle,
$$
$$
\Delta_1(x, y, z_1, \ldots, z_{k-1}), \ldots, \Delta_n(x, y, z_1, \ldots, z_{k-1})
$$
$$(24) \qquad \vdash_{\mathcal{S}} \langle z_1, \ldots, z_{i-1}, y, z_i, \ldots, z_{k-1} \rangle
$$

for all $i = 1, \ldots, k$.

Recall that the sequence $\langle z_1, \ldots, z_{k-1} \rangle$ is collectively denoted as $\mathbf{z}$.

Theorem 6. *A k-deductive system $\mathcal{S}$ is protoalgebraic if and only if it has a finite system of $k+1$-ary equivalence k-formulas with $k-1$ parameters.*

Proof. Assume that there is a system $\boldsymbol{\Delta}(x, y, z_1, \ldots, z_{k-1})$ of $k+1$-ary equivalence k-formulas. Let $\mathbf{M}$ be a Λ-algebra and let $F \subseteq G$ be two $\mathcal{S}$-filters on $\mathbf{M}$. We want to show that $\Omega(F)$ is compatible with G.

Suppose that $(a, b) \in \Omega(F)$ and that $\langle c_1, \ldots, c_{i-1}, a, c_i, \ldots, c_{k-1} \rangle \in G$, for some $a, b, c_1, \ldots, c_{k-1} \in A$. By (23), $\boldsymbol{\Delta}(a, a, c_1, \ldots, c_{k-1}) \subseteq F$ and since $(a, b) \in \Omega(F)$, also $\boldsymbol{\Delta}(a, b, c_1, \ldots, c_{k-1}) \in F \subseteq G$. By (24), $\langle c_1, \ldots, c_{i-1}, b, c_i, \ldots, c_{k-1} \rangle \in G$, finishing the proof that $\Omega(F)$ is compatible with G. By the definition of $\Omega(G)$ as the largest congruence on $\mathbf{M}$ compatible with G, we get the inclusion $\Omega(F) \subseteq \Omega(G)$, which finishes the proof of the "if" direction.

For the "only if" direction, assume that $\mathcal{S}$ is protoalgebraic. Consider the term algebra $\mathbf{Te}(x, y, z_1, \ldots, z_{k-1})$. Let T consist of all k-terms $\mathbf{t}(x, y, \mathbf{z})$ such that $\vdash_{\mathcal{S}} \mathbf{t}(x, x, \mathbf{z})$. Notice that then T is an $\mathcal{S}$-filter on $\mathbf{Te}(x, y, z_1, \ldots, z_{k-1})$. We first show that $(x, y) \in \Omega(T)$.

Assume that $\mathbf{s}(v, x, y, \mathbf{z})$ is a k-term such that $\mathbf{s}(x, x, y, \mathbf{z}) \in T$. By the definition of T, $\mathbf{s}(x, x, x, \mathbf{z}) \in \mathrm{Cn}_{\mathcal{S}}(\emptyset)$. Let $\hat{\mathbf{s}}(x, y, \mathbf{z}) := \mathbf{s}(y, x, y, \mathbf{z})$. Then $\hat{\mathbf{s}}(x, x, \mathbf{z}) := \mathbf{s}(x, x, x, \mathbf{z}) \in \mathrm{Cn}_{\mathcal{S}}(\emptyset)$. Hence $\hat{\mathbf{s}}(x, y, \mathbf{z}) \in T$, i.e., $\mathbf{s}(y, x, y, \mathbf{z}) \in T$. So indeed, $(x, y) \in \Omega(T)$.

Consider $i = 1, \ldots, k$ and let

$$
S_i := \mathrm{Cn}_{\mathcal{S}}\left(T \cup \{\langle z_1, \ldots, z_{i-1}, x, z_i, \ldots, z_{k-1} \rangle\}\right).
$$

Since $(x, y) \in \Omega(T)$ and $\mathcal{S}$ is protoalgebraic, $(x, y) \in \Omega(S_i)$. It follows that

$$
\langle z_1, \ldots, z_{i-1}, y, z_i, \ldots, z_{k-1} \rangle \in S_i.
$$

Therefore there exists a finite set $\boldsymbol{\Delta}^{(i)}(x, y, z_1, \ldots, z_{k-1}) \subseteq T$ such that

$$\langle z_1,\ldots,z_{i-1},x,z_i,\ldots,z_{k-1}\rangle, \mathbf{\Delta}^{(i)}(x,y,z_1,\ldots,z_{k-1})$$
$$(25) \qquad\qquad \vdash \langle z_1,\ldots,z_{i-1},y,z_i,\ldots,z_{k-1}\rangle.$$

Taking the union $\mathbf{\Delta} := \bigcup_{i=1}^{k} \mathbf{\Delta}^{(i)}$ we obtain a system satisfying conditions (23) and (24).

For $k=1$ this gives the well-known characterization of protoalgebraic logics by the existence of a finite set of binary equivalence formulas without parameters.

The following corollary finishes the discussion of Example 3.

Corollary 7. *The 2-deductive system of Example 3 is protoalgebraic.*

Proof. Let $\mathbf{\Delta}(x,y,z)$ be the set $\{\langle xz,yz\rangle\}$. By (10) and (11) $\mathbf{\Delta}(x,y,z)$ is a finite system of ternary equivalence formulas for $\mathcal{S}$. By Theorem 6, $\mathcal{S}$ is protoalgebraic.

Thus the *protoalgebraic* property of k-deductive systems cannot be characterized by the existence of a finite system of equivalence k-formulas without parameters. An alternative example can be found in (Pynko, 1999, section 5).

5.2 A comment of the original argument

The argument in (Blok and Pigozzi, 1992) makes use of a certain theory T on the term algebra. The term algebra is generated by the full set V of variables, but the theory T itself is generated by k-terms that do not involve variables $z_1,\ldots,z_k$ and have some additional property. Specifically,

$$T = \{\mathbf{t}(x,y,u_1,\ldots,u_m): \quad \vdash \mathbf{t}(x,x,u_1,\ldots,u_m)\},$$

with m ranging over natural numbers and u_i being arbitrary variables other than $x,y,z_1,\ldots,z_{k-1}$. At a crucial stage of the proof it is assumed that $\langle x,y\rangle \notin \Omega(T)$, with the aim of deriving a contradiction. It follows from this assumption that there is a k-formula $\mathbf{s} = \mathbf{s}(v,x,y,u_1,\ldots,u_m,z_1,\ldots,z_k)$ such that

$$T \vdash \mathbf{s}(x,x,y,u_1,\ldots,u_m,z_1,\ldots,z_k)$$

while

$$T \nvdash \mathbf{s}(y,x,y,u_1,\ldots,u_m,z_1,\ldots,z_k).$$

Consider the k-formula

$$\widetilde{\mathbf{s}}(v,x,y,u_1,\ldots,u_m)$$

resulting from $s(v,x,y,u_1,\ldots,u_m,\mathbf{z})$ by replacing $\mathbf{z}$'s by the variables $u_1,\ldots,u_m$ disjoint with them. By structurality,

$$T \vdash_{\mathcal{S}} \widetilde{\mathbf{s}}(x,x,y,u_1,\ldots,u_m).$$

It does not follow, however, that $T \nvdash \widetilde{\mathbf{s}}(y,x,y,u_1,\ldots,u_m)$, for the implication

$$T \nvdash \mathbf{s}(y,x,y,u_1,\ldots,u_m,z_1,\ldots,z_k) \Rightarrow T \nvdash \widetilde{\mathbf{s}}(y,x,y,u_1,\ldots,u_m),$$

does not hold in general. The authors' claim that without the loss of generality $\mathbf{s}$ may be assumed to be free from variables $\mathbf{z}$, cannot be justified.

As an illustration, consider again the 2-deductive system $\mathcal{S}$ of Example 3 and let $\mathbf{s}(v,x,y,z) = \langle v,z \rangle$. Then $\mathbf{s}(y,x,y,z) = \langle y,z \rangle \notin T$. Replacing z with y we get $\widetilde{\mathbf{s}}(v,y) = \langle v,y \rangle$. Then $T \vdash \widetilde{\mathbf{s}}(y,y)$.

6 Congruence formulas

The proofs of some results in (Blok and Pigozzi, 1992, Section 13) depending on Theorem 13.2. need tiny changes to make the application of this theorem right; such a change is obvious for Corollary 13.3. and Theorem 13.13. Below we discuss Theorems 13.5 and 13.10, and propose a notion of a system of congruence formulas with $k-1$ parameters $\mathbf{z}$, as a possible additional one to consider in addition to those of a *finite system of congruence formulas* and an *infinite system of congruence formulas with parameters* used in these theorems.

Definition 8. Let $\mathcal{S}$ be a k-deductive system and let $\boldsymbol{\Delta}(x,y,\mathbf{z})$ be a finite system of $k+1$-ary equivalence k-formulas with $k-1$ parameters $\mathbf{z}$. Then $\boldsymbol{\Delta}(x,y,\mathbf{z})$ is called a finite system of congruence formulas with $k-1$ parameters $\mathbf{z}$ if in addition

$$(26) \qquad \bigcup_{i=1}^{\rho(\Lambda)} \boldsymbol{\Delta}(x_i,y_i,\mathbf{z}) \vdash \boldsymbol{\Delta}\left(\Lambda(x_1,\ldots,x_{\rho(\Lambda)}),\Lambda(y_1,\ldots,y_{\rho(\Lambda)},\mathbf{z}\right).$$

If $\boldsymbol{\Delta}(x,y,\mathbf{z}) = \boldsymbol{\Delta}(x,y)$ does not depend on $\mathbf{z}$, then we call it a *finite system of congruence k-formulas*, following (Blok and Pigozzi, 1992, Definition 13.4)

Theorem 9. (Compare with (Blok and Pigozzi, 1992, Theorem 13.5)) *Let $\mathcal{S}$ be a protoalgebraic k-deductive system with a finite system of $k+1$-ary equivalence k-formulas $\boldsymbol{\Delta}(x,y,\mathbf{z})$ with $k-1$ parameters $\mathbf{z}$. Then for every Λ-algebra $\mathbf{M}$, every pair $a,b \in M$ and every $\mathcal{S}$-filter F on $\mathbf{M}$, we have*

$$(27) \qquad (a,b) \in \Omega(F) \quad \textit{iff} \quad \boldsymbol{\Delta}^{\mathbf{M}}\left(t(a,\bar{c}),t(b,\bar{c}),\bar{d}\right) \subseteq F,$$

for all $t \in Te$ such that $t = t(x, \bar{u})$ for some sequence $\bar{u}$ of variables from $V \setminus \{x\}$, for all sequences $\bar{c}$ of elements of M of the same length as $\bar{u}$ and for all $\bar{d} = \langle d_1, \ldots, d_{k-1} \rangle \in M^{k-1}$.

If $\mathbf{\Delta}(x, y, \mathbf{z})$ is a finite system of $k+1$-ary congruence k-formulas with $k-1$ parameters $\mathbf{z}$ then for every Λ-algebra $\mathbf{M}$, every pair of elements $a, b \in M$, every $\mathcal{S}$-filter F on $\mathbf{M}$, we have

$$(a, b) \in \Omega(F) \quad \text{iff} \quad \mathbf{\Delta}(a, b, \bar{d}) \subseteq F,$$
$$\text{(28)} \qquad \text{for all } \bar{d} = \langle d_1, \ldots, d_{k-1} \rangle \text{ of } k-1 \text{ elements of } M.$$

Finally, if $\mathbf{\Delta}(x, y)$ is a finite system of congruence k-formulas then for every Λ-algebra $\mathbf{M}$, every pair of elements $a, b \in M$, every $\mathcal{S}$-filter F on $\mathbf{M}$, we have

$$(a, b) \in \Omega(F) \quad \text{iff} \quad \mathbf{\Delta}(a, b) \subseteq F.$$

An example given at the end of this section shows a k-deductive system with a finite set of congruence k-formulas with parameters $\mathbf{z}$ that does not have such a system without parameters. A direct consequence is the following version of (Blok and Pigozzi, 1992, Corollary 13.6 (i)), where D is the symbol for the "truth" predicate in the language of the universal Horn theory interpreted as the filter F.

Corollary 10. (Compare with (Blok and Pigozzi, 1992, Corollary 13.6 (i)))
Let $\mathcal{S}$ be a k-deductive system with a finite system of $k+1$-ary congruence k-formulas with $k-1$ parameters. Let $\mathfrak{M}$ be a model of $\mathcal{S}$. Then $\mathfrak{M}$ is reduced iff

$$\mathfrak{M} \models \left[\forall_{z_1, \ldots, z_{k-1}} \bigwedge_{i=1}^{n} D\left(\Delta_i(x, y, \mathbf{z})\right) \right] \rightarrow x \approx y.$$

Corollary 13.6. in (Blok and Pigozzi, 1992) has a second part that states that the class of reduced models of $\mathcal{S}$ is closed under submodels and filtered products if $\mathcal{S}$ has a finite system of congruence formulas without parameters. Under the weaker assumption that congruence formulas may involve $k - 1$ parameters, the closure under filtered products continues to hold, but a submodel of a reduced model need not be reduced.

Corollary 11. (Compare with (Blok and Pigozzi, 1992, Corollary 13.6 (ii)))
Let $\mathcal{S}$ be a k-deductive system with a finite system of $k+1$-ary congruence k-formulas with $k-1$ parameters. Then the class $\mathrm{Mod}^*(\mathcal{S})$ of reduced models of $\mathcal{S}$ is closed under the formation of filtered products.

For the same reason, an analogue of Theorem 13.7 for does not hold under the weaker assumption.

Theorem 12. (Blok and Pigozzi, 1992, Theorem 13.7) *Let S be a k-deductive system with a finite system of congruence k-formulas. Then for all $\mathbf{K} \subseteq$ $\mathrm{Mod}^*(S)$ the reduced Horn class $\mathrm{Mod}^*(\mathbf{K})$ generated by $\mathbf{K}$ is* $\mathbf{ISPP_U K}$.

In the next theorem we consider a system of congruence k-formulas with arbitrary parameters. The system may be infinite and the parameters may form an infinite set, intersecting with the variables in the sequence $\mathbf{z}$. Of course in a single k-formula there are oly finitely many parameters at once.

Definition 13. (Blok and Pigozzi, 1992, Definition 13.9.) Let $\bar{w}$ be an infinite sequence of variables, possibly including some or all of $z_1, \ldots, z_{k-1}$. A (possibly infinite) set $\boldsymbol{\Delta}$ of k-formulas, $\boldsymbol{\Delta}(x, y, \bar{w})$ is called an infinite system of congruence k-formulas with parameters if for every Λ-algebra $\mathbf{M}$, every S-filter F on $\mathbf{M}$ and for every pair of elements $a, b \in M$, we have

$$(a, b) \in \Omega(F) \quad \text{iff} \quad \boldsymbol{\Delta}(a, b, \bar{c}) \subseteq F,$$

for all choices of sequences $\bar{c}$ of elements of M of appropriate length.

Finally, (Blok and Pigozzi, 1992, Theorem 13.10) characterizes protoalgebraic k-deductive systems by the existence of an infinite system of congruence formulas with parameters. The original proof is used, with small corrections concerning the use of structurality and equivalence k-formulas.

Theorem 14. (Blok and Pigozzi, 1992, Theorem 13.10) *A k-deductive system S is protoalgebraic iff S has a possibly infinite system of congruence k-formulas with parameters.*

Proof. Assume first that S is protoalgebraic. Let $\boldsymbol{\Delta}$ be a finite system of $k + 1$-ary equovalence k-formulas with $k - 1$ parameters $z_1, \ldots, z_{k-1}$ that exists by Theorem 6. Let $\boldsymbol{\Sigma}(x, y, \bar{w}, \mathbf{z})$ be the union of all sets

$$\boldsymbol{\Delta}(t(x, \bar{w}), t(y, \bar{w}), \mathbf{z}),$$

for all possible terms $t(x, \bar{w})$. Then $\boldsymbol{\Sigma}$ is an infinite set of congruence k-formulas with parameters for S.

Assume now that there is an infinite set $\boldsymbol{\Sigma}(x, y, \bar{w})$ of congruence k-formulas with parameters for S. In order to show that S is protoalgebraic we want to use Theorem 6 again. Let $l \in \{1, \ldots, k-1\}$. Define T_l to be the theory generated by the union of the sets

$$\boldsymbol{\Sigma}(x, y, \bar{t}(x, y, \bar{y})),$$

for all possible choices of sequence $\bar{t}$ of terms. together with the k-formula $\langle z_1, \ldots, z_{l-1}, x, z_l, \ldots, z_{k-1} \rangle$.

Since $\boldsymbol{\Sigma}$ is a system of congruence formulas, it follows by (29) that $(x, y) \in \Omega(T_l)$. So $\langle z_1, \ldots, z_{l-1}, y, z_l, \ldots, z_{k-1} \rangle \in T_l$, i.e.,

$$T_l \vdash \langle z_1, \ldots, z_{l-1}, y, z_l, \ldots, z_{k-1} \rangle.$$

It follows by structurality that there is a finite subset $\mathbf{\Delta}^{(l)}$ of the union of all sets $\mathbf{\Sigma}(x, y, \bar{t}(x, y, \bar{y}))$ for all sequences of terms $\bar{t}$, that together with the k-term $\langle z_1, \ldots, z_{l-1}, x, z_l, \ldots, z_{k-1} \rangle$ allows to deduce $\langle z_1, \ldots, z_{l-1}, y, z_l, \ldots, z_{k-1} \rangle$, i.e.,

$$\mathbf{\Delta}^{(l)}, \langle z_1, \ldots, z_{l-1}, x, z_l, \ldots, z_{k-1} \rangle \vdash \langle z_1, \ldots, z_{l-1}, y, z_l, \ldots, z_{k-1} \rangle.$$

There is a finite set of variables, say $\bar{w}$ occurring in the k-terms in $\mathbf{\Delta}^{(l)}$ other than x and y; some of the $\bar{w}$'s may be in $\mathbf{z}$. Also, since $\mathbf{\Delta}^{(l)} \subseteq \mathbf{\Sigma}$ and $\mathbf{\Sigma}$ is a congruence system, we have that

$$\vdash \mathbf{\Delta}^{(l)}(x, x, \bar{w}).$$

Finally, substitute in $\mathbf{\Delta}^{(l)}$ every variable from the sequence $\bar{w}$ other than x, y and $\mathbf{z}$ with x and obtain $\mathbf{\Gamma}^{(l)} = \mathbf{\Gamma}^{(l)}(x, y, \mathbf{z})$. By structurality,

$$\vdash \mathbf{\Gamma}^{(l)}(x, x, \mathbf{z}) \text{ and}$$

$$\mathbf{\Gamma}^{(l)}(x, y, \mathbf{z}), \langle z_1, \ldots, z_{l-1}, x, z_l, \ldots, z_{k-1} \rangle$$
$$\vdash \langle z_1, \ldots, z_{l-1}, y, z_l, \ldots, z_{k-1} \rangle.$$

Let $\mathbf{\Gamma} = \bigcup_{l=1}^{k-1} \mathbf{\Gamma}^l(x, y, \mathbf{z})$. Then $\mathbf{\Gamma}$ is a finite system of $k+1$-ary equivalence k-formulas with $k-1$ parameters $\mathbf{z}$ for $\mathcal{S}$. By Theorem 6, $\mathcal{S}$ is protoalgebraic.

Example 15. *We define a 2-deductive system $\mathcal{T}$ as an extension of $\mathcal{S}$ from Example 3 by rules. Here v denotes any variable from V other than x, y, z.*

Let $\Lambda = \{\cdot\}$ with $\rho(\cdot) = 2$. The operation symbol $\cdot$ is omitted when writing terms. Let $\mathcal{T}$ be the 2-deductive system defined by the following axiom and rules.

$$\vdash \langle x, x \rangle$$
$$\langle xz, yz \rangle, \langle x, z \rangle \quad \vdash \langle y, z \rangle$$
$$\langle xz, yz \rangle, \langle z, x \rangle \quad \vdash \langle z, y \rangle$$

(29) $$\langle xz, yz \rangle, \vdash \langle (xv)z, (yv)z \rangle$$

(30) $$\langle xz, yz \rangle \vdash \langle (vx)z, (vy)z \rangle$$

Clearly, the two-formula $\langle xz, yz \rangle$ forms by itself the system of congruence formulas for $\mathcal{T}$ with parameter z. However, we have the following

Proposition 16. *The 2-deductive systems $\mathcal{T}$ of Example 15 does not have a system of binary congruence formulas without parameters.*

Proof. For the proof by contradiction suppose that $\Sigma(x,y)$ is a congruence system without parameters. Let $\gamma = \langle \gamma_1, \gamma_2 \rangle$ be a 2 formula. The following claim is proved by induction on the length of the proof of γ.

Claim

$$\text{If } \Sigma(x,y), \langle x,z \rangle \vdash_{\mathcal{T}} \gamma$$

$$\text{then } \gamma = \langle x,z \rangle \text{ or } \mathrm{Var}(\gamma_1) \setminus \{x,y\} = \mathrm{Var}(\gamma_2) \setminus \{x,y\},$$

where for a term t, $\mathrm{Var}(t)$ denotes the set of all variables occurring in t.

It follows from the claim that

$$\Sigma(x,y), \langle x,z \rangle \nvdash \langle y,z \rangle.$$

Hence $\Sigma(x,y)$ is not even a system of equivalence formulas for $\mathcal{T}$.

Acknowledgment The author was a graduate student of professor Don Pigozzi at Iowa State in the years 1989-1994 and this is when she has learned from him about k-deductive systems. The influence of professor Pigozzi on the author's mathematical and academic growth cannot be overstated. Since the very first moment of her graduate program professor Pigozzi was extremely supportive, encouraging and generous in sharing his profound knowledge, mathematical insight and time. He also quickly became an irreplaceable friend of this student and her family who had arrived to the US from behind the Iron Curtain with three small children. The times were very different from the contemporary ones; the challenges of this arrival would not be overcome if not the dedication and help of Don Pigozzi. He has ever since remained a close friend of the family, in spite of the passing years and the distance between California and Poland. I would definitely not be the person I am if not for the enormous positive impact professor Pigozzi had on my academics and life. Thank you, Don.

References

Blok, W. and Pigozzi, D. (1986). Potoalgebraic logics, *Studia Logica* 45, 337–369.

Blok, W. and Pigozzi, D. (1989). *Algebraizable Logics*, Memoirs of the American Mathematical Society, 396, Providence.

Blok, W. and Pigozzi, D.(1992). Algebraic semantics for universal Horn logic without equality, in: *Universal Algebra and Quasigroup Theory*, A. Romanowska and J. D. H.Smith (eds.), Heldermann Verlag, Berlin, 1–56.

Czelakowski, J. (2001). *Protoalgebraic Logics*, Trends in Logic, Vol. 10, Kluwer, Dordrecht.

Elgueta, R. and Jansana, R. (1999). Definability of the Leibniz equality, *Studia Logica* 63, 223–243.

Font, J. M. (2016). *Abstract Algebraic Logic, An Introductory Textbook*, Studies in Logic, Mathematical Logic and Foundations, Vol. 60, College Publications, London.

Prucnal, T. and Wroński, A. (1974). An algebraic characterization of the notion of structural completeness, *Bulletin of the Section of Logic* 3, Polish Academy of Sciences, Institute of Philosophy and Sociology, 30–33.

Czelakowski, J. (1981). Equivalential logics (I), (II), *Studia Logica* 40, 227–236, 335–372.

Herrmann, B. (1997). Characterizing equivalential and algebraizable logics by the Leibniz operator, *Studia Logica* 58, 305–323.

Herrmann, B. (1993). *Equivalential Logics and Definability of Truth*, Ph.D. Dissertation, Freie Universität Berlin.

Pałasińska, K. (1994). *Deductive Systems and Finite Axiomatization Properties*, Ph.D. Dissertation, Iowa State Unversity.

Pałasińska, K. (2003). Finite Basis Theorem for fllter-distributive protoalgebraic deductive systems and strict universal Horn classes,, *Studia Logica* 74, 233–273.

Pynko, A. (1999). Implication systems for many-dimensional logics, *Reports on Mathematical Logic* 33, 11–27.

Diagrammatic duality

Anna B. Romanowska and Jonathan D.H. Smith*

Abstract If a class of algebras is characterized by diagrams in a category of dualizable algebras, diagrammatic duality provides dual representing objects in terms of corresponding dual diagrams appearing in the dual category of representation spaces for the dualizable algebras. The general technique is illustrated by a selection of examples, including quasigroups, bilattices, and Nelson algebras.

Key words: duality, bilattices, Nelson algebras, constructive logic, quasigroups, nets

1 Introduction

Dualities between algebras and representation spaces have now become a staple topic of interest at the interface of algebra and logic, and indeed in many areas of mathematics. While dualities are often produced with the aid of "schizophrenic objects" as codomains for both algebra and space homomorphisms, various other techniques are also available. As an addition to the palette of alternative techniques, diagrammatic duality is presented as a method for obtaining new dualities founded on existing ones. Whenever algebras of a certain class are equivalent to diagrams in a category of known

Anna B. Romanowska
Faculty of Mathematics and Information Science, Warsaw University of Technology, Warsaw, Poland, e-mail: `aroman@mini.pw.edu.pl`

Jonathan D.H. Smith
Department of Mathematics, Iowa State University, Ames, Iowa, 50011, USA, e-mail: `jdhsmith@iastate.edu`

* The first author's research was supported by the Warsaw University of Technology under grant number 504P 1120 0064 0001.

J. Czelakowski (eds.), *Don Pigozzi on Abstract Algebraic Logic,
Universal Algebra, and Computer Science*, Outstanding Contributions
to Logic 16, https://doi.org/10.1007/978-3-319-74772-9_11

dualizable algebras, diagrammatic duality furnishes representation spaces for the algebras in the class by examining dual diagrams in the category of representation spaces for the known dualizable algebras.

Sections 2–5 summarize the required background from categories and duality theory. Section 6 introduces diagrammatic algebras as objects of a category that is equivalent to a category of diagrams of algebras. Among the more immediate examples are the interlaced bilattices studied by Don Pigozzi and others (6.1). Further examples are given by the N-lattices or Nelson algebras that provide the algebraic semantics for Nelson's constructive logic with strong negation (6.2), and central piques (6.4). More generally, 6.3 interprets arbitrary classical universal algebras as diagrammatic algebras over sets.

Section 7 introduces diagrammatic duality in detail, the precise definition appearing in 7.2. Corresponding to the examples of diagrammatic algebras presented in Section 6, diagrammatic dualities for these algebras are given in Section 8. In particular, Lindenbaum-Tarski duality between sets and complete atomic Boolean algebras provides the basis for a so-called cabalistic (or CABAlistic) duality for each class of classical universal algebras. In the concluding section, this technique is used to show how quasigroups are dual to 3-nets.

Algebraic conventions generally follow those of (Smith and Romanowska, 1999). In particular, algebraic or "reverse-Polish" notation serves to limit the proliferation of brackets, and to avoid the awkward twisting caused by Eulerian notation, for example in the domain of the composition function of a category.

2 Diagram categories

2.1 Quivers

Definition 1. A *quiver* or "directed graph"

$$C = (C_0, C_1, \partial_0, \partial_1)$$

consists of two classes C_0, C_1 and two maps $\partial_0 \colon C_1 \to C_0$, $\partial_1 \colon C_1 \to C_0$.

(a) Elements of C_0 are called *vertices*, *points*, or *objects*.
(b) Elements of C_1 are called *edges*, *arrows*, or *morphisms*.
(c) The map ∂_0 is variously called the *tail* or *domain map*.
(d) The map ∂_1 is variously called the *head* or *codomain map*.

Edges f are depicted in the form $f \colon x \to y$ to indicate that $f^{\partial_0} = x$ and $f^{\partial_1} = y$. For a given pair (x, y) of vertices in a quiver C, set

$$(1) \qquad C(x, y) = \{ f \in C_1 \mid f^{\partial_0} = x, f^{\partial_1} = y \}.$$

A quiver $C = (C_0, C_1, \partial_0, \partial_1)$ is *small* if C_0 and C_1 are sets.

Definition 2. The *opposite* or *dual* of a quiver $C = (C_0, C_1, \partial_0, \partial_1)$ is the quiver $C^{\mathrm{op}} = (C_0, C_1, \partial_1, \partial_0)$.

In Definition 2, note that each edge $f\colon x \to y$ of C corresponds to an edge $x \leftarrow y\colon f$ of C^{op}. Thus $C^{\mathrm{op}}(y, x) = C(x, y)$.

2.2 Categories and their duals

Definition 3. A *category* $\mathbf{C} = (\mathbf{C}_0, \mathbf{C}_1)$ is defined as a quiver

$$\mathbf{C} = (\mathbf{C}_0, \mathbf{C}_1, \partial_0, \partial_1)$$

with an *identity function* $\epsilon\colon \mathbf{C}_0 \to \mathbf{C}_1; X \mapsto 1_X$ such that $\epsilon\partial_0 = 1_{\mathbf{C}_0} = \epsilon\partial_1$, and a *composition function*

$$(2) \qquad \mathbf{C}(U, V) \times \mathbf{C}(V, W) \to \mathbf{C}(U, W); (f, g) \mapsto fg$$

for $\{U, V, W, X\} \subseteq \mathbf{C}_0$, such that

$$1_U f = f = f 1_V$$

for f in $\mathbf{C}(U, V)$, and the *associative law*

$$(3) \qquad (fg)h = f(gh)$$

holds for f in $\mathbf{C}(U, V)$, g in $\mathbf{C}(V, W)$, and h in $\mathbf{C}(W, X)$. The category $\mathbf{C}$ is said to be *concrete* if its objects are sets (possibly equipped with additional structure) and its morphisms are functions, the composition function (2) being given by the usual mixed associative law $u^{fg} = (u^f)^g$ for $u \in U$.

Proposition 4. *Suppose that* $\mathbf{C} = (\mathbf{C}_0, \mathbf{C}_1)$ *is a category, as in Definition 3. Then the dual* $\mathbf{C}^{\mathrm{op}}$ *is a category, where the composition function*

$$\mathbf{C}^{\mathrm{op}}(U, V) \times \mathbf{C}^{\mathrm{op}}(V, W) \to \mathbf{C}^{\mathrm{op}}(U, W); (f, g) \mapsto f \circ g$$

is exactly the same function as the composition function

$$\mathbf{C}(V, U) \times \mathbf{C}(W, V) \to \mathbf{C}(W, U); (f, g) \mapsto gf$$

within $\mathbf{C}$. $\qquad\qquad\qquad\qquad\qquad\qquad\qquad\qquad\qquad\qquad\square$

2.3 Isomorphisms and inverses

Definition 5. Let $\mathbf{C}$ be a category.

(a) A morphism $f \colon X \to Y$ of $\mathbf{C}$ is an *isomorphism* (or *invertible*) if there is a morphism $g \colon Y \to X$ such that $fg = 1_X$ and $gf = 1_Y$.

(b) Two objects X and Y of $\mathbf{C}$ are said to be *isomorphic* if there is an isomorphism $f \colon X \to Y$.

In the situation of Definition 5, the morphism g is unique. It is written as the *inverse* f^{-1} of the morphism f.

2.4 Graph maps and diagrams

Definition 6. A *graph map* $F \colon D \to C$ from a quiver D to a quiver C consists of two functions, a *vertex map* or *object part* $F_0 \colon D_0 \to C_0$ and an *edge map* or *morphism part* $F_1 \colon D_1 \to C_1$, such that for each pair x, y of vertices of D, the map F_1 restricts to

$$(4) \qquad\qquad F_1 \colon D(x,y) \to C(x^{F_0}, y^{F_0}).$$

The respective suffices 0 and 1 on the object and morphism parts are usually suppressed.

Definition 7. A *diagram* in a category $\mathbf{C}$ is a graph map $F \colon D \to \mathbf{C}$ with codomain $\mathbf{C}$. The diagram is *proper* if its domain D is small.

Definition 8. Consider two diagrams $F \colon D \to \mathbf{C}$ and $G \colon D \to \mathbf{C}$ with common domain quiver D and codomain category $\mathbf{C}$. A *natural transformation* $\tau \colon F \to G$ is a vector having a component $\tau_x \colon x^F \to x^G$ in $\mathbf{C}(x^F, x^G)$ for each vertex x of D, such that the *naturality property* $f^F \tau_y = \tau_x f^G$ is satisfied for each edge $f \colon x \to y$ of D.

Proposition 9. *For a quiver D and category $\mathbf{C}$, let $\left(\mathbf{C}^D\right)_0$ be the class of diagrams $F \colon D \to \mathbf{C}$. For given diagrams $F \colon D \to \mathbf{C}$ and $G \colon D \to \mathbf{C}$, let $\left(\mathbf{C}^D\right)(F,G)$ be the class of natural transformations $\tau \colon F \to G$. Define $(1_F)_x = 1_{x^F}$ at each vertex x of D. For H in $\left(\mathbf{C}^D\right)_0$ and $\upsilon \colon G \to H$, define $(\tau\upsilon)_x = \tau_x \upsilon_x$ at each vertex x of D. Then $\mathbf{C}^D$ forms a category.* $\square$

If D is a proper diagram, then categories of the form $\mathbf{C}^D$ described in Proposition 9 are known as *diagram categories*.

3 Duality

3.1 Functors and natural isomorphisms

Definition 10. (a) A (*covariant*) *functor* $F\colon \mathbf{B} \to \mathbf{C}$ from a domain category $\mathbf{B}$ to a codomain category $\mathbf{C}$ is a graph map satisfying the two *functoriality properties*:

(i) For each object X of $\mathbf{B}$, one has $(1_X)^F = 1_{XF}$;
(ii) For $\{U,V,W\} \subseteq \mathbf{B}_0$, $f \in \mathbf{B}(U,V)$, and $g \in \mathbf{B}(V,W)$, one has $f^F g^F = (fg)^F$.

(b) A *contravariant functor* $F\colon \mathbf{B} \to \mathbf{C}$ is a functor $F\colon \mathbf{B}^{\mathsf{op}} \to \mathbf{C}$, or equivalently $F\colon \mathbf{B} \to \mathbf{C}^{\mathsf{op}}$.

(c) The *identity functor* $1_\mathbf{C}$ on a category $\mathbf{C}$ consists of the identity functions on the classes $\mathbf{C}_0$ and $\mathbf{C}_1$.

Two functors $F\colon \mathbf{B} \to \mathbf{C}$ and $G\colon \mathbf{B} \to \mathbf{C}$ are said to be *naturally isomorphic* if they are isomorphic within the category $\mathbf{C}^\mathbf{B}$ of Proposition 9. In that case, for an object X of $\mathbf{B}$, the objects XF and XG of $\mathbf{C}$ are also said to be *naturally isomorphic*. Thus if $\tau\colon F \to G$ is an isomorphism within $\mathbf{C}^\mathbf{B}$, the component $\tau_X\colon XF \to XG$ provides an isomorphism within $\mathbf{C}$.

3.2 Equivalence and dual equivalence

Definition 11. An *equivalence* $F : \mathbf{B} \cong \mathbf{C} : G$ between categories $\mathbf{B}$ and $\mathbf{C}$ consists of covariant functors $F\colon \mathbf{B} \to \mathbf{C}$ and $G\colon \mathbf{C} \to \mathbf{B}$ such that:

(a) the functors FG and $1_\mathbf{B}$ are naturally isomorphic, and
(b) the functors GF and $1_\mathbf{C}$ are naturally isomorphic.

For a *dual equivalence*, the same conditions (a), (b) are imposed on a pair of contravariant functors $F\colon \mathbf{B} \to \mathbf{C}$ and $G\colon \mathbf{C} \to \mathbf{B}$.

The notation

$$F : \mathbf{B} \rightleftarrows \mathbf{C} : G$$

will be used to denote a dual equivalence as in Definition 11.

3.3 Duality

A *duality* will denote a dual equivalence

(5) $$D : \mathfrak{A} \rightleftarrows \mathfrak{X} : E$$

in which $\mathfrak{A}$ is a category of *algebras* (in the sense of modern universal algebra) and homomorphisms, while $\mathfrak{X}$ is a concrete category of objects known as *spaces*. For an algebra A, the image AD is called the *representation space* of A. For a space X, the image XE is called the *algebra represented* by X. The functor D is called the *dual space functor*. The functor E is called the *represented algebra functor*.

Remark 12. In (5), the determination of which side has the algebras and which side has the spaces may be purely conventional. For example, in the Lindenbaum-Tarski duality (4.3 below), one might equally well take the sets as the spaces and the Boolean algebras as the algebras.

4 Examples of duality

4.1 Finite-dimensional vector spaces

For a field K, take both $\mathfrak{A}$ and $\mathfrak{X}$ to be the category $\underline{\underline{K}}^{<\omega}$ of finite-dimensional vector spaces and linear transformations over K. Then for a finite-dimensional vector space V, both V^D and V^E are defined as the space $\underline{\underline{K}}^{<\omega}(V, K)$ of linear functionals on K.

4.2 Pontryagin duality

Take $\mathfrak{A}$ to be the category **Ab** of abelian groups. Take $\mathfrak{X}$ to be the category **CHAb** of compact Hausdorff topological abelian groups and continuous homomorphisms. For a (discrete) abelian group A, the representation space is the group $A^D := \mathbf{Ab}(A, \mathbb{R}/\mathbb{Z})$ of *characters* or homomorphisms into the circle group $\mathbb{R}/\mathbb{Z}$, topologized as a subspace of the product space $(\mathbb{R}/\mathbb{Z})^A$. Then a compact Hausdorff abelian group X represents the discrete group $X^E := \mathbf{CHAb}(X, \mathbb{R}/\mathbb{Z})$ of *continuous characters*, continuous homomorphisms into the circle group. The duality between discrete and compact Hausdorff abelian groups is extended to a self-duality on the category of locally compact abelian groups (Dixmier, 1964),(Pontryagin, 1966, §37).

4.3 Lindenbaum-Tarski duality

Take $\mathfrak{A}$ to be the category **Set** of sets (algebras without operations). Take $\mathfrak{X}$ to be the category **CABA** of complete atomic Boolean algebras and homomorphisms preserving all joins and meets. Consider the set $2 = \{0,1\}$, possibly endowed with Boolean algebra structure. For a set A, the representation space A^D is defined to be the set 2^A or $\mathcal{P}(A)$ of (characteristic functions of) subsets of A, with the singletons as atoms. For a complete atomic Boolean algebra B, the represented algebra $B^E := \mathbf{CABA}(B,2)$ is naturally isomorphic to the set of atoms of B (Johnstone, 1982, p. xiv), (Tarski, 1935).

4.4 Priestley duality

Take $\mathfrak{A}$ to be the category **DL** of distributive lattices. Take $\mathfrak{X}$ to be the category **O′Stone** of partially ordered Stone spaces (a.k.a Priestley spaces) and monotone continuous maps. Consider the set $2 = \{0 < 1\}$ as a distributive lattice or as a partially ordered Stone space. For a distributive lattice L, the representation space $L^D = \mathbf{DL}(L,2)$ carries the induced order and subspace topology from the product 2^L. A partially ordered Stone space S represents the algebra $S^E = \mathbf{O′Stone}(S,2)$, a (distributive) sublattice of 2^S (Johnstone, 1982, pp. 66, 75), (Priestley, 1970).

4.5 Esakia duality

Take $\mathfrak{A}$ to be the category **Heyt** of Heyting algebras. Take $\mathfrak{X}$ to be the category **Esakia** of *Esakia spaces*, partially ordered Stone spaces where the downset $C^{\geq}$ of each clopen subset C is clopen. Morphisms in **Esakia** are **O′Stone**-morphisms f such that $(x^{\leq})f = (xf)^{\leq}$ for all elements x of the domain of f. Consider the set $2 = \{0 < 1\}$ as a Heyting algebra or as an Esakia space. For a Heyting algebra H, the representation space $H^D = \mathbf{Heyt}(H,2)$ carries the induced order and subspace topology from the product 2^H. An Esakia space S represents the algebra $S^E = \mathbf{Esakia}(S,2)$, a Heyting subalgebra of 2^S (Esakia, 1974, 1985).

4.6 Gel'fand duality

Take $\mathfrak{A}$ to be the category $\underline{C}^*$ of commutative C^*-algebras and norm-decreasing, involution-preserving algebra homomorphisms. Take $\mathfrak{X}$ to be the category **CH** of compact Hausdorff topological spaces. For a C^*-algebra A, the representation space A^D is the space of maximal ideals. A compact Hausdorff space X represents the C^*-algebra $C^*(X)$ of continuous complex-valued functions on X. Compare (Herrlich and Strecker, 1973, §38.4) to justify consideration of $\underline{C}^*$ as a category of algebras (Loomis, 1953, §26E).

4.7 Affine schemes

Take $\mathfrak{A}$ to be the category **CRing** of commutative, unital rings. Take $\mathfrak{X}$ to be the category **Aff** of affine schemes. For a ring A, the space A^D is the affine scheme $(\operatorname{Spec} A, \mathcal{O}_{\operatorname{Spec} A})$. For an affine scheme X or $(X, \mathcal{O}_X)$, the algebra X^E is the ring $\Gamma(X, \mathcal{O}_X)$ of global sections (Mumford, 1988, II.2 Cor. 1).

4.8 Schizophrenic objects

The examples 4.1–4.5 each arise from a *schizophrenic object* T appearing in an *algebraic* personality as an object of $\mathfrak{A}$, and in a *spatial* personality as an object of $\mathfrak{X}$. The dual space functor D is then naturally isomorphic with $\mathfrak{A}(\ ,T)$, while the represented algebra functor E is naturally isomorphic with $\mathfrak{X}(\ ,T)$:

- In 4.1, T appears as the one-dimensional vector space K in both its personalities.
- In 4.2, T is the one-dimensional Torus $\mathbb{R}/\mathbb{Z}$, discrete in its algebraic personality, but with the usual topology in its spatial personality.
- In 4.3, T is the Two-element set, with no additional structure in its algebraic personality, and appearing as a Boolean algebra in its spatial personality.
- In 4.4, T is the Two-element set, as a distributive lattice in its algebraic personality, and as a partially ordered discrete space in its spatial personality.
- In 4.5, T is the Two-element set, as a Heyting algebra in its algebraic personality, and as an Esakia space in its spatial personality.

The dualities 4.3–4.5 are "natural" in the sense of (Clark and Davey, 1998). Note that the duality 4.7 does not arise from a schizophrenic object.

4.9 Finiteness

In (5), each object of $\mathfrak{A}$ generally appears as a directed colimit of its finitely generated subalgebras. Since the functors D and E are mutually right and left adjoint, they preserve limits and colimits respectively (Smith and Romanowska, 1999, Th. III.3.4.1). To specify the duality (5), it then suffices to specify its restriction to finitely generated algebras. The somewhat pathological topologies used in axiomatizing $\mathfrak{X}$ (e.g., in Priestley duality) are often best understood merely as directed limits of the representation spaces of finitely generated algebras.

5 Applications

Dualities in the sense of (5) have a variety of applications, illustrated by the examples from 4.

5.1 Simplification

It may happen that the space A^D is much simpler than the original algebra A, so the duality may be regarded as a useful way of constructing A from the simple object A^D as A^{DE}. Thus in the finite case, Priestley duality (4.4) reconstructs a distributive lattice as the lattice of ideals of its poset of join-irreducibles.

5.2 Coproducts

The underlying set functor from $\mathfrak{A}$ to **Set** preserves products, so that products in $\mathfrak{A}$ are easy to understand, by virtue of their componentwise structure. On the other hand, coproducts in $\mathfrak{A}$ are often messy to construct. Duality constructs the coproduct $A_1 + A_2$ of two objects A_1, A_2 of $\mathfrak{A}$ as the algebra $(A_1^D \times A_2^D)E$ represented by the product of the spaces of the individual objects.

5.3 Logic

Many dualities provide backgrounds for logic. A basic example is Lindenbaum-Tarski duality $D : \mathbf{Set} \rightleftarrows \mathbf{CABA} : E$ (4.3). For a set S, the complete atomic Boolean algebra S^D is the home for propositions describing properties of S.

5.4 Coordinatization

Dualities such as 4.1 and 4.7 may be understood as coordinatizing geometries. In 4.1, a non-zero linear functional on a vector space V (or element of V^D) measures coordinates along a certain axis.

6 Diagrammatic algebras

Definition 13. Suppose that $\mathfrak{C}$ and $\mathfrak{A}$ are categories of algebras and homomorphisms. Suppose that there is an equivalence $\mathfrak{C} \cong \mathfrak{C}_{\mathfrak{A}}$ between $\mathfrak{C}$ and a subcategory $\mathfrak{C}_{\mathfrak{A}}$ of a diagram category $\mathfrak{A}^V$ with given domain diagram V. Then the objects of $\mathfrak{C}$ are known as *diagrammatic algebras* (relative to $\mathfrak{A}$).

In the context of Definition 13, it is often convenient to abuse notation and suppress the distinction between $\mathfrak{C}$ and $\mathfrak{C}_{\mathfrak{A}}$, merely stating that a $\mathfrak{C}$-algebra C is equivalent to a diagram $\gamma\colon V \to \mathfrak{A}$, or further that a $\mathfrak{C}$-homomorphism $f\colon C_1 \to C_2$ is equivalent to an $\mathfrak{A}^V$-morphism $\varphi\colon \gamma_1 \to \gamma_2$. For this reason, the functors producing the equivalence $\mathfrak{C} \cong \mathfrak{C}_{\mathfrak{A}}$ are not mentioned explicitly in the definition.

6.1 Interlaced and distributive bilattices

Among the simplest examples of diagrammatic algebras in the sense of Definition 13 are interlaced or distributive bilattices, as studied by many authors, including Pigozzi *et al.* (Mobasher et al., 2000).

Definition 14. Consider an algebra $(B, \vee, \wedge, +, \cdot)$ equipped with four binary operations.

(a) The algebra is a *bilattice* if both the reducts $B_1 = (B, \vee, \wedge)$ and $B_2 = (B, +, \cdot)$ are lattices.

(b) A bilattice B is *bounded* if both B_1 and B_2 are bounded lattices.

(c) A bilattice is *distributive* if for each pair of basic operations $\times_1, \times_2$, the identity

$$(6) \qquad ((x \times_1 z) \times_2 (y \times_1 z) = (x \times_2 y) \times_1 z$$

is satisfied.

(d) A bilattice is *interlaced* if each basic operation preserves the ordering relations in each of the lattices B_1 and B_2.

Proposition 15. *(Romanowska and Trakul, 1989) A bilattice is interlaced if and only if it satisfies the identities:*

$$(7) \qquad\qquad ((x \times_2 y) \times_1 z) \times_2 (y \times_1 z) = (x \times_2 y) \times_1 z$$

for each pair of operations $\times_1, \times_2 \in \{\vee, \wedge, +, \cdot\}$. $\qquad\qquad\qquad\qquad$ $\square$

Remark 16. The equations (6) and (7) may be considered as hyperidentities (remaining valid under substitution of basic binary operations as well as under substitution of arguments).

Corollary 17. *A distributive bilattice is interlaced.*

Proof. Within bilattices, note that the hyperidentity (6) implies the hyperidentity (7). $\qquad\qquad\qquad\qquad\qquad\qquad\qquad\qquad\qquad\qquad\qquad\qquad\qquad$ $\square$

Let $(L_1, \vee_1, \wedge_1)$ and $(L_2, \vee_2, \wedge_2)$ be lattices. The *superproduct* $B = L_1 \bowtie L_2$ of the lattices L_1 and L_2 is the algebra

$$B = (L_1 \times L_2, \wedge, \vee, \cdot, +)$$

with basic operations defined by

$$\wedge = (\wedge_1, \vee_2), \quad \vee = (\vee_1, \wedge_2), \quad \cdot = (\wedge_1, \wedge_2), \quad + = (\vee_1, \vee_2).$$

The reducts B_1 and B_2 of B are lattices. In fact $B_1 \cong L_1 \times L_2^d$, with L_2^d as the dual of L_2, while $B_2 \cong L_1 \times L_2$. For bounded lattices L_1 and L_2, their superproduct $L_1 \bowtie L_2$ is also a bounded bilattice, with the four bounds

$$\bot_1 = (0_1, 1_2), \quad \top_1 = (1_1, 0_2), \quad \bot_2 = (0_1, 0_2), \quad \top_2 = (1_1, 1_2)$$

written in terms of the respective bounds $0_1, 1_1$ of L_1 and $0_2, 1_2$ of L_2.

Proposition 18. *(Movsisyan et al., 2006) Consider an algebra* $(B, \vee, \wedge, +, \cdot)$ *equipped with four binary operations.*

(a) *The algebra B is an interlaced bilattice iff it is isomorphic to the superproduct $L_1 \bowtie L_2$ of two lattices L_1 and L_2.*
(b) *The algebra B is a bounded interlaced bilattice if and only if it is isomorphic to the superproduct $L_1 \bowtie L_2$ of two bounded lattices L_1 and L_2.*
(c) *The algebra B is a distributive bilattice iff it is isomorphic to the superproduct $L_1 \bowtie L_2$ of two distributive lattices L_1 and L_2.*
(d) *The algebra B is a bounded distributive bilattice iff it is isomorphic to the superproduct $L_1 \bowtie L_2$ of two bounded distributive lattices L_1 and L_2.* $\square$

Now define the following categories of algebras and homomorphisms:

- **BL**: the category of bounded lattices;
- **BIB**: the category of bounded interlaced bilattices;
- **BDL**: the category of bounded distributive lattices;
- **BDB**: the category of bounded distributive bilattices.

Suppose that V is the "discrete" quiver with two vertices and no edges. By virtue of Proposition 18(b), there is an equivalence $\mathbf{BIB} \cong \mathbf{BL}_{\mathbf{BIB}}$ (Mobasher et al., 2000, Th. 10). Similarly, by virtue of Proposition 18(d), there is an equivalence $\mathbf{BDB} \cong \mathbf{BDL}_{\mathbf{BDB}}$ (Mobasher et al., 2000, Cor. 11). Thus the results of Pigozzi *et al.* realize both bounded interlaced bilattices and bounded distributive bilattices as diagrammatic algebras, relative to the categories of bounded lattices and bounded distributive lattices respectively. Pigozzi *et al.* were only able to consider these bounded cases, since they did not have Proposition 18(a),(c) at their disposal. However, their methods extend to cover all the cases summarized as follows.

Theorem 19. (a) *Interlaced bilattices are diagrammatic relative to lattices.*

(b) *Distributive bilattices are diagrammatic relative to distributive lattices.*

(c) *Bounded interlaced bilattices are diagrammatic relative to bounded lattices.*

(d) *Bounded distributive bilattices are diagrammatic relative to bounded distributive lattices.* $\qquad\square$

6.2 Nelson algebras

These algebras (Sendlewski, 1990), also known as "N-lattices" (Rasiowa, 1958), provide the algebraic semantics for Nelson's *constructive logic with strong negation* (Nelson, 1949; Rasiowa, 1974).

Definition 20. (Odintsov, 2010, Defn. 2.1) Consider an algebra

$$(B, \vee, \wedge, \rightarrow, \sim, 0, 1)$$

equipped with three binary operations $\vee, \wedge, \rightarrow$, and with $\sim$ as a unary operation (*strong negation*). Suppose that $(B, \vee, \wedge, 0, 1)$ is a bounded distributive lattice, with $\leq$ as the lattice ordering. Then the algebra $(B, \vee, \wedge, \rightarrow, \sim, 0, 1)$ is a *Nelson algebra* if the following conditions are satisfied:

(i) The reduct $(B, \vee, \wedge, \sim, 0, 1)$ is a De Morgan algebra (i.e. the De Morgan identities $\sim (x \vee y) = \sim x \wedge \sim y$ and $\sim (x \wedge y) = \sim x \vee \sim y$ hold, along with $\sim\sim x = x$);

(ii) A reflexive, transitive relation $\preceq$ is defined on B by setting

$$x \preceq y \text{ iff } (x \rightarrow y) \rightarrow (x \rightarrow y) = x \rightarrow y;$$

(iii) The lattice order relation $x \leq y$ on B is equivalent to $x \preceq y$ and $\sim x \preceq \sim y$;

(iv) The equivalence relation χ, defined on B by $x \chi y$ iff $x \preceq y$ and $y \preceq x$, is a congruence on the reduct $(B, \vee, \wedge, \rightarrow)$ such that the quotient

$$(B, \vee, \wedge, \rightarrow, 0, 1)^\chi$$

is a Heyting algebra;

(v) For all $x, y \in B$, one has $(x \wedge \sim x, 0) \in \chi$ and $(\sim (x \to y), x \wedge \sim y) \in \chi$.

Remark 21. Nelson algebras form a variety (Odintsov, 2010).

Definition 22. A congruence α on a Heyting algebra H is *Boolean* if the quotient H^α is a Boolean algebra.

Theorem 23. *(Sendlewski, 1990, Th. 4.1) The category* **Nelson** *of Nelson algebras is equivalent to the category of pairs* (H, α)*, where* H *is a Heyting algebra and* α *is a Boolean congruence on* H*.* $\square$

Theorem 23 associates a pair $B^P = (H, \alpha_B)$ with a Nelson algebra B, including the Heyting algebra quotient $H = B^\chi$ from Definition 20(iv). The Boolean congruence α_B is obtained by identifying the quotient B^χ with a certain subset B^* of B (Sendlewski, 1990, Lemma 3.3), and then restricting a certain congruence on B (Sendlewski, 1990, Lemma 3.4) to the subset B^*.

Conversely, consider a Heyting algebra H with a Boolean congruence α. The Nelson algebra $(H, \alpha)^N$ associated with the pair (H, α) is built on the subset

$$(H, \alpha)^N = \{(x, y) \in H^2 \mid x \wedge y = 0 \text{ and } x \vee y \in 1^\alpha\}$$

of H^2, with operations

$$(x_1, y_1) \vee (x_2, y_2) = (x_1 \vee x_2, y_1 \wedge y_2),$$
$$(x_1, y_1) \wedge (x_2, y_2) = (x_1 \wedge x_2, y_1 \vee y_2),$$
$$(x_1, y_1) \to (x_2, y_2) = (x_1 \to x_2, y_1 \wedge y_2),$$
$$\sim (x_1, y_1) = (y_1, x_1)$$

for $(x_1, y_1), (x_2, y_2) \in (H, \alpha)^N$. (See (Sendlewski, 1990, Theorem 3.6(i)).)

One may now exhibit Nelson algebras as diagrammatic relative to (the category **Heyt** of) Heyting algebras. Consider the quiver V given as $a \colon h \to b$. Then by virtue of Theorem 23, a Nelson algebra B is equivalent to a diagram $\beta \colon V \to$ **Heyt** sending the arrow a to the natural projection of the Boolean congruence α_B from the Heyting algebra B^χ.

6.3 Classical universal algebras

Consider a *type* $\tau \colon \Omega \to \mathbb{N}$, with *operator domain* Ω. Then a *τ-algebra* (A, τ), or more informally, an *Ω-algebra* (A, Ω), is a set A with an *operation*

$$(8) \qquad\qquad\qquad \omega \colon A^{\omega\tau} \to A$$

corresponding to each *operator* or element ω of Ω. Let $\underline{\underline{\tau}}$ denote the category of τ-algebras and homomorphisms between them (Smith and Romanowska, 1999, §§IV1.1–2).

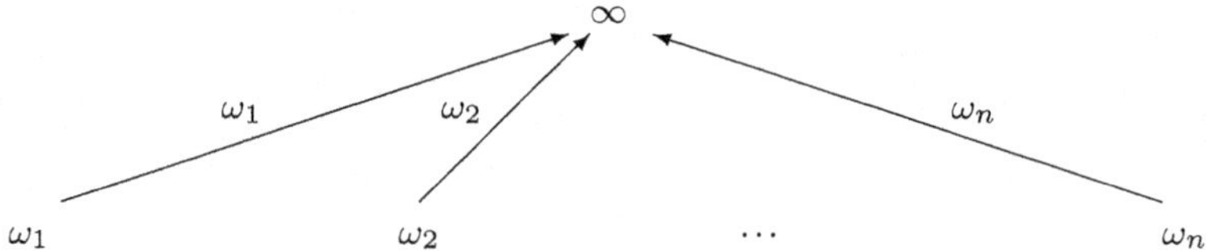

Fig. 1 The $\{\omega_1,\dots,\omega_n\}$-cospan.

The Ω-*cospan* is a quiver Ω_∞ with edge set Ω (compare Figure 1). Its vertex set is the disjoint union $\Omega + \top$ of Ω with a singleton $\top = \{\infty\}$. The tail map is the identity function on Ω, while the head map is the unique function $\Omega \to \top$. A τ-algebra A is equivalent to a diagram $\alpha\colon \Omega_\infty \to \mathbf{Set}$ with edge map $\alpha_1\colon \omega \mapsto (\omega\colon A^{\omega\tau} \to A)$. Thus the edge map sends each operator to the corresponding operation on the set A. Each τ-homomorphism

$$f\colon (A_1,\tau) \to (A_2,\tau)$$

is equivalent to a $\mathbf{Set}^{\Omega_\infty}$-morphism φ with component $f^{\omega\tau}\colon A_1^{\omega\tau} \to A_2^{\omega\tau}$ at a vertex ω, and component $f\colon A \to B$ at the unique element ∞ of $\top$. For the equivalence, consider the following.

Proposition 24. *Suppose that A_1 and A_2 are τ-algebras, with respective equivalent diagrams α_1 and α_2 in $\mathbf{Set}^{\Omega_\infty}$. Then a $\mathbf{Set}^{\Omega_\infty}$-morphism*

$$\varphi\colon \alpha_1 \to \alpha_2$$

is a $\underset{=\mathbf{Set}}{\tau}$-morphism if and only if $\varphi_\omega = \varphi_\infty^{\omega\tau}$ for each operator ω.

Proof. The naturality property of φ is the commuting of

$$(9) \qquad\qquad \begin{array}{ccc} A_1^{\omega\tau} & \xrightarrow{\ \varphi_\omega\ } & A_2^{\omega\tau} \\ {\scriptstyle\omega}\downarrow & & \downarrow{\scriptstyle\omega} \\ A_1 & \xrightarrow[\ \varphi_\infty\ ]{} & A_2 \end{array}$$

for each operator ω. If $\varphi_\omega = \varphi_\infty^{\omega\tau}$, the commuting of (9) shows that $\varphi_\infty\colon A_1 \to A_2$ is a τ-homomorphism. Conversely, if $\varphi_\infty = f$ and $\varphi_\omega = f^{\omega\tau}$ for operators ω and a τ-homomorphism $f\colon A_1 \to A_2$, the commuting of (9) is given, along with the condition $\varphi_\omega = \varphi_\infty^{\omega\tau}$. $\qquad\square$

One may summarize as follows.

Theorem 25. *Algebras from any variety are diagrammatic relative to sets.*

$$\qquad\qquad\qquad\qquad\qquad\qquad\qquad\qquad\qquad\qquad\qquad\qquad\square$$

6.4 Central piques

A *quasigroup* $(Q, \cdot)$ is a set Q equipped with a binary *multiplication* operation denoted by $\cdot$ or simple juxtaposition of the two arguments, in which specification of any two of x, y, z in the equation $x \cdot y = z$ determines the third uniquely. Equationally, a quasigroup $(Q, \cdot, /, \backslash)$ is a set Q equipped with three binary operations of multiplication, *right division* $/$ and *left division* $\backslash$, satisfying the identities:

$$\text{(IL)} \quad y \backslash (y \cdot x) = x;$$
$$\text{(IR)} \quad x = (x \cdot y)/y;$$
$$\text{(SL)} \quad y \cdot (y \backslash x) = x;$$
$$\text{(SR)} \quad x = (x/y) \cdot y.$$

An element e of a quasigroup Q is said to be *idempotent* if $\{e\}$ forms a singleton subquasigroup of Q. A *pique* or *pointed idempotent quasigroup* is a quasigroup P, containing an idempotent element 0, that has its quasigroup structure of multiplication and the divisions enriched by a nullary operation selecting the idempotent element 0 (Chein et al., 1990, §III.5),(Romanowska and Smith, 2006). Note that piques also form a variety.

On a pique $(P, \cdot, 0)$, there are permutations

$$R \colon P \to P; x \mapsto x \cdot 0$$

and

$$L \colon P \to P; x \mapsto 0 \cdot x$$

of the set P. A *loop B* is a pique in which the pointed idempotent element acts as an identity, so that $R = L = 1_B$. For a general pique $(P, \cdot, 0)$, the *cloop* or *corresponding loop* is the loop $B(P)$ or $(P, +, 0)$ in which the "multiplication" operation $+$ is defined by

$$(10) \qquad\qquad\qquad x + y = xR^{-1} \cdot yL^{-1}.$$

Inverting (10), the multiplication of a pique is recovered from the cloop by

$$(11) \qquad\qquad\qquad x \cdot y = xR + yL.$$

Definition 26. A pique $(P, \cdot, 0)$ is said to be *central* if:

1. The cloop $B(P)$ is an abelian group, and
2. The maps R and L are automorphisms of $B(P)$.

Remark 27. Central piques may also be characterized structurally as quasigroups Q where the diagonal $\widehat{Q} = \{(q, q) \mid q \in Q\}$ is a normal subquasigroup of the square Q^2, and $Q \cong Q^2/\widehat{Q}$ (Smith, 2007, (3.27)).

Take $\mathfrak{C}$ as the category of central piques. Take $\mathfrak{A}$ as the category of abelian groups and homomorphisms. Consider the quiver

$$V = {}_L \, \overset{\curvearrowright}{\bigcirc} \cdot \overset{\curvearrowleft}{\bigcirc} \, {}_R$$

Take $\mathfrak{C}_\mathfrak{A}$ as the subcategory of the diagram category $\mathfrak{A}^V$ given by those diagrams where the images of L and R are invertible. The preceding discussion may then be summarized as follows.

Proposition 28. *Central piques are diagrammatic relative to abelian groups.*

$\square$

Now recall that for a given type $\tau\colon \Omega \to \mathbb{N}$ (in the sense of 6.3), a τ-algebra (A, Ω) is *topological* if A is a topological space and the operations (8) are continuous.

Proposition 29. *(Romanowska and Smith, 2006, Prop. 3.3) A central pique P is topological if and only if:*

1. *The cloop $B(P)$ is a topological abelian group, and*
2. *The maps $R\colon P \to P$ and $L\colon P \to P$ are homeomorphisms.* $\square$

Proposition 28 then yields:

Proposition 30. (a) *Topological central piques are diagrammatic relative to topological abelian groups.*

(b) *Locally compact central piques are diagrammatic relative to locally compact topological abelian groups.* $\square$

7 Diagrammatic duality

7.1 Duality of diagram categories

Theorem 31. *Let V be a quiver, and let $F\colon \mathbf{B} \rightleftarrows \mathbf{C}\colon G$ be a dual equivalence. Define a contravariant functor*

$$F^V \colon \mathbf{B}^V \to \mathbf{C}^{(V^{\mathrm{op}})}$$

by

$$(F^V)_0 \colon \beta \mapsto \beta F$$

(compare Figure 2) and

$$(F^V)_1 \colon \varphi \mapsto \varphi^F$$

for $\varphi \in F^V(\beta_1, \beta_2)$, where the component of φ^F at a vertex x of V is given by $\left(\varphi^F\right)_x = (\varphi_x)^F$. Define a contravariant functor

$$G^V \colon \mathbf{C}^{(V^{\mathrm{op}})} \to \mathbf{B}^V$$

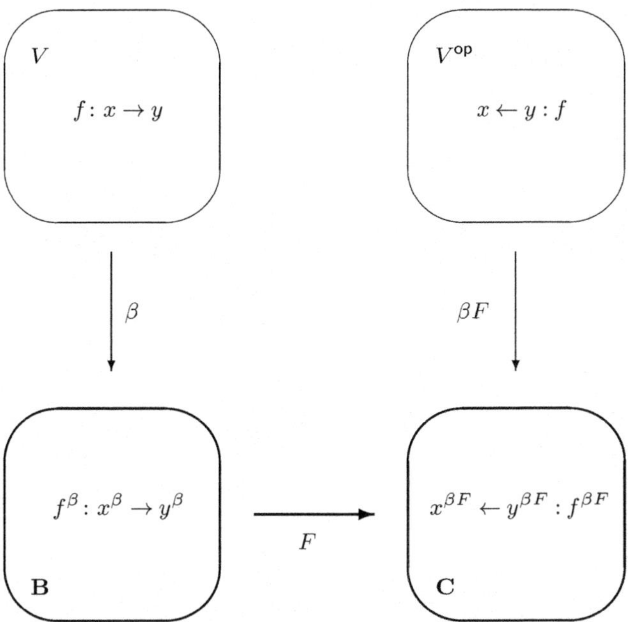

Fig. 2 The object part of F^V.

in similar fashion. Then

$$F^V : \mathbf{B}^V \rightleftarrows \mathbf{C}^{(V^{\mathrm{op}})} : G^V$$

is a dual equivalence.

Proof. Let $\tau \colon FG \to 1_{\mathbf{B}}$ be an isomorphism in $\mathbf{B}^{\mathbf{B}}$. Then a natural isomorphism $\tau^V \colon F^V G^V \to 1_{\mathbf{B}^V}$ will be defined, by giving its component

$$(12) \qquad\qquad (\tau^V)_\beta \colon \beta^{F^V G^V} \to \beta$$

at a diagram $\beta \colon V \to \mathbf{B}$. The component of (12) at a vertex x of V is

$$(13) \qquad\qquad \tau_{x\beta} \colon x^{\beta FG} \to x^\beta \,.$$

Note that (13) is an isomorphism in $\mathbf{B}$, with inverse $(\tau_{x\beta})^{-1}$. Thus (12) is an isomorphism in $\mathbf{B}^V$: its inverse has component $(\tau_{x\beta})^{-1}$ at the vertex x of V. In turn, this means that $\tau^V \colon F^V G^V \to 1_{\mathbf{B}^V}$ is a natural isomorphism. In similar fashion, it may be shown that $G^V F^V$ and $1_{\mathbf{C}^{V^{\mathrm{op}}}}$ are naturally isomorphic. $\qquad\square$

7.2 Diagrammatic duality

Consider a duality

$$(14) \qquad\qquad D : \mathfrak{A} \rightleftarrows \mathfrak{X} : E$$

as given by (5). In the current context, (14) is known as the *seed duality*. Let $\mathfrak{C}$ be a category of algebras that are diagrammatic relative to $\mathfrak{A}$, by an equivalence $\mathfrak{C} \cong \mathfrak{C}_{\mathfrak{A}}$ of $\mathfrak{C}$ with a subcategory $\mathfrak{C}_{\mathfrak{A}}$ of a diagram category $\mathfrak{A}^V$. Then the dual equivalence

$$D^V : \mathfrak{A}^V \rightleftarrows \mathfrak{X}^{(V^{\mathrm{op}})} : E^V$$

given by Theorem 31 restricts to a dual equivalence

$$(15) \qquad\qquad D^V : \mathfrak{C}_{\mathfrak{A}} \rightleftarrows \mathfrak{X}_{\mathfrak{A}} : E^V$$

of $\mathfrak{C}_{\mathfrak{A}}$ with a subcategory $\mathfrak{X}_{\mathfrak{A}}$ of $\mathfrak{X}^{(V^{\mathrm{op}})}$. The dual equivalence (15) becomes

$$(16) \qquad\qquad D^V : \mathfrak{C} \rightleftarrows \mathfrak{X}_{\mathfrak{A}} : E^V$$

when rewritten in terms of the category $\mathfrak{C}$ that is equivalent to $\mathfrak{C}_{\mathfrak{A}}$.

Definition 32. The duality (16) is known as *diagrammatic duality* of $\mathfrak{C}$, or between $\mathfrak{C}$ and $\mathfrak{X}_{\mathfrak{A}}$, relative to the seed duality (14).

8 Examples of diagrammatic duality

8.1 Duality for distributive bilattices

According to Theorem 19(d), bounded distributive bilattices are diagrammatic relative to bounded distributive lattices. The duality theory for bounded distributive bilattices given by Pigozzi *et al.* (Mobasher et al., 2000, Cor. 12) may then be construed as a diagrammatic duality for bounded distributive bilattices relative to Priestley duality for bounded distributive lattices. Using Theorem 19(b), one further obtains a diagrammatic duality for general distributive bilattices, relative to Priestley duality (4.4) for general distributive lattices. An alternative approach to duality for general distributive bilattices has recently been presented by Cabrer and Priestley (Cabrer and Priestley, 2015).

8.2 Duality for Nelson algebras

Duality for Nelson algebras has been discussed in (Cignoli, 1986; Sendlewski, 1990). The considerations of 6.2 yield a diagrammatic duality for Nelson algebras, based on Esakia duality for Heyting algebras.

8.3 Cabalistic duality

It was observed in 6.3 that for a given type $\tau\colon \Omega \to \mathbb{N}$, the τ-algebras are diagrammatic relative to sets. More specifically, the category $\underline{\underline{\tau}}$ of τ-algebras is equivalent to a subcategory $\underline{\underline{\tau}}_{\mathbf{Set}}$ of the diagram category $\mathbf{Set}^{\Omega_\infty}$ given by the Ω-cospan Ω_∞. Now consider Lindenbaum-Tarski duality

$$D : \mathbf{Set} \rightleftarrows \mathbf{CABA} : E$$

between sets and complete atomic Boolean algebras (4.3). Then with Lindenbaum-Tarski duality as the seed duality (14), the corresponding diagrammatic duality

$$(17) \qquad D^{\Omega_\infty} : \underline{\underline{\tau}} \rightleftarrows \mathbf{CABA}_{\mathbf{Set}} : E^{\Omega_\infty}$$

given by (16) is known as *CABAlistic* or *cabalistic duality* for (the class of) τ-algebras. This terminology carries over to restrictions of (17) to subcategories of $\underline{\underline{\tau}}$.

The dual of the Ω-cospan Ω_∞ is known as the Ω-*span* Ω^∞. Thus in the cabalistic duality (17), the representation spaces are given by objects in the diagram category $\mathbf{CABA}^{\Omega^\infty}$, or their equivalents. An important detailed case, interpreting the relationship between quasigroups and the 3-nets they coordinatize, is worked out in 9.

8.4 Locally compact central piques and Suvorov duality

The considerations of 6.4, and most notably Proposition 30(b), yield a diagrammatic duality for locally compact central piques, based on Pontryagin duality for locally compact abelian groups (Romanowska and Smith, 2006, Th. 5.5). This duality specializes to the duality obtained by Suvorov (Suvarov, 1969) for locally compact, idempotent, entropic quasigroups (locally compact *quasigroup modes* in the terminology of (Romanowska and Smith, 2002)).

9 Quasigroups and 3-nets

As observed in 5.4, one general application of duality is the coordinatization of a geometry. In this section, it will be shown how cabalistic duality for quasigroups leads to the well-known geometric or combinatorial structures, namely 3-nets or (discrete) 3-webs, that are coordinatized by quasigroups.

9.1 3-nets

For a set N, let $\widehat{N}$ denote the diagonal $\{(n,n) \mid n \in N\}$. For binary relations α, β on N, let $\alpha \circ \beta$ denote the relation product

$$\{(l,n) \in N^2 \mid \exists m \in N \,.\, (l,m) \in \alpha \text{ and } (m,n) \in \beta\}$$

of α and β.

Definition 33. (Bruck, 1963, p.73), (Chein et al., 1990, §II.8), (Smith and Romanowska, 1999, Defn. I.4.2) (a) A 3-*net* is a relational structure $(N, \alpha_1, \alpha_2, \alpha_3)$ with three equivalence relations $\alpha_1, \alpha_2, \alpha_3$ on N such that

$$(18) \qquad \forall 1 \leq i \neq j \leq 3, \ \alpha_i \cap \alpha_j = \widehat{N} \text{ and } \alpha_i \circ \alpha_j = N^2.$$

(b) For $1 \leq i \leq 3$, the α_i-classes are known as *i-lines*.

The condition (18) means that N is isomorphic to the direct product $N^{\alpha_i} \times N^{\alpha_j}$ for $1 \leq i \neq j \leq 3$.

Definition 34. A 3-net N as in Definition 33 is *labeled* by a set Q iff for $1 \leq i \leq 3$, there is a bijection $\theta_i \colon Q \to N^{\alpha_i}$. In this case, an element q of Q is described as the *label* of the i-line q^{θ_i}.

9.2 Labeled 3-nets from quasigroups

Each quasigroup $(Q, \cdot)$ determines a 3-net Q^2 labeled by Q: The 1-line labeled by q in the 3-net specified by the quasigroup $(Q, \cdot)$ is

$$(19) \qquad\qquad\qquad \{(x,y) \in Q^2 \mid x = q\}.$$

The 2-line labeled by q is

$$(20) \qquad\qquad\qquad \{(x,y) \in Q^2 \mid y = q\}.$$

The 3-line labeled by q is

$$(21) \qquad\qquad \{(x,y) \in Q^2 \mid x \cdot y = q\}.$$

9.3 Quasigroups from labeled 3-nets

A quasigroup Q may be recovered from its labeled 3-net. For elements x, y, z of Q, one has $x \cdot y = z$ if and only if the unique point of intersection of the 1-line labeled x and the 2-line labeled y lies on the 3-line labeled z. More generally, the same procedure yields a quasigroup structure $(Q, \cdot)$ on the label set Q of a labeled 3-net.

9.4 3-nets from cabalistic duality

For the cabalistic duality considered here, a quasigroup $(Q, \cdot)$ is construed as carrying three binary operations: a left zero band or left projection operation

$$(22) \qquad\qquad p_1 \colon Q^2 \to Q; (x_1, x_2) \mapsto x_1,$$

a right zero band or right projection operation

$$(23) \qquad\qquad p_2 \colon Q^2 \to Q; (x_1, x_2) \mapsto x_2,$$

and the quasigroup multiplication

$$(24) \qquad\qquad p_3 \colon Q^2 \to Q; (x_1, x_2) \mapsto x_1 \cdot x_2.$$

The cabalistic dual of (22) is the **CABA**-morphism (inverse image function)

$$(25) \qquad\qquad p_1^{-1} \colon \mathcal{P}(Q) \to \mathcal{P}(Q^2); X \mapsto p_1^{-1} X.$$

This morphism is specified by its action on the atoms of $\mathcal{P}(Q)$, namely the singletons. For such an atom $\{q\}$, the image under (25) is

$$p_1^{-1}\{q\} = \{(x,y) \in Q^2 \mid x = q\}.$$

This is the 1-line (19) labeled by q in the 3-net specified by the quasigroup $(Q, \cdot)$. The 2- and 3-lines appear in similar fashion. For example, the image of an atom $\{q\}$ under the cabalistic dual p_3^{-1} of (24) is

$$p_3^{-1}\{q\} = \{(x,y) \in Q^2 \mid x \cdot y = q\},$$

the 3-line labeled by q. In this way, the cabalistic dual of the quasigroup Q may be interpreted as the 3-net coordinatized by Q.

9.5 Covariant and contravariant passage to the 3-net

In 9.4, the association of a 3-net to a quasigroup provided by cabalistic duality is contravariant with respect to quasigroup homomorphisms. For contrast, a covariant association is given in (Smith, 2008, Th. 1).

References

Bruck, R.H. (1963). What is a loop?, in *Studies in Modern Algebra*, Albert, A.A., Ed., Prentice-Hall, Engelwood Cliffs, NJ, 59–99.

Cabrer, L.M. and Priestley, H.A., (2015). Distributive bilattices from the perspective of natural duality theory, *Algebra Universalis* **73**. 103–141. `arXiv:1308.4495v2 [math.RA]`

Chein, O., Pflugfelder, H.A., and Smith, J.D.H., Eds. (1990). *Quasigroups and Loops: Theory and Applications*, Heldermann, Berlin.

Cignoli, R. (1986). The class of Kleene algebras satisfying interpolation property and Nelson Algebras, *Algebra Universalis* **23**, 262–292.

Clark, D.M. and Davey, B.A. (1998). *Natural Dualities for the Working Algebraist*, Cambridge University Press, Cambridge.

Dixmier, J. (1964). *Les C^*-Algèbres et leurs Représentations*, Gauthier-Villars, Paris.

Esakia, L.L. (1974). Topological Kripke models, *Soviet Mathematics – Doklady* **15**, 147–151.

Esakia, L.L. (1985). *Heyting Algebras I, Duality Theory* (Russian), Metsniereba, Tbilisi.

Herrlich, H. and Strecker, G.E. (1973). *Category Theory*, Allyn and Bacon, Boston, MA.

Johnstone, P.T. (1982). *Stone Spaces*, Cambridge University Press, Cambridge.

Loomis, L.H. (1953). *Introduction to Abstract Harmonic Analysis*, Van Nostrand, New York, NY.

Mobasher, B., Pigozzi, D., Slutzki, G., and Voutsadakis, G. (2000). A duality theory for bilattices, *Algebra Universalis* **43**, 109–125.

Movsisyan, Yu.M., Romanowska, A.B., and Smith, J.D.H. (2006). Superproducts, hyperidentities and algebraic structures of logic programming' *Journal of Combinatorial Mathematics and Combinatorial Computing* **58**, 101–111.

Mumford, D. (1988). *The Red Book of Varieties and Schemes*, Springer, Berlin.

Nelson, D. (1949). Constructible falsity, *Journal of Symbolic Logic* **14**, 16–26.

Odintsov, S.P. (2010). Priestley duality for paraconsistent Nelson's logic, *Studia Logica* **96**, 65–93.

Priestley, H.A. (1970). Representation of distributive lattices by means of ordered Stone spaces, *Bulletin of the London Mathematical Society* **2**, 186–190.

Pontryagin, L.S. (1966). *Topological Groups* (2nd. ed.), Gordon and Breach, New York, NY.

Rasiowa, H. (1958). *N*-lattices and constructive logics with strong negation, *Fundamenta Mathematicae* **46**, 61–80.

Rasiowa, H. (1974). *An Algebraic Approach to Non-Classical Logics*, North-Holland, Amsterdam/PWN, Warsaw.

Romanowska, A.B. and Smith, J.D.H. (2002). *Modes*, World Scientific, Singapore.

Romanowska, A.B. and Smith, J.D.H. (2006). Duality for central piques, *Quasigroups Related Systems* **14**, 153–159.

Romanowska, A.B. and Trakul, A. (1989). On the structure of some bilattices, in *Universal and Applied Algebra*, Hałkowska, K. and Stawski, B., Eds., World Scientific, Singapore, 235–253.

Sendlewski, A. (1990). Nelson algebras through Heyting ones: I, *Studia Logica* **49**, 106–126.

Smith, J.D.H. (2008). Quasigroup homotopies, semisymmetrization, and reversible automata, *International Journal of Algebra and Computing* **18**, 1203–1221.

Smith, J.D.H. (2007). *An Introduction to Quasigroups and their Representations*, Chapman & Hall/CRC, Boca Raton, FL.

Smith, J.D.H. and Romanowska, A.B. (1999). *Post-Modern Algebra*, Wiley, New York, NY.

Suvarov, N.M. (1969). A duality theorem for distributive transitive topological quasigroups (Russian), *Matem. Issledovanija* **4**, 162–166.

Tarski, A. (1935). Zur Grundlegung der Boole'schen Algebra, *Fundamenta Mathematicae* **24**, 177–198.

Boolean product representations of algebras via binary polynomials

Antonino Salibra, Antonio Ledda, and Francesco Paoli

Dedicated to Don Pigozzi
on the occasion of his eightieth birthday

Abstract We mimick the construction of *guard algebras* and show how to extract a Church algebra out of the binary functions on an arbitrary algebra, containing a Church subalgebra of binary polynomial operations. We put to good use the weak Boolean product representations of these Church algebras to obtain weak Boolean product representations of the original algebras. Although we cannot, in general, say much about the factors in these products, we identify a number of sufficient conditions for the stalks to be directly indecomposable. As an application, we prove that every skew Boolean algebra is a weak Boolean product of directly indecomposable skew Boolean algebras.

Key words: Church algebras, Guard algebras, skew Boolean algebras, weak Boolean product.

2010 Mathematics Subject Classification: 08A05, 06E75

1 Introduction

According to one of the most celebrated theorems of universal algebra, proved by Birkhoff at the very dawn of the discipline, every algebra in a variety has

Antonino Salibra
Università Ca'Foscari Venezia, Italy, e-mail: `salibra@dsi.unive.it`

Antonio Ledda
Università di Cagliari, Italy, e-mail: `antonio.ledda@unica.it`

Francesco Paoli
Università di Cagliari, Italy, e-mail: `paoli@unica.it`

© Springer International Publishing AG 2018
J. Czelakowski (eds.), *Don Pigozzi on Abstract Algebraic Logic,*
Universal Algebra, and Computer Science, Outstanding Contributions
to Logic 16, https://doi.org/10.1007/978-3-319-74772-9_12

a subdirect representation with subdirectly irreducible factors that belong to the same variety. This result, whose scope is as wide as it can be, is not very informative in itself. In some special cases, though, qualitatively superior representations can be attained by imposing additional *desiderata* on the target structure. Boolean products, of course, are the prime example of this situation.

Recall that a *weak Boolean product* of a family $(\mathbf{A}_i)_{i \in I}$ of algebras is a subdirect product $\mathbf{A} \leq \prod_{i \in I} \mathbf{A}_i$, where I can be endowed with a Boolean space topology such that: (i) the set $\{i \in I : a_i = b_i\}$ is open for all $a, b \in A$, and (ii) if $a, b \in A$ and $N \subseteq I$ is clopen, then the element c, defined by $c_i = a_i$ for $i \in N$ and $c_i = b_i$ for $i \in I - N$, belongs to A. Also, recall that a weak Boolean product of a family $(\mathbf{A}_i)_{i \in I}$ of algebras is a *Boolean product* if the set $\{i \in I : a_i = b_i\}$ is clopen for all $a, b \in A$. The study of (weak) Boolean products is motivated by Stone's representation theorem: every Boolean algebra $\mathbf{B}$ is isomorphic to the algebra of clopen subsets of its Stone space $S(\mathbf{B})$. Since the 1970s, various researchers have sought to generalise Stone's result to ever-larger classes of algebras. Pierce (Peirce, 1967) proved that every commutative ring with unit is representable as a Boolean product of directly indecomposable rings. Subsequently, Burris and Werner (Burris and Werner, 1979, 1980) obtained Boolean product representations for algebras in discriminator varieties. The technique of Boolean products underwent remarkable developments over the following years (Burris and Sankappanavar, 1981, Ch. 4.8), giving rise to further generalisations of Stone's theorem by Comer (covering the case of algebras with Boolean factor congruences: (Comer, 1971)) and by Vaggione (who emphasised the importance of central elements in (weak) Boolean product-like constructions: (Vaggione, 1996)).

Contemporaneous with these developments, researchers in theoretical computer science have extensively pursued the study of the if-then-else construct. Focussing solely on algebraic developments, Bloom and Tindall (Bloom and Tindall, 1983) and Mekler and Nelson (Mekker and Nelson, 1987), among others, investigated a number of algebraic functions modeling if-then-else constructs, including the ternary discriminator. With applications to logic in mind, Pigozzi introduced the concept of an equality-test algebra, and published a number of papers on these structures, including (Pigozzi, 1990, 1991). In a different direction, in (Bergman, 1991) Bergman modelled the if-then-else construct by considering Boolean algebras acting on sets. If the Boolean algebra of actions is the 2-element algebra, simply set $1(a, b) = a$ and $0(a, b) = b$ to mimic the if-then-else construct. The approach followed by the first author and Manzonetto in (Manzonetto and Salibra, 2008) differs from Bergman's in that the if-then-else is treated as a proper algebraic ternary operation q on a double-pointed algebra $\mathbf{A}$, having the property that for every $a, b \in A$, $q(1, a, b) = a$ and $q(0, a, b) = b$. The resulting variety of Church algebras is one of the fundamental notions in the present work and is investigated in (Manzonetto and Salibra, 2008, 2010; Salibra et al., 2013).

Not all (weak) Boolean product representations are in the same league. At one extreme we have the optimal example of discriminator varieties, whose members are representable as Boolean products of simple algebras; yet, the weak Boolean product construction is so flexible that little can be said about the factors of the product (the *stalks* of the representation, as we will call them hereafter) in the general case. The situation improves if $\mathcal{V}$ is a Church variety. Using Vaggione's concept of central element in a double-pointed algebra, it is proved in (Salibra et al., 2013) (following the lead of (Comer, 1971) and (Vaggione, 1996)) that every algebra $\mathbf{A}$ in a Church variety $\mathcal{V}$ admits a weak Boolean product representation $f : A \to \prod_{I \in S} A/\theta_I$ (S the spectrum of maximal ideals), and that the stalks $\mathbf{A}/\theta_I$ are directly indecomposable whenever the class of directly indecomposable members of $\mathcal{V}$ is a universal class. Outside the borders of the double-pointed territory, however, universal algebra is of little avail and we often have to proceed case by case.

In this paper, with an eye to extending the above results to arbitrary algebras, we mimic the construction of *guard algebras* (Urbanik, 1965; Bloom et al., 1990). In (Bloom et al., 1990), Bloom, Esik, and Manes proved a Cayley-type theorem for Boolean algebras, which says that any Boolean algebra is isomorphic to a Boolean algebra of binary functions on a set; such a Boolean algebra of binary functions is called a guard algebra. Borrowing from this idea, we introduce the notion of a substitution Church algebra $F(\mathbf{A})$ of binary functions on an arbitrary algebra $\mathbf{A}$, and show that this algebra always contains a Church subalgebra that we denote by $\widehat{\mathbf{A}}$ of binary polynomial operations, which latter may then be used to recover a weak Boolean product representation $f_{|A} : A \to \prod_{I \in S} A/\theta_I$ of $\mathbf{A}$ (S the spectrum of maximal ideals). Although we cannot, in general, say much about the factors in these products, we identify a number of sufficient conditions for the stalks to be directly indecomposable. As an application, we prove that every skew Boolean algebra is a weak Boolean product of directly indecomposable skew Boolean algebras.

The topics we cover, and the approach we adopt, throughout the paper are consonant with Don Pigozzi's angle on universal algebra. Pigozzi consistenly paid a special attention to the cross-fertilisation potential inherent not only to the application of universal algebraic methods to theoretical computer science, but also to the construction of abstract algebraic models of computational structures, which have oftentimes delivered an unexpected payoff — the creation of new tools for addressing purely algebraic problems (Pigozzi, 1991; Pigozzi and Salibra, 1998; Martins and Pigozzi, 2007).

The article is structured as follows. In Section 2 we dispatch various preliminaries on factor congruences, decomposition operations, Church algebras, and guard algebras. In Section 3 we introduce the idea of a *substitution Church algebra*. Roughly speaking, a substitution Church algebra is an algebra $\mathbf{A}$ of given type ν expanded by a ternary operation $q^{\mathbf{A}}(a,b,c)$ and constants 0 and 1 such that $(\mathbf{A}, q^{\mathbf{A}}, 0, 1)$ is a Church algebra and, for each

n-ary $f \in \nu$, the operation $f^{\mathbf{A}}$ respects $q^{\mathbf{A}}(a,b,c)$. We also identify a certain subalgebra $\mathbf{A}_0 \leq \mathbf{A}$, called *zero-dimensional*, that plays an important role in subsequent developments. In Section 4, we show that any algebra $\mathbf{A}$ of given type ν is isomorphic to the zero-dimensional subreduct $F(\mathbf{A})_0$ of a substitution Church algebra $F(\mathbf{A})$ of binary functions on A, obtained by appropriately tweaking the guard algebra construction. We also prove that the central elements of any subalgebra $\mathbf{B} \leq F(\mathbf{A})$ such that $F(\mathbf{A})_0 \subseteq B$ correspond to decomposition operations enjoying certain commutation properties; such a subalgebra $\mathbf{B}$ is called a *functional Church algebra of value domain* $\mathbf{A}$.

In Section 5, we consider the situation in which the functional Church algebra of value domain $\mathbf{A}$ is the algebra $\widehat{\mathbf{A}}$ of binary polynomial operations on $\mathbf{A}$. We show that the central elements of $\widehat{\mathbf{A}}$ are exactly the operations on $\mathbf{A}$ that are simultaneously polynomial operations on $\mathbf{A}$ and decomposition operations on $\mathbf{A}$. From this observation it follows that, given an arbitrary algebra $\mathbf{A}$, the map $f : \widehat{A} \to \prod_{I \in S} \widehat{A/\theta_I}$ (S the spectrum of maximal ideals) yields a weak Boolean product representation of $\widehat{\mathbf{A}}$, the restriction of which to the constant polynomials provides a weak Boolean product representation $f_{|A} : A \to \prod_{I \in S} A/\theta_I$ of $\mathbf{A}$. Moreover, we identify sufficient conditions under which this representation has directly indecomposable stalks. In Section 6, we apply these results to a concrete setting: building on previous results from (Cvetko-Vah and Salibra, 2015), we show that every skew Boolean algebra (Leech, 1990) is a weak Boolean product of directly indecomposable skew Boolean algebras.

2 Preliminaries

If $\mathbf{A}$ is an algebra and $x, y \in A$, then $\theta(x, y)$ denotes the least congruence on $\mathbf{A}$ including the pair (x, y). We denote respectively by Δ, ∇ the least and the greatest congruence of the congruence lattice $\mathrm{Con}(\mathbf{A})$.

If $f : A^k \to A$ and $g_1, \ldots, g_k : A^n \to A$ are maps, then we denote by $f\langle g_1, \ldots, g_k \rangle : A^n \to A$ the function defined as follows:

$$(1) \qquad f\langle g_1, \ldots, g_k \rangle(x_1, \ldots, x_n) = f(g_1(x_1, \ldots, x_n), \ldots, g_k(x_1, \ldots, x_n)).$$

We recall from (McKenzie et al., 1987) that, if $f, g : A \times A \to A$ are binary maps, then f and g *commute*, and we write $f \, \mathrm{Cm} \, g$, if the following condition holds for all $x_{ij} \in A$:

$$(2) \qquad f(g(x_{11}, x_{12}), g(x_{21}, x_{22})) = g(f(x_{11}, x_{21}), f(x_{12}, x_{22})).$$

Equation (2) holds for f and g iff f is a homomorphism of $(A, g)^2$ into (A, g) iff g is a homomorphism of $(A, f)^2$ into (A, f).

2.1 Factor congruences and decomposition operations

Definition 1. *A congruence ϕ on an algebra* $\mathbf{A}$ *is a factor congruence if there exists a congruence $\bar{\phi}$ such that $\phi \cap \bar{\phi} = \Delta$ and $\phi \circ \bar{\phi} = \nabla$. In this case we call $(\phi, \bar{\phi})$* a pair of complementary factor congruences.

Under the hypotheses of Definition 1, the homomorphism $f : \mathbf{A} \to \mathbf{A}/\phi \times \mathbf{A}/\bar{\phi}$ defined by $f(x) = (x/\phi, x/\bar{\phi})$ is an isomorphism. Consequently, $(\phi, \bar{\phi})$ is a pair of complementary factor congruences of $\mathbf{A}$ if, and only if, $\mathbf{A} \cong \mathbf{A}/\phi \times \mathbf{A}/\bar{\phi}$ under the natural map $x \mapsto (x/\phi, x/\bar{\phi})$. Δ and ∇ are the *trivial* factor congruences, corresponding to $\mathbf{A} \cong \mathbf{A} \times \mathbf{B}$, where $\mathbf{B}$ is a trivial algebra; of course, $\mathbf{B}$ is isomorphic to $\mathbf{A}/\nabla$ and $\mathbf{A}$ is isomorphic to $\mathbf{A}/\Delta$.

We denote by $\mathcal{FC}(\mathbf{A})$ the set of factor congruences of an algebra $\mathbf{A}$. We recall that factor congruences in a generic algebra do not satisfy any particular condition. For example, the set of factor congruences is not in general a sublattice of the lattice of all congruences.

An algebra $\mathbf{A}$ is *directly indecomposable* if $\mathcal{FC}(\mathbf{A}) = \{\Delta, \nabla\}$. Clearly, every subdirectly irreducible algebra is directly indecomposable, while the converse need not hold.

Factor congruences can be characterised in terms of certain algebra homomorphisms called *decomposition operations* (see (McKenzie et al., 1987, Def. 4.32) for more details).

Definition 2. *Let $\mathbf{A}$ be an algebra of type ν. A* decomposition operation *on $\mathbf{A}$ is a function $f : A \times A \to A$ satisfying the following conditions:*

D1: $f(x,x) = x$;
D2: $f(f(x_{11}, x_{12}), f(x_{21}, x_{22})) = f(x_{11}, x_{22})$;
D3: f *is an algebra homomorphism from* $\mathbf{A} \times \mathbf{A}$ *into* $\mathbf{A}$.

We denote by $\mathcal{DE}(\mathbf{A})$ the set of all decomposition operations on $\mathbf{A}$.

There exists a bijective correspondence between pairs of complementary factor congruences and decomposition operations, and thus, between decomposition operations and factorisations of the form $\mathbf{A} \cong \mathbf{B} \times \mathbf{C}$.

Proposition 3. (McKenzie et al., 1987, Thm. 4.33) *Let $\mathbf{A}$ be an algebra of type ν. Given a decomposition operation f on $\mathbf{A}$, the binary relations θ_f and $\bar{\theta}_f$ defined by:*

$$x \; \theta_f \; y \quad \text{iff} \quad f(x,y) = x,$$
$$x \; \bar{\theta}_f \; y \quad \text{iff} \quad f(x,y) = y,$$

form a pair of complementary factor congruences. Conversely, given a pair $(\phi, \bar{\phi})$ of complementary factor congruences, the function f defined by:

$$(3) \qquad\qquad f(x,y) = u \; \text{iff} \; y \phi u \bar{\phi} x,$$

determines a decomposition operation on $\mathbf{A}$ such that $\phi = \theta_f$ and $\bar{\phi} = \bar{\theta}_f$.

Notice that if $(\phi, \bar{\phi})$ is a pair of complementary factor congruences, then for all x and y there is exactly one element u such that $y\phi u\bar{\phi}x$.

An algebra $\mathbf{A}$ has *Boolean factor congruences* (BFC, for short) if the factor congruences of $\mathbf{A}$ form a distributive sublattice of the congruence lattice $\mathrm{Con}(\mathbf{A})$ of $\mathbf{A}$. A class of algebras has BFC if each algebra in the class has BFC.

A congruence ϕ is said to be:

- *balanced*, if $\phi = (\phi \vee \theta) \cap (\phi \vee \bar{\theta})$ for all pairs $(\theta, \bar{\theta})$ of complementary factor congruences.
- *bi-balanced* if ϕ is a balanced factor congruence which admits a balanced factor complement.

We have that:

Lemma 4. (Swamy and Suryanarayana Murti, 1981, Theorems 1 and 2) *(i) A congruence ϕ is balanced if and only if $\phi \circ \theta = \theta \circ \phi$ and $(\phi \vee \theta) \cap \bar{\theta} \subseteq \phi$ for every factor congruence θ. (ii) The set $\mathcal{B}(\mathbf{A})$ of all bi-balanced factor congruences is the universe of a permutable Boolean sublattice of $\mathrm{Con}(\mathbf{A})$, which we also denote by $\mathcal{B}(\mathbf{A})$.*

2.2 Church Algebras

The key observation motivating the introduction of *Church algebras* (Manzonetto and Salibra, 2008) is that many algebras arising in completely different fields of mathematics — including Heyting algebras, rings with unit, or combinatory algebras — have a term operation q satisfying the fundamental properties of the if-then-else connective:

$$q(1, x, y) \approx x \text{ and } q(0, x, y) \approx y.$$

As simple as they may appear, these properties are enough to yield rather strong results. This motivates the next definition.

Definition 5. *An algebra $\mathbf{A}$ of type ν is a Church algebra if there are term definable elements $0^{\mathbf{A}}, 1^{\mathbf{A}} \in A$ and a term operation $q^{\mathbf{A}}$ such that, for all $a, b \in A$*

$$q^{\mathbf{A}}(1^{\mathbf{A}}, a, b) = a \text{ and } q^{\mathbf{A}}(0^{\mathbf{A}}, a, b) = b.$$

A variety $\mathcal{V}$ of type ν is a Church variety if every member of $\mathcal{V}$ is a Church algebra with respect to the same term $q(x, y, z)$ and the same constants $0, 1$.

Examples of Church algebras include FL_{ew}-algebras (commutative, integral and double-pointed residuated lattices, for which see (Galatos, 2007)) and, in particular, Heyting algebras and thus also Boolean algebras; ortholattices; rings with unit; combinatory algebras.

Expanding on an idea due to Vaggione (Vaggione, 1996), we also define:

Definition 6. *An element e of a Church algebra* $\mathbf{A}$ *is called* central *if the pair* $(\theta(e,0),\theta(e,1))$ *is a pair of complementary factor congruences on* $\mathbf{A}$*. A central element e is nontrivial if $e \notin \{0,1\}$. By* $\mathrm{Ce}(\mathbf{A})$ *we denote the* centre *of* $\mathbf{A}$*, i.e., the set of central elements of the algebra* $\mathbf{A}$*.*

Proposition 7. (Salibra et al., 2013, Prop. 3.6) *An element e of a Church algebra* $\mathbf{A}$ *of type ν is central if and only if it satisfies the following conditions for all $a, a_{ij}, \overline{b}, \overline{c}$ in A:*

A1: $q(e,a,a) = a$.
A2: $q(e,q(e,a_{11},a_{12}),q(e,a_{21},a_{22})) = q(e,a_{11},a_{22})$.
A3: $q(e,\sigma(\overline{b}),\sigma(\overline{c})) = \sigma(q(e,b_1,c_1),\ldots,q(e,b_n,c_n))$*, for every $\sigma \in \nu$ of arity n.*
A4: $q(a,1,0) = a$.

It is proved in (Salibra et al., 2013, Thm. 3.7) that Church algebras have BFC and that, by defining

$$(4) \qquad x \wedge y = q(x,y,0); \quad x \vee y = q(x,1,y); \quad x' = q(x,0,1),$$

we get:

Theorem 8. *Let* $\mathbf{A}$ *be a Church algebra. Then* $c[\mathbf{A}] = (\mathrm{Ce}(\mathbf{A}),\vee,\wedge,{}',0,1)$ *is a Boolean algebra which is isomorphic to the Boolean algebra of factor congruences of* $\mathbf{A}$*.*

It clearly follows that a Church algebra is directly indecomposable iff $\mathrm{Ce}(\mathbf{A}) = \{0,1\}$.

Corollary 9. *Let* $\mathbf{A}$ *be a Church algebra. For every decomposition operation f on* $\mathbf{A}$*, there exists a central element e such that $f(x,y) = q(e,y,x)$, $\theta_f = \theta(e,0)$ and $\overline{\theta}_f = \theta(e,1)$.*

Theorem 8, together with theorems by Comer (Comer, 1971) and Vaggione (Vaggione, 1996), implies:

Theorem 10. (Salibra et al., 2013, Thm. 3.8) *Let* $\mathbf{A}$ *be a Church algebra, S be the Boolean space of maximal ideals of $c[\mathbf{A}]$ and $f : A \to \prod_{I \in S} A/\theta_I$ be the map defined by*

$$f(a) = (a/\theta_I : I \in S),$$

where $\theta_I = \bigvee_{e \in I} \theta(0,e)$. Then:

(1) *f gives a weak Boolean representation of* $\mathbf{A}$*;*
(2) *f provides a Boolean representation of* $\mathbf{A}$ *iff, for all $a \neq b \in A$, there exists a least central element e such that $q(e,a,b) = a$ (i.e., $(a,b) \in \theta(0,e)$).*

In general, not much can be said about the factors in this representation for a generic Church variety $\mathcal{V}$. However, these factors are guaranteed to be directly indecomposable provided that the directly indecomposable members of $\mathcal{V}$ form a universal class. In fact, following (Vaggione, 1996), it is shown in (Salibra et al., 2013, Thm. 3.9) that:

Theorem 11. *Let $\mathcal{V}$ be a Church variety of type ν. Then, the following conditions are equivalent:*

(1) *For all $\mathbf{A} \in \mathcal{V}$, the stalks $\mathbf{A}/\theta_I$ ($I \in S$ a maximal ideal) are directly indecomposable.*

(2) *The class $\mathcal{V}_{\mathrm{di}}$ of directly indecomposable members of $\mathcal{V}$ is a universal class.*

2.3 Guard algebras

Let $\mathbf{A}$ be an algebra of a fixed type ν. We add to ν a symbol c_a of arity 0 for each $a \in A$, and call the new type ν_A. The binary terms of type ν_A are called the *binary polynomials* of $\mathbf{A}$. If $p = p(x,y)$ is a polynomial, we call *polynomial operation* the interpretation $p^{\mathbf{A}} : A \times A \to A$ of p in the algebra $\mathbf{A}$. Moreover, oftentimes we use the same symbol a for an element $a \in A$ and its realisation in the type ν_A. The set of all binary polynomial operations on $\mathbf{A}$ is noted as $P^2(\mathbf{A})$.

In 1965, K. Urbanik (Urbanik, 1965) defined an algebra of polynomial operations along the following lines. Given an algebra $\mathbf{A}$ of type ν, he set:

$$\mathbf{B}_{\mathbf{A}} = (P^2(\mathbf{A}), \vee, \wedge, ', 0, 1),$$

where:

- $(p_1 \vee p_2)(x,y) = p_1(x, p_2(x,y))$;
- $(p_1 \wedge p_2)(x,y) = p_1(p_2(x,y), y)$;
- $p'(x,y) = p(y,x)$;
- $1(x,y) = x$ and $0(x,y) = y$.

We reproduce hereafter the main result in his paper:

Theorem 12. *If $\mathbf{A}$ is an idempotent algebra that has an essentially binary operation and no essentially n-ary polynomial operation for some $n \geq 3$, then $\mathbf{B}_{\mathbf{A}}$ is a finite Boolean algebra.*

Later on, Urbanik's ideas were developed along several different directions. On the one hand, instead of focussing on polynomial operations on an algebra $\mathbf{A}$, some authors have considered more general sets of binary functions on $\mathbf{A}$ (or even on an unstructured set X) satisfying appropriate closure conditions. On the other hand, there have been attempts to replace the rather unwieldy assumptions of Theorem 12 by equational conditions that **A** must satisfy for the result to hold true. Finally, it has been investigated whether *every* Boolean algebra is so representable. Bloom et al. (Bloom et al., 1990) proved the following result.

Theorem 13.   (1) *Let X be a set, and let Y be any set of binary functions on X (i.e., functions from $X \times X$ into X) that is closed under the operations $\vee, \wedge, ', 0, 1$, defined as above. Then the algebra*

$$\mathbf{B}'_X = (Y, \vee, \wedge, ', 0, 1)$$

is a Boolean algebra if all functions in Y satisfy conditions (D1) and (D2) in Definition 2 and commute with each other. Such Boolean algebras of binary functions on X are called guard algebras *on X.*

(2) *Every Boolean algebra $\mathbf{A}$ is isomorphic to a guard algebra on an appropriate set X.*

The denomination "guard algebra" is clearly inspired by guard conditions in computer science, see e.g. (Manes and Arbib, 1986) — in fact, the target algebra in part (2) of Theorem 13 is an algebra on the set of all polynomial "if-then-else" operations $q(a,\text{-},\text{-})$, where $a \in A$ and q is the Church term for Boolean algebras. Observe that part (1) of the same theorem implies that the guard algebra of all polynomial decomposition functions on an algebra $\mathbf{A}$ is a Boolean algebra. For other results along these lines, see e.g. (Padmanabhan and Penner, 1967) or (Movsisyan, 2009).

3 Substitution Church algebras

As important as they are, the results of Section 2.3 are still somewhat unsatisfactory in that, given some algebra, one obtains a Boolean algebra of polynomial operations in the above-described manner only under rather restrictive conditions. We intend to generalise this approach in such a way as to construct *Church* algebras of functions out of arbitrary algebras. Since the resulting Church algebras will enjoy special properties, we need an abstract concept to accommodate them, which it is the aim of the present section to provide.

Let ν be a type of algebra and let $\nu' = \nu \cup \{q, 0, 1\}$ be the expansion of ν by the ternary operation symbol q and the constants $0, 1$.

Definition 14. *A* substitution Church algebra *is an algebra $\mathbf{S} = (S, \sigma, q, 0, 1)_{\sigma \in \nu}$ of type ν' satisfying the following identities:*

S1: *The Church algebra identities for $(S, q, 0, 1)$.*
S2: $q(x, 1, 0) \approx x$.
S3: $q(q(x, y, z), t, u) \approx q(x, q(y, t, u), q(z, t, u))$.

S4: $q(\sigma(\bar{x}),y,z) \approx \sigma(q(x_1,y,z),\ldots,q(x_k,y,z))$ *for every* $\sigma \in \nu$ *of arity* k.

As usual, define $x \wedge y = q(x,y,0)$, $x \vee y = q(x,1,y)$ and $x' = q(x,0,1)$.

Proposition 15. *Any substitution Church algebra* **S** *satisfies the following conditions for all* $a,b,c \in S$:

(1) $(S,\wedge,1)$ *and* $(S,\vee,0)$ *are monoids with respective absorbing elements* 0 *and* 1.
(2) $(a')' = a$.
(3) $(a \wedge b)' = a' \vee b'$ *and* $(a \vee b)' = a' \wedge b'$.
(4) $0' = 1; \quad 1' = 0$.
(5) $q(a',b,c) = q(a,c,b)$.

Proof. Let $a,b,c \in S$. Then:
(1) $(a \wedge b) \wedge c = q(q(a,b,0),c,0) =_{S2,S0} q(a,q(b,c,0),0) = a \wedge (b \wedge c)$ and $(a \vee b) \vee c = q(q(a,1,b),1,c) =_{S2,S0} q(a,1,q(b,1,c)) = a \vee (b \vee c)$.
(2) $(a')' = q(q(a,0,1),0,1) =_{S2} q(a,q(0,0,1),q(1,0,1)) =_{S0} q(a,1,0) =_{S1} a$.
(3) $(a \wedge b)' = q(q(a,b,0),0,1) =_{S2} q(a,q(b,0,1),1) =_{S2} q(q(a,0,1),1,q(b,0,1)) = a' \vee b'$. Similarly for $(a \vee b)'$.
(4) Trivial.
(5) follows by applying (S2) to $q(q(a,0,1),b,c)$.

If **S** is an arbitrary substitution Church algebra, then $b \in S$ is *zero-dimensional* if $q(b,x,y) = b$ for all $x,y \in S$. We denote by S_0 the set of all zero-dimensional elements of **S**.

Lemma 16. S_0 *is a subuniverse of the* ν*-reduct of* **S** *satisfying the following condition:*

$$x \in S \text{ and } y,z \in S_0 \;\Rightarrow\; q(x,y,z) \in S_0.$$

Proof. Let $\bar{b} \equiv b_1,\ldots,b_k \in S_0$, $t,u \in S$ and $\sigma \in \nu$ of arity k. Then we have:

$$q(\sigma(\bar{b}),t,u) = \sigma(q(b_1,t,u),\ldots q(b_k,t,u)) \text{ by (S3)}$$
$$= \sigma(\bar{b}) \qquad\qquad\qquad \text{by } b_i \in S_0.$$

Moreover, for every $x \in S$ and $y,z \in S_0$, we have:

$$q(q(x,y,z),t,u) =_{S2} q(x,q(y,t,u),q(z,t,u)) = q(x,y,z),$$

because $y,z \in S_0$.

We denote by $\mathbf{S}_0$ the ν-algebra of universe S_0.

Example 17. *Let* $\mathbf{F}_\nu$ *be the absolutely free algebra of type* ν *over a countable infinite set* X *of generators and let* $x_0,x_1 \in X$. *We define*

$$0^{\mathbf{F}_\nu} = x_0, \quad 1^{\mathbf{F}_\nu} = x_1, \quad q^{\mathbf{F}_\nu}(t,u_1,u_0) = t[u_1/x_1,u_0/x_0],$$

where $t[u_1/x_1, u_0/x_0]$ is the term obtained by substituting the term u_i for every occurrence of x_i in the term t $(i = 0, 1)$. The algebra $(F_\nu, \sigma, q^{\mathbf{F}\nu}, 1^{\mathbf{F}\nu}, 0^{\mathbf{F}\nu})$ is a substitution Church algebra. A term t is zero-dimensional if and only if the variables x_0 and x_1 do not occur in t.

4 Substitution Church algebras of binary functions

In what follows, we dovetail the results of Subsections 2.2 and 2.3. The fact that central elements in a Church algebra form a Boolean algebra isomorphic to the Boolean algebra of its factor congruences invites a conjecture to the effect that Theorem 13 can be appropriately generalised. Mimicking the construction of guard algebras, in fact, we construct a substitution Church algebra $F(\mathbf{A})$ out of the binary functions on an arbitrary algebra $\mathbf{A}$, which remains embedded therein as its subreduct of zero-dimensional elements. We show that the central elements of any subalgebra $\mathbf{B}$ of $F(\mathbf{A})$ containing $F(\mathbf{A})_0$ are decomposition operations on $\mathbf{A}$ that commute with every element of B. We also prove that the factor congruences corresponding to decomposition operations on $\mathbf{A}$ that commute with every other decomposition operation are bi-balanced and form a Boolean sublattice of the lattice of congruences of $\mathbf{A}$.

Let $\mathbf{A}$ be an algebra of type ν and $F(A)$ be the set of all functions from $A \times A$ into A. Consider the algebra of type ν'

$$F(\mathbf{A}) = (F(A), \sigma^{F(\mathbf{A})}, q^{F(\mathbf{A})}, \pi_0^{F(\mathbf{A})}, \pi_1^{F(\mathbf{A})})_{\sigma \in \nu},$$

whose operations are defined as follows (for all $f, g, h, f_1, \ldots, f_k \in F(A)$ and all $a, b \in A$):

(1) $\pi_0^{F(\mathbf{A})}(a, b) = b,$
(2) $\pi_1^{F(\mathbf{A})}(a, b) = a,$
(3) $q^{F(\mathbf{A})}(f, g, h) = f\langle g, h \rangle,$
(4) $\sigma^{F(\mathbf{A})}(f_1, \ldots, f_k) = \sigma^{\mathbf{A}}\langle f_1, \ldots, f_k \rangle,$

where the operation $-\langle -, \ldots, - \rangle$ is defined in Equation (1) on page 300.

Proposition 18. *(i) The algebra $F(\mathbf{A})$ is a substitution Church algebra.*

(ii) The algebra $\mathbf{A}$ is isomorphic to the ν-algebra $F(\mathbf{A})_0$ of all zero-dimensional elements of $F(\mathbf{A})$.

Proof. (i) It is immediate to see that $F(\mathbf{A})$ abides by the conditions of Definition 14. By way of example, we show Condition (S1); in fact, for $a, b \in A$, $f\langle \pi_1, \pi_0 \rangle(a, b) = f(\pi_1(a, b), \pi_0(a, b)) = f(a, b)$. (ii) The required map, for any $a \in A$, is $a \mapsto f_a$, where $f_a(x, y) = a$.

Any subalgebra $\mathbf{B}$ of $F(\mathbf{A})$ such that $F(\mathbf{A})_0 \subseteq B$ is called a *functional Church algebra of value domain* $\mathbf{A}$.

In the next proposition we give a representation theorem for substitution Church algebras, in a similar vein to Theorem 13(2).

Proposition 19. *Let* **S** *be a substitution Church algebra of type* ν'*. The map*

$$(5) \qquad\qquad a \in S \mapsto q^{\mathbf{S}}(a,\text{-},\text{-}) : S_0 \times S_0 \to S_0$$

is a homomorphism from **S** *to the functional Church algebra* $F(\mathbf{S}_0)$*, whose value domain is the* ν*-algebra of all zero-dimensional elements of* **S**.

Proof. If $a \in S$, then $q^{\mathbf{S}}(a,x,y) \in S_0$ for all $x,y \in S_0$, by Lemma 16. It follows that the map defined in (5) is well-defined. We now prove that it is a homomorphism.
Let $x,y \in S_0$.

$$
\begin{aligned}
q^{\mathbf{S}}(\sigma^{\mathbf{S}}(\bar{a}),x,y) &= \sigma^{\mathbf{S}}(q(a_1,x,y),\ldots,q(a_k,x,y)) && \text{by (S3)} \\
&= \sigma^{\mathbf{S}}\langle q(a_1,\text{-},\text{-}),\ldots,q(a_k,\text{-},\text{-})\rangle(x,y) \\
&= \sigma^{F(\mathbf{S}_0)}(q(a_1,\text{-},\text{-}),\ldots,q(a_k,\text{-},\text{-}))(x,y).
\end{aligned}
$$

$$
\begin{aligned}
q^{\mathbf{S}}(q^{\mathbf{S}}(\bar{a}),x,y) &= q^{\mathbf{S}}(a_1,q^{\mathbf{S}}(a_2,x,y),q^{\mathbf{S}}(a_3,x,y)) && \text{by (S2)} \\
&= q^{F(\mathbf{S}_0)}(q^{\mathbf{S}}(a_1,\text{-},\text{-}),q^{\mathbf{S}}(a_2,\text{-},\text{-}),q^{\mathbf{S}}(a_3,\text{-},\text{-}))(x,y).
\end{aligned}
$$

Moreover, $q^{\mathbf{S}}(0,\text{-},\text{-}) = \pi_0^{F(\mathbf{S}_0)}$ and $q^{\mathbf{S}}(1,\text{-},\text{-}) = \pi_1^{F(\mathbf{S}_0)}$. This concludes the proof that $a \mapsto q(a,\text{-},\text{-})$ is a homomorphism.

4.1 Commuting decomposition operations

Recall that two binary functions f and g on A commute, noted by $f \text{ Cm } g$, if the equation (2) of Section 2 holds. We denote by $\text{Cm}(f)$ the set $\{g \in F(A) : f \text{ Cm } g\}$. If $g,h \in \text{Cm}(f)$ it is not the case, in general, that g and h commute.

Proposition 20. *Let* $f : A \times A \to A$ *be a decomposition operation on* **A**. *Then the set* $\text{Cm}(f)$ *is a subuniverse of the functional Church algebra* $F(\mathbf{A})$.

Proof. Let $\sigma \in \nu$ of arity k. We show that, for every $g_1,\ldots,g_k \in \text{Cm}(f)$, f and $\sigma^{\mathbf{A}}\langle g_1,\ldots,g_k\rangle$ commute.
Let $H = f(\sigma^{\mathbf{A}}\langle g_1,\ldots,g_k\rangle(x_1,x_2),\sigma^{\mathbf{A}}\langle g_1,\ldots,g_k\rangle(x_3,x_4))$. Then we have:

$$
\begin{aligned}
H &= f(\sigma^{\mathbf{A}}(\ldots g_i(x_1,x_2)\ldots),\sigma^{\mathbf{A}}(\ldots g_i(x_3,x_4)\ldots)) \\
&= \sigma^{\mathbf{A}}(\ldots f(g_i(x_1,x_2),g_i(x_3,x_4))\ldots) && f \text{ homomorphism} \\
&= \sigma^{\mathbf{A}}(\ldots g_i(f(x_1,x_3),f(x_2,x_4))\ldots) && \text{by } f \text{ Cm } g_i \\
&= \sigma^{\mathbf{A}}\langle g_1,\ldots,g_k\rangle(f(x_1,x_3),f(x_2,x_4)).
\end{aligned}
$$

We show that, for every $g,h,u \in \text{Cm}(f)$, f and $q^{F(\mathbf{A})}(g,h,u)$ commute. Let $K = f(g\langle h,u\rangle(x_1,x_2),g\langle h,u\rangle(x_3,x_4))$. Then we have:

$$K = f(g(h(x_1,x_2),u(x_1,x_2)),g(h(x_3,x_4),u(x_3,x_4)))$$
$$= g(f(h(x_1,x_2),h(x_3,x_4)),f(u(x_1,x_2),u(x_3,x_4))) \text{ by } f \text{ Cm } g$$
$$= g(h(f(x_1,x_3),f(x_2,x_4)),u(f(x_1,x_3),f(x_2,x_4))) \text{ by } f \text{ Cm } h,u$$
$$= g\langle h,u\rangle(f(x_1,x_3),f(x_2,x_4)).$$

In the hypothesis of Proposition 20, $\mathrm{Cm}(f)$ contains all binary constant functions. We denote by $\mathbf{Cm}(f)$ the functional Church algebra of value domain $\mathbf{A}$ with universe $\mathrm{Cm}(f)$.

The following proposition characterises commuting decomposition operations in terms of factor congruences.

Recall that, if $g : A \times A \to A$ is a function, then g' is the function defined by

$$g'(x,y) = g(y,x).$$

Proposition 21. *Let $\mathbf{A}$ be an algebra of type ν and g,h be decomposition operations on $\mathbf{A}$. Then the following conditions are equivalent:*

(i) *g Cm h;*
(ii) *$\theta_x = (\theta_x \vee \theta_y) \cap (\theta_x \vee \theta_{y'})$ for every $x,y \in \{g,h,g',h'\}$;*
(iii) *$\theta_x \circ \theta_y = \theta_y \circ \theta_x$ and $(\theta_x \vee \theta_y) \cap \theta_{y'} \subseteq \theta_x$ for every $x,y \in \{g,h,g',h'\}$.*

Proof. The equivalence of (ii) and (iii) follows from the proof of (Swamy and Suryanarayana Murti, 1981, Theorem 1).

(i) $\Rightarrow$ (iii): We prove that g Cm h implies:

$$(6) \qquad \exists z(x\theta_h z\theta_g y) \Rightarrow x\theta_g h(x,y)\theta_h y.$$

Let $x\theta_h z\theta_g y$. Then $h(x,z) = x$ and $g(y,z) = y$. By this last equality and by g Cm h we derive that $g(h(x,y),h(x,z)) = h(g(x,x),g(y,z)) = h(x,y)$, whence $h(x,y)\theta_g h(x,z) = x$. Since $h(y,h(x,y)) = h(y,y) = y$, then the conclusion $x\theta_g h(x,y)\theta_h y$ follows and $\theta_g \circ \theta_h = \theta_h \circ \theta_g$ holds.

We show that $(\theta_g \vee \theta_h) \cap \theta_{h'} \subseteq \theta_g$. Let $x\theta_g u\theta_h y\bar{\theta}_h x$. We have to show that $g(x,y) = x$ (i.e., $x\theta_g y$) assuming $g(x,u) = x$, $h(y,x) = x$ and $h(y,u) = y$:

$$g(x,y) = g(h(y,x),h(y,u)) = h(g(y,y),g(x,u)) = h(y,x) = x,$$

where the second equality holds because by hypothesis g and h commute.

From the hypothesis g Cm h it follows that g' Cm h, g Cm h' and g' Cm h'. Then the other conditions of (iii) can be proved in a similar way.

(ii) $\Rightarrow$ (i): By (Swamy and Suryanarayana Murti, 1981, Lemma 3(2)) and by the hypothesis we derive that $\phi_1 \equiv \theta_h \vee \theta_g$, $\phi_2 \equiv \theta_h \vee \bar{\theta}_g$, $\phi_3 \equiv \bar{\theta}_h \vee \theta_g$ and $\phi_4 \equiv \bar{\theta}_h \vee \bar{\theta}_g$ are factor congruences. Then we have:

$$\mathbf{A} \cong \mathbf{A}/\theta_h \times \mathbf{A}/\bar{\theta}_h \cong [\mathbf{A}/(\theta_h \vee \theta_g) \times \mathbf{A}/(\theta_h \vee \bar{\theta}_g)] \times [\mathbf{A}/(\bar{\theta}_h \vee \theta_g) \times \mathbf{A}/(\bar{\theta}_h \vee \bar{\theta}_g)]$$

because by hypothesis $\theta_h = (\theta_h \vee \theta_g) \cap (\theta_h \vee \bar{\theta}_g)$ and $\bar{\theta}_h = (\bar{\theta}_h \vee \theta_g) \cap (\bar{\theta}_h \vee \bar{\theta}_g)$. It is easy to check that the map $t(x_1,x_2,x_3,x_4) = h(g(x_4,x_3),g(x_2,x_1))$ is the unique element $u \in A$ such that $u\phi_i x_i$ for every $i = 1,2,3,4$, and that t satisfies the following identities:

- $t(x,x,x,x) = x$;
- $t(t(x_{11},x_{12},x_{13},x_{14}),\ldots,t(x_{41},x_{42},x_{43},x_{44})) = t(x_{11},x_{22},x_{33},x_{44})$;
- t commutes with the operations of $\mathbf{A}$.

Then the conclusion that h and g commute follows from (McKenzie et al., 1987, Exercise 4.38(15)).

4.2 Central elements of functional Church algebras

Recall that, by Theorem 8, the algebra

$$c[\mathbf{S}] = (\mathrm{Ce}(\mathbf{S}), \vee, \wedge, ', 0, 1)$$

of central elements of a substitution Church algebra $\mathbf{S}$ is a Boolean algebra isomorphic to the Boolean algebra of factor congruences of $\mathbf{S}$. We now prove the main theorem of this section.

Theorem 22. *Let $\mathbf{A}$ be a ν-algebra and $\mathbf{B}$ be a functional Church ν'-algebra of value domain $\mathbf{A}$. Then the following conditions are equivalent, for every $e \in B$:*

- (i) *e is a central element of $\mathbf{B}$;*
- (ii) *e is a decomposition operation on $\mathbf{A}$ such that e Cm g for every $g \in B$ (in other words, $\mathbf{B}$ is a subalgebra of $\mathbf{Cm}(e)$).*

Proof. (ii)$\Rightarrow$(i). Let $e \in B$ be a decomposition operation on $\mathbf{A}$ such that e Cm g for every $g \in B$. We show that e is a central element of $\mathbf{B}$, i.e., it satisfies conditions (A1)-(A4) of Proposition 7. In the following, x,y range over A and g,h,t,r,s,u over B.

(A1): $q(e,g,g) = e\langle g,g \rangle = g$, because by (D1) in Definition 2 we have that $e(g(x,y),g(x,y)) = g(x,y)$ for every $x,y \in A$.

(A2):

$$
\begin{aligned}
q(e,q(e,g_{11},g_{12}),q(e,g_{21},g_{22})) &= e\langle e\langle g_{11},g_{12}\rangle, e\langle g_{21},g_{22}\rangle\rangle \\
&= e\langle g_{11},g_{22}\rangle \qquad\qquad \text{by (D2)} \\
&= q(e,g_{11},g_{22}).
\end{aligned}
$$

(A3): We recall that the type of the algebra $\mathbf{B}$ is $\nu \cup \{q,0,1\}$. Then, taking into account the fact that e is a decomposition operation on $\mathbf{A}$ and the fact that e Cm g and e Cm r, we have that:

$$
\begin{aligned}
q(e,q(g,h,t),q(r,s,u)) &= e\langle g\langle h,t\rangle, r\langle s,u\rangle\rangle \\
&= e\langle e\langle g\langle h,t\rangle, g\langle s,u\rangle\rangle, e\langle r\langle h,t\rangle, r\langle s,u\rangle\rangle\rangle \quad \text{by (A2)} \\
&\quad \text{where } g_{11} = g\langle h,t\rangle,\ g_{12} = g\langle s,u\rangle,\ \text{etc.} \\
&= e\langle g\langle e\langle h,s\rangle, e\langle t,u\rangle\rangle, r\langle e\langle h,s\rangle, e\langle t,u\rangle\rangle\rangle \\
&\quad \text{because } e\ \mathrm{Cm}\ g\ \text{and}\ e\ \mathrm{Cm}\ r \\
&= (e\langle g,r\rangle)\langle e\langle h,s\rangle, e\langle t,u\rangle\rangle \\
&= q(q(e,g,r),q(e,h,s),q(e,t,u)).
\end{aligned}
$$

Now, let $\sigma \in \nu$:

$$
\begin{aligned}
q(e,\sigma^{\mathbf{B}}(\bar{g}),\sigma^{\mathbf{B}}(\bar{h}))(x,y) &= e\langle \sigma^{\mathbf{B}}(\bar{g}), \sigma^{\mathbf{B}}(\bar{h})\rangle(x,y) \\
&= e(\sigma^{\mathbf{A}}(\bar{g}(x,y)), \sigma^{\mathbf{A}}(\bar{h}(x,y))) \\
&= \sigma^{\mathbf{A}}(\ldots, e(g_i(x,y), h_i(x,y)), \ldots) \quad\quad \text{by (D3)} \\
&= \sigma^{\mathbf{B}}(q(e,g_1,h_1), \ldots, q(e,g_n,h_n))(x,y).
\end{aligned}
$$

(A4): By definition of π_1, π_0 we easily obtain the conclusion: $q(e,\pi_1,\pi_0) = e\langle \pi_1, \pi_0\rangle = e$.

(i)$\Rightarrow$(ii). Let $e: A \times A \to A$ be a central element of $\mathbf{B}$. Then by definition of $q^{\mathbf{B}}$ and by Proposition 7, the map $f_e : B \times B \to B$, defined by

$$
f_e(g,h) = q^{\mathbf{B}}(e,g,h) = e\langle g,h\rangle,
$$

is a decomposition operation on $\mathbf{B}$ that satisfies the following conditions for all maps $g, g_{ij} \in B$:

(1) $e\langle g,g\rangle = g$;
(2) $e\langle e\langle g_{11}, g_{12}\rangle, e\langle g_{21}, g_{22}\rangle\rangle = e\langle g_{11}, g_{22}\rangle$;
(3) f_e is a homomorphism from $\mathbf{B} \times \mathbf{B}$ into $\mathbf{B}$.

The proof that e is a decomposition operation on $\mathbf{A}$ is now trivial, because, by Proposition 18(ii), $\mathbf{A}$ is isomorphic to $\mathbf{B}_0$ and f_e restricted to $\mathbf{B}_0 \times \mathbf{B}_0$ maps $\mathbf{B}_0 \times \mathbf{B}_0$ into $\mathbf{B}_0$. Moreover, $e\ \mathrm{Cm}\ g$ for every $g \in B$ follows from the fact that f_e is a homomorphism from $\mathbf{B} \times \mathbf{B}$ into $\mathbf{B}$.

Corollary 23. *Let f be a decomposition operation on $\mathbf{A}$. Then f is a central element of the algebra $\mathbf{Cm}(f)$.*

Proposition 24. *Let $\mathbf{A}$ be an algebra of type ν and $\mathbf{B}$ be an algebra in the similarity type of Boolean algebras such that $B \subseteq F(A)$. Then $\mathbf{B}$ is a guard algebra of decomposition operations on $\mathbf{A}$ if and only if $\mathbf{B}$ is a Boolean algebra of central elements of a functional Church algebra of value domain $\mathbf{A}$.*

Proof. ($\Leftarrow$) By Theorems 22 and 13(1).

($\Rightarrow$) By Theorem 13(1) B is a set of mutually commuting decomposition operations. Then from Theorem 22 it follows that $\mathbf{B}$ is a subalgebra of the Boolean algebra of central elements of the functional Church algebra $\bigcap_{f \in B} \mathbf{Cm}(f)$.

Let $\mathbf{A}$ be an algebra of type ν. The function

$$f \in \mathcal{DE}(\mathbf{A}) \mapsto \theta_f = \{(x,y) \in A \times A : f(x,y) = x\} \in \mathcal{FC}(\mathbf{A})$$

is a bijective correspondence between the set $\mathcal{DE}(\mathbf{A})$ of decomposition operations and the set $\mathcal{FC}(\mathbf{A})$ of factor congruences. If X is a set of decomposition operations we denote by Θ_X the set $\{\theta_f : f \in X\}$.

Proposition 25. *The map associating to any set of decomposition operations X the set Θ_X determines a bijective correspondence between universes of guard algebras of decomposition operations and universes of Boolean sublattices of $\mathrm{Con}(\mathbf{A})$ of permutable factor congruences.*

Proof. If g and h are mutually commuting decomposition operations on $\mathbf{A}$, then it is easy to show that $\theta_{g \wedge h} = \theta_g \cap \theta_h$ and $\theta_{g \vee h} = \theta_g \vee \theta_h$. Then the conclusion follows from Proposition 21.

Another proof of the above proposition can be found in (Knoebel, 2012, Proposition VI.2.2).

4.3 Totally commuting factor congruences

Let $\mathbf{A}$ be an algebra of type ν. A decomposition operation $f : A \times A \to A$ is called *totally commuting* if f Cm g for every decomposition operation g on $\mathbf{A}$. We denote by $\mathcal{TC}(\mathbf{A})$ the set of all totally commuting decomposition operations on $\mathbf{A}$. If f is totally commuting, then the factor congruence θ_f is also called totally commuting.

Proposition 26. *Let $\mathbf{A}$ be an algebra of type ν. A factor congruence of $\mathbf{A}$ is totally commuting if and only if it is bi-balanced.*

Proof. By Proposition 21.

The following proposition provides a proof of (Swamy and Suryanarayana Murti, 1981, Theorem 2).

Proposition 27. *Let $\mathbf{A}$ be an algebra of type ν. The set of all bi-balanced factor congruences is a Boolean sublattice of $\mathrm{Con}(\mathbf{A})$ of permutable factor congruences.*

Proof. If g is a totally commuting decomposition operation on $\mathbf{A}$, then by Proposition 20 $\mathbf{Cm}(g)$ is a functional Church algebra containing all decomposition operations on $\mathbf{A}$. The functional Church algebra $\mathbf{B} = \bigcap_{g \in \mathcal{TC}(\mathbf{A})} \mathbf{Cm}(g)$ contains all decomposition operations on $\mathbf{A}$ and satisfies $\mathcal{TC}(\mathbf{A}) \subseteq \mathrm{Ce}(\mathbf{B})$. By Theorem 22 a decomposition operation e is a central element of $\mathbf{B}$ if and only if e Cm g for every $g \in B$, that implies e Cm g for every decomposition operation g. In conclusion, $\mathcal{TC}(\mathbf{A}) = \mathrm{Ce}(\mathbf{B})$.

The following proposition partially solves (Knoebel, 2012, Problem 2.16(a)).

Proposition 28. *Let* **A** *be an algebra of type* ν. *The set of all bi-balanced factor congruences is the intersection of all maximal Boolean sublattices of* $\mathrm{Con}(\mathbf{A})$ *of permutable factor congruences.*

Proof. By Propositions 25 and 26 we can work on decomposition operations. Let **L** be a maximal Boolean algebra of mutually commuting decomposition operations on **A** and let $\mathbf{B} = \bigcap_{f \in L} \mathbf{Cm}(f)$. Since f Cm g for all $f, g \in L$, then by Theorem 22 and by the maximality of **L** we derive that $L = \mathrm{Ce}(\mathbf{B})$. We now show that $\mathcal{TC}(\mathbf{A}) \subseteq L$. Let h be a totally commuting decomposition operation. Since $h \in \mathrm{Cm}(f)$ and $f \in \mathrm{Cm}(h)$ for every $f \in L$, then $h \in B \cap \mathrm{Cm}(h)$, so that $L \cup \{h\} \subseteq \mathrm{Ce}(\mathbf{B} \cap \mathbf{Cm}(h))$. By maximality of **L** we derive that $h \in L = \mathrm{Ce}(\mathbf{B} \cap \mathbf{Cm}(h))$.

Let $h \in L$ for every maximal Boolean lattice **L** of mutually commuting decomposition operations on **A**. If there exists a decomposition operation f such that $h \notin \mathrm{Cm}(f)$, then by Zorn's Lemma there exists a maximal Boolean lattice **L** such that $\mathrm{Ce}(\mathbf{Cm}(f)) \subseteq L$ but $h \notin L$. This contradicts the hypothesis, so h is totally commuting.

5 Weak Boolean product representations via polynomials

A case of special interest as regards the construction of the foregoing section arises when the functional algebra of value domain **A** is the algebra $\hat{\mathbf{A}}$ of binary polynomial operations of **A**. Under this circumstance, the central elements of $\hat{\mathbf{A}}$ are exactly the polynomial decomposition operations of **A**. This allows us to take advantage of the results in Section 2.2 and obtain a weak Boolean decomposition of **A** out of the decomposition of $\hat{\mathbf{A}}$ provided by Theorem 10. Although we cannot, in general, say much about the factors in these products, we identify a number of sufficient conditions for the stalks to be directly indecomposable.

Definition 29. *Let* **A** *be an arbitrary algebra of type* ν.

    (1) *A map* $f : A \times A \to A$ *that is both a decomposition operation and a polynomial operation is called a* polynomial decomposition operation.

    (2) *If* f *is as in (1), the complementary factor congruences* θ_f *and* $\bar{\theta}_f$ *are called* polynomial factor congruences.

    (3) **A** *is* polynomially directly indecomposable *if* Δ *and* ∇ *are the unique polynomial factor congruences.*

The set $\hat{A}$ of all binary polynomial operations on **A** is the universe of the least subalgebra $\hat{\mathbf{A}}$ of the functional Church algebra $F(\mathbf{A})$ of value domain **A**. The next theorem sheds some light on the reasons behind Theorem 13(1) and the other results in Section 2.3.

Theorem 30. *A polynomial operation is a central element of* $\hat{\mathbf{A}}$ *if and only if it is a decomposition operation on* $\mathbf{A}$.

Proof. The conclusion follows from Theorem 22, because the decomposition operations commute with all polynomial operations.

Let $p \in \mathrm{Ce}(\hat{\mathbf{A}})$. We denote by $(\widehat{\theta_p}, \widehat{\theta'_p})$ the pair of complementary factor congruences on $\hat{\mathbf{A}}$ determined by the central element p. By Corollary 9 we have that $\widehat{\theta_p} = \theta(p, \pi_0^{\hat{\mathbf{A}}})$ and $\widehat{\theta'_p} = \theta(p, \pi_1^{\hat{\mathbf{A}}})$. Since $p \in \mathrm{Ce}(\hat{\mathbf{A}})$ is a decomposition operation on $\mathbf{A}$, we denote by (θ_p, θ'_p) the pair of complementary factor congruences on $\mathbf{A}$ determined by p. Clearly, for all $x, y \in A$ and $f, g \in \hat{A}$,

$$x\theta_p y \text{ iff } p(x,y) = x \text{ and } f\widehat{\theta_p}g \text{ iff } p\langle f,g \rangle = f.$$

Also,

$$f\widehat{\theta_p}g \text{ iff } f(x,y)\theta_p g(x,y) \text{ for all } x,y \in A,$$

and similarly for $\widehat{\theta'_p}$ and θ'_p. We now show that the Church algebra of polynomial operations on the quotient of $\mathbf{A}$ modulo θ_p is nothing but the quotient modulo $\widehat{\theta_p}$ of the Church algebra $\hat{\mathbf{A}}$ of polynomial operations on $\mathbf{A}$.

Proposition 31. $\hat{\mathbf{A}}/\widehat{\theta_p} \cong \widehat{\mathbf{A}/\theta_p}$.

Proof. Let $t(c_{\bar{a}}, y, z)$ be a polynomial. Define

$$\varphi(t^{\mathbf{A}}(c_{\bar{a}}, y, z)/\widehat{\theta_p}) = t^{\mathbf{A}/\theta_p}(c_{\bar{a}/\theta_p}, y, z).$$

It is not difficult to show that the map φ is a well-defined isomorphism.

Let X be the Boolean space of the maximal ideals of the Boolean algebra $c[\hat{\mathbf{A}}]$. For every maximal ideal I we define two congruences:

$$\theta_I = \{(a,b) \in A^2 : \exists p(p \in I \;\bar{\wedge}\; p(a,b) = a)\}$$

and

$$\widehat{\theta_I} = \{(f,g) \in \hat{A}^2 : \exists p(p \in I \;\bar{\wedge}\; p\langle f,g \rangle = f)\}.$$

Since $\theta_I = \bigcup_{p \in I} \theta_p$ and $\widehat{\theta_I} = \bigcup_{p \in I} \widehat{\theta_p}$, from Proposition 31 it follows that $\hat{\mathbf{A}}/\widehat{\theta_I} \cong \widehat{\mathbf{A}/\theta_I}$. We are now ready to state the main result of this section:

Theorem 32. *Let* $\mathbf{A}$ *be an arbitrary algebra and let* X *be the Boolean space of maximal ideals of the Boolean algebra* $c[\hat{\mathbf{A}}]$. *Then:*

(1) *The map*

$$F : \hat{A} \to \prod_{I \in X} \hat{A}/\widehat{\theta_I} \cong \prod_{I \in X} \widehat{A/\theta_I},$$

defined by

$$F(p) = (p/\widehat{\theta}_I : I \in X)$$

gives a weak Boolean representation of $\hat{\mathbf{A}}$.

(2) *The restriction* $F_{|A}$ *of* F *to the constant polynomials provides a weak Boolean product representation* $F_{|A} : A \to \prod_{I \in X} A/\theta_I$ *of* $\mathbf{A}$.

(3) *The stalks of the representation of* $\hat{\mathbf{A}}$ *are directly indecomposable if and only if the stalks of the representation of* $\mathbf{A}$ *are polynomially directly indecomposable.*

Proof. (1) By Theorem 10 and Proposition 31.

(2) The polynomial factor congruences constitute a Boolean algebra of permuting congruences.

(3) By Theorem 30.

If $\mathcal{V}$ is a variety of algebras of type ν, then we denote by $\mathrm{Ch}(\mathcal{V})$ the variety of algebras of type $\nu \cup \{q, 0, 1\}$ axiomatised by the equational theory $Eq(\mathcal{V})$ of $\mathcal{V}$ and the axioms of substitution Church algebra.

We say that a variety $\mathcal{V}$ has *polynomial factor congruences* (PFC, for short) if, for every $\mathbf{A} \in \mathcal{V}$, all factor congruences of $\mathbf{A}$ are polynomial factor congruences. By Theorem 30, PFC implies BFC. Now, Theorems 11 and 32 imply the following corollary.

Corollary 33.

(1) *If* $\mathrm{Ch}(\mathcal{V})_{\mathrm{di}}$ *is a universal class then every algebra* $\mathbf{A} \in \mathcal{V}$ *is representable as a weak Boolean product of polynomially directly indecomposable algebras.*

(2) *If* $\mathcal{V}$ *has* PFC *and* $\mathrm{Ch}(\mathcal{V})_{\mathrm{di}}$ *is a universal class then every algebra* $\mathbf{A} \in \mathcal{V}$ *is representable as a weak Boolean product of directly indecomposable algebras.*

If p is a (binary) polynomial on $\mathbf{A}$, then we write $p \equiv p(a_1, \ldots, a_k, x, y)$, where $a_i \in A$, to spell out in full all constants and variables occurring in p. For every homomorphism $g : \mathbf{A} \to \mathbf{B}$ of algebras of type ν, let $\hat{g} : \hat{\mathbf{A}} \to \hat{\mathbf{B}}$ be the homomorphism of algebras of type $\nu \cup \{q, 0, 1\}$, defined by

$$\hat{g}(p^{\mathbf{A}}(a_1, \ldots, a_k, x, y)) = p^{\mathbf{B}}(g(a_1), \ldots, g(a_k), x, y)$$

for every polynomial p on $\mathbf{A}$. We recall that, for every onto homomorphism $g : \mathbf{C} \to \mathbf{D}$ of Church algebras, the restriction of g to the central elements of $\mathbf{C}$ is a (non necessarily onto) Boolean homomorphism from $c[\mathbf{C}]$ into $c[\mathbf{D}]$.

Theorem 34. *Let* $\mathcal{V}$ *be a variety of algebras satisfying the following two conditions:*

(1) *The polynomially directly indecomposable members of* $\mathcal{V}$ *are directly indecomposable;*

(2) *For every homomorphism g from $\mathbf{A} \in \mathcal{V}$ onto a directly decomposable algebra $\mathbf{B}$, the codomain of $\hat{g}_{|c[\hat{\mathbf{A}}]} : c[\hat{\mathbf{A}}] \to c[\hat{\mathbf{B}}]$ properly includes $\{\pi_0^{\mathbf{B}}, \pi_1^{\mathbf{B}}\}$.*

Then the weak Boolean product representations, provided by Theorem 32, have directly indecomposable stalks.

Proof. Let $\mathbf{A} \in \mathcal{V}$, and let I be a maximal ideal of the Boolean algebra $c[\hat{\mathbf{A}}]$ of central elements of $\hat{\mathbf{A}}$. Let $f : \mathbf{A} \to \mathbf{A}/\theta_I$ be the onto homomorphism mapping a to a/θ_I. Assume $\mathbf{A}/\theta_I$ to be directly decomposable. Then by (1) $c[\widehat{\mathbf{A}/\theta_I}] \neq \{\pi_0, \pi_1\}$. By (2), there exists a polynomial operation $p \in c[\hat{\mathbf{A}}]$ such that $\hat{f}(p) \neq \pi_0, \pi_1$. To simplify notation, let $\hat{p} = \hat{f}(p)$. So, there are $a, b, c, d \in A$ such that $a/\theta_I \neq d/\theta_I$, $b/\theta_I \neq c/\theta_I$ and

$$\hat{p}(b/\theta_I, c/\theta_I) = p(b,c)/\theta_I = c/\theta_I; \qquad \hat{p}(a/\theta_I, d/\theta_I) = p(a,d)/\theta_I = a/\theta_I$$

that is, $p(b,c) \; \theta_I \; c$ and $p(a,d) \; \theta_I \; a$. Since I is a maximal ideal, then either $p \in I$ or $p' \in I$ (recall that $p'(x,y) = p(y,x)$). Assume w.l.g. that $p' \in I$. So $\bar{\theta}_p = \{(x,y) : p(x,y) = y\} \subseteq \theta_I$. Since

$$\theta_I = \{(a,b) \in A^2 : \exists r \in I \; (r(a,b) = a)\} = \{(a,b) \in A^2 : \exists r \in I \; (a\theta_r b)\} = \bigcup_{r \in I} \theta_r,$$

then by $p(b,c) \; \theta_I \; c$ there exists a polynomial decomposition operation $e \in I$ such that $p(b,c) \; \theta_e \; c$. In other words, the algebra $\mathbf{A}$ satisfies the equality $e(c, p(b,c)) = c$. Since e and p are polynomial decomposition operations, we have:

$$c = e(c, p(b,c)) = e(p(c,c), p(b,c)) = p(e(c,b), e(c,c)) = p(e(c,b), c).$$

This, together with $\{(x,y) : p(x,y) = y\} \subseteq \theta_I$, implies $c\theta_I e(c,b)$. Since $b\theta_e e(c,b)\bar{\theta}_e c$ and $\theta_e \subseteq \theta_I$, we obtain $b\theta_I e(c,b)$. In conclusion, from $c\theta_I e(c,b)$, $b\theta_I e(c,b)$ we get $b\theta_I c$, contradicting the hypothesis $b/\theta_I \neq c/\theta_I$. A similar reasoning works if $p \in I$.

6 An application

As a first application of the results in the previous section, we give a weak Boolean product representation of skew Boolean algebras.

Weakenings of lattices where the meet and join operations may fail to be commutative attracted from time to time the attention of various communities of scholars, including ordered algebra theorists (for their connection with preordered sets) and semigroup theorists (who viewed them as structurally enriched bands possessing a dual multiplication). Probably the most interesting and successful such generalisation is the concept of *skew lattice* (Leech,

1996), along with the related notion of *skew Boolean algebra* (Leech, 1990). Here we will just review some definitions needed in the sequel; the interested reader is referred to (Leech, 1996) or (Spinks, 2003) for far more comprehensive accounts and for an illustration of the importance of both notions, especially in light of their connection to discriminator varieties (Bignall and Leech, 1995; Cvetko-Vah and Salibra, 2015).

Definition 35. *A* band *is a semigroup* $(A, \cdot)$ *satisfying the identity* $xx \approx x$. *A band is* regular *if it satisfies* $xyxzx \approx xyzx$; *it is* left (right) regular *if it satisfies the identity* $xyx \approx xy$ $(xyx \approx yx)$.

Left and right regular bands are obviously regular. Observe that, given a band **A**, the relation

$$a \leq b \iff ab = a = ba$$

is a partial ordering on A.

Definition 36. *A* double band *is an algebra* $(A, +, \cdot)$ *of type* $(2, 2)$ *such that the reducts* $(A, \cdot)$ *and* $(A, +)$ *are both bands. A double band satisfying the absorption identities*

$$x(x + y) \approx x \approx x + xy;$$
$$(y + x)x \approx x \approx yx + x.$$

is called a skew lattice. *A* skew lattice *is called* left-handed (right-handed) *if the reduct* $(A, \cdot)$ *is left (right) regular and the reduct* $(A, +)$ *is right (left) regular.*

If we expand skew lattices by a subtraction operation and a constant 0, we get the following noncommutative variant of Boolean algebras.

Definition 37. *A* skew Boolean algebra *is an algebra* $\mathbf{A} = (A, +, \cdot, \backslash, 0)$ *of type* $(2, 2, 2, 0)$ *such that:*

- *its reduct* $(A, +, \cdot)$ *is a skew lattice satisfying the identities* $xyzx \approx xzyx$, $x(y + z) \approx xy + xz$ *and* $(y + z)x \approx yx + zx$;
- *0 is left and right absorbing w.r.t. multiplication;*
- *the operation* $\backslash$ *satisfies the identities*

$$xyx + (x \backslash y) \approx x \approx (x \backslash y) + xyx;$$
$$xyx(x \backslash y) \approx 0 \approx (x \backslash y)xyx.$$

We call right- (left-) handed any skew Boolean algebra that is right- (left-) handed as a skew lattice. In the interests of brevity, we write "right-handed SBA" for "right-handed skew Boolean algebra".

Let $\mathbf{A} = (A, +, \cdot, \backslash, 0)$ be a right-handed SBA. Define the following term:

$$t(x, y, z) = (xy) + (z \backslash x).$$

For every $a \in A$, t_a will denote the polynomial operation on **A** given by

$$t_a(x,y) = t^{\mathbf{A}}(a,x,y).$$

Cvetko-Vah and the first author of the present paper proved the following result.

Lemma 38. (Cvetko-Vah and Salibra, 2015) *If $\mathbf{A}$ is a right-handed SBA then, for every $a \in A$, the map t_a is a polynomial decomposition operation on $\mathbf{A}$. Moreover, the factor congruence θ_{t_a} associated with t_a is $\theta(a,0)$, the least congruence collapsing a and 0.*

It follows that the pair $(\theta_{t_a}, \bar{\theta}_{t_a})$, where $\theta_{t_a} = \theta(a,0) = \{(x,y) : t^{\mathbf{A}}(a,x,y) = x\}$ and $\bar{\theta}_{t_a} = \{(x,y) : t^{\mathbf{A}}(a,x,y) = y\}$, is the pair of complementary factor congruences determined by the polynomial decomposition operation t_a. Notice that $t_0 = \pi_0^{\mathbf{A}}$. We recall one more result from (Cvetko-Vah and Salibra, 2015).

Lemma 39. (Cvetko-Vah and Salibra, 2015, Lemma 4.5) *A right-handed SBA $\mathbf{A}$ is directly indecomposable iff $t_a = \pi_1^{\mathbf{A}}$ for every $a \neq 0$ (that is, $\theta_{t_a} = \theta(a,0) = \nabla$ for every $a \neq 0$).*

By Proposition 3, a decomposition operation f on an algebra $\mathbf{A}$ corresponds to a pair of trivial factor congruences if and only if either $f = \pi_0^{\mathbf{A}}$ or $f = \pi_1^{\mathbf{A}}$.

Lemma 40.   (i) *Every right-handed SBA $\mathbf{A}$ is isomorphic to a weak Boolean product of directly indecomposable right-handed SBAs[1].*

   (ii) *Every left-handed SBA $\mathbf{A}$ is isomorphic to a weak Boolean product of directly indecomposable left-handed SBAs.*

Proof. (i) We show that the assumptions (1) and (2) of Theorem 34 are satisfied.

(1) Assume that there exists $a \neq 0 \in A$ such that $t_a = \pi_0$. Then by Lemma 38 we have

$$\theta(a,0) = \{(x,y) : t^{\mathbf{A}}(a,x,y) = x\} = \Delta,$$

which contradicts $a \neq 0$. Then $t_a = \pi_1$ for all $a \neq 0$ and the conclusion follows from Lemma 39(ii).

(2) Let $f : \mathbf{A} \to \mathbf{B}$ be an onto homomorphism of right-handed SBAs. Assume $\mathbf{B}$ to be directly decomposable. Then, $\mathrm{Ce}(\hat{\mathbf{B}}) \neq \{\pi_0^{\mathbf{B}}, \pi_1^{\mathbf{B}}\}$. By Lemma 39 there exists $b \in B$ such that $t_b \neq \pi_1^{\mathbf{B}}$ defines a nontrivial pair of factor congruences. Since f is onto, then there exists $a \in A$ such that $f(a) = b$. The polynomial t_a defines a nontrivial pair of factor congruences on $\mathbf{A}$ and $\hat{f}(t_a) = t_b$. (ii) Follows from (i) by skew-lattice duality.

Theorem 41. *Every skew Boolean algebra is isomorphic to a weak Boolean product of directly indecomposable skew Boolean algebras.*

---

[1] This result has been part of the folklore on the subject for more than a decade (Matthew Spinks, personal communication). To the best of the authors' knowledge, it has never been explicitly written down in print.

Proof. Recall that by (Leech, 1989, Thm. 1.15) every skew Boolean algebra **A** is isomorphic to the fibred product

$$i : \mathbf{A} \cong \mathbf{A}/R \times_{\mathbf{A}/D} \mathbf{A}/L,$$

where $\mathbf{A}/R$ and $\mathbf{A}/L$ are the maximal left- and right-handed homomorphic images of **A**, respectively, and D is the Green's congruence (Leech, 1989, Sec. 1.6). By Lemma 40, $\mathbf{A}/L$ admits a weak Boolean product representation $f : \mathbf{A}/L \to \prod_{I \in S} \mathbf{A}/L/\theta_I$ (S the spectrum of maximal ideals), with directly indecomposable stalks, and similarly for $\mathbf{A}/R$, $g : \mathbf{A}/R \to \prod_{I \in T} \mathbf{A}/R/\theta_I$ (T the spectrum of maximal ideals). Consider the Boolean space $T \uplus S$, which is the disjoint union of the spaces T and S with the topology in which a subset U of $T \uplus S$ is open if $U \cap T$ is open in T and $U \cap S$ is open in S. By (Koppleberg, 1989, Sec. 3, Prop 8.7), the Boolean algebra $\mathrm{Clop}(T \uplus S)$ of clopen subsets of $T \uplus S$ is isomorphic to the product $\mathrm{Clop}(T) \times \mathrm{Clop}(S)$. Then $(g, f) :$ $\mathbf{A}/R \times \mathbf{A}/L \to \prod_{I \in T} \mathbf{A}/R/\theta_I \times \prod_{I \in S} \mathbf{A}/L/\theta_I \cong \prod_{I \in T \uplus S} \mathbf{A}/\phi_I$, where, for all $a, b \in A$, $a\phi_I b$ iff $a/R\theta_I b/R$ (resp. $a/L\theta_I b/L$) in the case $I \in T$ (resp. $I \in S$). Therefore it can be easily seen that the map $(g, f) \circ i$ provides a weak Boolean product representation of **A** in $\prod_{I \in (S \uplus T)} \mathbf{A}/\phi_I$, with directly indecomposable stalks.

Acknowledgement. We warmly thank an anonymous referee for her/his extremely valuable suggestions, and Matthew Spinks for the conversations we had on some topics addressed in the present paper. The second author gratefully acknowledges the support of the Italian Ministry of Scientific Research (MIUR) within the FIRB project "Structures and Dynamics of Knowledge and Cognition", Cagliari: F21J12000140001.

References

Bergman, G.M. (1991). Actions of Boolean rings on sets, *Algebra Universalis*, 28, pp. 153–187.

Bignall, R.J. and Leech, J. (1995). Skew Boolean algebras and discriminator varieties, *Algebra Universalis*, 33, pp. 387–398.

Bloom, S., Ésik, Z. and Manes, E.G. (1990). A Cayley theorem for Boolean algebras, *American Mathematical Monthly*, 97, pp. 831–833.

Bloom, S. and Tindall, R. (1983). Varieties of 'If-then-else" algebras, *SIAM Journal on Computing*, 12, pp. 677–707.

Burris, S.N. and Sankappanavar, H.P. (1981). *A Course in Universal Algebra*, Springer, Berlin.

Burris, S.N. and Werner, H. (1979). Sheaf constructions and their elementary properties, *Transactions of the American Mathematical Society*, 248, pp. 269–309.

Burris, S.N. and Werner, H. (1980). Remarks on Boolean products, *Algebra Universalis*, 10, pp. 333–344.

Comer, S. (1971). Representations by algebras of sections over Boolean spaces, *Pacific Journal of Mathematics*, 38, pp. 29–38.

Cvetko-Vah, K. and Salibra, A. (2015). The connection of skew Boolean algebras and discriminator varieties to Church algebras, *Algebra Universalis*, DOI 10.1007/s00012-015-0320-9

Galatos, N., Jipsen, P., Kowalski, T. and Ono, H. (2007). *Residuated Lattices: An Algebraic Glimpse on Substructural Logics*, Elsevier, Amsterdam.

Knoebel, A. (2012). *Sheaves of Algebras over Boolean Spaces*, Springer, Berlin.

Koppleberg, S. (1989). *General Theory of Boolean Algebras. Handbook of Boolean Algebras*, Part I, North Holland, Amsterdam.

Leech, J. (1989). Skew lattices in rings, *Algebra Universalis*, 26, pp. 48–72.

Leech, J. (1990). Skew Boolean algebras, *Algebra Universalis*, 27, pp. 497–506.

Leech, J. (1996). Recent developments in the theory of skew lattices, *Semigroup Forum*, 52, pp. 7–24.

Manes, E.G and, Arbib, M.A. (1986). *Algebraic Approaches to Program Semantics*, Springer, Berlin.

Manzonetto, G. and Salibra, A. (2008). From λ-calculus to universal algebra and back, in *Symposium on Mathematical Foundations of Computer Science MFCS 2008*, volume 5162 of LNCS, Springer, Berlin, pp. 479–490.

Manzonetto, G. and Salibra, A. (2010). Applying universal algebra to lambda calculus, *Journal of Logic and Computation*, 20, pp. 877–915.

Martins, M.A. and Pigozzi, D. (2007). Behavioural reasoning for conditional equations, *Mathematical Structures in Computer Science*, 17, pp. 1075–1113.

McKenzie, R.N., McNulty, G.F. and Taylor, W.F. (1987). *Algebras, Lattices, Varieties*, Vol. I, Wadsworth Brooks, Monterey, California.

Mekker, A.H. and Nelson, E.M. (1987). Equational bases for if-then-else, *SIAM Journal on Computing*, 16, pp. 465–485.

Movsisyan, Y.M. (2009). Binary representations of algebras with at most two binary operations. A Cayley theorem for distributive lattices, *International Journal of Algebra and Computation*, 19, pp. 97–106.

Padmanabhan, R. and Penner, P. (1998). Lattice ordered polynomial algebras, *Order*, 15, pp. 75–86.

Peirce, R.S. (1967). *Modules over Commutative Regular Rings*, Memoirs of the American Mathematical Society, Number 70.

Pigozzi, D. (1990). Data types over multiple-values logics, *Theoretical Computer Science*, 77, pp. 161–194.

Pigozzi, D. (1991). Equality-test and if-then-else algebras: Axiomatization and specification, *SIAM Journal on Computing*, 20, pp. 766–805.

Pigozzi, D. and Salibra, A. (1998). Lambda abstraction algebras: coordinatizing models of lambda calculus, *Fundamenta Informaticae*, 33, pp. 149–200.

Salibra, A., Ledda, A., Paoli, F. and Kowalski, T. (2013). Boolean-like algebras, *Algebra Universalis*, 69, pp. 113–138.

Spinks, M. (2003). *On the Theory of Pre-BCK Algebras*, PhD Thesis, Monash University.

Swamy, U.M. and Suryanarayana Murti, G. (1981). Boolean centre of a universal algebra, *Algebra Universalis*, 13, pp. 202–205.

Urbanik K. (1965). On algebraic operations in idempotent algebras, *Colloquium Mathematicum*, 13, pp. 129–157.

Vaggione, D. (1996). Varieties in which the Pierce stalks are directly indecomposable, *Journal of Algebra*, 184, pp. 424–434.

Paraconsistent constructive logic with strong negation as a contraction-free relevant logic

Matthew Spinks and Robert Veroff*

1 Introduction

Summary Logics with strong negation are a class of sentential calculi that originally arose from concerns about the non-constructive nature of negation in intuitionistic logic. Nelson's paraconsistent constructive logic with strong negation **N4** (Almukdad and Nelson, 1984; Odintsov, 2003, 2004, 2008), the most important member of this class, is an axiomatic expansion of the negation-free fragment of the intuitionistic propositional calculus (Rasiowa, 1974, Chapter X) by a unary logical connective $\sim$ of strong negation. It is well known that strong negation plays an important role as 'explicit negation' in logic programming (Akama, 1997; Eiter et al., 1999; Gelfond, 2002; Kamide and Wansing, 2012; Pearce, 1999; Wansing, 1993). Nelson's constructive logic with strong negation **N3** (Nelson, 1949; Rasiowa, 1974; Sendlewski, 1984; Vakarelov, 1977) is the axiomatic extension of **N4** by the *ex falso quodlibet* law $\vdash x \to (\sim x \to y)$. Recent work due to (Järvinen et al., 2013; Järvinen and Radeleczki, 2011a,b, 2014) establishes fundamental connections between the algebraic counterpart of **N3** and the rough sets of Pawlak (1982).

Relevant logics are a class of logics that were occasioned in the first instance by the desire to avoid the so-called paradoxes of material and strict implication. The basic 'logic of relevant implication' **R** of Anderson and Belnap (1975, §3) is the $\{\wedge, \vee, *, \Rightarrow, \sim\}$-fragment of the involutive distributive full Lambek calculus with exchange and contraction (Galatos et al., 2007, Sec-

Matthew Spinks
Department of Philosophy, University of Cagliari, 09123 Cagliari, Italy, e-mail: `mspinksau@yahoo.com.au`

Robert Veroff
Department of Computer Science, University of New Mexico, Albuquerque NM 87131, U.S.A., e-mail: `veroff@cs.unm.edu`

* Dedicated to Don Pigozzi on the occasion of his eightieth birthday.

© Springer International Publishing AG 2018
J. Czelakowski (eds.), *Don Pigozzi on Abstract Algebraic Logic,
Universal Algebra, and Computer Science*, Outstanding Contributions
to Logic 16, https://doi.org/10.1007/978-3-319-74772-9_13

tion 2.2.3). Brady's contraction-free relevant logic **RW** (Brady, 1990, 1991, 1996a,b) is the deductive system obtained from **R** upon dropping the contraction axiom $\vdash (x \Rightarrow (x \Rightarrow y)) \Rightarrow (x \Rightarrow y)$. Girard's linear logic **LL** (Girard, 1987) is closely related to **RW**: roughly speaking, **LL** + contraction = **R** − distribution + exponentials; *cf.* (Avron, 1988, Section 2, p. 173). Over the past three decades, linear logic has found extensive application in computer science as a logic of resources (Abramsky, 1993; Girard, 1987, 1995; Troelstra, 1992).

In the present work, we uncover connections between the previously disparate classes of relevant logics and logics with strong negation. The main result of the paper, Theorem 2.1, announces that **N4** is, up to definitional equivalence, the axiomatic extension of **RW** by the axioms

(Paraconsistent Nelson$_\vdash^{\geqslant}$)	$\vdash \big((x \twoheadrightarrow y) \wedge (\sim y \twoheadrightarrow \sim x)\big) \Rightarrow (x \Rightarrow y)$
(Internal weakening$_\vdash^{\geqslant}$)	$\vdash (x * y) \twoheadrightarrow x$

for a certain naturally arising formula-definable connective $\twoheadrightarrow$. (For an alternative approach to commingling **RW** and **N4**, see (Kamide, 2016).) This generalises an earlier theorem of the authors (Spinks and Veroff, 2008b, Theorem 1.1) showing that **N3** is, up to definitional equivalence, an axiomatic extension of **InFL**$_{ew}$, the involutive full Lambek calculus with exchange and weakening. The description of **N4** *qua* a contraction-free relevant logic is subsequently used in showing that numerous well-known and familiar paraconsistent and relevant logics, including the 3-valued relevant logic with mingle **RM3** and the 4-valued paraconsistent logic **BN4** of Brady and Dunn, arise, up to definitional equivalence, as naturally occurring axiomatic extensions of **N4**. For an omnibus (yet partial) statement of results connecting **N4** with the existing literature of non-classical logics, see Theorem 2.2 below.

Aims of the paper The present work meets a triple duty. First, it offers a conspectus of (some of) the main results of the series of papers (Spinks and Veroff, a,b,c), serving in particular to announce Theorems 2.1 and 2.2 below. Second, the paper provides an overview of the global structure of the proof of Theorem 2.1. The proof of this theorem is conceptually difficult, owing both to its length and to its intricate structure (the proof comprises many highly interlocking parts, almost all of which are established via lengthy first-order computations). The present work thus functions as a kind of high-level guide to the proof. We warn the reader that the arguments sketched in this paper, although indicative, do not do justice to the proof of Theorem 2.1; for complete proofs of all results reported herein, the series of papers (Spinks and Veroff, a,b,c) should instead be consulted. Finally, the paper promotes the thesis that the deductive system **N4** is of central import to the study of non-classical logics, in that it sits at the nexus of two distinct logical domains—*viz.*, the paraconsistent and relevant logics.

Algebraic and logical preliminaries The set of natural numbers is denoted $\mathbb{N}$. For typographical convenience, we sometimes denote the application of the function f to a by a^f. Given a set A and an equivalence relation θ on A, the equivalence class of $a \in A$ is denoted $[a]_\theta$. We assume familiarity with the rudiments of general algebra and model theory, especially that part of first-order logic known as equational logic. For general algebraic background, see (Burris and Sankappanavar, 1981; Grätzer, 2008; McKenzie et al., 1987); for particulars on equational logic, see (Burris, 1998; McNulty, 1992; Pigozzi, 1975; Tarski, 1968; Taylor, 1979). Algebras are denoted $\mathbf{A}, \mathbf{B}, \ldots$ Given an algebra $\mathbf{A}$, the set of all congruences on $\mathbf{A}$ is denoted $\mathrm{Con}\,\mathbf{A}$ and the principal congruence on $\mathbf{A}$ generated by $\{a, b\} \subseteq A$ is denoted $\Theta^{\mathbf{A}}(a, b)$. Given a join semilattice $\langle A; \vee \rangle$, the supremum of $a, b \in A$ is denoted $\mathrm{l.u.b.}\{a, b\}$. Classes of algebras are denoted $\mathsf{K}, \mathsf{V}, \ldots$ Standard use is made throughout of the class operators I, $\underline{\mathsf{H}}$, $\underline{\mathsf{S}}$, $\underline{\mathsf{P}}$, and $\underline{\mathsf{P}}\mathsf{s}$ (for subdirect products); class operators always yield abstract classes.

We fix a countably infinite set $\mathbf{X} := \{x_i : i \in \mathbb{N}\}$ of variables for use throughout the paper; we write $x, y, z, \ldots$ as metavariables ranging over $\mathbf{X}$. Arbitrary language types are denoted Λ, while applied language types are denoted $\Lambda[\ldots]$. All language types are algebraic unless stated otherwise. Throughout the paper we overload a profusion of operation symbols [resp. logical connectives], in particular the symbols $\wedge, \vee, *, \Rightarrow, \rightarrow$, and $\sim$, repeatedly and without warning. We assume familiarity with the fundamentals of abstract algebraic logic, especially that part of abstract algebraic logic that concerns logics that are (strongly) algebraisable in the sense of Blok and Pigozzi (1989). For background on abstract algebraic logic, see (Blok and Pigozzi, 1989; Czelakowski, 2001; Font et al., 2003); for particulars on Blok-Pigozzi algebraisable logics, see (Blok and Pigozzi, 1989, 2001) or (Czelakowski, 2001, Chapter 4§6). The absolutely free algebra of type Λ generated by $\mathbf{X}$ is denoted $\mathbf{Fm}_\Lambda$ and its universe is denoted Fm_Λ; a *substitution* of type Λ is an endomorphism of $\mathbf{Fm}_\Lambda$. Following (Blok and Pigozzi, 1989, Chapter 1), a *deductive system* is a pair $\langle \Lambda, \vdash \rangle$ where Λ is a language type and $\vdash$ is a finitary and substitution-invariant consequence relation over Λ. Deductive systems are denoted $\mathbf{S}, \mathbf{T}, \ldots$ Given a deductive system $\mathbf{S}$, the set of all theories of $\mathbf{S}$ is denoted $\mathrm{Th}\,\mathbf{S}$. Unless otherwise specified, all deductive systems considered in the sequel are Blok-Pigozzi (finitary and finitely) algebraisable. Logical matrices are denoted $\mathfrak{M}, \mathfrak{N}, \ldots$

Organisation The remainder of this paper is organised as follows. The main results of the paper, Theorem 2.1 and Theorem 2.2, are stated in Section 2. The proof of Theorem 2.1 is delineated in Section 3, while the proofs composing Theorem 2.2 are outlined in Section 4. Some remarks relating the main results of the paper to contemporary developments in abstract algebraic logic (Blok and Pigozzi, 1989; Czelakowski, 2001; Font and Jansana, 2009) in general and to the work of Czelakowski and Pigozzi on Fregean logics (Czelakowski, 2001; Czelakowski and Pigozzi, 2004) in particular are presented in Section 5.

2 Main results

The main theorem Let $\mathbf{IPC}^+$ denote the negation-free fragment of the intuitionistic propositional calculus, considered over the language $\Lambda[\mathbf{IPC}^+] := \{\wedge, \vee, \to\}$ where $\wedge$, $\vee$, and $\to$ are all binary logical connectives, and let $\Sigma[\mathbf{IPC}^+]$ denote the standard Hilbert-style presentation of $\mathbf{IPC}^+$ given in (Rasiowa, 1974, Chapter X§1). *Paraconsistent constructive logic with strong negation*, in symbols $\mathbf{N4}$, is the deductive system over the language $\Lambda[\mathbf{N4}] := \Lambda[\mathbf{IPC}^+] \cup \{\sim\}$, where $\sim$ is a unary logical connective (called the *strong negation*), determined by the presentation $\Sigma[\mathbf{N4}]$ consisting of the axioms and inference rules of $\Sigma[\mathbf{IPC}^+]$ together with the axioms (Odintsov, 2003, Section 2)

$$(1) \quad \begin{aligned} &\vdash \sim(x \vee y) \leftrightarrow (\sim x \wedge \sim y) & &\vdash \sim(x \to y) \leftrightarrow (x \wedge \sim y) \\ &\vdash \sim(x \wedge y) \leftrightarrow (\sim x \vee \sim y) & &\vdash \sim\sim x \leftrightarrow x. \end{aligned}$$

Here the expression $\varphi \leftrightarrow \psi$ abbreviates $(\varphi \to \psi) \wedge (\psi \to \varphi)$ for all formulas φ, ψ in the language of $\mathbf{N4}$. The deductive system $\mathbf{N4}$ was introduced in (Routley, 1974) (not (Almukdad and Nelson, 1984), as is commonly asserted) and has recently been the subject of intense investigation; witness for instance the works (Odintsov, 2003, 2004, 2005, 2007, 2008).

Constructive logic with strong negation, in symbols $\mathbf{N3}$, is the deductive system over the language $\Lambda[\mathbf{N4}]$ determined by the axioms and inference rules of $\Sigma[\mathbf{N4}]$ together with the *ex falso quodlibet* law $\vdash x \to (\sim x \to y)$. The deductive system $\mathbf{N3}$ was introduced in (Nelson, 1949) and has been considered extensively in the literature; a representative selection of works includes (Busaniche and Cignoli, 2010; Kracht, 1998; Rasiowa, 1974; Sendlewski, 1984, 1990; Spinks and Veroff, 2008a,b; Vakarelov, 1977).

The logic $\mathbf{N4}^\perp$, introduced recently by Odintsov in (Odintsov, 2005), is the deductive system over the language $\Lambda[\mathbf{N4}^\perp] := \Lambda[\mathbf{N4}] \cup \{\perp\}$, where $\perp$ is a nullary logical connective, determined by the axioms and inference rules of $\Sigma[\mathbf{N4}]$ together with the additional axioms $\vdash \perp \to x$ and $\vdash x \to \sim\perp$.

Let $\mathbf{RW}$ denote the contraction-free relevant logic obtained from Anderson and Belnap's 'logic of relevant implication' $\mathbf{R}$ (Anderson and Belnap, 1975, §3) upon omitting the contraction axiom $\vdash (x \Rightarrow (x \Rightarrow y)) \Rightarrow (x \Rightarrow y)$ from the standard Hilbert-style presentation of $\mathbf{R}$ given in (Anderson and Belnap, 1975, §27.1.1). Here and throughout we are considering $\mathbf{R}$, hence $\mathbf{RW}$, to have language $\Lambda[\mathbf{RW}] := \{\wedge, \vee, *, \Rightarrow, \sim\}$, where $\wedge$, $\vee$, $*$, and $\Rightarrow$ are binary logical connectives and $\sim$ is a unary logical connective. The study of the deductive system $\mathbf{RW}$ dates back to at least (Slaney, 1984), and has been extensively pursued by Brady (1990, 1991, 1996a,b) in the context of decidability questions.

Let Λ be a language type having a distinguished binary logical connective $\Rightarrow$. For all Λ-formulas φ, let $|\varphi|$ abbreviate $\varphi \Rightarrow \varphi$. Further, let Λ be a language type having distinguished binary logical connectives $\Rightarrow$ and $\wedge$. For all Λ-formulas φ, ψ, let

$$\varphi \stackrel{\mathbf{RM}}{\Rightarrow} \psi \quad \text{abbreviate} \quad (\varphi \wedge |\psi|) \Rightarrow \psi, \quad \text{and}$$

$$\varphi \rightarrowtail \psi \quad \text{abbreviate} \quad \varphi \stackrel{\mathbf{RM}}{\Rightarrow} (\varphi \stackrel{\mathbf{RM}}{\Rightarrow} \psi).$$

Notice $\varphi \stackrel{\mathbf{RM}}{\Rightarrow} \psi$ is the "enthymematic" implication of Dunn and McCall (Anderson and Belnap, 1975, §29). Let $\mathbf{DPNRW}^{\gg}$ denote the axiomatic extension of $\mathbf{RW}$ by the axioms

(Paraconsistent Nelson$_{\vdash}^{\gg}$) $\qquad \vdash \big((x \rightarrowtail y) \wedge (\sim y \rightarrowtail \sim x) \big) \Rightarrow (x \Rightarrow y)$

(Internal weakening$_{\vdash}^{\gg}$) $\qquad \vdash (x * y) \rightarrowtail x.$

The main result of the present work asserts:

Theorem 2.1.

1. The map $\alpha : \Lambda[\mathbf{RW}] \to \mathrm{Fm}_{\Lambda[\mathbf{N4}]}$ defined by

$$x \wedge y \mapsto x \wedge y$$
$$x \vee y \mapsto x \vee y$$
$$(*_{\mathrm{def}}) \qquad x * y \mapsto \sim(x \to \sim y) \vee \sim(y \to \sim x)$$
$$(\Rightarrow_{\mathrm{def}}) \qquad x \Rightarrow y \mapsto (x \to y) \wedge (\sim y \to \sim x)$$
$$\sim x \mapsto \sim x$$

is an interpretation of $\mathbf{DPNRW}^{\gg}$ in $\mathbf{N4}$.

2. The map $\beta : \Lambda[\mathbf{N4}] \to \mathrm{Fm}_{\Lambda[\mathbf{RW}]}$ defined by

$$x \wedge y \mapsto x \wedge y$$
$$x \vee y \mapsto x \vee y$$
$$(\to_{\mathrm{def}}) \qquad x \to y \mapsto \Big(x \wedge \big(((x \wedge (y \Rightarrow y)) \Rightarrow y) \Rightarrow$$
$$((x \wedge (y \Rightarrow y)) \Rightarrow y) \big) \Big) \Rightarrow ((x \wedge (y \Rightarrow y)) \Rightarrow y)$$
$$\sim x \mapsto \sim x$$

is an interpretation of $\mathbf{N4}$ in $\mathbf{DPNRW}^{\gg}$.

3. The interpretations α and β are mutually inverse.

Hence the deductive systems $\mathbf{N4}$ and $\mathbf{DPNRW}^{\gg}$ are definitionally equivalent.

For deductive systems that are algebraisable, the notion of definitional equivalence (Gyuris, 1999) used throughout this paper is an analogue of the well known notion of term equivalence (McKenzie et al., 1987) for varieties. For details, see Subsection 3.5 below. See also (Spinks and Veroff, 2008b, Section 4). For alternative notions of definitional equivalence with applicability to abstract algebraic logic see (Caleiro and Gonçalves, 2005; Pynko, 1999; Wójcicki, 1988).

Consequences of the main theorem Theorem 2.1 allows connections to be drawn with many deductive systems that have been considered previously in the literature, including (but not limited to) those of the following theorem.

Theorem 2.2.

1. *The axiomatic extension of* **N4** *by the weakening axiom* $\vdash x \Rightarrow (y \Rightarrow x)$ *is* **N3**.

2. *The axiomatic expansion of* **N4** *by the Ackermann constant* t *is, up to definitional equivalence, the deductive system* **BN** *of Slaney* (Restall, 1993; Slaney, 1991; Slaney et al., 1989).

3. *The axiomatic extension of* **N4** *by the contraction axiom* $\vdash \big(x \Rightarrow (x \Rightarrow y)\big) \Rightarrow (x \Rightarrow y)$ *is, up to definitional equivalence, the 3-valued relevant logic with mingle* **RM3** (Anderson and Belnap, 1975, p. 470 ff.).

4. *The axiomatic extension of* **N4** *by the weakening axiom* $\vdash x \Rightarrow (y \Rightarrow x)$ *and the contraction axiom* $\vdash \big(x \Rightarrow (x \Rightarrow y)\big) \Rightarrow (x \Rightarrow y)$ *is, up to definitional equivalence, the classical propositional calculus* **CPC**.

5. *The axiomatic extension of* **N4**$^{\perp}$ *by the contraction axiom* $\vdash \big(x \Rightarrow (x \Rightarrow y)\big) \Rightarrow (x \Rightarrow y)$ *is, up to definitional equivalence, the 3-valued logic* **J3** *of D'Ottaviano and da Costa* (Carnielli and Marcos, 2002; D'Ottaviano and da Costa, 1970; Epstein, 1995).

6. *The axiomatic extension of* **N4** *by the weakening axiom* $\vdash x \Rightarrow (y \Rightarrow x)$ *and the prelinearity axiom* $\vdash (x \rightarrow y) \vee (y \rightarrow x)$ *is, up to definitional equivalence, the nilpotent minimum logic* **NM** *of Esteva and Godo* (Bianchi, 2011; Esteva and Godo, 2001; Gispert, 2003; Noguera, 2007; Noguera et al., 2008).

7. *The axiomatic extension of* **N4** *by the Peirce law* $\vdash \big((x \rightarrow y) \rightarrow x\big) \rightarrow x$ *is the (Hilbert style) 'basic system'* **HBe** *of Avron* (Arieli, 1999; Arieli and Avron, 1994, 1996; Avron, 1991).

8. *The axiomatic extension of* **N4** *by the Peirce law* $\vdash \big((x \rightarrow y) \rightarrow x\big) \rightarrow x$ *is, up to definitional equivalence, the deductive system* **BN4** *of Brady* (Brady, 1982; Meyer et al., 1984; Restall, 1993; Slaney, 1991).

9. *The axiomatic extension of* **N4** *by the weakening axiom* $\vdash x \Rightarrow (y \Rightarrow x)$ *and the Peirce law* $\vdash \big((x \rightarrow y) \rightarrow x\big) \rightarrow x$ *is, up to definitional equivalence, the 3-valued logic* **Ł**$_3$ *of Łukasiewicz* (Łukasiewicz, 1970a,b).

10. *The axiomatic extension of* **N4**$^{\perp}$ *by the Peirce law* $\vdash \big((x \rightarrow y) \rightarrow x\big) \rightarrow x$ *is, up to definitional equivalence, the logic* **BS4** *of Omori and Waragai* (De and Omori, 2015; Omori and Sano, 2014; Omori and Waragai, 2011; Sano and Omori, 2014).

Theorem 2.2 is by no means exhaustive. For example, it follows from the results of the series of papers (Spinks and Veroff, a,b,c) that the axiomatic extension of $\mathbf{N4}^{\perp}$ by the Peirce law $\vdash \left((x \to y) \to x \right) \to x$ interprets the deductive system determined by Smiley's matrices for the logic $\mathbf{E}_{\mathrm{fde}}$ of first degree entailments (Anderson and Belnap, 1975, pp. 161–162); *cf.* (Muskens, 1995, Chapter 5) or (Villadsen, 2001, Section 2.1). Numerous other connections with the existing literature may be similarly established.

3 Definitional Equivalence

3.1 Proof strategy

The proof of Theorem 2.1 mirrors the proof of the authors' earlier result (Spinks and Veroff, 2008b, Theorem 1.1) showing that $\mathbf{N3}$ is definitionally equivalent to an axiomatic extension of $\mathbf{InFL}_{ew}$. First, we establish the term equivalence of the algebraic counterparts of $\mathbf{N4}$ and $\mathbf{DPNRW}^{\gg}$. Subsequently, this term equivalence result is lifted to the setting of deductive systems to establish the definitional equivalence of the logics $\mathbf{N4}$ and $\mathbf{DPNRW}^{\gg}$. In this subsection, the proof strategy used to establish the requisite term equivalence result is presented.

N4-lattices Following (Nemitz, 1965, Section 2), an algebra $\langle A; \wedge, \vee, \to \rangle$ of type $\langle 2,2,2 \rangle$ is an ***implicative lattice*** if: (i) $\langle A; \wedge, \vee \rangle$ is a lattice (with lattice ordering $\leq$); and (ii) for all $a,b,c \in A$, it holds that $a \wedge b \leq c$ iff $a \leq b \to c$. Let Λ be a language type having a distinguished binary operation symbol $\to$. By abuse of notation, let $|\varphi|$ abbreviate $\varphi \to \varphi$ for every Λ-term φ. Observe that if $\mathbf{A}$ is an implicative lattice, then the unary term $\top := |x|$ is constant over $\mathbf{A}$; we write $\top \in A$ for $\top^{\mathbf{A}}$.

Recall from (Odintsov, 2003, Definition 5.1) that an algebra $\mathbf{A} := \langle A; \wedge, \vee, \to, \sim \rangle$ of type $\langle 2,2,2,1 \rangle$ is an $\mathbf{N4}$-***lattice*** if:

(N1)   The reduct $\langle A; \wedge, \vee, \sim \rangle$ is a De Morgan lattice with lattice ordering $\leq$.

(N2)   The relation $\preceq$ defined for all $a,b \in A$ by $a \preceq b$ iff $a \to b = |a \to b|$ is a quasiorder on A.

(N3)   The relation $\Xi := \preceq \cap (\preceq)^{-1}$ is a congruence on the reduct $\langle A; \wedge, \vee, \to \rangle$ such that the quotient algebra $\mathbf{A}_{\bowtie} := \langle A; \wedge, \vee, \to \rangle / \Xi$ is an implicative lattice.

(N4)   For all $a,b \in A$, it holds that $\sim(a \to b) \equiv a \wedge \sim b \pmod{\Xi}$.

(N5)   For all $a,b \in A$, it holds that $a \leq b$ iff $a \preceq b$ and $\sim b \preceq \sim a$.

N4-lattices were introduced in (Odintsov, 2003) and they have been further studied in (Odintsov, 2004, 2005, 2007, 2008). By (Odintsov, 2008, Theorem 8.5.3), the class N4 of all **N4**-lattices forms a variety. Let Λ be a language type having distinguished binary operation symbols $\wedge$ and $\to$ and a distinguished unary operation symbol $\sim$. For all Λ-terms φ and ψ, let $\varphi \Leftrightarrow \psi$ abbreviate $\big((\varphi \to \psi) \wedge (\sim\psi \to \sim\varphi)\big) \wedge \big((\psi \to \varphi) \wedge (\sim\varphi \to \sim\psi)\big)$; notice $\varphi \Leftrightarrow \psi$ can also be written as $(\varphi \Rightarrow \psi) \wedge (\psi \Rightarrow \varphi)$, where the derived connective $\Rightarrow$ is as fixed by the map $(\Rightarrow_{\text{def}})$. By (Rivieccio, 2011, Theorem 2.6), **N4** is strongly algebraisable with system of equivalence formulas $\{\varphi \Leftrightarrow \psi\}$, system of defining equations $\big\{x \approx |x|\big\}$, and equivalent variety semantics N4.

Lemma 3.1. *The variety of* **N4**-*lattices satisfies the identities*

$$(2) \qquad \sim(\sim x \to \sim y) \to \sim x \approx \big|\sim(\sim x \to \sim y) \to \sim x\big|$$

$$(3) \qquad x \Rightarrow (y \to x) \approx \big|x \Rightarrow (y \to x)\big|$$

$$(4) \qquad (x \Leftrightarrow y) \to x \approx (x \Leftrightarrow y) \to y$$

$$(5) \qquad x \to y \approx x \to (x \to y)$$

$$(6) \qquad x \to (y \to z) \approx y \to (x \to z)$$

$$(7) \qquad x \to (y \to z) \approx (x \to y) \to (x \to z).$$

A ***Nelson algebra*** is an **N4**-lattice satisfying the *ex falso quodlibet* identity

$$(8) \qquad x \to (\sim x \to y) \approx \big|x \to (\sim x \to y)\big|.$$

Nelson algebras were introduced in (Białynicki-Birula and Rasiowa, 1958; Rasiowa, 1958, 1959) and they have been considered extensively in the literature (Kracht, 1998; Rasiowa, 1974; Sendlewski, 1984; Vakarelov, 1977; Viglizzo, 1999); for a recent survey and further references, see (Vakarelov, 2006). Because N4 is equationally definable, the class N3 of all Nelson algebras also forms a variety; this was first shown by (Brignole, 1969, Theorem 3, Theorem 4). By the algebraisability of **N4** and (Blok and Pigozzi, 1989, Corollary 4.9), the deductive system **N3** is strongly algebraisable with the same systems of equivalence formulas and defining equations as **N4** and with equivalent variety semantics N3.

An **N4**$^{\perp}$-*lattice* is an algebra $\mathbf{A} := \langle A; \wedge, \vee, \to, \sim, \perp \rangle$ of type $\langle 2, 2, 2, 1, 0 \rangle$ such that: (i) $\langle A; \wedge, \vee, \to, \sim \rangle$ is an N4-lattice; and (ii) for all $a \in A$, it holds that $\perp \preceq a$ and $a \preceq \sim\perp$. By (N5), condition (ii) is equivalent to asserting that $\perp \leq a$ for all $a \in A$; the class N4$^{\perp}$ of all **N4**$^{\perp}$-lattices thus forms a variety. By the algebraisability of **N4** and (Blok and Pigozzi, 1989, Theorem 4.7), the deductive system **N4**$^{\perp}$ is strongly algebraisable with the same systems of equivalence formulas and defining equations as **N4** and with equivalent variety semantics N4$^{\perp}$.

Dimorphic paraconsistent Nelson RW-algebras Let $\langle A; \leq \rangle$ be a poset. A binary operation $*$ on A is ***compatible*** with $\leq$ if, for all $a, b, c \in A$, it holds that

$$(\text{Compat}) \qquad \text{if } a \leq b, \text{ then } a * c \leq b * c \text{ and } c * a \leq c * b.$$

A structure $\langle A; *, \Rightarrow; \leq \rangle$ of type $\langle 2, 2; 2 \rangle$ is a ***residuated po-semigroup*** if: (i) $\leq$ is a partial order on A; (ii) $*$ is an associative commutative binary operation on A compatible with $\leq$; and (iii) for all $a, b, c \in A$, it holds that

(Res) $\qquad\qquad\qquad\qquad a * b \leq c$ iff $a \leq b \Rightarrow c$.

Observe that every residuated po-semigroup satisfies the identity

$$(9) \qquad\qquad x \Rightarrow (y \Rightarrow z) \approx (x * y) \Rightarrow z.$$

A ***residuated lo-semigroup*** is an algebra $\langle A; \wedge, \vee, *, \Rightarrow \rangle$ of type $\langle 2, 2, 2, 2 \rangle$ such that: (i) $\langle A; \wedge, \vee \rangle$ is a lattice (with lattice ordering $\leq$); and (ii) $\langle A; *, \Rightarrow; \leq \rangle$ is a residuated po-semigroup. A residuated lo-semigroup $\mathbf{A}$ is ***distributive*** if its lattice reduct is distributive; it is ***adjunctive*** (Blok and Raftery, 2008; Font and Pérez, 1992; Font and Rodríguez, 1990; Hsieh, 2008; Hsieh and Raftery, 2007) if the inequality $\big(|a| \wedge |b|\big) \Rightarrow c \leq c$ holds identically on $\mathbf{A}$. By a result implicit in (van Alten and Raftery, 2004, Section 7), a residuated lo-semigroup is adjunctive iff it is the $\langle \wedge, \vee, *, \Rightarrow \rangle$-subreduct of a residuated lattice; recall a ***residuated lattice*** (Hart et al., 2002) is an algebra $\langle A; \wedge, \vee, *, \Rightarrow, e \rangle$ where $\langle A; \wedge, \vee, *, \Rightarrow \rangle$ is a residuated lo-semigroup and $\langle A; *, e \rangle$ is a monoid. The study of residuated lattices has been pursued extensively in the literature; see (Jipsen and Tsinakis, 2002) for a survey and (Galatos et al., 2007) for a detailed treatment.

An algebra $\mathbf{A} := \langle A; \wedge, \vee, *, \Rightarrow, \sim \rangle$ of type $\langle 2, 2, 2, 2, 1 \rangle$ is a ***dimorphic paraconsistent Nelson $\mathbf{RW}^{\gg}$-algebra*** if:

(D1) The reduct $\langle A; \wedge, \vee, *, \Rightarrow \rangle$ is a residuated lo-semigroup (with lattice order $\leq$) that is adjunctive and distributive.

(D2) The operation $\sim$ is a (compatible) ***negation*** on $\mathbf{A}$. That is, for all $a, b \in A$ it holds that $\sim \sim a = a$ and $a \Rightarrow b = \sim b \Rightarrow \sim a$.

(D3$^{\gg}$) The algebra $\mathbf{A}$ satisfies the identities

(Paraconsistent Nelson$^{\gg}$) $\big((x \twoheadrightarrow y) \wedge (\sim y \twoheadrightarrow \sim x)\big) \vee (x \Rightarrow y) \approx x \Rightarrow y$

(Internal weakening$^{\gg}$) $\qquad (x * y) \twoheadrightarrow x \approx \big|(x * y) \twoheadrightarrow x\big|.$

Notice that a residuated lo-semigroup $\mathbf{A}$ satisfies (Paraconsistent Nelson$^{\gg}$) iff $(a \twoheadrightarrow b) \wedge (\sim b \twoheadrightarrow \sim a) \leq a \Rightarrow b$ holds identically on $\mathbf{A}$. Because of (van Alten and Raftery, 2004, Proposition 7.1), the class DPNRW$^{\gg}$ of all dimorphic paraconsistent Nelson $\mathbf{RW}^{\gg}$-algebras forms a variety.

Recall that, for a language type Λ having a distinguished binary operation symbol $\Rightarrow$, the expression $|\varphi|$ abbreviates $\varphi \Rightarrow \varphi$ for every Λ-term φ. The next result is almost immediate on combining (Hsieh and Raftery, 2007, Theorem 7.7) with (Blok and Pigozzi, 1989, Corollary 4.9).

Proposition 3.2. *The deductive system* $\mathbf{DPNRW}^{\gg}$ *is strongly algebraisable with system of equivalence formulas* $\{\varphi \Rightarrow \psi, \psi \Rightarrow \varphi\}$ *and system of defin-*

ing equations $\{x \approx |x|\}$. *The equivalent variety semantics of* **DPNRW**$^\gg$ *is* DPNRW$^\gg$.

An algebraic analogue of Theorem 2.1 As previously remarked, to prove Theorem 2.1, we establish an algebraic analogue of Theorem 2.1 and then lift this result to the setting of deductive systems. The algebraic analogue of Theorem 2.1 that we prove is:

Theorem 3.3.

1. *The map α of Theorem 2.1.(1) is an interpretation of* DPNRW$^\gg$ *in* N4.
2. *The map β of Theorem 2.1.(2) is an interpretation of* N4 *in* DPNRW$^\gg$.
3. *The interpretations α and β are mutually inverse.*

Hence the varieties N4 *and* DPNRW$^\gg$ *are term equivalent.*

Owing to its syntactic complexity, the right projection of the ordered pair ($\rightarrow_{\text{def}}$) is difficult to work with in practice. Thus, to prove Theorem 3.3, we establish a variant in which the right projection of the ordered pair ($\rightarrow_{\text{def}}$) is replaced with a simpler expression; Theorem 3.3 then follows easily from this variant result. To this end, let Λ be a language type having distinguished binary operation symbols $\wedge$ and $\Rightarrow$. For all Λ-terms φ and ψ, let

$$\varphi \rightarrow \psi \quad \text{abbreviate} \quad \big(\varphi \wedge |\psi|\big) \Rightarrow \big((\varphi \wedge |\psi|) \Rightarrow \psi\big).$$

An algebra $\mathbf{A} := \langle A; \wedge, \vee, *, \Rightarrow, \sim \rangle$ of type $\langle 2, 2, 2, 2, 1 \rangle$ is a ***dimorphic paraconsistent Nelson* RW-*algebra*** if (D1)–(D2) hold, and:

(D3)   The algebra $\mathbf{A}$ satisfies the identities

     (Paraconsistent Nelson) $\big((x \rightarrow y) \wedge (\sim y \rightarrow \sim x)\big) \vee (x \Rightarrow y) \approx x \Rightarrow y$

     (Internal weakening) $(x * y) \rightarrow x \approx \big|(x * y) \rightarrow x\big|.$

Because of (van Alten and Raftery, 2004, Proposition 7.1), the class DPNRW of all dimorphic paraconsistent Nelson RW-algebras forms a variety. The variant of Theorem 3.3 we establish is:

Theorem 3.4.

1. *The map α of Theorem 2.1.(1) is an interpretation of* DPNRW *in* N4.
2. *The map $\beta : \Lambda[\mathbf{N4}] \rightarrow \mathrm{Fm}_{\Lambda[\mathbf{RW}]}$ defined by*

$$x \wedge y \mapsto x \wedge y$$
$$x \vee y \mapsto x \vee y$$
$$(\rightarrow'_{\text{def}}) \qquad x \rightarrow y \mapsto \big(x \wedge (y \Rightarrow y)\big) \Rightarrow \big((x \wedge (y \Rightarrow y)) \Rightarrow y\big)$$
$$\sim x \mapsto \sim x$$

    is an interpretation of N4 *in* DPNRW.
3. *The interpretations α and β are mutually inverse.*

Hence the varieties N4 *and* DPNRW *are term equivalent.*

3.2 Proof of Theorem 3.4.(1)

In this subsection we give the proof of Theorem 3.4.(1). The proof is along the lines of the proof of (Busaniche and Cignoli, 2010, Theorem 3.1); the main tool used in the proof is the twist-structure representation for **N4**-lattices of Odintsov (2003, 2004, 2008).

Twist structures Let **A** be an implicative lattice. The *full twist-structure* over **A** is the algebra $\mathbf{A}^{\bowtie} := \langle A \times A; \wedge, \vee, \rightarrow, \sim \rangle$ of type $\langle 2,2,2,1 \rangle$ with operations defined for all $\langle a,b \rangle, \langle c,d \rangle \in A \times A$ by

$$(\wedge_{\mathrm{def}}^{\bowtie}) \qquad\qquad \langle a,b \rangle \wedge \langle c,d \rangle := \langle a \wedge c, b \vee d \rangle$$

$$\langle a,b \rangle \vee \langle c,d \rangle := \langle a \vee c, b \wedge d \rangle$$

$$(\rightarrow_{\mathrm{def}}^{\bowtie}) \qquad\qquad \langle a,b \rangle \rightarrow \langle c,d \rangle := \langle a \rightarrow c, a \wedge d \rangle$$

$$(\sim_{\mathrm{def}}^{\bowtie}) \qquad\qquad\qquad \sim \langle a,b \rangle := \langle b,a \rangle.$$

By (Odintsov, 2003, Proposition 5.2), $\mathbf{A}^{\bowtie}$ is an **N4**-lattice, and by Odintsov (2004, Corollary 3.2), every **N4**-lattice can be represented as a subalgebra of the full twist-structure $\mathbf{B}^{\bowtie}$, for a suitable implicative lattice **B**; we give the details.

Let **A** be an implicative lattice. The set $\nabla_d(\mathbf{A}) := \{ a \vee (a \rightarrow b) : a,b \in A \}$ is the *filter of dense elements* of **A** (Odintsov, 2004, Section 3, p. 395; Proposition 3.2); it is easy to see that $\nabla_d(\mathbf{A})$ is a lattice filter. Let $\nabla \subseteq A$ be a non-empty lattice filter such that $\nabla_d(\mathbf{A}) \subseteq \nabla$ and let $\Delta \subseteq A$ be an arbitrary non-empty lattice ideal. Then the set $B := \{ \langle a,b \rangle \in A \times A : a \vee b \in \nabla, a \wedge b \in \Delta \}$ is closed under the operations $\wedge$, $\vee$, $\rightarrow$, and $\sim$ of $\mathbf{A}^{\bowtie}$. Thus $\langle B; \wedge, \vee, \rightarrow, \sim \rangle$ is an **N4**-lattice, that, following (Odintsov, 2004), we denote by $\mathbf{Tw}(\mathbf{A}, \nabla, \Delta)$.

Given an arbitrary **N4**-lattice **B**, let

$$\nabla(\mathbf{B}) := \{ [a \vee \sim a]_{\Xi} : a \in B \} \quad \text{and} \quad \Delta(\mathbf{B}) := \{ [a \wedge \sim a]_{\Xi} : a \in B \}.$$

Then $\nabla(\mathbf{B})$ is a lattice filter of the implicative lattice $\mathbf{B}_{\bowtie}$ such that $\nabla_d(\mathbf{B}) \subseteq \nabla(\mathbf{B})$, and $\Delta(\mathbf{B})$ is an ideal of $\mathbf{B}_{\bowtie}$.

Theorem 3.5. (Jansana and Rivieccio, 2013, Proposition 2.2) *Every* **N4**-*lattice* **B** *is isomorphic to the algebra* $\mathbf{Tw}(\mathbf{B}_{\bowtie}, \nabla(\mathbf{B}), \Delta(\mathbf{B}))$, *with* $\mathbf{B}_{\bowtie}$ *an implicative lattice, via the map* $B \xrightarrow{j_{\mathbf{B}}} B/\Xi \times B/\Xi$ *defined for all* $b \in B$ *by* $j_{\mathbf{B}}(b) := \langle [b]_{\Xi}, [\sim b]_{\Xi} \rangle$.

Proof of Theorem 3.4.(1) With the ingredients of Section 3.2 to hand, we have everything in place needed to prove Theorem 3.4.(1). But first, let Λ be a language type having a distinguished binary logical connective $*$. For every Λ-term φ, let φ^2 [resp. φ^3] abbreviate $\varphi * \varphi$ [resp. $\varphi * (\varphi * \varphi)$].

Theorem 3.6. *The map* α *of Theorem 3.4.(1) is an interpretation of* DPNRW *in* N4.

Proof. (Sketch) Let $\mathbf{A}$ be an $\mathbf{N4}$-lattice. To prove the theorem, it suffices to verify that (D1)–(D3) hold over $\mathbf{A}^\alpha := \langle A; c^\alpha \rangle_{c \in \Lambda[\mathbf{RW}]}$.

By Theorem 3.5, we may assume without loss of generality that $\mathbf{A}$ is the form $\mathbf{Tw}(\mathbf{A}_{\bowtie}, \nabla(\mathbf{A}), \Delta(\mathbf{A}))$. In what follows we write $\mathbf{Tw}(\mathbf{A})$ for $\mathbf{Tw}(\mathbf{A}_{\bowtie}, \nabla(\mathbf{A}), \Delta(\mathbf{A}))$ and $\mathbf{Tw}(A)$ for the universe of $\mathbf{Tw}(\mathbf{A})$. To begin, observe that the derived operations $*$ and $\Rightarrow$ on $\mathbf{Tw}(\mathbf{A})$ induced by the maps ($*_{\mathrm{def}}$) and ($\Rightarrow_{\mathrm{def}}$) respectively take the following forms for all $\langle a, b \rangle, \langle c, d \rangle \in \mathbf{Tw}(A)$:

$$(10) \qquad \langle a, b \rangle * \langle c, d \rangle = \langle a \wedge c, (a \to d) \wedge (c \to b) \rangle$$

$$(11) \qquad \langle a, b \rangle \Rightarrow \langle c, d \rangle = \langle (a \to c) \wedge (d \to b), a \wedge d \rangle.$$

Because $*$ and $\Rightarrow$ are both term operations on $\mathbf{Tw}(\mathbf{A})$, both these operations are well-defined. Observe in particular that for all $\langle a, b \rangle \in \mathbf{Tw}(A)$,

$$(12) \qquad \langle a, b \rangle \Rightarrow \langle a, b \rangle = \langle \top, a \wedge b \rangle$$

$$(13) \qquad \langle a, b \rangle * \langle a, b \rangle = \langle a, a \to b \rangle.$$

To complete the proof of the theorem, it suffices to verify that (D1)–(D3) hold over $\mathbf{Tw}(\mathbf{A})^\alpha$. This reduces to a routine series of computations; we show $\mathbf{Tw}(\mathbf{A})^\alpha \models$ (Paraconsistent Nelson) by way of example. But first, let $\mathbf{B}$ be an implicative lattice. Note $\mathbf{B}$ satisfies the identity

$$(14) \qquad (x \to y) \wedge \big(z \to (x \to (w \vee (y \wedge z)))\big) \approx x \to y.$$

Indeed, let $e, f, g, h \in B$. From $f \wedge g \leq h \vee (f \wedge g)$ we have $f \leq g \to (h \vee (f \wedge g))$ by the theory of implicative lattices, whence

$$e \to f \leq e \to \big(g \to (h \vee (f \wedge g))\big) = g \to \big(e \to (h \vee (f \wedge g))\big),$$

again by the theory of implicative lattices. Hence $\mathbf{B} \models$ (14).

Observe next that for all $\langle a, b \rangle, \langle c, d \rangle \in \mathbf{Tw}(A)$,

$$(15) \qquad \langle a, b \rangle \to^{\mathbf{Tw}(\mathbf{A})^{\alpha\beta}} \langle c, d \rangle = \langle a \to c, a \wedge d \rangle = \langle a, b \rangle \to^{\mathbf{Tw}(\mathbf{A})} \langle c, d \rangle.$$

Indeed, we have:

$$\langle a,b\rangle \to^{\mathbf{Tw}(\mathbf{A})^{\alpha\beta}}\langle c,d\rangle = \big(\langle a,b\rangle \wedge |\langle c,d\rangle|\big) \Rightarrow \big((\langle a,b\rangle \wedge |\langle c,d\rangle|) \Rightarrow \langle c,d\rangle\big)$$

$$= \big(\langle a,b\rangle \wedge |\langle c,d\rangle|\big)^2 \Rightarrow \langle c,d\rangle \qquad \text{by (9)}$$

$$= \big(\langle a,b\rangle \wedge \langle \top, c\wedge d\rangle\big)^2 \Rightarrow \langle c,d\rangle \qquad \text{by (12)}$$

$$= \big\langle a, b\vee(c\wedge d)\big\rangle^2 \Rightarrow \langle c,d\rangle \qquad \text{by } (\wedge_{\mathrm{def}}^{\bowtie})$$

$$= \big\langle a, a\to (b\vee(c\wedge d))\big\rangle \Rightarrow \langle c,d\rangle \qquad \text{by (13)}$$

$$= \Big\langle (a\to c)\wedge\big(d\to(a\to(b\vee(c\wedge d)))\big), a\wedge d\Big\rangle \quad \text{by (11)}$$

$$= \langle a\to c, a\wedge d\rangle \qquad \text{by (14)}$$

$$= \langle a,b\rangle \to^{\mathbf{Tw}(\mathbf{A})}\langle c,d\rangle \qquad \text{by } (\to_{\mathrm{def}}^{\bowtie}),$$

vindicating the claim. To see $\mathbf{Tw}(\mathbf{A})^\alpha \models$ (Paraconsistent Nelson), simply note

$$\big(\langle a,b\rangle \to^{\mathbf{Tw}(\mathbf{A})^{\alpha\beta}}\langle c,d\rangle\big) \wedge \big(\sim\langle c,d\rangle \to^{\mathbf{Tw}^{\alpha\beta}}\sim\langle a,b\rangle\big)$$

$$= \big(\langle a,b\rangle \to \langle c,d\rangle\big) \wedge \big(\sim\langle c,d\rangle \to \sim\langle a,b\rangle\big) \quad \text{by (15)}$$

$$= \big(\langle a,b\rangle \to \langle c,d\rangle\big) \wedge \big(\langle d,c\rangle \to \langle b,a\rangle\big) \qquad \text{by } (\sim_{\mathrm{def}}^{\bowtie})$$

$$= \langle a\to c, a\wedge d\rangle \wedge \langle d\to b, d\wedge a\rangle \qquad \text{by } (\to_{\mathrm{def}}^{\bowtie})$$

$$= \big\langle (a\to c)\wedge(d\to b), a\wedge d\big\rangle \qquad \text{by } (\wedge_{\mathrm{def}}^{\bowtie})$$

$$= \langle a,b\rangle \Rightarrow \langle c,d\rangle \qquad \text{by (12).}$$

Thus $\mathbf{Tw}(\mathbf{A})^\alpha \models$ (Paraconsistent Nelson). A similar argument shows $\mathbf{Tw}(\mathbf{A})^\alpha \models$ (Internal weakening). Thus (D3) holds over $\mathbf{Tw}(\mathbf{A})^\alpha$; conditions (D1) and (D2) are readily established.

The next result obtains from the proof of Theorem 3.6; it generalises a result for Nelson algebras found in (Viglizzo, 1999, Chapter 1, p. 18). See also (Spinks and Veroff, 2008a, Proposition 3.2) and the accompanying historical footnote.

Corollary 3.7. *The variety of* $\mathbf{N}4$*-lattices satisfies the identity*

$$x \to y \approx (x\wedge|y|) \Rightarrow ((x\wedge|y|) \Rightarrow y).$$

3.3 Proof of Theorem 3.4.(2)

In this subsection we give the proof of Theorem 3.4.(2). The proof proceeds in the naïvest manner possible, by showing that if $\mathbf{A} \in \mathrm{DPNRW}$, then conditions (N1)–(N5) hold over $\mathbf{A}^\beta := \langle A; c^\beta\rangle_{c\in\Lambda[\mathbf{N4}]}$, where β is the map of Theorem 3.4.(2).

The proof that conditions (N1)–(N5) hold over $\mathbf{A}^\beta$ proceeds largely by equational reasoning. The first-order results composing the proof that conditions (N1)–(N5) hold over $\mathbf{A}^\beta$ were obtained, in large measure, with the assistance of the automated reasoning program PROVER9 (McCune). PROVER9 is a resolution-based theorem-prover for first-order logic with equality that has been shown to be particularly useful (Spinks and Veroff, 2008a,b; Veroff and Spinks, 2006) in the investigation of problems arising from algebraic logic. For a brief survey on automated theorem-proving in general algebra, see (Phillips and Stanovský, 2010, Section 6); for examples of the application of automated reasoning to a wide range of problems in equational logic, see (McCune and Padmanabhan, 1996).

Fundamental properties of dimorphic paraconsistent Nelson RW-algebras We begin the proof of Theorem 3.4.(2) by establishing some fundamental properties of dimorphic paraconsistent Nelson $\mathbf{RW}$-algebras. But first, let $\mathbf{A}$ be a dimorphic paraconsistent Nelson $\mathbf{RW}$-algebra. Observe that, by the theory of residuated lo-semigroups, the inequality

$$(16) \qquad\qquad a \leq b \to a$$

holds identically on $\mathbf{A}$. In addition, call an algebra $\mathbf{A} := \langle A; \wedge, \vee, *, \Rightarrow, \sim \rangle$ of type $\langle 2,2,2,2,1 \rangle$ a *residuated lo-semigroup with negation* if: (i) $\langle A; \wedge, \vee, *, \Rightarrow \rangle$ is a residuated lo-semigroup; and (ii) (D2) holds on $\mathbf{A}$, that is, $\sim$ is a (compatible) negation on $\mathbf{A}$. *Mutatis mutandis*, let the notion of a *residuated lattice with negation* be defined analogously.

Lemma 3.8 (Fundamental properties lemma). *Let $\mathbf{A}$ be a dimorphic paraconsistent Nelson $\mathbf{RW}$-algebra. The following statements hold for all $a, b \in A$:*

1. $|a| \Rightarrow b \leq b.$ (Implicativity)
2. $a \leq |a|.$ (Mingle)
3. $a^2 \leq a.$ (Square decreasingness)
4. $a \Rightarrow |a| \leq |a|.$ (Auto-contraction)
5. $a \Rightarrow |a| = |a|.$ (Strong auto-contraction)
6. $a^3 = a^2.$ (3-potence)

Proof.

(1) By adjunctivity, $|a| \Rightarrow b = (|a| \wedge |a|) \Rightarrow b \leq b.$

(2) By (16), $\sim a \leq a \to \sim a.$ Thus $\sim a \leq a \to \sim a = (a \to \sim a) \wedge (\sim(\sim a) \to \sim a) \leq a \Rightarrow \sim a$ by (Paraconsistent Nelson). That is, $\sim a \leq a \Rightarrow \sim a.$ But then $a \leq |\sim a| = |a|.$

(3) This follows from (2) by (Res).

(4) By (Mingle) and the theory of residuated lo-semigroups with negation, $\sim a \leq |a^2|.$ Also, $|a^2| = a \to a^2$ by (Mingle) and the theory of residuated lo-semigroups. Thus $\sim a \leq a \to a^2.$ Now by (16), $\sim a \leq \sim(a^2) \to \sim a,$ whence $\sim a \leq (a \to a^2) \wedge (\sim(a^2) \to \sim a) \leq a \Rightarrow a^2$ by (Paraconsistent Nelson). Thus $\sim a \leq a \Rightarrow a^2;$ over residuated lo-semigroups with negation, this inequality holds identically iff $a \Rightarrow |a| \leq |a|$ holds identically.

(5) This follows from (2) and (4).

(6) By (Square decreasingness), $a^2 \leq a$, so by (Compat), $a^3 \leq a^2$. For the converse, note $\sim(a^3) = \sim(a^2 * \sim \sim a) = a^2 \Rightarrow \sim a = (a \wedge |\sim a|)^2 \Rightarrow \sim a$ (by (Mingle)) $= a \to \sim a = (a \to \sim a) \wedge (\sim(\sim a) \to \sim a) \leq a \Rightarrow \sim a$ (by (Paraconsistent Nelson) $= \sim(a * \sim \sim a) = \sim(a^2)$. That is, $\sim(a^3) \leq \sim(a^2)$. Thus $a^2 \leq a^3$.

Let (P) be a fundamental property of dimorphic paraconsistent Nelson **RW**-algebras, as identified in Lemma 3.8. In the sequel, a residuated lo-semigroup having property (P) is called a ***(P) residuated lo-semigroup***. Thus, for example, a square decreasing residuated lo-semigroup is a residuated lo-semigroup such that $a^2 \leq a$ holds identically.

The following trivial lemma plays a fundamental role in the sequel.

Lemma 3.9 (Doubling construction). *Let* $\langle A; \leq \rangle$ *be a partially ordered set and let* $*$ *be a binary operation on A such that* (Compat) *holds for all* $a, b, c \in A$. *Then for all* $a, b \in A$, *if* $a \leq b$, *then* $a^2 \leq b^2$.

Proof. Suppose $a \leq b$. By (Compat), $a * a \leq a * b$, and by (Compat) again, $a * b \leq b * b$. By transitivity, $a * a \leq b * b$.

The following lemma, which is inspired by (Sendlewski, 1990, Lemma 3.1(viii)), also plays a crucial role in subsequent developments.

Lemma 3.10 (The centripetal lemma). *Let* **A** *be a dimorphic paraconsistent Nelson* **RW**-*algebra. Then* **A** *satisfies the identity*

$$(\wedge\text{-Centrip.}) \qquad\qquad (x \wedge y)^2 \approx (x^2 \wedge y^2)^2.$$

Proof. Let $a, b \in A$. By two applications of (Square decreasingness), $a^2 \leq a$ and $b^2 \leq b$; thus $a^2 \wedge b^2 \leq a \wedge b$. By the doubling construction, $(a^2 \wedge b^2)^2 \leq (a \wedge b)^2$. Conversely, from $a \wedge b \leq a, b$ we have $(a \wedge b)^2 \leq a^2, b^2$ by two applications of the doubling construction, whence $(a \wedge b)^2 \leq a^2 \wedge b^2$. By the doubling construction, $((a \wedge b)^2)^2 \leq (a^2 \wedge b^2)^2$, whence $(a \wedge b)^2 \leq (a^2 \wedge b^2)^2$ by (3-potence). Hence **A** $\models (\wedge\text{-Centrip.})$.

A quasiorder on dimorphic paraconsistent Nelson RW-algebras
Armed with the fundamental properties lemma, the doubling construction, and the centripetal lemma, we have all we need in place to establish the following result; the long proof is omitted.

Lemma 3.11. (*cf.* (Spinks, 2004, Lemma 3.2)) *Let* **A** *be a dimorphic paraconsistent Nelson* **RW**-*algebra. Then* **A** *satisfies the identities*

$$(17) \qquad x \to (y \wedge z) \approx (x \to y) \wedge (x \to z)$$

$$(18) \qquad x \to (y \vee z) \approx x \to ((x \to y) \vee (x \to z))$$

$$x \to (y \Rightarrow z) \approx (x \to y) \Rightarrow (x \to z)$$

$$x \to (y * z) \approx x \to ((x \to y) * (x \to z))$$

$$(19) \qquad x \to \sim y \approx x \to \sim(x \to y).$$

The next result exploits ideas due to (Idziak et al., 2009, Section 1) and (Aglianò, 2001, Section 1). In preparation for the proposition, observe that, by adjunctivity, every $\mathbf{A} \in \mathsf{DPNRW}$ satisfies

$$(20) \qquad\qquad\qquad |x| \to y \approx y.$$

Proposition 3.12. *Let $\mathbf{A}$ be a dimorphic paraconsistent Nelson $\mathbf{RW}$-algebra. The relation $\preceq$ on A defined for all $a,b \in A$ by $a \preceq b$ iff $a \to b = a \to |b|$ is a quasiorder on A. Moreover, for all $a,b \in A$, the following statements hold:*

1. $a \preceq b$ iff $a \to b = |a \to b|$.
2. $a \preceq b$ iff $(a \wedge |b|)^2 \leq b$.
3. $a \preceq b$ iff $(a \wedge |b|)^2 = (a \wedge b)^2$.

Proof. (Sketch) Let $\sqsubseteq$ be the relation defined on A by $a \sqsubseteq b$ iff $\Theta^{\mathbf{A}}(b, |b|) \subseteq \Theta^{\mathbf{A}}(a, |a|)$. By properties of $\subseteq$, the relation $\sqsubseteq$ is a quasiorder on A. To prove the first statement, therefore, it suffices to show $\sqsubseteq \; = \; \preceq$. For each $c \in A$, let θ_c be the relation on A defined for all $a,b \in A$ by $c \to a = c \to b$. Each θ_c is clearly an equivalence relation on A, which moreover is a congruence relation on $\mathbf{A}$ by Lemma 3.11. We claim that for every $c \in A$, we have $\theta_c = \Theta^{\mathbf{A}}(c, |c|)$. Indeed, $c \to |c| = |c|$ by (Strong auto-contraction), whence $\Theta^{\mathbf{A}}(c, |c|) \subseteq \theta_c$. For the opposite inclusion, let $a,b \in A$ and suppose $a \equiv b \pmod{\theta_c}$. Then

$$a \overset{(20)}{=} |c| \to a \; \Theta^{\mathbf{A}}\big(|c|,c\big) \; c \to a \overset{(\mathrm{Hyp.})}{=} c \to b \; \Theta^{\mathbf{A}}\big(|c|,c\big) \; |c| \to b \overset{(20)}{=} b.$$

Hence $\theta_c \subseteq \Theta^{\mathbf{A}}\big(|c|,c\big)$, vindicating the claim. Observe now that for all $a,b \in A$, we have $a \preceq b$ iff $a \to b = a \to |b|$ iff $b \equiv |b| \pmod{\theta_a}$ iff $b \equiv |b| \pmod{\Theta^{\mathbf{A}}(a,|a|)}$ iff $\Theta^{\mathbf{A}}\big(b,|b|\big) \subseteq \Theta^{\mathbf{A}}\big(a,|a|\big)$ iff $a \sqsubseteq b$. Thus $\preceq$ is a quasiorder on A.

(1) It suffices to show $a \to |b| = |a \to b|$ for all $a,b \in A$. Observe first that, over implicative residuated lo-semigroups, the inequality $a \Rightarrow |a| \leq |a|$ holds identically iff the inequality $a \Rightarrow |b| \leq |a \Rightarrow b|$ holds identically (the argument is non-trivial). Let $a,b \in A$. Then $a \to |b| = \big(a \wedge ||b||\big)^2 \Rightarrow |b| = \big(a \wedge |b|\big)^2 \Rightarrow |b|$ (by (Implicativity)) $\leq \big|(a \wedge |b|)^2 \Rightarrow b\big|$ (by (Auto-contraction) and preceding remarks) $= |a \to b|$. Conversely, $|a \to b| \leq a \to |b|$ by the theory of implicative residuated lo-semigroups.

(2) It suffices to show that for all $a,b \in A$, the equivalence $a \leq b$ iff $a \Rightarrow b = |a \Rightarrow b|$ holds; by (Implicativity) and (Square decreasingness), this follows from (Hsieh and Raftery, 2006, Proposition 5.6).

(3) This follows easily from (2).

Congruence properties of dimorphic paraconsistent Nelson RW-algebras For every dimorphic paraconsistent Nelson $\mathbf{RW}$-algebra $\mathbf{A}$, let $\Xi := \; \preceq \cap \, (\preceq)^{-1}$, where $\preceq$ is the quasiorder of Proposition 3.12. Our next order of business is to show that for each $\mathbf{A} \in \mathsf{DPNRW}$, the equivalence relation Ξ on A enjoys the substitution property with respect to the operations $\wedge, \vee,$ and $\to$.

Lemma 3.13. *Let* **A** *be a dimorphic paraconsistent Nelson* **RW**-*algebra. Then* **A** *satisfies the identity*

$$(21) \qquad \left(x \wedge |x \wedge y|\right)^2 \approx \left(x \wedge |y|\right)^2.$$

Proof. (Sketch) Let $a, b \in A$. By (Mingle), $a \wedge |b| \leq |a| \wedge |b|$. Also, $|a| \wedge |b| \leq |a \wedge b|$ by the theory of implicative residuated lo-semigroups. Thus $a \wedge |b| \leq |a \wedge b|$, whence $a \wedge \left(a \wedge |b|\right) \leq a \wedge |a \wedge b|$. By the doubling construction, $\left(a \wedge |b|\right)^2 \leq \left(a \wedge |a \wedge b|\right)^2$. For the converse, from $a \wedge |a \wedge b| \leq a$ we have $\left(a \wedge |a \wedge b|\right)^2 \leq a^2$ by the doubling construction; as $a^2 \leq a$ by (Square decreasingness), it holds that $\left(a \wedge |a \wedge b|\right)^2 \leq a$. On the other hand, $\left(a \wedge |a \wedge b|\right)^2 \leq b$ by the theory of 3-potent strongly auto-contractive adjunctive distributive residuated lo-semigroups with negation (the argument is non-trivial). Therefore $\left(a \wedge |a \wedge b|\right)^2 \leq a \wedge b$. By the doubling construction, $\left((a \wedge |a \wedge b|)^2\right)^2 \leq \left(a \wedge |b|\right)^2$, whence by (3-potence), $\left(a \wedge |a \wedge b|\right)^2 \leq \left(a \wedge |b|\right)^2$. Hence $\mathbf{A} \models$ (21). $\quad\blacksquare$

Lemma 3.14. *Let* **A** *be a dimorphic paraconsistent Nelson* **RW**-*algebra. The following statements hold for all* $a, b, c, d \in A$ *such that* $a \equiv c$ *and* $b \equiv d$:

1. $a \wedge b \equiv c \wedge d$.
2. $a \vee b \equiv c \vee d$.

Proof. (Sketch) We show only (1); the proof of (2) is non-trivial. Suppose $a \preceq b$. By Proposition 3.12.(3), $\left(a \wedge |b|\right)^2 = (a \wedge b)^2$. By Proposition 3.12.(3) again, to see $a \wedge c \preceq b \wedge c$ it suffices to show $\left(a \wedge c \wedge |b \wedge c|\right)^2 = (a \wedge c \wedge b)^2$. For this, just note

$$
\begin{aligned}
\left((a \wedge c) \wedge |b \wedge c|\right)^2 &= \left(a \wedge (c \wedge |b \wedge c|)\right)^2 \\
&= \left(a^2 \wedge (c \wedge |b \wedge c|)^2\right)^2 & \text{by } (\wedge\text{-Centrip.}) \\
&= \left(a^2 \wedge (c \wedge |b|)^2\right)^2 & \text{by (21)} \\
&= \left(a \wedge (c \wedge |b|)\right)^2 & \text{by } (\wedge\text{-Centrip.}) \\
&= \left((a \wedge |b|) \wedge c\right)^2 \\
&= \left((a \wedge |b|)^2 \wedge c^2\right)^2 & \text{by } (\wedge\text{-Centrip.}) \\
&= \left((a \wedge b)^2 \wedge c^2\right)^2 & \text{by hypothesis} \\
&= (a \wedge b \wedge c)^2. & \text{by } (\wedge\text{-Centrip.}).
\end{aligned}
$$

Hence $a \wedge c \preceq b \wedge c$. The opposite inequality is established similarly. $\quad\blacksquare$

It is not easy to show directly that the relation $\equiv$ enjoys the substitution property with respect to the operation $\rightarrow$. For this reason, we instead establish:

Lemma 3.15. *Let* **A** *be a dimorphic paraconsistent Nelson* **RW***-algebra. Then* **A** *satisfies the identity*

$$(22) \qquad\qquad (x \wedge y) \to z \approx x \to (y \to z).$$

Proof. (Sketch) The crux of the proof lies in showing that for all $a, b \in A$, it holds that $a \wedge b \,\Xi\, a * b$. By (Internal weakening) and Proposition 3.12.(1) we have $a * b \preceq a, b$, so by (the proof of) Lemma 3.14.(1), $(a * b) \wedge (a * b) \preceq a \wedge b$. Hence $a * b \preceq a \wedge b$. Conversely, $(a \wedge b)^2 \leq (a * b)^2$ by the theory of 3-potent residuated lo-semigroups, so certainly $(a \wedge b)^2 \wedge |a * b|^2 \leq (a * b)^2$. By the doubling construction, therefore, $\left((a \wedge b)^2 \wedge |a * b|^2 \right)^2 \leq \left((a * b)^2 \right)^2$, whence $\left((a \wedge b) \wedge |a * b| \right)^2 \leq \left((a * b)^2 \right)^2$ by ($\wedge$-Centrip.). By (3-potence), $\left((a \wedge b) \wedge |a * b| \right)^2 \leq (a * b)^2$; as $(a * b)^2 \leq a * b$ by (Square decreasingness), we get that $\left((a \wedge b) \wedge |a * b| \right)^2 \leq a * b$. By Proposition 3.12.(2), $a \wedge b \preceq a * b$. Thus $a \wedge b \,\Xi\, a * b$. Hence $\mathbf{A} \models (22)$.

Corollary 3.16. *Let* **A** *be a dimorphic paraconsistent Nelson* **RW***-algebra. For all $a, b, c \in A$,*

(Internal residuation) $\qquad\qquad c \preceq a \to b \quad iff \quad c \wedge a \preceq b.$

Proof. Suppose $c \preceq a \to b$. By Proposition 3.12.(1), $c \to (a \to b) = \big| c \to (a \to b) \big|$, whence

$$
\begin{aligned}
(c \wedge a) \to b &= c \to (a \to b) && \text{by (22)} \\
&= \big| c \to (a \to b) \big| \\
&= \big| (c \wedge a) \to b \big| && \text{by (22)}.
\end{aligned}
$$

By Proposition 3.12.(1) again, we have that $c \wedge a \preceq b$. The converse is analogous.

Now we can easily see that:

Lemma 3.17. *Let* **A** *be a dimorphic paraconsistent Nelson* **RW***-algebra. For all $a, b, c, d \in A$, if $a \,\Xi\, b$ and $c \,\Xi\, d$, then $a \to c \,\Xi\, b \to d$.*

Proof. Suppose $a \,\Xi\, b$ and $c \,\Xi\, d$. To see $a \to c \,\Xi\, b \to d$, note that from $b \preceq a$ we have that $b \wedge (a \to c) \preceq a \wedge (a \to c)$ by (the proof of) Lemma 3.14.(1). By (Internal residuation), $a \wedge (a \to c) \preceq c$, so by transitivity, $b \wedge (a \to c) \preceq c$. Since $c \preceq d$, by transitivity again we have that $b \wedge (a \to c) \preceq d$, which is to say $(a \to c) \wedge b \preceq d$. By (Internal residuation), we conclude that $a \to c \preceq b \to d$. An analogous argument shows $b \to d \preceq a \to c$. Hence $a \to c \,\Xi\, b \to d$ as desired.

Proof of Theorem 3.4.(2) It is now a (comparatively) routine exercise to complete the proof of Theorem 3.4.(2).

Lemma 3.18. *Let* $\mathbf{A}$ *be a dimorphic paraconsistent Nelson* $\mathbf{RW}$-*algebra. For all* $a, b \in A$,

$$\sim(a \to b) \equiv a \wedge \sim b \pmod{\Xi}.$$

Proof. (Sketch) The theory of 3-potent square decreasing residuated lo-semigroups with negation shows that $\big((a \wedge \sim b) \wedge |\sim(a \to b)|\big)^2 = (a \wedge \sim b)^2 \leq \sim(a \to b)$. By Proposition 3.12.(2), therefore, $a \wedge \sim b \preceq \sim(a \to b)$. For the opposite inequality, note $\big(a \wedge |b|\big)^2 * \sim b \preceq \big(a \wedge |b|\big)^2$ (by (Internal weakening)) $\preceq a \wedge |b|$ (by Square decreasingness) $\preceq a$, whence $\sim(a \to b) = \sim\big((a \wedge |b|)^2 \Rightarrow b\big) = \sim\sim\big((a \wedge |b|)^2 * \sim b\big) = \big(a \wedge |b|\big)^2 * \sim b \preceq a$. On the other hand, $\sim(a \to b) \preceq \sim b$, owing to (16). By (the proof of) Lemma 3.14.(1), we conclude that $\sim(a \to b) \preceq a \wedge \sim b$. Thus $\sim(a \to b) \ \Xi \ a \wedge \sim b$.

Lemma 3.19. *Let* $\mathbf{A}$ *be a dimorphic paraconsistent Nelson* $\mathbf{RW}$-*algebra. For all* $a, b \in A$,

$$a \leq b \quad \text{iff} \quad a \preceq b \quad \text{and} \quad \sim b \preceq \sim a.$$

Proof. (Sketch) Suppose $a \preceq b$ and $\sim b \preceq \sim a$. Then

$$
\begin{aligned}
a \Rightarrow b &= (a \to b) \wedge (\sim b \to \sim a) && \text{by (Paraconsistent Nelson)} \\
&= |a \to b| \wedge |\sim b \to \sim a| && \text{by Proposition 3.12.(1), as } a \preceq b \text{ and } \sim b \preceq \sim a \\
&= \big||a \to b| \wedge |\sim b \to \sim a|\big| && \text{by adjunctivity and} \\
& && \text{(Hsieh and Raftery, 2007, Theorem 4.2)} \\
&= \big|(a \to b) \wedge (\sim b \to \sim a)\big| && \text{by Proposition 3.12.(1), as } a \preceq b \text{ and } \sim b \preceq \sim a \\
&= |a \Rightarrow b| && \text{by (Paraconsistent Nelson).}
\end{aligned}
$$

By implicativity and (Hsieh and Raftery, 2006, Proposition 5.6), $a \leq b$. Conversely, suppose $a \leq b$. By (Paraconsistent Nelson), $(a \to b) \wedge (\sim b \to \sim a) = a \Rightarrow b = |a \Rightarrow b|$ (by implicativity and (Hsieh and Raftery, 2006, Proposition 5.6), since $a \leq b$) $= \big|(a \to b) \wedge (\sim b \to \sim a)\big|$ by (Paraconsistent Nelson) again. By the theory of square decreasing implicative residuated lo-semigroups, this is enough to conclude $a \to b = |a \to b|$ and $\sim b \to \sim a = |\sim b \to \sim a|$. By Proposition 3.12.(1), $a \preceq b$ and $\sim b \preceq \sim a$.

Theorem 3.20. *The map* β *of Theorem 3.4.(2) is an interpretation of* N4 *in* DPNRW.

Proof. Let $\mathbf{A} \in$ DPNRW. To prove the theorem, it suffices to verify that (N1)–(N5) hold over $\mathbf{A}^\beta$. Conditions (N2)–(N5) hold over $\mathbf{A}^\beta$ by virtue of Proposition 3.12, Lemmas 3.14 and 3.17, Lemma 3.18, and Lemma 3.19 respectively. By implicativity and (Hsieh and Raftery, 2006, Lemma 4.4), the negation $\sim$ is an involution of $\langle A; \leq \rangle$; it follows directly that condition (N1) also holds over $\mathbf{A}^\beta$.

Proof of Theorem 3.4 We are finally in a position to complete the proof of Theorem 3.4.

Lemma 3.21. *The interpretations α and β of Theorem 3.4 are mutually inverse.*

Proof. By Corollary 3.7, N4 satisfies $x \to y \approx (x \wedge |y|) \Rightarrow ((x \wedge |y|) \Rightarrow y)$, while by (Hsieh and Raftery, 2006, Lemma 4.6), DPNRW satisfies $x * y \approx \sim(x \Rightarrow \sim y)$. Collectively, these two identities guarantee that the maps α and β are mutually inverse.

The proof of Theorem 3.4 now follows on combining Theorem 3.6, Theorem 3.20, and Lemma 3.21.

3.4 Proof of Theorem 3.3

In this subsection we give the proof of Theorem 3.3. The proof proceeds by showing that Theorem 3.3 is simply a restatement of Theorem 3.4; once again, equational reasoning composes the key tool used in the proof.

Proof of Theorem 3.3 With Theorem 3.4 to hand, to establish Theorem 3.3 it suffices to show: (i) $\mathsf{DPNRW}^{\gg} = \mathsf{DPNRW}$; and (ii) $\mathsf{V} \models x \to y \approx x \to y$, for each $\mathsf{V} \in \{\mathsf{DPNRW}^{\gg}, \mathsf{DPNRW}\}$. Theorem 3.3 then follows directly from Theorem 3.4.

The hypotheses of the next proposition are somewhat artificial, but are nonetheless sufficiently general to meet our avowed purpose.

Proposition 3.22. *Let $\mathbf{A}$ be a strongly auto-contractive adjunctive distributive residuated lo-semigroup with negation. Then $\mathbf{A}$ satisfies the identity*

$$(23) \qquad\qquad x \to y \approx x \overset{\mathbf{RM}}{\Rightarrow} (x \overset{\mathbf{RM}}{\Rightarrow} y).$$

Proof. (Sketch) Let $a, b \in A$. Then $a \overset{\mathbf{RM}}{\Rightarrow} (a \overset{\mathbf{RM}}{\Rightarrow} b) \leq a \to b$ by the theory of residuated lo-semigroups. For the converse, the proof strategy is to show that if $\mathbf{B}$ is a strongly auto-contractive adjunctive residuated lo-semigroup with negation, then $\mathbf{B}$ satisfies

$$(24) \qquad (x \wedge \sim x)^2 \Rightarrow (x \overset{\mathbf{RM}}{\Rightarrow} y) \approx (x \wedge \sim x) \Rightarrow (x \overset{\mathbf{RM}}{\Rightarrow} y).$$

By (24), adjunctivity, and distributivity, $a \to b \leq a \overset{\mathbf{RM}}{\Rightarrow} (a \overset{\mathbf{RM}}{\Rightarrow} b)$ then follows. (Both steps of the converse are non-trivial.) Hence $\mathbf{A} \models (23)$.

Proposition 3.23. *For each $\mathsf{V} \in \{\mathsf{DPNRW}, \mathsf{DPNRW}^{\gg}\}$, it holds that $\mathsf{V} \models$* (23). *Hence $\mathsf{DPNRW} = \mathsf{DPNRW}^{\gg}$.*

Proof. (Sketch) We prove the second statement first. By the fundamental properties lemma, each $\mathbf{A} \in$ DPNRW is strongly auto-contractive; also, $\mathbf{A}$ is adjunctive, distributive, and has a negation by fiat. By Proposition 3.22, it holds that $\mathbf{A} \models (23)$. Thus $\mathbf{A} \in$ DPNRW$^{\gg}$. So DPNRW $\subseteq$ DPNRW$^{\gg}$. For the converse, an analogue of the fundamental properties lemma for DPNRW$^{\gg}$ shows that every $\mathbf{A} \in$ DPNRW$^{\gg}$ is strongly auto-contractive. The argument showing DPNRW $\subseteq$ DPNRW$^{\gg}$ therefore also establishes, *mutatis mutandis*, that DPNRW$^{\gg} \subseteq$ DPNRW. The first statement is now clear.

The proof of Theorem 3.3 now follows from Theorem 3.4 and Proposition 3.23.

Categorical isomorphism Let N4 [resp. $\mathbb{DPNRW}^{\gg}$] denote the finitary algebraic category whose objects are the algebras in N4 [resp. the algebras in DPNRW$^{\gg}$] and whose morphisms are all the homomorphisms $f : \mathbf{A} \to \mathbf{B}$, where $\mathbf{A}, \mathbf{B} \in$ N4 [resp. $\mathbf{A}, \mathbf{B} \in$ DPNRW$^{\gg}$]. Consider the following functors, where α and β denote the maps of Theorem 3.3.(1) and 3.3.(2) respectively:

- $R : \mathrm{N4} \to \mathbb{DPNRW}^{\gg}$, where the object function R_{obj} sends each $\mathbf{A} \in$ N4 to its associated algebra $\mathbf{A}^{\alpha}$, and the arrow function R_{arw} sends each homomorphism $h : \mathbf{A} \to \mathbf{B}$ $(\mathbf{A}, \mathbf{B} \in$ N4$)$ to the same homomorphism $h : \mathbf{A}^{\alpha} \to \mathbf{B}^{\alpha}$.
- $N : \mathbb{DPNRW}^{\gg} \to \mathrm{N4}$, where the object function N_{obj} sends each $\mathbf{A} \in$ DPNRW$^{\gg}$ to its associated algebra $\mathbf{A}^{\beta}$, and the arrow function N_{arw} sends each homomorphism $h : \mathbf{A} \to \mathbf{B}$ $(\mathbf{A}, \mathbf{B} \in$ DPNRW$^{\gg})$ to the same homomorphism $h : \mathbf{A}^{\beta} \to \mathbf{B}^{\beta}$.

The next theorem generalises a result of (Busaniche and Cignoli, 2010, Theorem 3.11).

Theorem 3.24. *The categories* N4 *and* $\mathbb{DPNRW}^{\gg}$ *are isomorphic.*

Proof. From the definitions, it is easy to see that the functor N is both a right and left adjoint of the functor R. Moreover, the composition NR coincides with the identity 1_{N4}, while the composition RN coincides with the identity $1_{\mathbb{DPNRW}^{\gg}}$. Thus N4 and $\mathbb{DPNRW}^{\gg}$ are categorically isomorphic.

3.5 Proof of Theorem 2.1

In this subsection we lift the term equivalence result of Theorem 3.3 to the setting of deductive systems to establish the definitional equivalence of the logics **N4** and **DPNRW**$^{\gg}$.

Definitional equivalence Let $\mathbf{A} := \langle A; c^{\mathbf{A}} \rangle_{c \in \Lambda}$ be an algebra of type Λ, and let $F \subseteq A$. A congruence θ on $\mathbf{A}$ is **compatible** with F if $a \in F$ and $a\,\theta\,b$ implies $b \in F$. The **Leibniz congruence on $\mathbf{A}$ over F**, in symbols $\Omega^{\mathbf{A}} F$, is the largest congruence on $\mathbf{A}$ compatible with F. Thus $\Omega^{\mathbf{A}} F = \bigvee \{\theta \in \mathrm{Con}\,\mathbf{A} : \theta$ is compatible with $F\}$; we write simply Ω for $\Omega^{\mathbf{Fm}_\Lambda}$. For a survey of the (induced) operator $\Omega^{\mathbf{A}} F$ in abstract algebraic logic, see (Font, 1993).

For a deductive system $\mathbf{S}$ over a language type Λ, the **Tarski congruence** $\widetilde{\Omega}(\mathbf{S})$ is the largest congruence on $\mathbf{Fm}_\Lambda$ that is compatible with every theory of $\mathbf{S}$. Thus $\widetilde{\Omega}(\mathbf{S}) = \bigcap \{\Omega T : T \in \mathrm{Th}\,\mathbf{S}\}$. For studies of the Tarski congruence in (second-order) abstract algebraic logic see (Font and Jansana, 2009; Czelakowski and Pigozzi, 2004).

Let Λ_1 and Λ_2 be two language types, and let α be a map from Λ_1 to $\mathbf{Fm}_{\Lambda_2}$. The **standard extension** of α is the function $\bar{\alpha} : \mathbf{Fm}_{\Lambda_1} \to \mathbf{Fm}_{\Lambda_2}$ defined recursively based on the complexity of terms by

$$(x_i)^{\bar{\alpha}} := x_i,$$
$$(c\varphi_0, \ldots, \varphi_{n-1})^{\bar{\alpha}} := [\![\varphi_0^{\bar{\alpha}}, \ldots, \varphi_{n-1}^{\bar{\alpha}}]\!] c^{\alpha}$$

where x_i is a variable, $c \in \Lambda_1$ is an n-ary connective, $\varphi_0, \ldots, \varphi_{n-1}$ are Λ_1-formulas, and $[\![\varphi_0, \ldots, \varphi_{n-1}]\!]$ is the surjective substitution that takes values φ_i on x_i for $i = 0, \ldots, n-1$, and takes value x_j on x_{j+n} for all $j \geq 0$ (Gyuris, 1999, Section 2.1.1, p. 48). The map $\bar{\alpha}$ extends to sets of formulas in the natural way upon defining $\Gamma^{\bar{\alpha}} := \{\varphi^{\bar{\alpha}} : \varphi \in \Gamma\}$ for all $\Gamma \subseteq \mathbf{Fm}_{\Lambda_1}$.

Let $\mathbf{S}_1 := \langle \Lambda_1, \vdash_{\mathbf{S}_1} \rangle$ and $\mathbf{S}_2 := \langle \Lambda_2, \vdash_{\mathbf{S}_2} \rangle$ be two deductive systems. A map $\alpha : \Lambda_1 \to \mathbf{Fm}_{\Lambda_2}$ is an **interpretation** of $\mathbf{S}_1$ in $\mathbf{S}_2$ if it satisfies the following two conditions (Gyuris, 1999, Definition 2.5):

1. $\langle c^{\alpha}, \mu c^{\alpha} \rangle \in \widetilde{\Omega}(\mathbf{S}_2)$ for all connectives c of Λ_1 with arity n and substitutions μ of Λ_2 that fix the first n variables;
2. If $\Gamma \vdash_{\mathbf{S}_1} \varphi$ then $\Gamma^{\bar{\alpha}} \vdash_{\mathbf{S}_2} \varphi^{\bar{\alpha}}$ for all $\Gamma \subseteq \mathbf{Fm}_{\Lambda_1}$ and $\varphi \in \mathbf{Fm}_{\Lambda_1}$.

By (Gyuris, 1999, Chapter 2§2.1, pp. 49–50), this notion of interpretation generalises the usual notion (McKenzie et al., 1987, Chapter 4§12) of interpretation for quasivarieties.

Let α be an interpretation of $\mathbf{S}_1$ in $\mathbf{S}_2$, and β an interpretation of $\mathbf{S}_2$ in $\mathbf{S}_1$. We say that α and β are **mutually inverse** if $\langle \varphi, \varphi^{\bar{\alpha}\bar{\beta}} \rangle \in \widetilde{\Omega}(\mathbf{S}_1)$ and $\langle \psi, \psi^{\bar{\beta}\bar{\alpha}} \rangle \in \widetilde{\Omega}(\mathbf{S}_2)$ for all $\varphi \in \mathbf{Fm}_{\Lambda_1}$ and $\psi \in \mathbf{Fm}_{\Lambda_2}$. The deductive systems $\mathbf{S}_1$ and $\mathbf{S}_2$ are **definitionally equivalent** if there are interpretations α of $\mathbf{S}_1$ in $\mathbf{S}_2$ and β of $\mathbf{S}_2$ in $\mathbf{S}_1$ that are mutually inverse (Gyuris, 1999, Definition 2.14). By (Gyuris, 1999, Proposition 2.17), this notion of definitional equivalence is a generalisation of the usual notion (McKenzie et al., 1987, Chapter 4§12) of term equivalence for quasivarieties.

Proof of Theorem 2.1 For each $i = 1, 2$, let $\mathbf{S}_i$ be a deductive system over a language type Λ_i such that K_i is an algebraic (quasivariety) semantics for $\mathbf{S}_i$ with *singleton* system of defining equations $\delta_i(x) \approx \epsilon_i(x)$. Let $\alpha : \Lambda_1 \to \mathbf{Fm}_{\Lambda_2}$

and $\beta : \Lambda_2 \to \mathrm{Fm}_{\Lambda_1}$ be interpretations of K_1 in K_2 and K_2 in K_1 respectively. Adapting terminology from (Czelakowski, 2001, Chapter 4§2), the defining equations $\delta_i(x) \approx \epsilon_i(x)$, $i = 1, 2$, are said to **commute with the interpretations** α and β if for all $\Gamma_i \subseteq \mathrm{Fm}_{\Lambda_i}$ $(i = 1, 2)$, it holds that

$$\mathsf{K}_1 \models \big(\delta_2(\Gamma_2)\big)^{\bar{\beta}} \approx \delta_1(\Gamma_2^{\bar{\beta}}) \quad \text{and} \quad \mathsf{K}_1 \models \big(\epsilon_2(\Gamma_2)\big)^{\bar{\beta}} \approx \epsilon_1(\Gamma_2^{\bar{\beta}}),$$
$$\mathsf{K}_2 \models \big(\delta_1(\Gamma_1)\big)^{\bar{\alpha}} \approx \delta_2(\Gamma_1^{\bar{\alpha}}) \quad \text{and} \quad \mathsf{K}_2 \models \big(\epsilon_1(\Gamma_1)\big)^{\bar{\alpha}} \approx \epsilon_2(\Gamma_1^{\bar{\alpha}}).$$

The proof of the next theorem is along the lines of the proof of (Spinks and Veroff, 2008b, Theorem 4.6); it calls for considerable technical machinery of abstract algebraic logic.

Theorem 3.25. *Let $\mathbf{S}_1$ and $\mathbf{S}_2$ be algebraisable deductive systems over languages types Λ_1 and Λ_2 respectively, let $\delta_1(x) \approx \epsilon_1(x)$ and $\delta_2(x) \approx \epsilon_2(x)$ be singleton systems of defining equations for $\mathbf{S}_1$ and $\mathbf{S}_2$ respectively, and let K_1 and K_2 be the equivalent quasivariety semantics of $\mathbf{S}_1$ and $\mathbf{S}_2$ respectively. Suppose K_1 and K_2 are term equivalent with mutually inverse interpretations $\alpha : \Lambda_1 \to \mathrm{Fm}_{\Lambda_2}$ and $\beta : \Lambda_2 \to \mathrm{Fm}_{\Lambda_1}$ such that the defining equations $\delta_1(x) \approx \epsilon_1(x)$ and $\delta_2(x) \approx \epsilon_2(x)$ commute with the interpretations α and β. Then the following statements hold:*

1. *The maps α and β are interpretations of $\mathbf{S}_1$ in $\mathbf{S}_2$ and $\mathbf{S}_2$ in $\mathbf{S}_1$ respectively.*
2. *The interpretations α and β are mutually inverse.*
3. *The deductive systems $\mathbf{S}_1$ and $\mathbf{S}_2$ are definitionally equivalent with mutually inverse interpretations $\alpha : \Lambda_1 \to \mathrm{Fm}_{\Lambda_2}$ and $\beta : \Lambda_2 \to \mathrm{Fm}_{\Lambda_1}$.*

The proof of Theorem 2.1 follows from Theorem 3.3 and Theorem 3.25.

Inasmuch as Theorem 3.25 is oriented towards applications, the restriction to singleton systems of defining equations in the statement of the theorem is quite natural. For a discussion of this point, see (Raftery, 2006, Section 10) or (Blok and Raftery, 2008, Sections 5, 9, 11, 13).

Necessary conditions to be satisfied by any reasonable notion of definitional equivalence are postulated in (Caleiro and Gonçalves, 2005). There, it is asserted that deductive systems $\mathbf{S}_1$ and $\mathbf{S}_2$ are definitionally equivalent if there exist:

- mutually inverse syntactic translations, uniform in the sense of (Caleiro and Gonçalves, 2005), mapping formulas of $\mathbf{S}_1$ to formulas of $\mathbf{S}_2$ and conversely, and
- mutually inverse lattice isomorphisms between the lattices of theories of $\mathbf{S}_1$ and $\mathbf{S}_2$.

These stipulations are met by the notion of definitional equivalence used in this work.

4 Extensions and expansions of N4

4.1 Extensions of N4 *by the weakening axiom* $\vdash x \Rightarrow (y \Rightarrow x)$

Let **S** be an extension of **RW**. By the theory of relevant logics (Restall, 2000, Section 2.3, p. 24 ff.), (Galatos et al., 2007, Section 2.1, p. 85 ff.) the extension of **S** by weakening may be identified with the axiomatic extension of **S** by the weakening axiom $\vdash x \Rightarrow (y \Rightarrow x)$. Since **DPNRW**$^{\gg}$ is an (axiomatic) extension of **RW**, the extension of **N4** by weakening may therefore be identified (in view of Theorem 2.1) with the axiomatic extension of **N4** by the *weakening axiom* $\vdash x \Rightarrow (y \Rightarrow x)$, where the derived connective $\Rightarrow$ of this latter axiom is as fixed by the map $(\Rightarrow_{\mathrm{def}})$. *Mutatis mutandis*, these remarks extend also to **N4**$^{\perp}$ and **N3**. For each $\mathbf{S} \in \{\mathbf{N4}, \mathbf{N4}^{\perp}, \mathbf{N3}\}$, the *extension of* **S** *by weakening*, in symbols $\mathbf{S}_w$, is thus the axiomatic extension of **S** by the weakening axiom $\vdash x \Rightarrow (y \Rightarrow x)$.

Constructive logic with strong negation Recall from Section 2 that **N3** denotes the axiomatic extension of **N4** by the axiom $\vdash x \rightarrow (\sim x \rightarrow y)$.

Theorem 4.1. *The axiomatic extension of* **N4** *by the weakening axiom* $\vdash x \Rightarrow (y \Rightarrow x)$ *is* **N3**.

Proof. Let $\mathsf{N4}_w$ denote the class of all **N4**-lattices satisfying the identity

$$(25) \qquad\qquad x \Rightarrow (y \Rightarrow x) \approx \big| x \Rightarrow (y \Rightarrow x) \big|.$$

To prove the theorem, it suffices to show $\mathsf{N4}_w = \mathsf{N3}$. To this end, observe first that an **N4**-lattice **B** satisfies (25) iff the inequality $a \leq b \Rightarrow a$ holds identically on **B**. Let **A** be an **N4**-lattice and let $a, b \in A$. Suppose $\mathbf{A} \in \mathsf{N3}$. Then $\mathbf{A} \models (8)$. By (8), $a \preceq \sim a \rightarrow \sim b$. Also, $\sim(\sim a \rightarrow \sim b) \preceq \sim a$ by (2), so $a \leq \sim a \rightarrow \sim b$ by (N5). On the other hand, $a \leq b \rightarrow a$ by (3). Thus $a \leq (b \rightarrow a) \wedge (\sim a \rightarrow \sim b)$, which is to say $a \leq b \Rightarrow a$. Hence $\mathbf{A} \models (25)$. Conversely, suppose $\mathbf{A} \models (25)$. By (25), $a \leq \sim b \Rightarrow a$, whence $a \leq \sim a \Rightarrow \sim\sim b = \sim a \Rightarrow b = (\sim a \rightarrow b) \wedge (\sim b \rightarrow a)$. Thus $a \leq \sim a \rightarrow b$. By (N5), $a \preceq \sim a \rightarrow b$. Hence $\mathbf{A} \models (8)$. Thus $\mathbf{A} \in \mathsf{N3}$.

Observe that Theorem 4.1 asserts that the extension of **N4** by weakening is **N3**.

Let **S** be a deductive system over a language type Λ. Following Smiley (1962, 1963), Λ-formulas φ, ψ are *synonymous* in **S**, in symbols $\varphi \equiv_{\mathbf{S}} \psi$, if φ and ψ are freely interreplaceable in all contexts in **S** (Humberstone, 2015, Section 3, p. 300), (Humberstone, 2005, Section 2). That is, $\varphi \equiv_{\mathbf{S}} \psi$ if, for every formula $\xi(\varphi)$ and every formula $\xi(\psi)$ obtained from $\xi(\varphi)$ upon replacing one or more occurrences of φ in $\xi(\varphi)$ by ψ, it holds that $\xi(\varphi) \vdash_{\mathbf{S}} \xi(\psi)$ and vice versa. In particular, for all **N4**-formulas φ, ψ, the formulas $|\varphi|$ and $|\psi|$ are synonymous in $\mathbf{N4}_w$; thus the unary term $|x|$ is constant over $\mathsf{N3}$. From this it follows that the extension of $\mathbf{N4}^{\perp}$ by weakening is, up to definitional equivalence, $\mathbf{N3}$.

In (Spinks and Veroff, 2008b, Theorem 1.1) (see also (Busaniche and Cignoli, 2010, Sections 2–3)) the present authors proved that $\mathbf{N3}$ is, up to definitional equivalence, the axiomatic extension $\mathbf{NInFL}_{ew}$ of the involutive full Lambek calculus with weakening $\mathbf{InFL}_{ew}$ (Galatos et al., 2007, Section 2.1, p. 90) by the axiom

$$\text{(Nelson)} \qquad \vdash \big((x \Rightarrow (x \Rightarrow y)) \wedge (\sim y \Rightarrow (\sim y \Rightarrow \sim x)) \big) \Rightarrow (x \Rightarrow y).$$

In the result of (Spinks and Veroff, 2008b, Theorem 1.1), the ordered pair $(\rightarrow_{\mathrm{def}})$ is replaced by the ordered pair $x \rightarrow y \mapsto x \Rightarrow (x \Rightarrow y)$. Because $|\varphi| \equiv_{\mathbf{N4}_w} |\psi|$ for all $\mathbf{N4}$-formulas φ, ψ, it holds that $\varphi \wedge (\psi \Rightarrow \psi) \equiv_{\mathbf{N4}_w} \varphi$ for all $\mathbf{N4}$-formulas φ, ψ. Therefore $\varphi \overset{\mathbf{RM}}{\Rightarrow} \psi \equiv_{\mathbf{N3}} \varphi \Rightarrow \psi$ for all $\mathbf{N4}$-formulas φ, ψ; since the right projection of the ordered pair $(\rightarrow_{\mathrm{def}})$ rewrites to $x \overset{\mathbf{RM}}{\Rightarrow} (x \overset{\mathbf{RM}}{\Rightarrow} y)$, it follows that the right projection of the ordered pair $(\rightarrow_{\mathrm{def}})$ and the formula $x \Rightarrow (x \Rightarrow y)$ are synonymous in $\mathbf{N3}$. Moreover, (Internal weakening$_{\vdash}^{\geqslant}$) is a theorem of $\mathbf{NInFL}_{ew}$, while its analogue is provable in $\mathbf{N3}$. Theorem 2.1 thus specialises, up to definitional equivalence, to (Spinks and Veroff, 2008b, Theorem 1.1). The next theorem, which sharpens (Busaniche and Cignoli, 2010, Corollary 3.8), is an algebraic analogue of this latter result; in the statement of the theorem, a residuated lattice $\mathbf{A}$ is ***integral*** if $a \leq e$ for all $a \in A$. The long proof is again omitted.

Theorem 4.2. *Up to term equivalence, a Nelson algebra is an integral residuated lattice with negation $\mathbf{A}$ such that* $\mathrm{l.u.b.}\{a^2, a \wedge \sim a\} = a$ *for all $a \in A$.*

Slaney's logic $\mathbf{F}^{**}$ The deductive system $\mathbf{F}^{**}$ is the axiomatic extension of $\mathbf{RW}$ by the axiom $\vdash x \Rightarrow (y \Rightarrow x)$ and the axiom

$$\text{(Nelson')} \qquad \vdash \big((x \Rightarrow (x \Rightarrow y)) \wedge (\sim y \Rightarrow (x \Rightarrow y)) \big) \Rightarrow (x \Rightarrow y).$$

The logic $\mathbf{F}^{**}$ was introduced in (Slaney et al., 1989) in connection with the study of future contingent propositions; it has been further studied in (Restall, 2005; Slaney, 2010; Spinks and Veroff, 2010).

Let $\mathbf{S} \in \{\mathbf{F}^{**}, \mathbf{N3}\}$. Because $\sim$ is contraposable in $\mathbf{S}$, for all $\mathbf{S}$-formulas φ, ψ it holds that $\varphi \Rightarrow \psi \equiv_{\mathbf{S}} \sim\psi \Rightarrow \sim\varphi$. From this, it follows directly that the formulas (Nelson) and (Nelson') are synonymous in $\mathbf{S}$. As $\mathbf{N3}$ is the axiomatic extension of $\mathbf{N4}$ by the weakening axiom $\vdash x \Rightarrow (y \Rightarrow x)$, we conclude:

Theorem 4.3. (Spinks and Veroff, 2010, Theorem 1.1) *The axiomatic extension of $\mathbf{N4}$ by the weakening axiom $\vdash x \Rightarrow (y \Rightarrow x)$ is, up to definitional equivalence, the logic $\mathbf{F}^{**}$ of future contingents of Slaney et al. (Restall, 2005; Slaney, 2010; Slaney et al., 1989; Spinks and Veroff, 2010).*

Slaney's logic $\mathbf{BN}$ Let $\mathbf{S}$ be an axiomatic extension of $\mathbf{RW}$. The ***axiomatic expansion of $\mathbf{S}$ by the Ackermann constant*** t, in symbols $\mathbf{S}^{\mathrm{t}}$, is the deductive system over the language $\Lambda[\mathbf{RW}] \cup \{\mathsf{t}\}$, where t is a nullary logical connective, axiomatised by a presentation of $\mathbf{S}$ together with the axioms $\vdash \mathsf{t}$ and $\vdash \mathsf{t} \Rightarrow |x|$ (Restall, 2006, Section 2.2, pp. 298–299). The intended interpretation of t in this context is 'the conjunction of all logical truths'.

For a language type Λ having distinguished binary logical connectives $\wedge$ and $\Rightarrow$ and a distinguished nullary logical connective t, let

$$\varphi \overset{\mathsf{t}}{\to} \psi \quad \text{abbreviate} \quad (\varphi \wedge \mathsf{t}) \Rightarrow ((\varphi \wedge \mathsf{t}) \Rightarrow \psi)$$

for all Λ-formulas φ, ψ. The deductive system $\mathbf{BN}$ is the axiomatic extension of $\mathbf{RW}^{\mathsf{t}}$ by the axioms

$$\vdash (x * y) \overset{\mathsf{t}}{\to} x \quad \text{and} \quad \vdash \left((x \overset{\mathsf{t}}{\to} y) \wedge (\sim y \overset{\mathsf{t}}{\to} \sim x) \right) \Rightarrow (x \Rightarrow y).$$

As with $\mathbf{F}^{**}$, the deductive system $\mathbf{BN}$ was introduced in (Slaney et al., 1989) in connection with the study of future contingent propositions. The logic $\mathbf{BN}$ has been further studied in (Restall, 1993; Slaney, 1991).

Let $\mathbf{RM}$ denote the relevant logic with mingle of McCall and Dunn (Anderson and Belnap, 1975, §§29.3–29.4), (Avron, 2016), that is, the axiomatic extension of $\mathbf{R}$ by the mingle axiom $\vdash x \Rightarrow |x|$. In (Blok and Raftery, 2004, Section 2.5, p. 74) Blok and Raftery observe that for all $\mathbf{RM}^{\mathsf{t}}$-formulas φ, ψ, it holds that $\varphi \overset{\mathbf{RM}}{\Rightarrow} \psi \equiv_{\mathbf{RM}^{\mathsf{t}}} (\varphi \wedge \mathsf{t}) \Rightarrow \psi$. Similarly, it can be observed that $\varphi \to \psi \equiv_{(\mathbf{DPNRW}^{\gg})^{\mathsf{t}}} \varphi \overset{\mathsf{t}}{\to} \psi$ for all $(\mathbf{DPNRW}^{\gg})^{\mathsf{t}}$-formulas φ, ψ. Identifying the axiomatic expansion of $\mathbf{N4}$ by the Ackermann constant t with the deductive system $(\mathbf{DPNRW}^{\gg})^{\mathsf{t}}$, we obtain:

Theorem 4.4. *The axiomatic expansion of $\mathbf{N4}$ by the Ackermann constant t is, up to definitional equivalence, the deductive system $\mathbf{BN}$ of Slaney (Restall, 1993; Slaney, 1991; Slaney et al., 1989).*

4.2 Extensions of $\mathbf{N4}$ by the contraction axiom $\vdash \left(x \Rightarrow (x \Rightarrow y) \right) \Rightarrow (x \Rightarrow y)$

Let $\mathbf{S}$ be an extension of $\mathbf{RW}$. By the theory of relevant logics (Restall, 2000, Section 2.3, p. 24 ff.), (Galatos et al., 2007, Section 2.1, p. 85 ff.) the extension of $\mathbf{S}$ by contraction may be identified with the axiomatic extension of $\mathbf{S}$ by the contraction axiom $\vdash \left(x \Rightarrow (x \Rightarrow y) \right) \Rightarrow (x \Rightarrow y)$. Since $\mathbf{DPNRW}^{\gg}$ is an (axiomatic) extension of $\mathbf{RW}$, the extension of $\mathbf{N4}$ by contraction may therefore be identified (in view of Theorem 2.1) with the axiomatic extension of $\mathbf{N4}$ by the ***contraction axiom*** $\vdash \left(x \Rightarrow (x \Rightarrow y) \right) \Rightarrow (x \Rightarrow y)$, where the derived connective $\Rightarrow$ of this latter axiom is as fixed by the map $(\Rightarrow_{\mathrm{def}})$. *Mutatis mutandis*, these remarks extend also to $\mathbf{N4}^{\perp}$ and $\mathbf{N3}$. For each

$\mathbf{S} \in \{\mathbf{N4}, \mathbf{N4}^{\perp}, \mathbf{N3}\}$, the ***extension of*** $\mathbf{S}$ ***by contraction***, in symbols $\mathbf{S}_c$, is thus the axiomatic extension of $\mathbf{S}$ by the axiom $\vdash (x \Rightarrow (x \Rightarrow y)) \Rightarrow (x \Rightarrow y)$.

The 3-valued relevant logic with mingle RM3 The logic $\mathbf{RM3}$ (Anderson and Belnap, 1975, §26.9, §29.12) is the deductive system over the language $\Lambda[\mathbf{RW}]$ determined by the matrix:

$\wedge$	$\top$	b	$\perp$
$\dagger\top$	$\top$	b	$\perp$
$\dagger b$	b	b	$\perp$
$\perp$	$\perp$	$\perp$	$\perp$

$\vee$	$\top$	b	$\perp$
$\top$	$\top$	$\top$	$\top$
b	$\top$	b	b
$\perp$	$\top$	b	$\perp$

$*$	$\top$	b	$\perp$
$\top$	$\top$	$\top$	$\perp$
b	$\top$	b	$\perp$
$\perp$	$\perp$	$\perp$	$\perp$

$\Rightarrow$	$\top$	b	$\perp$
$\top$	$\top$	$\perp$	$\perp$
b	$\top$	b	$\perp$
$\perp$	$\top$	$\top$	$\top$

$\sim$	
$\top$	$\perp$
b	b
$\perp$	$\top$

(Here and in the sequel the designated elements of a matrix are marked with the symbol '$\dagger$'.) Denote this matrix by $\mathfrak{RM}3$. The logic $\mathbf{RM3}$ is, in a sense, the strongest logic in the family of relevant logics (Avron, 1991, p. 277); it has been further investigated in (for instance) (Avron, 1991; Brady, 1982; Dunn, 1970; Parks, 1972; Tokarz, 1975). Axiomatically, $\mathbf{RM3}$ may be obtained from a standard presentation of $\mathbf{RM}$ (Blok and Raftery, 2004, Section 2, p. 65) upon adjoining the axiom $\vdash x \vee (x \Rightarrow y)$. Recall from (Font and Pérez, 1992, Corollary 9, Theorem 4) that $\mathbf{RM}$, hence $\mathbf{RM3}$, is strongly algebraisable with system of equivalence formulas $\{\varphi \Rightarrow \psi, \psi \Rightarrow \varphi\}$ and system of defining equations $\{x \approx |x|\}$; the equivalent variety semantics of $\mathbf{RM}$ is the variety of Sugihara algebras considered in (Blok and Dziobiak, 1986; Blok and Raftery, 2004; Font and Pérez, 1992).

Let $\mathbf{RM}_3$ denote the algebra reduct of $\mathfrak{RM}3$. Let $\mathrm{RM}_3 := \underline{\mathsf{HSP}}(\mathbf{RM}_3)$. Observe that, since the right projection of the ordered pair $(\rightarrow'_{\mathrm{def}})$ is synonymous with the the right projection of the ordered pair $(\rightarrow_{\mathrm{def}})$ in $\mathbf{N4}$, it follows that the right projection of the ordered pair $(\rightarrow_{\mathrm{def}})$ and the formula $x \overset{\mathbf{RM}}{\Rightarrow} y$ are synonymous in $\mathbf{N4}_c$. This simplifies computations in the proof of the next theorem.

Theorem 4.5. *The axiomatic extension of* $\mathbf{N4}$ *by the contraction axiom* $\vdash (x \Rightarrow (x \Rightarrow y)) \Rightarrow (x \Rightarrow y)$ *is, up to definitional equivalence, the 3-valued relevant logic with mingle* $\mathbf{RM3}$ (Anderson and Belnap, 1975, p. 470 ff.).

Proof. (Sketch) Let $\mathbf{B}$ be an adjunctive distributive residuated lo-semigroup with negation. Notice first that:

(i) If $\mathbf{B}$ is square increasing (that is, if $b \leq b * b$ holds identically on $\mathbf{B}$), then $\mathbf{B} \models ((x \overset{\mathbf{RM}}{\Rightarrow} y) \wedge (\sim y \overset{\mathbf{RM}}{\Rightarrow} \sim x)) \vee (x \Rightarrow y) \approx x \Rightarrow y$ iff $\mathbf{B}$ is a Sugihara algebra. (See also (Avron, 1986, Theorem 1.7) for this observation.)

(ii) If $\mathbf{B}$ is a Sugihara algebra, then $\mathbf{B} \models x \vee (x \Rightarrow y) \approx |x \vee (x \Rightarrow y)|$ iff $\mathbf{B} \models (x * y) \overset{\mathbf{RM}}{\Rightarrow} x \approx |(x * y) \overset{\mathbf{RM}}{\Rightarrow} x|$.

Let $\mathsf{DPNRW}_c^{\geqslant}$ denote the class of all members of $\mathsf{DPNRW}^{\geqslant}$ satisfying the identity

$$(26) \qquad \big(x \Rightarrow (x \Rightarrow y)\big) \Rightarrow (x \Rightarrow y) \approx \big|(x \Rightarrow (x \Rightarrow y)) \Rightarrow (x \Rightarrow y)\big|$$

and let $\mathbf{A}$ be an adjunctive distributive residuated lo-semigroup with negation. From (i) and (ii), it follows that $\mathbf{A} \in \mathsf{DPNRW}_c^{\geqslant}$ iff $\mathbf{A} \in \mathsf{RM}_3$. Thus $\mathsf{DPNRW}_c^{\geqslant} = \mathsf{RM}_3$. Let $\mathsf{N4}_c$ denote the class of all $\mathbf{N4}$-lattices satisfying (26). From Theorem 3.3, it follows that $\mathsf{N4}_c$ and $\mathsf{DPNRW}_c^{\geqslant}$ are term equivalent; thus $\mathsf{N4}_c$ and RM_3 are term equivalent. Applying Theorem 3.25, we conclude that $\mathbf{N4}_c$ and $\mathbf{RM3}$ are definitionally equivalent.

Observe that Theorem 4.5 asserts that the extension of $\mathbf{N4}$ by contraction is $\mathbf{RM3}$.

The classical propositional calculus Let $\mathbf{CPC}$ denote the classical propositional calculus, considered in the language $\Lambda[\mathbf{N4}]$. In (Spinks and Veroff, 2008a, Corollary 3.8), the present authors essentially observe that $\mathbf{CPC}$ is the axiomatic extension of $\mathbf{N3}$ by the contraction axiom $\vdash \big(x \Rightarrow (x \Rightarrow y)\big) \Rightarrow (x \Rightarrow y)$. As $\mathbf{N3}$ is the axiomatic extension of $\mathbf{N4}$ by the weakening axiom $\vdash x \Rightarrow (y \Rightarrow x)$, we conclude:

Theorem 4.6. *The axiomatic extension of* $\mathbf{N4}$ *by the weakening axiom* $\vdash x \Rightarrow (y \Rightarrow x)$ *and the contraction axiom* $\vdash \big(x \Rightarrow (x \Rightarrow y)\big) \Rightarrow (x \Rightarrow y)$ *is, up to definitional equivalence, the classical propositional calculus* $\mathbf{CPC}$.

The axiomatic expansion of RM3 by the Church constant $\top$ Following (Restall, 2006, Section 2.2, pp. 298–299), the *axiomatic expansion of* $\mathbf{RM3}$ *by the Church constant* $\top$, in symbols $\mathbf{RM3}^{\top}$, is the deductive system over the language $\Lambda[\mathbf{RW}^{\top}] := \Lambda[\mathbf{RW}] \cup \{\top\}$, where $\top$ is a nullary logical connective, axiomatised by a presentation of $\mathbf{RM3}$ together with the axiom $\vdash x \Rightarrow \top$. The intended interpretation of $\top$ in this context is 'the disjunction of all sentences'; *cf.* (Meyer, 2004, Section 5.2, p. 209). Modulo Theorem 4.5, the next result is clear.

Theorem 4.7. *The axiomatic extension of* $\mathbf{N4}^{\perp}$ *by the contraction axiom* $\vdash \big(x \Rightarrow (x \Rightarrow y)\big) \Rightarrow (x \Rightarrow y)$ *is, up to definitional equivalence, the logic* $\mathbf{RM3}^{\top}$.

The deductive system Pac Let $\Lambda[\mathbf{Pac}]$ be a language consisting of binary logical connectives $\wedge$, $\vee$, and $\supset$, and the unary logical connective $\sim$. The logic $\mathbf{Pac}$ is the deductive system over $\Lambda[\mathbf{Pac}]$ determined by the matrix:

$\wedge$	$\top$	b	$\perp$		$\vee$	$\top$	b	$\perp$		$\supset$	$\top$	b	$\perp$		$\sim$	
$\dagger\top$	$\top$	b	$\perp$		$\top$	$\top$	$\top$	$\top$		$\top$	$\top$	b	$\perp$		$\top$	$\perp$
$\dagger b$	b	b	$\perp$		b	$\top$	b	b		b	$\top$	b	$\perp$		b	b
$\perp$	$\perp$	$\perp$	$\perp$		$\perp$	$\top$	b	$\perp$		$\perp$	$\top$	$\top$	$\top$		$\perp$	$\top$

The deductive system $\mathbf{Pac}$ was introduced in (Batens, 1980, Section 6) (but see also (Schütte, 1960, Chapter II§7)), where it was denoted $\mathbf{PI}^s$, and independently in (Avron, 1986), where it was denoted $\mathbf{RM}_3^{\supset}$. The designation $\mathbf{Pac}$ is due to (Avron, 1991). According to (Carnielli, 2002, Section 2),

Priest's logic of paradox **LP** (Priest, 1989) is the $\{\supset\}$-free fragment of **Pac**; (Arieli et al., 2010, Example 40) show **Pac** is maximally paraconsistent. For studies of **Pac**, see in particular (Avron, 1991) or (Arieli and Avron, 2015, Section 5.4); for a discussion and further references, see (Carnielli and Marcos, 2002, Section 2.4).

The next result is essentially (Avron, 1986, Theorem 2.10(a)). See also (Avron, 1991, Proposition, Section 3, p. 287).

Theorem 4.8.

1. *The map* $\alpha : \Lambda[\mathbf{Pac}] \to \mathrm{Fm}_{\Lambda[\mathbf{RW}]}$ *defined by*

$$x \wedge y \mapsto x \wedge y \qquad\qquad x \supset y \mapsto (x \Rightarrow y) \vee y$$
$$x \vee y \mapsto x \vee y \qquad\qquad {\sim}x \mapsto {\sim}x$$

is an interpretation of **Pac** *in* **RM3**.

2. *The map* $\beta : \Lambda[\mathbf{RW}] \to \mathrm{Fm}_{\Lambda[\mathbf{Pac}]}$ *defined by*

$$x \wedge y \mapsto x \wedge y \qquad\qquad x \Rightarrow y \mapsto (x \supset y) \wedge ({\sim}y \supset {\sim}x)$$
$$x \vee y \mapsto x \vee y \qquad\qquad x * y \mapsto {\sim}(x \supset {\sim}y) \vee {\sim}(y \supset {\sim}x)$$
$${\sim}x \mapsto {\sim}x$$

is an interpretation of **RM3** *in* **Pac**.

3. *The interpretations* α *and* β *are mutually inverse.*

Hence the deductive systems **Pac** *and* **RM3** *are definitionally equivalent.*

Since definitional equivalence is an equivalence relation on deductive systems (*cf.* (McKenzie et al., 1987, Section 4.12, p. 246)), we conclude:

Corollary 4.9. *The axiomatic extension of* **N4** *by the contraction axiom* $\vdash (x \Rightarrow (x \Rightarrow y)) \Rightarrow (x \Rightarrow y)$ *is, up to definitional equivalence, Avron's logic* **Pac** *(Arieli and Avron, 2015; Avron, 1986, 1991; Batens, 1980).*

The deductive system J3 Let $\Lambda[\mathbf{J3}] := \Lambda[\mathbf{Pac}] \cup \{\nabla\}$, where ∇ is a unary logical connective. The logic **J3** is the deductive system over $\Lambda[\mathbf{J3}]$ determined by the matrix:

$\wedge$	$\top$	b	$\bot$
†$\top$	$\top$	b	$\bot$
†b	b	b	$\bot$
$\bot$	$\bot$	$\bot$	$\bot$

$\vee$	$\top$	b	$\bot$
$\top$	$\top$	$\top$	$\top$
b	$\top$	b	b
$\bot$	$\top$	b	$\bot$

$\supset$	$\top$	b	$\bot$
$\top$	$\top$	b	$\bot$
b	$\top$	b	$\bot$
$\bot$	$\top$	$\top$	$\top$

$\sim$	
$\top$	$\bot$
b	b
$\bot$	$\top$

∇	
$\top$	$\top$
b	$\top$
$\bot$	$\bot$

The deductive system $\mathbf{J}_3$ was introduced by D'Ottaviano and da Costa (1970) in connection with the solution to a problem of Jaśkowski. It has been further studied by Epstein (1995, Chapter IX); applications are given in (da Costa, 1974). See also (Avron, 1986, 1991). The formulation of $\mathbf{J}_3$ given here can be found in (Marcos, 2005, Section 2.2).

The next result is basically stated in (Carnielli and Marcos, 2002, Section 2.4, p. 26).

Theorem 4.10.

1. The map $\alpha : \Lambda[\mathbf{J3}] \to \mathrm{Fm}_{\Lambda[\mathbf{RW}^\top]}$ *defined by*

$$
\begin{aligned}
x \wedge y &\mapsto x \wedge y & x \supset y &\mapsto (x \Rightarrow y) \vee y \\
x \vee y &\mapsto x \vee y & {\sim}x &\mapsto {\sim}x \\
\nabla x &\mapsto x * \top
\end{aligned}
$$

is an interpretation of $\mathbf{J3}$ *in* $\mathbf{RM3}^\top$.

2. The map $\beta : \Lambda[\mathbf{RW}^\top] \to \mathrm{Fm}_{\Lambda[\mathbf{J3}]}$ *defined by*

$$
\begin{aligned}
x \wedge y &\mapsto x \wedge y & x * y &\mapsto {\sim}(x \supset {\sim}y) \vee {\sim}(y \supset {\sim}x) \\
x \vee y &\mapsto x \vee y & {\sim}x &\mapsto {\sim}x \\
x \Rightarrow y &\mapsto (x \supset y) \wedge ({\sim}y \supset {\sim}x) & \top &\mapsto \nabla(x \supset x)
\end{aligned}
$$

is an interpretation of $\mathbf{RM3}^\top$ *in* $\mathbf{J3}$.

3. The interpretations α *and* β *are mutually inverse.*

Hence the deductive systems $\mathbf{RM3}^\top$ *and* $\mathbf{J3}$ *are definitionally equivalent.*

Since definitional equivalence is an equivalence relation on deductive systems (*cf.* (McKenzie et al., 1987, Section 4.12, p. 246)), we conclude:

Corollary 4.11. *The axiomatic extension of* $\mathbf{N4}^\perp$ *by the contraction axiom* $\vdash (x \Rightarrow (x \Rightarrow y)) \Rightarrow (x \Rightarrow y)$ *is, up to definitional equivalence, the 3-valued logic* $\mathbf{J3}$ *of D'Ottaviano and da Costa* (Carnielli and Marcos, 2002; D'Ottaviano and da Costa, 1970; Epstein, 1995).

The deductive system LFI1 Let $\Lambda[\mathbf{LFI1}]$ be a language type having binary logical connectives $\wedge$, $\vee$, and $\supset$, and unary logical connectives $\sim$ and $\bullet$. The logic $\mathbf{LFI1}$ is the deductive system over $\Lambda[\mathbf{LFI1}]$ determined by the matrix:

$\wedge$	$\top$	b	$\perp$
†$\top$	$\top$	b	$\perp$
†b	b	b	$\perp$
$\perp$	$\perp$	$\perp$	$\perp$

$\vee$	$\top$	b	$\perp$
$\top$	$\top$	$\top$	$\top$
b	$\top$	b	b
$\perp$	$\top$	b	$\perp$

$\supset$	$\top$	b	$\perp$
$\top$	$\top$	b	$\perp$
b	$\top$	b	$\perp$
$\perp$	$\top$	$\top$	$\top$

$\sim$	
$\top$	$\perp$
b	b
$\perp$	$\top$

$\bullet$	
$\top$	$\perp$
b	$\top$
$\perp$	$\perp$

The deductive system $\mathbf{LFI1}$ was introduced in (Carnielli et al., 2000) *qua* a 'logic of formal inconsistency' (Carnielli et al., 2007; Carnielli and Marcos, 2002), and it has been further considered in (for example) (de Amo et al., 2002; Carnielli et al., 2007; Carnielli and Marcos, 2002; Carnielli, 2002). The next result is essentially (Carnielli et al., 2000, Remark 3.2).

Theorem 4.12.

1. The map $\alpha : \Lambda[\mathbf{LFI1}] \to \mathrm{Fm}_{\Lambda[\mathbf{J3}]}$ defined by

$$x \wedge y \mapsto x \wedge y \qquad\qquad x \supset y \mapsto x \supset y$$
$$x \vee y \mapsto x \vee y \qquad\qquad {\sim}x \mapsto {\sim}x$$
$$\bullet x \mapsto \nabla x \wedge \nabla {\sim} x$$

is an interpretation of **LFI1** *in* **J3**.

2. The map $\beta : \Lambda[\mathbf{J3}] \to \mathrm{Fm}_{\Lambda[\mathbf{LFI1}]}$ defined by

$$x \wedge y \mapsto x \wedge y \qquad\qquad x \supset y \mapsto x \supset y$$
$$x \vee y \mapsto x \vee y \qquad\qquad {\sim}x \mapsto {\sim}x$$
$$\nabla x \mapsto x \vee \bullet x$$

is an interpretation of **J3** *in* **LFI1**.

3. The interpretations α and β are mutually inverse.

Hence the deductive systems **J3** *and* **LFI1** *are definitionally equivalent.*

Since definitional equivalence is an equivalence relation on deductive systems (*cf.* (McKenzie et al., 1987, Section 4.12, p. 246)), we conclude:

Corollary 4.13. *The axiomatic extension of $\mathbf{N4}^{\perp}$ by the contraction axiom $\vdash \big(x \Rightarrow (x \Rightarrow y)\big) \Rightarrow (x \Rightarrow y)$ is, up to definitional equivalence, the logic* **LFI1** *of Carnielli et al.* (Carnielli et al., 2007; Carnielli and Marcos, 2002; Carnielli et al., 2000).

4.3 Extensions of N4 by the prelinearity axiom $\vdash (x \to y) \vee (y \to x)$

The nilpotent minimum logic NM Let **NM** denote the deductive system over the language $\{\wedge, *, \Rightarrow, \perp\}$, where $\wedge$, $*$, and $\Rightarrow$ are binary logical connectives and $\perp$ is a nullary logical connective, presented by the axioms and inference rules

$$\vdash (x \Rightarrow y) \Rightarrow \big((y \Rightarrow z) \Rightarrow (x \Rightarrow z)\big) \qquad \vdash \big((x * y) \Rightarrow z\big) \Rightarrow \big(x \Rightarrow (y \Rightarrow z)\big)$$
$$\vdash (x * y) \Rightarrow x \qquad\qquad \vdash \big((x \Rightarrow y) \Rightarrow z\big) \Rightarrow \big(((y \Rightarrow x) \Rightarrow z) \Rightarrow z\big)$$
$$\vdash (x * y) \Rightarrow (y * x) \qquad\qquad \vdash \perp \Rightarrow x$$
$$\vdash (x \wedge y) \Rightarrow x \qquad\qquad \vdash \big((x \Rightarrow \perp) \Rightarrow \perp\big) \Rightarrow x$$
$$\vdash (x \wedge y) \Rightarrow (y \wedge x) \qquad\qquad \vdash \big((x * y) \Rightarrow \perp\big) \vee \big((x \wedge y) \Rightarrow (x * y)\big)$$
$$\vdash \big(x * (x \Rightarrow y)\big) \Rightarrow (x \wedge y) \qquad\qquad x, x \Rightarrow y \vdash y,$$
$$\vdash \big(x \Rightarrow (y \Rightarrow z)\big) \Rightarrow \big((x * y) \Rightarrow z\big)$$

where for all formulas φ, ψ in the language of **NM**, the expression $\varphi \vee \psi$ abbreviates $((\varphi \Rightarrow \psi) \Rightarrow \psi) \wedge ((\psi \Rightarrow \varphi) \Rightarrow \varphi)$. The deductive system **NM** is known as Esteva and Godo's *nilpotent minimum logic* in the literature (Bianchi, 2011; Esteva and Godo, 2001; Gispert, 2003; Noguera, 2007; Noguera et al., 2008).

Let NM denote the class of nilpotent minimum algebras introduced in (Esteva and Godo, 2001) and further studied in (Bianchi, 2011; Noguera, 2007; Noguera et al., 2008). By (Noguera et al., 2008, Theorem 2.8), NM is the variety generated by the (involutive) left-continuous t-norm of Fodor (1995) on the real unit interval $[0, 1]$, while by (Noguera, 2007, Theorem 3.4), **NM** is strongly algebraisable with system of equivalence formulas $\{\varphi \Rightarrow \psi, \psi \Rightarrow \varphi\}$, system of defining equations $\{x \approx |x|\}$, and equivalent variety semantics NM. Let $N3_\ell$ denote the equational class of all Nelson algebras satisfying the prelinearity identity $(x \to y) \vee (y \to x) \approx |(x \to y) \vee (y \to x)|$. By remarks due to (Busaniche and Cignoli, 2010, Section 6.3), the varieties $N3_\ell$ and NM are term equivalent. Because **N3** is the axiomatic extension of **N4** by the weakening axiom $\vdash x \Rightarrow (y \Rightarrow x)$, from Theorem 3.25 we deduce:

Theorem 4.14. *The axiomatic extension of* **N4** *by the weakening axiom* $\vdash x \Rightarrow (y \Rightarrow x)$ *and the prelinearity axiom* $\vdash (x \to y) \vee (y \to x)$ *is, up to definitional equivalence, the nilpotent minimum logic* **NM** *of Esteva and Godo* (Bianchi, 2011; Esteva and Godo, 2001; Gispert, 2003; Noguera, 2007; Noguera et al., 2008).

The logic L* of fuzzy reasoning of Wang The logic **L*** of fuzzy reasoning of Wang (Pei and Wang, 2002; Wang, 1997, 1999, 2000; Wang and Wang, 2001) is the deductive system over the language $\{\vee, \Rightarrow, \sim\}$, where $\vee$ and $\Rightarrow$ are binary logical connectives and $\sim$ is a unary logical connective, presented by the axioms and inference rules

$$\vdash x \Rightarrow (y \Rightarrow x)$$
$$\vdash (\sim x \Rightarrow \sim y) \Rightarrow (y \Rightarrow x)$$
$$\vdash (x \Rightarrow (y \Rightarrow z)) \Rightarrow (y \Rightarrow (x \Rightarrow z))$$
$$\vdash (y \Rightarrow z) \Rightarrow ((x \Rightarrow y) \Rightarrow (x \Rightarrow z))$$
$$\vdash x \Rightarrow \sim \sim x$$
$$\vdash x \Rightarrow (x \vee y)$$
$$\vdash (x \vee y) \Rightarrow (y \vee x)$$

$$\vdash ((x \Rightarrow z) \wedge (y \Rightarrow z)) \Rightarrow ((x \vee y) \Rightarrow z)$$
$$\vdash ((x \wedge y) \Rightarrow z) \Rightarrow ((x \Rightarrow z) \vee (y \Rightarrow z))$$
$$\vdash (x \Rightarrow y) \vee ((x \Rightarrow y) \Rightarrow (\sim x \vee y))$$
$$\vdash x \Rightarrow (y \Rightarrow (x \wedge y))$$
$$\vdash ((x \Rightarrow y) \wedge (x \Rightarrow z)) \Rightarrow (x \Rightarrow (y \wedge z))$$
$$x, x \Rightarrow y \vdash y,$$

where for all formulas φ, ψ in the language of **L***, the expression $\varphi \wedge \psi$ abbreviates $\sim(\sim\varphi \vee \sim\psi)$. Routine arguments show that **L*** is strongly algebraisable with system of equivalence formulas $\{\varphi \Rightarrow \psi, \psi \Rightarrow \varphi\}$ and system of defining equations $\{x \approx |x|\}$; the equivalent variety semantics of **L*** is the variety R_0 of R_0-algebras introduced in (Wang, 2000) and studied further in (Cheng and Wang, 1999; Wang, 2002; Zhou and Li, 2010).

In (Pei, 2003, Theorem 5) Pei showed that the varieties of R_0-algebras and **NM**-algebras are term equivalent. Since term equivalence is an equivalence relation on quasivarieties (*cf.* (McKenzie et al., 1987, Section 4.12, p. 246)), it also holds that the varieties $\mathsf{N3}_\ell$ and $\mathsf{R_0}$ are term equivalent. We therefore have the following result, which may be understood as both strengthening and sharpening (Pei, 2003, Theorem 4).

Theorem 4.15. *The axiomatic extension of* **N4** *by the weakening axiom* $\vdash x \Rightarrow (y \Rightarrow x)$ *and the prelinearity axiom* $\vdash (x \to y) \vee (y \to x)$ *is, up to definitional equivalence, the logic* **L*** *of fuzzy reasoning of Wang* (Pei and Wang, 2002; Wang, 1997, 1999, 2000; Wang and Wang, 2001).

4.4 Extensions of **N4** *by the Peirce law*
$$\vdash \big((x \to y) \to x\big) \to x$$

The logic B of Brouwerian bilattices Let $\Sigma'[\mathbf{IPC}^+]$ be the set of axioms and rules got from $\Sigma[\mathbf{IPC}^+]$ by replacing the axiom $\vdash (x \to y) \to ((x \to z) \to (x \to (y \wedge z)))$ of $\Sigma[\mathbf{IPC}^+]$ with the axiom $\vdash x \to (y \to (x \wedge y))$, and let $\Sigma'[\mathbf{N4}]$ denote the presentation determined by the axioms and inference rules of $\Sigma'[\mathbf{IPC}^+]$ together with the axioms for the strong negation connective of (1). Bou and Rivieccio's ***logic of Brouwerian bilattices*** (Bou and Rivieccio, 2013, Definition 1.1) is the deductive system **B** over the language $\{\wedge, \vee, \oplus, \otimes, \to, \sim\}$, where $\wedge, \vee, \oplus, \otimes$, and $\to$ are binary logical connectives and $\sim$ is a unary logical connective, presented by the axioms and inference rules of $\Sigma'[\mathbf{N4}]$ together with the axioms

$$\vdash (x \otimes y) \to x \qquad\qquad \vdash x \to (x \oplus y)$$
$$\vdash (x \otimes y) \to y \qquad\qquad \vdash y \to (x \oplus y)$$
$$\vdash x \to \big(y \to (x \otimes y)\big) \qquad \vdash (x \to z) \to \big((y \to z) \to ((x \oplus y) \to z)\big)$$
$$\vdash \sim(x \otimes y) \leftrightarrow \sim x \otimes \sim y \qquad \vdash \sim(x \oplus y) \leftrightarrow \sim x \oplus \sim y.$$

(Here $\varphi \leftrightarrow \psi$ abbreviates $(\varphi \to \psi) \wedge (\psi \to \varphi)$ for all **B**-formulas φ, ψ.)

Theorem 4.16. (Bou and Rivieccio, 2013, Section 2.2) **N4** *is the* $\{\wedge, \vee, \to , \sim\}$*-fragment of the logic* **B** *of Brouwerian bilattices of Bou and Rivieccio* (Bou and Rivieccio, 2013).

The Hilbert-style 'basic system' HBe of Avron The (*Hilbert-style*) ***basic system***, in symbols **HBe**, is the deductive system over the language $\Lambda[\mathbf{N4}]$ presented by the set of axioms and inference rules $\Sigma'[\mathbf{N4}]$, together with the Peirce law $\vdash ((x \to y) \to x) \to x$. The deductive system **HBe** was introduced in (Avron, 1991, Section 4) in connection with the study of 'natural' 3-valued logics.

For each $\mathbf{S} \in \{\mathbf{N4}, \mathbf{N4}^{\perp}, \mathbf{N3}\}$, let $\mathbf{S}_p$ denote the axiomatic extension of $\mathbf{S}$ by the Peirce law $\vdash ((x \to y) \to x) \to x$.

Theorem 4.17. *The axiomatic extension of* $\mathbf{N4}$ *by the Peirce law* $\vdash ((x \to y) \to x) \to x$ *is the (Hilbert style) 'basic system'* $\mathbf{HBe}$ *of Avron (Arieli, 1999; Arieli and Avron, 1994, 1996; Avron, 1991).*

Proof. It is well known that the presentations $\Sigma[\mathbf{IPC}^+]$ and $\Sigma'[\mathbf{IPC}^+]$ both axiomatise $\mathbf{IPC}^+$. It follows that $\Sigma'[\mathbf{N4}]$ constitutes an axiomatisation of $\mathbf{N4}$. As $\mathbf{HBe}$ and $\mathbf{N4}_p$ have the same language type, the two deductive systems coincide.

By (Avron, 1991, Theorem on Extensions), $\mathbf{Pac}$ is the axiomatic extension of $\mathbf{HBe}$ by the law of the excluded middle $\vdash x \vee {\sim}x$. This yields the next result, which is essentially observed in (Brady, 1982, Section 6, p. 32).

Corollary 4.18. *The axiomatic extension of* $\mathbf{N4}$ *by the Peirce law* $\vdash ((x \to y) \to x) \to x$ *and the law of the excluded middle* $\vdash x \vee {\sim}x$ *is, up to definitional equivalence,* $\mathbf{RM3}$.

The logic HBL of implicative bilattices The deductive system $\mathbf{HBe}$ has been further considered by Arieli et al. in the context of the logic of implicative bilattices (Arieli, 1999; Arieli and Avron, 1994, 1996). The *logic of implicative bilattices* (Arieli and Avron, 1996, Section 3), in symbols $\mathbf{HBL}$, is the axiomatic extension of the logic $\mathbf{B}$ by the Peirce law $\vdash ((x \to y) \to x) \to x)$. For recent studies of the logic of implicative bilattices, see (Bou and Rivieccio, 2011; Rivieccio, 2010; Rivieccio et al., 2011).

The next result is unsurprising in view of Theorem 4.16.

Theorem 4.19. *The axiomatic extension of* $\mathbf{N4}$ *by the Peirce law* $\vdash ((x \to y) \to x) \to x$ *is the* $\{\wedge, \vee, \to, {\sim}\}$*-fragment of the logic* $\mathbf{HBL}$ *of implicative bilattices of Arieli and Avron (Arieli and Avron, 1996; Bou and Rivieccio, 2011; Rivieccio, 2010; Rivieccio et al., 2011).*

Proof. Let $\Sigma[\mathbf{HBL}]$ denote the presentation of $\mathbf{HBL}$ given in (Arieli and Avron, 1996, Section 3). By (Arieli and Avron, 1996, Corollary 3.24), the presentation $\Sigma[\mathbf{HBL}]$ is separable in the sense that for each subset $\Lambda \subseteq \Lambda[\mathbf{HBL}]$ such that $\to \in \Lambda$, the Λ-fragment of $\mathbf{HBL}$ is axiomatised by those axioms and inference rules of $\Sigma[\mathbf{HBL}]$ whose connectives are among those in Λ. The $\{\wedge, \vee, \to, {\sim}\}$-fragment of $\mathbf{HBL}$ is thus axiomatised by $\Lambda'[\mathbf{N4}]$ together with the Peirce law $\vdash ((x \to y) \to x) \to x$. The $\{\wedge, \vee, \to, {\sim}\}$-fragment of $\mathbf{HBL}$ is thus $\mathbf{HBe}$, which is to say, $\mathbf{N4}_p$.

The logic BN4 The logic $\mathbf{BN4}$ (Restall, 2006, Example 17), (Restall, 2000, Example 8.39) is the deductive system over the language $\Lambda[\mathbf{RW}]$ determined by the matrix:

$\wedge$	T	b	n	$\bot$
†T	T	b	n	$\bot$
†b	b	b	$\bot$	$\bot$
n	n	$\bot$	n	$\bot$
$\bot$	$\bot$	$\bot$	$\bot$	$\bot$

$\vee$	T	b	n	$\bot$
T	T	T	T	T
b	T	b	T	b
n	T	T	n	n
$\bot$	T	b	n	$\bot$

$*$	T	b	n	$\bot$
T	T	T	n	$\bot$
b	T	b	n	$\bot$
n	n	n	$\bot$	$\bot$
$\bot$	$\bot$	$\bot$	$\bot$	$\bot$

$\Rightarrow$	T	b	n	$\bot$
T	T	$\bot$	n	$\bot$
b	T	b	n	$\bot$
n	T	n	T	n
$\bot$	T	T	T	T

$\sim$	
T	$\bot$
b	b
n	n
$\bot$	T

Denote this matrix by $\mathfrak{BN}4$. The logic **BN4** was introduced in (Brady, 1982, Section 1) in connection with the study of Anderson and Belnap's logic $\mathbf{E}_{\mathrm{fde}}$ of first-degree entailments (Anderson and Belnap, 1975, pp. 161–162). See also (Belnap, 1977, 1976) and (Dunn, 1976). The deductive system **BN4** has been further studied as a logic permitting both truth value 'gaps' and truth value 'gluts' in (Meyer et al., 1984; Restall, 1993; Slaney, 1991, 2004); *cf.* (Kapsner, 2014, Chapter 4).

Let **4** denote the algebra with universe $\{\mathsf{T}, \mathsf{b}, \mathsf{n}, \bot\}$, and whose operations $\wedge$, $\vee$, $\rightarrow$, and $\sim$ are fixed by the Cayley tables

$\wedge$	T	b	n	$\bot$
T	T	b	n	$\bot$
b	b	b	$\bot$	$\bot$
n	n	$\bot$	n	$\bot$
$\bot$	$\bot$	$\bot$	$\bot$	$\bot$

$\vee$	T	b	n	$\bot$
T	T	T	T	T
b	T	b	T	b
n	T	T	n	n
$\bot$	T	b	n	$\bot$

$\rightarrow$	T	b	n	$\bot$
T	T	b	n	$\bot$
b	T	b	n	$\bot$
n	T	T	T	T
$\bot$	T	T	T	T

$\sim$	
T	$\bot$
b	b
n	n
$\bot$	T

Let $\mathbf{BN}_4$ denote the algebra reduct of $\mathfrak{BN}4$. Let $\mathsf{BN}_4 := \underline{\mathsf{HSP}}(\mathbf{BN}_4)$. Observe that: (i) $\mathbf{4}^\alpha = \mathbf{BN}_4$, where α is the map of Theorem 3.3.(1); (ii) $(\mathbf{BN}_4)^\beta = \mathbf{4}$, where β is the map of Theorem 3.3.(2); and (iii) $\mathbf{4}^{\alpha\beta} = \mathbf{4}$ and $(\mathbf{BN}_4)^{\beta\alpha} = \mathbf{BN}_4$. Thus $\mathbf{4}$ and $\mathbf{BN}_4$ have the same n-ary term operations for $n > 0$, and so are term equivalent. By (McKenzie et al., 1987, Theorem 4.140), we conclude that the varieties $\underline{\mathsf{HSP}}(\mathbf{4})$ and BN_4 are term equivalent.

Given a variety V, let V_{S} [resp. V_{SI}; resp. V_{SS}] denote the class of simple [resp. subdirectly irreducible; resp. semisimple] members of V.

Theorem 4.20. $\underline{\mathsf{HSP}}(\mathbf{4}) = \underline{\mathsf{HSP}}(\mathsf{N4}_{\mathrm{SS}})$.

Proof. By direct inspection, $\mathbf{4}$ is simple. Thus $\mathbf{4} \in \mathsf{N4}_{\mathrm{SS}}$ and $\underline{\mathsf{HSP}}(\mathbf{4}) \subseteq \underline{\mathsf{HSP}}(\mathsf{N4}_{\mathrm{SS}})$. For the converse, suppose first that $\mathbf{B} \in \mathsf{N4}$ is simple. Then $|B| \leq 4$. Indeed, $\mathrm{Con}\,\mathbf{B} = \mathrm{Con}\,\mathbf{B}_{\bowtie}$, by (Odintsov, 2004, Corollary 4.3). But to within isomorphism, the only simple implicative lattice is the 2-element chain $\mathbf{2}$ (considered as an implicative lattice). Thus $\mathbf{B}$ has at most $2 \times 2 = 4$ elements. Observe next that, by direct inspection, there are precisely four simple **N4**-lattices to within isomorphism, *viz.*:

$$\mathbf{4} := \mathbf{2}^{\bowtie} := \mathbf{Tw}\big(\mathbf{2}, \{0,1\}, \{0,1\}\big) \quad \mathbf{2}_{\circ}^{\bowtie} := \mathbf{Tw}\big(\mathbf{2}, \{1\}, \{0\}\big)$$
$$\mathbf{2}_R^{\bowtie} := \mathbf{Tw}\big(\mathbf{2}, \{1\}, \{0,1\}\big) \quad \mathbf{2}_L^{\bowtie} := \mathbf{Tw}\big(\mathbf{2}, \{0,1\}, \{0\}\big).$$

(*Cf.* (Odintsov, 2005, Section 5, p. 307).) Thus $\mathsf{N4}_{\mathrm{S}} \subseteq \underline{\mathsf{I}}(\{\mathbf{2}^{\bowtie}, \mathbf{2}_{\circ}^{\bowtie}, \mathbf{2}_R^{\bowtie}, \mathbf{2}_L^{\bowtie}\})$. By direct inspection again, $\underline{\mathsf{I}}(\{\mathbf{2}^{\bowtie}, \mathbf{2}_{\circ}^{\bowtie}, \mathbf{2}_R^{\bowtie}, \mathbf{2}_L^{\bowtie}\}) \subseteq \underline{\mathsf{IS}}(\mathbf{4}) \subseteq \underline{\mathsf{S}}(\mathbf{4})$; thus $\mathsf{N4}_{\mathrm{S}} \subseteq \underline{\mathsf{S}}(\mathbf{4})$. Now let K be the class of simple members of $\mathsf{N4}_{\mathrm{SS}}$ and let $\mathbf{A} \in \mathsf{N4}_{\mathrm{SS}}$. As $\mathsf{N4}_{\mathrm{SS}}$

is semisimple, $\mathbf{A} \in \underline{\mathsf{Ps}}(\mathsf{K}) \subseteq \underline{\mathsf{Ps}}(\mathsf{N4}_\mathrm{S})$ (because $\mathsf{K} \subseteq \mathsf{N4}_\mathrm{S}$) $\subseteq \underline{\mathsf{PsS}}(4) = \underline{\mathsf{SP}}(4)$. Thus $\mathsf{N4}_\mathrm{SS} \subseteq \underline{\mathsf{SP}}(4)$. But then $\underline{\mathsf{HSP}}(\mathsf{N4}_\mathrm{SS}) \subseteq \underline{\mathsf{HSPSP}}(4) \subseteq \underline{\mathsf{HSP}}(4)$.

The next lemma generalises results of (Monteiro, 1963, 1995) to the setting of **N4**-lattices.

Lemma 4.21. *For a variety* V *of* **N4**-*lattices the following are equivalent:*

1. V *is semisimple.*
2. V *satisfies the identity* $((x \to y) \to x) \to x \approx \big| ((x \to y) \to x) \to x \big|$. *(Peirce)*
3. V *satisfies the identity* $x \vee (x \to y) \approx \big| x \vee (x \to y) \big|$.   *(Peirce$_\vee$)*

Proof. (Sketch) $(1) \Rightarrow (2)$ Suppose V is semisimple. Then $\mathsf{V} \subseteq \underline{\mathsf{HSP}}(4)$. By direct inspection, $4 \models$ (Peirce). Thus $\underline{\mathsf{HSP}}(4) \models$ (Peirce) and therefore so too does V.

$(2) \Rightarrow (3)$ By equational reasoning.

$(3) \Rightarrow (1)$ Observe first that, if $\mathbf{B}$ is an implicative lattice and $\mathbf{B} \models$ (Peirce$_\vee$), then $\mathbf{B}$ is subdirectly irreducible iff $\mathbf{B}$ is simple. Suppose now that $\mathsf{V} \models$ (Peirce$_\vee$). Let $\mathbf{A} \in \mathsf{V}_\mathrm{SI}$. By (Odintsov, 2004, Corollary 4.3), $\mathrm{Con}\,\mathbf{A} = \mathrm{Con}\,\mathbf{A}_{\bowtie}$; thus $\mathbf{A}_{\bowtie}$ is subdirectly irreducible. As $\mathbf{A}_{\bowtie} \models$ (Peirce$_\vee$) (because $\mathbf{A} \models$ (Peirce$_\vee$)), it follows that $\mathbf{A}_{\bowtie}$ is simple. But by (Odintsov, 2004, Corollary 4.3) again, $\mathrm{Con}\,\mathbf{A} = \mathrm{Con}\,\mathbf{A}_{\bowtie}$; thus $\mathbf{A}$ is simple. Thus $\mathbf{A} \in \mathsf{V}_\mathrm{S}$. Therefore $\mathsf{V}_\mathrm{SI} \subseteq \mathsf{V}_\mathrm{S}$, and V is semisimple.

From the remarks preceding Theorem 4.20, Theorem 4.20 itself, Lemma 4.21, Theorem 3.3, and Theorem 3.25 we have the following theorem, a version of which is established in (Méndez and Robles, 2016, Section 6.5). See also (Odintsov and Speranski, 2016, Introduction, p. 4).

Theorem 4.22. *The axiomatic extension of* **N4** *by the Peirce law* $\vdash ((x \to y) \to x) \to x$ *is, up to definitional equivalence, the deductive system* **BN4** *of Brady* (Brady, 1982; Meyer et al., 1984; Restall, 1993; Slaney, 1991).

Łukasiewicz's 3-valued logic $\mathbf{Ł_3}$ Let $\Lambda[\mathbf{Ł_3}]$ be a language type having binary logical connectives $\wedge$, $\vee$, and $\Rightarrow$ together with a unary logical connective $\sim$. The well-known 3-valued logic of Łukasiewicz (Łukasiewicz, 1970a,b), in symbols $\mathbf{Ł_3}$, is the deductive system over $\Lambda[\mathbf{Ł_3}]$ determined by the matrix:

$\wedge$	$\top$	n	$\bot$
$^\dagger\top$	$\top$	n	$\bot$
n	n	n	$\bot$
$\bot$	$\bot$	$\bot$	$\bot$

$\vee$	$\top$	n	$\bot$
$\top$	$\top$	$\top$	$\top$
n	$\top$	n	n
$\bot$	$\top$	n	$\bot$

$\Rightarrow$	$\top$	n	$\bot$
$\top$	$\top$	n	$\bot$
n	$\top$	$\top$	n
$\bot$	$\top$	$\top$	$\top$

$\sim$	
$\top$	$\bot$
n	n
$\bot$	$\top$

Since the 1950s, the deductive system $\mathbf{Ł_3}$ has been the subject of enormous exegesis; for surveys and standard references, see for instance (Wójcicki, 1988, Section 4.3.1) or (Malinowski, 1993, Chapter 2, Chapter 5). In (Vakarelov, 1977, Theorem 11) Vakarelov essentially proved that the axiomatic extension of **N3** by the Peirce law $\vdash \big((x \to y) \to x\big) \to x$ is, up to definitional equivalence, the deductive system $\mathbf{L}_3$; in this connection, see also (Spinks and Veroff, 2008b, Example 1.3). As **N3** is the axiomatic extension of **N4** by the weakening axiom $\vdash x \Rightarrow (y \Rightarrow x)$, we conclude:

Theorem 4.23. *The axiomatic extension of* **N4** *by the weakening axiom* $\vdash x \Rightarrow (y \Rightarrow x)$ *and the Peirce law* $\vdash \big((x \to y) \to x\big) \to x$ *is, up to definitional equivalence, the 3-valued logic* $\mathbf{L}_3$ *of Łukasiewicz* (Łukasiewicz, 1970a,b).

4.5 Extensions of $\mathbf{N4}^{\perp}$ *by the Peirce law* $\vdash \big((x \to y) \to x\big) \to x$

In this subsection extensions of $\mathbf{N4}^{\perp}$ by the Peirce law $\vdash \big((x \to y) \to x\big) \to x$ are considered; the logic $\mathbf{N4}_p^{\perp}$ itself has been briefly considered (under the appellation $\mathbf{B}_4^{\to}$) in (Odintsov, 2005; Odintsov and Speranski, 2016). The key tool used throughout this subsection is the ternary discriminator construct (Burris and Sankappanavar, 1981; Jónsson, 1995; Werner, 1978) of general algebra; use of this construct calls in turn for the study of the congruence properties of **N4**-lattices.

Congruence permutability Because every **N4**-lattice has a lattice reduct, the variety of **N4**-lattices is congruence distributive. The next two results, which collectively extend and sharpen (Odintsov, 2008, Proposition 9.2.12) and (Spinks, 2004, Theorem 4.4, Corollary 4.5), affirm that congruences on **N4**-lattices are very well behaved in general.

Theorem 4.24. *The variety of* **N4**-*lattices is congruence permutable, with Mal'cev term* $p(x,y,z) := \big((x \Rightarrow y) \overset{\mathbf{RM}}{\Rightarrow} z\big) \wedge \big((z \Rightarrow y) \overset{\mathbf{RM}}{\Rightarrow} x\big)$.

Proof. Using adjunctivity, mimic the proof of (van Alten and Raftery, 2004, Proposition 8.3).

Corollary 4.25. *The variety of* **N4**-*lattices is arithmetical.*

EDPC A variety V has ***equationally definable principal congruences*** (EDPC) if there exists a finite set $\big\{ \langle p_i(x,y,z,w), q_i(x,y,z,w) \rangle : i < n \big\}$ of pairs of quaternary terms of V such that for every $\mathbf{A} \in \mathsf{V}$ and for all $a,b,c,d \in A$,

$$c \equiv d \quad (\mathrm{mod}\ \Theta^{\mathbf{A}}(a,b)) \quad \text{iff} \quad p_i^{\mathbf{A}}(a,b,c,d) = q_i^{\mathbf{A}}(a,b,c,d), \quad \text{for all } i < n.$$

The study of varieties with EDPC was initiated by Baldwin and Berman in (Baldwin and Berman, 1975) and further developed in (Fried et al., 1980; Köhler and Pigozzi, 1980). By (Köhler and Pigozzi, 1980, Corollary 6) every variety with EDPC is congruence distributive; this turned the study of such varieties into a major topic. In their series of papers (Blok et al., 1984;

Blok and Pigozzi, 1982, 1994a,b, 2001) Blok and Pigozzi extensively investigated EDPC for quasivarieties of logic (in the sense of (Barbour and Raftery, 2003, Section 1), (Font et al., 2003, Theorem 4.8)), showing in particular that a strongly algebraisable deductive system has the deduction-detachment theorem iff its equivalent variety has EDPC (Blok and Pigozzi, 2001, Theorem 5.5). By (Wójcicki, 1988, Theorem 2.4.2), **N4** has the deduction-detachment theorem; thus N4 has EDPC. Indeed, a modification of the proof of (Blok et al., 1984, Lemma 2.4) shows that for every **N4**-lattice **A** and for all $a, b, c, d \in A$, it holds that $c \equiv d \pmod{\Theta^{\mathbf{A}}(a,b)}$ iff $a \Leftrightarrow b \preceq c \Leftrightarrow d$. This observation is sharpened in Theorem 4.26 below.

TD terms Following (Blok and Pigozzi, 1994a, Definition 2.1), a ternary term $e(x, y, z)$ is a ***ternary deductive*** (***TD***) ***term*** for a class K of similar algebras if: (i) $\mathsf{K} \models e(x, x, z) \approx z$; and (ii) for all $\mathbf{A} \in \mathsf{K}$ and for all $a, b, c, d \in A$, $e^{\mathbf{A}}(a, b, c) = e^{\mathbf{A}}(a, b, d)$ if $c \equiv d \pmod{\Theta^{\mathbf{A}}(a,b)}$. By (Blok and Pigozzi, 1994a, Corollary 2.4(iii)), $e(x, y, z)$ is a TD term for the variety $\underline{\mathsf{HSP}}(\mathsf{K})$ generated by K iff K satisfies the identities

$$(27) \qquad\qquad\qquad e(x, x, y) \approx y$$
$$(28) \qquad\qquad\qquad e(x, y, x) \approx e(x, y, y)$$

together with the identities

$$(29) \qquad e\big(x, y, c(z_1, \ldots, z_n)\big) \approx e\big(x, y, c(e(x, y, z_1), \ldots, e(x, y, z_n))\big)$$

for every n-ary operation symbol c in the type of K. A TD term $e(x, y, z)$ on an algebra **A** is ***commutative*** if $\mathbf{A} \models e\big(x, y, e(x', y', z)\big) \approx e\big(x', y', e(x, y, z)\big)$ (Blok and Pigozzi, 1994a, Definition 3.1). A TD term $e(x, y, z)$ on a class K of similar algebras is ***commutative*** if it is commutative on every member of K.

The next result generalises (Spinks, 2004, Theorem 3.3) to **N4**-lattices.

Theorem 4.26. *The ternary term* $e(x, y, z) := (x \Leftrightarrow y) \to z$ *is a commutative TD term for the variety of* **N4**-*lattices.*

Proof. Let **A** be an **N4**-lattice and let $a, b, c, d \in A$. Note $e^{\mathbf{A}}(a, a, b) = (a \Leftrightarrow a) \to b = |a| \to b = b$ (by (20)), whence $\mathbf{A} \models (27)$. And, $e^{\mathbf{A}}(a, b, a) = (a \Leftrightarrow b) \to a = (a \Leftrightarrow b) \to b$ (by (4)) $= e^{\mathbf{A}}(a, b, b)$, whence $\mathbf{A} \models (28)$. Also,

$$
\begin{aligned}
e^{\mathbf{A}}(a, b, c \wedge d) &= (a \Leftrightarrow b) \to (c \wedge d) \\
&= (a \Leftrightarrow b) \to \big((a \Leftrightarrow b) \to (c \wedge d)\big) && \text{by (5)} \\
&= (a \Leftrightarrow b) \to \big(((a \Leftrightarrow b) \to c) \wedge ((a \Leftrightarrow b) \to d)\big) && \text{by (17)} \\
&= e^{\mathbf{A}}\big(a, b, e^{\mathbf{A}}(a, b, c) \wedge e^{\mathbf{A}}(a, b, d)\big)
\end{aligned}
$$

whence $\mathbf{A} \models (29)$ for the basic operation symbol $\wedge$. Similar arguments using the identities (18), (7), and (19) show $\mathbf{A} \models (29)$ for each of the fundamen-

tal operation symbols $\vee$, $\to$, and $\sim$. So $e(x,y,z)$ is a TD term for N4. To see $e(x,y,z)$ is commutative, note that for all $a,b,a',b',c \in A$, it holds that
$$e^{\mathbf{A}}\big(a,b,e^{\mathbf{A}}(a',b',c)\big) = (a \Leftrightarrow b) \to \big((a' \Leftrightarrow b') \to c\big) = (a' \Leftrightarrow b') \to \big((a \Leftrightarrow b) \to c\big)$$
(by (6)) $= e^{\mathbf{A}}\big(a',b',e^{\mathbf{A}}(a,b,c)\big)$.

From (Blok and Pigozzi, 1994a, Corollary 2.5), we infer:

Corollary 4.27. *For every* **N4**-*lattice* **A** *and for all* $a,b,c,d \in A$, *it holds that* $c \equiv d \pmod{\Theta^{\mathbf{A}}(a,b)}$ *iff* $e^{\mathbf{A}}(a,b,c) = e^{\mathbf{A}}(a,b,d)$. *Hence the variety of* **N4**-*lattices has EDPC.*

QD Terms Following (Blok et al., 1984, p. 359), a quaternary term $q(x,y,z,w)$ on an algebra **A** is a ***Quaternary Deductive (QD) term*** on **A** if, for all $a,b,c,d \in A$,

$$q^{\mathbf{A}}(a,b,c,d) = \begin{cases} c & \text{if } a = b \\ d & \text{if } c \equiv d \pmod{\Theta^{\mathbf{A}}(a,b)}. \end{cases}$$

A ***Quaternary Deductive (QD) term*** on a variety V is a quaternary term $q(x,y,z,w)$ of V such that q is a QD term on every member of V.

Lemma 4.28. (Spinks, 2004, Lemma 5.1) *Let* V *be a variety with a TD term* $e(x,y,z)$ *and a Mal'cev term* $p(x,y,z)$. *Then the term* $q(x,y,z,w) := p\big(e(x,y,z),e(x,y,w),w\big)$ *is a QD term for* V.

Combining Lemma 4.28 with Theorem 4.26 and Theorem 4.24 yields the following result, which generalises (Spinks, 2004, Theorem 5.2) to **N4**-lattices.

Theorem 4.29. *The variety of* **N4**-*lattices has a QD term. Given the commutative TD term* $e(x,y,z)$ *of Theorem 4.26 and the Mal'cev term* $p(x,y,z)$ *of Theorem 4.24, a QD term for* N4 *is* $q(x,y,z,w) := p\big(e(x,y,z),e(x,y,w),w\big)$.

Discriminator varieties of N4-lattices The (***ternary***) ***discriminator*** (Burris and Sankappanavar, 1981, Definition IV§9.1) on a set A is the function $t : A^3 \to A$ defined for all $a,b,c \in A$ by $t(a,b,c) := c$ if $a = b$ and a otherwise. A (***ternary***) ***discriminator variety*** is a variety V for which there exists a ternary term $t(x,y,z)$ of V that realises the discriminator on each subdirectly irreducible member of V. According to (Burris and Sankappanavar, 1981, Chapter IV§9,10), discriminator varieties constitute "…the most successful generalization of Boolean algebras to date." As such, discriminator varieties have been considered extensively in the literature; standard references include (Werner, 1978) and (Jónsson, 1995).

By a classic result of Blok, Köhler, and Pigozzi (1984, Corollary 3.4) (see also (Fried and Kiss, 1983)), an equational class is a discriminator variety iff it is congruence permutable, semisimple, and has EDPC. For each $\mathsf{K} \in \{\mathsf{N4},\mathsf{N4}^{\perp},\mathsf{N3}\}$, let K_p be the subvariety of K axiomatised (relative to K) by the identity $\big((x \to y) \to x\big) \to x \approx \big|((x \to y) \to x) \to x\big|$. Observe that, by Lemma 4.21, a variety V of **N4**-lattices is semisimple iff $\mathsf{V} \subseteq \mathsf{N4}_p$. Generalising (Spinks, 2004, Corollary 5.3), we have:

Corollary 4.30. *A variety* V *of* **N4**-*lattices is a discriminator variety iff* $\mathsf{V} \subseteq \mathsf{N4}_p$. *In this case a discriminator term for* V *is given by* $t(x,y,z) := q(x,y,z,x)$, *where* $q(x,y,z,w)$ *is the QD term of Theorem 4.29.*

Proof. The first assertion is clear in view of Theorem 4.24, Corollary 4.27, and the remarks preceding the corollary. If V is a discriminator variety, then $q(x,y,z,w)$ must coincide with the normal transform (see (Burris and Sankappanavar, 1981, Chapter IV§9)) on each subdirectly irreducible member of V, by virtue of the semisimplicity of V. From (Burris and Sankappanavar, 1981, Lemma IV§9.2(a)) it follows that $t(x,y,z)$ is a discriminator term for V.

For $\mathbf{RM}_3$, the first assertion of the next result may be deduced from Font and Pérez (1992, Theorem 12).

Corollary 4.31. *Each of the varieties* RM_3 *and* BN_4 *is a discriminator variety, with the discriminator term of Corollary 4.30.*

Proof. (Sketch) Let $\mathsf{V} \in \{\mathsf{RM}_3, \mathsf{BN}_4\}$. Then V is generated as a variety by either $\mathbf{RM}_3$ or $\mathbf{BN}_4$. Since each of these algebras is simple, V is semisimple (by (van Alten, 1998, Proposition 5.19)), and the result follows.

Given an algebra $\mathbf{A}$ and pairwise distinct elements $a_1, \dots, a_n \in A$, let $\mathbf{A}^{a_1, \dots, a_n}$ denote the algebra obtained from $\mathbf{A}$ upon distinguishing the elements $a_1, \dots, a_n$. Thus, for example, $\mathbf{RM}_3^{\perp}$ denotes the algebra obtained from $\mathbf{RM}_3$ upon distinguishing the element $\perp$. Let $\mathsf{RM}_3^{\perp} := \underline{\mathsf{HSP}}(\mathbf{RM}_3^{\perp})$ and $\mathsf{BN}_4^{\perp} := \underline{\mathsf{HSP}}(\mathbf{BN}_4^{\perp})$.

Corollary 4.32. *Each of the varieties* $\mathsf{RM}_3^{\perp}$ *and* $\mathsf{BN}_4^{\perp}$ *is a discriminator variety, with the discriminator term of Corollary 4.30.*

Primality Recall from (Burris and Sankappanavar, 1981, Chapter IV§7) that a finite algebra $\mathbf{A}$ is **primal** if every n-ary function on A, for every $n \geq 1$, is representable by a term. The next result is essentially (Avron, 1999, Theorem 3.8).

Corollary 4.33. *The algebra* $\mathbf{4}^{\mathsf{b},\mathsf{n}}$ *is primal.*

Proof. Observe that the term $t(x,y,z)$ of Corollary 4.30 realises the ternary discriminator on $\mathbf{4}^{\mathsf{b},\mathsf{n}}$. In view of this, to prove the theorem it suffices by the main result of (Werner, 1970) to show that every element of (the universe of) $\mathbf{4}^{\mathsf{b},\mathsf{n}}$ is realised by a constant. For this, just note $\mathsf{n}^{\mathbf{4}^{\mathsf{b},\mathsf{n}}} \to \mathsf{n}^{\mathbf{4}^{\mathsf{b},\mathsf{n}}} = \top$ and $\sim(\mathsf{n}^{\mathbf{4}^{\mathsf{b},\mathsf{n}}} \to \mathsf{n}^{\mathbf{4}^{\mathsf{b},\mathsf{n}}}) = \perp$.

Observe that $\mathbf{4}^{\perp}$ is not primal; $\{\perp, \top\}$ is the universe of a 2-element subalgebra.

The algebra $\mathbf{4}^{\top, \mathsf{b}, \mathsf{n}, \perp}$ serves as an algebraic 'base' for various of the modal bilattice logics studied in (Jung and Rivieccio, 2013; Rivieccio, 2014a,b);

by Corollary 4.33, the algebra $\mathbf{4}^{\top,\mathsf{b},\mathsf{n},\perp}$ is primal. The utility of Corollary 4.30 is that, when combined with (Burris, 1992, Lemma 3.5), other (bilattice) operations potentially of interest to (Jung and Rivieccio, 2013; Rivieccio, 2014a,b) can be realised readily on $\mathbf{4}^{\mathsf{b},\mathsf{n}}$, hence $\mathbf{4}^{\top,\mathsf{b},\mathsf{n},\perp}$. Indeed, let $q(x,y,z,w)$ be the QD term of Theorem 4.29. By (Burris, 1992, Lemma 3.5), $q\big(q(x,\top,x,\mathsf{b}),x,y,\mathsf{n}\big)$ realises Fitting's guard operation : (Fitting, 1994) on $\mathbf{4}^{\mathsf{b},\mathsf{n}}$ (*qua* $\mathbf{4}^{\top,\mathsf{b},\mathsf{n},\perp}$); the term $q\big(q(x,\top,x,\mathsf{b}),x,\top,\perp\big)$ realises Moore's autoepistemic operator L (Moore, 1985), as interpreted by Ginsberg (1990); and $q\big(q(x,\perp,x,\top),x,x,q(\mathsf{n},x,\mathsf{b},\mathsf{n})\big)$ realises Fitting's conflation operator $-$ (Fitting, 1989, 1991).

The deductive system BS4 Let $\Lambda[\mathbf{BS4}] := \Lambda[\mathbf{N4}] \cup \{\circ\}$, where $\circ$ is a unary logical connective. The logic **BS4** is the deductive system over $\Lambda[\mathbf{BS4}]$ determined by the matrix:

$\wedge$	$\top$	b	n	$\perp$		$\vee$	$\top$	b	n	$\perp$		$\rightarrow$	$\top$	b	n	$\perp$		$\sim$			$\circ$	
†$\top$	$\top$	b	n	$\perp$		$\top$	$\top$	$\top$	$\top$	$\top$		$\top$	$\top$	b	n	$\perp$		$\top$	$\perp$		$\top$	$\top$
†b	b	b	$\perp$	$\perp$		b	$\top$	b	$\top$	b		b	$\top$	b	n	$\perp$		b	b		b	$\perp$
n	n	$\perp$	n	$\perp$		n	$\top$	$\top$	n	n		n	$\top$	$\top$	$\top$	$\top$		n	n		n	$\perp$
$\perp$	$\perp$	$\perp$	$\perp$	$\perp$		$\perp$	$\top$	b	n	$\perp$		$\perp$	$\top$	$\top$	$\top$	$\top$		$\perp$	$\top$		$\perp$	$\top$

Denote this matrix by $\mathfrak{BG}4$. The deductive system **BS4** was introduced in (Omori and Waragai, 2011) as a generalisation of **LFI1** to the Belnapian four-valued setting; it has been further investigated in (De and Omori, 2015; Omori and Sano, 2014; Sano and Omori, 2014). In (Omori and Waragai, 2011, Theorem 7) Omori and Waragai establish the existence of translations $\tau_1 : \Lambda[\mathbf{BS4}] \to \mathrm{Fm}_{\Lambda[\mathbf{N4}^\perp]}$ and $\tau_2 : \Lambda[\mathbf{N4}^\perp] \to \mathrm{Fm}_{\Lambda[\mathbf{BS4}]}$ such that $\vdash_{\mathbf{BS4}} \varphi$ iff $\vdash_{\mathbf{N4}_p^\perp} \tau_1(\varphi)$ and $\vdash_{\mathbf{N4}_p^\perp} \varphi$ iff $\vdash_{\mathbf{BS4}} \tau_2(\varphi)$; because of (Omori and Sano, 2014, Remark 8), Omori and Sano's clause for mapping $\circ\varphi$ into $\mathbf{N4}^\perp$ can be given as $\tau_1(\varphi) := \big((\tau_1(\varphi) \wedge \tau_1(\sim\varphi)) \to \perp\big) \wedge \big(\tau_1(\varphi) \vee \tau_1(\sim\varphi)\big)$. In this connection, see also (Sano and Omori, 2014, Proposition 5.13).

In (Omori and Waragai, 2011, Remark 3, p. 321) Omori and Waragai notice that the unary formula $\circ x \wedge x \wedge \sim x$ acts as a falsum (bottom particle) in **BS4**; *cf.* the map α of Theorem 4.12.(1). Let $\mathbf{BS}_4$ denote the algebra reduct of $\mathfrak{BG}4$. Routine arguments show that $\mathbf{BS4}$ is strongly algebraisable with system of equivalence formulas $\{\varphi \Leftrightarrow \psi\}$ and system of defining equations $\{x \approx |x|\}$. Let V be the equivalent variety semantics V of **BS4**. Because each $\mathbf{A} \in \mathsf{V}$ is subdirectly irreducible iff its $\circ$-free reduct is subdirectly irreducible, $\mathsf{V} = \underline{\mathsf{HSP}}(\mathbf{BS}_4)$. Let $q(x,y,z,w)$ be the QD term of Theorem 4.29 and observe: (i) that $q(x,\sim x,\perp,\sim\perp)$ induces the operation $\circ$ on $\mathbf{4}^\perp$ (apply Corollary 4.32); and (ii) that $q(x,\sim x,\perp,\sim\perp)$ rewrites to the term $\big((x \to \sim x) \wedge (\sim x \to x)\big) \to \perp$ over $\mathbf{4}^\perp$. Notice also that

$$\mathbf{BS}_4 \models \circ x \approx \big((x \to {\sim}x) \wedge ({\sim}x \to x)\big) \to (\circ x \wedge x \wedge {\sim}x) \quad \text{and}$$

$$\mathbf{4}^{\perp} \models \perp \approx \big(((x \to {\sim}x) \wedge ({\sim}x \to x)) \to \perp\big) \wedge x \wedge {\sim}x.$$

Since the $\langle \wedge, \vee, \to, {\sim} \rangle$-reduct of $\mathbf{BS}_4$ is $\mathbf{4}$, we conclude that the algebras $\mathbf{BS}_4$ and $\mathbf{4}^{\perp}$ have the same n-ary term operations for $n > 0$, and so are term equivalent. By (McKenzie et al., 1987, Theorem 4.140), therefore, the varieties $\underline{\mathsf{HSP}}(\mathbf{BS}_4)$ and $\underline{\mathsf{HSP}}(\mathbf{4}^{\perp})$ are term equivalent. But $\underline{\mathsf{HSP}}(\mathbf{4}^{\perp})$ is $\mathsf{N4}_p^{\perp}$ (*cf.* (Odintsov and Speranski, 2016, Introduction, p. 4)), as can be seen from Theorem 4.20 and Lemma 4.21 (*mutatis mutandis*). Applying now standard techniques we have the following theorem, which sharpens the aforementioned translational equivalence of Omori and Sano; *cf.* (De and Omori, 2015, Remark 18), (Sano and Omori, 2014, Remark 5.12).

Theorem 4.34.

1. *The map* $\alpha : \Lambda[\mathbf{BS4}] \to \mathrm{Fm}_{\Lambda[\mathbf{N4}^{\perp}]}$ *defined by*

$$
\begin{aligned}
x \wedge y &\mapsto x \wedge y & \qquad\qquad {\sim}x &\mapsto {\sim}x \\
x \vee y &\mapsto x \vee y & \circ x &\mapsto \big((x \to {\sim}x) \wedge ({\sim}x \to x)\big) \to \perp \\
x \to y &\mapsto x \to y
\end{aligned}
$$

    is an interpretation of $\mathbf{BS4}$ *in* $\mathbf{N4}_p^{\perp}$.

2. *The map* $\beta : \Lambda[\mathbf{N4}^{\perp}] \to \mathrm{Fm}_{\Lambda[\mathbf{BS4}]}$ *defined by*

$$
\begin{aligned}
x \wedge y &\mapsto x \wedge y & \qquad\qquad {\sim}x &\mapsto {\sim}x \\
x \vee y &\mapsto x \vee y & \perp &\mapsto \circ x \wedge x \wedge {\sim}x \\
x \to y &\mapsto x \to y
\end{aligned}
$$

    is an interpretation of $\mathbf{N4}_p^{\perp}$ *in* $\mathbf{BS4}$.

3. *The interpretations* α *and* β *are mutually inverse.*

Hence the deductive systems $\mathbf{N4}_p^{\perp}$ *and* $\mathbf{BS4}$ *are definitionally equivalent.*

Corollary 4.35. *The axiomatic extension of* $\mathbf{N4}^{\perp}$ *by the Peirce law* $\vdash \big((x \to y) \to x\big) \to x$ *is, up to definitional equivalence, the logic* $\mathbf{BS4}$ *of Omori and Waragai* (De and Omori, 2015; Omori and Sano, 2014; Omori and Waragai, 2011; Sano and Omori, 2014).

Connections between $\mathbf{BS4}$ and other deductive systems considered previously in the literature—including Kachi's simple partial logic (Kachi, 2002, 2007) and Sano and Omori's logic $\mathbf{BD}\triangle$ (Sano and Omori, 2014)—are given in (De and Omori, 2015; Sano and Omori, 2014).

5 Concluding remarks

Fregean deductive systems Recall from (Czelakowski and Pigozzi, 2004, Definition 59) that a deductive system $\mathbf{S}$ over a language type Λ is ***Fregean*** if, for every theory T of $\mathbf{S}$, the relativised interderivability relation $\dashv\vdash^T_{\mathbf{S}}$ defined for all Λ-formulas φ, ψ by

$$\varphi \dashv\vdash^T_{\mathbf{S}} \psi \quad \text{iff} \quad T, \varphi \vdash_{\mathbf{S}} \psi \text{ and } T, \psi \vdash_{\mathbf{S}} \varphi$$

is a congruence relation on $\mathbf{Fm}_\Lambda$. A logic $\mathbf{S}$ is ***non-Fregean*** if it is not Fregean. Fregean deductive systems with the (uniterm) deduction-detachment theorem are essentially the intermediate logics, possibly with additional logical connectives (Czelakowski and Pigozzi, 2004, Theorem 63). For studies of Fregeanity in algebraic logic, see (Czelakowski, 2001; Czelakowski and Pigozzi, 2004; Font and Jansana, 2009; Idziak et al., 2009; Pigozzi, 1991).

In (Czelakowski and Pigozzi, 2004, Section 2.1) Czelakowski and Pigozzi observe that it is characteristic for the properties of protoalgebraicity and equivalentiality to coalesce in Fregean deductive systems. In particular, if $\mathbf{S}$ is a deductive system with the deduction-detachment theorem (with, say, $\Sigma(x,y)$ as a deduction-detachment system for $\mathbf{S}$), then $\mathbf{S}$ is Fregean iff $\Sigma(x,y) \cup \Sigma(y,x)$ is an equivalence system for $\mathbf{S}$ (Czelakowski and Pigozzi, 2004, Lemma 62). In consequence, if $\mathbf{S}$ is a non-Fregean equivalential deductive system with the deduction-detachment theorem, then the notions of equivalence and deduction necessarily diverge.

Strong and weak connectives As a rule, each non-Fregean deductive system $\mathbf{S}$ arises as the logical cognate of a Fregean deductive system $\mathbf{S}^F$; in the archetypical situation, $\mathbf{S}^F$ has the *uniterm* deduction-detachment theorem (and is thus strongly and regularly algebraisable (Czelakowski and Pigozzi, 2004, Theorem 66)). Non-Fregean deductive systems bearing witness to this phenomenon include, for example, Suzsko's sentential calculus with identity (Bloom and Suszko, 1972; Ishii, 2000; Suszko, 1968, 1971, 1975) and (various of) the substructural logics (Došen, 1993; Galatos et al., 2007; Restall, 2000; Spinks and Veroff, 2008b).

Let $\mathbf{S}$ be a non-Fregean equivalential logic with the deduction-detachment theorem arising as the logical cognate of some Fregean deductive system $\mathbf{S}^F$ with the deduction-detachment theorem; by (Czelakowski and Pigozzi, 2004, Theorem 61), $\mathbf{S}^F$ is regularly algebraisable. Suppose (prototypically) that $\{\supset\}$ is a uniterm deduction-detachment system for $\mathbf{S}^F$. Because the notions of equivalence and deduction necessarily diverge in $\mathbf{S}$, the connective $\supset$ of $\mathbf{S}^F$ ramifies on passage to $\mathbf{S}$. In the simplest (yet most common) situation, $\supset$ bifurcates, yielding, on the one hand, an implication-like connective $\Rightarrow$, whose symmetrisation $\{\varphi \Rightarrow \psi, \psi \Rightarrow \varphi\}$ acts as a system of equivalence formulas for $\mathbf{S}$; and, on the other hand, a separate, distinct implication connective $\rightarrow$ that witnesses the *uniterm* deduction-detachment system for $\mathbf{S}$

(and which is thus a conditional for $\mathbf{S}$ in the usual Tarskian sense). In general, $\varphi \Rightarrow \psi \vdash_{\mathbf{S}} \varphi \to \psi$ for all formulas φ, ψ, owing to properties of $\mathbf{S}^F$; the converse holds if $\mathbf{S}$ is Fregean. See for instance (Spinks et al., 2014, Proposition 3.1). The connectives $\Rightarrow$ and $\to$ are thus *strong* and *weak* respectively; *cf.* (Rasiowa, 1974, Chapter VIII§1, p. 279).

By the same token, if a nullary logical connective $\bot$ is available in the language of $\mathbf{S}^F$, hence $\mathbf{S}$, and negation is introduced in $\mathbf{S}^F$ via the definition $-\varphi := \varphi \supset \bot$, the derived connective $-$ also splits on passage to $\mathbf{S}$. (Here no postulates are assumed regarding $\bot$, as in the manner of Johansson (1936).) On the one hand, we obtain a negation $\sim$, defined as $\sim\varphi := \varphi \Rightarrow \bot$, which is typically contraposable; while on the other hand, we obtain a negation $\neg$, defined as $\sim\varphi := \varphi \to \bot$, for which it typically holds that $\Gamma \vdash_{\mathbf{S}} \neg\varphi$ iff $\Gamma \cup \{\varphi\} \vdash_{\mathbf{S}} \psi$ for all formulas φ, ψ; *cf.* (Wójcicki, 1988, Lemma 2.3.2). In general, $\sim\varphi \vdash_{\mathbf{S}} \neg\varphi$ for all formulas φ, owing to properties of $\mathbf{S}^F$; the converse holds if $\mathbf{S}$ is Fregean. The connectives $\sim$ and $\neg$ are thus *strong* and *weak* respectively; *cf.* (Vakarelov, 1977, p. 109). For more on the phenomenon of strong and weak connectives in the literature, see for example (Arieli and Avron, 1996; Avron, 1991; Jung and Rivieccio, 2013; Rivieccio, 2014a,b; Spinks et al., Spinks et al., 2014; Veroff and Spinks, 2006).

Constructive logic with strong negation Recall from the literature (Rasiowa, 1974, Chapter VIII) that Nelson's constructive logic with strong negation $\mathbf{N3}$ admits both a (primitive) conditional $\to$ and a (derived) implication $\Rightarrow$, where $\varphi \Rightarrow \psi$ abbreviates $(\varphi \to \psi) \wedge (\sim\psi \to \sim\varphi)$ for all $\varphi, \psi \in \mathrm{Fm}_{\Lambda[\mathbf{N4}]}$. Moreover, $\varphi \Rightarrow \psi \vdash_{\mathbf{N3}} \varphi \to \psi$ for all $\mathbf{N4}$-formulas φ, ψ. Further, $\mathbf{N3}$ admits both a (primitive) contraposable negation $\sim$ and a (derived) negation $\neg$ such that $\Gamma \vdash_{\mathbf{N3}} \neg\varphi$ iff $\Gamma \cup \{\varphi\} \vdash_{\mathbf{N3}} \psi$ for all $\varphi, \psi \in \mathrm{Fm}_{\Lambda[\mathbf{N4}]}$; because of (Spinks and Veroff, 2008b, Theorem 1.1), the negation $\sim$ [resp. $\neg$] may be realised up to synonymity as $\varphi \Rightarrow \bot$ [resp. $\varphi \to \bot$] for a suitably defined nullary logical connective $\bot$. Again, $\sim\varphi \vdash_{\mathbf{N3}} \neg\varphi$ for all $\mathbf{N4}$-formulas φ. Of course, $\mathbf{N3}$ is non-Fregean (Humberstone, 2011, Chapter 8§23).

The deductive system $\mathbf{N3}$ is thus the exemplar of an (integral) non-Fregean logic with the uniterm deduction-detachment theorem, in much the same way as Suszko's sentential calculus with identity is the paradigm of a non-Fregean logic; *cf.* (Pigozzi, 1991, Section 2, pp. 488–489) with respect to this latter. The results of the present work are informed by similarly viewing *paraconsistent* constructive logic with strong negation canonically as a non-Fregean logic with the uniterm deduction-detachment theorem. For consonant studies of a large class of non-Fregean logics generalising classical logic, see (Spinks et al., Spinks et al., 2014; Veroff and Spinks, 2006); for an alternative perspective on the remarks of this section, see (Caret and Weber, 2015, Section 2.3).

References

Abramsky, S. (1993). Computational intepretations of linear logic, *Theoretical Computer Science 111*, 3–57.

Aglianò, P. (2001). Congruence quasi-orderability in subtractive varieties, *Journal of the Australian Mathematical Society. Series A — Pure Mathematics and Statistics 71*, 421–445.

Akama, S. (1997). Tableaux for logic programming with strong negation, In D. Galmiche (Ed.), *Automated Reasoning with Analytical Tableaux and Related Methods. International Conference, TABLEAUX'97. Pont-á-Mousson, France, May, 1997. Proceedings*, Volume 1227 of *Lecture notes in Artificial Intelligence. Subseries of Lecture Notes in Computer Science*, Berlin, pp. 31–42. Springer.

Almukdad, A. and Nelson, D. (1984). Constructible falsity and inexact predicates, *Journal of Symbolic Logic 49*, 231–233.

van Alten, C. J. (1998). *An Algebraic Study of Residuated Ordered Monoids and Logics without Exchange and Contraction*, Ph. D. thesis, University of Natal, Durban, `http://hdl.handle.net/10413/3961`, accessed 12 March 2016.

van Alten, C. J. and Raftery, J. G. (2004). Rule separation and embedding theorems for logics without weakening, *Studia Logica 76*, 241–274.

de Amo, S. and Carnielli, W. A. and Marcos, J. (2002). A logical framework for integrating inconsistent information in multiple databases, In T. Eiter and K.-D. Schewe (Eds.), *Foundations of Information and Knowledge Systems. Second international Symposium (FoIKS 2002). Salzau Castle, Germany, February 2002. Proceedings*, Volume 2284 of *Lecture Notes in Computer Science*, Berlin, pp. 67–84. Springer.

Anderson, A. R. and Belnap, N. D. (1975). *Entailment. The Logic of Relevance and Necessity*, Volume 1, Princeton: Princeton University Press.

Arieli, O. (1999). *Multiple-valued Logics for Reasoning with Uncertainty*, Ph. D. thesis, Tel-Aviv University, `http://www2.mta.ac.il/~oarieli/Papers/phd.pdf`, accessed 12 March 2016.

Arieli, O. and Avron, A. (1994). Logical bilattices and inconsistent data, In *Ninth Annual IEEE Symposium on Logic in Computer Science (LICS'94). Proceedings*, Los Alamitos, pp. 468–476. IEEE Computer Society Press.

Arieli, O. and Avron, A. (1996). Reasoning with logical bilattices, *Journal of Logic, Language and Information 5*, 25–63.

Arieli, O. and Avron, A. (2015). Three-valued paraconsistent propositional logics, In J.-Y. Beziau, M. Chakraborty, and S. Dutta (Eds.), *New Directions in Paraconsistent Logic. 5th WCP, Kolkata, India, February 2014*, Volume 152 of *Springer Proceedings in Mathematics & Statistics*, New Delhi, pp. 91–129. Springer.

Arieli, O. and Avron, A. and Zamansky, A. (2010). Maximally paraconsistent three-valued logics, In F. Lin, U. Sattler, and M. Truszczyński (Eds.), *Twelfth International Conference on the Principles of Knowledge*

Representation and Reasoning (KR 2010). Toronto, Canada, May 2010. Proceedings, Massachusetts, pp. 310–318. AAAI Press.

Avron, A. (1986). On an implication connective of RM, *Notre Dame Journal of Formal Logic 27*, 201–209.

Avron, A. (1988). The semantics and proof theory of linear logic, *Theoretical Computer Science 57*, 161–184.

Avron, A. (1991). Natural 3-valued logics — characterization and proof theory, *Journal of Symbolic Logic 56*, 276–294.

Avron, A. (1999). On the expressive power of three-valued and four-valued languages, *Journal of Logic and Computation 9*, 977–994.

Avron, A. (2016). **RM** and its nice properties, In K. Bimbó (Ed.), *J. Michael Dunn on Information Based Logics*, Volume 8 of *Outstanding Contributions to Logic*, pp. 15–43. Switzerland: Springer International Publishing.

Baldwin, J. T. and Berman, J. (1975). The number of subdirectly irreducible algebras in a variety, *Algebra Universalis 5*, 379–389.

Barbour, G. D. and Raftery, J. G. (2003). Quasivarieties of logic, regularity conditions and parameterized algebraization, *Studia Logica 74*, 99–152.

Batens, D. (1980). Paraconsistent extensional propositional logics, *Logique et Analyse. Nouvelle Série 90–91*, 195–234.

Belnap, N. D. (1976). How a computer should think, In G. Ryle (Ed.), *Contemporary Aspects of Philosophy. Papers of the Oxford international Symposium of 29 September–4 October 1975, held in Christ Church College*, pp. 30–55. Stocksfield: Oriel Press.

Belnap, N. D. (1977). A useful four-valued logic, In J. M. Dunn and G. Epstein (Eds.), *Modern Uses of Multiple-Valued Logics. Invited papers from the fifth International Symposium on Multiple-Valued Logic held at Indiana University, Bloomington, Indiana, May 13–16, 1975, with a bibliography of many-valued logic by Robert. G. Wolf*, Volume 2 of *Episteme. A Series in the Foundational, Methodological, Philosophical, Psychological, Sociological and Political Aspects of the Sciences, Pure and Applied*, pp. 8–37. Dordrecht: D. Riedel Publishing Company.

Białynicki-Birula, A. and Rasiowa, H. (1958). On constructible falsity in the constructive logic with strong negation, *Colloquium Mathematicum 6*, 287–310.

Bianchi, M. (2011). *On some Axiomatic Extensions of the Monoidal T-norm Based Logic MTL: an Analysis in the Propositional and in the First-order Case*, Milan: Ledizioni LediPublishing.

Blok, W. J. and Dziobiak, W. (1986). On the lattice of quasivarieties of Sugihara algebras, *Studia Logica 45*, 275–280.

Blok, W. J. and Köhler, P. and Pigozzi, D. (1984). On the structure of varieties with equationally definable principal congruences II, *Algebra Universalis 18*, 334–379.

Blok, W. J. and Pigozzi, D. (1982). On the structure of varieties with equationally definable principal congruences I, *Algebra Universalis 15*, 195–227.

Blok, W. J. and Pigozzi, D. (1989). Algebraizable logics, *Memoirs of the American Mathematical Society* 77(396).

Blok, W. J. and Pigozzi, D. (1994a). On the structure of varieties with equationally definable principal congruences III, *Algebra Universalis 32*, 545–608.

Blok, W. J. and Pigozzi, D. (1994b). On the structure of varieties with equationally definable principal congruences IV, *Algebra Universalis 31*, 1–35.

Blok, W. J. and Pigozzi, D. (2001). Abstract algebraic logic and the deduction theorem, Manuscript, `http://orion.math.iastate.edu/dpigozzi/papers/aaldedth.pdf`, accessed 12 March 2016.

Blok, W. J. and Raftery, J. G. (2004). Fragments of R-Mingle, *Studia Logica 78*, 59–106.

Blok, W. J. and Raftery, J. G. (2008). Assertionally equivalent quasivarieties, *International Journal of Algebra and Computation 18*, 589–681.

Bloom, S. L. and Suszko, R. (1972). Investigations into the sentential calculus with identity, *Notre Dame Journal of Formal Logic 13*, 289–308.

Bou, F. and Rivieccio, U. (2011). The logic of distributive bilattices, *Logic Journal of the IGPL. Interest Group in Pure and Applied Logic 19*, 183–216.

Bou, F. and Rivieccio, U. (2013). Bilattices with implications, *Studia Logica 101*, 651–675.

Brady, R. T. (1982). Completeness proofs for the systems RM3 and BN4, *Logique et Analyse. Nouvelle Série 25*, 9–32.

Brady, R. T. (1990). The Gentzenization and decidability of RW, *Journal of Philosophical Logic 19*, 35–73.

Brady, R. T. (1991). Gentzenization and decidability of some contraction-less relevant logics, *Journal of Philosophical Logic 20*, 97–117.

Brady, R. T. (1996a). Gentzenization of relevant logics without distribution. I, *Journal of Symbolic Logic 61*, 353–378.

Brady, R. T. (1996b). Gentzenization of relevant logics without distribution. II, *Journal of Symbolic Logic 61*, 379–401.

Brignole, D. (1969). Equational characterisation of Nelson algebra, *Notre Dame Journal of Formal Logic 10*, 285–297.

Burris, S. (1992). Discriminator varieties and symbolic computation, *Journal of Symbolic Computation 13*, 175–207.

Burris, S. and Sankappanavar, H. P. (1981). *A Course in Universal Algebra*, Volume 78 of *Graduate Texts in Mathematics*, New York: Springer-Verlag.

Burris, S. N. (1998). *Logic for Mathematics and Computer Science*, New Jersey: Prentice-Hall.

Busaniche, M. and Cignoli, R. (2010). Constructive logic with strong negation as a substructural logic, *Journal of Logic and Computation 20*, 761–793.

Caleiro, C. and Gonçalves, R. (2005). Equipollent logical systems, In J.-Y. Beziau (Ed.), *Logica Universalis: Towards a General Theory of Logic*, pp. 99–111. Basel: Birkhäuser Verlag.

Caret, C. and Weber, Z. (2015). A note on contraction-free logic for validity, *Topoi. An international review of philosophy 34*, 63–74.

Carnielli, W. and Coniglio, M. E. and Marcos, J. (2007). Logics of formal inconsistency, In D. M. Gabbay and F. Guenther (Eds.), *The Handbook of Philosophical Logic* (2nd ed.), Volume 14, pp. 1–93. Dordrecht: Springer.

Carnielli, W. and Marcos, J. (2002). A taxonomy of C-systems, In W. Carnielli, M. E. Coniglio, and I. M. L. D'Ottaviano (Eds.), *Paraconsistency. The Logical Way to the Inconsistent. Second World Congress on Paraconsistency (WCP 2000). São Paulo, Brazil, May 2000. Proceedings*, Volume 228 of *Lecture Notes in Pure and Applied Mathematics*, New York, pp. 1–94. Marcel Dekker, Inc.

Carnielli, W. A. (2002). How to build your own paraconsistent logic: An introduction to the logics of formal (in)consistency, In J. Marcos, D. Batens, and W. A. Carnielli (Eds.), *Workshop on Paraconsistent Logic (WoPaLo). Held in Trento, Italy, as part of the 14th European Summer School on Logic, Language, and Information (ESSLLI'02), 5–9 August 2002. Proceedings, CLE e-Prints, vol. 2*, pp. 58–72.

Carnielli, W. A. and Marcos, J. and de Amo, S. (2000). Formal inconsistency and evolutionary databases, *Logic and Logical Philosophy 8*, 115–152.

Cheng, G. and Wang, G. J. (1999). R_0-algebras and their basic structure, *Acta Mathematica Scientia. Series A. Shuxue Wuli Xuebao. Chinese Edition 19*, 584–588, (Chinese).

da Costa, N. C. A. (1974). Theory of inconsistent formal systems, *Notre Dame Journal of Formal Logic 15*, 497–510.

Czelakowski, J. (2001). *Protoalgebraic Logics*, Volume 10 of *Trends in Logic. Studia Logica Library*, Dordrecht: Kluwer Academic Publishers.

Czelakowski, J. and Pigozzi, D. (2004). Fregean logics, *Annals of Pure and Applied Logic 127*, 17–76.

De, M. and Omori, H. (2015). Classical negation and expansions of Belnap-Dunn logic, *Studia Logica 103*, 825–851.

D'Ottaviano, I. M. L. and da Costa, N. C. A. (1970). Sur un problème de Jaśkowski, *Comptes Rendus de l'Académie des Sciences Paris. Serie A (Sciences mathématiques) 270A*, 1349–1353.

Došen, K. (1993). A historical introduction to substructural logics, In K. Došen and P. Schroeder-Heister (Eds.), *Substructural Logics*, Volume 2 of *Studies in Logic and Computation*, pp. 1–30. Oxford: Clarendon Press.

Dunn, J. M. (1970). Algebraic completeness results for R-mingle and its extensions, *Journal of Symbolic Logic 35*, 1–13.

Dunn, J. M. (1976). Intuitive semantics for first-degree entailments and 'coupled trees', *Philosophical Studies. An international journal of philosophy in the analytic tradition. 29*, 149–168.

Eiter, T. and Leone, N. and Pearce, D. (1999). Assumption sets for extended logic programs, In J. Gerbrandy, M. Marx, M. de Rijke, and Y. Venema (Eds.), *JFAK. Essays dedicated to Johan van Benthem on the occasion of his 50th birthday*, pp. 167–186. Amsterdam: Vossiuspers, Amsterdam University Press, `http://www.illc.uva.nl/j50/contribs/pearce/pearce.pdf`, accessed 12 March 2016.

Epstein, R. L. (1995). *The Semantic Foundations of Logic. Volume I. Propositional Logics*, Volume 35 of *Nijhoff International Philosophy Series*, Dordrecht: Kluwer Academic Publishers.

Esteva, F. and Godo, L. (2001). Monoidal t-norm based logic: towards a logic for left continuous t-norms, *Fuzzy Sets and Systems. An international journal in Information Science and Engineering 124*, 271–288.

Fitting, M. (1989). Bilattices and the theory of truth, *Journal of Philosophical Logic 18*, 225–256.

Fitting, M. (1991). Bilattices and the semantics of logic programming, *The Journal of Logic Programming 11*, 91–116.

Fitting, M. (1994). Kleene's three-valued logic and their children, *Fundamenta Informaticae 20*, 113–131.

Fodor, J. C. (1995). Contrapositive symmetry of fuzzy implications, *Fuzzy Sets and Systems. An international journal in Information Science and Engineering 69*, 142–156.

Font, J. M. (1993). On the Leibniz congruences, In C. Rauszer (Ed.), *Algebraic Methods in Logic and Computer Science*, Volume 28 of *Banach Center Publications*, pp. 17–36. Warszawa: Institute of Mathematics, Polish Academy of Sciences.

Font, J. M. and Jansana, R. (2009). *A General Algebraic Semantics for Sentential Logics* (2nd ed.), Volume 7 of *Lecture Notes in Logic*, New York: Association for Symbolic Logic.

Font, J. M. and Jansana, R. and Pigozzi, D. (2003). A survey of abstract algebraic logic, *Studia Logica 74*, 13–97.

Font, J. M. and Pérez, G. R. (1992). A note on Sugihara algebras, *Publicacions Matemàtiques 36*, 591–599.

Font, J. M. and Rodríguez, G. (1990). Note on algebraic models for relevance logic, *Zeitschrift für mathematische Logik und Grundlagen der Mathematik 36*, 535–540.

Fried, E. and Grätzer, G. and Quackenbush, R. (1980). Uniform congruence schemes, *Algebra Universalis 10*, 176–188.

Fried, E. and Kiss, E. W. (1983). Connections between congruence-lattices and polynomial properties, *Algebra Universalis 17*, 227–262.

Galatos, N. and Jipsen, P. and Kowalski, T. and Ono, H. (2007). *Residuated Lattices: An Algebraic Glimpse at Substructural Logics*, Volume 151 of *Studies in Logic and the Foundations of Mathematics*, Amsterdam: Elsevier.

Gelfond, M. (2002). Representing knowledge in A-Prolog, In A. C. Kakas and F. Sadri (Eds.), *Computational Logic: Logic Programming and Beyond. Essays in Honour of Robert A. Kowalski. Part II*, Volume 2408 of *Lecture notes in Artificial Intelligence. Subseries of Lecture Notes in Computer Science*, Berlin, pp. 413–451.

Ginsberg, M. (1990). Bilattices and modal operators, *Journal of Logic and Computation 1*, 41–69.

Girard, J.-Y. (1987). Linear logic, *Theoretical Computer Science 50*, 1–102.

Girard, J.-Y. (1995). Linear logic: its syntax and semantics, In J.-Y. Girard, Y. Lafont, and L. Regnier (Eds.), *Advances in Linear Logic*, Volume 222 of *London Mathematical Society Lecture Note Series*, pp. 1–42. Cambridge: Cambridge University Press.

Gispert, J. (2003). Axiomatic extensions of the nilpotent minimum logic, *Reports on Mathematical Logic 37*, 113–123.

Grätzer, G. (2008). *Universal Algebra* (2nd ed.), New York: Springer.

Gyuris, V. (1999). *Variations of Algebraizability*, Ph. D. thesis, The University of Illinois at Chicago.

Hart, J. B. and Rafter, L. and Tsinakis, C. (2002). The structure of commutative residuated lattices, *International Journal of Algebra and Computation 12*, 509–524.

Hsieh, A. (2008). *Embedding Theorems and Finiteness Properties for Residuated Structures and Substructural Logics*, Ph. D. thesis, University of KwaZulu-Natal, Durban, South Africa, `http://hdl.handle.net/10413/446`, accessed 12 March 2016.

Hsieh, A. and Raftery, J. G. (2006). A finite model property for $\mathbf{RMI}_{\min}$, *Mathematical Logic Quarterly 52*, 602–612.

Hsieh, A. and Raftery, J. G. (2007). Conserving involution in residuated structures, *Mathematical Logic Quarterly 53*, 583–609.

Humberstone, L. (2005). Logical discrimination, In J.-Y. Béziau (Ed.), *Logica Universalis : Towards a General Theory of Logic*, pp. 207–228. Basel: Birkhäuser Verlag.

Humberstone, L. (2011). *The Connectives*, Cambridge (Massachusetts): The MIT Press.

Humberstone, L. (2015). Béziau on *And* and *Or*, In A. Koslow and A. Buchsbaum (Eds.), *The Road to Universal Logic : Festschrift for 50th Birthday of Jean-Yves Béziau. Volume I*, Studies in Universal Logic, pp. 283–307. Basel: Birkhäuser.

Idziak, P. M. and Słomczyńska, K. and Wroński, A. (2009). Fregean varieties, *International Journal of Algebra and Computation 19*, 595–645.

Ishii, T. (2000). *Nonclassical Logics with Identity Connective and their Algebraic Characterization*, Ph. D. thesis, Japan Advanced Institute of Science and Technology, `http://hdl.handle.net/10119/898`, accessed 12 March 2016.

Jansana, R. and Rivieccio, U. (2013). Priestley duality for N4-lattices, In *Eighth Conference of the European Society for Fuzzy Logic and Technology (EUSFLAT 2013). Proceedings*, Volume 32 of *Advances in Intelligent Systems Research (AISR)*, Amsterdam, pp. 223–229. Atlantis Press.

Järvinen, J. and Pagliani, P. and Radeleczki, S. (2013). Information completeness in Nelson algebras of rough sets induced by quasiorders, *Studia Logica 101*, 1073–1092.

Järvinen, J. and Radeleczki, S. (2011a). Erratum to: Representation of Nelson algebras by rough sets determined by quasiorders, *Algebra Universalis 66*, 181.

Järvinen, J. and Radeleczki, S. (2011b). Representation of Nelson algebras by rough sets determined by quasiorders, *Algebra Universalis 66*, 163–179.

Järvinen, J. and Radeleczki, S. (2014). Monteiro spaces and rough sets determined by quasiorder relations: Models for Nelson algebras, *Fundamenta Informaticae 131*, 205–215.

Jipsen, P. and Tsinakis, C. (2002). A survey of residuated lattices, In J. Martínez (Ed.), *Ordered Algebraic Structures. Proceedings of the Gainesville Conference. Sponsored by the University of Florida. 28th February–3rd March, 2001*, Volume 7 of *Developments in Mathematics*, Dordrecht, pp. 19–56. Kluwer Academic Publishers.

Johansson, I. (1936). Der minimalkalkül, ein reduzierter intuitionistischer formalismus, *Compositio Mathematica 4*, 119–136.

Jónsson, B. (1995). Congruence distributive varieties, *Mathematica Japonica 42*, 353–401.

Jung, A. and Rivieccio, U. (2013). Kripke semantics for modal bilattice logics (extended abstract), In *Twenty-eighth Annual ACM/IEEE Symposium on Logic in Computer Science (LICS'13). Proceedings*, Los Alamitos, pp. 438–447. IEEE Computer Society Press.

Kachi, D. (2002). Validity in simple partial logic, *Annals of the Japan Association for Philosophy of Science 10*, 139–153.

Kachi, D. (2007). Partial logic as a logic of extensional alethic modality, *Journal of the Japan Association for Philosophy of Science 34*, 13–22, (Japanese).

Kamide, N. (2016). A decidable paraconsistent relevant logic: Gentzen system and Routley-Meyer semantics, *Mathematical Logic Quarterly 62*, 177–189.

Kamide, N. and Wansing, H. (2012). Proof theory of Nelson's paraconsistent logic: A uniform perspective, *Theoretical Computer Science 415*, 1–38.

Kapsner, A. (2014). *Logic and Falsifications. A New Perspective on Constructivist Semantics*, Volume 40 of *Trends in Logic. Studia Logica Library*, Basel: Springer.

Köhler, P. and Pigozzi, D. (1980). Varieties with equationally definable principal congruences, *Algebra Universalis 11*, 213–219.

Kracht, M. (1998). On extensions of intermediate logics by strong negation, *Journal of Philosophical Logic 27*, 49–73.

Łukasiewicz, J. (1970a). On three-valued logic, In L. Borkowski (Ed.), *Selected Works of Jan Łukasiewicz*, Studies in Logic and the Foundations of Mathematics, pp. 87–88. Warsaw: North-Holland Publishing Company.

Łukasiewicz, J. (1970b). Philosophical remarks on many-valued systems of propositional logic, In L. Borkowski (Ed.), *Selected Works of Jan Łukasiewicz*, Studies in Logic and the Foundation of Mathematics, pp. 153–178. Warsaw: North-Holland Publishing Company.

Malinowski, G. (1993). *Many-Valued Logics*, Volume 25 of *Oxford Logic Guides*, Oxford: Clarendon Press.

Marcos, J. (2005). On a problem of da Costa, In G. Sica (Ed.), *Essays on the Foundations of Mathematics and Logic*, Volume 2, pp. 53–69. Monza: Polimetrica.

McCune, W., Prover 9, `http://www.cs.unm.edu/~mccune/prover9/`, accessed 12 March 2016.

McCune, W. and Padmanabhan, R. (1996). *Automated Deduction in Equational Logic and Cubic Curves*, Volume 1095 of *Lecture notes in Artificial Intelligence. Subseries of Lecture Notes in Computer Science*, Berlin: Springer-Verlag.

McKenzie, R. N. and McNulty, G. F. and Taylor, W. F. (1987). *Algebras, Lattices, Varieties. Volume I*, The Wadsworth & Brooks/Cole Mathematics Series. Monterey: Wadsworth & Brooks/Cole Advanced Books & Software.

McNulty, G. (1992). A field guide to equational logic, *Journal of Symbolic Computation 14*, 371–397.

Méndez, J. M. and Robles, G. (2016). Strengthening Brady's paraconsistent 4-valued logic BN4 with truth-functional modal operators, *Journal of Logic, Language and Information 25*, 163–189.

Meyer, R. K. (2004). Ternary relations and relevant semantics, *Annals of Pure and Applied Logic 127*, 195–217.

Meyer, R. K. and Giambrone, S. and Brady, R. T. (1984). Where Gamma fails, *Studia Logica 43*, 247–256.

Monteiro, A. (1963). Algèbres de Nelson semi-simples. Résumé d'une communication présentée à la U.M.A. en octobre 1962, *Revista de la Unión Mathemática Argentina 21*, 145–146.

Monteiro, A. (1995). Les algèbres de Nelson semi-simples, Technical Report 50, Universidad Nacional del Sur, Bahía Blanca, Also available in *Unpublished papers I*, Notas de Lógica Matemática 40, Instituto de Matemática, Universidad Nacional del Sur, Bahía Blanca (1996).

Moore, R. (1985). Semantical considerations on nonmonotonic logic, *Artificial Intelligence. An International Journal 25*, 75–94.

Muskens, R. (1995). *Meaning and Partiality*, Studies in Logic, Language, and Information. Stanford: CSLI Publications and FoLLI.

Nelson, D. (1949). Constructible falsity, *Journal of Symbolic Logic 14*, 16–26.

Nemitz, W. (1965). Implicative semilattices, *Transactions of the American Mathematical Society 117*, 128–142.

Noguera, C. (2007). *Algebraic Study of Axiomatic Extensions of Triangular Norm Based Fuzzy Logics*, Number 27 in Monografies de l'Institut d'Investigació en Intel·ligència Artificial. Bellaterra: Institut d'Investigació en Intel·ligència Artificial.

Noguera, C. and Esteva, F. and Gispert, J. (2008). On triangular norm based axiomatic extensions of the weak nilpotent minimum logic, *Mathematical Logic Quarterly 54*, 403–425.

Odintsov, S. P. (2003). Algebraic semantics for paraconsistent Nelson's logic, *Journal of Logic and Computation 13*, 453–468.

Odintsov, S. P. (2004). On the representation of N4-lattices, *Studia Logica 76*, 385–405.

Odintsov, S. P. (2005). The class of extensions of Nelson's paraconsistent logic, *Studia Logica 80*, 291–320.

Odintsov, S. P. (2007). On extensions of Nelson's logic satisfying Dummett's axiom, *Siberian Mathematical Journal 48*, 232–247.

Odintsov, S. P. (2008). *Constructive Negations and Paraconsistency*, Volume 26 of *Trends in Logic. Studia Logica Library*, Berlin: Springer.

Odintsov, S. P. and Speranski, S. O. (2016). The lattice of Belnapian modal logics: Special extensions and counterparts, *Logic and Logical Philosophy 25*, 3–33.

Omori, H. and Sano, K. (2014). da Costa meets Belnap and Nelson, In R. Ciuni, H. Wansing, and C. Willkommen (Eds.), *Recent Trends in Philosophical Logic*, Volume 41 of *Trends in Logic. Studia Logica Library*, pp. 145–166. Heidelberg: Springer.

Omori, H. and Waragai, T. (2011). Some observations on the systems LFI and LFI*, In F. Morvan, A. M. Tjoa, and R. R. Wagner (Eds.), *Twenty-second International Workshop on Database and Expert System Applications (DEXA 2011). Toulouse, France, 29 August–2 September 2011. Proceedings*, Los Alamitos, pp. 320–324. IEEE Computer Society Press.

Parks, R. Z. (1972). A note on **R**-mingle and Sobociński's three-valued logic, *Notre Dame Journal of Formal Logic 13*, 227–228.

Pawlak, Z. (1982). Rough sets, *International Journal of Computer and Information Sciences 11*, 341–356.

Pearce, D. (1999). From here to there: Stable negation in logic programming, In D. M. Gabbay and H. Wansing (Eds.), *What is Negation?*, Volume 13 of *Applied Logic Series*, pp. 161–181. Dordrecht: Kluwer Academic Publishers.

Pei, D. (2003). On equivalent forms of fuzzy logic systems NM and IMTL, *Fuzzy Sets and Systems. An international journal in Information Science and Engineering 138*, 187–195.

Pei, D. and Wang, G. (2002). The completeness and applications of the formal system $\mathcal{L}^*$, *Science in China. Series F: Information Science 45*, 40–50.

Phillips, J. D. and Stanovský, D. (2010). Automated theorem proving in quasigroup and loop theory, *AI Communications 23*, 267–283.

Pigozzi, D. (1975). Equational logic and equational theories of algebras, Technical report CSD-135, Computer Science Department, Purdue University, Indiana, `http://docs.lib.purdue.edu/cstech/85`, accessed 12 March 2016.

Pigozzi, D. (1991). Fregean algebraic logic, In H. Andréka, J. D. Monk, and I. Németi (Eds.), *Algebraic Logic*, Volume 54 of *Colloquia Mathematica Societatis János Bolyai Budapest (Hungary)*, pp. 473–502. New York: North-Holland Publishing Company.

Priest, G. (1989). Reasoning about truth, *Artificial Intelligence. An International Journal 39*, 231–244.

Pynko, A. P. (1999). Definitional equivalence and algebraizability of generalised logical systems, *Annals of Pure and Applied Logic 98*, 1–68.

Raftery, J. G. (2006). The equational definability of truth predicates, *Reports on Mathematical Logic 41*, 95–149.

Rasiowa, H. (1958). 𝔑-lattices and constructive logic with strong negation, *Fundamenta Mathematicae 46*, 61–80.

Rasiowa, H. (1959). Algebraische Charakterisierung der intuitionistischen Logik mit starker Negation, In A. Heyting (Ed.), *Constructivity in Mathematics. Proceedings of the Colloquium held at Amsterdam, 1957*, Studies in Logic and the Foundations of Mathematics, Amsterdam, pp. 234–240. North-Holland Publishing Company.

Rasiowa, H. (1974). *An Algebraic Approach to Non-Classical Logics*, Volume 78 of *Studies in Logic and the Foundations of Mathematics*, Amsterdam: North-Holland Publishing Company.

Restall, G. (1993). How to be *really* contraction free, *Studia Logica 52*, 381–391.

Restall, G. (2000). *An Introduction to Substructual Logics*, London: Routledge.

Restall, G. (2005). Łukasiewicz, supervaluations, and the future, *L&PS — Logic and Philosophy of Science 3*, 1–10.

Restall, G. (2006). Relevant and substructural logics, In D. M. Gabbay and J. Woods (Eds.), *Logic and the Modalities in the Twentieth Century*, Volume 7 of *Handbook of the History of Logic*, pp. 289–398. Amsterdam: Elsevier.

Rivieccio, U. (2010). *An Algbraic Study of Bilattice-based Logics*, Ph. D. thesis, University of Barcelona, `arXiv:1010.2552 [math.LO]`, accessed 12 March 2016.

Rivieccio, U. (2011). Paraconsistent modal logics, *Electronic Notes in Theoretical Computer Science 278*, 173–186.

Rivieccio, U. (2014a). Algebraic semantics for bilattice public announcement logic, In A. Indrzejczak, J. Kaczmarek, and M. Zawidzki (Eds.), *Trends in Logic XIII. Łódź, Poland, July 2014. Proceedings*, Łódź, pp. 199–215. Łódź University Press.

Rivieccio, U. (2014b). Bilattice public announcement logic, In R. Goré, B. Kooi, and A. Kurucz (Eds.), *Advances in Modal Logic. Proceedings*, Volume 10, London, pp. 459–477. College Publications.

Rivieccio, U. and Bou, F. and Jansana, R. (2011). Varieties of interlaced bilattices, *Algebra Universalis 66*, 115–141.

Routley, R. (1974). Semantical analyses of propositional systems of Fitch and Nelson, *Studia Logica 33*, 283–298.

Sano, K. and Omori, H. (2014). An expansion of first-order Belnap-Dunn logic, *Logic Journal of the IGPL. Interest Group in Pure and Applied Logic 22*, 458–481.

Schütte, K. (1960). *Beweistheorie*, Berlin: Springer-Verlag.

Sendlewski, A. (1984). Some investigations of varieties of $\mathcal{N}$-lattices, *Studia Logica 43*, 257–280.

Sendlewski, A. (1990). Nelson algebras through Heyting ones: I, *Studia Logica 49*, 105–126.

Slaney, J. (1984). A metacompleteness theorem for contraction-free relevant logics, *Studia Logica 43*, 159–168.

Slaney, J. (1991). The implications of paraconsistency, In J. Mylopoulos and R. Reiter (Eds.), *Twelfth International Joint Conference on Artificial Intelligence (IJCAI-91). Sydney, Australia, August 1991. Proceedings*, San Francisco, pp. 1052–1057. Morgan Kaufmann.

Slaney, J. (2004). Relevant logic and paraconsistency, In L. Bertossi, A. Hunter, and T. Schaub (Eds.), *Inconsistency Tolerance*, Volume 3300 of *Lecture Notes in Computer Science*, pp. 270–293. Berlin: Springer-Verlag.

Slaney, J. (2010). A logic for vagueness, *Australasian Journal of Logic 8*, 100–134.

Slaney, J. and Surendonk, T. and Girle, R. (1989). Time, truth and logic, Technical Report TR-ARP-11/89, Automated Reasoning Project, Australian National University, Canberra, `http://ftp.rsise.anu.edu.au/papers/slaney/TTL/TTL.ps.gz`, accessed 12 March 2016.

Smiley, T. (1962). The independence of connectives, *Journal of Symbolic Logic 27*, 426–436.

Smiley, T. (1963). Relative necessity, *Journal of Symbolic Logic 28*, 113–134.

Spinks, M. (2004). Ternary and quaternary deductive terms for Nelson algebras, *Algebra Universalis 51*, 125–136.

Spinks, M. and Bignall, R. J. and Veroff, R., Discriminator logics, In preparation.

Spinks, M. and Bignall, R. J. and Veroff, R. (2014). Discriminator logics (Research announcement), *Australasian Journal of Logic 11*, 159–171.

Spinks, M. and Veroff, R. (a). Paraconsistent constructive logic with strong negation is a contraction-free relevant logic. I. Term equivalence, In preparation.

Spinks, M. and Veroff, R. (b). Paraconsistent constructive logic with strong negation is a contraction-free relevant logic. II. Definitional equivalence, In preparation.

Spinks, M. and Veroff, R. (c). Paraconsistent constructive logic with strong negation is a contraction-free relevant logic. III. Extensions and expansions, In preparation.

Spinks, M. and Veroff, R. (2008a). Constructive logic with strong negation is a substructural logic. I, *Studia Logica 88*, 325–348.

Spinks, M. and Veroff, R. (2008b). Constructive logic with strong negation is a substructural logic. II, *Studia Logica 89*, 401–425.

Spinks, M. and Veroff, R. (2010). Slaney's logic **F**** is constructive logic with strong negation, *Bulletin of the Section of Logic 39*, 161–174.

Suszko, R. (1968). Ontology in the *Tractatus* of L. Wittgenstein, *Notre Dame Journal of Formal Logic 9*, 7–33.

Suszko, R. (1971). Identity connective and modality, *Studia Logica 27*, 9–39.

Suszko, R. (1975). Abolition of the Fregean axiom, In R. Parikh (Ed.), *Logic Colloquium. Symposium on Logic held at Boston, 1972–73*, Volume 453 of *Lecture Notes in Mathematics*, pp. 169–236. Berlin: Springer-Verlag.

Tarski, A. (1968). Equational logic and equational theories of algebras, In H. A. Schmidt, K. Schütte, and H.-J. Thiele (Eds.), *Contributions to mathematical logic. Proceedings of the Logic Colloquium, Hannover 1966*, Volume 50 of *Studies in Logic and the Foundations of Mathematics*, pp. 275–288. Amsterdam: North-Holland Publishing Company.

Taylor, W. (Survey 1979). Equational logic, *Houston Journal of Mathematics*.

Tokarz, M. (1975). Functions definable in Sugihara algebras and their fragments (I), *Studia Logica 34*, 295–304.

Troelstra, A. S. (1992). *Lectures on linear logic*, Number 29 in CSLI Lecture Notes. Stanford: Center for the Study of Language and Information.

Vakarelov, D. (1977). Notes on $\mathcal{N}$-lattices and constructive logic with strong negation, *Studia Logica 36*, 109–125.

Vakarelov, D. (2006). Non-classical negation in the works of Helena Rasiowa and their impact on the theory of negation, *Studia Logica 84*, 105–127.

Veroff, R. and Spinks, M. (2006). Axiomatizing the skew Boolean propositional calculus, *Journal of Automated Reasoning 37*, 3–20.

Viglizzo, I. D. (1999). Álgebras de Nelson (Nelson Aglebras), Master's thesis, Universidad Nacional del Sur, Bahía Blanca, `http://sites.google.com/site/viglizzo/investigacion/viglizzo99nelson`, accessed 12 March 2016.

Villadsen, J. (2001). Combinators for paraconsistent attitudes, In P. de Groote, G. Morrill, and C. Retoré (Eds.), *Logical Aspects of Computational Linguistics. 4th International Conference, LACL 2001. Le Croisic, France, June 27–29, 2001. Proceedings*, Volume 2099 of *Lecture notes in Artificial Intelligence. Subseries of Lecture Notes in Computer Science*, pp. 91–102.

Wang, G. J. (1997). A formal deductive system for fuzzy propositional calculus, *Chinese Science Bulletin 42*, 1041–1045.

Wang, G. J. (1999). On the logic foundation of fuzzy reasoning, *Information Sciences 117*, 47–88.

Wang, G. J. (2000). *Non-classical Mathematical Logic and Approximate Reasoning*, Beijing: Science Press, (Chinese).

Wang, G. J. (2002). MV-algebras, BL-algebras, R_0-algebras, and multiple-valued logic, *Fuzzy Systems and Mathematics 15*, 1–15, (Chinese).

Wang, G. J. and Wang, H. (2001). Non-fuzzy versions of fuzzy reasoning in classical logics, *Information Sciences 138*, 211–236.

Wansing, H. (1993). *The Logic of Information Structures*, Volume 681 of *Lecture notes in Artificial Intelligence. Subseries of Lecture Notes in Computer Science*, Berlin: Springer-Verlag.

Werner, H. (1970). Eine charakterisierung funktional vollständiger algebren, *Archiv der Mathematik 21*, 381–385.

Werner, H. (1978). *Discriminator-Algebras*, Number 6 in Studien zur Algebra und ihre Anwendungen. Berlin: Akademie-Verlag.

Wójcicki, R. (1988). *Theory of Logical Calculi. Basic Theory of Consequence Operations*, Volume 199 of *Synthese Library. Studies in Epistemology, Logic, Methodology, and Philosophy of Science,* Dordrecht: Kluwer Academic Publishers.

Zhou, X. and Li, Q. (2010). Boolean products of R_0-algebras, *Mathematical Logic Quarterly 56*, 289–298.

Possible classification of finite-dimensional compact Hausdorff topological algebras

Walter Taylor

With gratitude to Professor Don Pigozzi

Abstract The time may be right to ask for an enumeration or classification of all the reasonably small topological algebras. Here the terrain is surveyed, and a program of investigation is proposed.

2010 Mathematics Subject Classification: 22A30

Introduction.

This paper is part of a continuing investigation—see the author's papers (Taylor, 1986), (Taylor, 2000), (Taylor, 2006), (Taylor, 2010), and (Taylor, 2011)—into the compatibility relation, which is described in (2) below. A. D. Wallace defined the inquiry succinctly in 1955, when he asked (Wallace, 1955, p. 96), "Which spaces admit what structures?" By "structure," he meant the existence of continuous operations identically satisfying certain equations: e.g., the structure of a topological group or a topological lattice, and so on. Here we survey the current state of knowledge in this area, especially for finite simplicial complexes, and ask some refined versions of Wallace's question.

Walter Taylor
Mathematics Department, University of Colorado, Boulder, Colorado 80309–0395, USA,
e-mail: `walter.taylor@colorado.edu`

© Springer International Publishing AG 2018
J. Czelakowski (eds.), *Don Pigozzi on Abstract Algebraic Logic,
Universal Algebra, and Computer Science*, Outstanding Contributions
to Logic 16, https://doi.org/10.1007/978-3-319-74772-9_14

0.1 Role of this investigation in mathematics.

We see such topological structure as fundamental to mathematics. Generally, there seem to be two ways to put an infinite number system on a firm logical and practical foundation. The first is through recursion, the method that underlies calculations with integers and rational numbers (for example). The second is through continuity of operations, as in the real number system $\mathbb{R}$, where we can meaningfully calculate values like $\sin(\pi/3)$ or $\sqrt[3]{2}$ by approximation. Here we are looking into the possibilities of calculation through continuity.

For these two modalities to be available in any practical way, we must, at the very least, be talking about a topological space that has a countable dense subset—the first axiom of countability. Thus, for example, discrete topological spaces are compatible with any consistent set of equations, but a discrete topological algebra is of no use in pursuing calculations through continuity. Discrete spaces play no role in the rest of this paper.

0.2 Limited focus of this investigation.

The investigation emphasizes topological algebras that satisfy some non-trivial (in the sense of §6.1) equations. We do not wish to diminish the importance of other algebras—for instance some topological semigroups are of paramount importance, and yet the associative law remains trivial in the sense of §6.1. But the main unknown, under the present focus, is the identical satisfaction of equations.

In keeping with the ideas of §0.1, we limit our attention to first-countable spaces. In fact we mostly limit ourselves to very simple spaces: finite simplicial complexes. First of all, much of the variation and mystery of the subject already lies in this seemingly elementary domain. Secondly, as soon as we admit infinite polyhedra, any consistent set of equations can be modeled (see §5 below).

Most of our attention will be to connected simplicial complexes. Every finite model of an equation-set Σ may be viewed as a topological model of Σ based on a zero-dimensional complex, which is of course disconnected. These models, and any other disconnected topological models, are generally not under consideration. This exclusion is most important, and will be re-iterated, in the open questions of §9.

0.3 Layout of the paper.

The beginning sections lay out concepts relevant to our results and problems. The reader who is proficient in these notions may proceed to §7, and examine the examples shown there, which are the heart of the paper. After that, one might read the comments and questions that arise in §§8–9. These sections comprise the main new material of this paper.

0.4 Acknowledgments.

I thank George M. Bergman, who read the manuscript very closely, and made many helpful suggestions. I want also to acknowledge and thank Don Pigozzi for his long-term encouragement of myself and others in the study of equational logic, the central focus of this article. This is a field where he clarified many issues; see for instance his articles with W. J. Blok (Blok and Pigozzi, 1988) and with K. Palasińska (Palasińska and Pigozzi, 1995).

1 Satisfaction of equations by operations.

Readers with some familiarity with logic or general algebra can easily skip §1, at least on a first reading.

1.1 Terms and equations.

A *similarity type* consists of a set T and a function $t \longmapsto n_t$ from T to natural numbers. A *term of type* $\langle n_t : t \in T \rangle$ is recursively either a variable or a formal expression of the form $F_t(\tau_1, \ldots, \tau_{n_t})$ for some $t \in T$ and some shorter terms τ_i of this type. An **equation** of this type is a formal expression $\tau \approx \sigma$ for terms τ and σ of this type. A formal equation makes no assertion, but merely presents two terms for consideration. The actual mathematical assertion of equality is made (in a given context) by the satisfaction relation $\models$ (see (1) below). We mostly work with a set Σ of equations, finite or infinite, and tacitly assume that there is a similarity type $\langle n_t : t \in T \rangle$ such that each equation in Σ is of this type.

Examples: In almost any concrete example of interest, the foregoing formality is not really necessary for comprehension. It suffices to give, for example, the familiar assertion that "Σ has two binary operations $\wedge$ and $\vee$," instead of insisting on e.g. "$\wedge = F_1$ and $\vee = F_2$, where $T = \{1, 2\}$ and $n_1 = n_2 = 2$." In such a simplified context, formal equations may be written like ordinary equations in standard lattice theory. (One should be careful, however, about writing informal terms such as "$x \wedge y \wedge z$," which is not meaningful in the absence of associativity.)

1.2 Satisfaction of equations.

Given a set A and for each $t \in T$ a function $\overline{F_t} : A^{n(t)} \longrightarrow A$ (called an *operation*), we say that the operations $\overline{F_t}$ *satisfy* Σ and write

$$(1) \qquad\qquad (A, \overline{F}_t)_{t \in T} \models \Sigma,$$

iff for each equation $\sigma \approx \tau$ in Σ, both σ and τ evaluate to the same function when the operations $\overline{F_t}$ are substituted for the symbols F_t appearing in σ and τ.

A structure of the form $(A, \overline{F}_t)_{t \in T}$ (as in (1)) is called an *algebra*; the set A is called the *universe* of $(A, \overline{F}_t)_{t \in T}$. Often, if the context permits, we denote $(A, \overline{F}_t)_{t \in T}$ by the bold letter corresponding to the letter denoting the universe, and so on. Then we can express (1) by saying that the algebra **A** *satisfies* (or *models*) Σ.

In discussing satisfaction of equations, it is standard (and helpful) to distinguish as we have done between an operation symbol F_t and an operation $\overline{F}_t$ interpreting the symbol.[1] Nevertheless in keeping with the last part of §1.1 above, we may sometimes omit the bar from familiar operations like $+$, $\wedge$ and so on.

2 Compatibility of a space with a set of equations.

Given a *topological space* A and a set of equations Σ, we write[2]

$$(2) \qquad\qquad A \models^{\mathrm{ctn}} \Sigma,$$

and say that A and Σ are *compatible*, iff there exist *continuous* operations $\overline{F}_t$ on A satisfying Σ, in other words iff (1) holds with continuous operations $\overline{F}_t$. (Here we mean that each function $\overline{F}_t : A^{n_t} \longrightarrow A$ should be continuous relative to the usual product topology formed on the direct power A^{n_t}.)

---

[1] Obviously the simple notation $\overline{F}_t$ will be inadequate if more than one operation interprets F_t in a given discussion.

[2] In a context that contains little possibility of confusing (1) and (2) one may omit the designation "ctn" for continuous modeling, and simply write $A \models \Sigma$ for (2). We have done this in (Taylor, 1986), (Taylor, 2000), (Taylor, 2006), (Taylor, 2010) and (Taylor, 2011).

One may also read (2) as "A topologically models Σ," or "A continuously models Σ."

Given operations $\overline{F}_t$ on a topological space A, we may of course form the algebra $\mathbf{A} = (A; \overline{F}_t)$; if in addition each $\overline{F}_t$ is continuous, we may say that this $\mathbf{A}$ is a *topological algebra based on the space A*. With this vocabulary, the compatibility relation (2) may be rephrased as follows: *there exists a topological algebra satisfying Σ that is based on the space A.*

Thus, for instance, A is compatible with group theory if and only if A is the underlying space of some topological group. If desired, one may skip to §7 on a first reading, for a much longer list of examples.

3 General results on compatibility.

While the definitions are simple, the relation (2) remains mysterious. Two results, one fifty years old, the other recent, point toward this mystery. First, the algebraic topologists have long known that the n-dimensional sphere S^n is compatible with H-space theory ($x \cdot e \approx x \approx e \cdot x$) if and only if $n = 1, 3$ or 7. (There is a large literature on this topic; one landmark paper was Adams (Adams, 1960).) Second, for $A = \mathbb{R}$, the relation (2) is algorithmically undecidable for Σ — see (Taylor, 2006); i.e. there is no algorithm that accepts as input an arbitrary finite Σ, and outputs the truth value of (2) for $A = \mathbb{R}$. In any case, (2) appears to hold only sporadically, and with no readily discernible pattern.

The mathematical literature contains numerous but scattered further examples of the truth or falsity of specific instances of (2). The author's earlier papers (Taylor, 1986), (Taylor, 2000), (Taylor, 2006), (Taylor, 2010) collectively refer to most of what is known, and in fact many of the earlier examples illustrating incompatibility are recapitulated throughout the long article (Taylor, 2010). The present article will cover most of the known compatibilities for finite simplicial complexes.

4 Compatibility and the interpretability lattice.

Here we review a notion introduced by W. D. Neumann in 1974 (see (Neumann, 1974)), and further studied by O. C. García and W. Taylor in 1981 (see (García and Taylor, 1984)). (In 1968 J. Isbell (Isbell, 1968) constructed quasi-orderings of arbitrary categories; the ordering we use—i.e. Neumann's—can be seen as arising from Isbell's ordering.)

4.1 Interpretability as an order.

We introduce an order on the class of all sets Σ, Γ ... of equations, as follows. Let us suppose that the operation symbols of Σ are F_s ($s \in S$), and the operation symbols of Γ are G_t ($t \in T$). We say that Σ is *interpretable in* Γ, and write $\Sigma \leq \Gamma$, iff there are terms α_s ($s \in S$) in the operation symbols G_t such that, if $(A, \overline{G}_t)_{t \in T}$ is any model of Γ, then $(A, \overline{\alpha_s})_{s \in S}$ is a model of Σ.

A typical example has Γ defining Boolean algebras and Σ defining Abelian groups with operations $+, -, 0$. Here the terms α_+ and α_- are both equal to the so-called *symmetric difference* $\alpha_+(x,y) = \alpha_-(x,y) = (x \wedge (\neg y)) \vee (y \wedge (\neg x))$. (It is worthwhile noticing that this interpretation is neither one-one on the class of all BA's nor onto the class of all Abelian groups.) For further concrete examples, see §7.2.5, §7.5.4 and §8.3 below.

Strictly speaking, we need to observe that, so far, our relation $\leq$ is not anti-symmetric. It is easy to find distinct sets Σ_1 and Σ_2 that are mutually related by $\leq$. It is however a quasi-order, and when we speak of an order, or a least upper bound, and so on, we are referring to the order formed in the usual way modulo the equivalence relation that includes the pair (Σ_1, Σ_2) whenever the two Σ_i are as above, i.e. $\Sigma_1 \leq \Sigma_2 \leq \Sigma_1$. We generally will leave this fine point unexpressed.

4.2 Interpretability defines a lattice.

Given sets Σ and Γ of equations, there is a set $\Sigma \wedge \Gamma$ that is a greatest lower bound of Σ and Γ in the $\leq$-ordering of §4. For a precise definition, including an axiomatization of $\Sigma \wedge \Gamma$, the reader may consult R. McKenzie (McKenzie, 1975) or O. García and W. Taylor (García and Taylor, 1984).

We describe here the (algebraic) models of $\Sigma \wedge \Gamma$. We make the inessential assumption that the operation symbols of Σ (resp. Γ) are F_s ($s \in S$) (resp. F_t ($t \in T$)), with S disjoint from T. The operation symbols of $\Sigma \wedge \Gamma$ are F_j ($j \in S \cup T$), together with a new binary operation symbol p. The models of $\Sigma \wedge \Gamma$ are precisely all algebras isomorphic to a product $\mathbf{A} \times \mathbf{B}$, where

   (i) $\mathbf{A} \models \Sigma$.
   (ii) $\mathbf{B} \models \Gamma$.
   (iii) For each $t \in T$, $\mathbf{A} \models F_t(x_1, \ldots, x_n) \approx x_1$ and $\mathbf{A} \models p(x_1, x_2) \approx x_1$.
   (iv) For each $s \in S$, $\mathbf{B} \models F_s(x_1, \ldots, x_n) \approx x_1$ and $\mathbf{A} \models p(x_1, x_2) \approx x_2$.

For instance, to see that $\Sigma \wedge \Gamma \leq \Sigma$, we define an interpretation as follows. For $s \in S$, the term α_s is $F_s(x_1, x_2, \ldots)$; for $t \in T$, the term α_t is x_1, and $\alpha_p(x_1, x_2)$ is x_1. For any $(A, \overline{F}_s)_{s \in S}$, the interpreted algebra $(A, \overline{F}_j)_{j \in S \cup T}$ clearly has the form $\mathbf{A} \times \mathbf{B}$ described above, with $\mathbf{B}$ a singleton. *Mutatis mutandis*, we have $\Sigma \wedge \Gamma \leq \Gamma$. For the fact that $\Sigma \wedge \Gamma$ is a *greatest* lower bound, let us suppose that Φ is a set of equations with operation symbols G_i

$(i \in I)$, and that there are terms α_i (resp. β_i) interpreting Φ in Σ (resp. Γ). It is not hard to see that the terms $p(\alpha_i, \beta_i)$ will interpret Φ in $\Sigma \wedge \Gamma$.

Continuing our inessential assumption that $S \cap T = \emptyset$, it is not hard to see that $\Sigma \cup \Gamma$ *is a least upper bound*[3] *of* Σ *and* Γ, which we may also denote $\Sigma \vee \Gamma$.

4.2.1 The spaces compatible with $\Delta \wedge \Gamma$.

For an arbitrary topological space C, *and for arbitrary sets* Σ *and* Γ *of equations,* $C \models^{\mathrm{ctn}} \Sigma \wedge \Gamma$ *if and only if* C *is homeomorphic to a product space* $A \times B$, *where* $A \models^{\mathrm{ctn}} \Sigma$ *and* $B \models^{\mathrm{ctn}} \Gamma$.

Proof. We continue the notation and assumptions of §4.2. Given C equal (or homeomorphic) to the product space $A \times B$, and topological algebras **A** and **B** modeling Σ and Γ, respectively, we extend **A** and **B** to $(S \cup T)$-algebras using clauses (iii) and (iv) of §4.2, and then take their product. This product has underlying space C and models $\Sigma \wedge \Gamma$. Thus C is compatible with $\Sigma \wedge \Gamma$.

Conversely, given a space C compatible with $\Sigma \wedge \Gamma$, the existence of the corresponding spaces A and B is proved in (García and Taylor, 1984, Proposition 5, p. 22).

4.3 For each space, compatibility defines an ideal of the lattice.

Let A be an arbitrary topological space. We will see that *the class of all* Σ *that are compatible with* A *forms an ideal in the interpretability lattice.* In this report we shall denote this ideal by $I(A)$.

First, let us suppose that $A \models^{\mathrm{ctn}} \Gamma$ and that $\Sigma \leq \Gamma$. By definition of $\models^{\mathrm{ctn}}$, there is a topological algebra $(A, \overline{G}_t)_{t \in T}$, that models Γ. By the definition of $\Sigma \leq \Gamma$, we have that $(A, \overline{\alpha}_s)_{s \in S}$ models Σ, with S and T disjoint. The operations $\overline{\alpha}_s$ are built using composition from the continuous operations $\overline{G}_t$, hence are continuous themselves. In other words, $(A, \overline{\alpha}_s)_{s \in S}$ is a topological algebra that models Σ. Therefore $A \models^{\mathrm{ctn}} \Sigma$, as desired.

Next, given Σ and Γ, each compatible with the space A, we must show that $\Sigma \vee \Gamma = \Sigma \cup \Gamma$ (described at the end of §4.2) is compatible with A. This result is immediate from the definitions involved.

Thus each space A yields an ideal in the interpretability lattice, which is denoted $I(A)$.

---

[3] For any *set* A of sets of equations (with all their types disjoint), the union $\bigcup A$ is a least upper bound of the family A. However the lattice is a proper class, and there may exist a subclass that has no join.

4.3.1 $I(A)$ is principal: the theory Σ_A.

Given a space A, we define a theory Σ_A as follows. For each continuous function $\mu\colon A^n \longrightarrow A$, there is an n-ary operation symbol F_μ. For $1 \leq i \leq n < \omega$ we let $\pi_i^n\colon A^n \longrightarrow A$ be the continuous function defined by $\pi_i^n(a_1,\ldots,a_n) = a_i$. For a continuous function $\lambda\colon A^n \longrightarrow A^m$, and for $1 \leq i \leq m$, we let λ_i denote the continuous function $\pi_i^m \circ \lambda$. Now we define Σ_A to consist of the equations

$$(3) \qquad\qquad F_{\pi_i^n}(x_1,\ldots,x_n) \approx x_i$$

$$(4) \qquad F_\mu(F_{\lambda_1}(x_1,\ldots,x_m),\ldots,F_{\lambda_n}(x_1,\ldots,x_m)) \approx F_{\mu\circ\lambda}(x_1,\ldots,x_m).$$

for all $1 \leq i \leq n$ and all pairs of continuous functions $A^m \xrightarrow{\lambda} A^n \xrightarrow{\mu} A$. We shall see that Σ_A generates the ideal $I(A)$. (This was asserted without proof in (García and Taylor, 1984, Proposition 11).)

It is not hard to see that $A \models^{\mathrm{ctn}} \Sigma_A$: for the requisite topological algebra on A, one simply takes $\overline{F_\mu} = \mu$, for all $A^n \xrightarrow{\mu} A$. Thus $\Sigma_A \in I(A)$, and so the principal ideal generated by Σ_A is a subset of $I(A)$. For the reverse inclusion, let us consider an arbitrary $\Sigma \in I(A)$. This means that $A \models^{\mathrm{ctn}} \Sigma$; i.e., there exists a topological algebra $\mathbf{A} = (A, \overline{G}_s)_{s \in S}$ satisfying Σ. We construct an interpretation of Σ in Σ_A as follows. For each $n = 1, 2, \ldots$ and each n-ary $s \in S$ we define the term α_s to be $F_\lambda(x_1,\ldots,x_n)$, where λ is the operation $\overline{G}_s\colon A^n \longrightarrow A$. It is not hard to see that the terms α_s form an interpretation of Σ in Σ_A. (The proof uses the given fact that $A \models^{\mathrm{ctn}} \Sigma$, together with an inductive argument on all the subterms of terms appearing in Σ.) Thus $\Sigma \leq \Sigma_A$.; i.e., Σ lies in the desired principal ideal. Thus the two sets are equal: $I(A)$ *is the principal ideal generated by* Σ_A.

Nevertheless, the equation-set Σ_A is large and unwieldy. In a few cases, we know a simple finite generator of $I(A)$. For example if A is any of the spaces mentioned in §6.1 below, then $I(A)$ is the principal ideal generated by $F(x) \approx F(y)$, as one may easily see from the results cited in §6.1. For such a space A and a finite exponent k, the ideal $I(A^k)$ is also principal, as is proved in (Taylor, 2000, Theorem 2 and §11.4).

If A is the one-sphere S^1, then $I(A)$ is the principal ideal generated by Abelian group theory (Taylor, 2000, Theorems 42–43). If A is the dyadic solenoid, then $I(A)$ is the principal ideal generated by the theory of $\mathbb{Z}[1/2]$-modules[4] (Taylor, 2000, Theorems 46–47). For both $I(S^1)$ and $I(S)$ (S the solenoid), the ideal generator can be taken as a finite set of equations.

For any given A, we generally do not know whether $I(A)$ is generated by a single finite set Σ of equations. Further speculations on the generators (e.g. whether there exists such a Σ that is a recursive set of equations) remain equally opaque.

---

[4] $\mathbb{Z}[1/2]$ is the ring of all rationals with denominator a power of 2.

4.3.2 Unions of chains.

If Λ_2 is a set of equations, and if Λ_1 is an arbitrary subset of Λ_2, then $\Lambda_1 \leq \Lambda_2$ in our lattice. The converse is far from true: even if $\Lambda_1 \leq \Lambda_2$, it may be true that the two Λ_i have disjoint similarity types. Thus the consideration of the union of a chain (under inclusion) is not central to our main topic. Nevertheless, we include one small observation.

The ideal $I(A)$ may not be closed under unions of chains. One may have $\Lambda_1 \subseteq \Lambda_2 \subseteq \Lambda_3 \subseteq \ldots$, with $A \models^{\mathrm{ctn}} \Lambda_k$ for each k, but $A \not\models^{\mathrm{ctn}} \Lambda = \bigcup \Lambda_k$. Such Λ_k—with A taken as a closed interval of the real line—may be seen in §7.2.2 below. (The example comes from (Taylor, 1977, p. 525).) In other words, every finite subset of Λ lies in $I(A)$, but Λ does not.

Incidentally, this example shows that while the union of a chain (under inclusion) is an upper bound of that chain, it need not be a *least* upper bound.

4.3.3 Sometimes $I(A)$ is a prime ideal.

If C is a product-indecomposable space, then Σ_C is meet-prime, which further implies that $I(C)$ is a prime ideal in the lattice.

Proof. Suppose that $\Sigma \wedge \Gamma \leq \Sigma_C$. By §4.2.1, C is homeomorphic to a product space $A \times B$ with $A \models^{\mathrm{ctn}} \Sigma$ and $B \models^{\mathrm{ctn}} \Gamma$. Since C is product-indecomposable, either A or B is a singleton. Thus C is homeomorphic to A or to B. In the former case $\Sigma \leq \Sigma_C$, and in the latter $\Gamma \leq \Sigma_C$.

4.3.4 The complementary filter.

If A is product-indecomposable, then by §4.3.3 the complement of $I(A)$ is a *filter* which we will denote $F(A)$. This consists of all Σ that are not compatible with A. By §4.3.2, when A is a closed interval of $\mathbb{R}$, there is a set $\Sigma \in F(A)$ such that no finite subset of Σ is in $F(A)$. Hence this filter is generally not Mal'tsev-definable (see (Taylor, 1973)). It is unknown whether it might be subject in some cases to a syntactic definition (such as a weak Mal'tsev condition). (Exception: in (Taylor, 1973) we gave a Mal'tsev condition describing $F(A)$ for A a two-element discrete space.)

4.4 The ideal of a product of two spaces.

Let A and B be topological spaces. We saw in §4.2.1 that *if A and B are compatible with Σ and Γ, respectively, then the product space $A \times B$ is compatible with the meet $\Sigma \wedge \Gamma$.*

Now if we have $\Delta \in I(A) \cap I(B)$, then Δ is compatible with both A and B. By the previous paragraph, $A \times B$ is compatible with $\Delta \wedge \Delta$, which is co-interpretable with Δ, hence equal to Δ in the lattice. In other words, we now have $I(A) \cap I(B) \subseteq I(A \times B)$. They are not generally equal. For instance, if A is not homeomorphic to a perfect square, then, though $I(A \times A)$ will contain the perfect-square equations (§7.5 below), $I(A) \cap I(A)$ will not.

5 Note on free topological algebras.

Let A be a metrizable space, and Σ a finite or countable set of equations that is consistent (does not entail $x \approx y$) . Considering A purely as a set, one of course has the free algebra $\mathbf{F}_\Sigma(A)$; it has A embedded as a subset, and satisfies the equations Σ. In 1964, S. Świerczkowski showed (Świerczkowski, 1964) how to topologize (even metrize) $\mathbf{F}_\Sigma(A)$ in such a way that A is embedded as a subspace, and each operation is continuous. Thus in particular, Σ is compatible with the topological space that underlies $\mathbf{F}_\Sigma(A)$.

We mention this example of compatibility to illustrate the fact that, beyond consistency, there is no apparent constraint on the Σ that can appear in the compatibility relation $A \models^{\mathrm{ctn}} \Sigma$, even when we require A to satisfy the first countability axiom, as described in §0.1.

The topological spaces defined by Świerczkowski are large and non-compact. If A is a CW-complex, then so is $\mathbf{F}_\Sigma(A)$ (see Bateson (Bateson, 1982)), but the construction of the algebra $\mathbf{F}_\Sigma(A)$ is inherently infinitary, and so the complex structure is, to our knowledge, almost always infinite. It is only for very special and somewhat trivial equation-sets Σ that $\mathbf{F}_\Sigma(A)$ turns out to be finitely triangulable.[5] By way of contrast, our main proposal in this report (see §6 below) will be to consider $I(A)$ when A is a finite simplicial complex. Here the compatible Σ appear to be more limited.

6 Restrictions on compatibility for a finite complex.

We turn our attention toward compact Hausdorff spaces, mostly limiting it to those connected spaces that have the form of a finite simplicial complex. The latter form the most down-to-earth geometric corner of topology, and hopefully our understanding could be rooted there. For simplicity, we will refer to a space as *finite* if it has a finite triangulation, and as *compact* if it is compact and Hausdorff.

Our starting point is the impression that the various Σ that have been observed on finite connected complexes often fall into several broad cate-

---

[5] For example, for Σ defining G-sets over a finite group G.

gories: lattice-related equations, group-related equations, $[k]$-th power equations, simple equations, and a few special equation-sets. In this section we review a few incompatibility results that make such a division slightly more plausible.

6.1 Undemanding sets of equations.

A set Σ is called *undemanding* if it can be satisfied on some set of more than one element—equivalently, on any set— by taking each operation to be either a projection function or a constant function. Such operations are continuous, and hence if Σ is undemanding, then Σ is compatible with every space A. Taylor proved (Taylor, 2000) a sort of converse result: that *many finite spaces A have the property that $A \models^{\mathrm{ctn}} \Sigma$ only for undemanding sets Σ.* (The proofs apply algebraic topology of the sort used in analyzing H-spaces, as mentioned in §3.)

In other words, such an A is compatible with *no* interesting Σ! The list of such A contains, for instance, all spheres other than S^1, S^3 and S^7, the Klein bottle, the projective plane, a one-point join of two 1-spheres, and several others. It appears that the proofs could be extended to many other finite A, but no one has carried out this job. From these considerations it appears that for many finite A, perhaps for most, the situation is totally arid.

For such a space A, the ideal $I(A)$ is the smallest ideal containing every undemanding set of equations. In fact this ideal is generated by the single undemanding equation $f(x) \approx f(y)$ (which postulates the existence of a constant function).

Among those Σ that have at least one constant function, any undemanding Σ is least in the interpretability ordering.

(For a k-dimensional counterpart of "undemanding," see §7.5.3 below.)

6.1.1 "Undemanding" is an algorithmic property.

There is an easy algorithm that accepts any finite set Σ of equations as input, and halts with output 1 or 0, depending whether Σ is undemanding. We will describe this algorithm informally.

Given Σ, it has a finite similarity type $n : T \longrightarrow \mathbb{Z}$. We now consider an arbitrary finite set K of equations of the form

$$(5) \qquad\qquad F_t(x_1, \ldots, x_{n(t)}) \approx x_j$$

or of the form

$$(6) \qquad\qquad F_t(x_1,\ldots,x_{n(t)}) \approx C,$$

where our formal language has been augmented to include a single new constant symbol C. For each $t \in T$, our K must include a single equation involving F_t; that one equation must be either Equation (6) or one instance of Equation (5)—thereby choosing a value of j in that equation.

If σ is any term in the language of Σ, the equations in K will immediately imply either $\sigma \approx x_j$ for some unique j, or $\sigma \approx C$. For each $\sigma \approx \tau$ occurring in Σ, we may check whether σ and τ both reduce to the same x_j or else both to C. If this happens for all equations in Σ, we say that K is consistent with Σ.

We now undertake to do this for all of the (finitely many) possibilities for K. If one K turns out to be consistent with Σ, we may say Σ is undemanding. Otherwise, all such K turn out to be inconsistent with Σ, in which case we conclude that Σ is demanding.

If all the operations are for instance binary, then the number of sets K is obviously $3^{|T|}$; we see therefore that the algorithm is exponential in $|T|$. Nevertheless, in many cases of interest $|T|$ is small, and the algorithm is easily carried out. The reader is invited to try his/her hand at equations (25–26) in §7.6.2.

6.2 Not both groups and semilattices.

The incompatibility of (non-trivial) compact Hausdorff spaces with lattice-ordered groups was proved by M. Ja. Antonovskiĭ and A. V. Mironov (Antonovskiĭand Mironov, 1967) in 1967. Therefore, of course, if Σ is an axiom-set for LO-groups, we will not have $A \models^{\mathrm{ctn}} \Sigma$ for any non-trivial finite space A. If A is also connected, we in fact have the stronger conclusion—a consequence of J. D. Lawson and B. Madison (Lawson and Madison, 1970)—that A is not compatible both with group theory and with lattice theory.

Of course, from the perspective of the present investigation, it would be very desirable to have a stronger version of this result, where group theory and lattice theory are replaced by lower elements of the interpretability lattice. In any case we will use §6.2 as a rough guide in organizing §7 which follows, separating group-like topological algebras from lattice-like ones. (In §7.3, however, we find some examples that lie on the overlap.)

6.2.1 Proof of the assertion in §6.2.

In §6.2.1, all citations refer to items in Lawson and Madison (Lawson and Madison, 1970) (1970).

Our proof (that no nontrivial connected compact Hausdorff space is compatible with both group theory and lattice theory) is by contradiction. We will assume that A is a non-trivial compact connected Hausdorff space, and that there are continuous operations $\wedge$, $\vee$, $\cdot$ and $^{-1}$ such that $(A, \wedge, \vee)$ is a topological lattice and $(A, \cdot, ^{-1})$ is a topological group.

By compactness, $(A, \wedge, \vee)$ has 0 and 1. Therefore $(A, \wedge)$ is certainly a non-trivial compact connected nontrivial idempotent semigroup with 0. By Corollary 2.12 on page 135, each maximal idempotent of $(A, \vee)$ is marginal. Therefore A has at least one marginal element.

Marginal elements are defined in Definition 1.1 on page 129. Peripheral elements are defined in Definition 1.2. On the bottom of page 129 we are told that every marginal element is peripheral, with a reference to another paper of Lawson and Madison. Also in Definition 1.2 we have that a point is *inner* if and only if it is not peripheral. Therefore our space A contains one point that is not inner.

On the other hand, Theorem 1.6 on page 130 tells us that if A is a finite dimensional locally compact Hausdorff space, then the set of inner points of A is dense in A. And hence non-empty. Thus our space A has one point that is inner and one point that is not inner.

On the other hand $(A, \cdot, ^{-1})$ is a topological group, and hence A is homogeneous. Therefore, either all points of A are inner or no points of A are inner. This contradiction completes the proof.

7 $A \models^{\mathrm{ctn}} \Sigma$ for Σ non-trivial and A given by a finite complex.

We present essentially all the examples that we know for sure. Our rough division into types of Σ is partly based on the results mentioned in §6.2.

7.1 Σ related to group theory.

7.1.1 Grouplike algebra on spheres.

We look at one strengthening of group theory (i.e. higher in the lattice), and two weakenings.

The one-dimensional sphere S^1 is compatible with Abelian group theory. (The Abelian group may be modeled as the set of unit-modulus complex numbers under multiplication, or as the set of orthogonal 2×2 real matrices of determinant 1.) On the other hand, S^3 is the space underlying the group of unit quaternions, which is not Abelian. (R. Bott proved in 1953 that S^3 is not compatible with Abelian group theory—see (Bott, 1953).) S^7 has the multiplication of unit octonians. With this multiplication, S^7 forms an H-space (see §3), which in fact satisfies the alternative laws (associativity on all two-generated subalgebras). S^7 does not, however, have a multiplication forming an associative H-space (monoid), as was proved by I. M. James in 1957 (see (James, 1957)). Thus, for k any positive integer with $k \neq 1, 3, 7$, we have the set inequalities

$$I(S^k) = I(S^k) \cap I(S^7) \cap I(S^3) \cap I(S^1) \subset$$
$$I(S^7) \cap I(S^3) \cap I(S^1) \subset I(S^3) \cap I(S^1) \subset I(S^1).$$

The four ideals are separated by H-space theory, associative H-space theory (monoids) (or by group theory), and Abelian group theory, using the results cited here and in §3. (Recall that $I(S^1)$ is described near the end of §4.3.1, and $I(S^k)$ is described in §6.1.)

7.1.2 Other groups.

There are various compact Lie groups (orthogonal, special orthogonal, and so on). Matrix multiplication (which is inherently continuous) is often the basic operation. Their various underlying spaces appear to be very sparse among the class of all compact manifolds. The underlying spaces of compact Lie groups may be finitely triangulated (see (Wikipedia, 2016) and references given there).

7.2 Σ derived from lattice theory.

7.2.1 Distributive lattices (with 0 and 1).

A real interval $[a, b]$ has a well-known distributive lattice structure. Therefore each simplex $[a, b]^n$ has compatible distributive lattice operations, as does any of its sublattices. In the compact realm, every compatible lattice has a zero (bottom) and a one (top). The compact subuniverses of $([0, 1], \overline{\wedge}, \overline{\vee}, 0, 1)^2$ appear to be limited in their possible shapes, although a full description of the limitations has not yet been discovered.

For the first such limitation, we note that in 1959 Dyer and Shields proved (Dyer and Shields, 1959) that every compact connected metric topological lattice is contractible and locally contractible. In particular a finite graph (one-dimensional complex) with a lattice structure must be acyclic.

For a further limitation on finite graphs, we note that if A is an acyclic non-linear finite graph, i.e., a one-dimensional compact connected simplicial complex that does not define a line segment—e.g. if A is a Y-shaped space—

then A is not compatible with lattice theory, and hence cannot be such a subuniverse of $[0,1]^2$. See §7.2.4 below for further incompatibilities.

The aforementioned result, of non-compatibility between lattice theory and an acyclic non-linear finite graph A, comes in essence from Wallace (Wallace, 1955), although only a weaker theorem is stated, and without proof. (See the "Alphabet Theorem" on page 107 of (Wallace, 1955).) We shall thus include a short proof here, by contradiction. So suppose that $\mathbf{A} = (A, \overline{\wedge}, \overline{\vee})$ is a topological lattice. Since our space A is compact, $\mathbf{A}$ has a 0 and a 1. Since A is not homeomorphic to a segment, there is a point E of A such that $A \setminus E$ has at least three components $S_0, S_1, S_2, \ldots S_N$. Clearly there is one S_k that contains neither 0 nor 1.

We now take $P \in S_k$; clearly $P \neq E$. We consider the map $X \longmapsto X \overline{\vee} P$. It maps 0 to P and 1 to 1. By connectedness, and the fact that every path from P to 1 must pass through E, we see that our map must have E in its range. In other words, for some X, $P \overline{\vee} X = E$. Thus $P \leq E$.

A dual argument shows that $P \geq E$; by anti-symmetry $P = E$. This is the contradiction that establishes our result.

It is worth mentioning, for future reference (§8.2) that the lattice operations on a real interval are *piecewise linear*:

$$(7) \qquad\qquad x \overline{\wedge} y \;=\; x \text{ if } x \leq y; \; y \text{ if } x \geq y,$$

and similarly for join.

7.2.2 One can go higher in $I([0,1])$.

For this section, we let Λ_0 be a finite equational axiom system for distributive lattice theory with zero and one. For each integer $n \geq 1$ we let Λ_n be Λ_0 augmented with a unary operation symbol f and constant symbols $a_1, \ldots, a_n$, and extended with the following axioms:

$$a_1 \wedge a_2 \approx a_1, \quad a_2 \wedge a_3 \approx a_2, \quad \cdots, \quad a_{n-1} \wedge a_n \approx a_{n-1}$$
$$f(0) \approx 0, \quad f(a_1) \approx 1, \quad f(a_2) \approx 0, \quad f(a_3) \approx 1, \quad \cdots$$
$$f(1) \approx 1 \text{ if } n \text{ is even, } 0 \text{ otherwise.}$$

One easily checks that, in the interpretability lattice

$$\Lambda_0 \;<\; \Lambda_1 \;<\; \cdots \;<\; \Lambda_n \;<\; \Lambda_{n+1} \;<\; \cdots.$$

(For non-interpretability of Λ_{n+1} in Λ_n, we note that, modulo equational deductions, Λ_n has only $n+2$ constant terms, whereas any interpretation of Λ_{n+1} will require $n+3$ logically distinct constant terms.)

Compatibility of Λ_n with a closed interval is easiest if we use the interval $[-1,1]$. Then the desired function $\overline{f}$ can be taken as the Chebyshev polynomial T_{n+1} of degree $n+1$. (Or one can simply take $\overline{f}$ to be piecewise linear as specified by our equations.)

We therefore have an ω-chain of sets in the ideal $I([0,1])$, going upward from the theory of distributive lattices with zero and one (§7.2.1).

7.2.3 Lattices (with 0 and 1).

Lattice theory lies strictly below modular lattice theory in the interpretability lattice. Nevertheless, we do not know any space B that is compatible with lattice theory (with or without zero and one), and yet is not compatible with modular lattice theory. (It is possible that, for $\Sigma =$ lattice theory, and for suitably chosen A, the space of the free algebra $\mathbf{F}_\Sigma(A)$ (see §5) might be such a B. Furthermore, in (Bergman, 2015), G. Bergman has suggested several finite spaces that are compatible with lattice theory, but may fail to be compatible with modular lattice theory.)

We do know a space B that separates modular lattice theory from distributive lattice theory in this manner, namely such that B is compatible with modular lattice theory but not with distributive lattice theory. If B is the union of three closed 2-simplices along one common edge, then B has the announced properties. The proof, by G. Bergman and W. Taylor, appears in (Bergman, 2015, §3.1). See also §9.4.7 for a more general question about the separation of one variety from another by the compatibility relation

In the nineteen-fifties A. D. Wallace conjectured that every compact, connected topological lattice $(L, \overline{\wedge}, \overline{\vee})$ is distributive. This was disproved in 1956 by D. E. Edmondson (Edmondson, 1956), who gave a non-modular example[6] with L homeomorphic to $[0,1]^3$. (Of course this space is compatible with distributive lattice theory.) Wallace's conjecture holds for $L = [0,1]^2$ (see (Anderson, 1959)) and for modular lattices with $L = [0,1]^3$ (see (Gierz and Stralka, 1989)).

7.2.4 Semilattices (with 0 and 1).

By contrast with §7.2.1, every finite tree (see §7.2.1) is compatible with semilattice theory—as may be seen in §3.7 of W. Taylor (Taylor, 1986)—even semilattice theory with **0** and **1**. (And it is not hard to see from the proof that the semilattice operation may be taken to be piecewise linear, i.e. simplicial.)

Taylor proved in 1977 (see (Taylor, 1977)) that if A is a topological semilattice, then the homotopy group $\pi_n(A, a)$ is trivial for every $n \geq 1$ and every $a \in A$. (In 1965 (see (Brown, 1965)) D. R. Brown had obtained the same conclusion for a different equation-set: $x \wedge x \approx x, x \wedge 0 \approx 0 \wedge x \approx 0$. In the compact

---

[6] G. Bergman has recently found a simpler construction of an example with these properties—see §4.2 of (Bergman, 2015).

case Brown's result already applies to a semilattice A, since A will then have a zero.)

(In 1959 Dyer and Shields had proved (Dyer and Shields, 1959) that every compact connected metric topological lattice is contractible and locally contractible.)

7.2.5 Majority operations and median algebras.

It is well known that if $(A, \wedge, \vee)$ is a lattice, then the derived operation defined by the term

$$(8) \qquad m(x,y,z) \;=\; (x \wedge y) \vee (x \wedge z) \vee (y \wedge z)$$

satisfies the *majority equations*

$$(9) \qquad m(x,x,y) \approx m(x,y,x) \approx m(y,x,x) \approx x.$$

Thus the majority equations lie below lattice theory in the interpretability lattice, and so are compatible with the space of any topological lattice (§7.2.3).

Moreover the majority equations are also compatible with the finite trees mentioned in §7.2.1 and §7.2.4. The idea (due to M. Sholander in 1954—see (Sholander, 1954)) is very simple. Given such a tree T, for any two points $a, b \in T$, there is a smallest connected subset containing the two, which will be denoted $[a,b]$. Moreover, Sholander proved, for any three points a, b and $c \in T$, the intersection $[a,b] \cap [b,c] \cap [c,a]$ is a singleton. Taking its lone member as the value of $\overline{m}(a,b,c)$, we obtain a symmetric, continuous operation $\overline{m} : T^3 \longrightarrow T$ that satisfies Equation (9). Finally, we remark here that the $\overline{m}$ so defined on a tree T satisfies a stronger set of equations, the axioms of *median algebra*— see e.g. the 1983 treatise by Bandelt and Hedlíková (Bandelt and Hedlíková, 1983), or the 1980 treatise by Isbell (Isbell, 1980).

In fact, it was proved in 1979–82 by J. van Mill and M. van de Vel (Mill and Vel, 1979, 1982) that, among finite-dimensional spaces, the ones compatible with the majority equations are precisely the absolute retracts. (They refer to a continuous majority operation as a "mixer.")

7.2.6 Multiplication with one-sided unit and zero.

One very weak consequence of semilattice theory with zero and one—or of ring theory—is the following set of two equations:

$$x \wedge 0 \approx 0, \qquad x \wedge 1 \approx x.$$

These equations lie quite low in the interpretability lattice; hence it is not hard to find contractible spaces that model them. (For example see e.g. §7.2.1 and §7.2.4.) On the other hand, as was mentioned in §3.6 of (Taylor, 1986), it is easy to see that if A is a path-connected finite space compatible with these equations, then A is contractible.

7.3 Below both groups and lattices: H-spaces.

H-spaces (multiplication with two-sided unit element), and *associative H-spaces* (otherwise known as *monoids*) were mentioned in §7.1.1; their theories lie well below group theory. It is interesting to note that both of these theories also lie below ∧-semilattices with **1** (§7.2.4).

For example, we may let $\mathbf{S}^1 = (S^1, \cdot, e)$ denote the circle group, with unit element e. We may let $\mathbf{I} = (I, \cdot, 1)$ denote the unit interval, where $\cdot$ is the usual semilattice operation, and 1 is the top element, and also the unit element for this algebra. Then $\mathbf{S}^1 \times \mathbf{I}$ is also an associative H-space, with two-side unit element $(e, 1)$. One may easily check that

$$P = \{(u, v) \in S^1 \times I : u = e \text{ or } v = 0\}$$

is a subuniverse of $\mathbf{S}^1 \times \mathbf{I}$. It is homeomorphic to the pointed union of the pointed spaces (S^1, e) and $(I, 0)$. (In other words, the space P is homeomorphic to the letter P.) Thus the space P is, for example, compatible with monoids. (This example appeared in (Taylor, 1986), and is derived from work of Wallace (Wallace, 1955).)

If B is any compact metrizable space that is an absolute retract among metric spaces, then B is compatible with H-spaces (see §3.2.3 of W. Taylor (Taylor, 2009)).

If A is compatible with the Mal'tsev equations, then A is compatible with Σ_H—see §7.4.3 below.

7.3.1 A mysterious theorem.

Algebraic topology has a lot to say about—and methods concerning—H-spaces. As one sample result, we mention this:

J. R. Harper proved in 1972 (*inter alia*, see (Harper, 1972)) that if $\mathbf{A}$ is a finite connected H-space, then the homotopy group $\pi_4(A)$ obeys the law $x^2 = 1$. ($A = S^3$ is an example of such an H-space with $\pi_4(A) \neq 0$.)

7.3.2 Digression on homotopy.

One may examine *satisfaction up to homotopy*. In the case of H-space theory, one asks for a continuous map $F\colon A^2 \longrightarrow A$, and an element $e \in A$, such that the maps $x \longmapsto F(x,e)$ and $x \longmapsto F(e,x)$ are each homotopic to the identity map $x \longmapsto x$.

Many of the spaces of interest in this investigation are contractible, and therefore model $x \approx y$ up to homotopy — which means that they model any Σ up to homotopy. We therefore will not pursue this notion here, except to report that if A is a CW-complex, and if A is compatible with H-space theory up to homotopy, then[7] in fact A is an H-space.

7.4 Σ consisting of simple equations.

If A is an absolute retract in the class of metric spaces, and if Σ is a consistent set of simple equations, then A is compatible with Σ (see Taylor (Taylor, 2009)). A term σ is *simple* iff there is at most one operation symbol F_t in σ, appearing at most once. An equation $\sigma \approx \tau$ is simple iff both terms σ and τ are simple. For example, the majority equations (9) are simple,

For absolute retracts, consult works by Borsuk (Borsuk, 1967) and Hu (Hu, 1965). For example, the finite trees defined in §7.2.1 are absolute retracts (among, e.g., metric spaces). Thus the result of this section extends the compatibility results of §7.2.5.

Moreover, if Σ is a consistent set of simple equations in a finite similarity type, and if A is an absolute extensor (see (Hu, 1965)) in the class of completely regular spaces, then there is a topological algebra $\mathbf{A} = (A, \ldots F_t \ldots)$ whose simple identities are precisely the simple consequences of Σ (see (Taylor, 2009, Theorem 7(b))). This is the rare case where we have some control over equations *not* holding in an algebra $\mathbf{A}$ constructed in this report.

If Σ is a finite (or recursive) set of simple equations, and if A is a finite (or recursive) tree, and if we know some computable (hence continuous) operations modeling Σ on a closed interval, then there are computable (hence continuous) operations modeling Σ on A. The method is described in §4.2 of (Taylor, 2009); it probably can be extended to an arbitrary absolute retract which is a finite complex. A special case of the method is given in detail in §7.4.2 below. (For computability of real functions, see (Pour-El and Richards, 1989).)

---

[7] See Whitehead (Whitehead, 1978, Theorem III..4.7, page 117). In fact Whitehead proves this under assumptions weaker than what we have stated here.

7.4.1 Minority equations on a closed interval.

As an example of simple equations, we consider the *ternary minority equations*

$$(10) \qquad q(x,x,y) \approx q(x,y,x) \approx q(y,x,x) \approx y.$$

A closed real interval $[a,b]$ is well known to be an absolute retract, so by §7.4 there exists a ternary operation $\overline{q}$ on $[a,b]$ satisfying (10). We can, however, define such an operation directly, without reference to §7.4. A minority operation $\overline{q}$ may be defined by the following two conditions:

(i) If $u \leq v \leq w$, then $\overline{q}(u,v,w) = u - v + w$.
(ii) $\overline{q}$ is completely symmetric in its three variables.

It is worth noting that there is a single formula defining this $\overline{q}$, namely

$$(11) \qquad \overline{q}(u,v,w) = u \wedge v \wedge w - \overline{m}(u,v,w) + u \vee v \vee w,$$

where $\overline{m}$ is the ternary majority operation defined in Equation (8).

If A is a space homeomorphic to an interval, then of course our definition of $\overline{q}$ may be transferred to A by laying down coordinates. Any non-linear change of coordinates will effect the values of the resulting $\overline{q}^A : A^3 \longrightarrow A$, but Equation (10) will not be affected. Linear changes of coordinates will not affect any values of $\overline{q}^A$.

(A very different—and more complicated—$\overline{q}$ was described in Equation (71) of §9.3 of (Taylor, 2006).)

7.4.2 Minority equations on a tree.

Here we will illustrate one way to satisfy the minority equations (10) on a simple tree—as mentioned in §7.2.1 and §7.2.4 and §7.2.5. Specifically let Y stand for the Y-shaped space that is formed by joining three closed intervals with the amalgamation of one endpoint each. Y is an absolute retract; hence compatible with the minority equations (10) by §7.4. We can, however, define such an operation directly, without reference to §7.4.

Let Y_1, Y_2, Y_3 be the three subsets of Y that can be formed by joining two out of three of the constituent intervals. The significant facts about the Y_i are these:

(i) Each element of Y belongs to at least two of the Y_i.
(ii) Each Y_i is homeomorphic to an interval, and hence has a minority operation $\overline{q}_i$ by §7.4.1.
(iii) For each i there is a continuous function $\overline{p}_i$ retracting Y onto Y_i.

Let $\overline{m}$ be a majority operation on Y—whose existence is assured by §7.2.5. We now define $\overline{Q} : Y^3 \longrightarrow Y$ as follows:

$$\overline{Q}(a,b,c) = \overline{m}(\overline{q}_1(\overline{p}_1(a),\overline{p}_1(b),\overline{p}_1(c)),$$
$$\overline{q}_2(\overline{p}_2(a),\overline{p}_2(b),\overline{p}_2(c)), \overline{q}_3(\overline{p}_3(a),\overline{p}_3(b),\overline{p}_3(c))).$$

From points (i)–(iii) it follows easily that $\overline{Q}$ is a minority operation on Y.

As mentioned at the end of §7.4, the methods of §4.2 of (Taylor, 2009)—a recursive invocation of the methods here—will allow one to construct a ternary majority operation on any finite tree.

7.4.3 Mal'tsev operations.

The *Mal'tsev equations* are

$$(12) \qquad\qquad p(x,x,y) \approx p(y,x,x) \approx y.$$

One may say that their study initiated the investigation of relative strengths of equation-sets, ultimately leading to the lattice of §4.2. Equations (12) obviously lie below the minority equations (10) in the lattice. Thus Mal'tsev operations are found on a closed interval and on any finite tree (by §7.4.1 and §7.4.2).

Moreover, in any group $(A,\cdot,^{-1})$, the formula

$$(13) \qquad\qquad \overline{p}(a,b,c) = a\cdot b^{-1}\cdot c$$

defines a Mal'tsev operation on A. Therefore S^1, S^3 have Mal'tsev operations.

As a sort of hybrid example, we look at the cylinder $[a,b]\times S^1$. It has a Mal'tsev operation as does any (necessarily closed) subset onto which the entire space $[a,b]\times S^1$ retracts. (E.g. a belt around the cylinder that is pinched so as to be one-dimensional in spots and two-dimensional in other spots.)

Notice that any space A that has a Mal'tsev operation is an H-space (§7.3): if $\overline{p}\colon A^3 \longrightarrow A$ satisfies (12), and if $e \in A$, we may then define a multiplication $x\cdot y = \overline{p}(x,e,y)$. This multiplication has e as a two-sided unit.

7.4.4 Two-thirds minority operations.

The *two-thirds minority equations* are

$$(14) \qquad\quad t(x,x,y) \approx t(y,x,x) \approx y; \qquad t(x,y,x) \approx x.$$

Clearly they lie higher in the lattice than the Mal'tsev equations (12). (Strictly higher because they are not interpretable in Abelian group theory—cf. §4.3.1.) Equations (14) also lie above the ternary majority equations (9): $\overline{p}(x,y,z) = \overline{t}(x,\overline{t}(x,y,z),z)$ defines a majority operation, as one may easily check. Equations (14) play a significant role in the study of *arith-*

metic varieties (varieties that are congruence-permutable and congruence-distributive)—see e.g. A. F. Pixley (Pixley, 1979).

Of course an interval $[a,b]$ or a tree has a continuous two-thirds minority operation by the general results of §7.4. One choice for $\bar{t}$ on $[a,b]$ is this:

$$\bar{t}(u,v,w) \;=\; u - \overline{m}(u,v,w) + w,$$

whose form has much in common with Equations (11) and (13). For the tree Y one may use the method of §7.4.2.

7.5 Σ *defining* $[k]$*-th powers.*

For each set Σ of equations, and for each $k = 2,3,\ldots$, there exists a set $\Sigma^{[k]}$ with the following property: an arbitrary topological space A is compatible with $\Sigma^{[k]}$ if and only if there exists a space B such that $B \models^{\mathrm{ctn}} \Sigma$ and A is homeomorphic to the direct power B^k. If Σ is finite (resp. recursive, resp. r.e., etc.), then $\Sigma^{[k]}$ may be taken as finite (resp. recursive, resp. r.e., etc.).

From the definition (which we have skipped) it is immediate that $\Sigma^{[k]} \geq \Sigma$ in our lattice (§4). The theory $\Sigma^{[k]}$ was developed in 1975 by R. McKenzie (McKenzie, 1975); see also (Taylor, 1975, pp. 268–269) or §10.1 of (Taylor, 2006). The connection of $\Sigma^{[k]}$ with topological spaces was perhaps first noted in (García and Taylor, 1984).

Obviously, if $\Gamma^{[k]} \in I(A)$, then $\Gamma \in I(A)$ and A is a k-th power. The converse is false,[8] even when $k = 2$: take A to be a four-element discrete space, and Γ to be the $\Sigma^{[2]}$ of §7.5.2 below. Then $\Gamma \in I(A)$ and A is a square, but $\Gamma^{[2]} \notin I(A)$ (for then, by §7.5.2, A would be the square of a square, which it is not).

In this context, of course, every example adduced so far in §7 yields further examples for each $k = 2,3,\ldots$. If B is known to be compatible with Σ, then $A = B^k$ is known to be compatible with $\Sigma^{[k]}$. In the opposite direction, we of course need to know all possible factorizations of A as homeomorphic to some B^k. If each such B is incompatible with Σ, then we know that A is not compatible with $\Sigma^{[k]}$. (This of course includes the case where no such factorization exists.)

7.5.1 The operations of $\Sigma^{[k]}$.

Given operations $\overline{F}_1, \cdots, \overline{F}_k$ on a set B, each of arity nk, we may define an n-ary operation $\overline{F}$ on the set B^k as follows:

---

[8] This observation thanks to G. M. Bergman.

$$\overline{F}((b_1^1,\cdots,b_1^k),\cdots,(b_n^1,\cdots,b_n^k))$$
$$(15) \qquad = (\overline{F}_1(b_1^1,\cdots,b_n^k),\cdots,\overline{F}_k(b_1^1,\cdots,b_n^k)).$$

Clearly, if B has a topology, and if each $\overline{F}_j$ is continuous, then $\overline{F}$ is continuous. One may think of $\Sigma^{[k]}$ as having one such n-ary operation symbol for each k-tuple of nk-ary term operations of Σ. More usually, we take only these special cases as fundamental operations of $\Sigma^{[k]}$:

$$(16) \quad \overline{H}((b_1^1,\cdots,b_1^k),\cdots,(b_k^1,\cdots,b_k^k)) = (b_1^1,\cdots,b_k^k);$$

$$(17) \qquad\qquad \overline{d}((b_1^1,\cdots,b_1^k)) = (b_1^2,\cdots,b_1^k,b_1^1);$$

$$(18) \quad \overline{G}_t((b_1^1,\cdots,b_1^k),\cdots,(b_n^1,\cdots,b_n^k)) = (\overline{F}_t(b_1^1,\cdots,b_n^1),\cdots,\overline{F}_t(b_1^k,\cdots,b_n^k)),$$

where $\overline{F}_t$ $(t \in T)$ are the fundamental operations of Σ. (The other operations (15) can formed from these.)

7.5.2 Squares—Σ empty and $k = 2$.

For Σ empty, $\Sigma^{[2]}$ may be axiomatized as:

$$H(x,x) \approx x$$
$$H(x,H(y,z)) \approx H(x,z) \approx H(H(x,y),z)$$
$$d(d(x)) \approx x$$
$$d(H(x,y)) \approx H(d(y),d(x)).$$

If A is the square of another space B, i.e. $A = B^2$ with the product topology, then A is compatible with $\Sigma^{[2]}$ in the following manner. We define operations $\overline{H}$ and $\overline{d}$ on B^2 via

$$(19) \qquad\qquad \overline{H}((b_1,b_2),(b_3,b_4)) = (b_1,b_4)$$
$$(20) \qquad\qquad \overline{d}((b_1,b_2)) = (b_2,b_1),$$

for all $b_1,\ldots,b_4 \in B$. These operations are obviously continuous, and it is easy to check by direct calculations that they obey $\Sigma^{[2]}$. Thus $B^2 \models^{\mathrm{ctn}} \Sigma^{[2]}$. (Equations (19–20) are special cases of Equations (16–17) above.)

Conversely, it is not hard to prove that if A is any space with continuous operations H' and d' modeling this $\Sigma^{[2]}$, then there exist a space B and a bijection $\phi : A \longrightarrow B^2$ that is both a homeomorphism of spaces and an isomorphism of (A,H',d') with $(B^2,\overline{H},\overline{d})$, with $\overline{H}$ and $\overline{d}$ defined as above. (One begins by defining B to be the subspace $\{a \in A : d'(a) = a\}$.)

Thus this $\Sigma^{[2]}$ is compatible with A if and only if A is homeomorphic to a square, as claimed.

7.5.3 Squares of spaces.

If B is any space and $\overline{F}_i$ is a $2n$-ary operation on B $(i = 1, 2)$, then—as a special case of (15)—one has an n-ary operation $\overline{F}$ defined on $A = B^2$ as follows:

$$(21) \qquad \overline{F}((b_1^1, b_1^2), \cdots, (b_n^1, b_n^2)) = (\overline{F}_1(b_1^1, \cdots, b_n^1), \overline{F}_2(b_1^2, \cdots, b_n^2)).$$

If each $\overline{F}_i$ is continuous, then $\overline{F}$ is continuous.

For most spaces B, there are many continuous operations on B^2 besides those described in Equation (21), and there is little or no real restriction on the equation-sets that may be compatible with B^2. But for certain spaces, notably those described at the start of §6.1, the compatible equation-sets are very limited.

In Theorem 2 of (Taylor, 2000) it was proved that if B is one of these spaces, such as a figure-eight or a sphere S^n $(n \neq 1, 3, 7)$, then a set Σ is compatible with B^2 only if Σ is interpretable in operations of type (21), where each $\overline{F}_i$ is either a coordinate projection function or a constant. Such a set Σ is called 2-undemanding. There is an algorithm to determine if a finite set is 2-undemanding.

The reader may easily imagine the corresponding definition for k-undemanding sets. Then a k-th power such as $(S^n)^k$ $(n \neq 1, 3, 7)$ is compatible with Σ only if Σ is k-undemanding.

7.5.4 Below squares in the interpretability lattice.

Let Γ consist of the single equation

$$(22) \qquad\qquad (x \star y) \star (y \star z) \approx y.$$

In the context of §7.5.2, if we define

$$(23) \qquad\qquad x \star y = d(H(y, x)),$$

then it is not hard to check that Equation (22) follows from the equations $\Sigma^{[2]}$ of §7.5.2. In other words, Γ is interpretable in $\Sigma^{[2]}$ (where Σ is empty). Therefore, by §7.5.2 and by §4.3, if A is the square of another space B, then $A = B^2 \models^{\text{ctn}} \Gamma$.

In fact, if we apply the definition (23) to our operations $\overline{d}$ and $\overline{H}$ of §7.5.2, we obtain the following concrete definition of a continuous $\overline{\star}$ modeling Γ on any square B^2:

$$(24) \qquad\qquad (b_1, b_2) \,\overline{\star}\, (b_3, b_4) = (b_2, b_3).$$

(And the fact that $(B^2, \overline{\star}) \models \Gamma$ can be reconfirmed by an easy calculation.)

Thus (22) is an example of an equation that is 2-undemanding (§7.5.3) but is not undemanding (§6.1).

(This discussion of Γ and $\bar{\mp}$ is due in part to T. Evans (Evans, 1967). Equation (22) was also discussed on pages 202–203 of (Taylor, 2000).)

7.5.5 A special case: $A = \mathbb{R}^k$.

We mentioned at the start of §7.5 that one might need to know all topological factorizations of A as a power B^k in order to assess the truth of $B \models^{\text{ctn}} \Sigma^{[k]}$. There is one case where all such factorizations are known, namely $A = \mathbb{R}^k$. If $\mathbb{R}^k$ is homeomorphic to B^k for some B, then B is homeomorphic to $\mathbb{R}$ (Taylor, 2006, Lemma 29) (bassed on F. B. Jones and G. S. Young (Jones and Young, 1959)). (In other words, topologically the space $\mathbb{R}^k$ has unique k-th roots.)

The remarks in §7.5 now yield that, *for any finite k and any set Σ of equations, $\Sigma^{[k]}$ is compatible with $\mathbb{R}^k$ if and only if Σ is compatible with $\mathbb{R}$.* (This result first appeared as (Taylor, 2006, Corollary 30).)

Few other k-th power spaces are known to have unique k-th roots, and so the result stated here cannot be generalized very far. It does, however, hold for powers $[0,1]^k$.

7.5.6 The $[k]$-th root of a theory.

It is possible to turn the tables and define a theory $\sqrt[k]{\Sigma}$ such that an arbitrary space A is compatible with $\sqrt[k]{\Sigma}$ if and only if the space A^k is compatible with Σ. The theory $\sqrt[k]{\Sigma}$ was defined by R. McKenzie in 1975 (see (McKenzie, 1975)); it is also briefly discussed on page 68 of (García and Taylor, 1984).

We will exhibit $\sqrt[k]{\Sigma}$ for $k = 2$ and Σ the theory of H-spaces (binary multiplication with two-sided unit element, §7.2.6). Here is $\sqrt[2]{\Sigma}$; it has two constants and two 4-ary operations:

$$f_1(x_1, x_2, c_1, c_2) \approx x_1$$
$$f_2(x_1, x_2, c_1, c_2) \approx x_2$$
$$f_1(c_1, c_2, x_1, x_2) \approx x_1$$
$$f_2(c_1, c_2, x_1, x_2) \approx x_2.$$

It should be clear that if operations $\overline{f}_i, \overline{c}_i$ $(i = 1, 2)$ satisfy these equations on A, then one may define an H-space operation on A^2 via

$$\overline{F}((a_1, a_2), (a_3, a_4)) = (\overline{f}_1(a_1, \ldots, a_4), \overline{f}_2(a_1, \ldots, a_4))$$

for all $a_1, \ldots, a_4 \in A$. The general method of defining $\sqrt[k]{\Sigma}$ should be clear from here.

Obviously in general $I(A) \subseteq I(A^k)$, and the reverse inclusion may fail; for example, if $\Sigma = \Delta^{[k]}$ for some Δ and if A is not homeomorphic to a k-th power, then $\Delta^{[k]} \in I(A^k)$ but $\Delta^{[k]} \notin I(A)$ (for Δ taken as, say, the empty theory). In terms of this section, we may equivalently say that if $A \models^{\mathrm{ctn}} \Sigma$, then $A \models^{\mathrm{ctn}} \sqrt[k]{\Sigma}$, but not always conversely.

J. van Mill exhibited (Mill, 1981) a space V such that V is not compatible with group theory, but V^2 is compatible. In other words group theory lies in $I(V^2)$ but not $I(V)$. Nevertheless, the space V seems far from being a finite space, and we do not expect examples of this type to play a big role in the analysis of compatibility for finite spaces.

If Σ is a set of *simple equations* (see §7.4), then $\sqrt[k]{\Sigma}$ is equivalent to Σ in the interpretability lattice, which entails that $I(A^k) = I(A)$ and that $A^k \models^{\mathrm{ctn}} \Sigma$ implies $A \models^{\mathrm{ctn}} \Sigma$. This theorem was proved in 1983 by B. Davey and H. Werner (Davey and Werner, 1983), and about the same time by R. McKenzie [unpublished]. A later proof appears in (García and Taylor, 1984, Prop. 39, p. 69).

7.6 Miscellaneous Σ.

7.6.1 Exclusion of fixed points.

We consider the equation-set

$$F(x,x,y) \approx y; \qquad F(\phi(x),x,y) \approx x.$$

If A is a space of more than one element that has the *fixed-point property* (each continuous self-map has a fixed point), then, applying that property to a given $\overline{\phi}$, we clearly see that these equations are not compatible with A. Such spaces include the closed simplex of each finite dimension (Brouwer fixed-point Theorem).

The equations also fail to be compatible with S^1—which obviously does not have the fixed-point property. Indeed, in §4.3.1, $S^1 \models^{\mathrm{ctn}} \Sigma$ if and only if, in our lattice, Σ lies below the theory of Abelian groups. Thus ϕ will be interpreted as a unary Abelian group operation. All such operations have 0 as a fixed point, and so the fixed-point argument may be applied again.

It is easy to find operations that show $\mathbb{R}$ to be compatible with the equations, but in fact I do not know of any finite complex that is compatible.

In the reverse direction, one may note that in 1959 E. Dyer and A. Shields proved (Dyer and Shields, 1959) that if A is a finite-dimensional compact connected space compatible with lattice theory, then A has the fixed-point property.

7.6.2 One-one but not onto.

We consider the equations

(25) $$F(x,y,0) \approx x, \quad F(x,y,1) \approx y,$$

(26) $$\psi(\theta(x)) \approx x, \quad \phi(\theta(x)) \approx 0, \quad \phi(1) \approx 1,$$

which first appeared in (Taylor, 1986, §3.17). Clearly this set is demanding (see §6.1.1). In a non-singleton model $\mathbf{A} = (A,\overline{F},\overline{\psi},\overline{\theta},\overline{\phi},\overline{0},\overline{1})$, Equations (25) imply that $\overline{0} \neq \overline{1}$. The next equation tells us that θ is one-to-one, and the last two tell us (using $\overline{0} \neq \overline{1}$) that the range of θ is not all of A. Every one-one continuous self-map of the sphere S^n ($n = 1,2,\ldots$) maps onto S^n (for example, by the Invariance of Domain Theorem). Therefore these equations are incompatible with spheres S^n. (For most spheres, we already knew this, by §6.1. For S^1, S^3 and S^7, the result is new in this section; for all spheres, the proof here is much easier than the proof referenced in §6.1.)

On the other hand, it is not hard to satisfy the equations with continuous operations on the closed interval $[0,1]$:

(27) $$\overline{0} = 0, \quad \overline{1} = 1, \quad \overline{F}(a,b,c) = (1-c)a + cb$$

(28) $$\overline{\theta}(a) = a/2, \quad \overline{\psi}(a) = 2a \wedge 1, \quad \overline{\phi}(a) = (2a-1) \vee 0.$$

We would also like to see that Equations (25–26) can be satisfied on $[0,1]$ with (continuous) piecewise linear operations. The operations in Line (28) are already piecewise linear; we need only add a piecewise linear definition for (a new) $\overline{F}$ that satisfies (25). The reader may check that the following definition suffices:

$$\overline{F}(a,b,c) = \begin{cases} a \vee 2c & \text{if } c \leq 1/2 \\ b \vee (2-2c) & \text{if } c \geq 1/2. \end{cases}$$

A slight variant of Equations (25–26) replaces Equations (25) with the equations of §7.2.6. These equations serve, again, to separate 0 from 1 in any algebra of more than one element. They are satisfied on $[0,1]$ by using (28) together with the ordinary meet operation on $[0,1]$.

7.6.3 Entropic operations on $[0,1]$.

In 1974 Fajtlowicz and Mycielski (see (Fajtlowicz and Mycielski, 1974)) considered continuous affine combinations on $[0,1]$, i.e. functions that have this form:

(29) $$\overline{F}_\alpha(a,b) = \alpha a + (1-\alpha)b,$$

one such operation for each $\alpha \in [0,1]$. Such an operation is easily seen to satisfy the equations

$$F_\alpha(x,x) \approx x, \qquad F_\alpha(F_\alpha(x,y),F_\alpha(u,v)) \approx F_\alpha(F_\alpha(x,u),F_\alpha(y,v))$$

The first of these is the idempotent law; the second is the *entropic law*. They also proved that if α is transcendental, then $([0,1],\overline{F}_\alpha)$ satisfies no equations other than the logical consequences of idempotence and entropicity. These equations are obviously undemanding (see the algorithm in §6.1.1), and hence not interesting for the present investigation.

On the other hand, they proved that if α is algebraic, then $([0,1],\overline{F}_\alpha)$ satisfies some equations beyond the logical consequences of idempotence and entropicity. Regrettably, I don't know which values of α yield an equation set that is demanding. (E.g. when $\alpha = 1/2$, we have the equation $F_\alpha(x,y) \approx F_\alpha(y,x)$, which renders the equations demanding. I don't know other examples.)

One may further consider two or more $\overline{F}_\alpha$ in the same term. For instance, for any α and β we clearly have the *mixed entropic law*

$$F_\alpha(F_\beta(x,y),F_\beta(u,v)) \approx F_\beta(F_\alpha(x,u),F_\alpha(y,v)).$$

Moreover, one can consider affine combinations with more than two variables. We do not emphasize such combinations, since each of them can be formed by concatenating binary affine combinations. For example, given positive reals μ, ν, λ that sum to 1, if we let $\alpha = \mu+\nu$ and $\beta = \mu/(\mu+\nu)$, then we have

$$\overline{F}_\alpha(\overline{F}_\beta(x,y),z) = \mu x + \nu y + \lambda z.$$

7.6.4 Some twisted ternary operations on $[0,1]$.

Let $\overline{R}_\theta : \mathbb{R}^3 \longrightarrow \mathbb{R}^3$ be the rotation of 3-space through angle θ, whose axis is the line joining $(0,0,0)$ to $(1,1,1)$. (For example, when $\theta = 2\pi/3$ this rotation cyclically permutes the three positive coordinate axes.) Further, let $\overline{m}$ be the ternary majority operation on $\mathbb{R}$ that is defined by Equation (8) of §7.2.5. Here we consider the composite ternary operation on $\mathbb{R}$, defined by

$$\overline{F}_\theta = \overline{m} \circ \overline{R}_\theta.$$

As established in (Taylor, 2006, §9.4), the interval $[0,1]$ is a subuniverse of $(\mathbb{R},\overline{F}_\theta)$, and moreover the operation $\overline{F}_\theta$ satisfies the equations

$$F_\theta(x,x,x) \approx x, \qquad F_\theta(x,y,z) \approx F_\theta(z,x,y).$$

Moreover, the derived operation

$$\overline{p}_\theta(a,b) \;=\; \overline{F}_\theta(a,a,b)$$

turns out to be an affine combination on $[0,1]$ (as defined in Equation (29)). Therefore p_θ obeys the idempotent and entropic equations of §7.6.3, plus further equations if the coefficients of p_θ are algebraic. These easily translate to further laws for $\overline{F}_\theta$.

We do not have a clear idea of how high in the lattice these examples might lie.

8 The operations needed for the examples in §7.

Somewhat surprisingly, the concrete examples of compatibility provided throughout §7 require operations only of a very unsophisticated design. (A few examples above, such as the P in §7.3, are originally formed as products. In such a case, the following analysis should be seen as applying to the two factors separately.)

8.1 Multilinear Operations.

Let A be a topological subspace of some power $\mathbb{R}^k$. For $\alpha \in \mathbb{R}$ and K a subset of $\{1,\ldots,n\} \times \{1,\ldots,k\}$ we consider continuous functions $\overline{F}\colon A^n \longrightarrow \mathbb{R}$ of the form

$$\overline{F}((x_{(1,1)},\ldots,x_{(1,k)}),\ldots,(x_{(n,1)},\ldots,x_{(n,k)})) \;=\; \alpha \prod_{(i,j)\in K} x_{(i,j)}$$

uniformly over A^n. We usually restrict our attention to sets K with the property that for each $i \in \{1,\ldots,n\}$ there is at most one j with $(i,j) \in K$. For such a set K, we will say that this $\overline{F}$, or any linear combination of such $\overline{F}$'s for varying α and K, is *multilinear*.

If $k = 1$ (in which case, each $j = 1$) we may say that $\overline{F}$ is *linear*, resp. *bilinear*, if each K is a singleton, resp. has two elements.

Then for an operation $\overline{F}\colon A^n \longrightarrow A \subseteq \mathbb{R}^k$, we say that $\overline{F}$ is multilinear (resp. linear) iff each of its components is multilinear (resp. linear).

Given a continuous operation $\overline{F}\colon A^n \longrightarrow A$, if the space A^n may be triangulated so that on each simplex $\overline{F}$ is multilinear, then we may say that $\overline{F}$ is *piecewise multilinear*. Similarly *piecewise linear*, *piecewise bilinear*, and so on.

8.2 Piecewise linear operations seem to suffice on $[0,1]$.

Let us first look at $I([0,1])$, the sets Σ known to be compatible with the interval $[0,1]$. In fact, *piecewise linear operations suffice* for all the concrete examples included in §7. The piecewise linearity is made explicit in Equation (7) of §7.2.1, in points (i) and (ii) of §7.4.1 and in §7.6.2; elsewhere it may be easily inferred from the context.

In detail, the operations in §7.2.1 are piecewise linear, by Equation (7). The equations in §7.2.2 can be modeled either with piecewise linear functions or with Chebyshev polynomials (among infinitely many possibilities). The equation-sets below lattice theory—semilattices in §7.2.4, majority operations in §7.2.5 and 0,1-multiplication in §7.2.6—are *a fortiori* satisfied with piecewise linear operations on $[0,1]$. And then the minority operation $\bar{q}$ defined in Equation (11) of (§7.4.1), the Mal'tsev operation of §7.4.3, and the two-thirds minority operation $\bar{t}$ of §7.4.4 are linear combinations of operations defined earlier, and hence still piecewise linear.

Finally, it is not hard to check that the entropic operations in §7.6.3, and the twisted operations in §7.6.4 are all piecewise linear. As for the composite ring-lattice operations in §7.6.2, we gave two ways to define $\overline{F}$, one piecewise linear, and one not.

In the first sentence of §7.4 we cited only an existence proof for operations on $[0,1]$ to make A compatible with Σ. To constructively provide such operations would require solving the word problem for free Σ-algebras, and then analyzing the topological structure of $\mathbf{F}_\Sigma([0,1])$.

$\Sigma_{[0,1]}$ obviously defines a huge and complicated mathematical structure; complete knowledge of it may be impossible (unless, for example, we are so lucky as to find a simple finite generator for $I([0,1])$). We do, however, know something about it. In several places—notably §7.2, §7.4 and §7.6—we have reported on positive findings of $[0,1] \models^{\mathrm{ctn}} \Sigma$ for various sets Σ. Each of these reports amounts to a description of a finite piece of $\Sigma_{[0,1]}$.

8.3 Some further piecewise bilinear operations on a closed interval.

In this speculative section we note the possibility that for A a closed interval of the real line, there may exist $\Sigma \in I(A)$ that is higher than any other such Σ that we have considered so far in this account.

In this context it works best to consider the interval $[-1,1]$. The operations we will consider, beyond the ordinary join $\vee$ and meet $\wedge$ and constants 0 and 1 that we have already considered, are these:

　(i) Ordinary multiplication, $x \cdot y$
　(ii) Truncated addition: $x \boxplus y$, to mean $[(x+y) \wedge 1] \vee (-1)$

(iii) Shrinking: $\overline{F}(x)$ to mean[9] $x/3$.

Besides the distributive-lattice equations for $\wedge, \vee$, and commutativity and associativity for $x \cdot y$, the equations satisfied by our operations include[10] these:

$$(30) \qquad x \boxplus y \approx y \boxplus x; \quad (F(x) \boxplus F(y)) \boxplus F(z) \approx F(x) \boxplus (F(y) \boxplus F(z))$$

$$(31) \qquad (F(x) \boxplus F(x)) \boxplus F(x) \approx x$$

$$(32) \qquad x \cdot (F(y) \boxplus F(z)) \approx (x \cdot F(y)) \boxplus (x \cdot F(z))$$

$$(33) \qquad (x \vee 0) \cdot (y \wedge z) \approx (x \vee 0) \cdot y \wedge (x \vee 0) \cdot z$$

$$(34) \qquad ((x \vee 0) \boxplus y) \boxplus (z \vee 0) \approx (x \vee 0) \boxplus (y \boxplus (z \vee 0))$$

$$(35) \qquad ((x \vee 0) \boxplus y) \boxplus (z \wedge 0) \preccurlyeq (x \vee 0) \boxplus (y \boxplus (z \wedge 0))$$

$$(36) \qquad F(x \wedge y) \approx F(x) \wedge F(y)$$

$$(37) \qquad (x \cdot x) \vee 0 \approx x \cdot x$$

$$(38) \qquad (x \wedge 0) \cdot (y \wedge z) \approx ((x \wedge 0) \cdot y) \vee ((x \wedge 0) \cdot z)$$

and the duals of (33–36). (The dual of (37) is **not** included.) For the notation in (35): $\alpha \preccurlyeq \beta$ means $\alpha \vee \beta \approx \beta$. Probably the careful reader can find further interesting examples.[11]

For our context, the question is whether the operations defined here on $[-1, 1]$ satisfy an equation-set that lies higher in the interpretability lattice than (or incomparable with), say, the equations already seen in §7.2.2. Equations (30–36) do *not* have this property: they are (jointly) interpretable in distributive lattice theory by defining $F(x)$ to be x, and defining both $x \boxplus y$ and $x \cdot y$ to be $x \wedge y$. This interpretation does not work for the set of Equations (30–38); we do not know the location in the lattice of this set.

8.4 Multilinear maps define many group operations.

The groups on S^1, S^3 and S^7 (see §7.1.1) all proceed from coordinate systems (pairs, quadruples or octuples of real numbers). The product in S^3, say, of (x_1, x_2, x_3, x_4) and (y_1, y_2, y_3, y_4) has four components, each of which is a bilinear function of the x_i and the y_j—a linear combination of the sixteen products $x_i y_j$. Products in S^1 and S^7 are calculated similarly. In all three groups, inverses are calculated by a form of conjugation, which is linear.

The matrix groups (§7.1.2) involve the ordinary product of two $N \times N$ matrices; in the product, each entry is a bilinear function of the entries in the two given matrices. In dealing with unitary matrices, the inverse is simply

---

[9] The 3 is somewhat arbitrary here.

[10] We thank George M. Bergman for Equations (35) and (38).

[11] For example the list could be extended by adding 1 (or any other constants), and adding laws satisfied by such constant(s).

conjugation, which is linear. For more general non-singular matrices, one will also require the operation of calculating inverses, which can be seen as the calculation of many determinants, followed by non-zero division. Each determinant may be calculated as a multilinear function of the columns.

8.5 Point operations.

Operations such as those defined in Equations (19–20) of §7.5.2 were termed *point operations* by Trevor Evans in (Evans, 1967). Another point operation may be seen in Equation (24). The definition is that our space is a direct power, and each coordinate of an $\overline{F}$-value is determined as one of the input coordinates. Obviously point operations are multilinear in the sense of §8.1.

More generally, if each coordinate of an $\overline{F}$-value is determined as one of the input coordinates *or a constant*, then we have operations that can model the k-undemanding equations in the third paragraph of §7.5.3.

The operation of $\Sigma^{[k]}$ defined in Equation (15) of §7.5.1 may be seen as a hybrid of Evans' pure point operations, and the basic operations of the root variety Σ. If the basic operations $\overline{F}_i$ of §7.5.1 can be taken as multilinear, then the constructed operation $\overline{F}$ defined in Equation (15) will be multilinear as well.

8.6 Operations of arity 4 and higher.

None of our concrete examples mentions an operation of arity 4 or higher. (Of course simple equations (§7.4) can involve operation symbols of any arity.) We therefore do not know of any significant role played by N-ary operations for $N \geq 4$. For example, we do not know whether, for each $N \geq 4$, there exists a (finite) space A such that any generator of the ideal $I(A))$ (§4.3) must include an operation symbol of arity $\geq N$. (In fact we do not even know whether this holds with $N = 3$; some of our examples involve ternary operations, but in some or all cases they might be dispensable.)

9 Outlook and questions.

From known examples of the compatibility relation $A \models^{\mathrm{ctn}} \Sigma$, and from the many instances in which the relation is known to fail, it may be possible to catalog or classify the possibilities, at least for some finite spaces A (i.e. spaces homeomorphic to the realization of a finite simplicial complex) and for some finite Σ. Or at least to formulate a conjecture as to what is possible.

9.1 Topological models of a given theory Σ.

It may be difficult to characterize or enumerate the finite models of a given Σ. The overall difficulty should be apparent from the surprising results surrounding H-spaces (§7.1.1).

Moreover, there seems to be little structure to the collection (among finite complexes) of all topological groups, say, or all topological semilattices, etc. Algebraically, the collection is a category and a variety, and products are of some use—e.g. the product of two finite complexes is a finite complex. But subalgebras and homomorphic images of finite complexes are not usually finite complexes.

There are, of course, a few exceptional cases where the topological spaces compatible with Σ can be expressly described or classified. Such are for example the squaring equations of §7.5.3 (and analogous k-th power equations), and also the majority operations of §7.2.5.

9.2 The theories compatible with a given space A.

In a few places—such as §7.5.3 and §4.3.1—we have seen a space A for which the compatible theories Σ can be described or enumerated, such as $A = S^1$. For general A, however, the task eludes us.

More precisely, we are asking for some description of the ideal $I(A)$ of §4.3 and §4.3.1. We thus have the lattice-theoretic structure to help formulate a description. In particular, we know (§4.3.1) that $I(A)$ is principal. The task here will be to find a generator, or generating family, that is (in some sense) small and easily understood.

For a relatively simple space like $[0,1]$ or its finite powers, it may be possible to refine our understanding of $I(A)$. It seems interesting that all the known theories compatible with $[0,1]$ are very simple (or lie below some simpler compatible theory). This points either to an inherent simplicity of $I([0,1])$, or to a large misapprehension on the part of those who have studied it. Hopefully, the former.

9.3 The theories compatible with any finite space.

Let I be the union of the ideals $I(A)$ for all finite complexes A. By §6.2 it is not an ideal, but it is down-closed. Remarkably, it again seems that everything we know to be in I is relatively simple, or at least lies below a fairly simple set of equations. The upper boundary of I may be easier to define than the boundaries of an individual $I(A)$. (We have no conjecture as to a possible form.)

9.4 Specific questions.

9.4.1 Thoroughness of §7.

Does §7 include, at least implicitly, all the **known** *examples of equation-sets Σ that hold on a finite topological space A?*

(In saying "implicitly," we allow for example that Σ might lie below some theory mentioned in our text, or that A might be a direct product or a finite power.) (And of course, this could change with time; again please let the author know of any new discoveries.)

9.4.2 Completeness of §7.

Does §7 include, at least implicitly, all equation-sets Σ that hold on a finite topological space A?

In other words, we are asking about the down-set I described in §9.3. The answer here may surely be "no," even after §9.4.1 may have been corrected. It may, however, be true that we are close to a full knowledge of I.

9.4.3 What is $I = \bigcup\{I(A) : A$ any finite complex$\}$?

For example, *Does there exist a recursive sequence Σ_0, $\Sigma_1 \ldots$ (with each Σ_n a finite set of equations) such that $\Sigma \in I$ if and only if for some n, $\Sigma \leq \Sigma_n$ in the interpretability lattice?*

9.4.4 What operations are needed for I?

For each $\Sigma \in I$ (as defined in §9.4.3), do there exist a finite complex A and continuous piecewise multilinear operations $\overline{F}_t$ on A such that $(A, \overline{F}_t)_{t \in T} \models \Sigma$?

We suspect the answer here is "no," but we have no counterexample.

If not, does there exist some reasonable enlargement of the category "piecewise multilinear" for which the answer is yes?

9.4.5 Algorithmic questions: fixed space.

Given a fixed finite space A, does there exist an algorithm that inputs a finite set Σ of equations, and outputs whether $A \models^{\mathrm{ctn}} \Sigma$?

Given a fixed finite space A, is the set of all finite Σ with $A \models^{\mathrm{ctn}} \Sigma$ recursively enumerable?

(We assume that one can work out a language to code a set of equations.)

In special cases, an algorithm for $A \models^{\mathrm{ctn}} \Sigma$ exists and is implicit in what we have already written. For A one of the spaces of §6.1, the algorithm would check whether Σ is undemanding. For a k-th power of one of those spaces, the algorithm would check whether Σ is k-undemanding (§7.5.3). For the sphere S^1, one would check whether Σ can be modeled by linear operations with integer coefficients (see §4.3.1). For the majority of finite spaces, however, the answer is unknown. In fact, we know of no finite A for which we can say that the answer to either question is negative. By contrast, for $A = \mathbb{R}$, we do know that there is no algorithm (see (Taylor, 2006)).

(The proof in (Taylor, 2006) of the algorithmic undecidability of $\mathbb{R} \models^{\mathrm{ctn}} \Sigma$ seems to require a non-compact space, where some periodic functions can be found to live.)

9.4.6 Algorithmic questions: fixed theory.

For a fixed set Σ of equations, does there exist an algorithm to decide, for a finite complex A, whether $A \models^{\mathrm{ctn}} \Sigma$?

Is the set of such A recursively enumerable?

We advise the reader that some very simple questions on finite complexes— such as the question of the simple connectedness of a 2-complex—can fail to have an algorithmic solution. (See (Markov, 1958) or (Barwise, 1977) for examples.)

9.4.7 How well can $A \models^{\mathrm{ctn}} \Sigma$ separate two theories?

We restate a question from §7.2.3:

Does there exist a space A that is compatible with lattice theory but not with modular lattice theory? Is there a finite space A with this property?

Obviously the corresponding question may be asked for any two Σ that are distinct in the interpretability lattice. The specific question posed here remains open, and may be taken as an indicator of how little we know in this area. For a similar question, that of separating modular lattice theory from distributive lattice theory, an example is known. If B is the topological space formed by joining three triangles (closed 2-simplices) along a single common edge, then B is compatible with modular lattice theory, but not with distributive lattice theory. The proof—to which the present author contributed—may be found in §3.1 of G. Bergman (Bergman, 2015); see especially Theorem 7 there.

9.4.8 Description of $I([0,1])$?

(i) Does §7 give a thorough description of all known Σ compatible with $[0,1]$?

(ii) Is there a finite Σ that generates the ideal $I([0,1])$ of all theories compatible with the interval $[0,1]$? If so, attempt to exhibit a specific finite generator Σ.

(iii) If so, will the Σ that is implicit in §7 suffice for this purpose? Would it help to include the operations shown in §8.3?

(iv) Can one recursively enumerate a set of finite Σ's that collectively generate the ideal $I([0,1])$?

(v) In question (ii) or (iv), can one find such a Σ (or Σ's) that can be modeled with piecewise linear operations on $[0,1]$?

(vi) In question (ii) or (iv), can one find such a Σ (or Σ's) whose operation symbols all have arity ≤ 3? What about ≤ 2?

9.4.9 Description of $F([0,1])$?

Describe the filter $F([0,1])$ of all theories not compatible with the space $[0,1]$. If possible, frame this description as a weak Mal'tsev condition (Taylor, 1973).

As mentioned in §4.3.4, $F([0,1])$ is not a Mal'tsev filter.

9.4.10 Other spaces A.

The questions in §9.4.8 may be asked for any space A, and we consider them to be on the table, especially for A a finite complex. ("Linearity" may require a specified coordinate system.) With a few exceptions (such as $A = S^1$), we do not expect them to be any easier than the questions for $A = [0,1]$.

If A is product-indecomposable, then the questions of §9.4.9 may also be asked for A.

9.4.11 How dense are the non-trivial finite complexes?

Among those complexes that have at most m simplices, of dimension at most n, what fraction are compatible with some demanding theory (§6.1)?

We expect a meaningful answer only in the limit as m, or as m and n together, approach infinity. The precise method of counting complexes (simply by raw data, or by isomorphism types of complex, or by homeomorphism types of space, for example), is certainly part of the problem. We would not be surprised if the limiting fraction turns out to be zero.

References

Adams, J. F. (1960). On the non-existence of elements of Hopf-invariant one, *Mathematische Annalen* (2) 72, 20–104.

Anderson, L. W. (1959). On the distributivity and simple connectivity of plane topological lattices. *Transactions of the American Mathematical Society* 91, 102–112.

Antonovskiĭ, M. Ja. and Mironov, A. V. (1967). On the theory of topological l-groups. (Russian. Uzbek summary) *Dokl. Akad. Nauk UzSSR*, no. 6, 6–8. MR 46 #5528.

Baker, K.A. and Stralka, A.R. (1970). Compact, distributive lattices of finite breadth, *Pacific Journal of Mathematics* 34, 311–320. http:// projecteuclid.org/euclid.pjm/1102976425. MR0282895

Bandelt, H.-J. and Hedlíková, J. (1983). Median algebras, *Discrete Mathematics* 45, 1–30.

Bateson, A. (1982). Fundamental groups of topological R-modules, *Transactions of the American Mathematical Society* 270, 525–536.

Bergman, G. (2015). Simplicial complexes with lattice structures, preprint; to appear. https://math.berkeley.edu/~gbergman/papers/geom_lat.pdf

Blok, W.J. and Pigozzi, D. (1988). Alfred Tarski's Work on General Metamathematics, *Journal of Symbolic Logic* 53 (03), 36–50.

Borsuk, K. (1967). *Theory of retracts*, Monografie Matematyczne, Tom 44, Państwowe Wydawnictwo Naukowe, Warsaw, 251 pages.

Bott, R. (1953). On symmetric products and the Steenrod squares, *Annals of Mathematics* 57, 579–590.

Brown, D. R. (1965). Topological semilattices on the two-cell, *Pacific Journal of Mathematics* 15, 35–46.

Choe, T.H. (1969). The breadth and dimension of a topological lattice, *Proceedings of the American Mathematical Society* 23, 82–84. MR0248760

Clifford, A. H. (1958). Connected ordered topological semigroups with idempotent endpoints, I, *Transactions of the American Mathematical Society* 88, 80–98.

Davey, B. and Werner, H. (1983). Dualities and equivalences for varieties of algebras, *Colloquia mathematica Societatis János Bolyai* 33, 181–275.

Dyer, E. and Shields, A. (1959). Connectivity of topological lattices, *Pacific Journal of Mathematics* 9, 443–448.

Edmondson, D. E. (1956). A nonmodular compact connected topological lattice, *Proceedings of the American Mathematical Society* 7, 1157–1158.

Evans, T. (1967). Products of points — some simple algebras and their identities, *American Mathematical Monthly* 74, 362–372.

Fajtlowicz, S. and Mycielski, J. (1974). On convex linear forms, *Algebra Universalis* 4, 244–249.

Faucett, W. M. (1955). Compact semigroups irreducibly connected between two points; Topological semigroups and continua with cutpoints, *Proceedings of the American Mathematical Society* 6 , 741–756.

García, O. C. and Taylor, W. (1984). *The lattice of interpretability types of varieties*, Memoirs of the American Mathematical Society, Number 305, iii+125 pages. MR 86e:08006a.

Gierz, G. and Stralka, A. R. (1989). Modular lattices on the 3-cell are distributive, *Algebra Universalis* 26, 1–6.

Harper, J. R. (1972). Homotopy groups of finite H-spaces, *Bulletin of the American Mathematical Society* 78, 532–534.

Hedegaard, R. and Weisstein, E. W. (2016). H-space, `http://mathworld. wolfram.com/H-Space.html`

Hu, S. T. (1965). *Theory of retracts*, Wayne State University Press, Detroit.

Isbell, J. R. (1968). Remarks on decompositions of categories, *Proceedings of the American Mathematical Society* 19, 899-904.

Isbell, J. R. (1980). Median algebra, *Transactions of the American Mathematical Society* (2) 260, 319–362.

James, I. M. (1957). Multiplication on spheres. I, II, *Proceedings of the American Mathematical Society* 13, 192–196 and *Transactions of the American Mathematical Society* 84 (1957), 545–558.

Jones, F. B. and Young, G. S. (1959). Product spaces in n-manifolds, *Proceedings of the American Mathematical Society* 10, 307-308.

Kaplansky, I. (1947). Topological Rings, *American Journal of Mathematics* 69, 153–183.

Koch, R. J. and Wallace, A. D. (1958). Admissibility of semigroup structures on continua, *Transactions of the American Mathematical Society* 88, 277–287.

Koch, R. J. and Wallace, A. D. (1970). Topological semilattices and their underlying spaces, *Semigroup Forum* 1, 209–223.

Kolmogorov, A. N. (1957). On the representation of continuous functions of many variables by superposition of continuous functions of one variable and addition (Russian), *Dokl. Akad. Nauk SSSR* 114, 953–956.

Lawson, J. D. and Madison, B. (1970). Peripherality in semigroups, *Semigroup Forum* 1, 128–142.

Lie Groups, Wikipedia article, (2016). `http://en.wikipedia.org/wiki/ Triangulation_(topology)`.

Mal'tsev, A. A. and Nurutdinov, B. S. (1975). Iterative algebras of continuous functions (Russian), *Dokl. Akad. Nauk SSSR* 224, 532–535.

Markov, A. (1958). The insolubility of the problem of homeomorphy (Russian), *Dokl. Akad. Nauk SSSR* 121, 218–220.

McKenzie, R. (1975). On spectra, and the negative solution of the decision problem for identities having a non-trivial finite model, *The Journal of Symbolic Logic* 41, 186–196.

Neumann, W. D. (1974). On Mal'cev conditions, *Journal of the Australian Mathematical Society* 17, 376–384.

Palasińska, K. and Pigozzi, D. (1995). Gentzen-style axiomatizations in equational logic, *Algebra Universalis* 34, 128–143.

Pixley, A. F. (1979). Characterizations of arithmetical varieties, *Algebra Universalis* 9, 87-?98.

Pour-El, M. B. and Richards, J. I. (1989). *Computability in analysis and physics,* Perspectives in Mathematical Logic, Springer-Verlag, Berlin. xii+206 pp.

Barwise, J. (ed.) (1977). *Recursion theory, Handbook of mathematical logic,* Part C, pp. 525-815. Studies in Logic and the Foundations of Math., Vol. 90, North-Holland, Amsterdam.

Sholander, M. (1954). Medians and betweenness, *Proceedings of the American Mathematical Society* 5, 801–807; Medians, lattices and trees, *ibid*, 808–812.

Świerczkowski, S. (1964). Topologies in free algebras, *Proceedings of the London Mathematical Society* (3) 14, 566–576.

Taylor, W. (1973). Characterizing Mal'cev conditions, *Algebra Universalis* 3, 351–397.

Taylor, W. (1975). The fine spectrum of a variety, *Algebra Universalis* 5, 263–303.

Taylor, W. (1977). Varieties obeying homotopy laws. *Canadian Journal of Mathematics* 29, 498–527.

Taylor, W. (1986). *The clone of a topological space,* Volume 13 of *Research and Exposition in Mathematics,* 95 pages. Heldermann Verlag, Berlin.

Taylor, W. (2000). Spaces and equations, *Fundamenta Mathematicae* 164, 193–240.

Taylor, W. (2006). Equations on real intervals, *Algebra Universalis* 55, 409–456.

Taylor, W. (2009). Simple equations on real intervals, *Algebra Universalis* 61, 213–226.

Taylor, W. (2010). Approximate satisfaction of identities. 98 pp., to appear, see `http://euclid.colorado.edu/~wtaylor/approx.pdf`, see also `http://arXiv.org/abs/1504.01165`

Taylor, W. (2011). Discontinuities in the identical satisfaction of equations. 75 pp., to appear, see `http://euclid.colorado.edu/~wtaylor/jumps.pdf`, see also `http://arXiv.org/abs/1504.01445`

Mill, J. van (1981). A rigid space X for which $X \times X$ is homogeneous, *Proceedings of the American Mathematical Society* 83, 597–600.

Mill, J. van (1983). A topological group having no homeomorphisms other than translations, *Transactions of the American Mathematical Society* 280, 491–498.

Mill, J. van and Vel, M. van de (1979). On an internal property of absolute retracts, *Topology Proceedings* 4, 193–200.

Mill, J. van and Vel, M. van de (1982). On an internal property of absolute retracts II, *Topology and its applications* 13, 59–68.

Wallace, A. D. (1955). The structure of topological semigroups, *Bulletin of the American Mathematical Society* 61, 95–112.

Whitehead, G. W. (1978). *Elements of homotopy theory*, Graduate Texts in
    Mathematics, Volume 61, Springer-Verlag, Berlin.

Categorical Abstract Algebraic Logic: Compatibility Operators and Correspondence Theorems

George Voutsadakis

To **Don Pigozzi** *this work is dedicated
on the occasion of his 80th Birthday.*

Abstract Very recently Albuquerque, Font and Jansana, based on preceding work of Czelakowski on compatibility operators, introduced coherent compatibility operators and used Galois connections, formed by these operators, to provide a unified framework for the study of the Leibniz, the Suszko and the Tarski operators of abstract algebraic logic. Based on this work, we present a unified treatment of the operator approach to the categorical abstract algebraic logic hierarchy of π-institutions. This approach encompasses previous work by the author in this area, started under Don Pigozzi's guidance, and provides resources for new results on the semantic, i.e., operator-based, side of the hierarchy.

Key words: Leibniz operator, Tarski operator, Suszko operator, logical matrix, full model, reduced model, Leibniz filter, protoalgebraic logic, equivalential logic, algebraizable logic

2010 Mathematics Subject Classification: 03B22, 18C50, 03G27

1 Introduction: The Three Operators of AAL

The operator approach in abstract algebraic logic (AAL) has born many fruits and forms the cornerstone of all three main directions in the field: The association of classes of algebras with logical systems, the correspondence

George Voutsadakis
School of Mathematics and Computer Science, Lake Superior State University, Sault Sainte Marie, MI 49783, USA, e-mail: gvoutsad@lssu.edu

© Springer International Publishing AG 2018
J. Czelakowski (eds.), *Don Pigozzi on Abstract Algebraic Logic,
Universal Algebra, and Computer Science*, Outstanding Contributions
to Logic 16, https://doi.org/10.1007/978-3-319-74772-9_15

between logical and algebraic properties and the study of specific classes of logical systems or specific classes of algebras per se in the context of algebraization and/or property correspondence. General surveys of the approach can be found in (Font and Jansana, 1996; Czelakowski, 2001; Font et al., 2003).

The operator approach has been extended by the author to a categorical framework starting with (Voutsadakis, 2015a) and in a bulk of subsequent work (Voutsadakis, 2005a,b, 2007a,b) and has provided equally intriguing results, while in the last few years, the levels of the AAL hierarchy of sentential logics have been extended, based on this approach, to the categorical abstract algebraic logic (CAAL) hierarchy of π-institutions (Voutsadakis, 2007a, 2006, 2008, 2007c, 2015b).

Since the overviews cited above contain remarkably inclusive and relatively up-to-date presentations of the work in the field, we restrict the introduction to some of the essentials needed in placing the present work in context and in outlining some of the contents of the paper.

The operator approach in AAL was initiated by Blok and Pigozzi in their seminal "Memoirs" monograph (Blok and Pigozzi, 1989), which was, to a certain extent, anticipated in the ground-breaking work of Czelakowski on equivalential logics (Czelakowski, 1981a,b). This was followed by vital and influential contributions by Hermann (Herrmann, 1993, 1996, 1997), Czelakowski and Pigozzi (Czelakowski and Pigozzi, 1999, 2004a,b), Font and Jansana (Font and Jansana, 1996), Czelakowski and Jansana (Czelakowski and Jansana, 2000), Czelakowski (Czelakowski, 2003) and Raftery (Raftery, 2006).

Blok and Pigozzi introduced the Leibniz operator associating with a theory of a sentential logic the largest congruence on the formula algebra that is compatible with the theory. This may be generalized to an association with an arbitrary filter of the logic on any algebra of the same similarity type as the logic of the largest congruence on the algebra that is compatible with the filter. One of the main results of (Blok and Pigozzi, 1989) was a characterization of algebraizability via a correspondence between theories and congruences. This was subsequently refined in many works, e.g., in (Font and Jansana, 1996), and, much more recently, in a unifying setting, encompassing many previously known results of this type, in (Albuquerque et al., 2016).

The bulk of Blok and Pigozzi's and subsequent work focuses on establishing the main classes of the AAL Leibniz hierarchy, consisting of the protoalgebraic (Blok and Pigozzi, 1986), equivalential (Czelakowski, 1981a,b), truth-equational (Raftery, 2006), weakly algebraizable (Czelakowski and Jansana, 2000) and algebraizable logics (Blok and Pigozzi, 1989; Herrmann, 1993, 1996, 1997), based on properties of the Leibniz operator, and on exploring various metalogical properties and related properties of their associated classes of algebras.

In the CAAL context, a categorical Leibniz operator, inspired by the work in AAL, was introduced in (Voutsadakis, 2007a) with the goal of obtaining abstractions of the various results obtained in the AAL context using the Leibniz operator of Blok and Pigozzi; among them, perhaps most importantly, developing a CAAL hierarchy of π-institutions based on their algebraic character, indicating the strength of ties between the structure of the lattice of their theory families and that of the congruence systems on algebraic systems.

The second operator that was introduced historically was the Tarski operator (Font and Jansana, 1996) in seminal work carried out by Font and Jansana and ushering in a period of intense and fruitful investigations by the Barcelona School of AAL. The Tarski operator associates to a collection of filters on a specific algebra the largest congruence on the algebra that is compatible with all filters in the collection. The Tarski operator served the purpose of lifting the model theory of sentential logics from the level of logical matrices, which had been at the focus of the work of Czelakowski and Blok and Pigozzi, to the level of generalized matrices and of abstract logics (see, e.g., (Font and Jansana, 1996; Hendriks and Malinowski, 2003)). A key aspect of this theory, playing a decisive role in the characterization of the classes in the Leibniz hierarchy, is the determination of full models of either single logical systems under consideration in a specific study or of classes of logical systems (Font and Jansana, 1995; Jansana, 2002; Babenyshev, 2003; Font et al., 2006). These are the models that include all filters that are compatible with the Tarski congruence of the model.

Perhaps surprisingly, but understandably, if one takes into account the nature of closure systems defining π-institutions, the categorical Tarski operator (Voutsadakis, 2015a) was introduced in CAAL before the corresponding abstraction of the Leibniz operator (Voutsadakis, 2007a, 2006). It served the purpose of abstracting the theory of abstract logics of Font and Jansana to the level of models of π-institutions. Very important for our present work was the establishment of a General Correspondence Theorem (Voutsadakis, 2005a,b), which parallels a celebrated Correspondence Theorem in the context of sentential logics (Font and Jansana, 1996), in turn generalizing one of the original results of Blok and Pigozzi (Blok and Pigozzi, 1989, 1992).

In (Czelakowski, 2003), Czelakowski introduced the last of the three major operators, the Suszko operator. Raftery (Raftery, 2006) has employed the Suszko operator in the study of truth equational sentential logics. More recently, it has played a potent role in providing alternative characterizations and in studying properties of various classes of the Leibniz hierarchy (Albuquerque et al., 2016), in addition to its critical role in the characterization of truth equationality. Following (Albuquerque et al., 2016), we use in a similar way the corresponding CAAL operator, also termed the Suszko operator, which was introduced in (Voutsadakis, 2007b), following (Czelakowski, 2003).

The three operators are closely related, the Leibniz being in a sense the fundamental one. The Leibniz operator and the Suszko operator are applied to single filters of a logic and the Suszko congruence associated with a given filter is the intersection of all Leibniz congruences associated with filters of the logic that include the given filter. On the other hand the Tarski operator is applied to collections of filters of a logic and it gives the intersection of all Leibniz congruences of the filters in the collection. Thus, both the Suszko and the Tarski operators can be expressed in terms of the Leibniz operator in a straightforward way. One of the elegant contributions of (Albuquerque et al., 2016) was the introduction of the unifying framework of compatibility operators in which all three operators can be treated uniformly to a far-reaching extent. We follow here (Albuquerque et al., 2016) in treating the categorical operators in a similar way. We are able, as a result, on the one hand to both unify and simplify already known results from CAAL, and, on the other, to establish many hitherto unknown ones, that generalize to π-institutions corresponding known results from the AAL domain.

2 Abstract Compatibility Operators

In (Albuquerque et al., 2016), Albuquerque, Font and Jansana developed the theory of $\mathcal{S}$-compatibility operators, encompassing and treating under a unified framework the three classical operators of AAL. We review briefly the basic components of the work in (Albuquerque et al., 2016) since it forms the foundation for the work developed in the present paper.

We fix a sentential logic $\mathcal{S} = \langle \mathcal{L}, C_{\mathcal{S}} \rangle$. $\mathcal{S}$-compatibility operators are mappings $\nabla^{\boldsymbol{A}}$ from the set of all $\mathcal{S}$-filters $\mathrm{Fi}_{\mathcal{S}}(\boldsymbol{A})$ on an arbitrary algebra $\boldsymbol{A}$, of the similarity type $\mathcal{L}$ of $\mathcal{S}$, to the set of congruences $\mathrm{Con}(\boldsymbol{A})$ on the algebra. Such an operator $\nabla^{\boldsymbol{A}}$ maps an $\mathcal{S}$-filter F on $\boldsymbol{A}$ to a congruence $\nabla^{\boldsymbol{A}}(F)$ that is compatible with the filter. Since, by definition, the Leibniz congruence $\Omega^{\boldsymbol{A}}(F)$ is the largest congruence on $\boldsymbol{A}$ compatible with F (Blok and Pigozzi, 1989), it follows that $\Omega^{\boldsymbol{A}}$ is the largest $\mathcal{S}$-compatibility operator on $\boldsymbol{A}$. Moreover, as shown by Czelakowski in (Czelakowski, 2003), the Suszko operator $\widetilde{\Omega}^{\boldsymbol{A}}$ is the largest order-preserving $\mathcal{S}$-compatibility operator.

In the abstract theory, the Leibniz and Suszko operators form an example of another type of relationship. Namely, given an $\mathcal{S}$-compatibility operator $\nabla^{\boldsymbol{A}}$, two more "companion" operators are defined from it (Albuquerque et al., 2016):

- The *lifting* $\widetilde{\nabla}^{\boldsymbol{A}}$ is applied to arbitrary collections of $\mathcal{S}$-filters on $\boldsymbol{A}$; it associates with such a collection, the largest congruence on $\boldsymbol{A}$ that is compatible with all filters in the collection.
- The *relativization* $\widetilde{\nabla}_{\mathcal{S}}^{\boldsymbol{A}}$ is applied to an $\mathcal{S}$-filter and associates with it the largest congruence on $\boldsymbol{A}$ that is compatible with all $\mathcal{S}$-filters on $\boldsymbol{A}$ containing the given filter. Thus, its action is that of the lifting applied on the upset of the lattice of all $\mathcal{S}$-filters generated by the given filter.

Clearly, the Tarski operator is the lifting of the Leibniz operator and the Suszko operator is its relativization, and they constitute the prototypical examples of operators that motivate the general theory.

The springboard of the theory in (Albuquerque et al., 2016) is the observation that $\widetilde{\nabla}^A$ is part of a Galois connection between the powerset $\mathcal{P}(\mathrm{Fi}_{\mathcal{S}}(A))$ of the collection of $\mathcal{S}$-filters on A and the collection $\mathrm{Con}(A)$ of congruences on A. The fixed points are the so-called ∇^A-*full sets of $\mathcal{S}$-filters* and the ∇^A-*full congruences*.

Another pair of important concepts consists of the ∇^A-*class* $[\![F]\!]^{\nabla^A}$ of an $\mathcal{S}$-filter F, which is composed of all filters with which $\nabla^A(F)$ is compatible, and the smallest element F^{∇^A} of this class. A filter F is termed a ∇^A-*filter* in (Albuquerque et al., 2016) if $F = F^{\nabla^A}$, i.e., if it is the smallest filter that is compatible with its ∇^A-associated congruence, again a concept that has been studied extensively in the traditional setting by Font and Jansana (Font and Jansana, 2001, 2011) and Jansana (Jansana, 2003).

If an $\mathcal{S}$-compatibility operator ∇^A is defined for every algebra A of the same similarity type $\mathcal{L}$ as that of the sentential logic $\mathcal{S}$, then a family $\nabla = \{\nabla^A\}_{A \in \mathsf{Alg}(\mathcal{L})}$ is assembled. To relate the members of ∇ the increasing in strength notions of coherence, commutativity with inverse images of surjective homomorphisms and commutativity with inverse images of arbitrary homomorphisms are introduced. The first is novel in (Albuquerque et al., 2016) whereas the latter two are well known in traditional AAL and play a critical role in the theory of protoalgebraic (Blok and Pigozzi, 1986), equivalential (Czelakowski, 1981a,b) and algebraizable (Blok and Pigozzi, 1989) logics (see also (Czelakowski, 2001; Font et al., 2003)).

Remarkably, taking advantage of coherence, a General Correspondence Theorem (Theorem 4.17 of (Albuquerque et al., 2016)) is obtained to the effect that, for every surjective homomorphism $h : A \to B$ and every $\mathcal{S}$-filter F on A, such that h is ∇^A-compatible with F, a technical condition, h induces an order isomorphism between the ∇^A-class of F and the ∇^B-class of $h(F)$, with inverse h^{-1}. Several well-known isomorphism theorems from the theory of protoalgebraic logics and beyond follow from this General Correspondence Theorem, including results of Blok and Pigozzi (Blok and Pigozzi, 1989, 1992), of Czelakowski (Czelakowski, 2003) and of Font and Jansana (Font and Jansana, 2001).

Following the lead from the classical theory of AAL, based on $\nabla^A, \widetilde{\nabla}^A$ and $\widetilde{\nabla}^A_{\mathcal{S}}$, classes of algebras are defined consisting of algebras that are reduced with respect to corresponding types of congruences. The abstract hypotheses of coherence and commutativity with inverse images of surjective homomorphisms imply various relationships between the classes analogous to those established in the traditional context in relation to the well-known classes $\mathsf{Alg}^*\mathcal{S}, \mathsf{Alg}^{\mathsf{Su}}\mathcal{S}$ and $\mathsf{Alg}\mathcal{S}$ (see Subsection 4.2 of (Albuquerque et al., 2016)).

Using the concepts of full generalized matrix models, of the Leibniz operator, of the Suszko operator and of the aforementioned classes of algebras associated with $\mathcal{S}$, a wealth of characterizations of the classes in the AAL hierarchy is obtained in Section 6 of (Albuquerque et al., 2016). Some of these have been well-known in the AAL literature, some less well-known and some are new. What is remarkable, however, and motivated the present exposition, is the fact that they are all obtained as consequences of the treatment of abstract S-compatibility operators and the basic Galois connection, as specialized in the context of the three main operators of AAL, essentially the Leibniz operator, since it is the fundamental among the three, and the Tarski and Suszko as the lifting and relativization of the Leibniz operator.

3 The Categorical Operators

Let **Sign** be a category, referred to as a **category of signatures**. Let, also, SEN : **Sign** $\rightarrow$ **Set** be a set-valued functor from the category of signatures, referred to as a **sentence functor**. A collection $T = \{T_\Sigma\}_{\Sigma \in |\mathbf{Sign}|}$, with $T_\Sigma \subseteq \mathrm{SEN}(\Sigma)$, for all $\Sigma \in |\mathbf{Sign}|$, is called a **sentence family** of SEN.[1] A sentence family is a **sentence system** if it is invariant under **Sign**-morphisms, i.e., for all $\Sigma, \Sigma' \in |\mathbf{Sign}|$ and all $f \in \mathbf{Sign}(\Sigma, \Sigma')$, $\mathrm{SEN}(f)(T_\Sigma) \subseteq T_{\Sigma'}$. An **equivalence family** $\theta = \{\theta_\Sigma\}_{\Sigma \in |\mathbf{Sign}|}$ on SEN is a $|\mathbf{Sign}|$-indexed family of equivalence relations $\theta_\Sigma \subseteq \mathrm{SEN}(\Sigma)^2$. It is called an **equivalence system** if it is invariant under **Sign**-morphisms, i.e., if, for all $\Sigma, \Sigma' \in |\mathbf{Sign}|$ and $f \in \mathbf{Sign}(\Sigma, \Sigma')$, $\mathrm{SEN}(f)^2(\theta_\Sigma) \subseteq \theta_{\Sigma'}$. Signature-wise inclusion of both sentence families and equivalence families is denoted by $\leq$, i.e.,

$$T \leq T' \quad \text{iff} \quad T_\Sigma \subseteq T'_\Sigma, \text{ for all } \Sigma \in |\mathbf{Sign}|,$$

and

$$\theta \leq \theta' \quad \text{iff} \quad \theta_\Sigma \subseteq \theta'_\Sigma, \text{ for all } \Sigma \in |\mathbf{Sign}|.$$

Consider, in addition to a sentence functor SEN : **Sign** $\rightarrow$ **Set**, a **category N of natural transformations** on SEN in the sense of, e.g., Section 2 of (Voutsadakis, 2007a). The triple $\boldsymbol{A} = \langle \mathbf{Sign}, \mathrm{SEN}, N \rangle$ is called an **algebraic system**. An equivalence family θ on SEN is called a **congruence family** on $\boldsymbol{A}$ if it is invariant under N-morphisms, i.e., if, for all $\sigma : \mathrm{SEN}^k \rightarrow \mathrm{SEN}$ in N, all $\Sigma \in |\mathbf{Sign}|$ and all $\varphi_0, \psi_0, \ldots, \varphi_{k-1}, \psi_{k-1} \in \mathrm{SEN}(\Sigma)$,

$$\langle \varphi_i, \psi_i \rangle \in \theta_\Sigma, \ i < k, \ \text{imply} \ \langle \sigma_\Sigma(\varphi_0, \ldots, \varphi_{k-1}), \sigma_\Sigma(\psi_0, \ldots, \psi_{k-1}) \rangle \in \theta_\Sigma.$$

A **congruence system** is a congruence family that is an equivalence system, i.e., an equivalence family that is invariant under both **Sign**-morphisms and N-morphisms. The collection of all congruence systems on $\boldsymbol{A}$ is denoted by ConSys($\boldsymbol{A}$). Ordered by signature-wise inclusion $\leq$, they form a complete lattice, denoted by $\mathbf{ConSys}(\boldsymbol{A}) = \langle \mathrm{ConSys}(\boldsymbol{A}), \leq \rangle$.

Let $\boldsymbol{F} = \langle \mathbf{Sign}, \mathrm{SEN}, N \rangle$ be an algebraic system, termed the **base algebraic system**. An algebraic system $\boldsymbol{A} = \langle \mathbf{Sign}', \mathrm{SEN}', N' \rangle$ is called an N-

---

[1] This was called an *axiom family* in CAAL before.

algebraic system if there exists a surjective functor $'': N \to N'$ that preserves all projection natural transformations and, therefore, preserves also the arities of all natural transformations in N. We write σ' in N' to indicate the image in N' of a σ in N under the functor $''$. Given two N-algebraic systems $\boldsymbol{A} = \langle \mathbf{Sign}', \mathrm{SEN}', N' \rangle$ and $\boldsymbol{B} = \langle \mathbf{Sign}'', \mathrm{SEN}'', N'' \rangle$, an N-(**algebraic system**) **morphism** $\langle H, \gamma \rangle : \boldsymbol{A} \to \boldsymbol{B}$ consists of

- a functor $H : \mathbf{Sign}' \to \mathbf{Sign}''$ and
- a natural transformation $\gamma : \mathrm{SEN}' \to \mathrm{SEN}'' \circ H$, such that, for all $\sigma : \mathrm{SEN}^k \to \mathrm{SEN}$ in N, all $\Sigma \in |\mathbf{Sign}'|$ and all $\varphi_0, \ldots, \varphi_{k-1} \in \mathrm{SEN}'(\Sigma)$,

$$\gamma_\Sigma(\sigma'_\Sigma(\varphi_0, \ldots, \varphi_{k-1})) = \sigma''_{H(\Sigma)}(\gamma_\Sigma(\varphi_0), \ldots, \gamma_\Sigma(\varphi_{k-1})).$$

Given an N-morphism $\langle H, \gamma \rangle : \boldsymbol{A} \to \boldsymbol{B}$, the **kernel** of $\langle H, \gamma \rangle$ is the congruence system $\mathrm{Ker}(\langle H, \gamma \rangle) = \{\mathrm{Ker}_\Sigma(\langle H, \gamma \rangle)\}_{\Sigma \in |\mathbf{Sign}'|}$, defined, for all $\Sigma \in |\mathbf{Sign}'|$, by

$$\mathrm{Ker}_\Sigma(\langle H, \gamma \rangle) = \{\langle \varphi, \psi \rangle \in \mathrm{SEN}'(\Sigma)^2 : \gamma_\Sigma(\varphi) = \gamma_\Sigma(\psi)\}.$$

Given an algebraic system $\boldsymbol{A} = \langle \mathbf{Sign}, \mathrm{SEN}, N \rangle$ and a congruence system θ on $\boldsymbol{A}$, one can define the **quotient algebraic system** $\boldsymbol{A}/\theta = \langle \mathbf{Sign}, \mathrm{SEN}^\theta, N^\theta \rangle$ of $\boldsymbol{A}$ by θ (see, e.g., (Voutsadakis, 2015a)). In this case $\langle I_{\mathbf{Sign}}, \pi^\theta \rangle : \boldsymbol{A} \to \boldsymbol{A}/\theta$ denotes the projection morphism from $\boldsymbol{A}$ onto $\boldsymbol{A}/\theta$. Thus, given a class K of algebraic systems, it makes sense to consider the K-**relative congruence systems on** $\boldsymbol{A}$, i.e., those $\theta \in \mathrm{ConSys}(\boldsymbol{A})$, such that $\boldsymbol{A}/\theta \in \mathsf{K}$. The class of all relative K-congruence systems on $\boldsymbol{A}$ is denoted by $\mathrm{ConSys}_{\mathsf{K}}(\boldsymbol{A})$.

Let $\boldsymbol{A} = \langle \mathbf{Sign}, \mathrm{SEN}, N \rangle$ be an algebraic system and $T = \{T_\Sigma\}_{\Sigma \in |\mathbf{Sign}|}$ a sentence family of SEN. A congruence system $\theta = \{\theta_\Sigma\}_{\Sigma \in |\mathbf{Sign}|}$ on $\boldsymbol{A}$ is **compatible with** T if, for all $\Sigma \in |\mathbf{Sign}|$ and all $\varphi, \psi \in \mathrm{SEN}(\Sigma)$,

$$\langle \varphi, \psi \rangle \in \theta_\Sigma \quad \text{and} \quad \varphi \in T_\Sigma \quad \text{imply} \quad \psi \in T_\Sigma.$$

This condition is denoted $T \mathrel{\mathsf{comp}} \theta$ and may be characterized in the following ways:

Lemma 1. *Let $\boldsymbol{A} = \langle \mathbf{Sign}, \mathrm{SEN}, N \rangle$ be an algebraic system, $\theta \in \mathrm{ConSys}(\boldsymbol{A})$ and T a sentence family of SEN. The following conditions are equivalent:*

(i) *θ is compatible with T.*
(ii) *$\varphi \in T_\Sigma$ iff $\varphi/\theta_\Sigma \in T_\Sigma/\theta_\Sigma$, for all $\Sigma \in |\mathbf{Sign}|$ and all $\varphi \in \mathrm{SEN}(\Sigma)$.*
(iii) *$T = \pi^{\theta^{-1}}(\pi^\theta(T))$ $(\pi^{\theta^{-1}} := (\pi^\theta)^{-1})$.*
(iv) *$T_\Sigma = \bigcup_{\varphi \in T_\Sigma} \varphi/\theta_\Sigma$, for all $\Sigma \in |\mathbf{Sign}|$, i.e., T_Σ is a union of θ_Σ-equivalence classes, for all $\Sigma \in |\mathbf{Sign}|$.*

As for the kernel of an N-morphism, we have:

Lemma 2. *Let $\boldsymbol{A} = \langle \mathbf{Sign}', \mathrm{SEN}', N' \rangle$ and $\boldsymbol{B} = \langle \mathbf{Sign}'', \mathrm{SEN}'', N'' \rangle$ be N-algebraic systems and $\langle H, \gamma \rangle : \boldsymbol{A} \to \boldsymbol{B}$ an N-morphism.*

(1) *For all sentence families T of SEN', $\mathrm{Ker}(\langle H, \gamma \rangle)$ is compatible with T iff $\gamma_\Sigma^{-1}(\gamma_\Sigma(T_\Sigma)) = T_\Sigma$, for all $\Sigma \in |\mathbf{Sign}'|$.*

(2) *For all $\theta \in \mathrm{ConSys}(\boldsymbol{A})$, if $\mathrm{Ker}(\langle H, \gamma \rangle) \leq \theta$, then $\gamma_\Sigma^{-1}(\gamma_\Sigma(\theta_\Sigma)) = \theta_\Sigma$, for all $\Sigma \in |\mathbf{Sign}'|$.*

Consider again an algebraic system $\boldsymbol{A} = \langle \mathbf{Sign}, \mathrm{SEN}, N \rangle$. Given a sentence family T of SEN, there always exists a largest congruence system on $\boldsymbol{A}$ that is compatible with T (Proposition 2.2. of (Voutsadakis, 2007a)). It is called the **Leibniz congruence system** of T on $\boldsymbol{A}$ and denoted $\Omega^{\boldsymbol{A}}(T) = \{\Omega_\Sigma^{\boldsymbol{A}}(T)\}_{\Sigma \in |\mathbf{Sign}|}$.

Given a collection $\mathcal{T}$ of sentence families of SEN, there always exists a largest congruence system on $\boldsymbol{A}$ that is compatible with every $T \in \mathcal{T}$. This is termed the **Tarski congruence system** of $\mathcal{T}$ on $\boldsymbol{A}$ and denoted by $\widetilde{\Omega}^{\boldsymbol{A}}(\mathcal{T})$.

A π-**institution**[2] $\mathcal{I} = \langle \boldsymbol{A}, C \rangle$ consists of

- an algebraic system $\boldsymbol{A} = \langle \mathbf{Sign}, \mathrm{SEN}, N \rangle$ and
- a **closure system** C on SEN, i.e., a family of closure operators $C = \{C_\Sigma\}_{\Sigma \in |\mathbf{Sign}|}$ that satisfy, for all $\Sigma, \Sigma' \in |\mathbf{Sign}|$ and all $f \in \mathbf{Sign}(\Sigma, \Sigma')$,

$$\mathrm{SEN}(f)(C_\Sigma(\Phi)) \subseteq C_{\Sigma'}(\mathrm{SEN}(f)(\Phi)), \text{ for all } \Phi \subseteq \mathrm{SEN}(\Sigma),$$

a condition known as **structurality**.

Given a π-institution $\mathcal{I} = \langle \boldsymbol{A}, C \rangle$, a sentence family (system) $T = \{T_\Sigma\}_{\Sigma \in |\mathbf{Sign}|}$ of SEN is called a **theory family (system)** if each $T_\Sigma \subseteq \mathrm{SEN}(\Sigma)$ is a Σ-**theory**, i.e., a closed set under C: $C_\Sigma(T_\Sigma) = T_\Sigma$. The collection of all theory families of $\mathcal{I}$ is denoted by $\mathrm{ThFam}(\mathcal{I})$. Ordered by signature wise inclusion $\leq$, the collection of all theory families forms a complete lattice that is denoted by $\mathbf{ThFam}(\mathcal{I}) = \langle \mathrm{ThFam}(\mathcal{I}), \leq \rangle$.

Let $\mathcal{I} = \langle \boldsymbol{A}, C \rangle$ be a π-institution. As a special case of the definition of the Tarski congruence system of a collection of sentence families, we obtain the **Tarski congruence system of** $\mathcal{I}$, i.e., the largest congruence system that is compatible with every theory family $T \in \mathrm{ThFam}(\mathcal{I})$. Ordinarily, instead of the notation $\widetilde{\Omega}^{\boldsymbol{A}}(\mathrm{ThFam}(\mathcal{I}))$, we use the notation $\widetilde{\Omega}^{\boldsymbol{A}}(C)$ or $\widetilde{\Omega}(\mathcal{I})$ for this congruence system.

Consider again a π-institution $\mathcal{I} = \langle \boldsymbol{A}, C \rangle$ and a theory family $T \in \mathrm{ThFam}(\mathcal{I})$. The **Suszko congruence system of** T **in** $\mathcal{I}$, denoted $\widetilde{\Omega}^{\mathcal{I}}(T)$, is the largest congruence system that is compatible with all $T' \in \mathrm{ThFam}(\mathcal{I})$, such that $T \leq T'$. Taking after similar notation in AAL, this set is usually denoted by

$$(\mathrm{ThFam}(\mathcal{I}))^T = \{T' \in \mathrm{ThFam}(\mathcal{I}) : T \leq T'\}.$$

Therefore, $\widetilde{\Omega}^{\mathcal{I}}(T) = \widetilde{\Omega}^{\boldsymbol{A}}((\mathrm{ThFam}(\mathcal{I}))^T)$.

In summary, the three congruence systems $\Omega^{\boldsymbol{A}}(T)$, $\widetilde{\Omega}^{\mathcal{I}}(T)$ and $\widetilde{\Omega}^{\boldsymbol{A}}(C)$ are related by

---

[2] This is the same as a π-institution $\mathcal{I} = \langle \mathbf{Sign}, \mathrm{SEN}, C \rangle$, augmented with a category N of natural transformations on its sentence functor SEN, in traditional CAAL.

$$\widetilde{\Omega}^{\mathcal{I}}(T) = \bigcap\{\Omega^{\boldsymbol{A}}(T') : T' \in \mathrm{ThFam}(\mathcal{I}), T \leq T'\}$$

and

$$\widetilde{\Omega}(\mathcal{I}) = \bigcap\{\Omega^{\boldsymbol{A}}(T) : T \in \mathrm{ThFam}(\mathcal{I})\}.$$

Let $\mathbf{F} = \langle\mathbf{Sign}, \mathrm{SEN}, N\rangle$ be a base algebraic system and $\boldsymbol{A} = \langle\mathbf{Sign}', \mathrm{SEN}',$ $N'\rangle$ an N-algebraic system. A pair $\mathcal{A} = \langle\boldsymbol{A}, \langle F, \alpha\rangle\rangle$ is an (**interpreted**) N-**algebraic system**[3] if $\boldsymbol{A}$ is an N-algebraic system and $\langle F, \alpha\rangle : \mathrm{SEN} \to \mathrm{SEN}'$ is an N-morphism.

Let $\mathcal{A} = \langle\boldsymbol{A}, \langle F, \alpha\rangle\rangle$ and $\mathcal{B} = \langle\boldsymbol{B}, \langle G, \beta\rangle\rangle$ be two interpreted N-algebraic systems. An N-morphism $\langle H, \gamma\rangle : \boldsymbol{A} \to \boldsymbol{B}$ is called an N-**morphism from** $\mathcal{A}$ **to** $\mathcal{B}$, denoted $\langle H, \gamma\rangle : \mathcal{A} \to \mathcal{B}$, if the following triangle commutes:

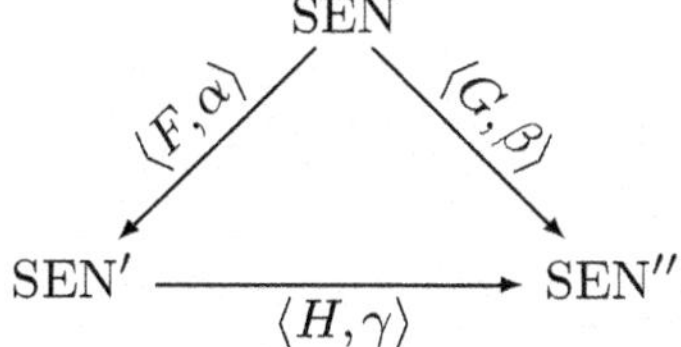

Such an N-morphism is said to be **surjective** if both $H : \mathbf{Sign}' \to \mathbf{Sign}''$ and all $\gamma_{\Sigma'} : \mathrm{SEN}'(\Sigma') \to \mathrm{SEN}''(H(\Sigma'))$, $\Sigma' \in |\mathbf{Sign}'|$, are surjective.

An N-**matrix system** $\mathfrak{A} = \langle\boldsymbol{A}, T'\rangle$ is a pair consisting of an N-algebraic system $\boldsymbol{A} = \langle\mathbf{Sign}', \mathrm{SEN}', N'\rangle$ and a sentence family $T' = \{T'_\Sigma\}_{\Sigma \in |\mathbf{Sign}'|}$ of SEN'. An (**interpreted**) N-**matrix system**[3] $\mathfrak{A} = \langle\mathcal{A}, T'\rangle$ is a pair consisting of an interpreted N-algebraic system $\mathcal{A} = \langle\boldsymbol{A}, \langle F, \alpha\rangle\rangle$ and a sentence family $T' = \{T'_\Sigma\}_{\Sigma \in |\mathbf{Sign}'|}$ of SEN'.

Fix a base algebraic system $\mathbf{F} = \langle\mathbf{Sign}, \mathrm{SEN}, N\rangle$ and a π-institution $\mathcal{I} = \langle\mathbf{F}, C\rangle$, referred to as the **base** π-**institution**.[4] Then an interpreted N-matrix system $\mathfrak{A} = \langle\mathcal{A}, T'\rangle$ is called an $\mathcal{I}$-**matrix system** if T' is an $\mathcal{I}$-**filter family** of $\mathcal{A}$, i.e., for all $\Sigma \in |\mathbf{Sign}|$, $\Phi \cup \{\varphi\} \subseteq \mathrm{SEN}(\Sigma)$, such that $\varphi \in C_\Sigma(\Phi)$, and all $f \in \mathbf{Sign}(\Sigma, \Sigma')$,

$$\alpha_{\Sigma'}(\mathrm{SEN}(f)(\Phi)) \subseteq T'_{F(\Sigma')} \quad \text{implies} \quad \alpha_{\Sigma'}(\mathrm{SEN}(f)(\varphi)) \in T'_{F(\Sigma')}.$$

We denote by $\mathrm{FiFam}^{\mathcal{I}}(\mathcal{A})$ the collection of all $\mathcal{I}$-filter families of $\mathcal{A}$. Ordered by signature-wise inclusion $\leq$, this collection becomes a complete lattice, denoted by $\mathbf{FiFam}^{\mathcal{I}}(\mathcal{A}) = \langle\mathrm{FiFam}^{\mathcal{I}}(\mathcal{A}), \leq\rangle$. Keeping in line with previously introduced notation, given $T' \in \mathrm{FiFam}^{\mathcal{I}}(\mathcal{A})$, we set

$$(\mathrm{FiFam}^{\mathcal{I}}(\mathcal{A}))^{T'} = \{T'' \in \mathrm{FiFam}^{\mathcal{I}}(\mathcal{A}) : T' \leq T''\}.$$

---

[3] Hopefully, the overloading of terminology will not cause any confusion.

[4] The qualifying "base" is omitted whenever $\mathcal{I}$ is considered fixed in a specific context.

The following lemma provides some preservation properties of $\mathcal{I}$-filter families under the application of N-morphisms between the underlying N-algebraic systems.

Lemma 3. *Let* $\mathcal{I} = \langle \mathbf{F}, C \rangle$ *be a* π-*institution,* $\mathcal{A} = \langle \mathbf{A}, \langle F, \alpha \rangle \rangle, \mathcal{B} = \langle \mathbf{B}, \langle G, \beta \rangle \rangle$ *be* N-*algebraic systems, with* $\mathbf{A} = \langle \mathbf{Sign}', \mathrm{SEN}', N' \rangle$ *and* $\mathbf{B} = \langle \mathbf{Sign}'', \mathrm{SEN}'',$ $N'' \rangle$, $\langle H, \gamma \rangle : \mathcal{A} \to \mathcal{B}$ *an* N-*morphism and* T'' *a sentence family of* $\mathcal{B}$.

1. *If* $T'' \in \mathrm{FiFam}^{\mathcal{I}}(\mathcal{B})$, *then* $\gamma^{-1}(T'') \in \mathrm{FiFam}^{\mathcal{I}}(\mathcal{A})$.
2. *If* $\gamma^{-1}(T'') \in \mathrm{FiFam}^{\mathcal{I}}(\mathcal{A})$, *then* $T'' \in \mathrm{FiFam}^{\mathcal{I}}(\mathcal{B})$.
3. *If* $\langle H, \gamma \rangle$ *is such that* H *is an isomorphism, and* $\mathrm{Ker}(\langle H, \gamma \rangle)$ *is compatible with* $T' \in \mathrm{FiFam}^{\mathcal{I}}(\mathcal{A})$, *then* $\gamma(T') \in \mathrm{FiFam}^{\mathcal{I}}(\mathcal{B})$.

Proof:

1. Suppose $\Sigma \in |\mathbf{Sign}|$, $\Phi \cup \{\varphi\} \subseteq \mathrm{SEN}(\Sigma)$, such that $\varphi \in C_{\Sigma}(\Phi)$ and
$\alpha_{\Sigma'}(\mathrm{SEN}(f)(\Phi)) \subseteq \gamma^{-1}_{F(\Sigma')}(T''_{H(F(\Sigma'))})$.

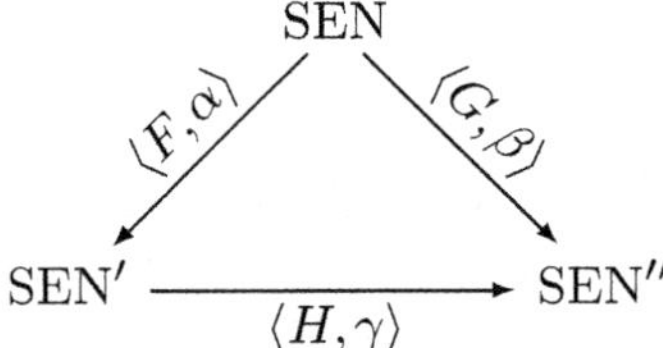

This holds iff

$$\gamma_{F(\Sigma')}(\alpha_{\Sigma'}(\mathrm{SEN}(f)(\Phi))) \subseteq T''_{H(F(\Sigma'))}$$
$$\text{iff} \quad \beta_{\Sigma'}(\mathrm{SEN}(f)(\Phi)) \subseteq T''_{G(\Sigma')}$$
$$\text{implies} \quad \beta_{\Sigma'}(\mathrm{SEN}(f)(\phi)) \in T''_{G(\Sigma')}$$
$$\text{iff} \quad \gamma_{F(\Sigma')}(\alpha_{\Sigma'}(\mathrm{SEN}(f)(\phi))) \in T''_{H(F(\Sigma'))}$$
$$\text{iff} \quad \alpha_{\Sigma'}(\mathrm{SEN}(f)(\phi)) \in \gamma^{-1}_{F(\Sigma')}(T''_{H(F(\Sigma'))}).$$

2. Suppose $\Sigma \in |\mathbf{Sign}|$, $\Phi \cup \{\varphi\} \subseteq \mathrm{SEN}(\Sigma)$, such that $\varphi \in C_{\Sigma}(\Phi)$ and $\beta_{\Sigma'}(\mathrm{SEN}(f)(\Phi)) \subseteq T''_{G(\Sigma')}$. This holds iff

$$\gamma_{F(\Sigma')}(\alpha_{\Sigma'}(\mathrm{SEN}(f)(\Phi))) \subseteq T''_{H(F(\Sigma'))}$$
$$\text{iff} \quad \alpha_{\Sigma'}(\mathrm{SEN}(f)(\Phi)) \subseteq \gamma^{-1}_{F(\Sigma')}(T''_{H(F(\Sigma'))})$$
$$\text{implies} \quad \alpha_{\Sigma'}(\mathrm{SEN}(f)(\phi)) \in \gamma^{-1}_{F(\Sigma')}(T''_{H(F(\Sigma'))})$$
$$\text{iff} \quad \gamma_{F(\Sigma')}(\alpha_{\Sigma'}(\mathrm{SEN}(f)(\phi))) \in T''_{H(F(\Sigma'))}$$
$$\text{iff} \quad \beta_{\Sigma'}(\mathrm{SEN}(f)(\phi)) \in T''_{G(\Sigma')}.$$

3. Note that compatibility of $\mathrm{Ker}(\langle H, \gamma \rangle)$ with $T' \in \mathrm{FiFam}^{\mathcal{I}}(\mathcal{A})$ implies that, for all $\Sigma \in |\mathbf{Sign}'|$, $\gamma^{-1}_{\Sigma}(\gamma_{\Sigma}(T'_{\Sigma})) = T'_{\Sigma}$, or, more compactly, $\gamma^{-1}(\gamma(T')) = T'$. Now assume $\Sigma \in |\mathbf{Sign}|$, $\Phi \cup \{\varphi\} \subseteq \mathrm{SEN}(\Sigma)$, such that $\varphi \in C_{\Sigma}(\Phi)$ and $\beta_{\Sigma'}(\mathrm{SEN}(f)(\Phi)) \subseteq \gamma_{F(\Sigma')}(T'_{F(\Sigma')})$. This holds iff

$$\gamma_{F(\Sigma')}(\alpha_{\Sigma'}(\mathrm{SEN}(f)(\Phi))) \subseteq \gamma_{F(\Sigma')}(T'_{F(\Sigma')})$$

$\qquad$ iff $\quad \alpha_{\Sigma'}(\mathrm{SEN}(f)(\Phi)) \subseteq \gamma_{F(\Sigma')}^{-1}(\gamma_{F(\Sigma')}(T'_{F(\Sigma')}))$

$\qquad$ iff $\quad \alpha_{\Sigma'}(\mathrm{SEN}(f)(\Phi)) \subseteq T'_{F(\Sigma')}$

$\qquad$ implies $\quad \alpha_{\Sigma'}(\mathrm{SEN}(f)(\phi)) \in T'_{F(\Sigma')}$

$\qquad$ iff $\quad \alpha_{\Sigma'}(\mathrm{SEN}(f)(\phi)) \in \gamma_{F(\Sigma')}^{-1}(\gamma_{F(\Sigma')}(T'_{F(\Sigma')}))$

$\qquad$ iff $\quad \gamma_{F(\Sigma')}(\alpha_{\Sigma'}(\mathrm{SEN}(f)(\phi))) \in \gamma_{F(\Sigma')}(T'_{F(\Sigma')})$

$\qquad$ iff $\quad \beta_{\Sigma'}(\mathrm{SEN}(f)(\phi)) \in \gamma_{F(\Sigma')}(T'_{F(\Sigma')}).$ $\qquad\square$

Similar concepts and terminology may be applied to the so-called generalized matrix systems or gmatrix systems for short. An N-**gmatrix system** $\mathbb{A} = \langle \boldsymbol{A}, \mathcal{T}' \rangle$ is a pair consisting of an N-algebraic system $\boldsymbol{A} = \langle \mathbf{Sign}', \mathrm{SEN}', N' \rangle$ and a collection of sentence families $\mathcal{T}'$ of SEN'. An (**interpreted**) N-**gmatrix system**[3] $\mathbb{A} = \langle \mathcal{A}, \mathcal{T}' \rangle$ is a pair consisting of an interpreted N-algebraic system $\mathcal{A} = \langle \boldsymbol{A}, \langle F, \alpha \rangle \rangle$ and a collection of sentence families $\mathcal{T}'$ of SEN'. An $\mathcal{I}$-**gmatrix system** $\mathbb{A} = \langle \mathcal{A}, \mathcal{T}' \rangle$ is a tuple, such that every sentence family in $\mathcal{T}'$ is an $\mathcal{I}$-filter family of $\mathcal{A}$.

Note that, given an interpreted N-algebraic system $\mathcal{A} = \langle \boldsymbol{A}, \langle F, \alpha \rangle \rangle$, the pair $\mathcal{I}' = \langle \boldsymbol{A}, \mathrm{FiFam}^{\mathcal{I}}(\mathcal{A}) \rangle$ is also a π-institution (in closure system form). In accordance, we define the **Suszko congruence of** $T' \in \mathrm{FiFam}^{\mathcal{I}}(\mathcal{A})$, denoted $\widetilde{\Omega}^{\mathcal{A},\mathcal{I}}(T')$ by

$$\widetilde{\Omega}^{\mathcal{A},\mathcal{I}}(T') = \widetilde{\Omega}^{\mathcal{I}'}(T') = \bigcap \{ \Omega^{\boldsymbol{A}}(T'') : T'' \in \mathrm{FiFam}^{\mathcal{I}}(\mathcal{A}), T' \leq T'' \}.$$

We also extend the notation $\Omega^{\boldsymbol{A}}(T')$ and $\Omega^{\boldsymbol{A}}(\mathcal{T}')$ to interpreted N-algebraic systems, writing $\Omega^{\mathcal{A}}(T')$ and $\Omega^{\mathcal{A}}(\mathcal{T}')$, with the meaning that these are identical to those applied to the underlying N-algebraic system $\boldsymbol{A}$ of $\mathcal{A}$. The restriction of $\Omega^{\mathcal{A}}$ to $\mathrm{FiFam}^{\mathcal{I}}(\mathcal{A})$ is the **Leibniz operator on** $\mathcal{A}$. The restriction of $\widetilde{\Omega}^{\mathcal{A},\mathcal{I}}$ to $\mathrm{ThFam}^{\mathcal{I}}(\mathcal{A})$ is the **Suszko operator on** $\mathcal{A}$ and the restriction of $\widetilde{\Omega}^{\mathcal{A}}$ on $\mathcal{P}(\mathrm{FiFam}^{\mathcal{I}}(\mathcal{A}))$ is the **Tarski operator on** $\mathcal{A}$. The families

$$\Omega = \{ \Omega^{\mathcal{A}} : \mathcal{A} \text{ an } N\text{-algebraic system} \}$$
$$\widetilde{\Omega}^{\mathcal{I}} := \widetilde{\Omega}^{\bullet,\mathcal{I}} = \{ \widetilde{\Omega}^{\mathcal{A},\mathcal{I}} : \mathcal{A} \text{ an } N\text{-algebraic system} \}$$
$$\widetilde{\Omega} = \{ \widetilde{\Omega}^{\mathcal{A}} : \mathcal{A} \text{ an } N\text{-algebraic system} \}$$

are termed the **Leibniz**, the **Suszko** and the **Tarski operator**, respectively. Saying that one of those **has a property** P **globally** means that property P holds for every member of the family. E.g., the Leibniz operator is globally order preserving if $\Omega^{\mathcal{A}} : \mathrm{FiFam}^{\mathcal{I}}(\mathcal{A}) \to \mathrm{ConSys}(\boldsymbol{A})$ is order preserving, for every N-algebraic system $\mathcal{A}$.

Concerning these operators, we have

Proposition 4. *Let* $\mathcal{I}$ *be a* π-*institution,* $\mathcal{A}, \mathcal{B}$ *two* N-*algebraic systems and* $\langle H, \gamma \rangle : \mathcal{A} \to \mathcal{B}$ *a surjective* N-*morphism. For all* $\mathcal{T}'' \cup \{T''\} \subseteq \mathrm{FiFam}^{\mathcal{I}}(\mathcal{B})$,

1. $\gamma^{-1}(\Omega^{\mathcal{B}}(T'')) = \Omega^{\mathcal{A}}(\gamma^{-1}(T''))$;
2. $\gamma^{-1}(\widetilde{\Omega}^{\mathcal{B}}(\mathcal{T}'')) = \widetilde{\Omega}^{\mathcal{A}}(\gamma^{-1}(\mathcal{T}''))$.

3. $\gamma^{-1}(\widetilde{\Omega}^{\mathcal{B},\mathcal{I}}(T'')) = \widetilde{\Omega}^{\mathcal{A},\mathcal{I}}((\gamma^{-1}(\mathrm{FiFam}^{\mathcal{I}}(\mathcal{B})))^{\gamma^{-1}(T'')})$.

Proof: Property 1 is a well-known property of the categorical Leibniz operator (see, e.g., Lemma 5.4 of (Voutsadakis, 2007a)). For Property 2,

$$\gamma^{-1}(\widetilde{\Omega}^{\mathcal{B}}(\mathcal{T}'')) = \gamma^{-1}(\bigcap_{T'' \in \mathcal{T}''} \Omega^{\mathcal{B}}(T'')) = \bigcap_{T'' \in \mathcal{T}''} \gamma^{-1}(\Omega^{\mathcal{B}}(T''))$$
$$= \bigcap_{T'' \in \mathcal{T}''} \Omega^{\mathcal{A}}(\gamma^{-1}(T'')) = \widetilde{\Omega}^{\mathcal{A}}(\gamma^{-1}(\mathcal{T}'')).$$

Finally, for Property 3, it suffices to notice that, because of surjectivity,

$$\gamma^{-1}((\mathrm{FiFam}^{\mathcal{I}}(\mathcal{B}))^{T''}) = (\gamma^{-1}(\mathrm{FiFam}^{\mathcal{I}}(\mathcal{B})))^{\gamma^{-1}(T'')}$$

and, then, take advantage of Property 2. $\qquad\qquad\square$

4 Full Models, Algebras and the Hierarchy

The original definition of a full model in AAL was given by Font and Jansana in (Font and Jansana, 1996) and, it was, subsequently, adapted in CAAL in (Voutsadakis, 2005a).

Let $\mathcal{I} = \langle \mathbf{F}, C \rangle$, with $\mathbf{F} = \langle \mathbf{Sign}, \mathrm{SEN}, N \rangle$, be a π-institution and $\mathcal{A} = \langle \mathbf{A}, \langle F, \alpha \rangle \rangle$, with $\mathbf{A} = \langle \mathbf{Sign}', \mathrm{SEN}', N' \rangle$, an N-algebraic system. A collection $\mathcal{T}' \subseteq \mathrm{FiFam}^{\mathcal{I}}(\mathcal{A})$ is **full** if

$$\mathcal{T}' = \{T' \in \mathrm{FiFam}^{\mathcal{I}}(\mathcal{A}) : \widetilde{\Omega}^{\mathcal{A}}(\mathcal{T}') \leq \Omega^{\mathcal{A}}(T')\},$$

i.e., $\mathcal{T}'$ consists of all $\mathcal{I}$-filter families on $\mathcal{A}$ with which the Tarski congruence system $\widetilde{\Omega}^{\mathcal{A}}(\mathcal{T}')$ of $\mathcal{T}'$ is compatible.

If $\mathcal{T}'$ is full, then $\mathcal{T}'$ is a closure system on $\mathcal{A}$, whence the pair $\mathcal{I}' = \langle \mathbf{A}, \mathcal{T}' \rangle$ is a π-institution. We use the terminology **full $\mathcal{I}$-gmatrix system** for $\mathbb{A} = \langle \mathcal{A}, \mathcal{T}' \rangle$ when $\mathcal{T}'$ is a full collection of $\mathcal{I}$-filter families.

Using the CAAL notion of a quotient algebraic system $\mathbf{A}/\theta = \mathbf{A}^{\theta} = \langle \mathbf{Sign}, \mathrm{SEN}^{\theta}, N^{\theta} \rangle$ of a given algebraic system $\mathbf{A} = \langle \mathbf{Sign}, \mathrm{SEN}, N \rangle$ modulo a congruence system θ on $\mathbf{A}$ (Voutsadakis, 2015a), we may give several characterizations of full $\mathcal{I}$-gmatrix systems that parallel results from AAL (Proposition 2.7 of (Albuquerque et al., 2016)).

Proposition 5. *Let $\mathcal{A} = \langle \mathbf{A}, \langle F, \alpha \rangle \rangle$ be an N-algebraic system, with $\mathbf{A} = \langle \mathbf{Sign}', \mathrm{SEN}', N' \rangle$, let $\mathcal{T}' \subseteq \mathrm{FiFam}^{\mathcal{I}}(\mathcal{A})$ and $\langle I_{\mathbf{Sign}'}, \pi \rangle : \mathbf{A} \to \mathbf{A}/\widetilde{\Omega}^{\mathcal{A}}(\mathcal{T}')$ be the canonical projection N-morphism. Then the following conditions are equivalent:*

    (i) *$\mathcal{T}'$ is full.*
    (ii) $\pi(\mathcal{T}') = \mathrm{FiFam}^{\mathcal{I}}(\mathcal{A}/\widetilde{\Omega}^{\mathcal{A}}(\mathcal{T}'))$.
    (iii) $\mathcal{T}' = \pi^{-1}(\mathrm{FiFam}^{\mathcal{I}}(\mathcal{A}/\widetilde{\Omega}^{\mathcal{A}}(\mathcal{T}')))$.

(iv) $\mathcal{T}' = \gamma^{-1}(\mathrm{FiFam}^{\mathcal{I}}(\mathcal{B}))$ *for some N-algebraic system $\mathcal{B}$ and some surjective N-morphism $\langle H, \gamma \rangle : \mathcal{A} \to \mathcal{B}$, with H an isomorphism.*

Proof:

(i)$\Rightarrow$(ii) Suppose that $\mathcal{T}'$ is full.

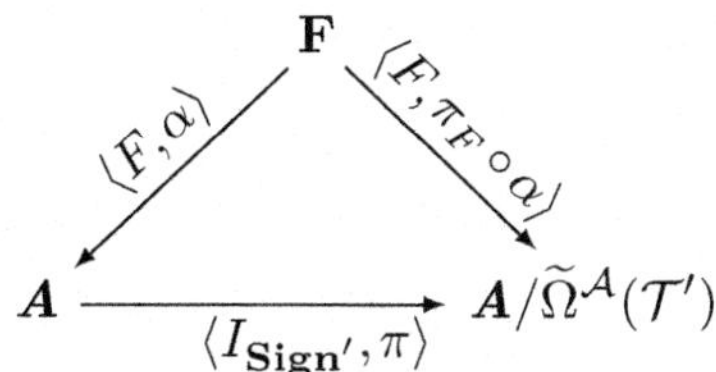

If $T' \in \mathcal{T}'$, then we have $\widetilde{\Omega}^{\mathcal{A}}(\mathcal{T}') \leq \Omega^{\mathcal{A}}(T')$. Thus, $\mathrm{Ker}(\langle I_{\mathbf{Sign'}}, \pi \rangle) = \widetilde{\Omega}^{\mathcal{A}}(\mathcal{T}')$ is compatible with T', and, hence, by Lemma 3, $\pi(T') \in \mathrm{FiFam}^{\mathcal{I}}(\mathcal{A}/\widetilde{\Omega}^{\mathcal{A}}(\mathcal{T}'))$. If, conversely, $T'' \in \mathrm{FiFam}^{\mathcal{I}}(\mathcal{A}/\widetilde{\Omega}^{\mathcal{A}}(\mathcal{T}'))$, consider $\pi^{-1}(T'')$. It is not difficult to see that $\widetilde{\Omega}^{\mathcal{A}}(\mathcal{T}')$ is compatible with $\pi^{-1}(T'')$, whence, since $\mathcal{T}'$ is full, $\pi^{-1}(T'') \in \mathcal{T}'$. Moreover, $\pi(\pi^{-1}(T'')) = T''$ by surjectivity and, therefore,
$\mathrm{FiFam}^{\mathcal{I}}(\mathcal{A}/\widetilde{\Omega}^{\mathcal{A}}(\mathcal{T}')) \subseteq \pi(\mathcal{T}')$.

(ii)$\Rightarrow$(iii) Since every filter family $T' \in \mathcal{T}'$ is compatible with $\widetilde{\Omega}^{\mathcal{A}}(\mathcal{T}')$, it follows that $\pi^{-1}(\pi(\mathcal{T}')) = \mathcal{T}'$, whence the hypothesis yields the conclusion.

(iii)$\Rightarrow$(iv) Obvious.

(iv)$\Rightarrow$(i) The inclusion $\mathcal{T}' \subseteq \{T' : \widetilde{\Omega}^{\mathcal{A}}(\mathcal{T}') \leq \Omega^{\mathcal{A}}(T')\}$ is universally valid, since $\widetilde{\Omega}^{\mathcal{A}}(\mathcal{T}')$ is compatible with every $T' \in \mathcal{T}'$. For the converse, we note that the hypothesis that $\widetilde{\Omega}^{\mathcal{A}}(\mathcal{T}')$ is compatible with every $T' \in \gamma^{-1}(\mathrm{FiFam}^{\mathcal{I}}(\mathcal{B}))$ implies that there exists $\langle H, \widetilde{\gamma} \rangle : \mathcal{A}/\widetilde{\Omega}^{\mathcal{A}}(\mathcal{T}') \to \mathcal{B}/\widetilde{\Omega}^{\mathcal{B}}(\mathrm{FiFam}^{\mathcal{I}}(\mathcal{B}))$ that makes the following diagram commute:

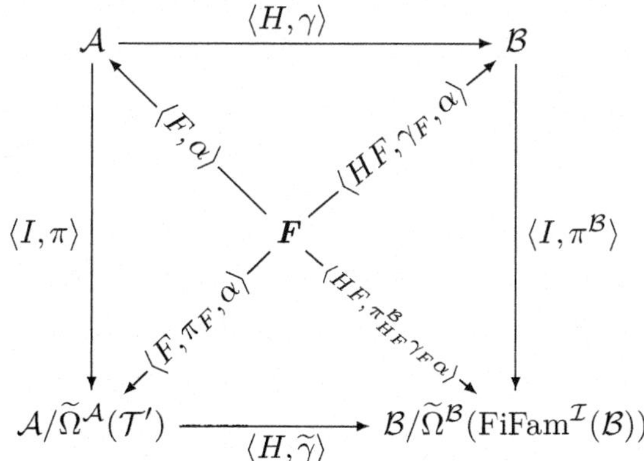

Now diagram chasing gives that, if $\widetilde{\Omega}^{\mathcal{A}}(\mathcal{T}')$ is compatible with T', then $T' \in \gamma^{-1}(\mathrm{FiFam}^{\mathcal{I}}(\mathcal{B})) = \mathcal{T}'$ and, hence, $\mathcal{T}'$ is full.

$\square$

Given two N-matrix systems $\mathfrak{A} = \langle \boldsymbol{A}, T' \rangle$ and $\mathfrak{B} = \langle \boldsymbol{B}, T'' \rangle$, an N-**matrix system morphism** $\langle H, \gamma \rangle : \mathfrak{A} \to \mathfrak{B}$ is a N-morphism $\langle H, \gamma \rangle : \boldsymbol{A} \to \boldsymbol{B}$, such that $\gamma^{-1}(T'') \leq T'$. It is called **strict** if $\gamma^{-1}(T'') = T'$. These definitions extend to interpreted systems with the proviso that N-morphisms must be replaced by morphisms between interpreted systems, i.e., algebraic morphisms commuting with the interpretations.

A N-matrix system $\mathfrak{A} = \langle \boldsymbol{A}, T' \rangle$, with $\boldsymbol{A} = \langle \mathbf{Sign}', \mathrm{SEN}', N' \rangle$ is said to be **Leibniz reduced** or simply **reduced** if $\Omega^{\boldsymbol{A}}(T') = \Delta^{\mathrm{SEN}'}$, where $\Delta^{\mathrm{SEN}'}$ is the identity congruence system of $\boldsymbol{A}$. This terminology applies also to interpreted N-matrix systems and to $\mathcal{I}$-matrix systems.

A gmatrix system $\mathbb{A} = \langle \boldsymbol{A}, \mathcal{T}' \rangle$ is **Tarski reduced** or simply **reduced** if $\widetilde{\Omega}^{\boldsymbol{A}}(\mathcal{T}') = \Delta^{\mathrm{SEN}'}$. This terminology also extends to interpreted N-gmatrix systems and to $\mathcal{I}$-gmatrix systems.

Finally, we call an $\mathcal{I}$-matrix system $\mathbb{A} = \langle \mathcal{A}, T' \rangle$ **Suszko reduced** if $\widetilde{\Omega}^{\mathcal{A}, \mathcal{I}}(T') = \Delta^{\mathrm{SEN}'}$.

By analogy with the universal algebraic framework, reduced $\mathcal{I}$-matrix systems, Suszko reduced $\mathcal{I}$-matrix systems and Tarski reduced $\mathcal{I}$-gmatrix systems give rise to natural classes of N-algebraic systems that are associated to a given base π-institution $\mathcal{I}$.

$$\mathsf{AlgSys}^*(\mathcal{I}) = \{\mathcal{A} : (\exists T' \in \mathrm{FiFam}^{\mathcal{I}}(\mathcal{A}))(\Omega^{\mathcal{A}}(T') = \Delta^{\mathrm{SEN}'})\}$$
$$\mathsf{AlgSys}^{\mathsf{Su}}(\mathcal{I}) = \{\mathcal{A} : (\exists T' \in \mathrm{FiFam}^{\mathcal{I}}(\mathcal{A}))(\widetilde{\Omega}^{\mathcal{A}, \mathcal{I}}(T') = \Delta^{\mathrm{SEN}'})\}$$
$$\mathsf{AlgSys}(\mathcal{I}) = \{\mathcal{A} : (\exists \mathcal{T}' \subseteq \mathrm{FiFam}^{\mathcal{I}}(\mathcal{A}))(\widetilde{\Omega}^{\mathcal{A}}(\mathcal{T}') = \Delta^{\mathrm{SEN}'})\}$$
$$= \{\mathcal{A} : \widetilde{\Omega}^{\mathcal{A}}(\mathrm{FiFam}^{\mathcal{I}}(\mathcal{A})) = \Delta^{\mathrm{SEN}'}\}.$$

Analogously with the corresponding AAL classes and accompanying results, established in (Blok and Pigozzi, 1989; Czelakowski, 2003; Font and Jansana, 1996), we may obtain the following characterizations of these classes (**I** denotes the isomorphic copies operator for interpreted N-algebraic systems):

Lemma 6. *Let $\mathcal{I}$ be a π-institution.*

1. $\mathsf{AlgSys}^*(\mathcal{I}) = \mathbf{I}(\{\mathcal{A}/\Omega^{\mathcal{A}}(T) : \mathcal{A} \; N\text{-alg system}, T \in \mathrm{FiFam}^{\mathcal{I}}(\mathcal{A})\}).$
2. $\mathsf{AlgSys}^{\mathsf{Su}}(\mathcal{I}) = \mathbf{I}(\{\mathcal{A}/\widetilde{\Omega}^{\mathcal{A}, \mathcal{I}}(T) : \mathcal{A} \; N\text{-alg system}, T \in \mathrm{FiFam}^{\mathcal{I}}(\mathcal{A})\}).$
3. $\mathsf{AlgSys}(\mathcal{I}) = \mathbf{I}(\{\mathcal{A}/\widetilde{\Omega}^{\mathcal{A}}(\mathcal{T}) : \mathcal{A} \; N\text{-alg system}, \mathcal{T} \subseteq \mathrm{FiFam}^{\mathcal{I}}(\mathcal{A}) \; full\}).$
4. $\mathsf{AlgSys}(\mathcal{I}) = \mathbf{I}(\{\mathcal{A}/\widetilde{\Omega}^{\mathcal{A}}(\mathcal{T}) : \mathcal{A} \; N\text{-alg system}, \mathcal{T} \subseteq \mathrm{FiFam}^{\mathcal{I}}(\mathcal{A})\}).$
5. $\mathsf{AlgSys}(\mathcal{I}) = \mathsf{AlgSys}^{\mathsf{Su}}(\mathcal{I}).$

Adopting the operator approach in defining the main classes of a categorical abstract algebraic hierarchy of π-institutions, we have:

Definition 7. *Let $\boldsymbol{F} = \langle \mathbf{Sign}, \mathrm{SEN}, N \rangle$ be a base algebraic system and $\mathcal{I} = \langle \boldsymbol{F}, C \rangle$ a π-institution based on $\boldsymbol{F}$.*

- $\mathcal{I}$ *is* **protoalgebraic** *((Blok and Pigozzi, 1986) in AAL and (Voutsadakis, 2007a) in CAAL) if Ω is globally order-preserving.*

- $\mathcal{I}$ *is* **equivalential** ((Czelakowski, 1981a,b) in AAL and (Voutsadakis, 2008) in CAAL) *if Ω is globally order preserving and commutes with inverse images of N-morphisms.*
- $\mathcal{I}$ *is* **truth-equational** ((Raftery, 2006) in AAL and (Voutsadakis, 2015b) in CAAL) *if Ω is globally completely order reflecting.*
- $\mathcal{I}$ *is* **weakly algebraizable** ((Czelakowski and Jansana, 2000) in AAL and (Voutsadakis, 2007c) in CAAL) *if it is protoalgebraic and truth-equational.*
- $\mathcal{I}$ *is* **algebraizable** ((Blok and Pigozzi, 1989; Herrmann, 1993) in AAL and (Voutsadakis, 2002) in CAAL) *if it is equivalential and truth-equational.*

These definitions preserved the structure of the AAL Leibniz hierarchy:

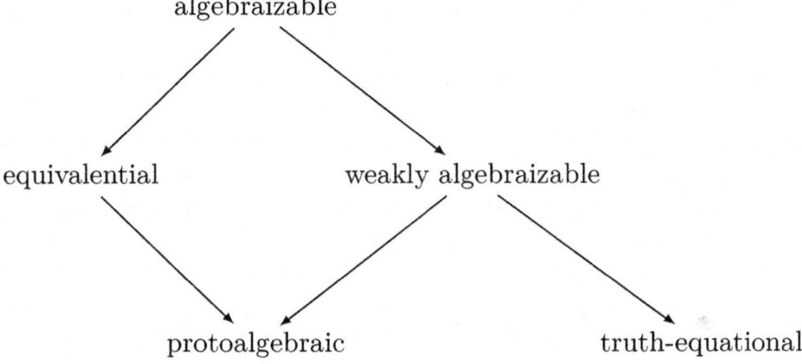

5 $\mathcal{I}$-Operators

Taking after (Albuquerque et al., 2016), we define and study arbitrary $\mathcal{I}$-operators, which correspond in the CAAL framework to arbitrary $\mathcal{S}$-operators in AAL.

Definition 8. *Let $\mathcal{I} = \langle \boldsymbol{F}, C \rangle$ be a base π-institution and $\mathcal{A} = \langle \boldsymbol{A}, \langle F, \alpha \rangle \rangle$ an N-algebraic system.*

- *An $\mathcal{I}$-**operator** on $\mathcal{A}$ is a map $\nabla^{\mathcal{A}} : \mathrm{FiFam}^{\mathcal{I}}(\mathcal{A}) \to \mathrm{ConSys}(\boldsymbol{A})$. The $\mathcal{I}$-operator $\nabla^{\mathcal{A}}$ is called **order-preserving** if, for all $T', T'' \in \mathrm{FiFam}^{\mathcal{I}}(\mathcal{A})$,*

$$T' \leq T'' \quad \text{implies} \quad \nabla^{\mathcal{A}}(T') \leq \nabla^{\mathcal{A}}(T'').$$

- *The **lifting** $\widetilde{\nabla}^{\mathcal{A}} : \mathcal{P}(\mathrm{FiFam}^{\mathcal{I}}(\mathcal{A})) \to \mathrm{ConSys}(\boldsymbol{A})$ of $\nabla^{\mathcal{A}}$ is defined by*

$$\widetilde{\nabla}^{\mathcal{A}}(\mathcal{T}') = \bigcap \{\nabla^{\mathcal{A}}(T') : T' \in \mathcal{T}'\}, \text{ for all } \mathcal{T}' \subseteq \mathrm{FiFam}^{\mathcal{I}}(\mathcal{A}).$$

- *The **relativization** $\widetilde{\nabla}^{\mathcal{A}, \mathcal{I}} : \mathrm{FiFam}^{\mathcal{I}}(\mathcal{A}) \to \mathrm{ConSys}(\boldsymbol{A})$ of $\nabla^{\mathcal{A}}$ is defined, for all $T' \in \mathrm{FiFam}^{\mathcal{I}}(\mathcal{A})$, by*

$$\widetilde{\nabla}^{\mathcal{A},\mathcal{I}}(T') = \bigcap \{\nabla^{\mathcal{A}}(T'') : T'' \in \mathrm{FiFam}^{\mathcal{I}}(\mathcal{A}), T' \leq T''\}$$
$$= \widetilde{\nabla}^{\mathcal{A}}((\mathrm{FiFam}^{\mathcal{I}}(\mathcal{A}))^{T'}).$$

- *The map* $\nabla^{\mathcal{A}^{-1}} : \mathrm{ConSys}(\boldsymbol{A}) \to \mathcal{P}(\mathrm{FiFam}^{\mathcal{I}}(\mathcal{A}))$ *is defined by*

$$\nabla^{\mathcal{A}^{-1}}(\theta) = \{T' \in \mathrm{FiFam}^{\mathcal{I}}(\mathcal{A}) : \theta \leq \nabla^{\mathcal{A}}(T')\}, \textit{ for all } \theta \in \mathrm{ConSys}(\boldsymbol{A}).$$

Directly from these definitions we obtain

Lemma 9. *Let* $\mathcal{I} = \langle \boldsymbol{F}, C \rangle$ *be a base* π*-institution,* $\mathcal{A} = \langle \boldsymbol{A}, \langle F, \alpha \rangle \rangle$ *an* N*-algebraic system and* $\nabla^{\mathcal{A}}$ *an* $\mathcal{I}$*-operator on* $\mathcal{A}$.

1. $\widetilde{\nabla}^{\mathcal{A},\mathcal{I}}$ *is also an* $\mathcal{I}$*-operator.*
2. $\widetilde{\nabla}^{\mathcal{A},\mathcal{I}}(T') \leq \nabla^{\mathcal{A}}(T')$, *for all* $T' \in \mathrm{FiFam}^{\mathcal{I}}(\mathcal{A})$.
3. $\widetilde{\nabla}^{\mathcal{A},\mathcal{I}}$ *is order-preserving.*
4. $\widetilde{\nabla}^{\mathcal{A}}(\mathcal{T}') \leq \nabla^{\mathcal{A}}(T')$, *for all* $T' \in \mathcal{T}'$.

By analogy with (Albuquerque et al., 2016), the categorical Leibniz and Suszko operators are the prototypical examples of the general notion of $\mathcal{I}$-operator. The Suszko operator is the relativization of the Leibniz operator and is order-preserving. Finally, the Tarski operator is the lifting of the Leibniz operator.

In all subsequent results, when we say "let $\nabla^{\mathcal{A}}$ be an $\mathcal{I}$-operator on $\mathcal{A}$" or quantify "for all $\mathcal{A}$", we implicitly make the assumption that $\boldsymbol{F} = \langle \mathbf{Sign}, \mathrm{SEN}, N \rangle$ is a base algebraic system, $\mathcal{I} = \langle \boldsymbol{F}, C \rangle$ is a π-institution based on $\boldsymbol{F}$ and $\mathcal{A} = \langle \boldsymbol{A}, \langle F, \alpha \rangle \rangle$ an N-algebraic system.

Proposition 10. *Let* $\nabla^{\mathcal{A}}$ *be an* $\mathcal{I}$*-operator on* $\mathcal{A}$. *The maps* $\widetilde{\nabla}^{\mathcal{A}}$ *and* $\nabla^{\mathcal{A}^{-1}}$ *establish a Galois connection between* $\mathcal{P}(\mathrm{FiFam}^{\mathcal{I}}(\mathcal{A}))$ *and* $\mathrm{ConSys}(\boldsymbol{A})$ *with the first ordering being the subset relation* $\subseteq$ *and the second the signature-wise inclusion relation* $\leq$.

Proof: Suppose $\mathcal{T} \subseteq \mathrm{FiFam}^{\mathcal{I}}(\mathcal{A})$ and $\theta \in \mathrm{ConSys}(\boldsymbol{A})$.

- Assume $\theta \leq \widetilde{\nabla}^{\mathcal{A}}(\mathcal{T})$. If $T \in \mathcal{T}$, then $\widetilde{\nabla}^{\mathcal{A}}(\mathcal{T}) \leq \nabla^{\mathcal{A}}(T)$. Thus, $\theta \leq \nabla^{\mathcal{A}}(T)$, whence $T \in \nabla^{\mathcal{A}^{-1}}(\theta)$. This proves that $\mathcal{T} \subseteq \nabla^{\mathcal{A}^{-1}}(\theta)$.
- If $\mathcal{T} \subseteq \nabla^{\mathcal{A}^{-1}}(\theta)$, then $\theta \leq \nabla^{\mathcal{A}}(T)$, for all $T \in \mathcal{T}$. Thus, $\theta \leq \widetilde{\nabla}^{\mathcal{A}}(\mathcal{T})$. $\quad\square$

Applying now general results pertaining to Galois connections (see, e.g., p. 55 onwards of (Davey and Priestley, 2002)), we may obtain the following statements as direct consequences of Proposition 10.

Corollary 11. *Let* $\nabla^{\mathcal{A}}$ *be an* $\mathcal{I}$*-operator on* $\mathcal{A}$.

1. *The maps* $\widetilde{\nabla}^{\mathcal{A}}$ *and* $\nabla^{\mathcal{A}^{-1}}$ *are order-reversing.*
2. *The map* $\nabla^{\mathcal{A}^{-1}} \circ \widetilde{\nabla}^{\mathcal{A}}$ *is a closure operator over* $\mathrm{FiFam}^{\mathcal{I}}(\mathcal{A})$.
3. *The map* $\widetilde{\nabla}^{\mathcal{A}} \circ \nabla^{\mathcal{A}^{-1}}$ *is a closure operator on* $\mathrm{ConSys}(\boldsymbol{A})$.

4. *The set of fixed points of $\nabla^{\mathcal{A}^{-1}} \circ \widetilde{\nabla}^{\mathcal{A}}$ is $\mathrm{Ran}(\nabla^{\mathcal{A}^{-1}})$.*
5. *The set of fixed points of $\widetilde{\nabla}^{\mathcal{A}} \circ \nabla^{\mathcal{A}^{-1}}$ is $\mathrm{Ran}(\widetilde{\nabla}^{\mathcal{A}})$.*
6. *The maps $\widetilde{\nabla}^{\mathcal{A}}$ and $\nabla^{\mathcal{A}^{-1}}$ restrict to mutually inverse dual order isomorphisms between the set of fixed points of $\nabla^{\mathcal{A}^{-1}} \circ \widetilde{\nabla}^{\mathcal{A}}$ and the set of fixed points of $\widetilde{\nabla}^{\mathcal{A}} \circ \nabla^{\mathcal{A}^{-1}}$.*

We assign special names to the fixed points of the closure operators of Parts 2 and 3 of the preceding corollary. Both will be central to our subsequent analysis and to many of our results.

Definition 12. *Given an $\mathcal{I}$-operator $\nabla^{\mathcal{A}}$ on $\mathcal{A}$,*

- *a family $\mathcal{T} \subseteq \mathrm{FiFam}^{\mathcal{I}}(\mathcal{A})$ is $\nabla^{\mathcal{A}}$-full if $\mathcal{T} = \nabla^{\mathcal{A}^{-1}}(\widetilde{\nabla}^{\mathcal{A}}(\mathcal{T}))$;*
- *a congruence system $\theta \in \mathrm{ConSys}(\boldsymbol{A})$ is $\nabla^{\mathcal{A}}$-full if $\theta = \widetilde{\nabla}^{\mathcal{A}}(\nabla^{\mathcal{A}^{-1}}(\theta))$.*

Then, Part 6 of the corollary asserts that $\widetilde{\nabla}^{\mathcal{A}}$ and $\nabla^{\mathcal{A}^{-1}}$ restrict to mutually inverse dual order isomorphisms between the sets of $\nabla^{\mathcal{A}}$-full $\mathcal{I}$-gmatrices on $\mathcal{A}$ and $\nabla^{\mathcal{A}}$-full congruence systems on $\boldsymbol{A}$.

Another consequence of the previously described Galois connection is the following

Proposition 13. *Let $\nabla^{\mathcal{A}}$ be an $\mathcal{I}$-operator on $\mathcal{A}$.*

1. *$\mathcal{T} \subseteq \mathrm{FiFam}^{\mathcal{I}}(\mathcal{A})$ is $\nabla^{\mathcal{A}}$-full iff it is the largest $\mathcal{U} \subseteq \mathrm{FiFam}^{\mathcal{I}}(\mathcal{A})$, such that $\widetilde{\nabla}^{\mathcal{A}}(\mathcal{U}) = \widetilde{\nabla}^{\mathcal{A}}(\mathcal{T})$.*
2. *$\theta \in \mathrm{ConSys}(\boldsymbol{A})$ is $\nabla^{\mathcal{A}}$-full iff it is the largest $\eta \in \mathrm{ConSys}(\boldsymbol{A})$, such that $\nabla^{\mathcal{A}^{-1}}(\eta) = \nabla^{\mathcal{A}^{-1}}(\theta)$.*

Focusing, next, on the Leibniz operator $\Omega^{\mathcal{A}}$, whose lifting is the Tarski operator $\widetilde{\Omega}^{\mathcal{A}}$, we note, first, that, if $\theta \in \mathrm{ConSys}(\boldsymbol{A})$,

$$
\begin{aligned}
\Omega^{\mathcal{A}^{-1}}(\theta) &= \{T \in \mathrm{FiFam}^{\mathcal{I}}(\mathcal{A}) : \theta \leq \Omega^{\mathcal{A}}(T)\} \\
&= \{T \in \mathrm{FiFam}^{\mathcal{I}}(\mathcal{A}) : T \text{ comp } \theta\} \\
&\subseteq \mathrm{FiFam}^{\mathcal{I}}(\mathcal{A}).
\end{aligned}
$$

We obtain

Proposition 14. *Let $\mathcal{A} = \langle \boldsymbol{A}, \langle F, \alpha \rangle \rangle$ be an N-algebraic system, with $\boldsymbol{A} = \langle \mathbf{Sign}', \mathrm{SEN}', N' \rangle$, $\theta \in \mathrm{ConSys}(\boldsymbol{A})$ and $\langle I_{\mathbf{Sign}}, \pi \rangle := \langle I_{\mathbf{Sign}}, \pi^{\theta} \rangle : \mathrm{SEN} \to \mathrm{SEN}^{\theta}$ the corresponding projection N-morphism.*

1. *$\Omega^{\mathcal{A}^{-1}}(\theta) = \pi^{-1}(\mathrm{FiFam}^{\mathcal{I}}(\mathcal{A}/\theta))$ and $\mathrm{FiFam}^{\mathcal{I}}(\mathcal{A}/\theta) = \pi(\Omega^{\mathcal{A}^{-1}}(\theta))$.*
2. *The natural transformations $\pi : \mathcal{P}\mathrm{SEN} \to \mathcal{P}\mathrm{SEN}^{\theta}$ and $\pi^{-1} : \mathcal{P}\mathrm{SEN}^{\theta} \to \mathcal{P}\mathrm{SEN}$ restrict to order-isomorphisms between the sets $\Omega^{\mathcal{A}^{-1}}(\theta)$ and $\mathrm{FiFam}^{\mathcal{I}}(\mathcal{A}/\theta)$.*

Proof:

1. Suppose that $T \in \Omega^{\mathcal{A}^{-1}}(\theta)$. Then θ is compatible with T. By Lemma 3, $\pi(T) \in \mathrm{FiFam}^{\mathcal{I}}(\mathcal{A}/\theta)$. Since, by compatibility, $T = \pi^{-1}(\pi(T))$, we obtain $T \in \pi^{-1}(\mathrm{FiFam}^{\mathcal{I}}(\mathcal{A}/\theta))$.

   If, conversely, $T' \in \mathrm{FiFam}^{\mathcal{I}}(\mathcal{A}/\theta)$, we get $\pi^{-1}(T') \in \mathrm{FiFam}^{\mathcal{I}}(\mathcal{A})$, and, by the surjectivity of π, $T' = \pi(\pi^{-1}(T'))$. This implies $\pi^{-1}(T') = \pi^{-1}(\pi(\pi^{-1}(T')))$, showing that θ is compatible with $\pi^{-1}(T')$, or, equivalently, $\pi^{-1}(T') \in \Omega^{\mathcal{A}^{-1}}(\theta)$.

   Taking into account the surjectivity of π, we get the second equality.

2. By Part 1, both π and π^{-1} are onto their respective codomains. Note, in addition, that

   - by the surjectivity of π, $\pi\pi^{-1} = \mathrm{I}_{\mathrm{FiFam}^{\mathcal{I}}(\mathcal{A}/\theta)}$ and
   - by the definition of $\Omega^{\mathcal{A}^{-1}}(\theta)$, $\pi^{-1}\pi = \mathrm{I}_{\Omega^{\mathcal{A}^{-1}}(\theta)}$.

   These show that π and π^{-1} are mutually inverse bijections and, therefore, being order preserving, must be order isomorphisms. $\square$

Taking into account that isomorphisms preserve least elements, we get

Corollary 15. *Let* $T \in \mathrm{FiFam}^{\mathcal{I}}(\mathcal{A})$ *and* $\theta \in \mathrm{ConSys}(\boldsymbol{A})$, *such that* θ *is compatible with* T. *Then* T *is the least element of* $\Omega^{\mathcal{A}^{-1}}(\theta)$ *iff* T/θ *is the least element of* $\mathrm{FiFam}^{\mathcal{I}}(\mathcal{A}/\theta)$.

Using the characterization of full $\mathcal{I}$-gmatrix systems of Proposition 5, we get

Corollary 16. *For all* $\theta \in \mathrm{ConSys}(\boldsymbol{A})$, *the set* $\Omega^{\mathcal{A}^{-1}}(\theta)$ *is full and, hence, a closure system.*

Specifically for the $\Omega^{\mathcal{A}}$-full sets of $\mathcal{I}$-filter families and the $\Omega^{\mathcal{A}}$-full congruence systems on $\boldsymbol{A}$, we have the following characterizations, which form an abstraction to the categorical level of Propositions 3.10 and 3.11 of (Albuquerque et al., 2016):

Proposition 17. *Let* $\mathcal{T} \subseteq \mathrm{FiFam}^{\mathcal{I}}(\mathcal{A})$ *and* $\theta \in \mathrm{ConSys}(\boldsymbol{A})$.

- $\mathcal{T}$ *is* $\Omega^{\mathcal{A}}$*-full iff it is full.*
- θ *is* $\Omega^{\mathcal{A}}$*-full iff* $\theta \in \mathrm{ConSys}_{\mathsf{AlgSys}(\mathcal{I})}(\boldsymbol{A})$.

Proof:

- In general, $\Omega^{\mathcal{A}^{-1}}(\widetilde{\Omega}^{\mathcal{A}}(\mathcal{T})) = \{T \in \mathrm{FiFam}^{\mathcal{I}}(\mathcal{A}) : \widetilde{\Omega}^{\mathcal{A}}(\mathcal{T}) \leq \Omega^{\mathcal{A}}(\mathrm{T})\}$. On the other hand, $\mathcal{T}$ is $\Omega^{\mathcal{A}}$-full iff $\mathcal{T} = \Omega^{\mathcal{A}^{-1}}(\widetilde{\Omega}^{\mathcal{A}}(\mathcal{T}))$ and it is full iff $\mathcal{T} = \{T \in \mathrm{FiFam}^{\mathcal{I}}(\mathcal{A}) : \widetilde{\Omega}^{\mathcal{A}}(\mathcal{T}) \leq \Omega^{\mathcal{A}}(\mathrm{T})\}$. Thus, the two notions coincide.
- Assume θ is $\Omega^{\mathcal{A}}$-full. Then, $\mathcal{A}/\theta = \mathcal{A}/\widetilde{\Omega}^{\mathcal{A}}(\Omega^{\mathcal{A}^{-1}}(\theta))$. By Corollary 16, $\Omega^{\mathcal{A}^{-1}}(\theta)$ is full. By Part 3 of Lemma 6, $\mathcal{A}/\widetilde{\Omega}^{\mathcal{A}}(\Omega^{\mathcal{A}^{-1}}(\theta)) \in \mathsf{AlgSys}(\mathcal{I})$. Therefore, we conclude that $\theta \in \mathrm{ConSys}_{\mathsf{AlgSys}(\mathcal{A})}(\mathcal{A})$.

If, conversely, $\mathcal{A}/\theta \in \mathsf{AlgSys}(\mathcal{I})$, then $\widetilde{\Omega}^{\mathcal{A}/\theta}(\mathrm{FiFam}^{\mathcal{I}}(\mathcal{A}/\theta)) = \Delta^{\mathrm{SEN}'^{\theta}}$, whence

$$
\begin{aligned}
\theta &= \mathrm{Ker}(\langle I_{\mathbf{Sign}'}, \pi^{\theta}\rangle) = \pi^{\theta^{-1}}(\Delta^{\mathrm{SEN}'^{\theta}})\\
&= \pi^{\theta^{-1}}(\widetilde{\Omega}^{\mathcal{A}/\theta}(\mathrm{FiFam}^{\mathcal{I}}(\mathcal{A}/\theta)))\\
&\overset{\mathrm{Prop.\ 4}}{=} \widetilde{\Omega}^{\mathcal{A}}(\pi^{\theta^{-1}}(\mathrm{FiFam}^{\mathcal{I}}(\mathcal{A}/\theta))) \overset{\mathrm{Prop.\ 14}}{=} \widetilde{\Omega}(\Omega^{\mathcal{A}^{-1}}(\theta)).
\end{aligned}
$$

Hence θ is $\Omega^{\mathcal{A}}$-full.

$\square$

Using Proposition 17 we obtain the following statement on the Galois connection established in Proposition 10 and Corollary 11 as pertaining to the special case of the Tarksi operator, viewed as the lifting of the Leibniz operator:

Corollary 18. *The maps $\widetilde{\Omega}^{\mathcal{A}}$ and $\Omega^{\mathcal{A}^{-1}}$ establish a Galois connection between $\mathcal{P}(\mathrm{FiFam}^{\mathcal{I}}(\mathcal{A}))$ and $\mathrm{ConSys}(\mathbf{A})$, and they restrict to mutually inverse dual order isomorphisms between the poset of all full $\mathcal{I}$-gmatrix systems on $\mathcal{A}$ and the poset $\mathrm{ConSys}_{\mathsf{AlgSys}(\mathcal{I})}(\mathbf{A})$.*

The isomorphism of Corollary 18 is actually the one established a decade ago as Theorem 13 of (Voutsadakis, 2005b), taking after the Isomorphism Theorem 2.30 of (Font and Jansana, 1996). The form in Corollary 18, expressed in terms of Galois connections in the context of $\mathcal{I}$-operators, is Corollary 3.12 of (Albuquerque et al., 2016).

Finally, putting together the equivalence between fullness of $\mathcal{I}$-gmatrix systems and $\nabla^{\mathcal{A}}$-fullness given in Proposition 17 and the general characterization of $\nabla^{\mathcal{A}}$-fullness given in Proposition 13, we obtain

Proposition 19. *A subset $\mathcal{T} \subseteq \mathrm{FiFam}^{\mathcal{I}}(\mathcal{A})$ is full iff $\mathcal{T}$ is the largest $\mathcal{U} \subseteq \mathrm{FiFam}^{\mathcal{I}}(\mathcal{A})$, such that $\widetilde{\Omega}^{\mathcal{A}}(\mathcal{T}) = \widetilde{\Omega}^{\mathcal{A}}(\mathcal{U})$.*

Leaving, once more, aside the special case of the Leibniz operator and returning to arbitrary $\mathcal{I}$-operators, and still following the ideas in (Albuquerque et al., 2016), we introduce the concept of a $\nabla^{\mathcal{A}}$-class of a theory family T and, based on it, that of a $\nabla^{\mathcal{A}}$-filter family (see Subsection 3.3 of (Albuquerque et al., 2016)).

Definition 20. *Let $\nabla^{\mathcal{A}}$ be an $\mathcal{I}$-operator on $\mathcal{A}$ and $T \in \mathrm{FiFam}^{\mathcal{I}}(\mathcal{A})$. The $\nabla^{\mathcal{A}}$-class of T is the set*

$$
[\![T]\!]^{\nabla^{\mathcal{A}}} = \Omega^{\mathcal{A}^{-1}}(\nabla^{\mathcal{A}}(T)) = \{T' \in \mathrm{FiFam}^{\mathcal{I}}(\mathcal{A}) : \nabla^{\mathcal{A}}(T) \leq \Omega^{\mathcal{A}}(T')\}.
$$

In other words, the $\nabla^{\mathcal{A}}$-class of a filter family T of $\mathcal{A}$ consists of all those filter families of $\mathcal{A}$ with which the $\nabla^{\mathcal{A}}$-congruence system of T is compatible.

Exploiting Corollary 16, with $\theta = \nabla^{\mathcal{A}}(T)$, we get

Proposition 21. *Let $\nabla^{\mathcal{A}}$ be an $\mathcal{I}$-operator on $\mathcal{A}$ and $T \in \mathrm{FiFam}^{\mathcal{I}}(\mathcal{A})$. The $\nabla^{\mathcal{A}}$-class $[\![T]\!]^{\nabla^{\mathcal{A}}}$ of T is full. Thus, it is a closure system and $[\![T]\!]^{\nabla^{\mathcal{A}}} = \Omega^{\mathcal{A}^{-1}}(\widetilde{\Omega}^{\mathcal{A}}([\![T]\!]^{\nabla^{\mathcal{A}}}))$.*

As a consequence it makes sense to consider the $\leq$-smallest $\mathcal{I}$-filter family in the $\nabla^{\mathcal{A}}$-class of T:

Definition 22. *Given an $\mathcal{I}$-operator $\nabla^{\mathcal{A}}$ on $\mathcal{A}$ and $T \in \mathrm{FiFam}^{\mathcal{I}}(\mathcal{A})$, the smallest element of the $\nabla^{\mathcal{A}}$-class $[\![T]\!]^{\nabla^{\mathcal{A}}}$ is denoted by $T^{\nabla^{\mathcal{A}}} = \bigcap [\![T]\!]^{\nabla^{\mathcal{A}}}$. We call T a $\nabla^{\mathcal{A}}$-**filter family** if $T = T^{\nabla^{\mathcal{A}}}$ and we denote the set of all $\nabla^{\mathcal{A}}$-filter families of $\mathcal{A}$ by $\mathrm{FiFam}^{\nabla^{\mathcal{A}}}(\mathcal{A})$.*

The first result asserts the injectivity of the $\mathcal{I}$-operator $\nabla^{\mathcal{A}}$ on the collection of $\nabla^{\mathcal{A}}$-filter families:

Proposition 23. *Every $\mathcal{I}$-operator $\nabla^{\mathcal{A}}$ on $\mathcal{A}$ is order-reflecting and, thus, injective, on $\mathrm{FiFam}^{\nabla^{\mathcal{A}}}(\mathcal{A})$.*

Proof: Suppose $T', T'' \in \mathrm{FiFam}^{\nabla^{\mathcal{A}}}(\mathcal{A})$, with $\nabla^{\mathcal{A}}(T') \leq \nabla^{\mathcal{A}}(T'')$. Then, clearly, $[\![T'']\!]^{\nabla^{\mathcal{A}}} \subseteq [\![T']\!]^{\nabla^{\mathcal{A}}}$, whence $T' = \bigcap [\![T']\!]^{\nabla^{\mathcal{A}}} \leq \bigcap [\![T'']\!]^{\nabla^{\mathcal{A}}} = T''$. $\qquad\square$

One can now state the following properties relating to $\nabla^{\mathcal{A}}$-filter families:

Lemma 24. *Let $\nabla^{\mathcal{A}}$ be an $\mathcal{I}$-operator on $\mathcal{A}$. For all $T, T' \in \mathrm{FiFam}^{\mathcal{I}}(\mathcal{A})$,*

1. $[\![T]\!]^{\nabla^{\mathcal{A}}} \subseteq (\mathrm{FiFam}^{\mathcal{I}}(\mathcal{A}))^{T^{\nabla^{\mathcal{A}}}}$;
2. *If $[\![T]\!]^{\nabla^{\mathcal{A}}} \subseteq [\![T']\!]^{\nabla^{\mathcal{A}}}$, then $T'^{\nabla^{\mathcal{A}}} \leq T^{\nabla^{\mathcal{A}}}$.*

If, moreover, $\nabla^{\mathcal{A}}$ is order-preserving, then:

3. *If $T \leq T'$, then $[\![T']\!]^{\nabla^{\mathcal{A}}} \subseteq [\![T]\!]^{\nabla^{\mathcal{A}}}$ and $T^{\nabla^{\mathcal{A}}} \leq T'^{\nabla^{\mathcal{A}}}$.*
4. $(\mathrm{FiFam}^{\nabla^{\mathcal{A}}}(\mathcal{A}))^{T} \subseteq [\![T]\!]^{\nabla^{\mathcal{A}}}$;

Proof: Part 1 follows from $T^{\nabla^{\mathcal{A}}} = \bigcap [\![T]\!]^{\nabla^{\mathcal{A}}}$. For Part 2, we have $T'^{\nabla^{\mathcal{A}}} = \bigcap [\![T']\!]^{\nabla^{\mathcal{A}}} \leq \bigcap [\![T]\!]^{\nabla^{\mathcal{A}}} = T^{\nabla^{\mathcal{A}}}$. For Part 3, taking into account order preservation, if $T \leq T'$, then $\nabla^{\mathcal{A}}(T) \leq \nabla^{\mathcal{A}}(T')$, whence $[\![T']\!]^{\nabla^{\mathcal{A}}} \subseteq [\![T]\!]^{\nabla^{\mathcal{A}}}$ and $T^{\nabla^{\mathcal{A}}} \leq T'^{\nabla^{\mathcal{A}}}$. For Part 4, suppose $T \leq T' = \bigcap [\![T']\!]^{\nabla^{\mathcal{A}}}$. By order preservation and by Proposition 21, $\nabla^{\mathcal{A}}(T) \leq \nabla^{\mathcal{A}}(T') \leq \Omega^{\mathcal{A}}(T')$, whence $T' \in [\![T]\!]^{\nabla^{\mathcal{A}}}$. $\qquad\square$

In concluding this section on $\mathcal{I}$-operators and their properties, we prove a lemma, relating the property of a filter family being a $\nabla^{\mathcal{A}}$-filter family with the form of its $\nabla^{\mathcal{A}}$-class, for an order preserving $\mathcal{I}$-operator $\nabla^{\mathcal{A}}$ that is dominated by the Leibniz operator on $\mathcal{A}$. Lemma 25 also helps usher in the material of Section 6.

Lemma 25. *Let $\nabla^{\mathcal{A}}$ be an order-preserving $\mathcal{I}$-operator, such that $\nabla^{\mathcal{A}}(T) \leq \Omega^{\mathcal{A}}(T)$, for all $T \in \mathrm{FiFam}^{\mathcal{I}}(\mathcal{A})$. Then $[\![T]\!]^{\nabla^{\mathcal{A}}} = (\mathrm{FiFam}^{\mathcal{I}}(\mathcal{A}))^{T}$ iff $T = T^{\nabla^{\mathcal{A}}}$, i.e., iff T is a $\nabla^{\mathcal{A}}$-filter family.*

Proof: Suppose, first, that $[\![T]\!]^{\nabla^{\mathcal{A}}} = (\mathrm{FiFam}^{\mathcal{I}}(\mathcal{A}))^T$. Then, we have $T^{\nabla^{\mathcal{A}}} = \bigcap [\![T]\!]^{\nabla^{\mathcal{A}}} = \bigcap (\mathrm{FiFam}^{\mathcal{I}}(\mathcal{A}))^T = T$.

Conversely, if $T = T^{\nabla^{\mathcal{A}}}$, then by Part 1 of Lemma 24, we have $[\![T]\!]^{\nabla^{\mathcal{A}}} \subseteq (\mathrm{FiFam}^{\mathcal{I}}(\mathcal{A}))^T$. On the other hand, if $T' \in \mathrm{FiFam}^{\mathcal{I}}(\mathcal{A})$, with $T \leq T'$, then, by the hypotheses, $\nabla^{\mathcal{A}}(T) \leq \nabla^{\mathcal{A}}(T') \leq \Omega^{\mathcal{A}}(T')$, whence $T' \in [\![T]\!]^{\nabla^{\mathcal{A}}}$. $\square$

6 $\mathcal{I}$-Compatibility Operators and Coherence

We focus next on $\mathcal{I}$-operators that associate to a given filter family on an algebraic system $\mathcal{A}$ a congruence system that is compatible with the filter family. In the AAL context of (Albuquerque et al., 2016), such operators are termed $\mathcal{S}$-compatibility operators.

Definition 26. *An $\mathcal{I}$-compatibility operator on $\mathcal{A}$ is an $\mathcal{I}$-operator $\nabla^{\mathcal{A}}$ on $\mathcal{A}$, such that, for all $T \in \mathrm{FiFam}^{\mathcal{I}}(\mathcal{A})$, the congruence system $\nabla^{\mathcal{A}}(T)$ is compatible with T, i.e., $\nabla^{\mathcal{A}}(T) \leq \Omega^{\mathcal{A}}(T)$.*

This is equivalent to saying that an $\mathcal{I}$-operator is an $\mathcal{I}$-compatibility operator iff $T \in [\![T]\!]^{\nabla^{\mathcal{A}}}$, for all $T \in \mathrm{FiFam}^{\mathcal{I}}(\mathcal{A})$. By definition, the largest $\mathcal{I}$-compatibility operator is $\Omega^{\mathcal{A}}$ and the smallest one is the one sending every $\mathcal{I}$-filter family to the identity congruence system $\Delta^{\mathrm{SEN}'}$ on SEN'. As has been shown in Theorem 4 of (Voutsadakis, 2007a) (see, also, (Voutsadakis, 2007b) and Theorem 1.6 of (Czelakowski, 2003) for the progenitor in AAL), the Suszko operator $\widetilde{\Omega}^{\mathcal{A},\mathcal{I}}$ has the distinction of being the largest order-preserving $\mathcal{I}$-compatibility operator on $\mathcal{A}$.

Some easy properties of $\mathcal{I}$-compatibility operators, refining those properties of $\mathcal{I}$-operators enumerated in Lemma 24, follow. Note, also, that Lemma 25 dealt with an $\mathcal{I}$-compatibility operator.

Lemma 27. *Let $\nabla^{\mathcal{A}}$ be an $\mathcal{I}$-compatibility operator on $\mathcal{A}$. For all $T \in \mathrm{FiFam}^{\mathcal{I}}(\mathcal{A})$,*

1. $T \in [\![T]\!]^{\nabla^{\mathcal{A}}}$;
2. $T^{\nabla^{\mathcal{A}}} \leq T$.

If $\nabla^{\mathcal{A}}$ is order-preserving, then:

3. $[\![T]\!]^{\nabla^{\mathcal{A}}} \subseteq [\![T^{\nabla^{\mathcal{A}}}]\!]^{\nabla^{\mathcal{A}}}$;
4. *Every $\nabla^{\mathcal{A}}$-full class of $\mathcal{I}$-filter families is an upset of $\mathrm{FiFam}^{\mathcal{I}}(\mathcal{A})$.*

Proof: Part 1 follows by the remark following Definition 26. Part 2 follows by Part 1 and the definition of $T^{\nabla^{\mathcal{A}}}$. Part 3 follows by Part 2 and Part 3 of Lemma 24. Finally, Part 4 follows by the definition $\nabla^{\mathcal{A}}$-fullness and the order preservation of $\nabla^{\mathcal{A}}$. $\square$

Compatibility of the $\mathcal{I}$-operators allows the following rewriting of Corollary 15 characterizing $\nabla^{\mathcal{A}}$-filter families:

Corollary 28. *Let $\nabla^{\mathcal{A}}$ be an $\mathcal{I}$-compatibility operator on $\mathcal{A}$. Then, for all $T \in \mathrm{FiFam}^{\mathcal{I}}(\mathcal{A})$, T is a $\nabla^{\mathcal{A}}$-filter family of $\mathcal{A}$ iff $T/\nabla^{\mathcal{A}}(T)$ is the least $\mathcal{I}$-filter family of $\mathcal{A}/\nabla^{\mathcal{A}}(T)$.*

Proof: Set $\theta = \nabla^{\mathcal{A}}(T)$ in Corollary 15. $\qquad\qquad\qquad\qquad\qquad\qquad\square$

The following corollary characterizes the property of an N-algebraic system having all filter families being $\nabla^{\mathcal{A}}$-filter families:

Corollary 29. *Let $\nabla^{\mathcal{A}}$ be an $\mathcal{I}$-compatibility operator on $\mathcal{A}$. The the following are equivalent:*

(i) *Every $\mathcal{I}$-filter family of $\mathcal{A}$ is a $\nabla^{\mathcal{A}}$-filter family.*
(ii) *For all $T, T' \in \mathrm{FiFam}^{\mathcal{I}}(\mathcal{A})$, $\nabla^{\mathcal{A}}(T) \leq \Omega^{\mathcal{A}}(T')$ implies $T \leq T'$.*

Proof:

(i)$\Rightarrow$(ii) Suppose $\nabla^{\mathcal{A}}(T) \leq \Omega^{\mathcal{A}}(T')$. Then $T' \in [\![T]\!]^{\nabla^{\mathcal{A}}}$. But, by hypothesis, T is the smallest theory family in $[\![T]\!]^{\nabla^{\mathcal{A}}}$, whence $T \leq T'$.

(ii)$\Rightarrow$(i) By Lemma 27, Part 2, we have $T^{\nabla^{\mathcal{A}}} \leq T$. On the other hand, by the definition of $T^{\nabla^{\mathcal{A}}}$, we get that $\nabla^{\mathcal{A}}(T)$ is compatible with $T^{\nabla^{\mathcal{A}}}$, whence $\nabla^{\mathcal{A}}(T) \leq \Omega^{\mathcal{A}}(T^{\nabla^{\mathcal{A}}})$. But, then, by hypothesis, $T \leq T^{\nabla^{\mathcal{A}}}$. So $T = T^{\nabla^{\mathcal{A}}}$. $\qquad\qquad\square$

A family of $\mathcal{I}$-compatibility operators (see Subsection 4.1 of (Albuquerque et al., 2016))

$$\nabla := \{\nabla^{\mathcal{A}} : \mathcal{A} \text{ an } N\text{-algebraic system}\}$$

is a collection, where $\nabla^{\mathcal{A}}$ is an $\mathcal{I}$-compatibility operator on $\mathcal{A}$, for every N-algebraic system $\mathcal{A}$.

Definition 30. *Let $\nabla^{\mathcal{A}}$ and $\nabla^{\mathcal{B}}$ be $\mathcal{I}$-compatibility operators on $\mathcal{A}$ and $\mathcal{B}$. The pair $\langle \nabla^{\mathcal{A}}, \nabla^{\mathcal{B}} \rangle$ **commutes with inverse (surjective) N-morphisms** if, for all (surjective) $\langle H, \gamma \rangle : \mathcal{A} \to \mathcal{B}$ and all $T'' \in \mathrm{FiFam}^{\mathcal{I}}(\mathcal{B})$,*

$$\nabla^{\mathcal{A}}(\gamma^{-1}(T'')) = \gamma^{-1}(\nabla^{\mathcal{B}}(T'')).$$

*A family ∇ of $\mathcal{I}$-compatibility operators **commutes with inverse (surjective) N-morphisms** if, for all N-algebraic systems $\mathcal{A}$ and $\mathcal{B}$, the pair $\langle \nabla^{\mathcal{A}}, \nabla^{\mathcal{B}} \rangle$ commutes with inverse (surjective) N-morphisms.*

The following notions of compatibility of morphisms with filter families and with collections of filter families will prove helpful. It abstracts to the categorical context Definition 4.7 of (Albuquerque et al., 2016).

Definition 31. *Let $\nabla^{\mathcal{A}}$ be an $\mathcal{I}$-compatibility operator on $\mathcal{A} = \langle \boldsymbol{A}, \langle F, \alpha \rangle \rangle$, with $\boldsymbol{A} = \langle \mathbf{Sign}', \mathrm{SEN}', N' \rangle$, and assume T is a $\mathcal{I}$-filter family on SEN' and $\mathcal{T}$ is a collection of $\mathcal{I}$-filter families on SEN'.*

- *An N-morphism $\langle H, \gamma \rangle : \mathcal{A} \to \mathcal{B}$ is $\nabla^{\mathcal{A}}$-**compatible with** T if*

$$\mathrm{Ker}(\langle H, \gamma \rangle) \leq \nabla^{\mathcal{A}}(T).$$

- *An N-morphism $\langle H, \gamma \rangle : \mathcal{A} \to \mathcal{B}$ is $\nabla^{\mathcal{A}}$-**compatible with** $\mathcal{T}$ if it is $\nabla^{\mathcal{A}}$-compatible with every $T \in \mathcal{T}$.*

Note that $\langle H, \gamma \rangle : \mathcal{A} \to \mathcal{B}$ is $\Omega^{\mathcal{A}}$-compatible with a filter family T on SEN' iff the congruence $\mathrm{Ker}(\langle H, \gamma \rangle)$ is compatible with T and, in case H is an isomorphism, this happens if and only if the matrix system morphism $\langle \mathcal{A}, T \rangle \to \langle \mathcal{B}, \gamma(T) \rangle$ is strict. Also note that, in case H is an isomorphism, $\langle \mathcal{A}, T \rangle \overset{\langle H, \gamma \rangle}{\to} \langle \mathcal{B}, \gamma(T) \rangle$ is a deductive matrix system morphism if and only if $\langle H, \gamma \rangle : \mathcal{A} \to \mathcal{B}$ is $\widetilde{\Omega}^{\mathcal{A}, \mathcal{I}}$-compatible with T. Czelakowski used the corresponding sentential concept in his study of the Suszko operator in (Czelakowski, 2003) to obtain a general Correspondence Theorem that was generalized in Theorem 4.17 of (Albuquerque et al., 2016). Using the abstract version encapsulated in Definition 31, we will obtain a similar general correspondence result in Theorem 40 as an analog of Theorem 4.17 of (Albuquerque et al., 2016).

For all $T \in \mathrm{FiFam}^{\mathcal{I}}(\mathcal{A})$, the projection N-morphism $\langle I_{\mathbf{Sign}'}, \pi \rangle : \mathcal{A} \to \mathcal{A}/\nabla^{\mathcal{A}}(T)$ is always $\nabla^{\mathcal{A}}$-compatible with T. In addition, since $\nabla^{\mathcal{A}}$ is an $\mathcal{I}$-compatibility operator, if $\langle H, \gamma \rangle : \mathcal{A} \to \mathcal{B}$ is $\nabla^{\mathcal{A}}$-compatible with T, then it is also $\Omega^{\mathcal{A}}$-compatible with T, i.e., $\mathrm{Ker}(\langle H, \gamma \rangle)$ is compatible with T. In case H is an isomorphism, this implies that $T = \gamma^{-1}(\gamma(T))$ and $\nabla^{\mathcal{A}}(T) = \gamma^{-1}(\gamma(\nabla^{\mathcal{A}}(T)))$.

Definition 32. *A family ∇ of $\mathcal{I}$-compatibility operators is called (**weakly**) **coherent** if, for all surjective N-morphisms $\langle H, \gamma \rangle : \mathcal{A} \to \mathcal{B}$ (with H an isomorphism) and all $T'' \in \mathrm{FiFam}^{\mathcal{I}}(\mathcal{B})$,*

$$\langle H, \gamma \rangle \ \nabla^{\mathcal{A}}\text{-compatible with } \gamma^{-1}(T'')$$
$$\text{implies} \quad \nabla^{\mathcal{A}}(\gamma^{-1}(T'')) = \gamma^{-1}(\nabla^{\mathcal{B}}(T'')).$$

Since the reverse implication of the defining condition is universally valid, a family ∇ of $\mathcal{I}$-compatibility operators is (weakly) coherent iff, for every surjective N-morphism $\langle H, \gamma \rangle : \mathcal{A} \to \mathcal{B}$ (with H an isomorphism) and every $T'' \in \mathrm{FiFam}^{\mathcal{I}}(\mathcal{B})$, $\langle H, \gamma \rangle$ is $\nabla^{\mathcal{A}}$-compatible with $\gamma^{-1}(T'')$ if and only if $\nabla^{\mathcal{A}}(\gamma^{-1}(T'')) = \gamma^{-1}(\nabla^{\mathcal{B}}(T''))$.

Lemma 33. *Let ∇ be a weakly coherent family of $\mathcal{I}$-compatibility operators. Then, for every surjective N-morphism $\langle H, \gamma \rangle : \mathcal{A} \to \mathcal{B}$, with H an isomorphism, and every $T' \in \mathrm{FiFam}^{\mathcal{I}}(\mathcal{A})$, if $\langle H, \gamma \rangle$ is $\nabla^{\mathcal{A}}$-compatible with T', then*

$$\gamma(\nabla^{\mathcal{A}}(T')) = \nabla^{\mathcal{B}}(\gamma(T')).$$

Proof: Suppose ∇ is weakly coherent, $T' \in \mathrm{FiFam}^{\mathcal{I}}(\mathcal{A})$, and $\langle H, \gamma \rangle : \mathcal{A} \to \mathcal{B}$ surjective, with H an isomorphism, and compatible with T'. By compatibility of ∇ and Lemma 3, we get $T' = \gamma^{-1}(\gamma(T'))$ and $\gamma(T') \in \mathrm{FiFam}^{\mathcal{I}}(\mathcal{B})$. Thus, by weak coherence $\nabla^{\mathcal{A}}(T') = \nabla^{\mathcal{A}}(\gamma^{-1}(\gamma(T'))) = \gamma^{-1}(\nabla^{\mathcal{B}}(\gamma(T')))$. Finally, by surjectivity, $\gamma(\nabla^{\mathcal{A}}(T')) = \nabla^{\mathcal{B}}(\gamma(T'))$. $\qquad\square$

Corollary 34. *Let ∇ be a weakly coherent family of $\mathcal{I}$-compatibility operators and $\langle H, \gamma \rangle : \mathcal{A} \to \mathcal{B}$ an N-isomorphism. Then, for all $T' \in \mathrm{FiFam}^{\mathcal{I}}(\mathcal{A})$ and all $T'' \in \mathrm{FiFam}^{\mathcal{I}}(\mathcal{B})$,*

$$\gamma(\nabla^{\mathcal{A}}(T')) = \nabla^{\mathcal{B}}(\gamma(T')) \quad and \quad \nabla^{\mathcal{A}}(\gamma^{-1}(T'')) = \gamma^{-1}(\nabla^{\mathcal{B}}(T'')).$$

Proof: The first property follows immediately by Lemma 33. For the second property, note that the kernel of an isomorphism is the identity congruence system, whence every isomorphism is $\nabla^{\mathcal{A}}$-compatible with all sentence families of $\mathcal{A}$, for any $\mathcal{I}$-operator $\nabla^{\mathcal{A}}$. Therefore, the property holds by weak coherence. $\qquad\square$

Putting together Definitions 30 and 32, we obtain

Proposition 35. *If ∇ is a family of $\mathcal{I}$-compatibility operators that commutes with inverse surjective N-morphisms, then it is coherent.*

Since Ω satisfies this property (see (Voutsadakis, 2007a)), we obtain that the Leibniz operator (viewed as a family of operators) is indeed a coherent family of $\mathcal{I}$-compatibility operators.

For weakly coherent families ∇ of $\mathcal{I}$-compatibility operators, the $\nabla^{\mathcal{A}}$-full congruence systems on $\mathcal{A}$ and the $\nabla^{\mathcal{A}}$-full collections of filter families, introduced in Definition 12 as the fixed-points of $\widetilde{\nabla}^{\mathcal{A}} \circ \nabla^{\mathcal{A}^{-1}}$ and $\nabla^{\mathcal{A}^{-1}} \circ \widetilde{\nabla}^{\mathcal{A}}$, respectively, can be characterized more elegantly (see Corollary 4.14 and Proposition 4.16 of (Albuquerque et al., 2016) for the original characterizations in the AAL context).

Proposition 36. *If ∇ is a weakly coherent family of $\mathcal{I}$-compatibility operators, for all $\theta \in \mathrm{ConSys}(\boldsymbol{A})$ (denoting $\langle I_{\mathbf{Sign}'}, \pi \rangle := \langle I_{\mathbf{Sign}'}, \pi^{\theta} \rangle : \mathcal{A} \to \mathcal{A}/\theta$),*

$$\nabla^{\mathcal{A}^{-1}}(\theta) = \pi^{-1}(\{T' \in \mathrm{FiFam}^{\mathcal{I}}(\mathcal{A}/\theta) : \pi^{-1}(\nabla^{\mathcal{A}/\theta}(T')) = \nabla^{\mathcal{A}}(\pi^{-1}(T'))\}).$$

Proof: If $T \in \nabla^{\mathcal{A}^{-1}}(\theta)$, then, by definition, $T \in \mathrm{FiFam}^{\mathcal{I}}(\mathcal{A})$ and $\theta \leq \nabla^{\mathcal{A}}(T)$. Thus, $T = \pi^{-1}(\pi(T))$, whence, by Lemma 33, $\pi(\nabla^{\mathcal{A}}(T)) = \nabla^{\mathcal{A}/\theta}(\pi(T))$. Set $T' = \pi(T) \in \mathrm{FiFam}^{\mathcal{I}}(\mathcal{A}/\theta)$. Then $T = \pi^{-1}(T')$ and $\nabla^{\mathcal{A}}(\pi^{-1}(T')) = \nabla^{\mathcal{A}}(T) = \pi^{-1}(\pi(\nabla^{\mathcal{A}}(T))) = \pi^{-1}(\nabla^{\mathcal{A}/\theta}(\pi(T))) = \pi^{-1}(\nabla^{\mathcal{A}/\theta}(T'))$.

Conversely, suppose that $T' \in \mathrm{FiFam}^{\mathcal{I}}(\mathcal{A}/\theta)$, such that $\pi^{-1}(\nabla^{\mathcal{A}/\theta}(T')) = \nabla^{\mathcal{A}}(\pi^{-1}(T'))$. Then $\langle I_{\mathbf{Sign}'}, \pi \rangle$ is $\nabla^{\mathcal{A}}$-compatible with $\pi^{-1}(T')$, or, equivalently, $\pi^{-1}(T') \in \nabla^{\mathcal{A}^{-1}}(\theta)$. $\qquad\square$

Now, Definition 12 immediately yields

Corollary 37. *Let ∇ be a weakly coherent family of $\mathcal{I}$-compatibility operators and $\mathcal{T} \subseteq \mathrm{FiFam}^{\mathcal{I}}(\mathcal{A})$. Then $\mathcal{T}$ is $\nabla^{\mathcal{A}}$-full iff*

$$\mathcal{T} = \pi^{\theta^{-1}}(\{T' \in \mathrm{FiFam}^{\mathcal{I}}(\mathcal{A}/\theta) : \pi^{\theta^{-1}}(\nabla^{\mathcal{A}/\theta}(T')) = \nabla^{\mathcal{A}}(\pi^{\theta^{-1}}(T'))\}),$$

for some $\theta \in \mathrm{ConSys}(\boldsymbol{A})$, which can be taken equal to $\widetilde{\nabla}^{\mathcal{A}}(\mathcal{T})$.

Recall that, by Proposition 14, $\Omega^{\mathcal{A}}$-full filter families are of the form $\pi^{\theta^{-1}}(\mathrm{FiFam}^{\mathcal{I}}(\mathcal{A}/\theta))$, for some $\theta \in \mathrm{ConSys}(\boldsymbol{A})$. But since, given an $\mathcal{I}$-compatibility operator $\nabla^{\mathcal{A}}$, for every filter family T, $\nabla^{\mathcal{A}}(T) \leq \Omega^{\mathcal{A}}(T)$, we get $\nabla^{\mathcal{A}^{-1}}(\theta) \subseteq \Omega^{\mathcal{A}^{-1}}(\theta) = \pi^{\theta^{-1}}(\mathrm{FiFam}^{\mathcal{I}}(\mathcal{A}/\theta))$. Thus, $\nabla^{\mathcal{A}^{-1}}(\theta)$ must be of the form $\pi^{\theta^{-1}}(\mathcal{T}')$, for some $\mathcal{T}' \subseteq \mathrm{FiFam}^{\mathcal{I}}(\mathcal{A}/\theta)$.

Lemma 38. *Let ∇ be a weakly coherent family of $\mathcal{I}$-compatibility operators and $\langle H, \gamma \rangle : \mathcal{A} \to \mathcal{B}$ be a surjective N-morphism, with H an isomorphism.*

1. *For all $\mathcal{T}'' \subseteq \mathrm{FiFam}^{\mathcal{I}}(\mathcal{B})$, if $\langle H, \gamma \rangle$ is $\nabla^{\mathcal{A}}$-compatible with $\gamma^{-1}(\mathcal{T}'')$, then $\widetilde{\nabla}^{\mathcal{A}}(\gamma^{-1}(\mathcal{T}'')) = \gamma^{-1}(\widetilde{\nabla}^{\mathcal{B}}(\mathcal{T}''))$.*
2. *For all $\mathcal{T}' \subseteq \mathrm{FiFam}^{\mathcal{I}}(\mathcal{A})$, if $\langle H, \gamma \rangle$ is $\nabla^{\mathcal{A}}$-compatible with $\mathcal{T}'$, then $\gamma(\widetilde{\nabla}^{\mathcal{A}}(\mathcal{T}')) = \widetilde{\nabla}^{\mathcal{B}}(\gamma(\mathcal{T}'))$.*

Proof:

1. If $\langle H, \gamma \rangle$ is $\nabla^{\mathcal{A}}$-compatible with $\gamma^{-1}(\mathcal{T}'')$, then, by definition, it is $\nabla^{\mathcal{A}}$-compatible with every $\gamma^{-1}(T'')$, for $T'' \in \mathcal{T}''$. Using weak coherence, we get $\widetilde{\nabla}^{\mathcal{A}}(\gamma^{-1}(\mathcal{T}'')) = \bigcap_{T'' \in \mathcal{T}''} \nabla^{\mathcal{A}}(\gamma^{-1}(T'')) = \bigcap_{T'' \in \mathcal{T}''} \gamma^{-1}(\nabla^{\mathcal{B}}(T'')) = \gamma^{-1}(\bigcap_{T'' \in \mathcal{T}''} \nabla^{\mathcal{B}}(T'')) = \gamma^{-1}(\widetilde{\nabla}^{\mathcal{B}}(\mathcal{T}''))$.
2. If $\langle H, \gamma \rangle$ is $\nabla^{\mathcal{A}}$-compatible with $\mathcal{T}'$, then $\langle H, \gamma \rangle$ is $\nabla^{\mathcal{A}}$-compatible with all $T' \in \mathcal{T}'$, by definition of $\nabla^{\mathcal{A}}$-compatibility, whence $\gamma^{-1}(\gamma(T')) = T'$, for all $T' \in \mathcal{T}'$, i.e., $\gamma^{-1}(\gamma(\mathcal{T}')) = \mathcal{T}'$. This implies that $\langle H, \gamma \rangle$ is $\nabla^{\mathcal{A}}$-compatible with $\gamma^{-1}(\gamma(\mathcal{T}'))$. Now using Part 1, we get

$$\widetilde{\nabla}^{\mathcal{A}}(\mathcal{T}') = \widetilde{\nabla}^{\mathcal{A}}(\gamma^{-1}(\gamma(\mathcal{T}'))) = \gamma^{-1}(\widetilde{\nabla}^{\mathcal{B}}(\gamma(\mathcal{T}')))$$

and, finally, using surjectivity, $\gamma(\widetilde{\nabla}^{\mathcal{A}}(\mathcal{T}')) = \widetilde{\nabla}^{\mathcal{B}}(\gamma(\mathcal{T}'))$. $\qquad\square$

A characterization of $\nabla^{\mathcal{A}}$-full congruence systems analogous to that of a $\nabla^{\mathcal{A}}$-full collection of filter families, given in Corollary 37, is as follows:

Proposition 39. *Let ∇ be a weakly coherent family of $\mathcal{I}$-compatibility operators and $\theta \in \mathrm{ConSys}(\boldsymbol{A})$. Then θ is $\nabla^{\mathcal{A}}$-full iff (denoting $\langle I_{\mathbf{Sign}'}, \pi \rangle := \langle I_{\mathbf{Sign}'}, \pi^{\theta} \rangle : \mathcal{A} \to \mathcal{A}/\theta)$*

$$\widetilde{\nabla}^{\mathcal{A}/\theta}(\{T'' \in \mathrm{FiFam}^{\mathcal{I}}(\mathcal{A}/\theta) : \pi^{-1}(\nabla^{\mathcal{A}/\theta}(T'')) = \nabla^{\mathcal{A}}(\pi^{-1}(T''))\}) = \Delta^{\mathrm{SEN}'/\theta}.$$

Proof: Let $\mathcal{T}'' = \{T'' \in \mathrm{FiFam}^{\mathcal{I}}(\mathcal{A}/\theta) : \pi^{-1}(\nabla^{\mathcal{A}/\theta}(T'')) = \nabla^{\mathcal{A}}(\pi^{-1}(T''))\}$. Then $\langle I_{\mathbf{Sign'}}, \pi \rangle$ is $\nabla^{\mathcal{A}}$-compatible with $\pi^{-1}(\mathcal{T}'')$, whence, by Proposition 36 and Lemma 38, we get θ $\nabla^{\mathcal{A}}$-full iff

$$(1) \qquad \theta = \widetilde{\nabla}^{\mathcal{A}}(\nabla^{\mathcal{A}^{-1}}(\theta)) = \widetilde{\nabla}^{\mathcal{A}}(\pi^{-1}(\mathcal{T}'')) = \pi^{-1}(\widetilde{\nabla}^{\mathcal{A}/\theta}(\mathcal{T}'')).$$

If θ $\nabla^{\mathcal{A}}$-full, then, by Condition (1) and the surjectivity of π, $\widetilde{\nabla}^{\mathcal{A}/\theta}(\mathcal{T}'') = \pi(\pi^{-1}(\widetilde{\nabla}^{\mathcal{A}/\theta}(\mathcal{T}''))) = \pi(\theta) = \Delta^{\mathrm{SEN'}/\theta}$.

If, on the other hand, $\widetilde{\nabla}^{\mathcal{A}/\theta}(\mathcal{T}'') = \Delta^{\mathrm{SEN'}/\theta}$, then $\theta = \pi^{-1}(\Delta^{\mathrm{SEN'}/\theta}) = \pi^{-1}(\widetilde{\nabla}^{\mathcal{A}/\theta}(\mathcal{T}''))$ and, therefore, θ is $\nabla^{\mathcal{A}}$-full, by Condition (1). $\qquad\square$

Since the Leibniz operator commutes with all surjective N-morphisms, when $\nabla \equiv \Omega$ in Proposition 39, the family $\mathcal{T}'' = \mathrm{FiFam}^{\mathcal{I}}(\mathcal{A}/\theta)$, whence $\Omega^{\mathcal{A}^{-1}}(\theta) = \mathrm{FiFam}^{\mathcal{I}}(\mathcal{A}/\theta)$, as was shown in Proposition 14.

Since $\widetilde{\Omega}^{\mathcal{A}/\theta}(\mathcal{T}'') = \Delta^{\mathrm{SEN'}/\theta}$ is equivalent to $\theta \in \mathrm{ConSys}_{\mathsf{AlgSys}(\mathcal{I})}(\mathcal{A})$, we also obtain the result proven in Proposition 17.

We are now ready to lift the General Correspondence Theorem 4.17 of (Albuquerque et al., 2016) to CAAL.

Theorem 40 (General Correspondence Theorem). *Let ∇ be a weakly coherent family of $\mathcal{I}$-compatibility operators. For every surjective N-morphism $\langle H, \gamma \rangle : \mathcal{A} \to \mathcal{B}$, with H an isomorphism, and every $T \in \mathrm{FiFam}^{\mathcal{I}}(\mathcal{A})$, if $\langle H, \gamma \rangle$ is $\nabla^{\mathcal{A}}$-compatible with T, then $\langle H, \gamma \rangle$ induces an order isomorphism between $[\![T]\!]^{\nabla^{\mathcal{A}}}$ and $[\![\gamma(T)]\!]^{\nabla^{\mathcal{B}}}$, whose inverse is given by γ^{-1}.*

Proof: Since $\langle H, \gamma \rangle$ is $\nabla^{\mathcal{A}}$-compatible with T, by Lemma 1, $\gamma^{-1}(\gamma(T)) = T$ and, in addition, by Lemma 3, taking into account the fact that H is postulated to be an isomorphism, $\gamma(T) \in \mathrm{FiFam}^{\mathcal{I}}(\mathcal{B})$.

Let, first, $U \in [\![T]\!]^{\nabla^{\mathcal{A}}}$. Then $\mathrm{Ker}(\langle H, \gamma \rangle) \leq \nabla^{\mathcal{A}}(T) \leq \Omega^{\mathcal{A}}(U)$. Thus, by Lemma 3, $\gamma^{-1}(\gamma(U)) = U$ and $\gamma(U) \in \mathrm{FiFam}^{\mathcal{I}}(\mathcal{B})$. Since $\langle H, \gamma \rangle$ is $\Omega^{\mathcal{A}}$-compatible with U and $\nabla^{\mathcal{A}}$-compatible with T, and both operators are weakly coherent, Lemma 33 yields $\nabla^{\mathcal{B}}(\gamma(T)) = \gamma(\nabla^{\mathcal{A}}(T)) \leq \gamma(\Omega^{\mathcal{A}}(U)) = \Omega^{\mathcal{B}}(\gamma(U))$. Therefore, $\gamma(U) \in [\![\gamma(T)]\!]^{\nabla^{\mathcal{B}}}$.

Next, assume $U' \in [\![\gamma(T)]\!]^{\nabla^{\mathcal{B}}}$. Thus, $\nabla^{\mathcal{B}}(\gamma(T)) \leq \Omega^{\mathcal{B}}(U')$. It is the case, by Lemma 3, that $\gamma^{-1}(U') \in \mathrm{FiFam}^{\mathcal{I}}(\mathcal{A})$ and, by surjectivity, that $\gamma(\gamma^{-1}(U')) = U'$. Moreover, $\langle H, \gamma \rangle$ is $\nabla^{\mathcal{A}}$-compatible with $T = \gamma^{-1}(\gamma(T))$. Therefore, by weak coherence,

$$\nabla^{\mathcal{A}}(T) = \nabla^{\mathcal{A}}(\gamma^{-1}(\gamma(T))) = \gamma^{-1}(\nabla^{\mathcal{B}}(\gamma(T))) \leq \gamma^{-1}(\Omega^{\mathcal{B}}(U')) = \Omega^{\mathcal{A}}(\gamma^{-1}(U')),$$

proving that $\gamma^{-1}(U') \in [\![T]\!]^{\nabla^{\mathcal{A}}}$.

Thus, $\langle H, \gamma \rangle$ induces a bijection between $[\![T]\!]^{\nabla^{\mathcal{A}}}$ and $[\![\gamma(T)]\!]^{\nabla^{\mathcal{B}}}$ with inverse γ^{-1}. Both mappings are order-preserving, whence we obtain the asserted order isomorphism. $\qquad\square$

This isomorphism also shows that the least elements of the corresponding isomorphic complete lattices correspond.

Corollary 41. *Let ∇ be a weakly coherent family of $\mathcal{I}$-compatibility operators. For every surjective N-morphism $\langle H, \gamma \rangle : \mathcal{A} \to \mathcal{B}$, with H an isomorphism, and every $T \in \mathrm{FiFam}^{\mathcal{I}}(\mathcal{A})$, if $\langle H, \gamma \rangle$ is $\nabla^{\mathcal{A}}$-compatible with T, then*

$$T \in \mathrm{FiFam}^{\nabla^{\mathcal{A}}}(\mathcal{A}) \quad \textit{iff} \quad \gamma(T) \in \mathrm{FiFam}^{\nabla^{\mathcal{B}}}(\mathcal{B}).$$

To obtain an analogous correspondence theorem for the relativized operators $\widetilde{\nabla}^{\mathcal{A},\mathcal{I}}$ (see Theorem 4.20 of (Albuquerque et al., 2016)), we first show that relativization preserves weak coherence:

Proposition 42. *If ∇ is a weakly coherent family of $\mathcal{I}$-compatibility operators, then the family*

$$\widetilde{\nabla}^{\bullet,\mathcal{I}} = \{\widetilde{\nabla}^{\mathcal{A},\mathcal{I}} : \mathcal{A} \text{ an } N\text{-algebraic system}\}$$

is also a weakly coherent family of $\mathcal{I}$-compatibility operators.

Proof: By definition of $\widetilde{\nabla}^{\mathcal{A},\mathcal{I}}$, if ∇ is a family of $\mathcal{I}$-compatibility operators, then $\nabla^{\bullet,\mathcal{I}}$ is one also. To show that it is also weakly coherent, let $T'' \in \mathrm{FiFam}^{\mathcal{I}}(\mathcal{B})$ and $\langle H, \gamma \rangle : \mathcal{A} \to \mathcal{B}$ surjective, with H an isomorphism, $\widetilde{\nabla}^{\mathcal{A},\mathcal{I}}$-compatible with $\gamma^{-1}(T'')$. Then $\mathrm{Ker}(\langle H, \gamma \rangle) \leq \widetilde{\nabla}^{\mathcal{A},\mathcal{I}}(\gamma^{-1}(T''))$. Let $T' \in (\mathrm{FiFam}^{\mathcal{I}}(\mathcal{A}))^{\gamma^{-1}(T'')}$, i.e., $\gamma^{-1}(T'') \leq T'$. Then $\mathrm{Ker}(\langle H, \gamma \rangle) \leq \widetilde{\nabla}^{\mathcal{A},\mathcal{I}}(\gamma^{-1}(T'')) \leq \widetilde{\nabla}^{\mathcal{A},\mathcal{I}}(T') \leq \nabla^{\mathcal{A}}(T')$. Therefore, $\langle H, \gamma \rangle$ is $\nabla^{\mathcal{A}}$-compatible with T' and, hence, $T' = \gamma^{-1}(\gamma(T'))$ and $\gamma(T') \in \mathrm{FiFam}^{\mathcal{I}}(\mathcal{B})$. Now we get

$$(2) \qquad \nabla^{\mathcal{A}}(T') = \nabla^{\mathcal{A}}(\gamma^{-1}(\gamma(T'))) = \gamma^{-1}(\nabla^{\mathcal{B}}(\gamma(T'))).$$

Claim: $\gamma((\mathrm{FiFam}^{\mathcal{I}}(\mathcal{A}))^{\gamma^{-1}(T'')}) = (\mathrm{FiFam}^{\mathcal{I}}(\mathcal{B}))^{T''}$.
If $T' \in \mathrm{FiFam}^{\mathcal{I}}(\mathcal{A})$, with $\gamma^{-1}(T'') \leq T'$, then we have already shown that $\gamma(T') \in \mathrm{FiFam}^{\mathcal{I}}(\mathcal{B})$ and $T'' = \gamma(\gamma^{-1}(T'')) \leq \gamma(T')$. Conversely, suppose that $U'' \in \mathrm{FiFam}^{\mathcal{I}}(\mathcal{B})$, with $T'' \leq U''$. Then $U'' = \gamma(\gamma^{-1}(U''))$, $\gamma^{-1}(U'') \in \mathrm{FiFam}^{\mathcal{I}}(\mathcal{A})$ and $\gamma^{-1}(T'') \leq \gamma^{-1}(U'')$. This finishes the proof of the claim.
▲

Using Equation (2), we now get

$$
\begin{aligned}
\widetilde{\nabla}^{\mathcal{A},\mathcal{I}}(\gamma^{-1}(T'')) &= \bigcap\{\nabla^{\mathcal{A}}(T') : T' \in (\mathrm{FiFam}^{\mathcal{I}}(\mathcal{A}))^{\gamma^{-1}(T'')}\} \\
&\overset{\mathrm{Eq.}\,(2)}{=} \bigcap\{\gamma^{-1}(\nabla^{\mathcal{B}}(\gamma(T'))) : T' \in (\mathrm{FiFam}^{\mathcal{I}}(\mathcal{A}))^{\gamma^{-1}(T'')}\} \\
&= \gamma^{-1}(\bigcap\{\nabla^{\mathcal{B}}(\gamma(T')) : T' \in (\mathrm{FiFam}^{\mathcal{I}}(\mathcal{A}))^{\gamma^{-1}(T'')}\}) \\
&\overset{\mathrm{Claim}}{=} \gamma^{-1}(\bigcap\{\nabla^{\mathcal{B}}(U'') : U'' \in (\mathrm{FiFam}^{\mathcal{I}}(\mathcal{B}))^{T''}\}) \\
&= \gamma^{-1}(\widetilde{\nabla}^{\mathcal{B},\mathcal{I}}(T'')).
\end{aligned}
$$

Therefore, the family $\widetilde{\nabla}^{\bullet,\mathcal{I}}$ is weakly coherent, as claimed. $\qquad\square$

Recall that to simplify notation, we sometimes use $\widetilde{\nabla}^{\mathcal{I}} := \widetilde{\nabla}^{\bullet,\mathcal{I}}$ for the family of the relativized operators corresponding to the $\mathcal{I}$-operator ∇.

Theorem 43 (Relativized Correspondence). *Let ∇ be a weakly coherent family of $\mathcal{I}$-compatibility operators. For every surjective N-morphism $\langle H, \gamma \rangle : \mathcal{A} \to \mathcal{B}$, with H an isomorphism, and every $T \in \mathrm{FiFam}^{\mathcal{I}}(\mathcal{A})$, if $\langle H, \gamma \rangle$ is $\widetilde{\nabla}^{\mathcal{A},\mathcal{I}}$-compatible with T, then $\langle H, \gamma \rangle$ induces an order isomorphism between $[\![T]\!]^{\widetilde{\nabla}^{\mathcal{A},\mathcal{I}}}$ and $[\![\gamma(T)]\!]^{\widetilde{\nabla}^{\mathcal{B},\mathcal{I}}}$, whose inverse is given by γ^{-1}.*

Proof: Immediately follows by Theorem 40, taking into account the weak coherence property of $\widetilde{\nabla}^{\bullet,\mathcal{I}}$, established in Proposition 42. $\qquad\square$

The ordinary reduction processes of AAL, using the Leibniz and Suszko operators, were abstracted to the case of an arbitrary family of $\mathcal{S}$-operators in Definition 4.21 of (Albuquerque et al., 2016). In a parallel treatment, the reductions with respect to the categorical Leibniz and Suszko operators, which give rise to the CAAL algebraic system classes, can be lifted to arbitrary $\mathcal{I}$-operators.

Definition 44. *Let ∇ be a family of $\mathcal{I}$-operators. Define*

$\mathsf{AlgSys}^{\nabla}(\mathcal{I}) = \mathbf{I}(\{\mathcal{A}/\nabla^{\mathcal{A}}(T) : \mathcal{A} \text{ an } N\text{-algebraic system}, \ T \in \mathrm{FiFam}^{\mathcal{I}}(\mathcal{A})\})$

$\mathsf{AlgSys}_{\nabla}(\mathcal{I}) = \mathbf{I}(\{\mathcal{A} : \text{exists } T \in \mathrm{FiFam}^{\mathcal{I}}(\mathcal{A}) \text{ such that } \nabla^{\mathcal{A}}(T) = \Delta^{\mathrm{SEN}'}\})$

$\mathsf{AlgSys}^{\widetilde{\nabla}^{\mathcal{I}}}(\mathcal{I}) = \mathbf{I}(\{\mathcal{A}/\widetilde{\nabla}^{\mathcal{A},\mathcal{I}}(T) : \mathcal{A} \text{ an } N\text{-algebraic system}, \ T \in \mathrm{FiFam}^{\mathcal{I}}(\mathcal{A})\})$

$\mathsf{AlgSys}_{\widetilde{\nabla}^{\mathcal{I}}}(\mathcal{I}) = \mathbf{I}(\{\mathcal{A} : \text{exists } T \in \mathrm{FiFam}^{\mathcal{I}}(\mathcal{A}) \text{ such that } \widetilde{\nabla}^{\mathcal{A},\mathcal{I}}(T) = \Delta^{\mathrm{SEN}'}\}).$

We undertake, first, the task of showing that each pair of identically sup- and sub-scripted classes of algebraic systems, i.e., classes referring to the same weakly coherent family of $\mathcal{I}$-compatibility operators, consists of identical classes of N-algebraic systems. The key observation is the well-known (in both AAL and CAAL) fact that the congruence system corresponding to a reduced matrix system is the identity congruence system, i.e., "reduction always produces a reduced system".

Lemma 45. *If ∇ is a weakly coherent family of $\mathcal{I}$-compatibility operators, then, for all $T \in \mathrm{FiFam}^{\mathcal{I}}(\mathcal{A})$ and all $\theta \in \mathrm{ConSys}(\mathbf{A})$, if $\theta \leq \nabla^{\mathcal{A}}(T)$, then $\nabla^{\mathcal{A}/\theta}(T/\theta) = \nabla^{\mathcal{A}}(T)/\theta$. In particular, $\nabla^{\mathcal{A}/\nabla^{\mathcal{A}}(T)}(T/\nabla^{\mathcal{A}}(T)) = \Delta^{\mathrm{SEN}'}/\nabla^{\mathcal{A}}(T)$.*

Proof: For the first equality, noting that $\langle I_{\mathbf{Sign}'}, \pi^{\theta} \rangle$ is $\nabla^{\mathcal{A}}$-compatible with T, by the hypothesis, and using weak coherence and Lemma 33, we get that

$$\nabla^{\mathcal{A}/\theta}(T/\theta) = \nabla^{\mathcal{A}/\theta}(\pi^{\theta}(T)) = \pi^{\theta}(\nabla^{\mathcal{A}}(T)) = \nabla^{\mathcal{A}}(T)/\theta.$$

For $\theta = \nabla^{\mathcal{A}}(T)$, then, we obtain $\nabla^{\mathcal{A}/\nabla^{\mathcal{A}}(T)}(T/\nabla^{\mathcal{A}}(T)) = \nabla^{\mathcal{A}}(T)/\nabla^{\mathcal{A}}(T) = \Delta^{\mathrm{SEN}'}/\nabla^{\mathcal{A}}(T)$. $\qquad\square$

Proposition 46. *If ∇ is a weakly coherent family of $\mathcal{I}$-compatibility operators, then* $\mathsf{AlgSys}^{\nabla}(\mathcal{I}) = \mathsf{AlgSys}_{\nabla}(\mathcal{I})$. *Moreover, the class*

$$\{\mathcal{A} : exists \; T \in \mathrm{FiFam}^{\mathcal{I}}(\mathcal{A}) \; such \; that \; \nabla^{\mathcal{A}}(T) = \Delta^{\mathrm{SEN}'}\}$$

is closed under isomorphic copies.

Proof: If $\mathcal{A} \in \mathsf{AlgSys}_{\nabla}(\mathcal{I})$, then, there exists $T \in \mathrm{FiFam}^{\mathcal{I}}(\mathcal{A})$, such that $\nabla^{\mathcal{A}}(T) = \Delta^{\mathrm{SEN}'}$. Thus, $\mathcal{A}/\nabla^{\mathcal{A}}(T) \cong \mathcal{A}$ and $\mathcal{A} \in \mathsf{AlgSys}^{\nabla}(\mathcal{I})$.

If, conversely, $\mathcal{A} \in \mathsf{AlgSys}^{\nabla}(\mathcal{I})$, then, there exists $\mathcal{B}$ and $T \in \mathrm{FiFam}^{\mathcal{I}}(\mathcal{B})$, such that $\mathcal{A} \cong \mathcal{B}/\nabla^{\mathcal{B}}(T)$. But then, by Lemma 45, $\nabla^{\mathcal{B}/\nabla^{\mathcal{B}}(T)}(T/\nabla^{\mathcal{B}}(T)) = \Delta^{\mathrm{SEN}''}/\nabla^{\mathcal{B}}(T)$, whence, $\mathcal{B}/\nabla^{\mathcal{B}}(T) \in \mathsf{AlgSys}_{\nabla}(\mathcal{I})$. Therefore, $\mathcal{A} \in \mathsf{AlgSys}_{\nabla}(\mathcal{I})$. $\square$

Corollary 47. *If ∇ is a weakly coherent family of $\mathcal{I}$-compatibility operators, then* $\mathsf{AlgSys}^{\widetilde{\nabla}^{\mathcal{I}}}(\mathcal{I}) = \mathsf{AlgSys}_{\widetilde{\nabla}^{\mathcal{I}}}(\mathcal{I})$. *Moreover, the class*

$$\{\mathcal{A} : exists \; T \in \mathrm{FiFam}^{\mathcal{I}}(\mathcal{A}) \; such \; that \; \widetilde{\nabla}^{\mathcal{A},\mathcal{I}}(T) = \Delta^{\mathrm{SEN}'}\}$$

is closed under isomorphic copies.

Proof: By putting together Proposition 42, asserting that $\widetilde{\nabla}^{\bullet,\mathcal{I}}$ is also a weakly coherent family of compatibility operators, and Proposition 46. $\square$

As special cases of Proposition 46 and Corollary 47, we get $\mathsf{AlgSys}^{\Omega}(\mathcal{I}) = \mathsf{AlgSys}_{\Omega}(\mathcal{I}) = \mathsf{AlgSys}^{*}(\mathcal{I})$ and $\mathsf{AlgSys}^{\widetilde{\Omega}^{\bullet,\mathcal{I}}}(\mathcal{I}) = \mathsf{AlgSys}_{\widetilde{\Omega}^{\bullet,\mathcal{I}}}(\mathcal{I}) = \mathsf{AlgSys}^{\mathrm{Su}}(\mathcal{I})$, equalities that were asserted in Lemma 6.

Relating to the lifting of an $\mathcal{I}$-operator ∇, we consider the following corresponding classes of N-algebraic systems.

Definition 48. *For a π-institution $\mathcal{I}$ and family ∇ of $\mathcal{I}$-operators, define*

$$\mathsf{AlgSys}^{\widetilde{\nabla}}(\mathcal{I}) = \mathbf{I}(\{\mathcal{A}/\widetilde{\nabla}^{\mathcal{A}}(\mathcal{T}) : \mathcal{A} \; an \; N\text{-}algebraic \; system, \; \mathcal{T} \subseteq \mathrm{FiFam}^{\mathcal{I}}(\mathcal{A})\})$$
$$\mathsf{AlgSys}_{\widetilde{\nabla}}(\mathcal{I}) = \mathbf{I}(\{\mathcal{A} : exists \; \mathcal{T} \subseteq \mathrm{FiFam}^{\mathcal{I}}(\mathcal{A}) \; such \; that \; \widetilde{\nabla}^{\mathcal{A}}(\mathcal{T}) = \Delta^{\mathrm{SEN}'}\}).$$

Like before, each of these two classes may be obtained by considering exclusively the ∇-full $\mathcal{I}$-gmatrix systems. Moreover, in the case of $\mathsf{AlgSys}_{\widetilde{\nabla}}(\mathcal{I})$, we may consider the largest $\mathcal{I}$-gmatrix system, which is always ∇-full.

Lemma 49. *Let ∇ be a family of $\mathcal{I}$-operators. The following hold:*

1. $\mathsf{AlgSys}^{\widetilde{\nabla}}(\mathcal{I}) = \mathbf{I}(\{\mathcal{A}/\widetilde{\nabla}^{\mathcal{A}}(\mathcal{T}) : \mathcal{A} \; an \; N\text{-}algebraic \; system,$
$$\mathcal{T} \subseteq \mathrm{FiFam}^{\mathcal{I}}(\mathcal{A}) \; \nabla\text{-}full\})$$

2. $\mathsf{AlgSys}_{\widetilde{\nabla}}(\mathcal{I}) = \mathbf{I}(\{\mathcal{A} : exists \; \nabla\text{-}full \; \mathcal{T} \subseteq \mathrm{FiFam}^{\mathcal{I}}(\mathcal{A})$
$$such \; that \; \widetilde{\nabla}^{\mathcal{A}}(\mathcal{T}) = \Delta^{\mathrm{SEN}'}\})$$
$$= \mathbf{I}(\{\mathcal{A} : \widetilde{\nabla}^{\mathcal{A}}(\mathrm{FiFam}^{\mathcal{I}}(\mathcal{A})) = \Delta^{\mathrm{SEN}'}\}).$$

Proof:

1. The right-to-left inclusion is obvious. Suppose that $\mathcal{T} \subseteq \mathrm{FiFam}^{\mathcal{I}}(\mathcal{A})$. By Corollary 11 and Definition 12, the congruence system $\widetilde{\nabla}^{\mathcal{A}}(\mathcal{T})$ is a ∇-full congruence system. Thus, there exists a ∇-full $\mathcal{T}' \subseteq \mathrm{FiFam}^{\mathcal{I}}(\mathcal{A})$, such that $\widetilde{\nabla}^{\mathcal{A}}(\mathcal{T}') = \widetilde{\nabla}^{\mathcal{A}}(\mathcal{T})$. Therefore, $\mathcal{A}/\widetilde{\nabla}^{\mathcal{A}}(\mathcal{T}) = \mathcal{A}/\widetilde{\nabla}^{\mathcal{A}}(\mathcal{T}') \in \mathsf{AlgSys}^{\nabla}(\mathcal{I})$.

2. The first equality repeats the argument in Part 1. For the second, the right-to-left inclusion is obvious and, for the reverse, if $\widetilde{\nabla}^{\mathcal{A}}(\mathcal{T}) = \Delta^{\mathrm{SEN}'}$, for some $\mathcal{T} \subseteq \mathrm{FiFam}^{\mathcal{I}}(\mathcal{A})$, then $\widetilde{\nabla}^{\mathcal{A}}(\mathrm{FiFam}^{\mathcal{I}}(\mathcal{A})) \leq \widetilde{\nabla}^{\mathcal{A}}(\mathcal{T}) = \Delta^{\mathrm{SEN}'}$, which yields the conclusion. $\qquad\square$

We establish some connections between the algebraic system classes associated with the lifting and those associated with the relativization of a family of $\mathcal{I}$-operators.

Proposition 50. *Let ∇ be a family of $\mathcal{I}$-operators. Then*

$$\mathsf{AlgSys}_{\widetilde{\nabla}}(\mathcal{I}) = \mathsf{AlgSys}_{\widetilde{\nabla}^{\mathcal{I}}}(\mathcal{I}) \quad and \quad \mathsf{AlgSys}^{\widetilde{\nabla}^{\mathcal{I}}}(\mathcal{I}) \subseteq \mathsf{AlgSys}^{\widetilde{\nabla}}(\mathcal{I}).$$

Proof: Since, for all $T \in \mathrm{FiFam}^{\mathcal{I}}(\mathcal{A})$, $\widetilde{\nabla}^{\mathcal{A},\mathcal{I}}(T) = \widetilde{\nabla}^{\mathcal{A}}((\mathrm{FiFam}^{\mathcal{I}}(\mathcal{A}))^{T})$, we obtain that $\mathsf{AlgSys}_{\widetilde{\nabla}^{\mathcal{I}}}(\mathcal{I}) \subseteq \mathsf{AlgSys}_{\widetilde{\nabla}}(\mathcal{I})$ and, also, that $\mathsf{AlgSys}^{\widetilde{\nabla}^{\mathcal{I}}}(\mathcal{I}) \subseteq \mathsf{AlgSys}^{\widetilde{\nabla}}(\mathcal{I})$. To show equality in the first case, suppose that $\mathcal{A} \in \mathsf{AlgSys}_{\widetilde{\nabla}}(\mathcal{I})$. By Lemma 49, we get $\widetilde{\nabla}^{\mathcal{A}}(\mathrm{FiFam}^{\mathcal{I}}(\mathcal{A})) = \Delta^{\mathrm{SEN}'}$. Setting $T' = \bigcap \mathrm{FiFam}^{\mathcal{I}}(\mathcal{A})$, we get

$$\widetilde{\nabla}^{\mathcal{A},\mathcal{I}}(T') = \widetilde{\nabla}^{\mathcal{A}}((\mathrm{FiFam}^{\mathcal{I}}(\mathcal{A}))^{T'}) = \widetilde{\nabla}^{\mathcal{A}}(\mathrm{FiFam}^{\mathcal{I}}(\mathcal{A})) = \Delta^{\mathrm{SEN}'}.$$

Therefore, $\mathcal{A} \in \mathsf{AlgSys}_{\widetilde{\nabla}^{\mathcal{I}}}(\mathcal{I})$. $\qquad\square$

Lemma 51. *If ∇ is a weakly coherent family of $\mathcal{I}$-compatibility operators, then, for all $\mathcal{T} \subseteq \mathrm{FiFam}^{\mathcal{I}}(\mathcal{A})$,*

$$\widetilde{\nabla}^{\mathcal{A}/\widetilde{\nabla}^{\mathcal{A}}(\mathcal{T})}(\mathcal{T}/\widetilde{\nabla}^{\mathcal{A}}(\mathcal{T})) = \Delta^{\mathrm{SEN}'/\widetilde{\nabla}^{\mathcal{A}}(\mathcal{T})}.$$

Proof: Let $\theta = \widetilde{\nabla}^{\mathcal{A}}(\mathcal{T})$. Note that $\langle I_{\mathbf{Sign}'}, \pi^{\theta} \rangle$ is $\nabla^{\mathcal{A}}$-compatible with $\mathcal{T}$, by the hypothesis. Thus, using weak coherence and Lemma 38, we get

$$\widetilde{\nabla}^{\mathcal{A}/\theta}(\mathcal{T}/\theta) = \widetilde{\nabla}^{\mathcal{A}/\theta}(\pi^{\theta}(\mathcal{T})) = \pi^{\theta}(\widetilde{\nabla}^{\mathcal{A}}(\mathcal{T})) = \widetilde{\nabla}^{\mathcal{A}}(\mathcal{T})/\theta = \Delta^{\mathrm{SEN}'/\widetilde{\nabla}^{\mathcal{A}}(\mathcal{T})}. \qquad\square$$

Proposition 52. *If ∇ is a weakly coherent family of $\mathcal{I}$-compatibility operators, then $\mathsf{AlgSys}^{\widetilde{\nabla}}(\mathcal{I}) = \mathsf{AlgSys}_{\widetilde{\nabla}}(\mathcal{I})$.*

Moreover, the class $\{\mathcal{A} : \widetilde{\nabla}^{\mathcal{A}}(\mathrm{FiFam}^{\mathcal{I}}(\mathcal{A})) = \Delta^{\mathrm{SEN}'}\}$ is closed under isomorphic images and

$$\mathsf{AlgSys}^{\widetilde{\nabla}}(\mathcal{I}) = \mathbf{I}(\{\mathcal{A}/\widetilde{\nabla}^{\mathcal{A}}(\mathrm{FiFam}^{\mathcal{I}}(\mathcal{A})) : \mathcal{A} \text{ an } N\text{-algebraic system}\}).$$

Proof: If $\mathcal{A} \in \mathsf{AlgSys}_{\widetilde{\nabla}}(\mathcal{I})$, then, there exists $\mathcal{T} \subseteq \mathrm{FiFam}^{\mathcal{I}}(\mathcal{A})$, such that $\widetilde{\nabla}^{\mathcal{A}}(\mathcal{T}) = \Delta^{\mathrm{SEN}'}$. Thus, $\mathcal{A}/\widetilde{\nabla}^{\mathcal{A}}(\mathcal{T}) \cong \mathcal{A}$ and $\mathcal{A} \in \mathsf{AlgSys}^{\widetilde{\nabla}}(\mathcal{I})$. If, conversely, $\mathcal{A} \in \mathsf{AlgSys}^{\widetilde{\nabla}}(\mathcal{I})$, then, $\mathcal{A} \cong \mathcal{B}/\widetilde{\nabla}^{\mathcal{B}}(\mathcal{T})$, for some $\mathcal{T} \subseteq \mathrm{FiFam}^{\mathcal{I}}(\mathcal{B})$. But then, by Lemma 51, $\widetilde{\nabla}^{\mathcal{B}/\widetilde{\nabla}^{\mathcal{B}}(\mathcal{T})}(\mathcal{T}/\widetilde{\nabla}^{\mathcal{B}}(\mathcal{T})) = \Delta^{\mathrm{SEN}''}/\widetilde{\nabla}^{\mathcal{B}}(\mathcal{T})$, whence, $\mathcal{B}/\widetilde{\nabla}^{\mathcal{B}}(\mathcal{T}) \in \mathsf{AlgSys}_{\widetilde{\nabla}}(\mathcal{I})$. Therefore, $\mathcal{A} \in \mathsf{AlgSys}_{\widetilde{\nabla}}(\mathcal{I})$.

The displayed equality now follows by Lemma 49. $\qquad\square$

Taking into account Corollary 47 and Proposition 50, Proposition 52 yields that, under weak coherence, four of our six classes of N-algebraic systems actually coincide.

Corollary 53. *If ∇ is a weakly coherent family of $\mathcal{I}$-compatibility operators, then*

$$\mathsf{AlgSys}^{\widetilde{\nabla}}(\mathcal{I}) = \mathsf{AlgSys}_{\widetilde{\nabla}}(\mathcal{I}) = \mathsf{AlgSys}^{\widetilde{\nabla}^{\mathcal{I}}}(\mathcal{I}) = \mathsf{AlgSys}_{\widetilde{\nabla}^{\mathcal{I}}}(\mathcal{I}).$$

Proposition 54. *Let ∇ be a family of $\mathcal{I}$-compatibility operators that commutes with inverse surjective N-morphisms. For every N-algebraic system $\mathcal{A}$ and $\theta \in \mathrm{ConSys}(\mathbf{A})$,*

$$\theta \text{ is } \nabla^{\mathcal{A}}\text{-full} \quad \text{iff} \quad \theta \in \mathrm{ConSys}_{\mathsf{AlgSys}^{\widetilde{\nabla}}(\mathcal{I})}(\mathcal{A}).$$

Proof: Suppose $\theta \in \mathrm{ConSys}(\mathbf{A})$ is $\nabla^{\mathcal{A}}$-full. By Corollary 11, $\theta = \widetilde{\nabla}^{\mathcal{A}}(\mathcal{T})$, for some $\mathcal{T} \subseteq \mathrm{FiFam}^{\mathcal{I}}(\mathcal{A})$. Thus, $\mathcal{A}/\theta \in \mathsf{AlgSys}^{\widetilde{\nabla}}(\mathcal{I})$.

Conversely, if $\theta \in \mathrm{ConSys}_{\mathsf{AlgSys}^{\widetilde{\nabla}}(\mathcal{I})}(\mathcal{A})$, then $\mathcal{A}/\theta \in \mathsf{AlgSys}^{\widetilde{\nabla}}(\mathcal{I})$. By Proposition 52, there exists $\mathcal{T}' \subseteq \mathrm{FiFam}^{\mathcal{I}}(\mathcal{A}/\theta)$, such that $\widetilde{\nabla}^{\mathcal{A}/\theta}(\mathcal{T}') = \Delta^{\mathrm{SEN}'}/\theta$. Let $\langle I_{\mathbf{Sign}'}, \pi \rangle := \langle I_{\mathbf{Sign}'}, \pi^{\theta} \rangle : \mathcal{A} \to \mathcal{A}/\theta$. Then $\pi^{-1}(\mathcal{T}) \subseteq \mathrm{FiFam}^{\mathcal{I}}(\mathcal{A})$ and, by commutativity,

$$\theta = \mathrm{Ker}(\langle I_{\mathbf{Sign}'}, \pi \rangle) = \pi^{-1}(\Delta^{\mathrm{SEN}'/\theta}) = \pi^{-1}(\widetilde{\nabla}^{\mathcal{A}/\theta}(\mathcal{T}')) = \widetilde{\nabla}^{\mathcal{A}}(\pi^{-1}(\mathcal{T}')).$$

Thus, by Corollary 11, θ is $\nabla^{\mathcal{A}}$-full. $\qquad\square$

Proposition 54, taking into account the isomorphism in Corollary 11, gives a natural generalization of the isomorphism of Corollary 18.

Corollary 55. *Let ∇ be a family of $\mathcal{I}$-compatibility operators that commutes with inverse surjective N-morphisms. For every $\mathcal{A}$, the maps $\widetilde{\nabla}^{\mathcal{A}}$ and $\nabla^{\mathcal{A}^{-1}}$ are mutually inverse dual order isomorphisms betwen the lattice of $\nabla^{\mathcal{A}}$-full $\mathcal{I}$-gmatrix systems and the lattice $\mathrm{ConSys}_{\mathsf{AlgSys}^{\widetilde{\nabla}}(\mathcal{I})}(\mathcal{A}).$*

We will continue our developments along the line of the general theory presented here, establishing more analogs of results obtained in the AAL framework in (Albuquerque et al., 2016), pertaining to characterizations of classes in the Leibniz hierarchy, in a forthcoming companion to the present work.

Acknowledgements

The author is heavily indebted to many people for their pioneering work in this field, which has inspired and guided his own work, as well as for having offered support and encouragement for the best part of almost two decades of work in the field. **Don Pigozzi** has made the beginning, the end and everything in between possible, since, without him, I would not have entered the field in the first place and, without his ongoing support, I would not have had the opportunities that I was afforded to have kept working in it, both physically and mentally. *Janusz Czelakowski* has performed pioneering work and, both during his visits in Ames, Iowa, and in subsequent years through his work has been a source of major inspiration. The Barcelona group, under the tireless guidance of *Josep Maria Font* and *Ramon Jansana*, based on the work of *Wim Blok*, Don and Janusz have contributed some of the most fascinating and beautiful recent developments in this area, including work on which the developments in the present paper are based, and have also been a source of inspiration and support for my own work. Last, but not less importantly, I would like to mention as other sources of motivation and stimulus the work of *James Raftery* and of *Manuel António Martins*.

Finally, outside the field, *Charles Wells* has provided critical support and encouragement and *Giora Slutzki* has been with me from the very beginning and, recently, became an "official advisor" of mine as well.

References

Albuquerque, H., Font, J.M. and Jansana, R. (2016). Compatibility Operators in Abstract Algebraic Logic, *The Journal of Symbolic Logic*, Vol. 81, No. 2, pp. 417–462.

Babenyshev, S. (2003). Fully Fregean Logics, *Reports on Mathematical Logic*, Vol. 37, pp. 59–77

Blok, W.J. and Pigozzi, D. (1986). Protoalgebraic Logics, *Studia Logica*, Vol. 45, pp. 337–369

Blok, W.J. and Pigozzi, D. (1989). *Algebraizable Logics*, Memoirs of the American Mathematical Society, Vol. 77, No. 396.

Blok, W.J. and Pigozzi, D. (1992). Algebraic Semantics for Universal Horn Logic Without Equality, in *Universal Algebra and Quasigroup Theory*, A. Romanowska and J.D.H. Smith, Eds., Heldermann Verlag, Berlin.

Czelakowski, J. (1981a). Equivalential Logics I, *Studia Logica*, Vol. 40, pp. 227–236

Czelakowski, J. (1981b). Equivalential Logics II, *Studia Logica*, Vol. 40, pp. 355–372

Czelakowski, J. (2001). *Protoalgebraic Logics*, Trends in Logic-Studia Logica Library 10, Kluwer, Dordrecht.

Czelakowski, J. (2003). The Suszko Operator Part I, *Studia Logica*, Vol. 74, pp. 181–231.

Czelakowski, J. and Jansana, R. (2000). Weakly Algebraizable Logics, *Journal of Symbolic Logic*, Vol. 64, pp. 641–668.

Czelakowski, J. and Pigozzi, D. (1999). Amalgamation and Interpolation in Abstract Algebraic Logic, Models, Algebras and Proofs, in *Lecture Notes in Pure and Applied Mathematics*, Vol. 203, Dekker, New York, pp. 187–265.

Czelakowski, J. and Pigozzi, D. (2004). Fregean Logics, *Annals of Pure and Applied Logic*, Vol. 127, pp. 17–76.

Czelakowski, J. and Pigozzi, D. (2004). Fregean Logics with the Multiterm Deduction Theorem and Their Algebraization, *Studia Logica*, Vol. 78, pp. 171–212.

Davey, B.A. and Priestley, H.A. (2002). *Introduction to Lattices and Order, Second Edition*, Cambridge University Press, Cambridge.

Font, J.M. 2003. Generalized Matrices in Abstract Algebraic Logic, in V. F. Hendriks and J. Malinowski, (eds.), *Trends in Logic. 50 years of Studia Logica, Trends in Logic - Studia Logica Library*, Vol. 21, Kluwer, Dordrecht, pp. 57–86.

Font, J.M. and Jansana, R. (1995). Full Models for Sentential Logics, *Bulletin of the Section of Logic*, Vol. 24, No. 3, pp. 123–131.

Font, J.M. and Jansana, R. (1996). A General Algebraic Semantics for Sentential Logics, *Lecture Notes in Logic*, Vol. 332, No. 7 (1996), Springer-Verlag, Berlin Heidelberg.

Font, J.M. and Jansana, R. (2001). Leibniz Filters and the Strong Version of a Protoalgebraic Logic, *Archive for Mathematical Logic*, Vol. 40, pp. 437–465.

Font, J.M. and Jansana, R. (2011). Leibniz-linked Pairs of Deductive Systems, *Studia Logica*, Vol. 99, No. 1–3, pp. 171–20.

Font, J.M., Jansana, R. and Pigozzi, D. (2003). A Survey of Abstract Algebraic Logic, *Studia Logica*, Vol. 74, No. 1/2, pp. 13–97.

Font, J.M., Jansana, R. and Pigozzi, D. (2006). On the Closure Properties of the Class of Full G-models of a Deductive System, *Studia Logica*, Vol. 83, pp. 215–278.

Herrmann, B. (1993). *Equivalential Logics and Definability of Truth*, Ph.D. Dissertation, Freie Universität Berlin, Berlin.

Herrmann, B. (1996). Equivalential and Algebraizable Logics, Studia Logica, Vol. 57, pp. 419–436.

Herrmann, B. (1997). Characterizing Equivalential and Algebraizable Logics by the Leibniz Operator, *Studia Logica*, Vol. 58, pp. 305–323.

Jansana, R. (2002). Full Models for Positive Modal Logic, *Mathematical Logic Quarterly*, Vol. 48, No. 3, pp. 427–445.

Jansana, R. (2003). Leibniz Filters Revisited, *Studia Logica*, Vol. 75, pp. 305–317.

Raftery, J.G. (2006). The Equational Definability of Truth Predicates, *Reports on Mathematical Logic*, Vol. 41, pp. 95–149.

Voutsadakis, G. (2002). Categorical Abstract Algebraic Logic: Algebraizable Institutions, *Applied Categorical Structures*, Vol. 10, No. 6, pp. 531–568.

Voutsadakis, G. (2005). Categorical Abstract Algebraic Logic: Models of π-institutions, *Notre Dame Journal of Formal Logic*, Vol. 46, No. 4, pp. 439–460.

Voutsadakis, G. (2005). Categorical Abstract Algebraic Logic: $(\mathcal{I}, N)$-Algebraic Systems, *Applied Categorical Structures*, Vol. 13, No. 3, pp. 265–280.

Voutsadakis, G. (2006). Categorical Abstract Algebraic Logic: More on Protoalgebraicity, *Notre Dame Journal of Formal Logic*, Vol. 47, No. 4, pp. 487–514.

Voutsadakis, G. (2007a). Categorical Abstract Algebraic Logic: Prealgebraicity and Protoalgebraicity, *Studia Logica*, Vol. 85, No. 2, pp. 217–251.

Voutsadakis, G. (2007b). Categorical Abstract Algebraic Logic: The Categorical Suszko Operator, *Mathematical Logic Quarterly*, Vol. 53, No. 6, pp. 616–635.

Voutsadakis, G. (2007c). Categorical Abstract Algebraic Logic: Weakly Algebraizable π-Institutions, preprint available at `http://www.voutsadakis.com/RESEARCH/papers.html`.

Voutsadakis, G. (2008). Categorical Abstract Algebraic Logic: Equivalential π-Institutions, *Australasian Journal of Logic*, Vol. 6, 24pp.

Voutsadakis, G. (2015). Categorical Abstract Algebraic Logic: Tarski Congruence Systems, Logical Morphisms and Logical Quotients, *Journal of Pure and Applied Mathematics: Advances and Applications*, Vol. 13, No. 1, pp. 27–73.

Voutsadakis, G. (2015). Categorical Abstract Algebraic Logic: Truth Equational π-Institutions, *Notre Dame Journal of Formal Logic*, Vol. 56, No. 2, pp 351–378.

Printed by Printforce, the Netherlands